# Mineral Matter and Trace Elements in Coal

## Special Issue Editors

Shifeng Dai

Xibo Wang

Lei Zhao

MDPI • Basel • Beijing • Wuhan • Barcelona • Belgrade

**MDPI**

*Special Issue Editors*

Shifeng Dai
China University of Mining
and Technology (Beijing)
China

Xibo Wang
China University of Mining
and Technology (Beijing)
China

Lei Zhao
China University of Mining
and Technology (Beijing)
China

*Editorial Office*
MDPI AG
St. Alban-Anlage 66
Basel, Switzerland

This edition is a reprint of the Special Issue published online in the open access journal *Minerals* (ISSN 2075-163X) from 2015–2016 (available at: http://www.mdpi.com/journal/minerals/special_issues/minerals_in_coal).

For citation purposes, cite each article independently as indicated on the article page online and as indicated below:

Author 1; Author 2. Article title. *Journal Name* **Year**, *Article number*, page range.

**First Edition 2017**

**ISBN 978-3-03842-622-6 (Pbk)**
**ISBN 978-3-03842-623-3 (PDF)**

# Table of Contents

# About the Special Issue Editors

**Shifeng Dai** is a professor at the China University of Mining and Technology. He is the Dean of the School of Resources and Geosciences of the China University of Mining and Technology, and the Deputy Director of the State Key Laboratory of Coal Resources and Safe Mining. He received his Ph.D. (2002) from the China University of Mining and Technology. His research topics include coal mineralogy, coal geochemistry, and coal geology. He is the Editor-in-Chief of the International Journal of Coal Geology (2007 to date) and the President of The Society For Organic Petrology (2015 to date). He is the Chief Scientist of the National Key Basic Research Program of China and Changjiang and a Scholar Professor of the Ministry of Education (China). He has published over 100 research papers, co-authored three books, and edited several Special Issues for international journals. He is a recipient of the Distinguished Service Award granted by The Society for Organic Petrology, the Dal Swaine Award, and the National Science Fund for Distinguished Young Scholars of China.

**Xibo Wang** is an associate professor at the Key Laboratory of Coal Resources and Safe Mining, China University of Mining and Technology (Beijing). His research topics include mineralogy, and trace-element geochemistry of coal and shale, and coal-hosted ore metal deposits. He received his M.Sc degree from the China University of Mining and Technology (Beijing) in 2008 and his Ph.D. from the same university in 2011. He has published over 20 research papers in international journals. His co-authored paper "Mineralogical and Geochemical Compositions of the Pennsylvanian Coal in the Adaohai Mine, Daqingshan Coal-field, Inner Mongolia, China: Modes of Occurrence and Origin of Diaspore, Gorceixite, and Ammonian Illite" (International Journal of Coal Geology, 2012, 94, 250–270) was also recognized as being among "The Most Influential 100 International Papers in 2012 in China," selected from 190,100 published international papers in 2012.

**Lei Zhao** is a lecturer at the State Key Laboratory of Coal Resources and Safe Mining, China University of Mining and Technology (Beijing). Her main area of research is the characterization, formation and behavior of mineral matter in coal, trace element geochemistry, and the environmental impact of coal utilization. She received her MSc degree from the China University of Mining and Technology (Beijing) in 2007 and her PhD from the University of New South Wales, Australia in 2012. Lei Zhao has co-authored over 20 refereed journal articles and presented her papers at various academic conferences. She is an editorial board member of the International Journal of Coal Geology.

# Preface to "Mineral Matter and Trace Elements in Coal"

Minerals are highly significant components of coal from both academic and practical perspectives. Minerals may react when the coal is burned, either forming an ash residue, or, in many cases, releasing volatile components, or as they need to be removed as slag from the blast furnace during metallurgical processing. Minerals in coal can also be a source of unwanted abrasion, stickiness, corrosion, or pollution associated with coal handling and use. Minerals in coal, in some cases, are major carriers of highly-evaluated critical metals, such as Ga, Al, Nb, Zr, and rare earth elements, and these coals have the potential to be sources of raw material for industrial use. From the genetic point of view, minerals in coal are products of the processes associated with peat accumulation and rank advance, as well as other aspects of epigenetic processes, and, thus, the minerals in coal can provide information on the depositional conditions and geologic history of individual coal beds, coal-bearing sequences, and regional tectonic evolution.

This Special Issue book "Mineral Matter and Trace Elements in Coal" includes 23 chapters, providing up-to-date research and technological developments in the nature, origin, and significance of the minerals and trace elements in coal, coal-mining wastes, and various byproducts derived from combustion, gasification, and pyrolysis and other related processes.

Coal and coal combustion byproducts can have significant concentrations of rare earth elements and Y (REY), as well as a number of other critical elements (e.g., Ga, Ge, Nb, Ta, Zr, Hf, etc.). This book begins with a comprehensive review chapter by Hower et al. (2016) who discuss the origin and enrichment mechanisms of REY in coal and associated strata in China, US, and other countries. The authors also comment on classification systems used to evaluate the relative value of the rare earth concentrations and the distribution of the elements within the coals and coal combustion byproducts. This is followed by a research chapter which investigates the clay minerals in Nb–Zr–REE–Ga mineralized beds in southwestern (SW) China (Zhao et al., 2016), and noted that these clay minerals can absorb a large amount of critical elements in the studied mineralized beds.

The book encompasses a series of studies on mineralogy and geochemistry of coals from different coal deposits (Xibo Wang et al., 2015; Yang et al., 2015; Xie et al., 2016; Xibo Wang et al., 2016; Ruixue Wang, 2016; Luo and Zheng, 2016). Not only coal, but also host strata and other non-coal strata can be potential sources of critical elements. Zou et al. (2016) studied the geochemistry and mineralogy of a tuff from SW China, and found that such tuff is a potential polymetallic ore and discusses the opportunity for recovery of these critical elements.

This is followed by two comprehensive studies on the petrology, palynology, and geochemistry of coal from Eastern Kentucky (Hower et al., 2015), and petrology and geochemistry of a number of coals from Kentucky, USA (Johnston et al., 2015).

Several chapters discuss abundance and modes of occurrence of trace elements in coals from various coal deposits in China (Zhao et al. 2015; Xiao et al., 2016; Yang et al., 2016). The naturally occurring radionuclides in coals might exhibit high radioactivity, and could also be a risk to the surrounding environment due to coal combustion and other processes. Xin Wang et al. (2015) assess radioactivity of natural nuclides ($^{40}$K, $^{238}$U, $^{232}$Th, $^{226}$Ra) in coals from different areas in the Yunnan Province, China.

Following on from these, are three chapters investigate compositions of coal combustion byproducts. The chapter by Liu et al. 2016, focuses on morphology and compositions of microspheres in fly ash from a coal-combustion power plant in SW China. Some trace elements are of particular concern due to their potential detrimental environmental impact. Guangmeng Wang et al. (2015) investigate the modes of occurrence of fluorine in a coal-fired power plant in Inner Mongolia, China by extraction and the SEM method. Liu et al. (2016) describe mineralogy of ash and slag from underground coal gasification.

The following two chapters are concerned with the potential release and environmental impact of trace elements on the environment. Yang et al. (2016) investigate leaching behavior and potential environmental effects of trace elements in coal gangue of an open-cast coal mine area in Inner Mongolia, China. Toxic elements can also be a potential risk to the health of workers and residents in coal mining areas. Jia et al. (2015) carry out a human health risk assessment of toxic elements in the soil environment of an open-cast coal mine in China.

The cement industry has the potential to become a major consumer of industrial by-products, including coal-mining wastes. The chapter by Giménez-García et al. (2016) examines the mineralogical transformations of coal waste across a range of temperatures, for the establishment of optimum calcination conditions that yield products with sufficient pozzolanic properties to be used as additives in the manufacture of cements and related materials. The chapter by Devasahayam et al. (2015) evaluates the water absorption potential of super absorbent polymers and low-rank coal in Australia.

Pyrolysis is an important coal-cleaning technology, producing fuel or basic chemical materials. Understanding the behavior of trace elements in this process is also significant from the environment point of view. The final chapter by Dang et al. (2016) studies the behavior of toxic elements in coal during fast pyrolysis.

The Guest Editors sincerely thank all the authors who contributed chapters to this SI book, including those whose papers for various reasons did not actually proceed to the publication stage. Sincere thanks are also expressed to colleagues who served as reviewers for the chapters that were submitted to this book. These high-profile reviewers provided numerous valuable comments and constructive suggestions that helped many of the authors improve the quality of their chapters, and generally reinforced the high standard of the work submitted.

We would also like to thank the National Key Basic Research Program of China (No. 2014CB238902), the National Natural Science Foundation of China (No. 41420104001), and the "111" Project (No. B17042), which financially supported guest editors' and some authors' travels for discussion on various aspects of this book and, in part, supported a number of papers included in this SI book.

**Shifeng Dai, Xibo Wang and Lei Zhao**
*Special Issue Editors*

*minerals*
MDPI

*Commentary*

# Notes on Contributions to the Science of Rare Earth Element Enrichment in Coal and Coal Combustion Byproducts

James C. Hower [1,*], Evan J. Granite [2,†], David B. Mayfield [3,†], Ari S. Lewis [4,†] and Robert B. Finkelman [5,†]

1   Center for Applied Energy Research, University of Kentucky, Lexington, KY 40511, USA
2   National Energy Technology Laboratory, U.S. Department of Energy, Pittsburgh, PA 15236-0940, USA; evan.granite@netl.doe.gov
3   Gradient, 600 Stewart Street, Suite 1900, Seattle, WA 98101, USA; dmayfield@gradientcorp.com
4   Gradient, 20 University Road, Cambridge, MA 02138, USA; alewis@gradientcorp.com
5   Department of Geosciences, The University of Texas at Dallas, ROC 21, 800 West Campbell Road, Richardson, TX 75080-3021, USA; bobf@utdallas.edu
*   Correspondence: james.hower@uky.edu; Tel.: +1-859-257-0261
†   These authors contributed equally to this work.

Academic Editor: Mostafa Fayek
Received: 29 January 2016; Accepted: 25 March 2016; Published: 31 March 2016

**Abstract:** Coal and coal combustion byproducts can have significant concentrations of lanthanides (rare earth elements). Rare earths are vital in the production of modern electronics and optics, among other uses. Enrichment in coals may have been a function of a number of processes, with contributions from volcanic ash falls being among the most significant mechanisms. In this paper, we discuss some of the important coal-based deposits in China and the US and critique classification systems used to evaluate the relative value of the rare earth concentrations and the distribution of the elements within the coals and coal combustion byproducts.

**Keywords:** lanthanide; yttrium; critical materials; coal; coal combustion by-products

## 1. Introduction

Coal is a precious resource, both in the United States and around the world. The United States has a 250-year supply of coal, and generates between 30%–40% of its electricity through coal combustion. Approximately 1 Gt of coal has been mined annually in the US, although the 2015 total will likely be closer to 900 Mt [1]. Most of the coal is burned for power generation, but substantial quantities are also employed in the manufacture of steel, chemicals, and activated carbons. Coal has a positive impact upon many industries, including mining, power, rail transportation, manufacturing, chemical, steel, activated carbon, and fuels. Everything that is in the Earth's crust is also present within coal to some extent, and the challenge is always to utilize abundant domestic coal in clean and environmentally friendly manners. In the case of the rare earth elements, these valuable and extraordinarily useful elements are present within the abundant coal and coal byproducts produced domestically and world-wide. These materials include the coals, as well as the combustion by-products such as ashes, coal preparation wastes, gasification slags, and mining byproducts. All of these materials can be viewed as potential sources of rare earth elements. Most of the common inorganic lanthanide compounds, such as the phosphates found in coal, have very high melting, boiling, and thermal decomposition temperatures, allowing them to concentrate in combustion and gasification by-products. Furthermore, rare earths have been found in interesting concentrations in the strata above and below certain coal seams.

The National Energy Technology Laboratory (NETL) initiated research for the determination and recovery of rare earths from abundant domestic coal by-products in 2014. The NETL Rare Earth EDX Database [2] is a resource for rare earth information as related to coal and byproducts. Many other research organizations have also initiated efforts for the determination and recovery of rare earths from unconventional sources, such as coal byproducts.

Fifty years ago, the rare earth elements (REE) were little more than an interesting diversion from the study of more commercially and environmentally important elements. As stated by Gschneidner [3] *"we know what we know about the Fraternal Fifteen [the rare earth elements, REE] essentially because of scientific curiosity, and this is still one of the most important reasons for studying the rare earths."* While he did anticipate wider applications of the niche uses at the time of his pamphlet, some applications were still decades away [4].

Today, numerous technologies and devices rely upon rare earth elements. Important commercial uses of REEs include automotive catalytic converters, petroleum refining catalysts, metallurgical additives and alloys, permanent magnets and rechargeable batteries (for hybrid vehicles, wind turbines, and mobile phones), phosphors (for lighting and flat panel displays), glass polishing and ceramics, and medical devices [5]. In short, modern society has become increasingly dependent on the REEs (Figure 1). Due to the growing application of REEs in modern technology (particularly sustainable energy), many countries are developing strategies to obtain or develop additional sources of REE materials [5–7]. While traditional mining has typically provided the majority of REEs, current limitations with developing new mines has resulted in the search for alternative sources, including coal and coal combustion byproducts [8].

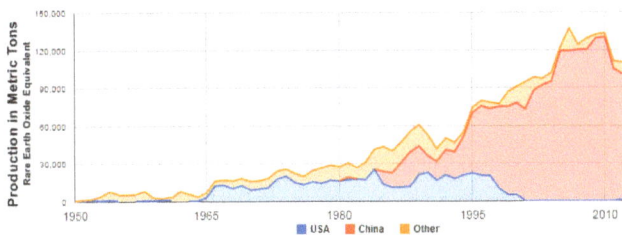

**Figure 1.** Production of REE oxides from 1950 in USA, China, and other countries.

Within the context of the expanded use of the rare earths and the widening search for economic sources of the elements, Seredin and Dai [9] made a fundamental theory in understanding of the origin and distribution of REEs or REE + yttrium (REY) in coal and, by extension, in coal combustion byproducts such as fly ash and bottom ash. While their paper was developed largely in the context of Chinese and former Soviet deposits, the background was built on knowledge of occurrences in Bulgaria [10–18]; Kentucky [19,20]; Utah [21]; Wyoming [22]; the Russian Far East [23]; China [24–29]; and elsewhere.

## 2. Rare Earth Elements in Coals

Several studies [30–33] have addressed the origin of rare earth elements in coal. Eskenazy [10–16] discussed the complications of REY enrichment in coals. Dealing primarily with lignites, Eskenazy [11,15] was able to observe organic associations not nearly as evident in the bituminous coals studied elsewhere. The loosely bound REE on clays could be desorbed by acidic waters, with heavy REE (HREE) preferentially desorbed and the subsequent increase in HREE in solution would lead to enrichment in HREE bound to organics [11,13]. As a supplemental or alternative source, the high organic-bound HREE could have resulted from high HREE in the waters feeding the swamp [13]. HREE generally have a stronger organic affinity than light REE (LREE) and HREE complexes are more stable than LREE complexes. Independent of peat or coal associations, soil studies by Aide and Aide [34] confirmed that HREE-organic complexes are more stable than LREE-organic complexes. Decreases in pH cause a decrease in the stability of the REE-organic complexes [35,36]. In testing of humic acids extracted from a Bulgarian lignite [15], Eskenazy found that $Na^+$, $K^+$, $Ca^{2+}$, and $Mg^{2+}$ bound to –COOH and –OH were replaced by REE cations.

Using a suite of bench samples from a Texas lignite strip mine Finkelman [37] demonstrated that the chondrite-normalized REE distribution pattern changed systematically with the ash yield. The high-ash bench (77 wt % ash) had a REE distribution pattern similar to those of North American shales and high-ash bituminous coals. With lower ash yields (3–51 wt %), the patterns were progressively flatter, indicating a higher proportion of heavier REE elements in the organic-rich benches. He interpreted this trend to indicate that the heavy REE (Eu to Lu) are preferentially complexed with the organics. Finkelman [37] estimated that no more than 10% of the total REE in the lignite had an organic association; the remaining 90% of the REE were associated with REE-bearing minerals.

Finkelman and Palmer (U.S. Geological Survey, unpublished data) used selective leaching on 14 bituminous coals, five subbituminous coals, and one lignite to determine the modes of occurrence of 37 elements including Y, Ce, La, Lu, Nd, Sm, and Yb. Based on the response of the elements to ammonium acetate, hydrochloric acid, hydrofluoric acid, and nitric acid leaches they concluded that in the bituminous samples approximately 70% of the light rare earths (Y, Ce, La, and Nd) were associated with phosphate minerals, about 20% were associated with clays, and about 10% were in carbonate minerals. A smaller proportion was organically associated. The heavier rare earths (Sm, Yb) were primarily associated with phosphates (50%), clays (20%), organics (30%), and carbonates. In contrast, the light rare earth elements in the lower rank coals were associated with clays (60%), phosphates (20%), carbonates (20%), and organics. The heavier rare earths were also associated with clays (50%), phosphates (25%), and carbonates, but had a much larger (25%) proportion associated with organics.

In contrast to Eskenazy's findings of strong HREE-organic associations [11,13,14], Seredin *et al.* [38], in their study of an Eocene subbituminous coal and a Miocene lignite from the Russian Far East, could not universally verify the association. Consequently, they noted that some high HREE concentrations in coal could not be explained by the higher sorption capacity or by higher HREE chelate stability, but rather by elevated HREE in waters which interacted with the organics.

Some Kentucky, Utah, and Wyoming REY occurrences are largely the result of volcanic ash falls. Crowley *et al.* [21] noted three enrichment mechanisms:

(1)   Leaching of volcanic ash with subsequent concentration by organic matter;
(2)   Leaching of volcanic ash with subsequent incorporation into secondary minerals; and
(3)   Incorporation of volcanic minerals into the peat.

Hower *et al.* [19] found that the coal immediately underlying the Fire Clay coal tonstein had 1965–4198 ppm (ash basis) REY, with REE-rich monazite and Y-bearing crandallite as the detectable REY minerals. The 4198-ppm REY lithotype contains thin lenses of the volcanic ash. They noted that, while volcanic glass may not have been stable in organic acids, zircons survived in the lithotype, as indicated by the 4540 ppm Zr (ash basis). Similarly, the Fire Clay-correlative Dean coal section in southern Knox County, Kentucky, has an REY enrichment (based on comparisons to REE levels in other coals in the region) but does not contain a tonstein [20], just as Crowley *et al.* [21] found in their study of Wyoming coals. In the central Eastern Kentucky Fire Clay coal locations, Hower *et al.* [19] noted the following enrichment mechanisms:

(1)   The highest LREE/HREE occurs in the tonstein and in the coal or illitic shale immediately underlying the tonstein;
(2)   The other lithotypes, in particular in the basal and uppermost lithotypes, have a lower LREE/HREE, suggesting concentration in secondary minerals.

Seredin [23] studied a complex assemblage of coals and volcanics in the Russian Far East. The REY entered the peat in a dissolved form. The bulk of the REY in the low-rank coals was sorbed onto the organics. The mineral assemblages included Eu-rich LREE phosphates with no Th or Y; HREE phosphates deposited on kaolinite; (Ca, Ba, Sr)-bearing aluminophosphates (crandallite) with LREE deposited on kaolinite; LREE-bearing F and Cl carbonates; REE-carbonates, -oxides, and -hydroxides; and other unknown REE mineral species. Based on the high concentrations of REY, he encouraged the recovery of REY from coal combustion by-products, something only considered for U and Ge at that time. Mardon and Hower [20], examining the path of the REY-enriched Dean coal from the mine to the boiler to the ash-collection system at a utility power plant, found that the REY were in concentrations exceeding 1600 ppm in some of the electrostatic-precipitator fly ashes.

The fundamental contributions of Dai and his colleagues [24–26,28,29] were based on deposits in the Jungar and Daqingshan coalfields, Inner Mongolia, and in host rocks in the Late Permian coal-bearing strata from Eastern Yunnan [39].

The interest in the Jungar coals has been driven by the prospects for commercial recovery of gallium, which substitutes for Al in boehmite [24–26], and Al from the coal combustion byproducts [29,40,41]. The REY is low in the partings and relatively high in the coal [25,26], attributed to leaching and incorporation in Al-hydrate minerals, goyazite, and organic matter [25,26]. For example, the REE content of one parting and its underlying coal bench were 231 and 1006 ppm. The LREE are both occur in Sr- and Ba-bearing minerals and have an organic affinity while the HREE are enriched in Sc-, Zr-, and Hf-bearing minerals [26]. The relative organic affinity of LREE *versus* HREE was found to vary between mines within the Jungar coalfield, perhaps indicative of different REE sources [28]. Examining the Light REY (LREY; La, Ce, Pr, Nd, and Sm), Medium REY (MREY; Eu, Gd, Tb, Dy, and Y), and Heavy REY (HREY; Ho, Er, Tm, Yb, and Lu), and the L-type ($La_N/Lu_N > 1$), M-type ($La_N/Sm_N < 1$; $Gd_N/Lu_N > 1$), and H-type ($La_N/Lu_N < 1$) distributions, Dai *et al.* [29] found LREY associations in goyazite and gorceixite, MREY and HREY in boehmite, and some indications of MREY and HREY associations in accessory minerals. L-type REE distributions are found in the upper portion of the Pennsylvanian No. 6 coal, Guanbanwusu mine, due to REE-rich colloidal input from weathered bauxite [29]. The H-type enrichment in the lower portion of the same coal is attributed to natural water influences. In both cases, mixed influences were evident.

The coal-bearing strata of Late Permian Xuanwei Formation in eastern Yunnan (Southwestern China) have (Nb, Ta)$_2$O$_5$–(Zr, Hf)O$_2$–(REY)$_2$O$_3$–Ga in 1–10-m-thick alkalic ore beds of pyroclastic origin [39]. Dai *et al.* [39] identified four types of ore lithologies: clay altered volcanic ash, tuffaceous clay, tuff, and volcanic breccia. The minerals associated with the above elevated concentrations of rare metals (e.g., the most common REY-bearing minerals monazite and xenotime) are rare, suggesting that the rare elements occur as adsorbed ions. Although the mineralization of (Nb, Ta)$_2$O$_5$–(Zr, Hf)O$_2$–(REY)$_2$O$_3$–Ga assemblage has been identified in felsic and alkalic tonsteins in many coal deposits for many years [19,21,42–47], this mineralization anomaly has never caused particular interest as raw materials for rare metals, owing to the low thickness (from 1–20 cm, mostly 3–6 cm) of the tonsteins. However, the occurrence of such thin tonsteins provides a basis for predicting the possibility of thick horizons of Nb–Zr–REY-bearing tuffs outside of coal seams [44]. This forecast has been successfully realized in China by discovery of such thick alkalic ore beds in Yunnan Province by Dai *et al.* [39], and thus, previous skeptical views in relation to this mineralization in coal-bearing strata should be reconsidered. As pointed out by Spears [46], *"Linked to the tonstein studies, Dai et al. (2010) found a new rare metal deposit comprised of several Nb–Zr–REE–Ga bearing tuffaceous horizons with thicknesses up to 10 m in Yunnan province,"* and to the best of our knowledge, this the first successful case of the application from the tonstein academic theory to discovery of rare-metal ore deposits.

## 3. Seredin and Dai Synthesis

Seredin and Dai [9] reinforced some of the basic principles outlined above, as shown in Table 1 (Table 2 as cited in [9]). The introduction of REY into a peat or coal falls into four basic paths. As we saw above, few coals are likely to have one dominant source of REY. Indeed, the Jungar coals were noted to have multiple modes of REY emplacement [25,26]. Similarly, while the Dean (Fire Clay) coal

REY is dominated by the REY-rich tonstein and the enrichment of adjacent coal lithologies through the leaching of REY from the tonstein, the coal bed had a depositional history prior to and following the ash fall [19]. In particular, Eastern Kentucky coals typically have $TiO_2$- and Zr-enriched basal lithologies which can also have REY enrichment [19,20]. For example, a section of the Fire Clay coal has 1358-ppm-REY (ash basis) basal lithotype, significantly less than the 4251-ppm-REY (ash basis) lithotype immediately underlying the tonstein, but double the 680-ppm-REY (ash basis) concentration in the lithotypes between those two portions of the coal [19]. Basically, on a whole-seam basis, and probably also for most lithotypes, mixed modes of REY emplacement are to be expected but it is also important to understand the end members in order to fully understand the continuum.

**Table 1.** The main genetic types of high REY accumulation in coals. After Seredin and Dai [9].

| Type | REO Content in Ash, % | Associated Elements | Typical Example |
|------|----------------------|---------------------|-----------------|
| Terrigenous | 0.1–0.4 | Al, Ga, Ba, Sr, | Jungar, China [25,26] |
| Tuffaceous | 0.1–0.5 | Zr, Hf, Nb, Ta, Ga | Dean, USA [20] |
| Infiltrational | 0.1–1.2 | U, Mo, Se, Re | Aduunchulun, Mongolia [48] |
| Hydrothermal | 0.1–1.5 | As, Sb, Hg, Ag, Au, *etc.* | Rettikhovka, Russia [49] |

REO, oxides of rare earth elements and yttrium.

In addition to the classification of light-, medium-, and heavy-REY, as well as the corresponding enrichment types (L-, M-, and H-types), Seredin and Dai [9] set a criterion of REY concentration evaluation of REO content $\geqslant$ 1000 ppm in coal ash, or 800–900 ppm in coal ash for coal seams with thicknesses of > 5 m, as the cut-off grade or beneficial recovery of the REY. The second criterion they set in their work for the evaluation of coal ash as REY raw materials is the individual composition of the elements. Seredin and Dai's [9] Figure 6 (Figure 2 in this paper) is a synthesis of the REY concentration in coal ashes and non-coal REY-enriched deposits compared to an expression of the current commercial need as weighted by the availability of the individual elements. The *x* axis, the outlook coefficient, is calculated as [9,50]:

$$C_{outl} = ((Nd + Eu + Tb + Dy + Er + Y/\Sigma REY)/((Ce + Ho + Tm + Yb + Lu)/\Sigma REY)$$

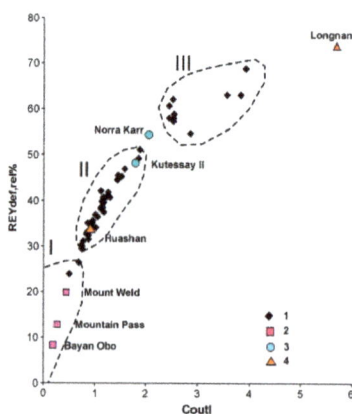

**Figure 2.** Classification of REY-rich coal ashes by outlook for individual REY composition in comparison with selected deposits of conventional types. 1, REE-rich coal ashes; 2, carbonatite deposits; 3, hydrothermal deposits; and 4, weathered crust elution-deposited (ion-adsorbed) deposits. Clusters of REE-rich coal ashes distinguished by outlook for REY composition (numerals in figure): I, unpromising; II, promising; and III, highly promising. From Seredin and Dai [9].

The $y$ axis ($REY_{def, rel\%}$) is the percentage of critical REY (Figure 3) in the total REY. Cluster I, which includes some of the mined REE ores, is not as promising as Cluster II. Seredin and Dai [9] noted that mining of Cluster I *"will neither mitigate the crisis in REY resources nor eliminate the shortage of the most critical REY, but will only result in overproduction of excessive Ce (p. 75)."* Cluster II, with a variety of L-, M-, and H-type distributions, contains many of the known coal ashes, including the Dean (Fire Clay) ash. Given concentrations exceeding the economic threshold, a variable, coal ashes in the Cluster II concentration and $C_{outl}$ range would be promising resources. Cluster III contains H-type REY's with hydrothermal origins. Seredin-Dai's classification and evaluation criteria of REY in coal deposits have been adopted and used by a number of researchers ([51–58], among others).

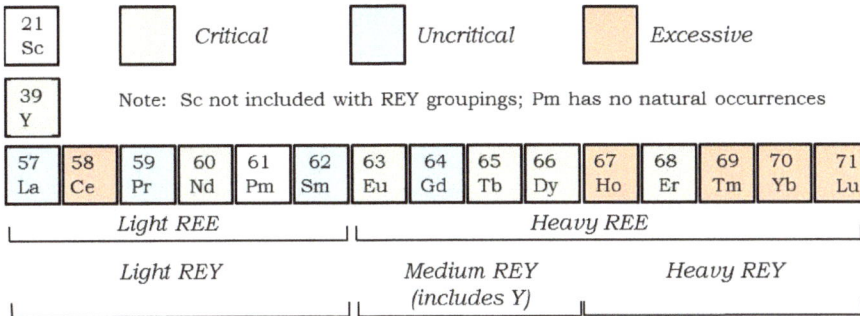

**Figure 3.** Divisions of lanthanides and yttrium into light, medium, and heavy REY; light and heavy REE; and critical, uncritical, and excessive groups (after Seredin [50] and Seredin and Dai [9]).

Coal scientists have identified coals that were successfully utilized as raw materials for rare-metal recovery during periods of raw material crises [41]. The first time the coal deposits were used as the major source of uranium was for the incipient nuclear industries in the former USSR and the United States following World War II. The second time was that the coal deposits are one of the major Ge sources for world industry. The third time, Al and Ga extraction was expanded from Jungar coal ashes of Northern China [41]. It is now time to address the coal-hosted rare earth elements and yttrium from coal deposits as a byproduct not only because of the REY supply crisis in recent years, but also because the distinct benefits of REY extraction from coal ash, such as the relatively low cost for the necessary infrastructure as compared to developing new mining projects, as well as the benefits associated with recycling of a waste product.

## 4. Conclusions

Much of the recent research on coal utilization in the United States has focused upon the capture of pollutants such as acid gases, particulates, and mercury, and the greenhouse gas carbon dioxide. The possible recovery of rare earth elements from abundant coal and byproducts is an exciting new research area. Additional data is needed on the rare earth contents of coals and byproducts in order to determine the most promising potential feed materials for extraction processes. Future work will likely focus on the characterization of coals and byproducts, as well as on separation methods for rare earth recovery.

**Acknowledgments:** We thank our respective employers for granting the time for us to contribute to this work.

**Author Contributions:** All authors contributed to the writing of the manuscript.

**Conflicts of Interest:** This paper/information was prepared as an account of work sponsored by an agency of the United States Government. Neither the United States Government nor any agency thereof, nor any of their employees, makes any warranty, express or implied, or assumes any legal liability or responsibility for the accuracy, completeness, or usefulness of any information, apparatus, product, or process disclosed, or represents that its use would not infringe privately owned rights. Reference herein to any specific commercial product, process, or service by trade name, trademark, manufacturer, or otherwise does not necessarily constitute or imply its endorsement, recommendation, or favoring by the United States Government or any agency thereof. The views and opinions of authors expressed herein do not necessarily state or reflect those of the United States Government or any agency thereof.

## References

1. US Department of Energy Coal Production Statistics. Available online: http://www.eia.gov/coal/production/quarterly/ (accessed on 29 January 2016).
2. US Department of Energy Rare Earth EDX Database. Available online: https://edx.netl.doe.gov/ree/ (accessed on 29 January 2016).
3. Gschneidner, K.A., Jr. *Rare Earths: The Fraternal Fifteen*; United States Atomic Energy Commission. Division of Technical Information: Washington, DC, USA, 1964; p. 46.
4. Dent, P.C. Rare Earth Future. Available online: http://www.magneticsmagazine.com/main/channels/materials-channels/rare-earth-future/ (accessed on 25 March 2016).
5. Pecht, M.G.; Kaczmarek, R.E.; Song, X.; Hazelwood, D.A.; Kavetsky, R.A.; Anand, D.K. *Rare Earth Materials: Insights and Concerns*; CALCE EPSC Press: College Park, MD, USA, 2012; p. 194.
6. U.S. Department of Energy. *Critical Materials Strategy*; U.S. Department of Energy: Washington, DC, USA, 2011; p. 197. Available online: http://energy.gov/sites/prod/files/DOE_CMS2011_FINAL_Full.pdf (accessed on 25 March 2016).
7. European Commission. Report on Critical Raw Materials for the EU. In *Report of the Ad Hoc Working Group on Defining Critical Raw Materials*; European Commission: Bruxelles, Belgium, 2014; p. 41. Available online: http://ec.europa.eu/DocsRoom/documents/10010/attachments/1/translations/en/renditions/native (accessed on 25 March 2016).
8. Mayfield, D.B.; Lewis, A.S. Environmental review of coal ash as a resource for rare earth and strategic elements. In Proceedings of the 2013 World of Coal Ash (WOCA) Conference, Lexington, KY, USA, 22–25 April 2013; pp. 22–25. Available online: http://www.flyash.info/2013/051-mayfield-2013.pdf (accessed on 25 March 2016).
9. Seredin, V.V.; Dai, S. Coal deposits as a potential alternative source for lanthanides and yttrium. *Int. J. Coal Geol.* **2012**, *94*, 67–93. [CrossRef]
10. Eskenazy, G. Rare-earth elements in some coal basins of Bulgaria. *Geol. Balc.* **1978**, *8*, 81–88.
11. Eskenazy, G.M. Rare earth elements and yttrium in lithotypes of Bulgarian coals. *Org. Geochem.* **1987**, *11*, 83–89. [CrossRef]
12. Eskenazy, G.M. Zirconium and hafnium in Bulgarian coals. *Fuel* **1987**, *66*, 1652–1657. [CrossRef]
13. Eskenazy, G.M. Rare earth elements in a sampled coal from the Pirin Deposit, Bulgaria. *Int. J. Coal Geol.* **1987**, *7*, 301–314. [CrossRef]
14. Eskenazy, G. Geochemistry of rare earth elements in Bulgarian coals. *Annuaire de l'Universite´ de Sofia "St. Kliment Ohridski", Faculte´ de Geologie et Geographie Livre 1-Geologie* **1995**, *88*, 39–65.
15. Eskenazy, G.M. Aspects of the geochemistry of rare earth elements in coal: An experimental approach. *Int. J. Coal Geol.* **1999**, *38*, 285–295. [CrossRef]
16. Eskenazy, G.M. Sorption of trace elements on xylain: An experimental study. *Int. J. Coal Geol.* **2015**, *150–151*, 166–169.
17. Eskenazy, G.M.; Mincheva, E.I.; Rousseva, D.P. Trace elements in lignite lithotypes from the Elhovo coal basin. *Comptes rendus del'Académie bulgare des Sciences* **1986**, *39*, 99–101.
18. Eskenazy, G.M.; Brinkin, K. Trace elements in the Karlovo coal deposit. *Comptes rendus del'Académie bulgare des Sciences* **1991**, *44*, 67–70.
19. Hower, J.C.; Ruppert, L.F.; Eble, C.F. Lanthanide, Yttrium, and Zirconium anomalies in the Fire Clay coal bed, Eastern Kentucky. *Int. J. Coal Geol.* **1999**, *39*, 141–153. [CrossRef]
20. Mardon, S.M.; Hower, J.C. Impact of coal properties on coal combustion by-product quality: Examples from a Kentucky power plant. *Int. J. Coal Geol.* **2004**, *59*, 153–169. [CrossRef]
21. Crowley, S.S.; Stanton, R.W.; Ryer, T.A. The effects of volcanic ash on the maceral and chemical composition of the C coal bed, Emery Coal Field, Utah. *Org. Geochem.* **1989**, *14*, 315–331. [CrossRef]

22. Crowley, S.S.; Ruppert, L.F.; Belkin, H.F.; Stanton, R.W.; Moore, T.A. Factors affecting the geochemistry of a thick, subbituminous coal bed in the Powder River Basin: Volcanic, detrital, and peat-forming processes. *Org. Geochem.* **1993**, *20*, 843–854. [CrossRef]

23. Seredin, V.V. Rare earth element-bearing coals from the Russian Far East deposits. *Int. J. Coal Geol.* **1996**, *30*, 101–129. [CrossRef]

24. Dai, S.; Ren, D.; Li, S. Discovery of the superlarge gallium ore deposit in Junger, Inner Mongolia, North China. *Chin. Sci. Bull.* **2006**, *51*, 2243–2252. [CrossRef]

25. Dai, S.; Ren, D.; Chou, C.L.; Li, S.; Jiang, Y. Mineralogy and geochemistry of the No. 6 coal (Pennsylvanian) in the Jungar Coalfield, Ordos Basin, China. *Int. J. Coal Geol.* **2006**, *66*, 253–270. [CrossRef]

26. Dai, S.; Li, D.; Chou, C.L.; Zhao, L.; Zhang, Y.; Ren, D.; Ma, Y.; Sun, Y. Mineralogy and geochemistry of boehmite-rich coals: New insights from the Haerwusu Surface Mine, Jungar Coalfield, Inner Mongolia, China. *Int. J. Coal Geol.* **2008**, *74*, 185–202. [CrossRef]

27. Dai, S.; Wang, X.; Chen, W.; Li, D.; Chou, C.L.; Zhou, Y.; Zhu, C.; Li, H.; Zhu, X.; Xing, Y.; et al. A high-pyrite semianthracite of Late Permian age in the Songzao Coalfield, southwestern China: Mineralogical and geochemical relations with underlying mafic tuffs. *Int. J. Coal Geol.* **2010**, *83*, 430–445. [CrossRef]

28. Dai, S.; Zou, J.; Jiang, Y.; Ward, C.R.; Wang, X.; Li, T.; Xue, W.; Liu, S.; Tian, H.; Sun, X.; et al. Mineralogical and geochemical compositions of the Pennsylvanian coal in the Adaohai Mine, Daqingshan Coalfield, Inner Mongolia, China: Modes of occurrence and origin of diaspore, gorceixite, and ammonian illite. *Int. J. Coal Geol.* **2012**, *94*, 250–270. [CrossRef]

29. Dai, S.; Jiang, Y.; Ward, C.R.; Gu, L.; Seredin, V.V.; Liu, H.; Zhou, D.; Wang, X.; Sun, Y.; Zou, J.; et al. Mineralogical and geochemical compositions of the coal in the Guanbanwusu Mine, Inner Mongolia, China: Further evidence for the existence of an Al (Ga and REE) ore deposit in the Jungar Coalfield. *Int. J. Coal Geol.* **2012**, *98*, 10–40. [CrossRef]

30. Birk, D.; White, J.C. Rare earth elements in bituminous coals and underclays of the Sydney Basin, Nova Scotia: Element sites, distribution, mineralogy. *Int. J. Coal Geol.* **1991**, *19*, 219–251. [CrossRef]

31. Kortenski, J.; Bakardjiev, S. Rare earth and radioactive elements in some coals from the Sofia, Svoge and Pernik Basins, Bulgaria. *Int. J. Coal Geol.* **1993**, *22*, 237–246. [CrossRef]

32. Pollock, S.M.; Goodarzi, F.; Riediger, C.L. Mineralogical and elemental variation of coal from Alberta, Canada: An example from the No.2 seam, Genesee Mine. *Int. J. Coal Geol.* **2000**, *43*, 259–286. [CrossRef]

33. Schatzel, S.J.; Stewart, B.W. Rare earth element sources and modification in the Lower Kittanning coal bed, Pennsylvania: Implications for the origin of coal mineral matter and rare earth element exposure in underground mines. *Int. J. Coal Geol.* **2003**, *54*, 223–251. [CrossRef]

34. Aide, M.T.; Aide, C. Rare earth elements: Their importance in understanding soil genesis. *ISRN Soil Sci.* **2012**, *2012*. [CrossRef]

35. Pédrot, M.; Dia, A.; Davranche, M. Dynamic stricture of humic substances: Rare earth elements as a fingerprint. *J. Colloid Interface Sci.* **2010**, *345*, 206–213. [CrossRef] [PubMed]

36. Davranche, M.; Grybos, M.; Gruau, G.; Pédrot, M.; Dia, A.; Marsac, R. Rare earth element patterns: A tool for identifying trace metal sources during wetland soil reduction. *Chem. Geol.* **2011**, *284*, 127–137. [CrossRef]

37. Finkelman, R.B. The origin, occurrence, and distribution of the inorganic constituents in low-rank coals. In Proceedings of the Basic Coal Science Workshop, Houston, TX, USA, 8–9 December 1981; Schobert, H.H., Ed.; Grand Forks Energy Technology Center: Grand Forks, ND, USA, 1981; pp. 70–90.

38. Seredin, V.; Shpirt, M.; Vassyanovich, A. REE contents and distribution in humic matter of REE-rich coals. In *Mineral Deposits: Processes to Processing*; Stanley, C.J., Ed.; A.A. Balkema: Rotterdam, The Netherlands, 1999; pp. 267–269.

39. Dai, S.; Zhou, Y.; Zhang, M.; Wang, X.; Wang, J.; Song, X.; Jiang, Y.; Luo, Y.; Song, Z.; Yang, Z.; et al. A new type of Nb (Ta)–Zr(Hf)–REE–Ga polymetallic deposit in the late Permian coal-bearing strata, eastern Yunnan, southwestern China: Possible economic significance and genetic implications. *Int. J. Coal Geol.* **2010**, *83*, 55–63. [CrossRef]

40. Dai, S.; Zhao, L.; Peng, S.; Chou, C.L.; Wang, X.; Zhang, Y.; Li, D.; Sun, Y. Abundances and distribution of minerals and elements in high-alumina coal fly ash from the Jungar Power Plant, Inner Mongolia, China. *Int. J. Coal Geol.* **2010**, *81*, 320–332. [CrossRef]

41. Seredin, V.V. From coal science to metal production and environmental protection: A new story of success. *Int. J. Coal Geol.* **2012**, *90–91*, 1–3. [CrossRef]

42. Arbuzov, S.I.; Ershov, V.V.; Potseluev, A.A.; Rikhvanov, L.P. *Rare Elements in Coals of the Kuznetsk Basin*; Publisher House KPK: Kemerovo, Russia, 2000; p. 248. (In Russian)

43. Arbuzov, S.I.; Mezhibor, A.M.; Spears, D.A.; Ilenok, S.S.; Shaldybin, M.V.; Belaya, E.V. Nature of tonsteins in the Azeisk deposit of the Irkutsk Coal Basin (Siberia, Russia). *Int. J. Coal Geol.* **2016**, *153*, 99–111. [CrossRef]

44. Seredin, V.V. The first data on abnormal Niobium content in Russian coals. *Dokl. Earth Sci.* **1994**, *335*, 634–636.

45. Seredin, V.V.; Finkelman, R.B. Metalliferous coals: A review of the main genetic and geochemical types. *Int. J. Coal Geol.* **2008**, *76*, 253–289. [CrossRef]

46. Spears, D.A. The origin of tonsteins, an overview, and links with seatearths, fireclays and fragmental clay rocks. *Int. J. Coal Geol.* **2012**, *94*, 22–31. [CrossRef]

47. Zhou, Y.; Bohor, B.F.; Ren, Y. Trace element geochemistry of altered volcanic ash layers (tonsteins) in late Permian coal-bearing formations of eastern Yunnan and western Guizhou Provinces, China. *Int. J. Coal Geol.* **2000**, *44*, 305–324. [CrossRef]

48. Arbuzov, S.I.; Mashen'kin, V.S. Oxidation zone of coalfields as a promising source of noble and rare metals: A case of coalfields in Central Asia. In *Problems and Outlook of Development of Mineral Resources and Fuel-Energetic Enterprises of Siberia*; Tomsk Polytechnic University: Tomsk, Russia, 2007; pp. 26–31. (In Russian)

49. Seredin, V.V. Metalliferous coals: Formation conditions and outlooks for development. In *Coal Resources of Russia*; Geoinformmark: Moscow, Russia, 2004; Volume 6, pp. 452–519. (In Russian)

50. Seredin, V.V. A new method for primary evaluation of the outlook for rare earth element ores. *Geol. Ore Depos.* **2010**, *52*, 428–433. [CrossRef]

51. Blissett, R.S.; Smalley, N.; Rowson, N.A. An investigation into six coal fly ashes from the United Kingdom and Poland to evaluate rare earth element content. *Fuel* **2014**, *119*, 236–239. [CrossRef]

52. Saikia, B.K.; Ward, C.R.; Oliveira, M.L.S.; Hower, J.C.; Leao, F.D.; Johnston, M.N.; O'Bryan, A.; Sharma, A.; Baruah, B.P.; Silva, L.F.O. Geochemistry and nano-mineralogy of feed coals, mine overburden, and coal-derived fly ashes from Assam (North-east India): A multi-faceted analytical approach. *Int. J. Coal Geol.* **2015**, *137*, 19–37. [CrossRef]

53. Chen, J.; Chen, P.; Yao, D.; Liu, Z.; Wu, Y.; Liu, W.; Hu, Y. Mineralogy and geochemistry of Late Permian coals from the Donglin Coal Mine in the Nantong coalfield in Chongqing, southwestern China. *Int. J. Coal Geol.* **2015**, *149*, 24–40. [CrossRef]

54. Franus, W.; Wiatros-Motyka, M.M.; Wdowin, M. Coal fly ash as a resource for rare earth elements. *Environ. Sci. Pollut. Res.* **2015**, *22*, 9464–9474. [CrossRef] [PubMed]

55. Zhao, L.; Ward, C.R.; French, D.; Graham, I.T. Major and Trace Element Geochemistry of Coals and Intra-Seam Claystones from the Songzao Coalfield, SW China. *Minerals* **2015**, *5*, 870–893. [CrossRef]

56. Hower, J.C.; Eble, C.F.; O'Keefe, J.M.K.; Dai, S.; Wang, P.; Xie, P.; Liu, J.; Ward, C.R.; French, D. Petrology, Palynology, and Geochemistry of Gray Hawk Coal (Early Pennsylvanian, Langsettian) in Eastern Kentucky, USA. *Minerals* **2015**, *5*, 894–918. [CrossRef]

57. Zhang, W.; Rezaee, M.; Bhagavatula, A.; Li, Y.; Groppo, J.; Honaker, R. A review of the occurrence and promising recovery methods of rare earth elements from coal and coal by-products. *Int. J. Coal Prep. Util.* **2015**, *35*, 295–330. [CrossRef]

58. Li, B.; Zhuang, X.; Li, J.; Querol, X.; Font, O.; Moreno, N. Geological controls on mineralogy and geochemistry of the Late Permian coals in the Liulong Mine of the Liuzhi Coalfield, Guizhou Province, Southwest China. *Int. J. Coal Geol.* **2016**, *154–155*, 1–15. [CrossRef]

*Article*

# Clay Mineralogy of Coal-Hosted Nb-Zr-REE-Ga Mineralized Beds from Late Permian Strata, Eastern Yunnan, SW China: Implications for Paleotemperature and Origin of the Micro-Quartz

**Lixin Zhao** [1,†], **Shifeng Dai** [1,2,*], **Ian T. Graham** [3,†] **and Peipei Wang** [1,†]

1    College of Geoscience and Surveying Engineering, China University of Mining and Technology (Beijing), Beijing 100083, China; TBP120201007@student.cumtb.edu.cn (L.Z.); wangpeipei1100@gmail.com (P.W.)
2    State Key Laboratory of Coal Resources and Safe Mining, China University of Mining and Technology (Beijing), Beijing 100083, China
3    School of Biological, Earth and Environmental Sciences, University of New South Wales, Sydney 2052, Australia; i.graham@unsw.edu.au
*    Correspondence: daishifeng@gmail.com; Tel.: +86-10-6234-1868
†    These authors contributed equally to this work.

Academic Editor: Dimitrina Dimitrova
Received: 24 January 2016; Accepted: 11 May 2016; Published: 17 May 2016

**Abstract:** The clay mineralogy of pyroclastic Nb(Ta)-Zr(Hf)-REE-Ga mineralization in Late Permian coal-bearing strata from eastern Yunnan Province; southwest China was investigated in this study. Samples from XW and LK drill holes in this area were analyzed using XRD (X-ray diffraction) and SEM (scanning electronic microscope). Results show that clay minerals in the Nb-Zr-REE-Ga mineralized samples are composed of mixed layer illite/smectite (I/S); kaolinite and berthierine. I/S is the major component among the clay assemblages. The source volcanic ashes controlled the modes of occurrence of the clay minerals. Volcanic ash-originated kaolinite and berthierine occur as vermicular and angular particles, respectively. I/S is confined to the matrix and is derived from illitization of smectite which was derived from the original volcanic ashes. Other types of clay minerals including I/S and berthierine precipitated from hydrothermal solutions were found within plant cells; and coexisting with angular berthierine and vermicular kaolinite. Inferred from the fact that most of the I/S is R1 ordered with one case of the R3 I/S; the paleo-diagenetic temperature could be up to 180 °C but mostly 100–160 °C. The micro-crystalline quartz grains (<10 μm) closely associated with I/S were observed under SEM and were most likely the product of desiliconization during illitization of smectite.

**Keywords:** coal-hosted Nb-Zr-REE-Ga mineralization; clay minerals; paleotemperature; microcrystalline quartz

## 1. Introduction

Polymetallic Nb(Ta)-Zr(Hf)-REE-Ga mineralization within Late Permian coal-bearing strata of eastern Yunnan Province, southwest China was reported by Dai, *et al.* [1]. The Nb(Ta)-Zr(Hf)-REE-Ga mineralization has anomalous response on natural gamma log curves and is widespread at the base of the Wuchiapingian of Late Permian age (*i.e.*, terrestrial Xuanwei and terrestrial-marine transitional Longtan Formations) in southwest China [1–3]. The mineralization is believed to be derived from alkali volcanic ashes and occurs as thick beds (up to 10 m but mostly 2–5 m) interbedded in the sedimentary rocks of terrigenous origin [1]. In most cases, the Nb-Zr-REE-Ga mineralized horizon has been argillized, but tuffaceous textures, volcanic breccia, and hematitization can be observed as well [1,2]. The Nb-Zr-REE-Ga mineralization occurs within coal-bearing strata and within coal beds directly [3], and accordingly, this Nb(Ta)-Zr(Hf)-REE-Ga polymetallic mineralization was generalized as a coal-hosted rare metal deposit [4].

This Nb-Zr-REE-Ga-enriched mineralization is characterized by significant enrichment in Nb, Ta, Zr, Hf, REE (rare earth elements and Y) and Ga, for example, $(Nb,Ta)_2O_5$, 0.0302–0.0627 wt %; $(Zr,Hf)O_2$, 0.3805–0.8468 wt %; REO (oxides of REE), 0.1216–0.1358 wt %; and Ga, 52.4–81.3 ppm [1,3]. Notably, the content of $(Nb,Ta)_2O_5$ is much higher than the required industrial $(Nb,Ta)_2O_5$ grade of weathered crust niobium deposits (0.016–0.02 wt %) [5] while the concentration of Ga is also higher than the required Ga industrial grade in coal (30 ppm) and in bauxite (20 ppm) [6]. In most cases, the contents of REO and $(Zr,Hf)_2O_5$ have also been up to their corresponding industrial utilization grades [5,7].

Although the mineralization of Nb(Ta)-Zr(Hf)-REE-Ga in the study area is notably significant, their hosted minerals such as zircon, pyrochlore, columbite *etc.* are rarely observed within the Nb-Zr-REE-Ga mineralized beds under both the microscope and X-ray diffraction (XRD) [1]. While using scanning electron microscopy (SEM), Dai *et al.* [2] identified several rare metal-bearing minerals including REE-bearing minerals (rhabdophane, silico-rhabdophane, florencite, parasite and xenotime), zircon, and Nb-bearing anatase within the Nb-Zr-REE-Ga mineralization. These rare metal-bearing minerals mainly occur within pores and cavities of clay minerals as very fine dispersed grains (in most cases <5 μm) indicating that they are probably of authigenic origin derived from re-deposition of rare metals leached from the Nb-Zr-REE-Ga-enriched tuff by hydrothermal solutions [2]. However, these rare-metal bearing minerals are rare to be observed under SEM and not in sufficient concentration to explain the high contents of rare metals found in the geochemical analyses, for example, zircon as the only Zr-bearing mineral phase identified by Dai *et al.* [2] was rarely observed in the samples which contain up to thousands of ppm zirconium. On the other hand, the amount of zircon within the Nb(Ta)-Zr(Hf)-REE-Ga-mineralized samples (with Zr in thousands ppm level and Nb in hundreds ppm level) in the lower Xuanwei Formation (the mineralization-bearing strata) is 10–100 times that of the pyroclastic tonsteins which only contain hundreds of ppm Zr and tens of ppm Nb from the upper part of the Xuanwei Formation [8]. Therefore, the majority of rare metals do not occur within discrete mineral phases such as zircon, Nb-anatase, and REE-phosphate/carbonate, and must therefore be inferred to occur as absorbed ions within the clay minerals [1,8,9]. In fact, our unpublished results of leaching experiments showed that the ammonium sulfate solutions could extract a large amount of rare metals from the Nb-Zr-REE-Ga-mineralized samples (though not the specifically studied samples) providing indirect evidence for a certain amount of rare metals being absorbed within clay minerals. Therefore, a detailed study of the clay minerals is required to understand the modes of occurrence and industrial utilization of these rare metal elements. In this paper, we report on the clay mineralogy (species, abundances, modes of occurrence, and ordering of mixed layer illite/smectite) in the Nb-Zr-REE-Ga mineralized beds by an investigation of samples collected from two drill holes (XW and LK) in eastern Yunnan Province, SW China. This paper also provides an insight into the paleo-diagenetic temperature and origin of the ultrafine quartz particles (<10 μm) found in the studied samples.

## 2. Geological Setting

The ~260 Ma Emeishan Large Igneous Province (ELIP) in SW China is considered to be the result of mantle plume activity and mainly comprises massive flood basalts and contemporary felsic, mafic and ultramafic intrusions [10–15]. Basalts as the predominant component of ELIP could be divided into two groups: high-Ti (Ti/Y > 500) and low-Ti (Ti/Y < 500) basalts [13]. Three ELIP zones including inner, intermediate, and outer zones were recognized (Figure 1A) based on the extent of erosion of the pre-ELIP eruption Maokou Formation (with limestone-dominated compositions) of Middle Permian age in SW China [11].

Within the inner zone of the ELIP is the Kangdian Oldland comprising a sequence of Emeishan basalts, which existed until the Middle Triassic [11]. In the eastern ELIP, the Emeishan basalts unconformably overlie the Maokou Formation of Middle Permian age while the ELIP basalts are, in turn, overlain by the Late Permian Xuanwei and Longtan formations (Figure 1B) [11]. During the Late Permian, the ELIP and associated volcanism and hydrothermal activity controlled the development of coal-bearing strata in SW China, not only in serving the source for the peat-accumulation process, but also the distribution of peat-mire sites (*i.e.*, peat-mires located in the middle and outer zones of ELIP) [2].

The Nb(Ta)-Zr(Hf)-REE-Ga polymetallic mineralization is found in the intermediate ELIP zone and is located in the lowest Xuanwei Formation in eastern Yunnan Province (Figures 1 and 2) [1]. The Xuanwei Formation is a continental formation containing the major coal-bearing strata of Late Permian age in eastern Yunnan Province, southwest China [1,16–19] and is mainly derived from erosion of the Kangdian Oldland in the central ELIP [2,15,20–22].

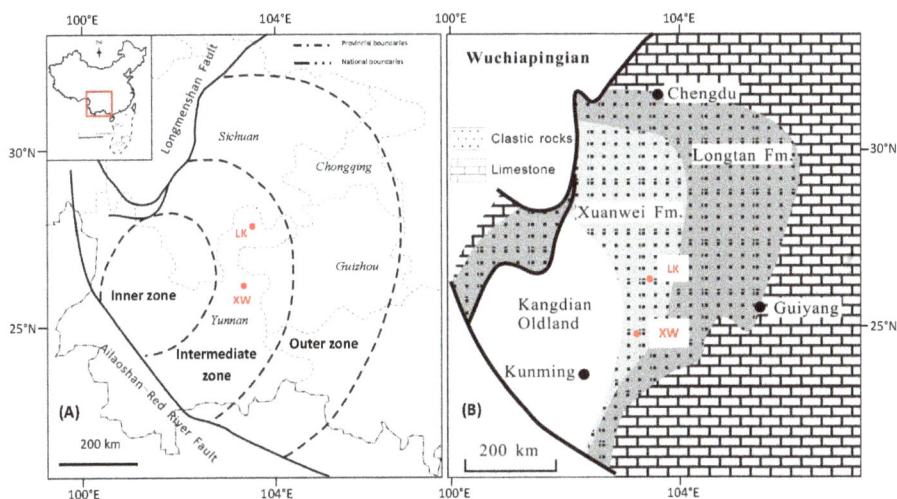

**Figure 1.** Geological setting during Late Permian in southwest China, showing the location of Nb-Zr-REE-Ga mineralization. (**A**) Schematic map showing the inner, intermediate and outer zones of the Emeishan Large Igneous Province. (**B**) Paleogeography map showing the distribution of terrestrial Xuanwei Formation and transitional Longtan Formation during the Wuchiapingian in SW China. The red spots indicate the localities of the two studied drill holes (XW and LK). Figure 1 is modified from [20].

## 3. Samples and Analytical Procedures

A total of 39 samples, corresponding to the high anomalies on natural gamma log curves, were collected from Nos. XW and LK drill holes in eastern Yunnan Province, SW China. These samples were identified as X-1 to X-17, and L-1 to L-22 from top to bottom, respectively (Figure 2). The samples are mainly mudstone, and in a few cases sandstones. Calcite veins, pyrite grains, hematite, and plant fragments are commonly observed in hand-specimens.

Polished thin-sections and polished block samples were prepared from selected samples for optical and scanning electron microscopic observations. All samples were then crushed and milled to pass 200-mesh for bulk X-ray diffraction (XRD) analysis (Rigaku, Tokyo, Japan). Bulk-XRD analysis was performed using a Rigaku D/max 2500 pc powder diffractometer equipped with Ni-filtered Cu-K$\alpha$ radiation and scintillation detector in China University of Mining and Technology (Beijing). Each XRD pattern was recorded over a 2$\theta$ interval of 2.6°–70°, with a step size of 0.02°. Quantitative analysis for each mineral phase was carried out by a commercial XRD interpretation software Siroquant™ (Sietronics Pty Ltd., Belconnen, ACT, Australia).

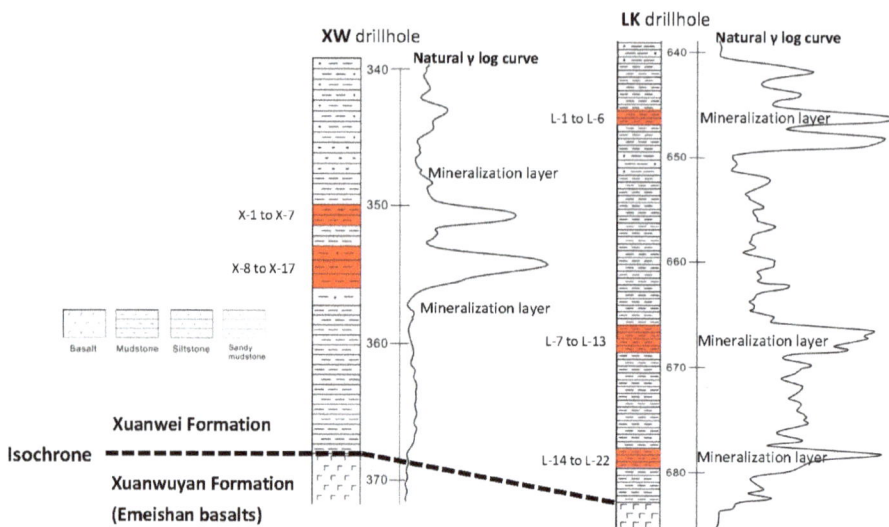

**Figure 2.** Stratigraphic sections of the XW and LK drill holes. The red areas indicate the sampling locations.

Modes of occurrence of minerals were investigated using a FEI Quanta™ 650 FEG scanning electron microscope (SEM, EDAX Inc., Mahwah, NJ, USA) in China University of Mining and Technology (Beijing), China and a Hitachi S3400-X/I SEM (Hitachi, Tokyo, Japan) at the University of New South Wales, Australia. The selected polished thin-sections and sample blocks were carbon-coated before SEM observation. The Quanta SEM worked with a beam voltage of 20.0 kV, working distance ~10 mm, and a spot-size of 5.5 while, for the Hitachi-S3400 X/I, the accelerating voltage was 20 kV, and the beam current was 40–60 mA during SEM operation.

Sample powders were mixed with water and then settled for approximately 2 h for acquiring clay-bearing suspensions. Suspended clay particles were concentrated through centrifuging (8 min at 2500 rpm). The concentrated clay fractions were placed evenly on glass slides. Three XRD runs were performed on each slide after air-drying, exposure to ethylene-glycol vapor (more than 24 h) and heating to 400 °C for 1 h using a PANalytical Empyrean II XRD (Co-Kα; PANalytical Ltd., Almelo, The Netherlands) at the University of New South Wales, Australia, with tube voltage of 45 kV and current of 40 mA.

## 4. Results

### 4.1. Mineral Phases and Clay Species in the Studied Sample

Mineralogical compositions based on powder XRD and Siroquant analyses in the studied samples are given in Table 1. Mineral phases in this study include clay minerals, quartz, anatase, calcite, siderite, hematite, albite, and florencite (Table 1; Figure 3). Pyrite occurring either as discrete grains or as veinlets was commonly observed but its content was generally below the detection limit of XRD or Siroquant techniques.

**Table 1.** Mineral contents (%) in the studied samples, and the estimated percentages of smectite layers *i.e.*, S (%) and Reichweite values for I/S. "-" means below the detection limit of Siroquant analysis.

| Samples | I/S | Kaolinite | Berthierine | Quartz | Anatase | Calcite | Florencite | Siderite | Albite | Hematite | Total Clay | S (%) | Reichweite Value |
|---|---|---|---|---|---|---|---|---|---|---|---|---|---|
| X-1 | 15.2 | 42.6 | 17.5 | 22.8 | 1.9 | - | - | - | - | - | 75.3 | 35 | R1 |
| X-2 | 23.5 | 57.7 | 12.3 | 5.2 | 1.4 | - | - | - | - | - | 93.5 | 35 | R1 |
| X-3 | 27.0 | 56.5 | 4.7 | 10.3 | 1.6 | - | - | - | - | - | 88.2 | 35 | R1 |
| X-4 | 51.9 | 28.4 | 4.3 | 12.8 | 1.6 | 0.9 | - | - | - | - | 84.6 | 30 | R1 |
| X-5 | 35.0 | 35.4 | 18.4 | 8.8 | 2.0 | 0.5 | - | - | - | - | 88.8 | 30 | R1 |
| X-6 | 22.8 | 39.7 | 22.9 | 13.6 | 1.0 | - | - | - | - | - | 85.4 | 30 | R1 |
| X-7 | 54.2 | 12.9 | - | 28.0 | 1.3 | 3.7 | - | - | - | - | 67.1 | 25 | R1 |
| X-8 | 45.5 | 29.6 | 11.6 | 11.8 | 0.6 | 0.9 | - | - | - | - | 86.7 | 30 | R1 |
| X-9 | 54.5 | 24.2 | 5.1 | 15.7 | 0.5 | - | - | - | - | - | 83.8 | 30 | R1 |
| X-10 | 43.9 | 28.2 | 10.1 | 16.9 | 0.9 | - | - | - | - | - | 82.2 | 30 | R1 |
| X-11 | 78.4 | 3.9 | 2.1 | 15.5 | 0.2 | - | - | - | - | - | 84.4 | 25 | R1 |
| X-12 | 75.9 | 1.2 | 1.5 | 19.8 | - | 1.5 | - | - | - | - | 78.6 | 25 | R1 |
| X-13 | 69.6 | 2.6 | 0.7 | 26.2 | 0.8 | - | - | - | - | - | 72.9 | 25 | R1 |
| X-14 | 75.7 | 1.2 | 3.9 | 19.2 | - | - | - | - | - | - | 80.8 | 25 | R1 |
| X-15 | 77.6 | 4.4 | 2.0 | 16.0 | - | - | - | - | - | - | 84.0 | 25 | R1 |
| X-16 | 57.8 | 9.0 | 6.0 | 14.1 | 0.2 | 12.3 | - | 0.7 | - | - | 72.7 | 25 | R1 |
| X-17 | 45.1 | 19.2 | 1.4 | 32.3 | 1.1 | 0.8 | - | - | - | - | 65.7 | 20 | R1 |
| L-1 | 31.1 | 13.4 | 19.1 | 33.5 | 2.3 | - | 0.5 | - | - | - | 63.6 | 20 | R1 |
| L-2 | 52.9 | 0.9 | 16.6 | 25.3 | 1.5 | 2.3 | 0.5 | - | - | - | 70.4 | 20 | R1 |
| L-3 | 56.1 | 1.5 | 6.1 | 28.0 | 0.9 | 7.0 | 0.3 | - | - | - | 63.7 | 20 | R1 |
| L-4 | 66.3 | 5.3 | 1.4 | 15.2 | 1.1 | 10.0 | 0.6 | - | - | - | 73.0 | 20 | R1 |
| L-5 | 20.6 | 8.1 | 10.5 | 52.2 | - | 7.9 | 0.7 | - | - | - | 39.2 | 20 | R1 |
| L-6 | 74.0 | 10.0 | 9.9 | 0.8 | 5.3 | - | - | - | - | - | 93.9 | 20 | R1 |
| L-7 | 28.0 | 4.1 | 31.9 | 23.6 | 1.4 | - | - | 11.1 | - | - | 64.0 | 20 | R1 |
| L-8 | 65.9 | 4.9 | 15.5 | 12.2 | 1.5 | - | - | - | - | - | 86.3 | 20 | R1 |
| L-9 | 58.7 | 5.3 | 4.3 | 30.8 | 0.9 | - | - | - | - | - | 68.3 | 20 | R1 |
| L-10 | 66.6 | 7.7 | 4.9 | 19.1 | 1.8 | - | - | - | - | - | 79.2 | 20 | R1 |
| L-11 | 33.3 | - | 51.9 | 0.4 | 4.5 | - | 0.1 | 6.4 | - | 3.3 | 85.2 | 20 | R1 |
| L-12 | 54.1 | - | 32.3 | 0.8 | 6.0 | - | - | 3.7 | - | 3.1 | 86.4 | 20 | R1 |
| L-13 | 69.0 | 11.0 | 6.4 | 7.4 | 2.3 | - | 0.5 | - | - | 3.5 | 86.4 | 25 | R1 |
| L-14 | 53.6 | 17.3 | 16.8 | 4.1 | 4.7 | - | 0.7 | - | - | 2.7 | 87.7 | 20 | R1 |
| L-15 | 71.7 | 5.9 | 8.8 | 9.1 | 1.8 | - | - | - | - | 2.7 | 86.4 | 20 | R1 |
| L-16 | 69.5 | 8.7 | 3.3 | 13.8 | 0.8 | - | 0.7 | - | - | 3.1 | 81.5 | 20 | R1 |
| L-17 | 65.5 | 9.8 | 0.4 | 23.3 | 1.0 | - | - | - | - | - | 75.7 | 20 | R1 |
| L-18 | 53.2 | 8.0 | 15.8 | 20.3 | 2.7 | - | - | - | - | - | 77.0 | 20 | R1 |
| L-19 | 41.9 | 10.4 | 9.2 | 30.6 | 0.1 | - | 0.5 | - | - | - | 61.5 | 20 | R1 |
| L-20 | 64.9 | 3.2 | 13.1 | 15.1 | 3.1 | - | 0.7 | - | 7.3 | - | 81.2 | 20 | R1 |
| L-21 | 44.1 | 8.6 | 17.9 | 28.1 | 0.8 | - | 0.5 | - | - | - | 70.6 | 20 | R1 |
| L-22 | 59.0 | 8.3 | 14.2 | 17.4 | 1.1 | - | - | - | - | - | 81.5 | 15 | R3 |

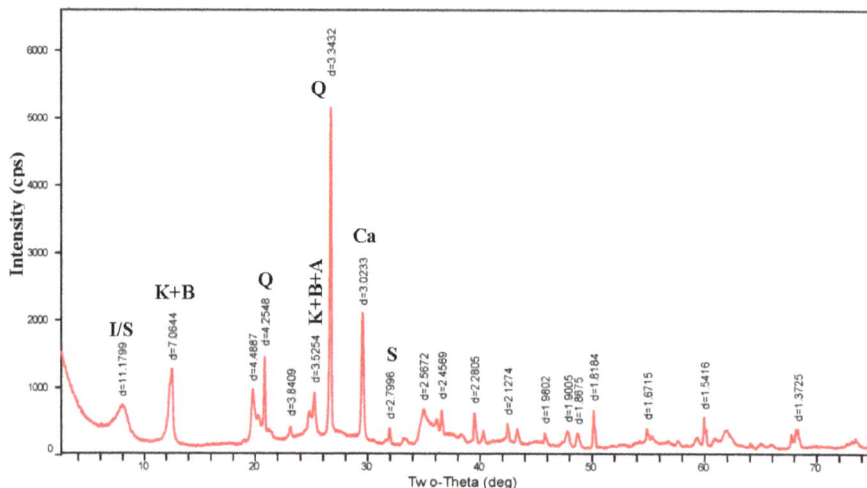

**Figure 3.** XRD of a selected sample (X-16). Abbreviations in the figure indicate the minerals identified, such as I/S: mixed layer illite/smectite; K: kaolinite; B: berthierine; Q: quartz; A: anatase; Ca: calcite; and S: siderite.

The clay species were further identified by the analyses of the three XRD runs on air-dried, Ethylene Glycol (EG)-solvation and heated specimens. Clay minerals identified in the studied samples comprise mixed layer illite/smectite (I/S), kaolinite, and berthierine (Figures 3 and 4). I/S has been recognized by comparing the bulk-XRD pattern (and the air-dried patterns) to XRD patterns of the ethylene-glycol treated and heated specimens. The characteristic broad peaks of mixed layer I/S are located at 11 Å± in the bulk and air-dried oriented patterns, split into two peaks at 12 and 9 Å± after EG solvation, and move towards 10 Å for the heated sample (Figures 3 and 4) [23]. Kaolinite is distinguished by the 7.2 and 3.58 Å peaks in the bulk and air-dried patterns which do not change in the patterns for the EG specimen while in the heated specimen, these peaks reduce or disappear [23]. Berthierine, which has a similar structure to kaolinite and similar chemical composition to chlorite, is distinguished by a lack of 14 Å reflections and a 7 Å basal spacing in the XRD patterns [24–26]. In the XRD pattern, the d(001) intensity of berthierine is slightly lower than that of kaolinite (Figure 4). Kaolinite may mask berthierine in the bulk XRD patterns owing to the proximate (001) and (002) reflections of these two minerals; however, they can be identified from each other in the air-dried oriented pattern (Figure 4). The d(001) reflection of berthierine will considerably reduce after being heated (Figure 4) [27].

I/S in all the studied samples was identified as ordered according to the position of split peaks, as well as the presence of the basal peak at 27 Å in the EG-solvation specimens [23]. The percentages of smectite layers of I/S *i.e.*, S (%) in the studied samples were estimated based on the d-spacing values of I/S in EG-XRD patterns, as listed in Table 1. Apparently, S (%) in samples from XW drill hole (25%–35%) is higher than those from LK drill hole (15%–20%). I/S in most of the samples are identified as R1 ordered with only one R3 I/S in L-22 (Table 1; R: Reichweite parameters; [28]).

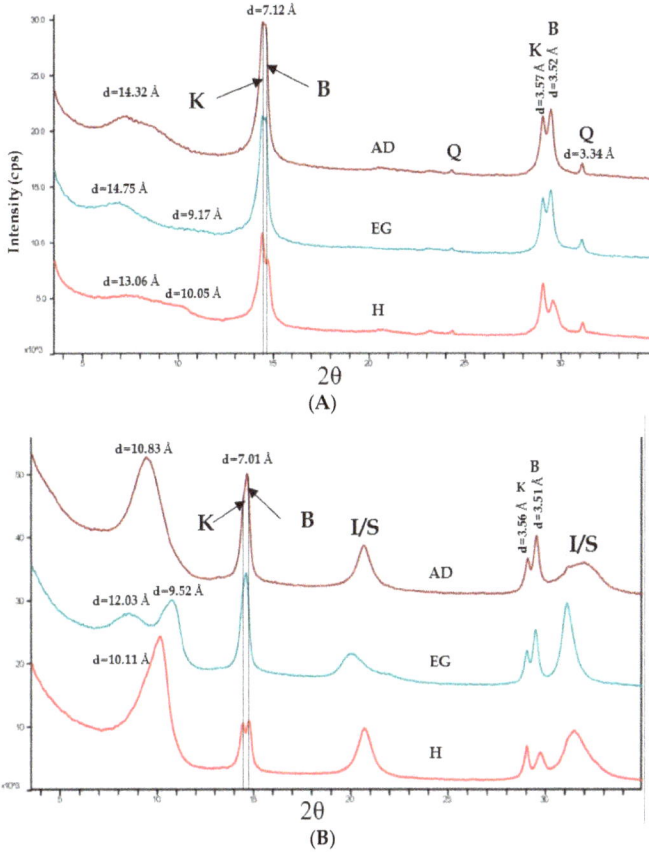

**Figure 4.** XRD patterns for air-dried, EG-solvation, and heated clay fractions of selected samples (**A**) X-2 and (**B**) L-19. Abbreviations are same as in Figure 3. AD: air-dried; EG: ethylene glycol saturated; H: heated.

## 4.2. Modes of Occurrence of Clay Minerals

Clay minerals in this study show various modes of occurrence. Mixed layer illite/smectite mainly occurs as groundmass for other minerals (not only non-clay minerals, but also kaolinite and berthierine; Figure 5). In some cases, I/S has a needle-/lath-like shape (Figure 6A) and is also found in plant cells (Figure 6B) probably indicating an authigenic origin.

Berthierine in this study shows pale red-yellowish discrete particles within I/S matrix under the microscope (Figure 5A). The berthierine particles are usually sharp-cornered and elongated in shape and vary in length from 50 μm to more than 300 μm (Figures 5A and 6C,D). Under SEM, berthierine not only occurs as angular particles (Figure 6C,D), but also colloidal infillings in plant cells coexisting with I/S or quartz (Figure 6B,E). In a few cases, berthierine precipitated along the cracks of vermicular kaolinite (Figure 6F). In Figure 6C,D, berthierine grains were eroded with I/S filling within the cavities or surrounding the remnant berthierine particles.

Vermicular kaolinite is commonly found under both the microscope and SEM. Vermicular kaolinite shows yellowish color under the cross-polarized light under SEM, and in some cases, kaolinite was altered to colloidal berthierine along the margins of kaolinite crystal (Figures 5B and 6F).

16

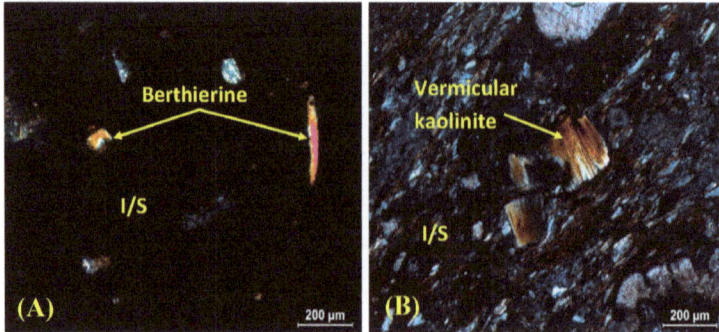

**Figure 5.** Microscopic observations of the studied samples (cross-polarized light). (**A**) Volcanic shard-like berthierine particles (X-17); (**B**) Vermicular kaolinite (L-11).

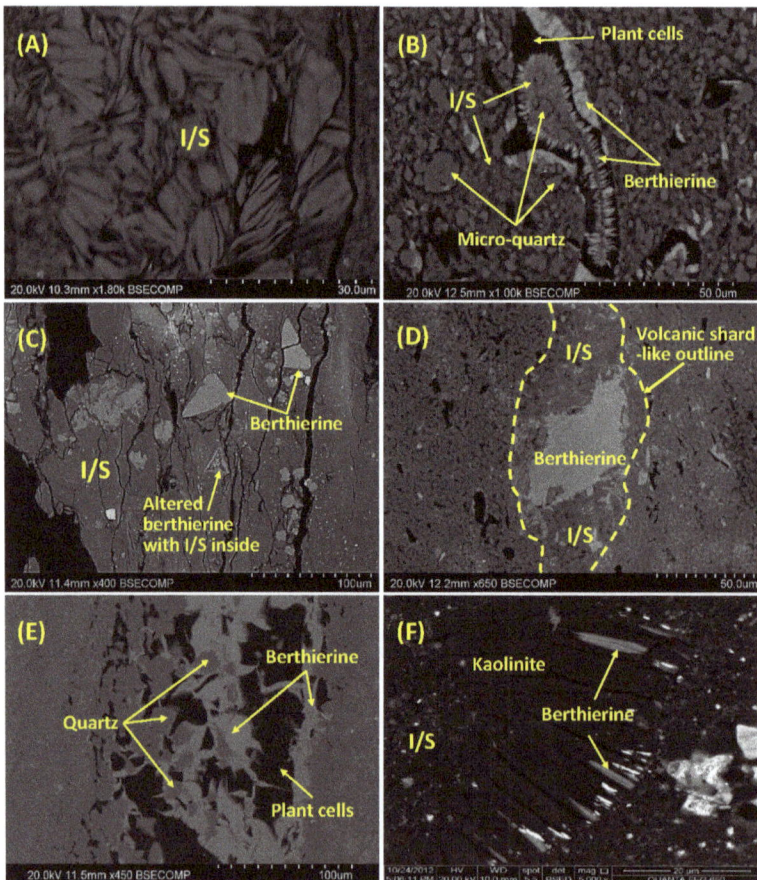

**Figure 6.** Back-scattered electron images of clay minerals. (**A**) Authigenic lath-like I/S (X-2); (**B**) Berthierine and I/S within plant cell (black areas), and micro-quartz particles (X-1); (**C**) Berthierine particles (X-2); (**D**) I/S surrounding berthierine (L-18); (**E**) Quartz coexisting with berthierine within plant cells (L-5); (**F**) berthierine within fractures of vermicular kaolinite (X-10).

*4.3. Abundances of Clay Minerals*

Clay minerals are dominant in almost all the studied samples (total clay: 65.7%–93.5%, mean 80.9% in XW#; 39.2%–93.9%, mean 75.6% in LK#; Table 1), followed by quartz (XW#: 5.2%–32.3% and LK#: 0.4%–52.2%; Table 1) and anatase. Calcite is commonly found under both micro- and macro-observations and in a few cases, the content of calcite is up to 12.3% (XW-16; Table 1). Siderite, albite, and hematite are only rarely found in LK drill holes. Trace REE-bearing phosphate florencite was also discovered in some samples from LK drill holes.

Regarding the clay minerals, I/S is more abundant than kaolinite and berthierine, for example the content of I/S in XW drill holes ranges from 15.2% to 78.4%, averaging 50.6% while in LK drill holes, this value is from 20.6% to 74%, with an average of 54% (Table 1). In general, samples from XW drill hole contain more kaolinite (1.2%–57.7%; Table 1) than those from LK drill hole (bdl-17.3%; bdl: below detection limit; Table 1). In contrast, berthierine seems more abundant in LK drill hole (0.4%–51.9%; Table 1) than in XW drill hole (bdl-22.9%; Table 1).

## 5. Discussions

*5.1. Volcanic Ash Control on Modes of Occurrence of Clay Assemblages*

Based on the petrologic, mineralogical, geochemical, and geophysical studies by Dai, *et al.* [1], the Nb-Zr-REE-Ga-mineralized horizons represent argillized tuffs originated from alkaline volcanic ashes. This is mainly because we have found typical tuffaceous instead of sedimentary textures, and shard-like and euhedral magmatic high-temperature minerals within the Nb-Zr-REE-Ga-mineralized samples under both the microscope and SEM [1,2]. These high-temperature magmatic mineral phases (such as beta-quartz, euhedral apatite, and zircon *etc.*) with high-T cracks, embayments, and sharp-edged outlines were not hydraulically sorted debris which usually have rounded morphologies. It is believed that the Nb-Zr-REE-Ga-mineralized samples and the contained elevated rare metals were derived from alkaline volcanic ash [1]. Natural gamma log data from more than 300 drill holes have shown that the Nb-Zr-REE-Ga-mineralized beds with high positive natural gamma anomalies have a continuous lateral extent across the lowest Upper Permian strata of SW China [1,2]. The widespread Nb-Zr-REE-Ga-mineralized beds with a uniform geochemistry and mineralogy are also indicative of a volcanic tuff deposition [1,2]. In some cases, the abrupt contact between the Nb-Zr-REE-Ga-mineralized rocks and the wall rocks may also be caused by the volcanic-ash origin of the former [1]. These Nb-Zr-REE-Ga-mineralized beds in the lowest Upper Permian strata, have a temporal link to the ~260 Ma Emeishan large igneous province and are the results of waning activity of Emeishan mantle plume [2,29]. The glass-rich volcanic ashes would be an ideal precursor for the clay minerals [30]. In this study, the volcanic origin of the studied samples is also reflected by the modes of occurrence of the clay minerals.

Berthierine is one of the dominant minerals in the Nb(Ta)-Zr(Hf)-REE-Ga ore deposit [1] and has been found within Paleogene and Late Triassic coals of Japan [31]; however, in the Late Permian coals from southwest China, berthierine is not widely reported while an Fe-rich chlorite (*i.e.*, chamosite which is characterized by the 14 Å on XRD pattern; cf. Dai and Chou [32]) is commonly found [32–35]. Berthierine in this study is mainly found occurring as individual angular particles although a small proportion of authigenetic colloidal berthierine can be observed as well. Such berthierine with various irregular shapes (*cf.* Figures 5 and 6) rather than the rounded shape or lumps following the bedding planes indicates that it had not been sorted by weathering and transportation process but is most likely to be transformed from the volcanic glass shards transported by air [36]. Under a microscope, shard-like berthierine occurring discretely within the matrix is also similar to the texture of a volcanic tuff (Figure 5A).

Kaolinite is a common mineral phase in coals and the intra-seam parting tonstein, generally a dominant clay species within clay assemblages in coal, probably due to its stability in low-pH peat mire which contains humic acid released from organic matter at early stage of coal-forming process [16,30,37]. Deconinck *et al.* [38] generated that kaolinite may be derived either from volcanic ashes or from detrital materials. Additionally, Ward [39] suggested that authigenic kaolinite occurring in pores and cavities in coal may be precipitated from solutions. The well-crystallized vermicular texture of kaolinite crystal is believed as volcanic ash-altered product in oxygen-depleted conditions and would occur in the volcanic horizons within coal-forming peat mires [30,34,40,41] or marine environments [38]. In this study, the microscopic and SEM observations revealed the presence of widespread vermicular aggregates of kaolinite is widespread (Figures 5B and 6F), indicating these kaolinite crystals were of authigenic origin and formed *in situ* through alteration of volcanic glass [38].

I/S is also a common mineral in coal [39] and in general, I/S is not the dominant minerals [42]. In fact, the clay fractions in this study, especially kaolinite and I/S have been described within volcanic-ash originated depositions of coal-bearing strata [30]. I/S has been reported in most Mesozoic bentonites [30], the lower Cretaceous bentonites, British Columbia [43–45], Oxfordian bentonites from the Subalpine Basin, Turonian bentonites of France [46], and Ordovician Kinnekulle K-bentonites, France [47]. Furthermore, I/S-enriched Late Permian coals were reported in the Changxing Mine, eastern Yunnan, southwest China (closely located to the drill holes present in this study) [42]. Volcanic ashes falling into the peat mire would transform to smectite; I/S was probably derived from illitisation of volcanic ashes originated smectite during burial diagenesis [30]. Based on the occurrence that I/S coexists with berthierine within the outline of volcanic glass (Figure 6D), it is inferred that both I/S and berthierine were derived from alteration of volcanic glass.

In addition to the volcanic ashes, hydrothermal fluids may have also participated in the formation of clay minerals. The minerals in the plant cells (Figure 6B,E), along with the rare metal-bearing minerals (including hydrothermal REE-bearing phosphate and carbonate, and Nb-bearing anatase *etc.*) occurring within pores and cavities of clay minerals [2] indicate that these minerals are authigenic origin derived from re-deposition of free-ions leached from volcanic ashes by hydrothermal fluids [2]. Clay minerals may have grown from hydrothermal solutions enriched in Si, Al, Fe, K, *etc.* by dissolution of volcanic ashes (including volcanic minerals and glasses).

*5.2. Implications for Paleo-Diagenetic Temperature*

The clay minerals in this study, particularly I/S, are sensitive to thermal conditions, thus their characteristics can be used to estimate the paleo-diagenetic temperatures during the burial process [31,38,48–50]. Factors affecting formation of ordered I/S include temperature, fluid chemistry, time, source material composition, and permeability of the host rock [48,51,52]. In the present study, the latter three factors are unlikely to be the main controls on the ordering of I/S because most of the studied samples are mudstones derived from the same phase of alkaline volcanism with same magma source in the earliest Late Permian [1,2]. Temperature and fluid chemistry thus should be the dominant factors that would have influenced the formation of I/S in this study. Many studies have focused on the diagenetic temperature during the burial process, using the evolution of smectite illitization (especially the percentages of smectite layers within I/S) [48,50,53–55]. In the XRD patterns for the EG-saturated clay fractions, as smectite layers decrease in I/S, the peaks at 9 Å, as well as peaks at 5 Å, become sharp and narrow showing a trend towards illite and indicating a progressive increment in temperature (Figure 7). It has been suggested that the temperature that reflects the appearance of R1 I/S during the smectite illitization process is generally around 100 °C while the transition temperature from R1 to R3 I/S can be up to *ca.* 180 °C [48,50,53–55]. In this study, most of the studied samples have the R1 ordered I/S, except for only one sample that has the R3 ordered I/S (Table 1) indicating that the paleo-diagenetic temperature for the studied samples could have been up to 180 °C but for most of the samples, this value ranges from 100 to 180 °C.

Considering the influence of fluids, it is interesting to note that in the samples from XW drill hole, especially in the lower bed (X-11 to X-16), the contents of kaolinite and berthierine tend to decrease from X-10 to (X-11 to X-16), and then increase in X-17 (Table 1). In addition to the sharp decrease of contents of kaolinite and berthierine, the ordering of their structure also starts to become poor (reflected by the broad peaks at 7 and 3.5 Å), along with an increase in the contents and ordering of I/S in X-11 to X-16 (Table 1; Figures 3, 4 and 7). The changes in contents and crystal ordering may reflect the re-forming of clay minerals under conditions that allow for the poorly-ordered kaolinite and berthierine to form [51]. While in the presence of K released from volcanic ashes, the conversion of well-ordered kaolinite to disordered kaolinite and I/S would have happened under hydrothermal conditions [51]. The colloidal/infilling minerals which were precipitated from hydrothermal fluids revealed in part 4 and a previous study [2] suggest that hydrothermal fluids influenced the studied samples during diagenesis. Accordingly, a reaction mechanism that the kaolinite, as well as berthierine (which has a similar structure to kaolinite [26]) had transformed into I/S under hydrothermal conditions is possible. Such a transformation was estimated to have occurred in a thermal metamorphism environment at a temperature around 225–250 °C [51]; however, as for this study, if the temperature had reached 225 °C, the corresponding I/S in these samples should be R3 ordered. In contrast, as revealed in Table 1, I/S in these samples R1 is still ordered with a slight decrease of 5% smectite layers within I/S. Therefore, such a high temperature seems improbable for the present study.

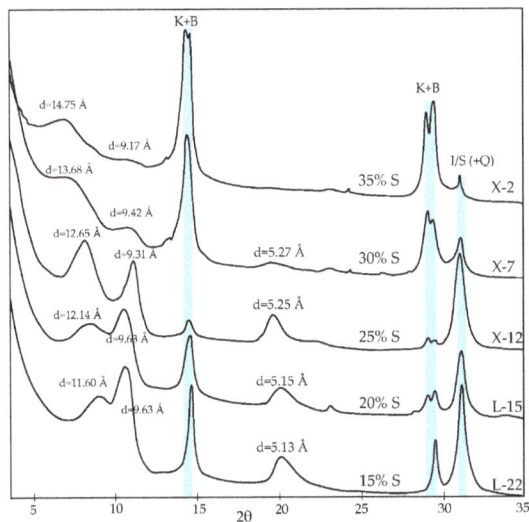

**Figure 7.** XRD patterns for EG-solvation slides with S (%) = 15, 20, 25, 30, 35, respectively. Blue bars indicate the synthetic peaks at 7, 3.5 and 3.3 Å. Abbreviations are the same as Figure 3.

Berthierine is another temperature-sensitive mineral which remains stable under a wide range of temperatures but generally lower than 200 °C [56]. Although in some cases, the transformation of berthierine to chamosite during diagenesis at lower temperatures was reported (below 70 °C when accompanied by organic matter) [31,56,57], that berthierine could have formed at higher temperature was revealed by Iijima and Matsumoto [31], who suggested the formation temperature of berthierine in coal to be 65–150 °C while the alteration temperature of berthierine higher than 160 °C. A similar crystallization temperature of berthierine (150 °C) was also reported by Rivas-Sanchez, *et al.* [25]. If we adopt Iijima and Matsumoto's [31] viewpoint to estimate the paleotemperature of the coal-hosted Nb-Zr-REE-Ga mineralization, since no traces of berthierine transforming to chamosite were found (*i.e.*, no 14 Å were found in XRD patterns) in this study, the paleotemperature should be below 160 °C.

Overall, temperature should be the primary control on the formation of I/S with in some cases, the hydrothermal activity as a secondary control. It is reasonable to estimate that the paleo-diagenetic temperature for the studied samples was 100–160 °C, with one case up to 180 °C. The distribution patterns of rare earth elements (e.g., negative Eu anomalies) in the ore beds also showed that the temperature is lower than 200 °C [58].

*5.3. Origin of the Micro-Crystalline Quartz Associated with Mixed Layer I/S*

Microcrystalline quartz with sizes generally less than 10 μm (mostly <5 μm; Figure 6B) was observed in this study. These very fine quartz grains/cements were found isolated and surrounded by a clay matrix (mainly I/S; Figure 6B). It is unlikely that this micro-quartz formed via mechanical transportation as quartz of a detrital origin is generally in silt- to sand-size particles [59]. Additionally, the pyroclastic quartz in the Nb-Zr-REE-Ga mineralization was reported as large angular particles (>50 μm) [1]. These individual micro-crystalline quartz particles are, therefore, most likely authigenic rather than detrital or pyroclastic in origin. The authigenic ultrafine quartz grains (around 10 μm) were also found in Late Permian coal from Xuanwei, east Yunnan Province, SW China [32] and the Early Cretaceous Wulantuga coal, Inner Mongolia, North China [59].

It has been suggested that illitization of smectite releases Si and involves addition of K [51,60–63]. The excess released Si in solutions (e.g., pore water) may precipitate locally as authigenic micro-crystalline quartz coexisting with the neoformed I/S [62,63]. A simplified reaction process would be:

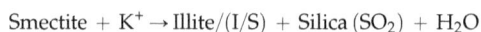

$$\text{Smectite} + \text{K}^+ \rightarrow \text{Illite}/(\text{I}/\text{S}) + \text{Silica} (\text{SO}_2) + \text{H}_2\text{O}$$

Peltonen *et al.* [62] suggested that the sources of smectite and potassium will govern the amount of Si released in such a reaction. In this study, as the studied samples are of pyroclastic origin [1], the smectite in the above equation would be from the alteration of volcanic glasses while volcanic ash-accompanied K-feldspar could have provided K for the illitization of smectite. Another factor limiting the precipitation of quartz from Si-solutions is the permeability of the host rocks, for example, Si-fluids would migrate easier from sandstone (high permeability) than from mudstone (low permeability) [63]. In this study, the low permeability of the studied samples (*i.e.*, mostly mudstones), restricting the diffusion of fluids that contain Si released from illitization of smectite, could also have favored the *in situ* deposition of quartz.

With all the above in mind, we assume that the illitization of smectite during diagenesis may have released significant amounts of $SiO_2$ into solutions first; then due to the low permeability of the studied mudstones, the $SiO_2$-rich solutions resulted in the formation of the micro-crystalline quartz *in situ* as grains near the I/S. As the progressive illitization of smectite proceeded during diagenesis, $SiO_2$ was continuously released to form the high silica saturated solutions favoring the further continuous crystallization and growth of the micro-quartz to become macro-crystalline quartz cements or aggregates [63]. Such a process may also be the cause of high quartz contents in the studied samples (Table 1).

## 6. Conclusions

(1) The clay minerals in the Nb(Ta)-Zr(Hf)-REE-Ga mineralized beds mainly comprise I/S, kaolinite, and berthierine. Generally, I/S is the most abundant species among the clay minerals while the contents of kaolinite and berthierine vary greatly. Angular berthierine particles and vermicular kaolinite occur within the I/S groundmass, while a small proportion of berthierine occurs as colloidal infillings coexisting with I/S in plant cells or in the fractures of vermicular kaolinite.

(2) The modes of occurrence of kaolinite and berthierine verify a volcanic origin for the studied samples. Vermicular kaolinite and the angular berthierine are probably *in situ* alteration products of volcanic ashes. I/S is the product of illitization of volcanic-ash originated smectite.

(3) Indicated by the presence of berthierine and the ordering of the I/S, the paleo-diagenetic temperature reached *ca.* 180 °C, but was generally within 100–160 °C.

(4) The authigenic micro-crystalline quartz coexisting with I/S is probably the result of illitization of smectite during the diagenetic process.

**Acknowledgments:** This research was supported by the National Key Basic Research Program of China (No. 2014CB238902), the National Natural Science Foundation of China (Nos. 41420104001 and 41272182), and the Program for Changjiang Scholars and Innovative Research Team in University (IRT13099). We would like to thank Colin Ward and David French from the University of New South Wales, Australia for their help in clay mineral identification and quantification, and Xisheng Lin from the Research Institute of Petroleum Exploration & Development, China for analysis of mixed layer illite/smectite. Joanne Wilde from the University of New South Wales, Australia is thanked for preparing the polished thin-sections for this study.

**Author Contributions:** All co-authors participated in the work of this study. Lixin Zhao carried-out the clay mineralogy analyses and interpreted mineralogical data under the supervision of Ian Graham. Shifeng Dai collected the samples in the field and has carried-out the field investigation. Ian Graham and Shifeng Dai helped to design the research and structure of the manuscript, as well as the correcting the English language of this manuscript. Peipei Wang conceived and determined the ordering of mixed layer I/S.

**Conflicts of Interest:** The authors declare no conflict of interest.

# References

1. Dai, S.; Zhou, Y.; Zhang, M.; Wang, X.; Wang, J.; Song, X.; Jiang, Y.; Luo, Y.; Song, Z.; Yang, Z.; Ren, D. A new type of Nb(Ta)-Zr(Hf)-REE-Ga polymetallic deposit in the Late Permian coal-bearing strata, eastern Yunnan, southwestern China: Possible economic significance and genetic implications. *Int. J. Coal Geol.* **2010**, *83*, 55–63. [CrossRef]

2. Dai, S.; Chekryzhov, I.Y.; Seredin, V.V.; Nechaev, V.P.; Graham, I.T.; Hower, J.C.; Ward, C.R.; Ren, D.; Wang, X. Metalliferous coal deposits in East Asia (Primorye of Russia and South China): A review of geodynamic controls and styles of mineralization. *Gondwana Res.* **2016**, *29*, 60–82. [CrossRef]

3. Dai, S.; Ren, D.; Chou, C.-L.; Finkelman, R.B.; Seredin, V.V.; Zhou, Y. Geochemistry of trace elements in Chinese coals: A review of abundances, genetic types, impacts on human health, and industrial utilization. *Int. J. Coal Geol.* **2012**, *94*, 3–21. [CrossRef]

4. Dai, S.; Ren, D.; Zhou, Y.; Seredin, V.V.; Li, D.; Zhang, M.; Hower, J.C.; Ward, C.R.; Wang, X.; Zhao, L.; *et al.* Coal-hosted rare metal deposits: Genetic types, modes of occurrence, and utilization evaluation. *J. China Coal Soc.* **2014**, *39*, 1707–1715. (In Chinese)

5. *Geology Mineral Industry Standard of P.R. China: Specifications for Rare Metal Mineral Exploration*; DZ/T 0203-2002; Geological Press: Beijing, China, 2002. (In Chinese)

6. Committee of National Resources. *Reference Handbook for Mineral Industry Requirements*; Geological Press: Beijing, China, 2010. (In Chinese)

7. *Geology Mineral Industry Standard of P.R. China: Specifications for Rare Earth Mineral Exploration*; DZ/T 0204-2002; Geological Press: Beijing, China, 2002. (In Chinese)

8. Zhou, Y. The synsedimentary alkalinity-volcanic ash derived tonsteins of early Longtan age in southwestern China. *Coal Geol. Explor.* **1999**, *27*, 5–9. (In Chinese)

9. Zhou, Y.; Bohor, B.F.; Ren, Y. Trace element geochemistry of altered volcanic ash layers (tonsteins) in Late Permian coal-bearing formations of eastern Yunnan and western Guizhou provinces, China. *Int. J. Coal Geol.* **2000**, *44*, 305–324. [CrossRef]

10. Ali, J.R.; Fitton, J.G.; Herzberg, C. Emeishan large igneous province (SW china) and the mantle-plume up-doming hypothesis. *J. Geol. Soc.* **2010**, *167*, 953–959. [CrossRef]

11. He, B.; Xu, Y.-G.; Chung, S.-L.; Xiao, L.; Wang, Y. Sedimentary evidence for a rapid, kilometer-scale crustal doming prior to the eruption of the Emeishan flood basalts. *Earth Planet. Sci. Lett.* **2003**, *213*, 391–405. [CrossRef]

12. Shellnutt, J.G. The Emeishan large igneous province: A synthesis. *Geosci. Front.* **2014**, *5*, 369–394. [CrossRef]

13. Xu, Y.; Chung, S.-L.; Jahn, B.-M.; Wu, G. Petrologic and geochemical constraints on the petrogenesis of Permian-Triassic Emeishan flood basalt in southwestern China. *Lithos* **2001**, *58*, 145–168. [CrossRef]

14. Chung, S.L.; Jahn, B.M. Plume-lithosphere interaction in generation of the Emeishan flood basalts at the Permian-Triassic boundary. *Geology* **1995**, *23*, 889–892. [CrossRef]

15. Xu, Y.; He, B.; Luo, Z.; Liu, H. Large igneous provinces in China and mantle plume: An overview and perspective. *Bull. Mineral. Petrol. Geochem.* **2013**, *32*, 25–39. (In Chinese)

16. China Coal Geology Bureau. *Sedimentary Environments and Coal Accumulation of Late Permian Coal Formation in Western Guizhou, Southern Sichuan and Eastern Yunnan, China*; Chongqing University Press: Chongqing, China, 1996. (In Chinese)

17. Feng, Z.; Yang, Y.; Jin, Z.; Li, S.; Bao, Z. *Lithofacies Paleogeograohy of Permian of South China*; China University of Petroleum Press: Dongying, China, 1997. (In Chinese)

18. Wang, S.; Yin, H. *Study in Terrestrial Permian-Triassic Boundary in Eastern Yunnan and Western Guizhou*; China University of Geoscience Press: Wuhan, China, 2001. (In Chinese)

19. Zhang, Z.; Yang, X.; Li, S.; Zhang, Z. Geochemical characteristics of the Xuanwei Formation in West Guizhou: Significance of sedimentary environment and mineralization. *Chin. J. Geochem.* **2010**, *29*, 355–364. [CrossRef]

20. He, B.; Xu, Y.-G.; Huang, X.-L.; Luo, Z.-Y.; Shi, Y.-R.; Yang, Q.-J.; Yu, S.-Y. Age and duration of the Emeishan flood volcanism, SW China: Geochemistry and SHRIMP zircon U–Pb dating of silicic ignimbrites, post-volcanic Xuanwei Formation and clay tuff at the Chaotian section. *Earth Planet. Sci. Lett.* **2007**, *255*, 306–323. [CrossRef]

21. He, B.; Xu, Y.-G.; Zhong, Y.-T.; Guan, J.-P. The Guadalupian-Lopingian boundary mudstones at Chaotian (SW China) are clastic rocks rather than acidic tuffs: Implication for a temporal coincidence between the end-Guadalupian mass extinction and the Emeishan volcanism. *Lithos* **2010**, *119*, 10–19. [CrossRef]

22. Xu, Y.; He, B.; Chung, S.-L.; Menzies, M.A.; Frey, F.A. Geologic, geochemical, and geophysical consequences of plume involvement in the Emeishan flood-basalt province. *Geology* **2004**, *32*, 917–920. [CrossRef]

23. *Analysis Method for Clay Minerals and Ordinary Non-Clay Minerals in Sedimentary Rocks by the X-ray Diffraction*; SY/T 5163-2010; Petroleum Industrial Publishing House: Beijing, China, 2010. (In Chinese)

24. Brindley, G.W. Chemical compositions of berthierines-a review. *Clays Clay Miner.* **1982**, *30*, 153–155. [CrossRef]

25. Rivas-Sanchez, M.; Alva-Valdivia, L.; Arenas-Alatorre, J.; Urrutia-Fucugauchi, J.; Ruiz-Sandoval, M.; Ramos-Molina, M. Berthierine and chamosite hydrothermal: Genetic guides in the Pena Colorada magnetite-bearing ore deposit, Mexico. *Earth Planet. Space* **2006**, *58*, 1389–1400. [CrossRef]

26. Toth, T.A.; Fritz, S.J. An Fe-berthierine from a Cretaceous laterite: Part I. Characterization. *Clays Clay Miner.* **1997**, *45*, 564–579. [CrossRef]

27. Moore, D.; Hughes, R.E. Ordovician and Pennsylvanian berthierine-bearing flint clays. *Clays Clay Miner.* **2000**, *48*, 145–149. [CrossRef]

28. Moore, D.; Reynolds, R.C. *X-ray Diffraction and the Identification and Analysis of Clay Minerals*, 2nd ed.; Oxford University Press: Oxford, UK, 1997.

29. Zhao, L.; Graham, I. Origin of the alkali tonsteins from southwest china: Implications for alkaline magmatism associated with the waning stages of the emeishan large igneous province. *Aust. J. Earth Sci.* **2016**, *63*, 123–128. [CrossRef]

30. Spears, D.A. The origin of tonsteins, an overview, and links with seatearths, fireclays and fragmental clay rocks. *Int. J. Coal Geol.* **2012**, *94*, 22–31. [CrossRef]

31. Iijima, A.; Matsumoto, R. Berthierine and chamosite in coal measures of Japan. *Clays Clay Miner.* **1982**, *30*, 264–274. [CrossRef]

32. Dai, S.; Chou, C.-L. Occurrence and origin of minerals in a chamosite-bearing coal of Late Permian age, Zhaotong, Yunnan, China. *Am. Mineral.* **2007**, *92*, 1253–1261. [CrossRef]

33. Wang, X.; Dai, S.; Chou, C.-L.; Zhang, M.; Wang, J.; Song, X.; Wang, W.; Jiang, Y.; Zhou, Y.; Ren, D. Mineralogy and geochemistry of Late Permian coals from the Taoshuping Mine, Yunnan province, China: Evidences for the sources of minerals. *Int. J. Coal Geol.* **2012**, *96–97*, 49–59. [CrossRef]

34. Dai, S.; Li, T.; Seredin, V.V.; Ward, C.R.; Hower, J.C.; Zhou, Y.; Zhang, M.; Song, X.; Song, W.; Zhao, C. Origin of minerals and elements in the Late Permian coals, tonsteins, and host rocks of the Xinde Mine, Xuanwei, eastern Yunnan, China. *Int. J. Coal Geol.* **2014**, *121*, 53–78. [CrossRef]

35. Dai, S.; Tian, L.; Chou, C.-L.; Zhou, Y.; Zhang, M.; Zhao, L.; Wang, J.; Yang, Z.; Cao, H.; Ren, D. Mineralogical and compositional characteristics of Late Permian coals from an area of high lung cancer rate in Xuanwei, Yunnan, China: Occurrence and origin of quartz and chamosite. *Int. J. Coal Geol.* **2008**, *76*, 318–327. [CrossRef]

36. Dai, S.; Luo, Y.; Seredin, V.V.; Ward, C.R.; Hower, J.C.; Zhao, L.; Liu, S.; Zhao, C.; Tian, H.; Zou, J. Revisiting the Late Permian coal from the Huayingshan, Sichuan, southwestern China: Enrichment and occurrence modes of minerals and trace elements. *Int. J. Coal Geol.* **2014**, *122*, 110–128. [CrossRef]

37. Bohor, B.F.; Triplehorn, D.M. Tonsteins: Altered volcanic-ash layers in coal-bearing sequences. *Geol. Soc. Am. Spec. Pap.* **1993**, *285*, 1–44.

38. Deconinck, J.-F.; Crasquin, S.; Bruneau, L.; Pellenard, P.; Baudin, F.; Feng, Q. Diagenesis of clay minerals and K-bentonites in Late Permian/Early Triassic sediments of the Sichuan Basin (Chaotian section, central China). *J. Asian Earth Sci.* **2014**, *81*, 28–37. [CrossRef]

39. Ward, C.R. Analysis and significance of mineral matter in coal seams. *Int. J. Coal Geol.* **2002**, *50*, 135–168. [CrossRef]

40. Zhao, L.; Ward, C.R.; French, D.; Graham, I.T. Mineralogy of the volcanic-influenced Great Northern coal seam in the Sydney Basin, Australia. *Int. J. Coal Geol.* **2012**, *94*, 94–110. [CrossRef]

41. Ren, D. *Coal Petrology of China*; China University of Mining and Technology Press: Xuzhou, China, 1996. (In Chinese)

42. Wang, X.; Zhang, M.; Zhang, W.; Wang, J.; Zhou, Y.; Song, X.; Li, T.; Li, X.; Liu, H.; Zhao, L. Occurrence and origins of minerals in mixed-layer illite/smectite-rich coals of the Late Permian age from the Changxing Mine, eastern Yunnan, China. *Int. J. Coal Geol.* **2012**, *102*, 26–34. [CrossRef]

43. Spears, D.A.; Duff, P.M.D. Kaolinite and mixed-layer illite–smectite in lower cretaceous bentonites from the peace river coalfield, british columbia. *Can. J. Earth Sci.* **1984**, *21*, 465–476. [CrossRef]

44. Pevear, D.R.; Williams, V.E.; Mustoe, G.E. Kaolinite, smectite, and k-rectorite in bentonites: Relation to coal rank at tulameen, british columbia. *Clays Clay Miner.* **1980**, *28*, 241–254. [CrossRef]

45. Reinink-Smith, L.M. Mineral assemblages of volcanic and detrital partings in tertiary coal beds. *Clays Clay Miner.* **1990**, *38*, 97–108. [CrossRef]

46. Pellenard, P.; Deconinck, J.-F.; Huff, W.D.; Thierry, J.; Marchand, D.; Fortwengler, D.; Trouiller, A. Characterization and correlation of Upper Jurassic (Oxfordian) bentonite deposits in the Paris Basin and the Subalpine Basin, France. *Sedimentology* **2003**, *50*, 1035–1060. [CrossRef]

47. Deconinck, J.-F.; Amédro, F.; Baudin, F.; Godet, A.; Pellenard, P.; Robaszynski, F.; Zimmerlin, I. Late cretaceous paleoenvironments expressed by the clay mineralogy of cenomanian-campanian chalks from the east of the paris basin. *Cretac. Res.* **2005**, *26*, 171–179. [CrossRef]

48. Vázquez, M.; Nieto, F.; Morata, D.; Droguett, B.; Carrillo-Rosua, F.J.; Morales, S. Evolution of clay mineral assemblages in the Tinguiririca geothermal field, Andean Cordillera of central Chile: An XRD and HRTEM-AEM study. *J. Volcanol. Geotherm. Res.* **2014**, *282*, 43–59. [CrossRef]

49. Somelar, P.; Kirsimäe, K.; Środoń, J. Mixed-layer illite-smectite in the Kinnekulle K-bentonite, northern Baltic Basin. *Clay Miner.* **2009**, *44*, 455–468. [CrossRef]

50. Hoffman, J.; Hower, J. Clay mineral assemblages as low grade metamorphic geothermometers: Application to the thrust faulted disturbed belt of Montana, USA. *SEPM Spec. Publ.* **1979**, *26*, 55–79.

51. Susilawati, R.; Ward, C.R. Metamorphism of mineral matter in coal from the Bukit Asam deposit, South Sumatra, Indonesia. *Int. J. Coal Geol.* **2006**, *68*, 171–195. [CrossRef]

52. Uysal, I.T.; Glikson, M.; Golding, S.D.; Audsley, F. The thermal history of the Bowen Basin, Queensland, Australia: Vitrinite reflectance and clay mineralogy of Late Permian coal measures. *Tectonophysics* **2000**, *323*, 105–129. [CrossRef]

53. Abid, I.A.; Hesse, R.; Harper, J.D. Variations in mixed-layer illite/smectite diagenesis in the rift and post-rift sediments of the Jeanne d'Arc Basin, Grand Banks offshore Newfoundland, Canada. *Can. J. Earth Sci.* **2004**, *41*, 401–429. [CrossRef]

54. Schegg, R.; Leu, W. Clay mineral diagenesis and thermal history of the Thonex Well, western Swiss Molasse Basin. *Clays Clay Miner.* **1996**, *44*, 693–705. [CrossRef]

55. Środoń, J.; Clauer, N.; Huff, W.; Dudek, T.; Banaś, M. K-Ar dating of the Lower Paleozoic K-bentonites from the Baltic Basin and the Baltic Shield: Implications for the role of temperature and time in the illitization of smectite. *Clay Miner.* **2009**, *44*, 361–387. [CrossRef]

56. Hornibrook, E.R.; Longstaffe, F.J. Berthierine from the Lower Cretaceous Clearwater Formation, Alberta, Canada. *Clays Clay Miner.* **1996**, *44*, 1–21. [CrossRef]

57. Jahren, J.S.; Aagaard, P. Compositional variations in diagenetic chlorites and illites, and relationships with formation-water chemistry. *Clay Miner.* **1989**, *24*, 157–170. [CrossRef]

58. Dai, S.; Graham, I.T.; Ward, C.R. A review of anomalous rare earth elements and yttrium in coal. *Int. J. Coal Geol.* **2016**, *159*, 82–95. [CrossRef]

59. Dai, S.; Wang, X.; Seredin, V.V.; Hower, J.C.; Ward, C.R.; O'Keefe, J.M.K.; Huang, W.; Li, T.; Li, X.; Liu, H.; et al. Petrology, mineralogy, and geochemistry of the Ge-rich coal from the Wulantuga Ge ore deposit, Inner Mongolia, China: New data and genetic implications. *Int. J. Coal Geol.* **2012**, *90–91*, 72–99. [CrossRef]

60. Środoń, J. Nature of mixed-layer clays and mechanisms of their formation and alteration. *Annu. Rev. Earth and Planet. Sci.* **1999**, *27*, 19–53. [CrossRef]
61. Hower, J.; Eslinger, E.V.; Hower, M.E.; Perry, E.A. Mechanism of burial metamorphism of argillaceous sediment: 1. Mineralogical and chemical evidence. *Geol. Soc. Am. Bull.* **1976**, *87*, 725–737. [CrossRef]
62. Peltonen, C.; Marcussen, Ø.; Bjørlykke, K.; Jahren, J. Clay mineral diagenesis and quartz cementation in mudstones: The effects of smectite to illite reaction on rock properties. *Mar. Pet. Geol.* **2009**, *26*, 887–898. [CrossRef]
63. Thyberg, B.; Jahren, J.; Winje, T.; Bjørlykke, K.; Faleide, J.I.; Marcussen, Ø. Quartz cementation in Late Cretaceous mudstones, northern North Sea: Changes in rock properties due to dissolution of smectite and precipitation of micro-quartz crystals. *Mar. Pet. Geol.* **2010**, *27*, 1752–1764. [CrossRef]

minerals

MDPI

*Article*

# Mineralogical and Geochemical Characteristics of Late Permian Coals from the Mahe Mine, Zhaotong Coalfield, Northeastern Yunnan, China

**Xibo Wang \*, Ruixue Wang, Qiang Wei, Peipei Wang and Jianpeng Wei**

State Key Laboratory of Coal Resources and Safe Mining, China University of Mining and Technology, Beijing 100083, China; wangruixue504@gmail.com (R.W.); tonyweiq@gmail.com (Q.W.); wangpeipei1100@gmail.com (P.W.); weijianpeng15@gmail.com (J.W.)

\* Author to whom correspondence should be addressed; xibowang@gmail.com; Tel./Fax: +86-10-6234-1868.

Academic Editor: Kota Hanumantha Rao

Received: 8 June 2015; Accepted: 26 June 2015; Published: 2 July 2015

**Abstract:** This paper reports the mineralogical and geochemical compositions of the Late Permian C2, C5a, C5b, C6a, and C6b semianthracite coals from the Mahe mine, northeastern Yunnan, China. Minerals in the coals are mainly made up of quartz, chamosite, kaolinite, mixed-layer illite/smectite (I/S), pyrite, and calcite; followed by anatase, dolomite, siderite, illite and marcasite. Similar to the Late Permian coals from eastern Yunnan, the authigenic quartz and chamosite were precipitated from the weathering solution of Emeishan basalt, while kaolinite and mixed-layer I/S occurring as lenses or thin beds were related to the weathering residual detrital of Emeishan basalt. However, the euhedral quartz and apatite particles in the Mahe coals were attributed to silicic-rock detrital input. It further indicates that there has been silicic igneous eruption in the northeastern Yunnan. Due to the silicic rock detrital input, the Eu/Eu\* value of the Mahe coals is lower than that of the Late Permian coals from eastern Yunnan, where the detrital particles were mainly derived from the basalt. The high contents of Sc, V, Cr, Co, Ni, Cu, Ga, and Sn in the Mahe coals were mainly derived from the Kangdian Upland.

**Keywords:** Late Permian coal; minerals; trace elements; Emeishan basalt/silicic rock; Mahe mine

## 1. Introduction

Late Permian coals from the eastern Yunnan Province have recently attracted much attention, because of both the high female lung cancer rate caused by the indoor coal burning and the geological implication of mineral matter in the coals to the origin and evolution of Emeishan mantle plume. For the first aspect, Tian [1] and Tian *et al.* [2] found that the lung cancer risk was associated with crystalline silica released from the indoor coal burning. Dai *et al.* [3,4], Large *et al.* [5], and Wang *et al.* [6] observed high concentration of authigenic quartz (from nanometer to less than 20 μm in size) in the Xuanwei coals. For the second aspect, Dai *et al.* [3,4,7], Zhou *et al.* [8], and Wang *et al.* [6] conducted a systematic mineralogical and geochemical study of the Late Permian coals and tonsteins from the southwest Chongqing and eastern Yunnan and suggested that, after massive flood basalt eruption, the magma evolved from mafic to silicic or alkalinity. The previous studies mostly focused on the coals in the eastern Yunnan, only Dai and Chou investigated mineral compositions in the coals from the northeastern Yunnan [9]. In this paper, we report the new data on the mineralogy and elemental geochemistry of the 5 coals in the Mahe mine, Zhaotong coalfield, northeastern Yunnan, China.

## 2. Geological Setting

The geological setting of the Late Permian coal basin from eastern Yunnan Province has been described in detail by several authors [9,10]. The Emeishan mantle plume uplift and extensive flood basalt eruption resulted in the formation of the Kangdian Upland [11,12]. Due to the existence of the Qinling sea trench to the north and Songpan basin to the west, the Kangdian Upland is the only possible source for the Xuanwei and Longtan/Changxing Formations in eastern Yunnan [12].

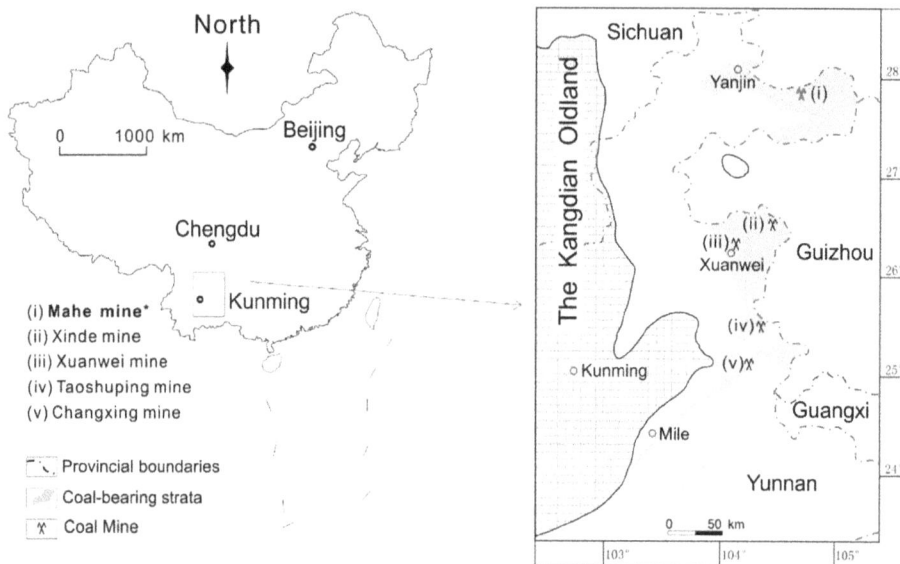

**Figure 1.** Location of the Mahe mine, Zhaotong coalfield, northeastern Yunnan Province, China, as well as locations of the Xinde, Xuanwei, Taoshuping, and Changxing mines, eastern Yunnan Province, China.

The Mahe mine from the Zhaotong coalfield is situated in the northeastern Yunnan Province (Figure 1). The coal-bearing strata are mainly the Changxing Formation and the Longtan Formation of Late Permian age. The Changxing Formation, with a thickness of 47 m, is mainly made up of clastic sediments, including sandstone, mudstone, coal or limestone. It was deposited in a continent-marine transitional environment. There are four coal seams in Changxing Formation, named C1, C2, C3, and C4 in order from up to bottom, and most of them are too thin to be mined. The Longtan Formation has a thickness of 197 m and overlies the Middle Permian Xuanwuyan Formation (Figure 2). The Longtan Formation is comprised of mudstone, sandstone, gravel, and coal seams including C5, C6, and C7 coals. The No.5 coal is the major minable seam. Due to the less continuity of C6 and C7 coals in thickness, they could only be locally mined.

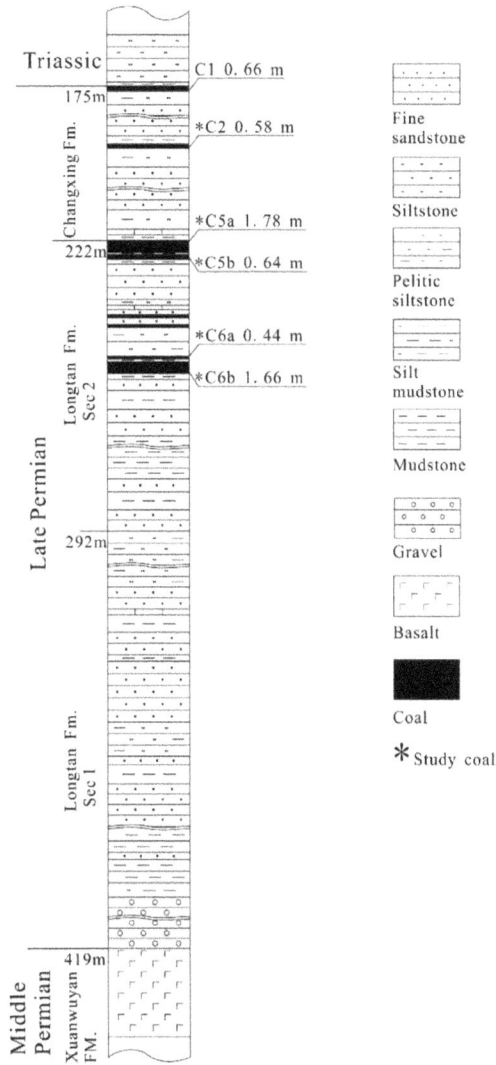

**Figure 2.** Stratigraphic column of the drill core from the Mahe mine, Zhaotong coalfield, northeastern Yunnan Province, China.

## 3. Sample Collection and Methods

The C5 and C6 coals in the Mahe mine are divided into two benches based on a stable parting, respectively. Thus, they were named as C5a, C5b, C6a, and C6b, respectively. A total of five coal samples named MhC2 (0.78 m), MhC5a (1.78 m), MhC5b (0.64 m), MhC6a (0.44 m), and MhC6b (1.66 m) were collected from a drill core from the Mahe mine, northeastern Yunnan, China (Figures 1 and 2; Table 1). The coal samples were pulverized to <1 mm fractions by hammer for briquette preparation.

**Table 1.** Random vitrinite reflectance, proximate analyses, and total sulfur in coals from the Mahe mine (%).

| Samples | $M_{ad}$ | $A_d$ | $V_{daf}$ | $R_{o,ran}$ | $S_{t,d}$ |
|---------|------|-------|-------|--------|--------|
| MhC2    | 0.97 | 42.24 | 14.82 | 2.21 | 3.17 |
| MhC5a   | 0.7  | 25.75 | 12.86 | 2.27 | 2.31 |
| MhC5b   | 0.96 | 33.67 | 13.83 | 2.24 | 3.86 |
| MhC6a   | 0.7  | 36.72 | 16.29 | 2.45 | 2.45 |
| MhC6b   | 0.66 | 29.62 | 13.67 | 2.31 | 2.49 |
| Average | 0.80 | 33.60 | 14.29 | 2.30 | 2.86 |
| Maximum | 0.97 | 42.24 | 16.29 | 2.45 | 3.86 |
| Minimum | 0.66 | 25.75 | 12.86 | 2.21 | 2.31 |

$R_{o,ran}$, random vitrinite reflectance; $M$, moisture; $A$, ash yield; $V$, volatile matter; $S_t$, total sulfur; ad, air-dry basis; d, dry basis; daf, dry and ash-free basis

Proximate analysis, covering moisture, volatile matter, and ash yield, was measured following the ASTMD3173-11 [13], ASTM D3175-11 [14], and ASTMD3174-11 [15], respectively. Total sulfur was determined in accordance with ASTMD 3177-02 [16]. Mean random vitrinite reflectance were measured using a Leica DM 4500P microscopy (Leica, Wetzlar, Germany) (at a magnification of $500\times$) equipped with a Craic QDI 302TM spectrophotometer (Craic, San Dimas, CA, USA). The standard reference for vitrinite reflectance measurement was gadolinium gallium garnet (Klein and Becker, Idar-Oberstein, Germany) with a calculated standard reflectance of 1.722% for $\lambda = 546$ nm under oil immersion.

Mineral phases were investigated in coal briquette by optical microscopy and by a scanning electron microscope equipped with an Oxford energy-dispersive X-ray spectrometer (SEM-EDX, Hitachi, Tokyo, Japan), confirmed by X-ray diffraction (XRD, D/max 2500, Rigaku, Tokyo, Japan) on low-temperature ashes of coal (LTA). Low-temperature ashes of coal was performed using an EMITECH K1050 Plasma Asher (Quorum, Lewes, UK). XRD analysis of low temperature ashes was performed on a powder diffractometer with a Ni-filtered Cu-K$\alpha$ radiation with a scintillation detector. The XRD pattern was recorded over a 2$\theta$ interval of 2.6°–70°, with a step size of 0.02°, and a 0.3 mm receiving slit. The X-ray diffractograms of LTAs of five coal samples were subjected to quantitative mineralogical analysis using Siroquant™ (Sietronics, Mitchell, Australia). Further details demonstrating the use of this technique for coal-related materials are given by Ward *et al.* [17,18] and Ruan and Ward [19].

Samples were crushed and ground by a shatterbox to pass 200 mesh for geochemical analysis. X-ray fluorescence (XRF) spectrometry (ARL Advant'XP+, ThermoFisher, Waltham, MA, USA) was used to determine the major element oxides including $Al_2O_3$, $SiO_2$, $Fe_2O_3$, CaO, $Na_2O$, MgO, $K_2O$, $TiO_2$, $P_2O_5$, and MnO as outlined by Dai *et al.* [20]. Inductively coupled plasma mass spectrometry (X series II ICP-MS, ThermoFisher) was used to determine elements in coal samples as outlined by Dai *et al.* [7], except for Hg and F. Mercury was determined by DMA-80 (Milestone, Sorisole, Italy) in which samples are heated and the evolved Hg is selectively captured as an amalgam and measured by atomic absorption spectrophotometry. The detection limit of Hg is 0.005 ng, the relative standard deviation (RSD) from eleven runs on Hg standard reference is 1.5%, and the linearity of the calibration is in the range 0–1000 ng [21]. Fluorine was determined by the ion-selective electrode method (ISE) following the procedures described in ASTM method D5987-96 [22]. For quality control in fluorine determination, the standard reference materials GBW 11121 and GBW 11123 were analyzed with each batch of samples.

## 4. Results

### 4.1. Coal Chemistry

Table 1 lists the data of chemical analysis (including proximate analysis, and total sulfur), and random vitrinite reflectance. Based on ASTM standard D388-12 [23], the five Mahe coals are semianthracite in rank, with an average volatile yield of 14.29% (ranging from 12.86% to 16.29%) on the ash free basis and an average random vitrinite reflectance of 2.30% (ranging from 2.21% to 2.45%). The Mahe coals have moisture of 0.66% to 0.97% (average 0.80%), ash yield of 25.75% to 42.24% (average 33.60%), and total sulfur of 2.31% to 3.86% (average 2.86%).

The C5a and C6b seams are classified as medium-ash coal, the C5b and C6a seams are medium-high-ash coal, and the C2 seam is high-ash coal according to Chinese Standards GB/T 15224.1-2010 [24]; coals with ash yield 20.01%–30.00% are medium-ash coals, coals with ash yield 30.01%–40.00% are medium-high-ash coals, and coals with 40.01%–50.00% ash are high-ash coals. The C5a, C6a, and C6b seam are medium-high sulfur (2.01%–3.00%) coals, while the C2 and C5b seams are high-sulfur (>3.00%) coals according to Chinese Standards GB/T 15224.2-2010 [25].

The rank of Late Permian coal from eastern Yunnan is different from that of northeastern area. The coals from the northeastern Zhaotong coalfield (155 channel samples) are mainly semiathracite ($R_{o,ran}$ = 1.98%–2.52%, mean 2.25%; $V_{daf}$ = 7.64%–20.6%, mean 11.9%) [9]. The Mahe coals are also mainly semianthracite. However, the coals from eastern Yunnan are mainly medium volatile bituminous, such as the Xinde coals ($R_{o,ran}$ = 1.21%–1.23%, $V_{daf}$ = 25.69%–26.74%) [4], the Xuanwei coals ($V_{daf}$ = 27.1%–32.68%) [3], and the Taoshuping coals ($R_{o,ran}$ = 0.94%–1.37%, $V_{daf}$ = 23%–29.3%) [6]. Different from the most Late Permian coals from eastern Yunnan, the Changxing coals are high in rank, which is semianthracite ($R_{o,ran}$ = 2.46%–2.63%, $V_{daf}$ = 7%–11.48%) [26]. Metamorphism of the Late Permian coals in eastern Yunnan is mainly attributed to the burial geothermal gradient. For example, in the Taoshuping profile the decreasing trend of volatile yield is coupled with random vitrinite reflectance increasing with increasing depth [6]. However, the elevated rank of the Changxing coal was related to the thermal alteration induced by the igneous intrusion [26]. No massive igneous intrusion was observed in the Zhaotong coalfield in the previous study by Dai and Chou [9] and the present study as discussed below, it is inferred the metamorphism of Zhaotong coals mainly resulted from a burial geothermal gradient which is higher than that of Taoshuping mine.

*4.2. Minerals in the Mahe Coals*

Minerals in the Mahe coals were observed by optical microscopy and SEM-EDX, confirmed by LTA + XRD. The quantitative XRD results from siroquant$^{TM}$ show that minerals in the Mahe coals are mainly comprised of quartz, chamosite, kaolinite, mixed-layer I/S, pyrite, and calcite, and followed by anatase, dolomite, siderite, illite and marcasite (Table 2).

**Table 2.** Mineral compositions of low-temperature ashes (LTA) of the Mahe coals by XRD analysis and Siroquant (%; on organic matter-free basis).

| Samples | Qu | Ana | Cal | Dol | Sid | Py | Mar | Kao | Cha | Illite | I/S |
|---------|------|-----|------|-----|-----|-----|-----|------|------|--------|-----|
| Mh C2   | 43.4 | -   | 21   | -   | 1.2 | 8.2 | -   | -    | 14.4 | 6.8    | 5   |
| Mh C5a  | 49.2 | -   | 8    | 1.3 | -   | -   | -   | 26.6 | 12.9 | -      | 2.1 |
| Mh C5b  | 50.4 | 4.9 | 11.1 | -   | -   | 10.2| -   | 13.6 | 2.3  | -      | 7.5 |
| Mh C6a  | 8.9  | 4.5 | 5.1  | 6.1 | 0.6 | 5.5 | 3.4 | 54   | 8.6  | -      | 3.3 |
| Mh C6b  | 27.8 | -   | 4.9  | -   | 1   | 10  | 3.3 | 36.6 | 13.8 | -      | 2.7 |

Qu, quartz; Ana, Anatase; Cal, Calcite; Dol, Dolomite; Sid, Siderite; Py, Pyrite; Mar, Marcasite; Kao, Kaolinite; Cha, Chamosite; I/S, mixed-layer illite/smectite (I/S).

Quartz is the most common mineral in coal. Its concentration in the LTA of the Mahe coals varies from 8.9% to 50.4%, with an average of 38.2% (Table 2). Quartz in the Mahe coals has two origins: (i) authigenic and (ii) terrigenous or silicic volcanic ashes. The authigenic quartz is mainly present as disseminated irregular particles distributed in collodetrinite (Figure 3A), and is usually less than 20 μm in size. The microscopy observation shows that the authigenic quartz accounts for more than 90% of the total quartz. However, quartz of terrigenous origin, preserving well edges and angles and a completely euhedral crystal form (Figure 3B), occurs mainly as assemblages or discrete particles. It is larger in size (100 to 500 μm) than the authigenic quartz (Figure 3A) and the quartz of volcanic origin

(20 to 100 μm) (Figure 3C). Based on the modes of occurrences, it is suggested that the source area of the terrigenous quartz is not far from the peat mire, because the frequently collision and friction suffered from a long distance transport would have resulted in a more rounded shape. β-form quartz is common in the Late Permian coals from eastern Yunnan, indicating a silicic volcanic ashes input [3,4,6]. This is also the case in the Mahe coal samples. The β-form quartz in the Mahe coals is similar in shape and size with that in the Xinde and Taoshuping coals [3,6]. It has a typical hexagonal form and a small size (less than 100 μm), and some of them occur as a doubly terminated bipyramidal form with an additional prism face (Figure 3C). Compared with common silicic volcanic ashes input indicated by the β-form quartz in the eastern Yunnan, mafic volcanic ashes input was only identified in the k21b coal from the Taoshuping mine [6].

Although chlorite is not common in coals [27], it was also observed in some high rank coals and formed by epigenetic processes [28,29]. Mixed-layer I/S in coal could be altered to chlorite [27,30,31]. Chamosite $[Fe_3^{2+}Mg_{1.5}AlFe_{0.5}^{3+}Si_3AlO_{12}(OH)_6]$, a type of chlorite, is common in the Late Permian coals of eastern Yunnan [3,4,6,9]. Chamosite is also present in all the Mahe coal samples (Figure 4), with a concentration from 2.3% to 14.4% (average 10.4%) (Table 2). The chamosite occurs mainly as cell-infillings alone (Figure 3D), sometimes exists with quartz and/or calcite. The coexisting of authigenic quartz and chamosite in the coals were mainly precipitated from the Fe-Mg-rich siliceous solution derived from the weathering of basalt, similar to the mechanism reported by Dai *et al.* [3] and Wang *et al.* [6]. Dai and Chou [9] has shown that some chamosite is closely related to quartz and kaolinite in the Zhaotong coals, and was suggested to be derived from the interaction of kaolinite with Fe-Mg rich fluids during early diagenesis.

Kaolinite and mixed-layer I/S are common in coal [32,33]. Kaolinite was identified in the LTA of the samples MhC5a, MhC5b, MhC6a, and MhC6b, with concentrations of 26.6%, 13.6%, 54%, and 36.6%, respectively (Table 2). Kaolinite in the sample MhC2 is below detection limit of XRD and Siroquant. Mixed-layer I/S is present in all the LTA of the coal samples, ranging from 2.1% to 7.5% (average 4.1%), and much lower than those of kaolinite and chamosite (Table 2). Illite is only detected in the MhC2 sample. Kaolinite and mixed-layer I/S in the Mahe coals occur mainly as recrystallization particles, lenses, and thin beds, indicating a terrigenous origin. The kaolinite and mixed-layer I/S were derived from the weathering products of basalt from the Kangdian Upland [6,9]. In addition, kaolinite as infillings of cells or fractures (Figure 3E) is similar to that in the Xinde coals, suggesting a authigenic origin [4]. Mixed-layer I/S in the Changxing coals occurs not only as lenses or thin beds but also as infillings of maceral fractures. The mixed-layer infillings of fractures is of epigenetic origin and precipitated from hydrothermal fluids of a igneous intrusion [6].

Calcite is common carbonate in the Late Permian coals from eastern Yunnan. It occurs mainly as fracture/pores infillings, indicating an epigenetic origin of hydrothermal fluids [3,4,9]. Although detrital calcite is very rare in coal because calcite can be easily-decomposed under acid conditions in the peat bog; however, syngenetic deposition of calcite (aragonite) is possible if a sediment source region mainly made up of carbonate rocks is located close to the peat mire [34,35]. Detrital calcite was identified in the Taoshuping coals [6]. The detrital calcite was probably blown by winds to the peat mire from the limestone of Middle Permian Maokou Formation [6], which underlies below the Xuanwuyan Formation. The content of calcite in the LTA of the Mahe coals varies from 4.9% to 21% (average 10%) (Table 2). It is mainly present as veins in macerals and display twin-striation characteristics under crossed polarized light (Figure 3F). In some cases, calcite fills the cells with quartz and/or chamosite.

Pyrite in the Mahe coals is dominated by disseminated fine or framboidal particles in macerals, and followed by massive particles of several micrometers in size (Figure 3G). Their modes of occurrences suggest a syngenetic origin [36].

Due to its low concentration, apatite was below the detection limit of XRD technique. It was only observed in the sample MhC2 under SEM-EDX. The apatite preserved distinct edges and angles, and has a big size of 220 μm in length and 60 μm in width (Figure 3H). However, apatite of silicic volcanic ashes origin in the Taoshuping coals is less than 5 μm in size [6].

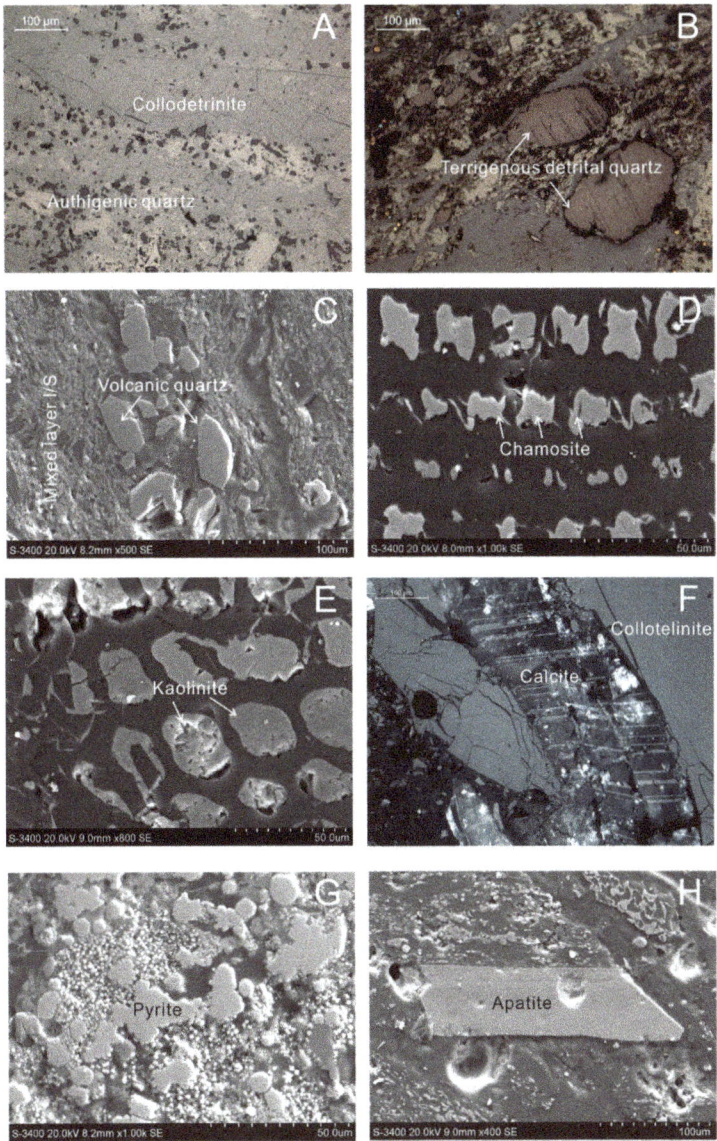

**Figure 3.** Minerals in the Mahe coals. (**A**) Disseminated irregular quartz particles in collodetrinite in the sample MhC2 (reflected light); (**B**) euhedral quartz in the sample MhC2 (reflected light); (**C**) β-form quartz of silicic volcanic ashes origin in the sample MhC2 (SEM, secondary electron image); (**D**) chamosite as cell-infillings in the sample MhC2 (SEM, secondary electron image); (**E**) kaolinite as cell-infillings in the sample MhC5a (SEM, secondary electron image); (**F**) calcite veins in the fractures of collotelinite with the twin-striation characteristics under crossed polarized light in the sample MhC2 (reflected light, oil immersion); (**G**) pyrite as disseminated or framboidal particles in the sample MhC2 (SEM, secondary electron image); (**H**) euhedral apatite in the sample MhC2 (SEM, secondary electron image).

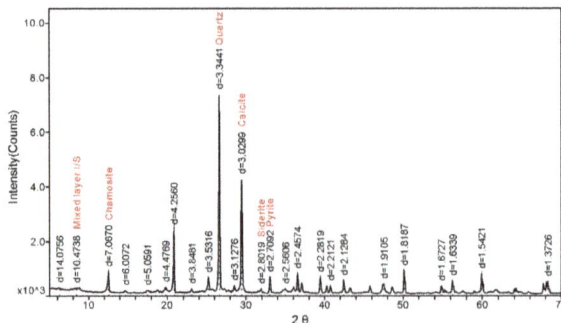

**Figure 4.** Identification of minerals in the XRD pattern of the low temperature ashes (LTA) of the MhC2.

*4.3. Major and Trace Elements*

4.3.1. Major Elements

As compared with Chinese average coals [37], the mean concentrations of major element oxides including $SiO_2$ (17.31%), $Al_2O_3$ (8.26%), CaO (1.57%), $K_2O$ (0.32%), MnO (0.03%), and $TiO_2$ (0.50%) are enriched with concentration coefficients (CC, the ratio of the average elements concentration of the Mahe coals to that of Chinese average coals) of 2.04, 1.38, 1.28, 1.68, 1.50, and 1.52, respectively (Table 3). However, the mean concentrations of $Fe_2O_3$ (3.4%), MgO (0.21%), $Na_2O$ (0.14%), and $P_2O_5$ (0.07%) are lower than or close to that of Chinese average coals [37].

The ratio of $SiO_2/Al_2O_3$ in the Mahe mine ranges from 1.57 to 4.49, with an average of 2.24. It is much higher than that of Chinese average coals 1.42 [37] and the theoretic value of kaolinite (1.18). The higher value of $SiO_2/Al_2O_3$ is attributed to the presence of abundant authigenic quartz. The CaO has a content of 0.61% to 3.14%, and the MnO has a content of 0.01% to 0.07%. CaO in the coals is mainly related to veins of calcite (Figure 3F). The significant relation coefficient between CaO and MnO (0.88) supports that they have a similar mode of occurrence. The concentration of $Fe_2O_3$ varies from 2.3% to 5.37%, with an average of 3.4%. The low positive relation coefficient between $Fe_2O_3$ and $S_{p,d}$ ($r_{Fe_2O_3-S_{p,d}} = 0.61$) suggest that Fe is not only associated with pyrite, but also associated with chamosite. The average contents of $K_2O$ and $Na_2O$ are 0.32% and 0.14%, respectively. $K_2O$ and $Na_2O$ are probably attributed to mixed-layer I/S, and illite. Anatase is primarily responsible for $TiO_2$ in the Mahe coals.

4.3.2. Trace Elements

As compared with Chinese average coals [37], the Mahe coals are high in Sc (4.38 µg/g), V (105 µg/g), Cr (45.7 µg/g), Co (19.0 µg/g), Ni (29.8 µg/g), Cu (70.4 µg/g), Ga (14.9 µg/g), and Sn (4.75 µg/g), with a concentration coefficient (CC) higher than 2 (Table 3). This is similar to the Late Permian coals from Taoshuping [6], Xinde [4], and Changxing [27], Xuanwei [3] mines, eastern Yunnan. The high concentrations of Sc, V, Cr, Co, and Ni in the Mahe coals are mainly attributed to the Emeishan basalt from Kangdian Upland, which is located to west of the basin and is the only source region of the Late Permian coals in eastern Yunnan [10]. The Emeishan basalt is high in Sc (29.8 µg/g), V (317 µg/g), Cr (176 µg/g), Co (43.1 µg/g), and Ni (104 µg/g) [38].

Although the average contents of potentially toxic elements F and Hg are lower than that of Chinese average coals [37], they are enriched in some coal samples. The MhC2 has a 246 µg/g fluorine, which is probably related to the P-bearing mineral apatite (Figure 3H). Fluorine in the Taoshuping coals also shows a significant positive relationship with phosphorous [6]. Mercury in the Mahe coals varies from 254 to 320 ng/g (average 277 ng/g), which is much higher than that of the Xinde coals (average 44 ng/g). Because mercury in coal is usually related to sulfur-bearing minerals [39–41], the high total sulfur of Mahe coals ($S_{t,d} = 2.86\%$, Table 1) is probably responsible for the elevated Hg, while the total sulfur of Xinde coal is only 0.16% [4].

Table 3. Contents of oxides of major elements and trace elements in the Mahe coals from northeastern Yunnan, China (in µg/g unless as indicated).

| Samples | $SiO_2$ | $Al_2O_3$ | CaO | $Fe_2O_3$ | $K_2O$ | MgO | MnO | $Na_2O$ | $P_2O_5$ | $TiO_2$ | Li | Be | F | Sc | V | Cr | Co | Ni |
|---|---|---|---|---|---|---|---|---|---|---|---|---|---|---|---|---|---|---|
| MhC2 | 22.92 | 5.11 | 3.14 | 5.37 | 0.68 | 0.48 | 0.07 | 0.08 | 0.17 | 0.23 | 17.5 | 1.36 | 246 | 5.54 | 115 | 31.5 | 14.1 | 29.1 |
| MhC5a | 13 | 5.66 | 2.24 | 2.3 | 0.11 | 0.14 | 0.03 | 0.07 | 0.06 | 0.17 | 26.8 | 1 | 79 | 2.9 | 20.8 | 18 | 12 | 15.9 |
| MhC5b | 18.3 | 7.6 | 1.15 | 3.88 | 0.33 | 0.15 | 0.02 | 0.2 | 0.02 | 0.76 | 23 | 0.93 | 77 | 9.76 | 67.5 | 55 | 22.3 | 20.4 |
| MhC6a | 17.56 | 13.53 | 0.73 | 2.4 | 0.29 | 0.15 | 0.02 | 0.21 | 0.05 | 0.84 | 45.2 | 2.66 | 104 | 13.3 | 182 | 68.4 | 17.5 | 37 |
| MhC6b | 14.78 | 9.39 | 0.61 | 3.04 | 0.19 | 0.15 | 0.01 | 0.16 | 0.04 | 0.5 | 30 | 2.01 | 68 | 13.3 | 139 | 55.4 | 28.9 | 46.5 |
| Average[a] | 17.31 | 8.26 | 1.57 | 3.40 | 0.32 | 0.21 | 0.03 | 0.14 | 0.07 | 0.50 | 28.50 | 1.59 | 115 | 8.96 | 105 | 45.66 | 18.96 | 29.78 |
| Coal[b] | 8.47 | 5.98 | 1.23 | 4.85 | 0.19 | 0.22 | 0.02 | 0.16 | 0.09 | 0.33 | 31.8 | 2.11 | 130 | 4.38 | 35.1 | 15.4 | 7.08 | 13.7 |
| CC[c] | 2.04 | 1.38 | 1.28 | 0.70 | 1.68 | 0.97 | 1.50 | 0.90 | 0.76 | 1.52 | 0.90 | 0.75 | 0.88 | 2.05 | 2.99 | 2.96 | 2.68 | 2.17 |

| Samples | Cu | Zn | Ga | Ge | Rb | Sr | Y | Zr | Nb | Mo | Sn | Sb | Cs | Ba | La | Ce | Pr | Nd |
|---|---|---|---|---|---|---|---|---|---|---|---|---|---|---|---|---|---|---|
| MhC2 | 80.1 | 58.9 | 13.1 | 2.04 | 30.7 | 132 | 27.2 | 77.6 | 6.12 | 2.95 | 3.43 | 0.37 | 1.95 | 87.2 | 35.3 | 70.6 | 9.28 | 40.5 |
| MhC5a | 18.2 | 11.8 | 5.72 | 0.9 | 3.98 | 112 | 19.5 | 53.5 | 4.18 | 0.68 | 3.77 | 0.41 | 0.4 | 30 | 16.1 | 31.8 | 3.76 | 13.9 |
| MhC5b | 45.5 | 32.2 | 10.4 | 1.17 | 15 | 94 | 16.6 | 120 | 16.4 | 1.13 | 4.55 | 0.31 | 0.66 | 98.8 | 23.3 | 46.8 | 5.48 | 20.3 |
| MhC6a | 124 | 40 | 24.3 | 3.44 | 12.7 | 87.8 | 38.7 | 191 | 21.9 | 1.36 | 7.1 | 0.82 | 1.45 | 51.7 | 69.5 | 133 | 16 | 62.9 |
| MhC6b | 84.2 | 36.8 | 21 | 3.45 | 8.52 | 73.5 | 55.9 | 118 | 13.4 | 2.1 | 4.88 | 0.91 | 0.85 | 54.9 | 61.7 | 126 | 15.5 | 59.9 |
| Average[a] | 70.40 | 35.94 | 14.90 | 2.20 | 14.18 | 99.86 | 31.58 | 112 | 12.40 | 1.64 | 4.75 | 0.56 | 1.06 | 64.52 | 41.18 | 81.64 | 10.00 | 39.50 |
| Coal[b] | 17.5 | 41.4 | 6.55 | 2.78 | 9.25 | 140 | 18.2 | 89.5 | 9.44 | 3.08 | 2.11 | 0.84 | 1.13 | 159 | 22.5 | 46.7 | 6.42 | 22.3 |
| CC[c] | 4.02 | 0.87 | 2.28 | 0.79 | 1.53 | 0.71 | 1.74 | 1.25 | 1.31 | 0.53 | 2.25 | 0.67 | 0.94 | 0.41 | 1.83 | 1.75 | 1.56 | 1.77 |

| Samples | Sm | Eu | Gd | Tb | Dy | Ho | Er | Tm | Yb | Lu | Hf | Ta | $Hg^d$ | Tl | Pb | Bi | Th | U |
|---|---|---|---|---|---|---|---|---|---|---|---|---|---|---|---|---|---|---|
| MhC2 | 7.95 | 1.7 | 7.34 | 1.09 | 5.06 | 0.94 | 2.45 | 0.36 | 2.15 | 0.34 | 3.44 | 0.44 | 258 | 0.9 | 9.76 | 0.17 | 4.72 | 3.28 |
| MhC5a | 2.83 | 0.53 | 2.79 | 0.5 | 3.07 | 0.61 | 2.01 | 0.29 | 1.99 | 0.3 | 2.49 | 0.33 | 254 | 0.11 | 7.55 | 0.21 | 5.2 | 1.49 |
| MhC5b | 3.59 | 0.61 | 3.55 | 0.53 | 3.21 | 0.56 | 1.69 | 0.24 | 1.72 | 0.22 | 5.11 | 1.14 | 320 | 0.17 | 11.4 | 0.18 | 6.56 | 1.37 |
| MhC6a | 9.24 | 1.73 | 7.73 | 1.28 | 7.23 | 1.35 | 4.25 | 0.6 | 4.03 | 0.63 | 8.99 | 1.57 | 264 | 0.17 | 20.4 | 0.34 | 11.1 | 2.84 |
| MhC6b | 10.1 | 2.04 | 9.54 | 1.62 | 9.59 | 1.82 | 5.53 | 0.78 | 4.91 | 0.71 | 5.46 | 0.97 | 289 | 0.21 | 34.7 | 0.28 | 7.73 | 2.1 |
| Average[a] | 6.74 | 1.32 | 6.19 | 1.00 | 5.63 | 1.06 | 3.19 | 0.45 | 2.96 | 0.44 | 5.10 | 0.89 | 277 | 0.31 | 16.76 | 0.24 | 7.06 | 2.22 |
| Coal[b] | 4.07 | 0.84 | 4.65 | 0.62 | 3.74 | 0.96 | 1.79 | 0.64 | 2.08 | 0.38 | 3.71 | 0.62 | 163 | 0.47 | 15.1 | 0.79 | 5.84 | 2.43 |
| CC[c] | 1.66 | 1.57 | 1.33 | 1.62 | 1.51 | 1.10 | 1.78 | 0.71 | 1.42 | 1.16 | 1.37 | 1.44 | 1.70 | 0.66 | 1.11 | 0.30 | 1.21 | 0.91 |

a, Average, arithmetic mean; b, Coal, Chinese average coals value by Dai *et al.* [37]; c, CC, the ratio of average elements in the Mahe coals to Chinese average coals [37]; d, Hg in ng/g.

### 4.3.3. Rare Earth Elements and Yttrium (REY)

A three-fold classification of rare earth elements and yttrium (REY) was used for this study: light (LREY: La, Ce, Pr, Nd, and Sm), medium (MREY: Eu, Gd, Tb, Dy, and Y), and heavy (HREY: Ho, Er, Tm, Yb, and Lu) [42]. After the normalization to the Upper Continental Crust (UCC) [43], three distribution types are identified: L-type (light-REY; $La_N/Lu_N > 1$), M-type (medium-REY; $La_N/Sm_N < 1$, $Gd_N/Lu_N > 1$), and H-type (heavy REY; $La_N/Lu_N < 1$) [42].

The REY contents of the Mahe coals range from 100 µg/g to 336 µg/g as whole coal basis, with an average of 233 µg/g (Table 4). The Eu/Eu* value is from 0.78 to 1.02, with an average of 0.91 (Table 4). Three distribution types of REY are also present in the Mahe coal samples: L-type of MhC6a, M-type of McC2 and Mc5b, and H-type of MhC5a, respectively (Figure 5). The MhC6a shows an M-H type REY distribution pattern. In coal, L-type REY pattern is of terrigenous origin; M-type REY pattern is probably attributed to REY supply by hydrothermal solutions; and H-type REY pattern was resulted from the circulation of HREE-bearing natural water and finally adsorbed by organic matter [42]. The L-type REY pattern of the sample MhC6a is also attributed to the terrigenous detrital input. Among the LTAs of the five coal samples, the McC6a has the highest content of kaolinite and mixed-layer I/S (illite) (57.3%), following by Mh6b (39.3%), Mh5a (28.7%), Mh5b (21.1%), and MhC2 (11.8%) (Table 2). Kaolinite and mixed-layer I/S as lenses and thin beds were derived from the weathering products of the basalt from the Kangdian Upland [6]. The higher content of kaolinite and mixed-layer I/S (or illite) means that more terrigenous detrital input were received by the MhC6 coal during the peat mire accumulation than other coal seams. However, the lowest content of the kaolinite and mixed-layer I/S (illite) in McC2 suggests that the terrigenous detrital input has a less contribution to the mineral matter than the hydrothermal solution precipitation indicated by authigenic quartz and chamosite. Thus, the M-type REY pattern of MhC2 is mainly associated with the precipitation of the weathering Fe-Mg-rich solution of basalt. This is the similar situation for the sample MhC5a. The H-type REY pattern of MhC5a is attributed to its high content of organic matter (indicated by lowest ash yield 25.75%, Table 1). Because organic matter in coal favors HREE brought by the circulation of natural water [44].

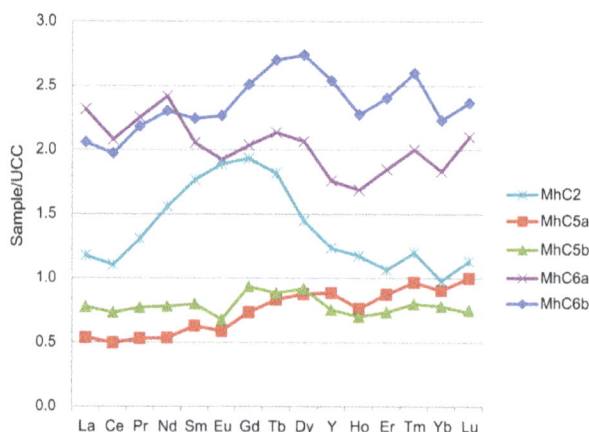

**Figure 5.** Distribution patterns of REY in the Mahe coal samples. REY are normalized by Upper Continental Crust (UCC) [43].

**Table 4.** Rare earth elements and yttrium (REY) concentrations and parameters in the Mahe coals, normalized to Upper Continental Crust [43].

| Samples | REY (µg/g) | $La_N/Lu_N$ | $La_N/Sm_N$ | $Gd_N/Lu_N$ | Ce/Ce* | Eu/Eu* |
|---------|-----------|-------------|-------------|-------------|--------|--------|
| MhC2 | 212 | 1.04 | 0.67 | 1.70 | 0.89 | 1.02 |
| MhC5a | 100 | 0.54 | 0.85 | 0.73 | 0.93 | 0.86 |
| MhC5b | 128 | 1.06 | 0.97 | 1.27 | 0.94 | 0.78 |
| MhC6a | 358 | 1.10 | 1.13 | 0.97 | 0.91 | 0.94 |
| MhC6b | 366 | 0.87 | 0.92 | 1.06 | 0.93 | 0.95 |

Ce/Ce* = $2Ce_N/(La_N + Pr_N)$; Eu/Eu* = $2Eu_N/(Sm_N + Gd_N)$.

## 5. Discussion: The Sources of Minerals in the Mahe Coals

For the Late Permian coals from the eastern Yunnan (Figure 1) including the Xuanwei [3], Xinde [4], Taoshuping [6], and Changxing [26] mines, four mineral sources were identified: (i) Emeishan basalt (the weathering detrital residues or the weathering Fe-Mg-rich silicon solution); (ii) silicic/mafic volcanic ashes; (iii) Maokou limestone; and (iii) hydrothermal fluids/igneous invasion.

However, in the northeastern Yunnan (Figure 1), mineral sources of the Mahe coals are attributed to not only (i) Emeishan basalt; (ii) silicic volcanic ashes; and (iii) hydrothermal fluids, but also Emeishan silicic rock. The kaolinite and mixed-layer I/S occurring as lenses and thin beds were derived from the weathering detrital residues of the basalt. It is also supported by the elevated concentrations of Sc, V, Cr, Co, and Ni in the Mahe coals, because Emeishan basalt are high in these elements [38]. The abundant authigenic quartz and chamosite were precipitated from the weathering Fe-Mg-rich silicon solution of Emeishan basalt [6,45]. The presence of β-form quartz in the Mahe coals suggests an input of silicic volcanic ashes during the peat mire formation. However, calcite as veins in the Mahe coals may have been precipitated by circulation of Ca-bearing meteoric fluids [46].

The euhedral quartz (Figure 3B) and apatite (Figure 3H) in the Mahe coals could not be attributed to the three sources discussed above. Firstly, quartz and apatite could not be the terrigenous detrital input of the weathering basalt residues. Because no quartz and apatite exists in the Emeishan basalt of the Kangdian Upland, which is mainly comprised of clinopyroxene, olivine, plagioclase, basaltic glasses or microlite, and magnetite [47]. Secondly, the euhedral quartz is not precipitated by the weathering Fe-Mg-rich silicon solution of the basalt. The authigenic quartz is irregular in shape rather than a euhedral form. In addition, the authigenic quartz is smaller in size than the euhedral quartz (Figure 3A,B). Thirdly, the euhedral quartz and apatite are not attributed to the silicic volcanic ashes input. β-form quartz and apatite induced by the silicic volcanic ashes in the Late Permian coal from eastern Yunnan are much smaller in size (about 20–100 µm and less than 5 µm, respectively) [3,4,6] than that of the euhedral quartz and apatite in the Mahe coals (100–500 µm and 220 um in size, respectively) (Figure 3C,H).

In the Binchuan from the western Yunnan, the silicic rock occurring at the uppermost part of the basalt sequence means there were also small scale silicic igneous eruption after massive flood basalt eruption [48]. Since quartz and apatite are common in silicic rock rather than basalt, it is inferred that euhedral quartz and apatite in the Mahe coals are probably related to the Emeishan silicic rock. However, the silicic sources of the Mahe coals is close to the peat mire in the eastern Yunnan rather than the faraway Binchuan area (more than 400 km between Binchuan and Mahe mine) from western Yunnan. Because the frequent collision and friction suffered from the the long-distance transport from western Yunnan would have resulted in more rounded shape of quartz and apatite. Thus, the euhedral quartz and apatite indicate there was silicic rock input during the Mahe coal accumulation. This further implies that there has been silicic igneous eruption in eastern Yunnan Province.

Implication of Eu/Eu* value also supports a silicic rock detrital input in the Mahe coals. As shown in Table 5, the Mahe coals show a slight negative Eu/Eu* anomaly, whereas other coals from the Xinde, Xuanwei, and Taoshuping mines have no or slight positive Eu/Eu* anomaly. The Eu/Eu* anomaly in coal, which is usually inherited from the parent rocks, is frequently used as an indicator of the terrigenous detrital sources [44,49]. The Emeishan silicic rock has a strong negative Eu/Eu* anomaly with the value of 0.56, whereas the basalt shows a positive Eu/Eu* anomaly with the value of 1.41 (Table 5). The detrital particles of silicic rock input in the Mahe coals resulted in a lower Eu/Eu* value (0.91) of the Mahe coals than that of other Late Permian coals (1.04 of the Xinde coals, 0.99 of the Xuanwei coals, and 1.06 of the Taoshuping coals), in which the terrigenous detrital input was mainly derived from the weathering residue of basalt [3,4,6].

**Table 5.** Eu/Eu* value, normalized to Upper Continental Crust (UCC) [43], in Late Permian coals from eastern Yunnan, and the Emeishan silicic rock and basalt.

| Eu/Eu* | Mahe | Xinde* | Xuanwei* | Taoshuping* | Silicic Rock* | Basalt* |
|---|---|---|---|---|---|---|
| Range | 0.78–1.92 | 0.80–1.19 | 0.73–1.27 | 0.77–1.25 | - | - |
| Average | 0.91 | 1.04 | 0.99 | 1.06 | 0.56 | 1.41 |
| Number * | 5 | 9 | 6 | 17 | - | - |

* Xinde from Dai *et al.* [4]; Xuanwei from Dai *et al.* [3]; Taoshuping from Wang *et al.* [6]; Silicic rock is rhyolite from Xu *et al.* [48]; basalt from Xiao *et al.* [38]. Number, sample number.

## 6. Conclusions

The Late Permian C2, C5a, C5b, C6a, and C6b coals from the Mahe mine are semianthracite with medium to high-ash yield and medium-high- to high-sulfur content. Minerals in the coals are mainly made up of quartz, chamosite, kaolinite, mixed-layer I/S, pyrite, and calcite, followed by anatase, dolomite, siderite, illite and marcasite. Compared with the Chinese average coals, the Mahe coals are high in Sc (4.38 µg/g), V (105 µg/g), Cr (45.7 µg/g), Co (19.0 µg/g), Ni (29.8 µg/g), Cu (70.4 µg/g), Ga (14.9 µg/g), and Sn (4.75 µg/g). The mineralogical and geochemical characteristics of the Mahe coals are attributed to four factors including Emeishan basalt, Emeishan silicic rock, silicic volcanic ashes, and hydrothermal fluid. The euhedral quartz and apatite particles in the Mahe coals mean that they were derived from silicic rocks. In addition, this further implies that there has been silicic igneous eruption in the northeast of Yunnan.

**Acknowledgments:** This research was supported by the National Key Basic Research Program of China (No. 2014CB238900), the National Natural Science Foundation of China (No. 41202121), the Program for Changjiang Scholars and Innovative Research Team in University (IRT13099). The authors wish to thank Yiping Zhou and Xiaolin Song for their great supports to this study.

**Author Contributions:** Xibo Wang conceived the overall experimental strategy and performed major elements measurement and microscopic experiments. Ruixue Wang, Jiangpeng Wei and Peipei Wang did the XRD and trace elements determination. Qiang Wei observed these samples using SEM-EDX. All authors participated in writing the manuscript.

**Conflicts of Interest:** The authors declare no conflict of interest.

## References

1. Tian, L. Coal Combustion Emissions and Lung Cancer in Xuan Wei, China. Ph.D. Thesis, University of California, Berkeley, CA, USA, 2005.
2. Tian, L.; Dai, S.; Wang, J.; Huang, Y.; Ho, S.C.; Zhou, Y.; Lucas, D.; Koshland, C.P. Nanoquartz in Late Permian C1 coal and the high incidence of female lung cancer in the Pearl River Origin area: A retrospective cohort study. *BMC Public Health* **2008**, *8*, 398. [CrossRef] [PubMed]

3.  Dai, S.F.; Tian, L.W.; Chou, C.L.; Zhou, Y.P.; Zhang, M.Q.; Zhao, L.; Wang, J.M; Yang, Z.; Cao, H.Z.; Ren, D.Y. Mineralogical and compositional characteristics of Late Permian coals from an area of high lung cancer rate in Xuan Wei, Yunnan, China: Occurrence and origin of quartz and chamosite. *Int. J. Coal Geol.* **2008**, *76*, 318–327. [CrossRef]
4.  Dai, S.F.; Li, T.; Seredin, V.V.; Ward, R.C.; Hower, J.C.; Zhou, Y.P.; Zhang, M.Q.; Song, X.L.; Song, W.J.; Zhao, C.L. Origin of minerals and elements in the Late Permian coals, tonsteins, and host rocks of the Xinde Mine, Xuanwei, eastern Yunnan, China. *Int. J. Coal Geol.* **2014**, *121*, 53–78. [CrossRef]
5.  Large, D.J.; Kelly, S.; Spiro, B.; Tian, L.; Shao, L.; Finkelman, R.; Zhang, M.; Somerfield, C.; Plint, S.; Ali, Y.; *et al.* Silica-volatile interaction and the geological cause of the Xuan Wei lung cancer epidemic. *Environ. Sci. Technol.* **2009**, *43*, 9016–9021. [CrossRef] [PubMed]
6.  Wang, X.; Dai, S.; Chou, C.-L.; Zhang, M.; Wang, J.; Song, X.; Wang, W.; Jiang, Y.; Zhou, Y.; Ren, D. Mineralogy and geochemistry of Late Permian coals from the Taoshuping Mine, Yunnan Province, China: Evidences for the sources of minerals. *Int. J. Coal Geol.* **2012**, *96–97*, 49–59. [CrossRef]
7.  Dai, S.; Wang, X.; Zhou, Y.; Hower, J.C.; Li, D.; Chen, W.M.; Zhu, X.W. Chemical and mineralogical compositions of silicic, mafic, and alkali tonsteins in the late Permian coals from the Songzao Coalfield, Chongqing. Southwest China. *Chem. Geol.* **2011**, *282*, 29–44. [CrossRef]
8.  Zhou, Y.; Bohor, B.F.; Ren, Y.L. Trace element geochemistry of altered volcanic ash layers (tonsteins) in Late Permian coal-bearing formations of eastern Yunnan and western Guizhou Province, China. *Int. J. Coal Geol.* **2000**, *44*, 305–324. [CrossRef]
9.  Dai, S.F.; Chou, C.L. Occurrence and origin of minerals in a chamosite-bearing coal of Late Permian age, Zhaotong, Yunnan, China. *Am. Mineral.* **2007**, *92*, 1253–1261. [CrossRef]
10. Coal Geology Bureau of China. *Sedimentary Environments and Coal Accumulation of Late Permian Coal Formation in Western Guizhou., Southern Sichuan and Eastern Yunnan, China*; Chongqing University Press: Chongqing, China, 1996. (In Chinese)
11. He, B.; Xu, Y.G.; Huang, X.L.; Luo, Z.Y.; Shi, Y.R.; Yang, Q.J.; Yu, S.Y. Age and duration of the Emeishan Flood volcanism, SW China: Geochemistry and SHRIMP zircon U-Pb dating of silicic ignimbrites, post volcanic Xuanwei Formation and clay tuff at Chaotian section. *Earth Planet. Sci. Lett.* **2007**, *255*, 306–323. [CrossRef]
12. He, B.; Xu, Y.G.; Guan, J.P.; Zhong, Y.T. Paleokarst on the top of the Maokou Formation: Further evidence for domal crustal uplift prior to the Emeishan flood volcanism. *Lithos* **2010**, *119*, 1–9. [CrossRef]
13. ASTM International. *Test Method for Moisture in the Analysis Sample of Coal and Coke*; ASTM D3173-11; ASTM International: West Conshohocken, PA, USA, 2011.
14. ASTM International. *Test Method for Volatile Matter in the Analysis Sample of Coal and Coke*; ASTM D3175-11; ASTM International: West Conshohocken, PA, USA, 2011.
15. ASTM International. *Test Method for Ash in the Analysis Sample of Coal and Coke from Coal*; ASTM D3174-11; ASTM International: West Conshohocken, PA, USA, 2011.
16. ASTM International. *Test Methods for Total Sulfur in the Analysis Sample of Coal and Coke*; ASTM D3177-02; ASTM International: West Conshohocken, PA, USA, 2011.
17. Ward, C.R.; Spears, D.A.; Booth, C.A.; Staton, I.; Gurba, L.W. Mineral matter and trace elements in coals of the Gunnedah Basin, NewSouth Wales, Australia. *Int. J. Coal Geol.* **1999**, *40*, 281–308. [CrossRef]
18. Ward, C.R.; Matulis, C.E.; Taylor, J.C.; Dale, L.S. Quantification of mineral matter in the Argonne Premium coals using interactive Rietveld-based X-ray diffraction. *Int. J. Coal Geol.* **2001**, *46*, 67–82. [CrossRef]
19. Ruan, C.-D.; Ward, C.R. Quantitative X-ray powder diffraction analysis of clay minerals in Australian coals using Rietveld methods. *Appl. Clay Sci.* **2002**, *21*, 227–240. [CrossRef]
20. Dai, S.F.; Li, T.J.; Jiang, Y.F.; Ward, C.R.; Hower, J.C.; Sun, J.H.; Liu, J.J.; Song, H.J.; Wei, J.P.; Li, Q.Q.; *et al.* Mineralogical and geochemical compositions of the Pennsylvanian coal in the Hailiushu Mine, Daqingshan Coalfield, Inner Mongolia, China: Implications of sediment-source region and acid hydrothermal solutions. *Int. J. Coal Geol.* **2015**, *137*, 92–110. [CrossRef]
21. Dai, S.F.; Hower, J.C.; Ward, C.R.; Guo, W.M.; Song, H.J.; O'Keefe, J.M.K.; Xie, P.P.; Hood, M.M.; Yan, X.Y. Elements and phosphorus minerals in the middle Jurassic inertinite-rich coals of the Muli Coalfield on the Tibetan Plateau. *Int. J. Coal Geol.* **2015**, *144–145*, 23–47. [CrossRef]
22. ASTM International. *Standard Test Method for Total Fluorine in Coal and Coke by Pyrohydrolytic Extraction and Ion Selective Electrode or Ion Chromatograph Methods*; ASTM Standard D5987-96 (2002) (Reapproved 2007); ASTM International: West Conshohocken, PA, USA, 2011.

23. ASTM International. *Standard Classification of Coals by Rank*; ASTM D388-12; ASTM International: West Conshohocken, PA, USA, 2012.

24. Standardization Administration of the People's Republic of China. *Classification for Quality of Coal. Part 1: Ash, 2010*; Chinese Standard GB/T 15224, 1-2010; Standardization Administration of the People's Republic of China: Beijing, China. (In Chinese)

25. Standardization Administration of the People's Republic of China. *Classification for Quality of Coal. Part 2: Sulfur, 2010*; Chinese Standard GB/T 15224, 2–2010; Standardization Administration of the People's Republic of China: Beijing, China. (In Chinese)

26. Wang, X.; Zhang, M.; Zhang, W.; Wang, J.; Zhou, Y.; Song, X.; Li, T.; Li, X.; Liu, H.; Zhao, L. Occurrence and origins of minerals in mixed-layer illite/smectite-rich coals of the Late Permian age from the Changxing Mine, eastern Yunnan, China. *Int. J. Coal Geol.* **2012**, *102*, 26–34. [CrossRef]

27. Tang, X.Y.; Huang, W.H. *Trace Elements in Chinese Coal*; Business Press: Beijing, China, 2004. (In Chinese)

28. Faraj, B.S.M.; Fielding, C.R.; Mackinnon, D.R. Cleat mineralization of Upper Permian Baralaba/Rangal Coal Measures, Bowen Basin, Australia. In *Coalbed Methane and Coal Geology*; Gayer, R., Harris, I., Eds.; Geological Society Special Publication: London, UK, 1996; Volume 109, pp. 151–164.

29. Permana, A.; Ward, C.R.; Li, Z.; Gurba, L.W. Distribution and origin of minerals in high-rank coals of the South Walker Creek area, Bowen Basin, Australia. *Int. J. Coal Geol.* **2013**, *116–167*, 185–207. [CrossRef]

30. Vassilev, S.V.; Kitano, K.; Vassileva, C.G. Some relationships between coal rank and chemical and mineral composition. *Fuel* **1996**, *75*, 1537–1542. [CrossRef]

31. Wang, X.B.; Jiang, Y.F.; Zhou, G.Q.; Wang, P.P.; Wang, R.X.; Zhao, L.; Chou, C.-L. Behavior of minerals and trace elements during natural coking: A case study of an intruded bituminous coal in the Shuoli mine, Anhui Province, China. *Energy Fuels* **2015**. [CrossRef]

32. Ward, C.R. Analysis and significance of mineral matter in coal seams. *Int. J. Coal Geol.* **2002**, *50*, 135–168. [CrossRef]

33. Dai, S.F.; Seredin, V.V.; Ward, C.R.; Hower, J.C.; Xing, Y.W.; Zhang, W.G.; Song, W.J.; Wang, P.P. Enrichment of U–Se–Mo–Re–V in coals preserved within marine carbonate successions: Geochemical and mineralogical data from the Late Permian Guiding Coalfield, Guizhou, China. *Miner. Deposita* **2015**, *50*, 159–186. [CrossRef]

34. Bouška, V.; Pešek, J.; Sýkorová, I. Probable modes of occurrence of chemical elements in coal. *Acta Mont. Ser. B Fuel Carbon Miner. Process.* **2000**, *10*, 53–90.

35. Dai, S.F.; Yang, J.Y.; Ward, C.R.; Hower, J.C.; Liu, H.D.; Garrison, T.M.; French, D.; O'Keefe, J.M.K. Geochemical and mineralogical evidence for a coal-hosted uranium deposit in the Yili Basin, Xinjiang, northwestern China. *Ore Geol. Rev.* **2015**, *70*, 1–30. [CrossRef]

36. Chou, C.-L. Sulfur in coals: A review of geochemistry and origins. *Int. J. Coal Geol.* **2012**, *100*, 1–13. [CrossRef]

37. Dai, S.F.; Ren, D.Y.; Chou, C.L.; Finkelman, R.B.; Seredin, V.V.; Zhou, Y.P. Geochemistry of trace elements in Chinese coals: A review of abundances, genetic types, impacts on human health, and industrial utilization. *Int. J. Coal Geol.* **2012**, *94*, 3–21. [CrossRef]

38. Xiao, L.; Xu, Y.; Mei, H.; Zheng, Y.; He, B.; Pirajno, F. Distinct mantle sources of low-Ti and high-Ti basalts from the western Emeishan large igneous province, SW China: Implications for plume-lithosphere interaction. *Earth Planet. Sci. Lett.* **2004**, *228*, 525–546. [CrossRef]

39. Dai, S.F; Zeng, R.S; Sun, Y.Z. Enrichment of arsenic, antimony, mercury, and thallium in a late Permian anthracite from Xingren, Guizhou, southwest China. *Int. J. Coal Geol.* **2006**, *66*, 217–226. [CrossRef]

40. Hower, J.C.; Campbell, J.L.; Teesdale, W.J.; Nejedly, Z.; Robertson, J.D. Scanning proton microprobe analysis of mercury and other trace elements in Fe-sulfides from a Kentucky coal. *Int. J. Coal Geol.* **2008**, *75*, 88–92. [CrossRef]

41. Yudovich, Ya.E.; Ketris, M.P. Mercury in coal: A review. Part 1. Geochemistry. *Int. J. Coal Geol.* **2005**, *62*, 107–134. [CrossRef]

42. Seredin, V.V.; Dai, S. Coal deposits as a potential alternative source for lanthanides and yttrium. *Int. J. Coal Geol.* **2012**, *94*, 67–93. [CrossRef]

43. Taylor, S.R.; McLennan, S.M. *The Continental Crust: Its Composition and Evolution*; Blackwell: Oxford, UK, 1985; p. 312.

44. Seredin, V.; Finkelman, R. Metalliferous coals: A review of the main genetic and geochemical types. *Int. J. Coal Geol.* **2008**, *76*, 253–289. [CrossRef]

45. Ren, D.Y. Mineral matter in coal. In *Coal Petrology of China*; Han, D.X., Ed.; Publishing House of China University of Mining and Technology: Xuzhou, China, 1996; pp. 67–78.

46. Kolker, A.; Chou, C.-L. Cleat-filling calcite in Illinois Basin coals: Trace element evidence for meteoric fluid migration in a coal basin. *J. Geol.* **1994**, *102*, 111–116. [CrossRef]

47. Xu, Y.G.; Chung, S.L.; Jahn, B.M.; Wu, G.Y. Petrologic and geochemical constraints on the petrogenesis of Permian-Triassic Emeishan flood basalts in southwestern China. *Lithos* **2001**, *58*, 145–168. [CrossRef]

48. Xu, Y.G.; Chung, S.L.; Shao, H.; He, B. Silicic magmas from the Emeishan large igneous province, southwest China: Petrogenesis and their link with the end-Guadalupian biological crisis. *Lithos* **2010**, *119*, 47–60. [CrossRef]

49. Goodarzi, F.; Sanei, H.; Stasiuk, L.D.; Bagheri-Sadeghi, H.; Reyes, J. A preliminary study of mineralogy and geochemistry of four coal samples from northern Iran. *Int. J. Coal Geol.* **2006**, *65*, 35–50. [CrossRef]

*minerals*

Article

# Mineralogical and Geochemical Compositions of the No. 5 Coal in Chuancaogedan Mine, Junger Coalfield, China

Ning Yang, Shuheng Tang *, Songhang Zhang and Yunyun Chen

School of Energy Resources, China University of Geosciences, Beijing 100083, China;
yangning@cugb.edu.cn (N.Y.); zhangsh@cugb.edu.cn (S.Z.); chenyy@cugb.edu.cn (Y.C.)
* Correspondence: tangsh@cugb.edu.cn; Tel./Fax: +86-10-8232-2005

Academic Editor: Shifeng Dai
Received: 21 September 2015; Accepted: 17 November 2015; Published: 25 November 2015

**Abstract:** This paper reports the mineralogy and geochemistry of the Early Permian No. 5 coal from the Chuancaogedan Mine, Junger Coalfield, China, using optical microscopy, scanning electron microscopy (SEM), Low-temperature ashing X-ray diffraction (LTA-XRD) in combination with Siroquant software, X-ray fluorescence (XRF), and inductively coupled plasma mass spectrometry (ICP-MS). The minerals in the No. 5 coal from the Chuancaogedan Mine dominantly consist of kaolinite, with minor amounts of quartz, pyrite, magnetite, gypsum, calcite, jarosite and mixed-layer illite/smectite (I/S). The most abundant species within high-temperature plasma-derived coals were $SiO_2$ (averaging 16.90%), $Al_2O_3$ (13.87%), $TiO_2$ (0.55%) and $P_2O_5$ (0.05%). Notable minor and trace elements of the coal include Zr (245.89 mg/kg), Li (78.54 mg/kg), Hg (65.42 mg/kg), Pb (38.95 mg/kg), U (7.85 mg/kg) and Se (6.69 mg/kg). The coal has an ultra-low sulfur content (0.40%). Lithium, Ga, Se, Zr and Hf present strongly positive correlation with ash yield, Si and Al, suggesting they are associated with aluminosilicate minerals in the No. 5 coal. Arsenic is only weakly associated with mineral matter and Ge in the No. 5 coals might be of organic and/or sulfide affinity.

**Keywords:** early Permian coals; minerals; trace elements; Junger Coalfield

## 1. Introduction

Coal is responsible for about 65% of electricity generation in China. The large abundance of coal makes it a reliable, long-term fuel source for both in China and in other coal-rich countries like Australia, Turkey and South Africa. With the increasing use of coal, a large amount of pollutants are produced, not only gas emissions ($SOx$, $NOx$ and $CO_2$) but also ash residues. Environmental impact of coal and coal combustion are generally associated with the minerals and the trace elements in coal. Studies on the mineralogy and geochemistry of coal are the basic work for researching environmental impact of coal and coal combustion. Dai *et al.* [1,2], Gürdal [3], Yang [4] Wang [5], Kolker [6], Finkelman [7] and Tang *et al.* [8] have done much research on mineralogical and geochemical characteristics of the coal in many areas. The Ordos basin is the most important energy base in China. Late Paleozoic coals from the Ordos basin have attracted much attention. Dai *et al.* [9–11], and Wang *et al.* [12] have studied the geochemistry and mineralogy of the coal and its coal combustion products from the Heidaigou, Guanbanwusu, and Haerwusu Surface Mines in the Junger Coalfield. The previous studies mostly focused on the No. 6 coals in Junger Coalfield. In this paper, we report the data on the mineralogy and elemental geochemistry of the No. 5 Coals in the Chuancaogedan mine, Junger coalfield, China.

## 2. Geological Setting

The Junger Coalfield is located on the northeastern margin of the Ordos Basin. The coalfield is 65-km long (N–S) and 26-km wide (W–E), with a total area of 1700 $km^2$. The geological setting of the area has been described in detail by Dai *et al.* [9]. The Chuancaogedan Mine is situated in the southeastern part of the Junger Coalfield (Figure 1).

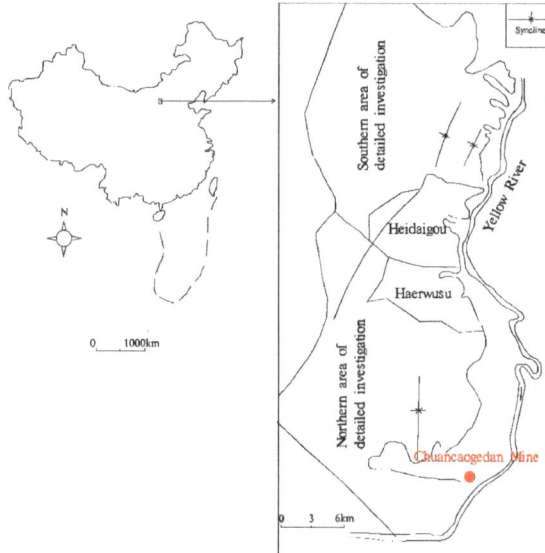

**Figure 1.** Location of the Chuancaogedan Mine in the Junger Coalfield, northern China (modified after Dai *et al.* [10]).

The coal-bearing sequences include Benxi Formation and Taiyuan Formation (both Pennsylvanian) and the Shanxi Formation (Lower Permian) with a collective thickness of 134 m; 110–150 m of which is mainly the Taiyuan and Shanxi formation (Figure 2). The Taiyuan Formation, with a thickness of 52 m, is mainly made up of sandstone, mudstone and coals. In the Shanxi Formation, which has a thickness of 67 m, there are five coal seams, named No. 1, No. 2, No. 3, No. 4 and No. 5 Coals in order from top to bottom.

## 3. Samples and Analytical Procedures

Fifteen bench samples of the No. 5 Coal were collected from the Chuancaogedan Mine, Junger Coalfield following the Chinese Standard Method GB 482-2008 [13], the cumulative thickness of the No. 5 Coal is about 4.0 m. From bottom to top, the fifteen bench samples are ZG501 to ZG517. All samples were air-dried, sealed in polyethylene bags to prevent oxidation, and parts of them were ground to pass 200 mesh, and stored in brown glass bottles for chemical analyses.

Proximate analyses were measured in accordance with ASTM standards (ASTM D3173-11 [14], ASTM D3175-11 [15], and ASTM D3174-11 [16], respectively). Total sulfur was determined following the ASTMD 3177-02 [17].

Mineralogical analyses of the coal samples were performed by means of Powder X-ray diffraction (XRD), optical microscopy and scanning electron microscopy (SEM).

Low-temperature ashing of the powdered coal samples was carried out using an EMITECH K1050X plasma asher (Quorum, Lewes, UK) prior to XRD analysis. XRD analysis of the low-temperature ashes was performed on a D/max-2500/PC powder diffractometer (Rigaku, Tokyo, Japan) with Ni-filtered Cu-K$\alpha$ radiation and a scintillation detector. Each XRD pattern was recorded over a 2$\theta$ interval of 2.6°–70°, with a step size of 0.01°. X-ray diffractograms of the Low-temperature ashings (LTAs) and non-coal samples were subjected to quantitative mineralogical analysis using the Siroquant™ interpretation software system (Sietronics, Mitchell, Australia). More analytical details are given by Dai *et al.* [18,19] and Wang *et al.* [20].

X-ray fluorescence (XRF) spectrometry (ARL ADVANT'XP+, ThermoFisher, Waltham, MA, USA) was used to determine the major element oxides in high-temperature ashed coal samples, including $SiO_2$, $Al_2O_3$, $CaO$, $K_2O$, $Na_2O$, $Fe_2O_3$, $MnO$, $MgO$, $TiO_2$ and $P_2O_5$. Trace elements within acid-digested ashed coal samples, except for As, Se, Hg and F, were determined by conventional inductively coupled plasma mass spectrometry (ICP-MS). For its analysis, samples were digested using an UltraClave Microwave High Pressure Reactor (Milestone, Sorisole, Italy). The basic load for the digestion tank was composed of 330-mL distilled $H_2O$, 30-mL 30% $H_2O_2$, and 2-mL 98% $H_2SO_4$. Initial nitrogen pressure was set at 50 bars and the highest temperature was set at 240 °C that lasted for 75 min. The reagents for 50-mg sample digestion were 5 mL 40% HF, 2 mL 65% $HNO_3$ and 1 mL 30% $H_2O_2$. Multi-element standards were used for calibration of trace element concentrations. More details are given by Dai *et al.* [21] Arsenic and Sewere analyzed by more advanced ICP-MS which utilized collision/reaction cell technology (ICP-CCT-MS) as outlined by Li *et al.* [22]. Fluorine was determined by an ion-selective electrode (ISE) method. Mercury was determined using a Milestone DMA-80 Hg analyzer (Milestone, Sorisole, Italy).

The quantitative analysis of minerals and determinations of elements were completed at the State Key Laboratory of Coal Resources and Safe Mining of China University of Mining and Technology (Beijing, China).

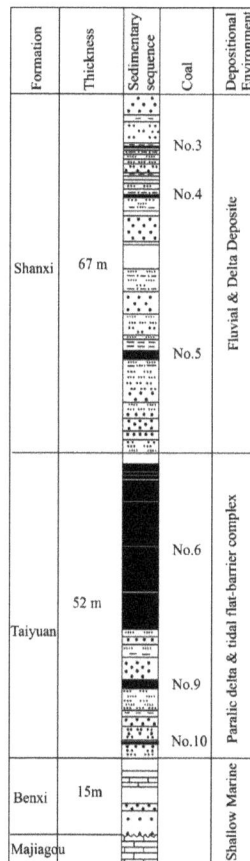

**Figure 2.** Stratigraphic sequence of the Junger Coalfield [9].

## 4. Results and Discussion

### 4.1. Coal Chemistry

The results of the total sulfur and proximate analysis of samples from the No. 5 coal are presented in Table 1. Ash yields of the Chuancaogedan No. 5 coal range from 5.95% to 60.70% (Figure 3), with an average of 32.69%, indicating a high ash coal according to Chinese National Standard (GB/T 15224.1-2004, 10.01% to 16.00% for low ash coal, 16.01% to 29.00% for medium ash coal, and >29.00% for high ash coal) [23]. The ash yields tend to increase from the bottom to the top in the coal seam.

The contents of volatile matter of the No. 5 coal varies from 32.57% to 50.30% through the coal-seam section (Figure 3), with a mean of 37.22%, suggesting that the Chuancaogedan coals are medium-high volatile bituminous coals based on MT/T 849-2000 (28.01% to 37.00% for medium-high volatile coal, 37.01% to 50.00% for high volatile coal and >29.00% for super high volatile coal) [24].

The No. 5 coals have a moisture content of 2.22% to 5.61% (Figure 3), with an average of 3.81%, indicating a low-medium rank coal in accordance of MT/T 850-2000 (≤5% for low moisture coal, 5% to 15% for medium moisture coal, and >15% for high moisture coal) [25].

The total sulfur of No. 5 coals changes from 0.12% to 0.83% (Figure 3), averaging 0.40%, which corresponds to ultra-low-sulfur coal according to Chinese National Standard (GB/T 15224.2-2010) (<0.5% for super low sulfur coal, 0.51% to 0.9% for low sulfur coal and 0.9% to 1.50% for medium sulfur coal) [26].

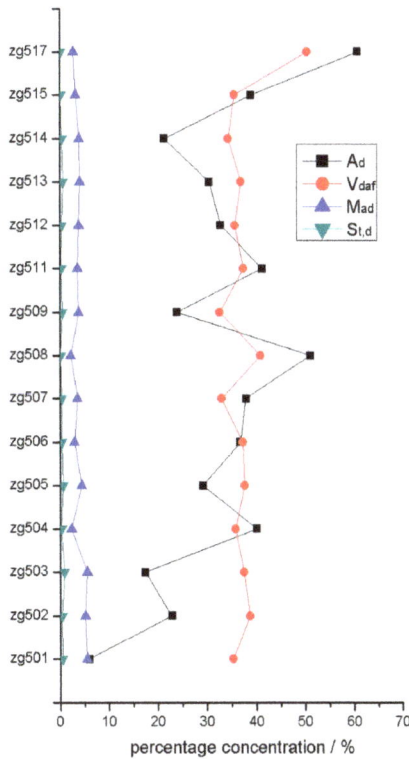

**Figure 3.** Variation of total sulfur and proximate analysis through the No. 5 Coal section.

**Table 1.** Proximate analysis and total sulfur in the No. 5 Coal (%).

| Sample | Proximate Analysis | | | $S_{t,d}$ |
|--------|--------|--------|--------|--------|
| | $M_{ad}$ | $V_{daf}$ | $A_d$ | |
| ZG517 | 2.78 | 50.3 | 60.7 | 0.12 |
| ZG515 | 3.19 | 35.52 | 38.9 | 0.18 |
| ZG514 | 3.91 | 34.3 | 21.38 | 0.40 |
| ZG513 | 4.14 | 36.9 | 30.46 | 0.61 |
| ZG512 | 3.86 | 35.59 | 32.79 | 0.35 |
| ZG511 | 3.64 | 37.39 | 41.16 | 0.30 |
| ZG509 | 3.82 | 32.57 | 23.9 | 0.47 |
| ZG508 | 2.22 | 40.84 | 51.02 | 0.21 |
| ZG507 | 3.54 | 32.91 | 37.88 | 0.36 |
| ZG506 | 2.94 | 37.23 | 36.66 | 0.36 |
| ZG505 | 4.39 | 37.61 | 29.17 | 0.64 |
| ZG504 | 2.42 | 35.75 | 40.04 | 0.29 |
| ZG503 | 5.61 | 37.47 | 17.41 | 0.83 |
| ZG502 | 5.2 | 38.63 | 22.89 | 0.46 |
| ZG501 | 5.52 | 35.29 | 5.95 | 0.48 |
| Average | 3.81 | 37.22 | 32.69 | 0.40 |

M, moisture; V, volatile matter; A, ash yield; $S_t$, total sulfur; ad, air-dry basis; d, dry basis; daf, dry and ash-free basis.

### 4.2. Minerals in the No. 5 Coal

The mineral phase percentages were calculated to a coal ash basis from the XRD results obtained on the low temperature ashes and are reported in Table 2. The results show that minerals in the No. 5 coal are mainly made up of kaolinite, followed by gypsum (averaging 0.99%), magnetite (0.85%), calcite (0.33%), quartz (0.31%), pyrite (0.26%) and mixed-layer I/S (0.01%).

**Table 2.** Mineral contents in coal samples from the Chuancaogedan Mine measured by Low-temperature ashing X-ray diffraction (LTA-XRD) (%).

| Samples | Kaolinite | Quartz | Magnetite | Pyrite | Gypsum | Calcite | Jarosite | I/S |
|---------|-----------|--------|-----------|--------|--------|---------|----------|-----|
| ZG517 | 55.42 | 0.97 | 4.31 | - | - | - | - | - |
| ZG515 | 38.32 | - | 0.58 | - | - | - | - | - |
| ZG514 | 21.38 | - | - | - | - | - | - | - |
| ZG513 | 30.46 | - | - | - | - | - | - | - |
| ZG512 | 32.79 | - | - | - | - | - | - | - |
| ZG511 | 41.16 | - | - | - | - | - | - | - |
| ZG509 | 23.57 | 0.07 | - | 0.26 | - | - | - | - |
| ZG508 | 50.76 | - | 0.26 | - | - | - | - | - |
| ZG507 | 37.77 | 0.11 | | - | - | - | - | - |
| ZG506 | 36.22 | - | 0.44 | - | - | - | - | - |
| ZG505 | 29.11 | - | 0.06 | - | - | - | - | - |
| ZG504 | 39.72 | 0.08 | 0.24 | - | - | - | - | - |
| ZG503 | 15.32 | - | - | - | 1.15 | 0.33 | 0.61 | - |
| ZG502 | 22.07 | - | - | - | 0.82 | - | - | - |
| ZG501 | 5.89 | - | 0.05 | - | - | - | - | 0.01 |

I/S: mixed-layer illite/smectite.

Kaolinite is common in coal [27,28]. As presented in Table 2, kaolinite is the most abundant mineral in the Chuancaogedan coal seam, with abundance within the ash varying from 5.89% to 55.42% (average 32.00%). Kaolinite occurs as infillings of cells or fractures (Figure 4A–C). In addition, kaolinite presents as thin-layered or flocculent forms (Figure 5A,B) in the No. 5 Coal.

Pyrite is only observed in ZG509 (0.26 wt %) (Figure 6), occurring as fracture-fillings (Figure 4D) or as pyrite aggregates (Figure 5C).

Magnetite presents in seven samples; the content varies from 0.05% to 4.31%. Other minerals, such as quartz, calcite, jarosite, mixed-layer illite/smectite (I/S) and gypsum, are only present in a few samples. Gypsum occurs in columnar form as shown by SEM scans (Figure 5D).

**Figure 4.** Minerals in the No. 5 Coal (reflected light): (**A**) kaolinite in dispersed form; (**B**) kaolinite in-filling cells; (**C**) kaolinite with organic matter; and (**D**) pyrite in vitrinite.

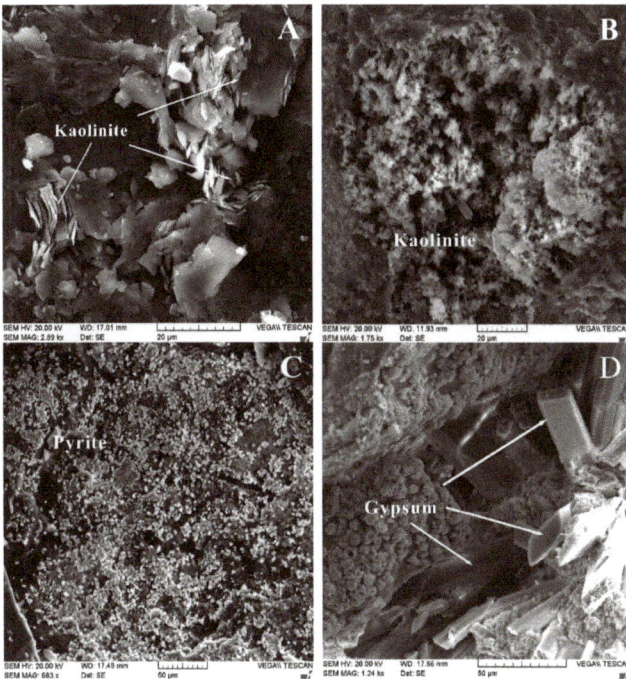

**Figure 5.** Minerals in the No. 5 Coal (SEM, secondary electron images): (**A**) kaolinite as thin-layered forms; (**B**) flocculent kaolinite; (**C**) pyrite aggregates; and (**D**) columnar gypsum.

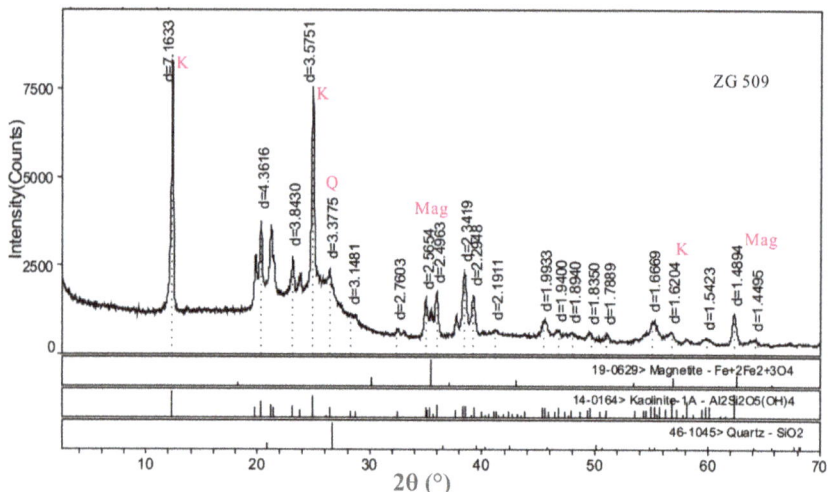

**Figure 6.** X-ray diffraction (XRD) patterns of coal samples (ZG509).

### 4.3. Geochemistry of the No. 5 Coals

#### 4.3.1. Major Elements

The major elements in coals from the Chuancaogedan Mine are dominated by $SiO_2$, $Al_2O_3$, and $Fe_2O_3$ (Table 3), which conform to major mineral compositions of the coals (kaolinite, magnetite and pyrite). Average values for high-temperature plasma No. 5 coal samples are as follows: $SiO_2$ (16.90 wt %), $Al_2O_3$ (13.87 wt %), $Fe_2O_3$ (0.70 wt %), $TiO_2$ (0.55 wt %), CaO (0.26 wt %), $K_2O$ (0.06 wt %), MgO (0.04 wt %), $Na_2O$ (0.02 wt %), and $P_2O_5$ (0.05 wt %). Coals from Chuancaogedan Mine contain higher proportions of $SiO_2$, $Al_2O_3$, $TiO_2$, $P_2O_5$, and lower proportions of $Fe_2O_3$, $Na_2O$ than the average values for Chinese coals reported by Dai *et al.* [29].

The $SiO_2/Al_2O_3$ ratios range from 1.17 to 1.27, with an average of 1.22 for the No.5 coal. This is higher than those of other Chinese coals (1.42) [29] and also higher than the theoretical $SiO_2/Al_2O_3$ ratio of kaolinite (1.18), suggesting quartz or amorphous silica occurs in the mineral matter portion of the coal. The ash has a $TiO_2$ content of 0.88% to 2.93%, much higher than the proportion within ash of other Chinese coals, and this is mainly affiliated with magnetite in No. 5 coal. Iron may be isomorphic replaced by Ti in magnetite ($Fe_3O_4$). The average contents of $K_2O$ and $Na_2O$ are 0.18% and 0.05%, respectively. $K_2O$ and $Na_2O$ are probably attributed to mixed-layer I/S. The concentration of $Fe_2O_3$ varies from 0.27% to 1.66%, with an average of 0.70%. The positive relation coefficient between $Fe_2O_3$ and $S_{t,d}$ ($rFe_2O_3$-$S_{t,d}$ = 0.66) suggest that Fe is mainly associated with sulfide (pyrite).

#### 4.3.2. Trace Elements

In contrast with the common Chinese coals [29], the No. 5 coals are slightly enriched in Li (averaging 78.54 mg/kg), Se (6.69 mg/kg), Zr (245.89 mg/kg), Hg (65.42 mg/kg), Pb (38.95 mg/kg), and U (7.85 mg/kg), with CC between 2 and 5 (CC, concentration coefficient, is the ratio of element concentration in investigated coals *vs.* Chinese coals or world hard coals [30]), while As (averaging 0.28 mg/kg), Co (3.44 mg/kg), Sr (93.10 mg/kg), Sb (0.36 mg/kg), and Tl (0.16 mg/kg) are depleted (with CC lower than 0.5), and the remaining elements (CC are between 0.5 and 2) are close to the average values for Chinese coals [29].

As stated above, elements including Li, Se, Zr and Hf are higher than that for Chinese average coals [29], and F and Ga are close to the average values. The correlation coefficients between F,

Ga and ash are 0.81 and 0.78, respectively, and the main mineral in coals is kaolinite, so they are probably related to the kaolinite (Figures 4 and 5). The high trace elements and boehmite in the No. 6 coals were derived from the weathered and oxidized bauxite in the exposed crust of the older Benxi Formation (Missisippian) situated to the northeast of the coal basin [11]. Benxi Formation bauxite; was an important terrigenous source for most Late Paleozoic coals in Junger coalfield, China [9]. During peat accumulation, the Junger Coalfield was in the low lying area between the Yinshan Oldland to the N and W and the upwarped Benxi Formation to the N and E. The paleo rivers ran dominantly in the N and E directions from these sediment-source regions to the Junger Coalfield [31].

### 4.3.3. Evaluated Li, Ga, Se, Zr, Hf, As, and Ge in the No. 5 Coal

*Lithium*: The content of Li in the No. 5 coals varies from 17.76 to 157.83 mg/kg (average 78.54 mg/kg), which is much higher than that of the No. 6 coals (average 37.80 mg/kg) [9] and Chinese coals (average 14 mg/kg) [29]. Lithium in coal samples is positively correlated with ash yield, Si, and Al, with correlation coefficients of 0.88, 0.69 and 0.62, respectively (Table 4), suggesting that Li is associated with aluminosilicate minerals.

*Gallium*: The Chuancaogedan coals have a Ga content close to the Chinese coal average [26], ranging from 6.34 to 27.10 mg/kg, with an average of 13.98 mg/kg. Gallium is generally related to clay minerals in coal [1,32]. The correlation coefficient between Ga and ash yield, Si and Al are 0.78, 0.51 and 0.24, respectively (Table 4). This strongly suggests that kaolinite may contain (but is not high in) Ga, and Ga mainly occurs in inorganic association.

*Selenium*: The concentration of Se in the No. 5 coals ranges from 2.02 to 19.07 mg/kg, with a mean of 6.69 mg/kg. The correlation between Se and ash yield, Si, and Al (correlation coefficient = 0.60, 0.37, 0.11 (Table 4)) suggest that only part of total Se exists in minerals.

*Zirconium and Hafnium*: Zr and Hf are enriched in the No. 5 coals, with average concentration of 245.89 mg/kg and 6.93 mg/kg, respectively. The correlation coefficient of Zr-Hf is 0.99 (Table 4), showing that they have similar occurrence. They are both positively correlated with ash yield, Si, and Al (rZr-ash = 0.76, rZr-Si = 0.59, rZr-Al = 0.62, rHf-ash = 0.81, rHf-Si = 0.64, rHf-Al = 0.67 (Table 4)), identifying the occurrence of Zr and Hf in association with aluminosilicate minerals. Zircon is the most common zirconium mineral, therefore the Zr is believed to be at least partly due to the probable presence of this heavy mineral these samples [10].

*Arsenic*: The content of As in the No. 5 coals was below the ICP-MS detection limit for three samples, but otherwise varies from 0.15 up to 0.64 mg/kg (average 0.28 mg/kg), which is lower than that of both the No. 6 coals (average 0.56 mg/kg) [9] and Chinese coals (average 5.00 mg/kg) [29]. A wide variety of As-bearing phases has been observed in high-As coals from southwestern Guizhou; for example: pyrite; Fe–As oxide; K–Fe sulfate; and As-bearing clays [33,34]. Occurrences of organically associated As have also been reported in Guizhou coal [34]. Arsenic in the Chongqing coal correlates with $Fe_2O_3$, suggesting a pyrite affinity [35]. The correlation coefficient between As and ash yield, Si, and Al in Chuancaogedan coals are 0.34, 0.54 and 0.22, respectively, which indicates that only a small part of the total As occurs in minerals. Arsenic has a negative correlation with $Fe_2O_3$ (correlation coefficient of −0.36), which suggests that As may not be affiliated with pyrite occurrence in the No. 5 coals.

*Germanium*: The Chuancaogedan coals have a Ge content of close to the average for Chinese coals [29], ranging from 0.35 to 4.21 mg/kg, with an average of 1.74 mg/kg. In the Tongda coal mine, Yimin coalfield, Ge occurs with major organic affinity, and partial sulfide affinity was observed also. As, Fe, and S show similar trends to Ge, though with a markedly higher sulfide affinity (mainly in pyrite) [36]. The correlation coefficients of Ge and ash yield, major elements and selected trace elements in the No. 5 coals range from −0.53 to 0.40, which means Ge may presents organic and/or sulfide affinity in these coals.

**Table 3.** Elemental concentrations in the No. 5 Coal from Chuancaogedan Mine (oxides in %, elements in mg/kg, Hg in ng/g).

| Elemental Concentrations | Sample | | | | | | | | | | | | | | | Average | Coal [a] |
|---|---|---|---|---|---|---|---|---|---|---|---|---|---|---|---|---|---|
| | ZG517 | ZG515 | ZG514 | ZG513 | ZG512 | ZG511 | ZG509 | ZG508 | ZG507 | ZG506 | ZG505 | ZG504 | ZG503 | ZG502 | ZG501 | | |
| $SiO_2$ | 32.55 | 20.7 | 11.13 | 15.98 | 17.15 | 21.82 | 12.32 | 27.33 | 19.57 | 19.55 | 15.1 | 20.24 | 7.94 | 11.63 | 2.98 | 16.9 | 8.47 |
| $Al_2O_3$ | 25.71 | 16.72 | 9.08 | 13 | 13.97 | 17.8 | 10.2 | 22.47 | 16.28 | 16.03 | 12.37 | 17.3 | 6.55 | 9.45 | 2.5 | 13.87 | 5.98 |
| $Fe_2O_3$ | 1.66 | 0.33 | 0.27 | 0.56 | 0.31 | 0.34 | 0.46 | 0.37 | 0.51 | 0.34 | 0.75 | 0.42 | 1.3 | 0.86 | 0.22 | 0.7 | 4.85 |
| $TiO_2$ | 1.11 | 0.73 | 0.52 | 0.45 | 0.96 | 0.76 | 0.5 | 0.47 | 0.73 | 0.36 | 0.45 | 0.77 | 0.24 | 0.2 | 0.07 | 0.55 | 0.33 |
| $CaO$ | 0.33 | 0.14 | 0.12 | 0.13 | 0.11 | 0.16 | 0.12 | 0.09 | 0.17 | 0.1 | 0.12 | 0.2 | 0.69 | 0.37 | 0.08 | 0.26 | 1.23 |
| $K_2O$ | 0.14 | 0.05 | 0.02 | 0.04 | 0.03 | 0.04 | 0.02 | 0.08 | 0.09 | 0.06 | 0.09 | 0.17 | 0.04 | 0.03 | 0.01 | 0.06 | 0.19 |
| $MgO$ | 0.1 | 0.04 | 0.02 | 0.03 | 0.03 | 0.02 | 0.03 | 0.04 | 0.05 | 0.03 | 0.04 | 0.06 | 0.06 | 0.04 | 0.01 | 0.04 | 0.22 |
| $Na_2O$ | 0.02 | 0.01 | 0.01 | 0.01 | 0.01 | 0.01 | 0.01 | 0.02 | 0.02 | 0.01 | 0.01 | 0.03 | 0.03 | 0.02 | 0 | 0.02 | 0.16 |
| $P_2O_5$ | 0.02 | 0.02 | 0.01 | 0.01 | 0.02 | 0.02 | 0.02 | 0.02 | 0.18 | 0.01 | 0.02 | 0.47 | 0.01 | 0.01 | 0 | 0.05 | 0.09 |
| Li | 114.2 | 83.99 | 46.99 | 72.25 | 82.51 | 120.42 | 75.33 | 157.83 | 83.09 | 103.42 | 56.04 | 79.12 | 35.33 | 49.82 | 17.76 | 78.54 | 14 |
| Be | 7.75 | 12.28 | 6.43 | 4.16 | 2.75 | 1.7 | 1.97 | 1.56 | 3.42 | 1.41 | 1.07 | 2.79 | 1.78 | 2.31 | 3.65 | 3.67 | 2 |
| F | 345.88 | 251.06 | 156.83 | 193.14 | 214.15 | 263.93 | 159.81 | 279.3 | 291.42 | 208.39 | 204.84 | 385.22 | 137.27 | 124.31 | 59.85 | 218.37 | 140 |
| Sc | 12.79 | 7.62 | 4.61 | 4.57 | 6.87 | 9.09 | 3.62 | 12.95 | 14.82 | 11.47 | 9.23 | 10.36 | 6.51 | 5.74 | 3.75 | 8.27 | 3 |
| V | 64.73 | 32.79 | 31.86 | 28.11 | 37.3 | 39.77 | 44.09 | 19.05 | 30.68 | 27.24 | 23.07 | 31.78 | 11.74 | 10.8 | 11.35 | 29.63 | 21 |
| Cr | 18.61 | 8.41 | 12.87 | 7.89 | 10.26 | 7.7 | 10.48 | 3.83 | 8.86 | 6.49 | 7.66 | 13.05 | 4.43 | 2.64 | 1.65 | 8.32 | 12 |
| Co | 4.98 | 2.9 | 4.86 | 5.4 | 2.69 | 2.03 | 3.23 | 1.23 | 1.13 | 1.7 | 1.89 | 0.86 | 5.99 | 5.61 | 7.05 | 3.44 | 7 |
| Ni | 10.37 | 10.74 | 11.54 | 12.64 | 6.9 | 5.13 | 7.35 | 4.35 | 4.63 | 3.71 | 5.21 | 4.16 | 13.59 | 17.47 | 16.9 | 8.98 | 14 |
| Cu | 18.84 | 22.08 | 19.63 | 20.61 | 17.23 | 10.95 | 17.78 | 9.13 | 13.36 | 16.73 | 12.42 | 22.44 | 7.06 | 8.36 | 8 | 14.97 | 13 |
| Zn | 14.85 | 7.12 | 16.37 | 19.96 | 17.08 | 16.39 | 14.62 | 11.81 | 19.36 | 35.83 | 30.03 | 25.94 | 54.71 | 35.57 | 13.6 | 22.22 | 35 |
| Ga | 27.1 | 15.62 | 9.51 | 13.58 | 18.24 | 16.19 | 13.34 | 12.89 | 14.7 | 13.36 | 17.16 | 12.3 | 10.34 | 9.05 | 6.34 | 13.98 | 9 |
| As | 0.64 | 0.42 | 0.36 | 0.62 | 0.23 | 0.18 | 0.42 | 0.15 | 0.24 | 0.19 | 0.41 | 0 | 0 | 0.31 | 6.34 | 0.28 | 5 |
| Se | 19.07 | 4.41 | 5.64 | 5.94 | 10.35 | 11.07 | 8.83 | 3.83 | 6.51 | 4.31 | 5.28 | 5.09 | 4.45 | 3.62 | 2.02 | 6.69 | 2 |
| Rb | 5.81 | 1.92 | 0.37 | 1.41 | 0.73 | 1 | 0.25 | 2.65 | 2.59 | 1.84 | 2.21 | 4.04 | 0.28 | 1.01 | 0.2 | 1.75 | 8 |
| Sr | 20.8 | 12.34 | 15.2 | 14.31 | 11.25 | 11.54 | 14.68 | 14.97 | 321.07 | 14.08 | 24.89 | 849.94 | 35.48 | 18.81 | 17.13 | 93.1 | 423 |
| Y | 0.19 | 0.22 | 0.29 | 0.2 | 0.27 | 0.18 | 0.38 | 0.21 | 0.28 | 0.08 | 0.12 | 0.3 | 3.77 | 0.27 | 5.43 | 0.81 | 20.76 |
| Zr | 450.76 | 241.43 | 165.49 | 292.81 | 303.95 | 354.05 | 272.13 | 221.16 | 270.52 | 262.09 | 326.43 | 202.89 | 139.76 | 150.53 | 34.28 | 245.89 | 52 |
| Ge | 1.41 | 1.15 | 3.49 | 4.21 | 2.74 | 1.72 | 1.66 | 0.77 | 0.43 | 1.39 | 1.67 | 0.35 | 1.33 | 1.4 | 2.4 | 1.74 | 2.78 |
| Mo | 1.45 | 1.77 | 2.11 | 2.31 | 2.72 | 1.6 | 1.77 | 0.72 | 1.52 | 1.69 | 3.14 | 1.69 | 3.05 | 1.9 | 3.13 | 2.04 | 4 |
| Cd | 0.35 | 0.17 | 0.12 | 0.21 | 0.19 | 0.22 | 0.18 | 0.32 | 0.37 | 0.39 | 0.63 | 0.29 | 0.22 | 0.21 | 0.06 | 0.26 | 0.2 |
| Sn | 5.21 | 1.11 | 0.06 | 0.77 | 1.02 | 1.75 | 0 | 2.98 | 3.01 | 2.29 | 1.94 | 3.08 | 1.17 | 2.84 | 0.63 | 1.86 | 2 |
| Sb | 0.34 | 0.28 | 0.49 | 0.52 | 0.38 | 0.31 | 0.4 | 0.22 | 0.15 | 0.3 | 0.64 | 0.13 | 0.2 | 0.57 | 0.43 | 0.36 | 2 |
| Cs | 0.72 | 0.25 | 0.07 | 0.31 | 0.13 | 0.15 | 0.06 | 0.31 | 0.25 | 0.14 | 0.23 | 0.3 | 0.04 | 0.09 | 0.02 | 0.2 | 1 |
| Ba | 424.52 | 13.3 | 11.1 | 18.23 | 17.36 | 9.69 | 14.36 | 11.76 | 38.57 | 7.68 | 23.79 | 235.07 | 20.29 | 10.21 | 12.67 | 57.91 | 56.03 |
| La | 0.11 | 0.16 | 0.73 | 0.64 | 0.57 | 0.22 | 0.63 | 0.28 | 0.76 | 0.05 | 0.08 | 2.7 | 4.87 | 1.22 | 4.6 | 1.17 | 25.78 |
| Ce | 1.5 | 3.96 | 20.77 | 17.15 | 8.1 | 2.46 | 8.88 | 2.57 | 12.33 | 1.25 | 2.33 | 41.7 | 25.28 | 13.79 | 20.11 | 12.15 | 49.11 |
| Nd | 0.11 | 0.2 | 1.1 | 0.66 | 0.58 | 0.15 | 0.75 | 0.19 | 0.5 | 0.06 | 0.15 | 1.75 | 6.42 | 1.03 | 4.95 | 1.24 | 21.5 |

**Table 3.** *Cont.*

| Elemental Concentrations | Sample | | | | | | | | | | | | | | | Average | Coal [a] |
|---|---|---|---|---|---|---|---|---|---|---|---|---|---|---|---|---|---|
| | ZG517 | ZG515 | ZG514 | ZG513 | ZG512 | ZG511 | ZG509 | ZG508 | ZG507 | ZG506 | ZG505 | ZG504 | ZG503 | ZG502 | ZG501 | | |
| Sm | 0.02 | 0.04 | 0.2 | 0.1 | 0.09 | 0.03 | 0.14 | 0.03 | 0.09 | 0.01 | 0.03 | 0.36 | 1.33 | 0.13 | 1.02 | 0.24 | 4.3 |
| Eu | 0.05 | 0 | 0.02 | 0.01 | 0.01 | 0 | 0.02 | 0.01 | 0.02 | 0 | 0.01 | 0.07 | 0.24 | 0.02 | 0.21 | 0.05 | 0.87 |
| Yb | 0.04 | 0.03 | 0.05 | 0.02 | 0.03 | 0.02 | 0.05 | 0.03 | 0.04 | 0.02 | 0.02 | 0.04 | 0.45 | 0.04 | 0.59 | 0.1 | 2.12 |
| Hf | 13.5 | 6.96 | 4.41 | 7.82 | 8.16 | 10.04 | 7.51 | 6.61 | 7.54 | 8.11 | 8.94 | 5.88 | 3.43 | 4.03 | 1.01 | 6.93 | 2.4 |
| Ta | 3.86 | 0.95 | 0.5 | 0.73 | 0.97 | 1.21 | 0.48 | 0.77 | 0.86 | 0.61 | 0.38 | 0.72 | 0.46 | 0.82 | 0.11 | 0.89 | 0.7 |
| W | 2.5 | 1.42 | 0.7 | 0.65 | 1.69 | 1.43 | 0.67 | 0.9 | 1.16 | 0.42 | 0.02 | 1.21 | 0.66 | 0.61 | 1.2 | 1.02 | 2 |
| Hg | 29 | 20 | 44 | 54 | 81 | 17 | 129 | 38 | 45 | 90 | 145 | 52 | 87 | 83 | 66 | 65.42 | 15 |
| Tl | 0.37 | 0.36 | 0.45 | 0.28 | 0.02 | 0.03 | 0.03 | 0.02 | 0.03 | 0.11 | 0.27 | 0.05 | 0.13 | 0.14 | 0.14 | 0.16 | 0.4 |
| Pb | 55.82 | 52.08 | 42.78 | 40.99 | 57.31 | 55.41 | 54.81 | 36.74 | 38.99 | 32.3 | 37.3 | 30.22 | 20.5 | 20.03 | 8.95 | 38.95 | 13 |
| Bi | 0.77 | 0.66 | 0.36 | 0.44 | 0.5 | 0.51 | 0.39 | 0.51 | 0.74 | 0.42 | 0.37 | 0.56 | 0.36 | 0.33 | 0.1 | 0.47 | 0.8 |
| Th | 1.71 | 1.32 | 1.54 | 1.06 | 1.09 | 0.79 | 2.02 | 1.17 | 1.1 | 0.81 | 1.41 | 0.67 | 2.51 | 0.65 | 0.29 | 1.21 | 6 |
| U | 5.93 | 18.41 | 17.64 | 22.3 | 8.55 | 4.8 | 6.16 | 4.92 | 8.73 | 5.32 | 5.15 | 5.1 | 1.75 | 2.1 | 0.91 | 7.85 | 3 |

[a] Coal, Chinese average coals value by Dai *et al.* [29] or world hard coals [37].

**Table 4.** Correlation coefficients between the content of each element in coal and ash yield, major elements.

| | Ad | SiO$_2$ | Al$_2$O$_3$ | TiO$_2$ | Fe$_2$O$_3$ | CaO | K$_2$O | MgO | Na$_2$O | P$_2$O$_5$ | Li | Ga | Se | Zr | Hf | As | Ge |
|---|---|---|---|---|---|---|---|---|---|---|---|---|---|---|---|---|---|
| Ad | 1 | | | | | | | | | | | | | | | | |
| SiO$_2$ | 0.66 ** | 1 | | | | | | | | | | | | | | | |
| Al$_2$O$_3$ | 0.51 | 0.89 ** | 1 | | | | | | | | | | | | | | |
| TiO$_2$ | 0.09 | 0.15 | 0.12 | 1 | | | | | | | | | | | | | |
| Fe$_2$O$_3$ | −0.66 ** | −0.92 ** | −0.93 ** | −0.37 | 1 | | | | | | | | | | | | |
| CaO | −0.52 * | −0.92 ** | −0.96 ** | −0.29 | 0.96 ** | 1 | | | | | | | | | | | |
| K$_2$O | 0.22 | −0.30 | −0.13 | −0.15 | 0.14 | 0.14 | 1 | | | | | | | | | | |
| MgO | −0.51 | −0.91 ** | −0.91 ** | −0.27 | 0.92 ** | 0.92 ** | 0.33 | 1 | | | | | | | | | |
| Na$_2$O | −0.47 | −0.95 ** | −0.91 ** | −0.34 | 0.93 ** | 0.96 ** | 0.34 | 0.91 ** | 1 | | | | | | | | |
| P$_2$O$_5$ | 0.15 | −0.19 | 0.15 | 0.16 | −0.14 | −0.09 | 0.74 ** | 0.05 | 0.11 | 1 | | | | | | | |
| Li | 0.88 ** | 0.69 ** | 0.62 * | −0.02 | −0.67 ** | −0.56 * | −0.07 | −0.66 ** | −0.51 * | −0.01 | 1 | | | | | | |
| Ga | 0.78 ** | 0.51 | 0.24 | 0.37 | −0.47 | −0.39 | 0.13 | −0.30 | −0.41 | −0.10 | 0.55 * | 1 | | | | | |
| Se | 0.60 * | 0.37 | 0.11 | 0.48 | −0.37 | −0.24 | −0.07 | −0.18 | −0.30 | −0.12 | 0.41 | 0.87 ** | 1 | | | | |
| Zr | 0.76 ** | 0.59 * | 0.36 | 0.35 | −0.55 * | −0.51 | 0.05 | −0.47 | −0.50 | −0.11 | 0.62 * | 0.93 ** | 0.81 ** | 1 | | | |
| Hf | 0.81 ** | 0.64 * | 0.41 | 0.29 | −0.59 * | −0.53 * | 0.07 | −0.49 | −0.52 * | −0.10 | 0.67 ** | 0.94 ** | 0.81 ** | 0.99 ** | 1 | | |
| As | 0.34 | 0.54 * | 0.22 | 0.20 | −0.36 | −0.41 | −0.23 | −0.32 | −0.51 | −0.41 | 0.16 | 0.58 * | 0.51 | 0.63 * | 0.61 * | 1 | |
| Ge | −0.40 | 0.09 | −0.06 | 0.26 | 0.01 | −0.06 | −0.53 * | −0.09 | −0.22 | −0.47 | −0.35 | −0.13 | 0.03 | −0.04 | −0.10 | 0.40 | 1 |

** Correlation is significant at the 0.01 level (two-tailed); * Correlation is significant at the 0.05 level (two-tailed).

## 5. Conclusions

Based on mineralogical and geochemical investigation of the No. 5 coal from Chuancaogedan Mine, Junger Coalfield, the conclusions are summarized below.

The No. 5 coal at the Chuancaogedan Mine has a high-ash yield (averages of 32.69%) and an ultra-low-sulfur content (0.40%), while the mean contents of volatile matter and moisture are 37.22% and 3.81%, respectively.

The mineral component of the No. 5 coal mainly consists of kaolinite, followed by magnetite, quartz, gypsum, mixed-layer I/S, pyrite, and calcite. Kaolinite is characteristically abundant and may have been derived from the weathered surface of the Benxi Formation bauxite during peat accumulation in the coal swamp.

Compared with common Chinese coals, the No. 5 coal is slightly enriched in $SiO_2$ (averaging 16.90%), $Al_2O_3$ (13.87%), $TiO_2$ (0.55%), $P_2O_5$ (0.55%), Li (78.54 mg/kg), Se (6.69 mg/kg), Zr (245.89 mg/kg), Hg (65.42 mg/kg), Pb (38.95 mg/kg) and U (7.85 mg/kg), and has a lower concentration of $Fe_2O_3$, $Na_2O$, As, Co, Sr, Sb and Tl, while others are close to averages for Chinese coals. The $SiO_2/Al_2O_3$ ratios (average of 1.22) are higher than that of the Chinese coals (1.42) and the theoretical $SiO_2/Al_2O_3$ ratio of kaolinite (1.18), suggesting quartz occurs in the mineral matter.

The modes of occurrence of Li, Ga, Se, Zr, Hf, As and Ge in the No. 5 coal were preliminarily investigated by correlation analysis. The correlation coefficients of Li, Ga, Se, Zr and Hf and ash yield are 0.88, 0.78, 0.60, 0.76 and 0.81, respectively, suggesting they occur in inorganic association. Li, Zr and Hf present positive correlation with Si and Al (rLi-Si = 0.69, rLi-Al = 0.62, rZr-Si = 0.59, rZr-Al = 0.62, rHf-Si = 0.64, rHf-Al = 0.67), indicating they are associated with aluminosilicate minerals in the No. 5 coal. Arsenic may be associated with organic and/or inorganic components of the tested coal samples, given that it is only moderately correlated with ash yield, Si, Al, and $Fe_2O_3$. Germanium may have organic and/or sulfide affinity in the No. 5 coals.

**Acknowledgments:** This work was supported by the National Key Basic Research Program of China (No. 2014CB238901) and the Key Program of National Natural Science Foundation of China (No. 41330317). The authors are grateful to Shifeng Dai for his experimental and technical assistance. Special thanks are given to anonymous reviewers for their useful suggestions and comments.

**Author Contributions:** Ning Yang conceived the overall experimental strategy and performed optical microscopy. Shuheng Tang guided all experiments. Songhang Zhang and Yunyun Chen observed these samples using SEM. All authors participated in writing the manuscript.

**Conflicts of Interest:** The authors declare no conflict of interest.

## References

1. Dai, S.F.; Zou, J.H.; Jiang, Y.F.; Ward, C.R.; Wang, X.B.; Li, T.; Xue, W.F.; Liu, S.D.; Tian, H.M.; Sun, X.H.; *et al.* Mineralogical and geochemical compositions of the Pennsylvanian coal in the Adaohai Mine, Daqingshan Coalfield, Inner Mongolia, China: Modes of occurrence and origin of diaspore, gorceixite, and ammonian illite. *Int. J. Coal Geol.* **2012**, *94*, 250–270. [CrossRef]

2. Dai, S.F.; Luo, Y.B.; Seredin, V.V.; Ward, C.R.; Hower, J.C.; Zhao, L.; Liu, S.D.; Zhao, C.L.; Tian, H.M.; Zou, J.H. Revisiting the late Permian coal from the Huayingshan, Sichuan, southwestern China: Enrichment and occurrence modes of minerals and trace elements. *Int. J. Coal Geol.* **2014**, *122*, 110–128. [CrossRef]

3. Gürdal, G. Abundances and modes of occurrence of trace elements in the Çan coals (Miocene), Çanakkale-Turkey. *Int. J. Coal Geol.* **2011**, *87*, 157–173. [CrossRef]

4. Yang, J.Y. Concentrations and modes of occurrence of trace elements in the Late Permian coals from the Puan Coalfield, southwestern Guizhou, China. *Environ. Geochem. Health* **2006**, *28*, 567–576. [CrossRef] [PubMed]

5. Wang, X.B. Geochemistry of Late Triassic coals in the Changhe Mine, Sichuan Basin, southwestern China: Evidence for authigenic lanthanide enrichment. *Int. J. Coal Geol.* **2009**, *80*, 167–174. [CrossRef]

6. Kolker, A. Minor element distribution in iron disulfides in coal: A geochemical review. *Int. J. Coal Geol.* **2012**, *94*, 32–43. [CrossRef]

7.  Finkelman, R.B. Modes of occurrence of potentially hazardous elements in coals: Levels of confidence. *Fuel Process. Technol.* **1994**, *39*, 21–34. [CrossRef]
8.  Tang, S.S.; Sun, S.L.; Qin, Y.; Jiang, Y.F.; Wang, W.F. Distribution characteristics of sulfur and the main harmful trace elements in China's coal. *Acta Geol. Sin. Engl. Ed.* **2008**, *82*, 722–730.
9.  Dai, S.F.; Ren, D.Y.; Chou, C.L.; Li, S.S.; Jiang, Y.F. Mineralogy and geochemistry of the No. 6 Coal (Pennsylvanian) in the Junger Coalfield, Ordos Basin, China. *Int. J. Coal Geol.* **2006**, *66*, 253–270. [CrossRef]
10. Dai, S.F.; Li, D.; Chou, C.L.; Zhao, L.; Zhang, Y.; Ren, D.Y.; Ma, Y.W.; Sun, Y.Y. Mineralogy and geochemistry of boehmite-rich coals: New insights from the Haerwusu Surface Mine, Jungar Coalfield, Inner Mongolia, China. *Int. J. Coal Geol.* **2008**, *74*, 185–202. [CrossRef]
11. Dai, S.F.; Jiang, Y.F.; Ward, C.R.; Gu, L.; Seredin, V.V.; Liu, H.D.; Zhou, D.; Wang, X.B.; Sun, Y.Z.; Zou, J.H.; *et al.* Mineralogical and geochemical compositions of the coal in the Guanbanwusu Mine, Inner Mongolia, China: Further evidence for the existence of an Al (Ga and REE) ore deposit in the Jungar Coalfield. *Int. J. Coal Geol.* **2012**, *98*, 10–40. [CrossRef]
12. Wang, X.B.; Dai, S.F.; Sun, Y.Y.; Li, D.; Zhang, W.G.; Zhang, Y.; Luo, Y.B. Modes of occurrence of fluorine in the Late Paleozoic No. 6 coal from the Haerwusu Surface Mine, Inner Mongolia, China. *Fuel* **2011**, *90*, 248–254. [CrossRef]
13. China Coal Research Institute. *GB/T 482-2008. Sampling of Coal Seams*; Chinese National Standard; General Administration of Quality Supervision, Inspection and Quarantine of the People's Republic of China: Beijing, China, 2008. (In Chinese)
14. American Society for Testing and Materials (ASTM) International. *Test Method for Moisture in the Analysis Sample of Coal and Coke*; ASTM D3173-11; ASTM International: West Conshohocken, PA, USA, 2011.
15. American Society for Testing and Materials (ASTM) International. *Test Method for Volatile Matter in the Analysis Sample of Coal and Coke*; ASTM D3175-11; ASTM International: West Conshohocken, PA, USA, 2011.
16. American Society for Testing and Materials (ASTM) International. *Test Method for Ash in the Analysis Sample of Coal and Coke from Coal*; ASTM D3174-11; ASTM International: West Conshohocken, PA, USA, 2011.
17. American Society for Testing and Materials (ASTM) International. *Test Methods for Total Sulfur in the Analysis Sample of Coal and Coke*; ASTM D3177-02; ASTM International: West Conshohocken, PA, USA, 2011.
18. Dai, S.F.; Yang, J.Y.; Ward, C.R.; Hower, J.C.; Liu, H.D.; Garrison, T.M.; French, D.; O'Keefe, J.M.K. Geochemical and mineralogical evidence for a coal-hosted uranium deposit in the Yili Basin, Xinjiang, northwestern China. *Ore Geol. Rev.* **2015**, *70*, 1–30. [CrossRef]
19. Dai, S.F.; Li, T.J.; Jiang, Y.F.; Ward, C.R.; Hower, J.C.; Sun, J.H.; Liu, J.J.; Song, H.J.; Wei, J.P.; Li, Q.Q.; *et al.* Mineralogical and geochemical compositions of the Pennsylvanian coal in the Hailiushu Mine, Daqingshan Coalfield, Inner Mongolia, China: Implications of sediment-source region and acid hydrothermal solutions. *Int. J. Coal Geol.* **2015**, *137*, 92–110. [CrossRef]
20. Wang, X.B.; Wang, R.X.; Wei, Q.; Wang, P.P.; Wei, J.P. Mineralogical and geochemical characteristics of late Permian coals from the Mahe Mine, Zhaotong Coalfield, Northeastern Yunnan, China. *Minerals* **2015**, *5*, 380–396. [CrossRef]
21. Dai, S.F.; Wang, X.B.; Zhou, Y.P.; Hower, J.C.; Li, D.H.; Chen, W.M.; Zhu, X.W.; Zou, J.H. Chemical and mineralogical compositions of silicic, mafic, and alkali tonsteins in the late Permian coals from the Songzao Coalfield, Chongqing, Southwest China. *Chem. Geol.* **2011**, *282*, 29–44. [CrossRef]
22. Li, X.; Dai, S.F.; Zhang, W.G.; Li, T.; Zheng, X.; Chen, W. Determination of As and Se in coal and coal combustion products using closed vessel microwave digestion and collision/reaction cell technology (CCT) of inductively coupled plasma mass spectrometry (ICP-MS). *Int. J. Coal Geol.* **2014**, *124*, 1–4. [CrossRef]
23. China Coal Research Institute. *GB/T 15224.1-2004, Classification for Quality of Coal—Part 1: Ash*; Chinese National Standard; General Administration of Quality Supervision, Inspection and Quarantine of the People's Republic of China: Beijing, China, 2004. (In Chinese)
24. China Coal Research Institute. *MT/T849-2000, Classification for Volatile Matter of Coal*; Chinese National Standard; General Administration of Quality Supervision, Inspection and Quarantine of the People's Republic of China: Beijing, China, 2000. (In Chinese)
25. China Coal Science Research Institute Beijing Coal Chemical Research Branch. *MT/T850-2000, Classification for Total Moisture in Coal*; Chinese National Standard; General Administration of Quality Supervision, Inspection and Quarantine of the People's Republic of China: Beijing, China, 2000. (In Chinese)

26. China Coal Research Institute Beijing Coal Chemical Research Branch. *GB/T 15224.2-2004, Classification for Coal Quality—Part 2: Sulfur Content*; Chinese National Standard; General Administration of Quality Supervision, Inspection and Quarantine of the People's Republic of China: Beijing, China, 2004. (In Chinese)

27. Dai, S.F.; Wang, X.B.; Seredin, V.V.; Hower, J.C.; Ward, C.R.; O'Keefe, J.M.K.; Huang, W.H.; Li, T.; Li, X.; Liu, H.D.; *et al.* Petrology, mineralogy, and geochemistry of the Ge-rich coal from the Wulantuga Ge ore deposit, Inner Mongolia, China: New data and genetic implications. *Int. J. Coal Geol.* **2012**, *90–91*, 72–99. [CrossRef]

28. Dai, S.F.; Li, T.; Seredin, V.V.; Ward, C.R.; Hower, J.C.; Zhou, Y.P.; Zhang, M.Q.; Song, X.L.; Song, W.J.; Zhao, C.L. Origin of minerals and elements in the late Permian coals, tonsteins, and host rocks of the Xinde Mine, Xuanwei, eastern Yunnan, China. *Int. J. Coal Geol.* **2014**, *121*, 53–78. [CrossRef]

29. Dai, S.F.; Ren, D.Y.; Chou, C.-L.; Finkelman, R.B.; Seredin, V.V.; Zhou, Y.P. Geochemistry of trace elements in Chinese coals: A review of abundances, genetic types, impacts on human health, and industrial utilization. *Int. J. Coal Geol.* **2012**, *94*, 3–21. [CrossRef]

30. Dai, S.F.; Seredin, V.V.; Ward, C.R.; Hower, J.C.; Xing, Y.W.; Zhang, W.G.; Song, W.J.; Wang, P.P. Enrichment of U–Se–Mo–Re–V in coals preserved within marine carbonate successions: Geochemical and mineralogical data from the Late Permian Guiding Coalfield, Guizhou, China. *Min. Deposita* **2015**, *50*, 159–186. [CrossRef]

31. Wang, S. (Ed.) *Coal Accumulation and Coal Resources Evaluation of Ordos Basin, China*; China Coal Industry Publishing House: Beijing, China, 1996; p. 437. (In Chinese)

32. Chou, C.-L. Abundances of sulfur, chlorine, and trace elements in Illinois Basin coals, USA. In Proceedings of the 14th Annual International Pittsburgh Coal Conference & Workshop, Taiyuan, China, 23–27 September 1997; Section 1. pp. 76–87.

33. Belkin, H.E.; Zheng, B.S.; Finkelman, R.B. Geochemistry of Coals Causing Arsenismin Southwest China. In *4th International Symposium on Environmental Geochemistry*; US Geological Survey Open-File Report; U.S. Geological Survey: Reston, VA, USA, 1997.

34. Ding, Z.H.; Zheng, B.S.; Long, J.P.; Belkin, H.E.; Finkelman, R.B.; Chen, C.G.; Zhou, D.; Zhou, Y. Geological and geochemical characteristics of high arsenic coals from endemic arsenosis areas in southwestern Guizhou Province. *Appl. Geochem.* **2001**, *16*, 1353–1360. [CrossRef]

35. Chen, J.; Chen, P.; Yao, D.X.; Liu, Z.; Wu, Y.S.; Liu, W.Z.; Hu, Y.B. Mineralogy and geochemistry of late Permian coals from the Donglin Coal Mine in the Nantong coalfield in Chongqing, southwestern China. *Int. J. Coal Geol.* **2015**, *149*, 24–40. [CrossRef]

36. Li, J.; Zhuang, X.G.; Querol, X.; Font, O.; Izquierdo, M.; Wang, Z.M. New data on mineralogy and geochemistry of high-Ge coals in the Yimin coalfield, Inner Mongolia, China. *Int. J. Coal Geol.* **2014**, *125*, 10–21. [CrossRef]

37. Ketris, M.P.; Yudovich, Y.E. Estimations of clarkes for carbonaceous biolithes: World average for trace element contents in black shales and coals. *Int. J. Coal Geol.* **2009**, *78*, 135–148. [CrossRef]

*Article*

# *minerals*

MDPI

# Mineralogical Characteristics of Late Permian Coals from the Yueliangtian Coal Mine, Guizhou, Southwestern China

Panpan Xie *, Hongjian Song, Jianpeng Wei and Qingqian Li

College of Geoscience and Surveying Engineering, China University of Mining and Technology, Beijing 100083, China; songhongjian90@gmail.com (H.S.); weijianpeng15@gmail.com (J.W.); liqingqian7@gmail.com (Q.L.)
* Correspondence: xiepanpan91@gmail.com; Tel.: +86-10-6234-1868

Academic Editor: Thomas Kerestedjian
Received: 6 December 2015; Accepted: 16 March 2016; Published: 31 March 2016

**Abstract:** This paper reports the mineralogical compositions of super-low-sulfur (Yueliangtian 6-upper (YLT6U)) and high-sulfur (Yueliangtian 6-lower (YLT6L)) coals of the Late Permian No. 6 coal seam from the Yueliangtian coal mine, Guizhou, southwestern China. The mineral assemblages and morphology were detected and observed by X-ray diffractogram (XRD), optical microscopy and field-emission scanning electron microscope (FE-SEM) in conjunction with an energy-dispersive X-ray spectrometer. Major minerals in the coal samples, partings and host rocks (roof and floor strata) include calcite, quartz, kaolinite, mixed-layer illite/smectite, chlorite and pyrite and, to a lesser extent, chamosite, anatase and apatite. The Emeishan basalt and silicic rocks in the Kangdian Upland are the sediment source for the Yueliangtian coals. It was found that there are several modes of chamosite occurrence, and precursor minerals, such as anatase, had been corroded by Ti-rich hydrothermal solutions. The modes of occurrence of minerals present in the coal were controlled by the injection of different types of hydrothermal fluids during different deposition stages. The presence of abundant pyrite and extremely high total sulfur contents in the YLT6L coal are in sharp contrast to those in the YLT6U coal, suggesting that seawater invaded the peat swamp of the YLT6L coal and terminated at the YLT6U-9p sampling interval. High-temperature quartz, vermicular kaolinite and chloritized biotite were observed in the partings and roof strata. The three partings and floor strata of the No. 6 coal seam from the Yueliangtian coal mine appear to have been derived from felsic volcanic ash. Four factors, including sediment-source region, multi-stage injections of hydrothermal fluids, seawater influence and volcanic ash input, were responsible for the mineralogical characteristics of the Yueliangtian coals.

**Keywords:** minerals; coal; hydrothermal fluids; seawater influence; tonstein

## 1. Introduction

Guizhou province in southwestern (SW) China contains abundant coal resources. The Late Permian coals from western Guizhou province have attracted much attention [1–5], not only because of the coal-hosted rare-metal ore deposits found in this area [6–8], but also due to their mineralogical and geochemical indications for the reginal geology evolution, such as the mantle plume formation located to the west of the coal basin [1,3]. Practically, although Panxian county is closely located to the high-incidence area of the endemic arsenosis and fluorosis in SW China, the relation between mineralogical compositions of the coals in this area and the endemic disease occurring in the surrounding areas is unknown. The Yueliangtian coal mine in Panxian county provides large-scale energy resources for its region, and the coals are directly used for combustion, despite high mineral contents in these coals, as described below. Thus, it is necessary to determine the contents and modes of occurrence of minerals in coals for the environmental issues caused by coal combustion in the area. Theoretically, mineral matter in coal may indicate depositional environments during peat accumulation, as well as geological processes during diagenetic and epigenetic stages [9]. Although the coals in Panxian county of southwestern Guizhou province have been reported by a few researchers [4,5], geochemical characteristics, modes of mineral occurrence and their controlling geological factors in these coals, however, have not been well addressed. In this paper, we report the mineralogical characteristics of the Late Permian coals from the No. 6 coal seam at the Yueliangtian coal mine, Guizhou, SW China.

## 2. Geological Setting

The Yueliangtian coal mine is located in Panxian county, western Guizhou, southwestern China, covering a total area of around 15 km$^2$, 6 km N–S long and 2.5 km W–E wide (Figure 1). Tectonically, the Yueliangtian deposit belongs to a monocline structure with an approximate east dip [6]. It is limited by well-developed normal faults within the epsilon-type structure in the Puan tectonic zone [6,10], covering the area between latitudes 25°54′22″ and 25°57′44″ N and longitudes 104°30′36″ and 104°31′59″ E. The Kangdian Upland is the dominant sediment-source region for this coal deposit in western Guizhou [11].

The sedimentary sequences in the Yueliangtian coal mine are the Late Permian and Early Triassic (Figure 2) strata. The Late Permian strata, with major coal resources in the area, consist of the Longtan and Emeishan Basalt Formations. The upper and lower portions of the Emeishan Basalt Formation are dominated by grey-lilac tuffs and grey-dark basalts, respectively, similar to those in the surrounding areas such as the Zhijing Coalfiled in western Guizhou [12].

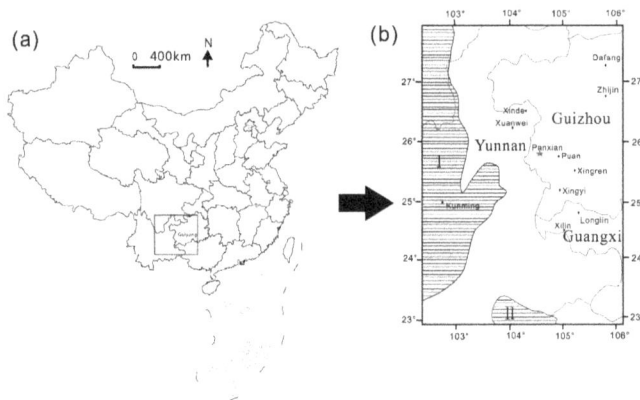

**Figure 1.** Location of the Yueliangtian coal mine. (**a**) China map and the location of study area; (**b**) Depositional environments during the Late Permian in Guizhou province, China. I, Kangdian Upland. II, Northern Vietnam Upland. (**b**) The enlargement of the red area in (**a**), modified from Dai *et al.* [15].

The Longtan Formation (P$_{2l}$) is the major coal-bearing strata of the coal mine. The sedimentary environment of the Longtan Formation varies greatly from lower delta plains, through tidal flats, to carbonate subtidal flats [11–14]. As shown in Figure 2, the upper portion of the Longtan Formation (81.97 m) is made up of siltstone, siderite layers and twelve coal seams. Its middle portion (89.75 m), intercalated with siderite layers and coal beds, is mainly composed of siltstone and silty mudstone. A total of 16 coal beds occur in the middle portion. The lower portion is composed of siltstone, silty mudstone, pelitic siltstone, siderite and ten coal seams. The Yueliangtian 6-upper (YLT6U) and 6-lower (YLT6L) coals, separated by a gray siltstone (2.94 m) that contains siderite layers and plenty of plant-root fossils, occur in the upper portion of the Longtan Formation. Currently, the YLT6U coal seam is the only mineable seam in this area (Figure 2).

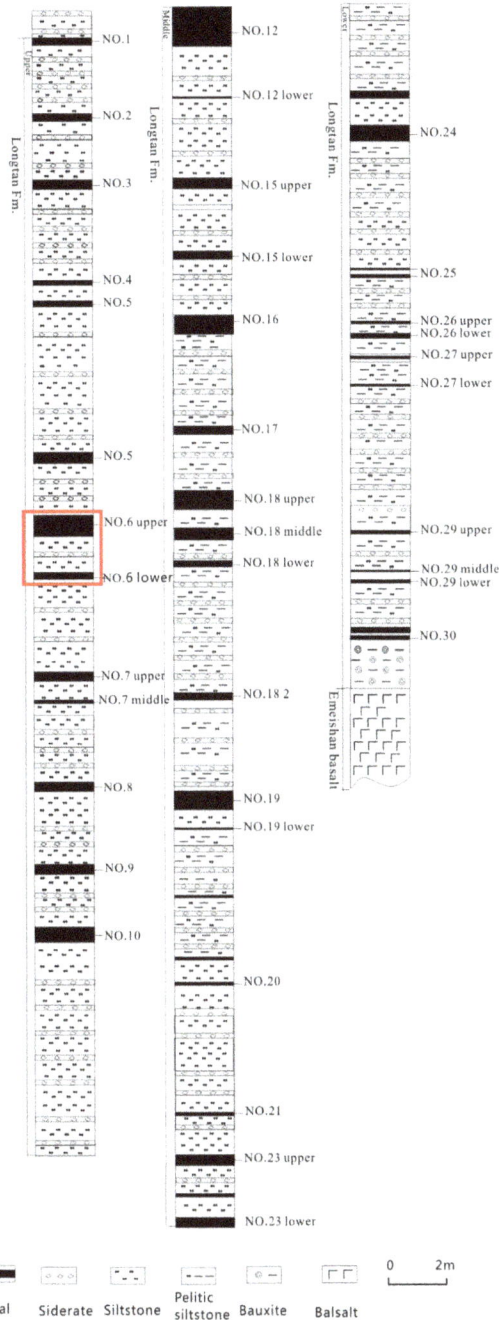

**Figure 2.** Sedimentary sequences and coal seams (red area) in the Yueliangtian coal mine.

### 3. Sample Collection and Analytical Procedures

According to Chinese Standard Method GB/T 482-2008 [16], a total of nineteen bench samples were collected from the No. 6 coal seam (YLT6) in the underground workings of the Yueliangtian coal mine in Panxian county, western Guizhou. The floor of the YLT6U coal and the roof of the YLT6L coal are unavailable, due to the complex structure, as well as limited capacity in sample collection. From top to bottom, the roof strata (with suffix-r), coal benches, partings (with suffix-p) and floor-stratum samples (with suffix-f) are identified, with the coal seams numbered in increasing order from top to bottom (Figure 3). Each bench sample was immediately stored in a clean and uncontaminated plastic bag, to ensure as little contamination and oxidation as possible. The thickness of each sample is given in Table 1.

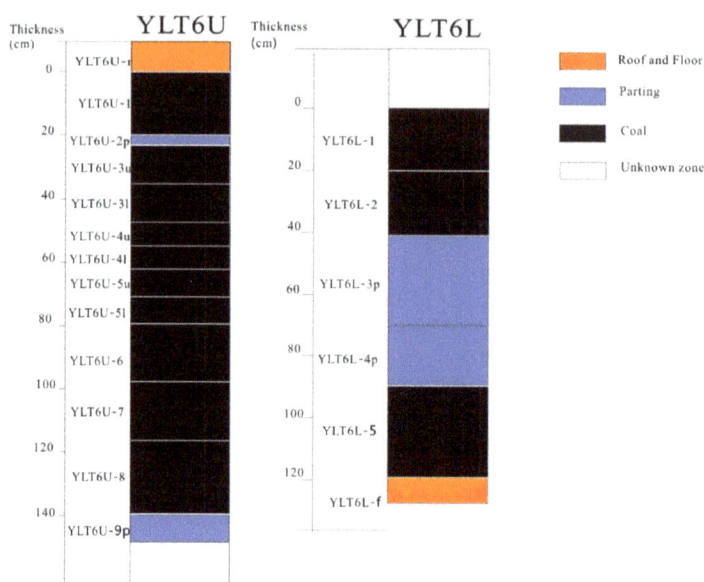

**Figure 3.** Lithologic column sections of the Yueliangtian 6-upper (YLT6U) and 6-lower (YLT6L) coal seams.

The samples were crushed and ground to pass the 75-μm sieve prior to analysis. In accordance with ASTM Standards D3173-11, D3175-11 and D3174-11 [17–19], proximate analysis, covering moisture and volatile matter percentages and ash yield, was conducted. Total sulfur was determined based on ASTM Standard D3177-02 [20]. The percentages of C, H and N in the coals were determined by an elemental analyzer (vario MACRO). Coarse-crushed samples of each coal were prepared as grain mounts and examined with 50× oil immersion for microscopic analysis. Following ASTM Standards D2797/D2797M-11a and D2798M-11a [21,22], vitrinite random reflectance (Rr, %) was determined using a Leica DM-4500P microscope (at a magnification of 500×) equipped with a Craic QDI 302™ spectrophotometer (Leica Inc., Wetzlar, Germany). An X-ray diffractogram (XRD) was used to determine the mineralogical compositions. Prior to XRD analysis, the low-temperature ashing (LTA) of coal was performed using an EMITECH K1050X plasma asher (Quorum Inc., Lewes, UK). A commercial interpretation software Siroquant™ [23,24] was used to obtain mineral contents in the LTAs and non-coal samples based on XRD. More information illustrating the use of this technique for coal-related materials was given by Ward *et al.* [25,26], Ruan and Ward [27] and Dai *et al.* [28,29]. Following the procedures described by Dai *et al.* [15,30], a field-emission scanning electron microscope

(FE-SEM) in conjunction with an energy-dispersive X-ray spectrometer (EDAX Inc., Mahwah, NJ, USA) was used to observe modes of mineral occurrence, and also to determine the distribution of some selected elements. Samples were carbon-coated using a Quorum Q150T ES sputtering coater (Quorum Inc.) or were not coated for low-vacuum SEM working conditions (60 bar). The working distance, beam voltage, aperture and spot size of the FE-SEM-EDS was 10 mm, 20.0 kV, 6 and 4.5–5.0, respectively. The images were captured by a back-scattered electron detector (BSE).

**Table 1.** Coal bench thickness (cm), proximate and ultimate analyses (%), and vitrinite random reflectance (%) of the No. 6 coals in the Yueliangtian coal mine, Guizhou, China.

| Sample | Thickness (cm) | Proximate Analyses (%) | | | Ultimate Analyses (%) | | | | Reflectance |
|---|---|---|---|---|---|---|---|---|---|
| | | $M_{ad}$ | $A_d$ | $V_{daf}$ | $S_{t,d}$ | $C_{daf}$ | $H_{daf}$ | $N_{daf}$ | $R_r$ |
| YLT6U Coal Seam | | | | | | | | | |
| YLT6U-1 | 20 | 0.57 | 36.62 | 41.59 | 0.65 | 83.38 | 5.06 | 1.56 | 1.01 |
| YLT6U-3u | 12.5 | 1.45 | 42.36 | 39.87 | 0.28 | 82.91 | 5.52 | 1.68 | 1.02 |
| YLT6U-3l | 12.5 | 0.72 | 18.71 | 34.63 | 0.61 | 88.37 | 5.35 | 1.76 | 0.95 |
| YLT6U-4u | 7.5 | 0.87 | 34.14 | 34.1 | 0.35 | 87.82 | 7.3 | 1.74 | 0.96 |
| YLT6U-4l | 7.5 | 1.02 | 10.97 | 35.75 | 0.27 | 87.73 | 4.71 | 1.87 | 0.91 |
| YLT6U-5u | 9 | 0.69 | 16 | 35.53 | 0.23 | 86.21 | 4.93 | 1.49 | 0.94 |
| YLT6U-5l | 9 | 0.83 | 10.6 | 36.43 | 0.33 | 86.55 | 4.95 | 1.71 | 0.91 |
| YLT6U-6 | 19 | 0.7 | 19.09 | 36.19 | 0.36 | 86.42 | 4.17 | 1.44 | 0.91 |
| YLT6U-7 | 19 | 0.59 | 26.88 | 37.96 | 0.75 | 92.54 | 4.39 | 1.48 | 0.94 |
| YLT6U-8 | 24 | 0.84 | 10.07 | 36.29 | 0.49 | 87.93 | 4.78 | 1.77 | 0.8 |
| Wa | 140 * | 0.79 | 22.78 | 37.24 | 0.48 | 87.04 | 4.97 | 1.61 | 0.93 |
| YLT6L Coal Seam | | | | | | | | | |
| YLT6L-1 | 20 | 0.89 | 40.6 | 44.15 | 13.34 | 58.71 | 5.19 | 1.11 | 0.77 |
| YLT6L-2 | 21 | 1.01 | 27.75 | 37.85 | 8.99 | 80.9 | 4.47 | 1.39 | 0.77 |
| YLT6L-5 | 30 | 0.45 | 36.37 | 35.44 | 3.61 | 86.15 | 4.54 | 1.24 | 0.75 |
| Wa | 71 * | 0.74 | 35.01 | 38.61 | 7.94 | 76.87 | 4.7 | 1.25 | 0.76 |

M, moisture; A, ash yield; V, volatile matter; $S_t$, total sulfur; C, carbon; H, hydrogen; N, nitrogen; ad, air-dried basis; d, dry basis; daf, dry and ash-free basis; $R_r$, random reflectance of vitrinite; Wa, weighted average for bench sample (weighted by thickness of sample interval); * total thickness.

The clay (<2 μm) fraction of rock samples was conducted based on the Chinese Industry Standard Method [31]. The sample was dissolved in 80 mL ultrapure water and isolated by ultrasonic dispersion. After 4h standing, the natural-oriented aggregate was analyzed in an air-dried state. The glycol-saturated (50 °C, 8 h) and heated (450 °C, 2.5 h) oriented aggregates were analyzed subsequently. The mineralogy of this fraction was also analyzed by XRD.

## 4. Result

### 4.1. Coal Chemistry and Vitrinite Reflectance

The vitrinite random reflectance values and the volatile matter yields of the YLT6U and YLT6L coals (Table 1) indicate a high volatile A bituminous coal in rank [32]. The ash yield of the YLT6U coal samples varies from 10.07%–42.36%, with the weighted average much lower than that of the YLT6L coal. Total sulfur contents vary greatly between the YLT6U and YLT6L coal seams (Table 1). As shown in Tables 1 and 2 the high-sulfur samples also have high percentages of pyrite (described in more detail below). Based on Chinese Standards GB/T 15224.1/2-2010 [33], the YLT6U coal is classified as a medium-ash and super-low-sulfur coal; the YLT6L coal, with its total sulfur content higher than other coals in Guizhou province [14,34], is a medium-high-ash and high-sulfur coal.

**Table 2.** Low-temperature ash yields (%) of coals and mineral compositions (%) of coal low-temperature ashing (LTA), partings and host rocks (%) determined by X-ray diffractogram (XRD) and Siroquant. I/S, illite/smectite.

| Sample | LTA | Quartz | Kaolinite | Chlorite | Illite | I/S | Calcite | Anatase | Pyrite | Bassanite | Ankerite | Rutile | Xenotime | Muscovite | Siderite | Stibnite | Apatite |
|---|---|---|---|---|---|---|---|---|---|---|---|---|---|---|---|---|---|
| YLT6U Coal Seam | | | | | | | | | | | | | | | | | |
| YLT6U-r | | 24.4 | | 17.5 | | | 1.9 | 6.2 | | | | | | 12 | 12 | | |
| YLT6U-1 | 43.8 | 38.1 | 14.8 | 8.5 | | 26 | 28 | | | | 10.5 | | | | | | |
| YLT6U-2p | | 5.4 | 53.7 | | | 34.7 | 6.2 | | | | | | | | | | |
| YLT6U-3u | 48.49 | 30.3 | 53.1 | | | | 16.6 | | | | | | | | | | |
| YLT6U-3l | 19.95 | 51.2 | 30.8 | 5.6 | | | 11.1 | 1.3 | | | | | | | | | |
| YLT6U-4u | 36.76 | 64.9 | 25.8 | 6.2 | | | 3.1 | | | | | | | | | | |
| YLT6U-4l | 11.76 | 23.5 | 20.6 | | | | 56 | | | | | | | | | | |
| YLT6U-5u | 19.35 | 37.3 | 49.4 | | | | 13.4 | | | | | | | | | | |
| YLT6U-5l | 12.46 | 6.3 | 17.4 | | | | 76.3 | | | | | | | | | | |
| YLT6U-6 | 22.89 | 8.6 | 14.3 | | | | 76.8 | 0.3 | | | | | | | | | |
| YLT6U-7 | 61.43 | 6.7 | 63.9 | | 11.3 | 3.5 | | 10.7 | 3.9 | | | | | | | | |
| YLT6U-8 | 10.2 | 24.6 | 22.4 | | | | 48.6 | 9.4 | | 4.4 | | | | | | | |
| YLT6U-9p | | 1.5 | 66.7 | | | 22.4 | | 1.6 | | | | | | | | | |
| Wa | 43.82 | 26.6 | 30.8 | 2 | 1.5 | 0.5 | 34.2 | | 0.5 | 0.8 | 1.5 | | | | | | |
| YLT6L Coal Seam | | | | | | | | | | | | | | | | | |
| YLT6L-1 | 59.35 | 5.6 | 10.8 | | | | 6.8 | | 65 | | | 11.7 | | | | | |
| YLT6L-2 | 35.11 | 25.4 | 9 | | | | 15 | | 43.6 | 4.2 | | 1 | 1.7 | 8.1 | | 1.5 | |
| YLT6L-3p | | 51.8 | | 20.4 | | | 2.4 | | 15.9 | | | | | | | | |
| YLT6L-4p | | 4.1 | 35.9 | | | 44.1 | | 10.1 | 5.8 | | | | | | | | |
| YLT6L-5 | 39.79 | 65.3 | | 10.8 | | | 10.5 | | 12.3 | 1.1 | | | | | | | |
| YLT6L-f | | 19 | 33.4 | | | 34.2 | 3.9 | 8.2 | | | | | | | | | 1.3 |
| Wa | 43.91 | 36.7 | 5.7 | 4.6 | | | 10.8 | | 36.4 | 1.7 | | 3.6 | 0.5 | | | | |

Wa, weighted average for bench sample (weighted by thickness of sample interval).

## 4.2. Minerals in the No. 6 Coals in the Yueliangtian Coal Mine

### 4.2.1. Minerals in Coal Benches

The proportions of each crystalline phase in the coal-LTA and non-coal samples identified from XRD and Siroquant are given in Table 2. The vertical concentration (%) variations of minerals, and low-temperature ash yields (%) through the No. 6 coals in the Yueliangtian coal mine are described in Figure 4. Like other Late Permian coals in western Guizhou described by Dai *et al.* [14,34,35] and Zhuang *et al.* [36], the major minerals of the YLT6U coal LTAs are mainly calcite, kaolinite and quartz, with trace amounts of chlorite and anatase. Various contents of illite, mixed-layer illite/smectite, pyrite, bassanite and ankerite also occur in a few coal benches (Table 2). The minerals in the LTA of the YLT6L coal are mainly represented by calcite, quartz and pyrite, with a lesser proportion of kaolinite, chlorite, rutile and bassanite. The LTA of sample YLT6L-2 also contains a small proportion of xenotime.

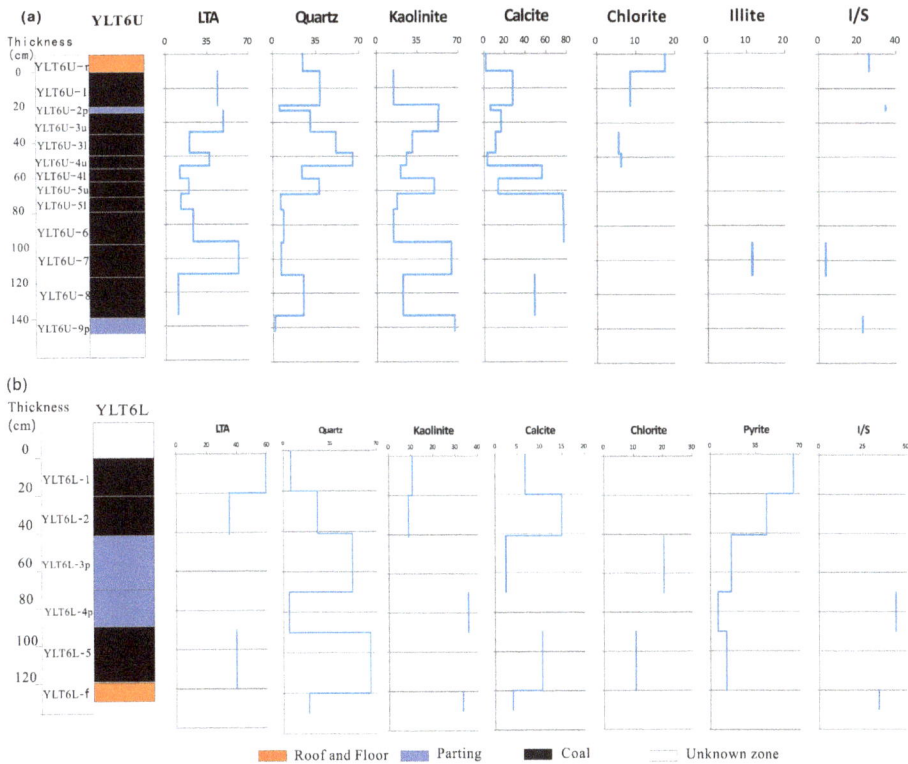

**Figure 4.** Concentration (%) variations of minerals and coal LTAs (%) through the No. 6 coal seam sections in the Yueliangtian coal mine. (**a**) YLT6U coal; (**b**) YLT6L coal.

Quartz in the YLT6U and YLT6L coals occurs in three forms: (1) as cell-fillings of structured inertinite macerals, suggesting an authigenic origin (Figure 5A); (2) as discrete particles embedded in collodetrinite (Figure 5B,C); and (3) coexisting with clay minerals (Figure 5D). Quartz with the latter two forms was probably derived from detrital materials of terrigenous origin.

As shown in Table 2 and Figure 4, the host rocks (roof and floor strata) and partings (YLT6U-9p and YLT6L-4p) of the YLT6U and YLT6L coals contain minor calcite. By contrast, the YLT6U coal contains abundant calcite, accounting for 76.8% of the total minerals in the LTA sample. Calcite layering has been macroscopically observed in sample YLT6U-5l during sampling in the field. Microscopically, it occurs mainly as cell-fillings (Figure 6A), as isolated particles with various forms (Figure 6B,C) and as vein-fillings in the vitrinite (Figure 6D). A platy calcite (Figure 6B) with a large size up to 570 μm in length and 100 μm in width has distinct edges and angles.

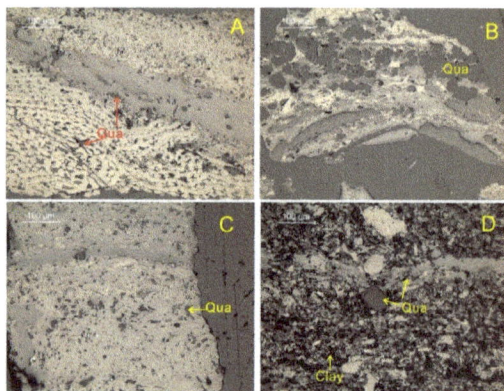

**Figure 5.** Quartz in the YLT6 coal (reflected white light microscopy). (**A**) Fusinite- and semifusinite-cell filling quartz in sample YLT6U-4u; (**B**) Quartz distributed along the bedding planes of sample YLT6U-4l; (**C**) Quartz embedded in collodetrinite in sample YLT6L-2; (**D**) Quartz coexisting with clay minerals in sample YLT6U-4u. Qua, quartz; Clay, clay minerals.

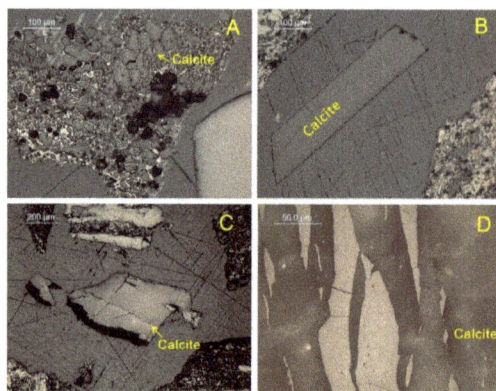

**Figure 6.** Calcite in the YLT6U coal (reflected white light microscope). (**A**) Cell-filling calcite in sample YLT6U-6; (**B**) calcite embedded in collodetrinite in sample YLT6U-4u; (**C**) calcite occurring as irregular shapes in sample YLT6U-1; (**D**) calcite as fracture-filling in vitrinite of sample YLT6U-1.

Kaolinite in the YLT6 coal has two modes of occurrence: as cell fillings (Figure 7A,B) and as detrital particles (Figure 5D). The former is of authigenic origin, and the latter is probably of terrigenous origin. In contrast to the minerals described above, pyrite is distributed completely differently between the YLT6U and YLT6L coals. Although below the detection limit of the XRD technique, cell-filling pyrite was observed under the optical microscope in sample YLT6U-8 (Figure 8A). Pyrites of framboidal (Figure 8B), cell-fillings (Figures 7A,B and 8C), cubic (Figure 8D) and nodular (Figure 8E) forms occur in the YLT6L coal. It is probably of syngenetic or early diagenetic origin [37–41]. Pyrite also occurs as cleat- and fracture-fillings (Figures 7A and 8C), suggesting an epigenetic origin of hydrothermal solutions with a formation temperature around 150–200 °C [5,41,42]. Fe-S compounds with different levels of brightness (Figure 8F) deserve a special note: the atomic ratios of Fe against S of Spots 1 and 2 (Figure 8G,H) are 1:2.4 and 1:2, respectively, probably suggesting an oxidation product of pyrite in Spot 1.

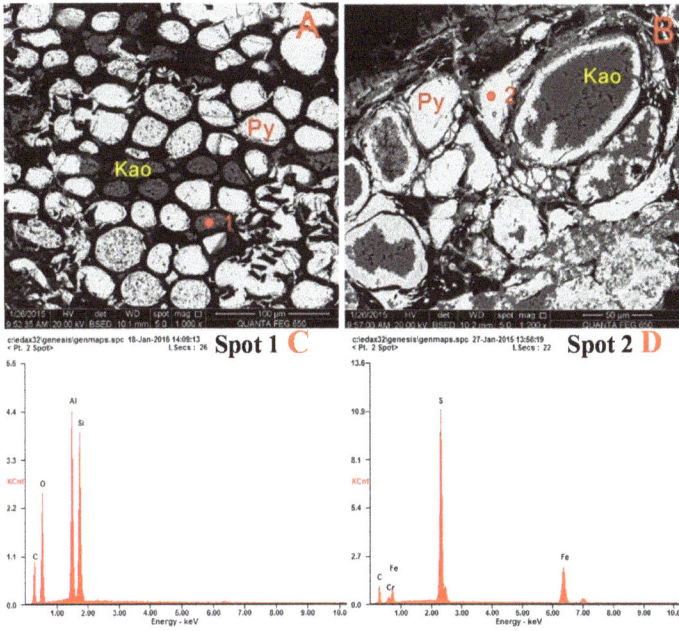

**Figure 7.** SEM back-scattered electron images and selected Energy Dispersive X-ray Spectroscopy (EDS) spectra of kaolinite and pyrite in YLT6L-1 coal. (**A**) Cell- and cleat-filling kaolinite and pyrite; (**B**) kaolinite and pyrite; (**C**) EDS spectrum of kaolinite; (**D**) EDS spectrum of pyrite. Kao, kaolinite; Py, pyrite.

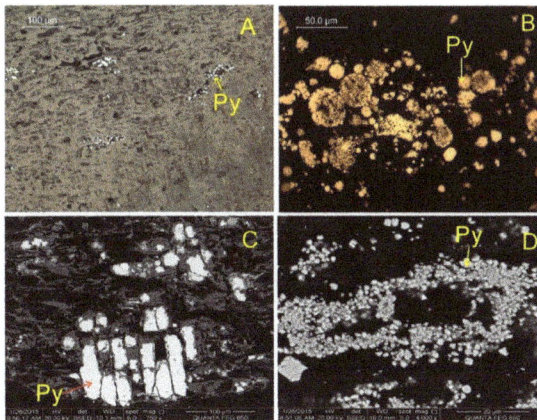

**Figure 8.** *Cont.*

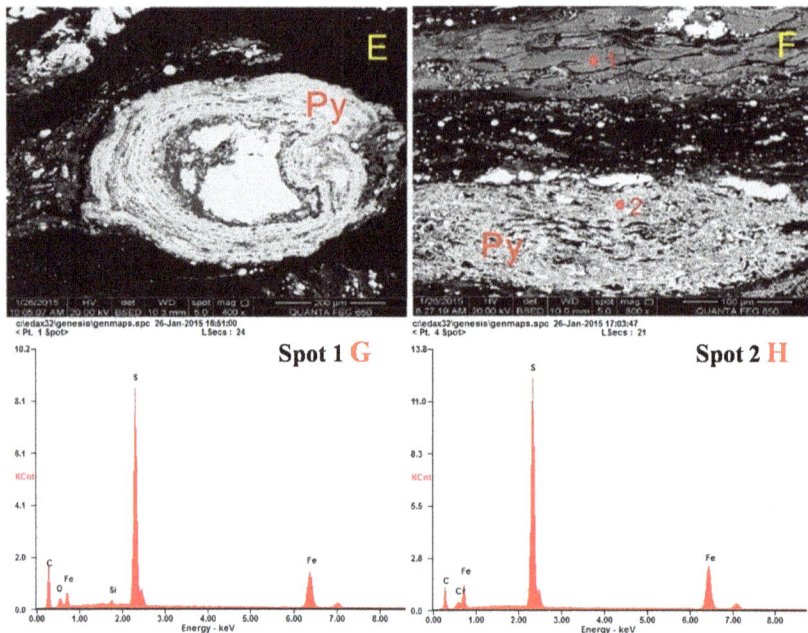

**Figure 8.** Reflected white light optical microscope, SEM back-scattered electron images and selected EDS data in YLT6U and YLT6L coals. (**A**) Fusinite- and semifusinite-cell filling pyrite in sample YLT6U-8 (optical microscope); (**B**) framboidal pyrite in sample YLT6L-2 (optical microscope, oil immersion); (**C**) cell- and fracture-filling pyrite in sample YLT6L-1; (**D**) cubic pyrite in sample YLT6L-1; (**E**) nodular pyrite in sample YLT6L-1; (**F**) cell-filling pyrite in sample YLT6L-1; (**G,H**) EDS spectra. Py, pyrite.

### 4.2.2. Minerals in the Partings

In comparison with the mineral matter in the coal benches, the four partings (YLT6U-2p, YLT6U-9p, YLT6L-3p and YLT6L-4p) have lower calcite and higher kaolinite and mixed-layer illite/smectite contents. In addition, the partings within the YLT6U coal have higher anatase and lower quartz contents. The partings within the YLT6L coal have higher chlorite and lower pyrite contents relative to those in the respective coal benches (Table 2, Figure 4). XRD analysis of the fractions of the partings shows relatively more abundant kaolinite and I/S, with small proportion of quartz, chamosite and berthierine in the clay mineral assemblage (Figure 9A–D). The chamosite is represented by a peak at 14.4 Å under air-dried conditions in sample YLT6L-4p, and its peak of 7.1 Å appears on heating (Figure 9D). Authigenic kaolinite and mixed-layer illite/smectite in the partings occur as fusinite- and semifusinite-cell fillings (Figure 10A), as vermicular form (Figure 10B,C) and as a matrix (Figure 10D). Kaolinite and mixed-layer illite/smectite (I/S) also occur as discrete particles (Figure 10E,F) and, in some cases, are distributed along bedding planes (Figure 10G).

Chlorite in the partings occurs generally as a cryptocrystalline matrix coexisting with anatase and pyrite (Figure 11A,C). Anatase provides Ti peaks in some of the chlorite EDS spectrum (Figure 10K). A small proportion of epigenetic chlorite occurs as fracture-filling in sample YLT6L-4p (Figure 11B). Chlorite with sub-angular forms (Figure 10F,J) formed because earlier formed apatite was chloritized by Fe-Mg fluids.

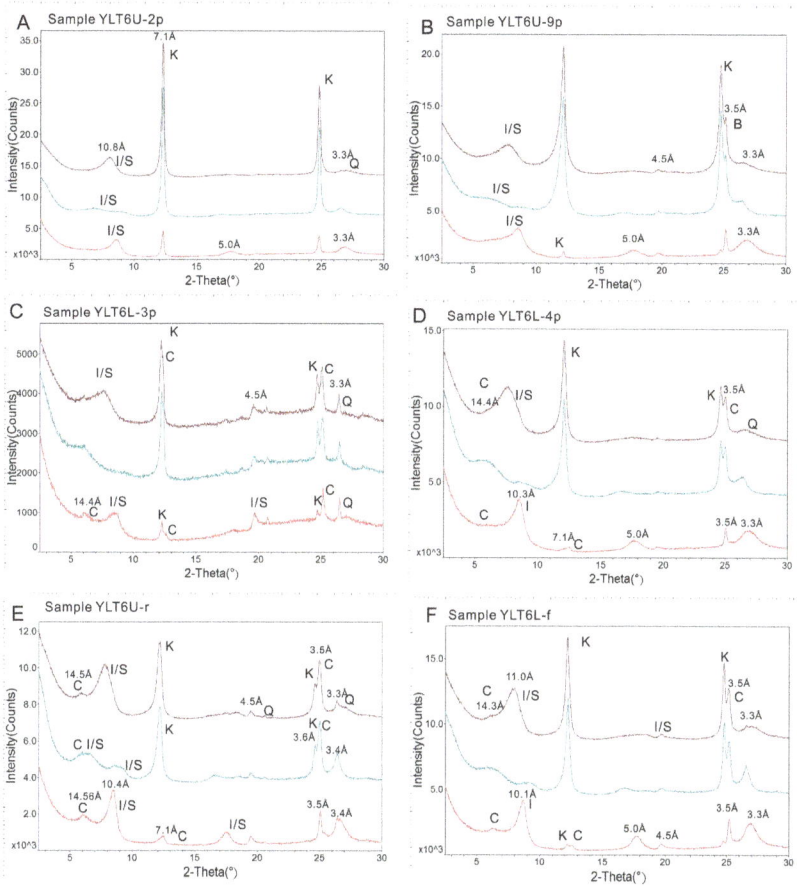

**Figure 9.** X-ray diffractogram (XRD) patterns of clay-fractions. (**A**) Sample YLT6U-2p; (**B**) sample YLT6U-9p; (**C**) sample YLT6L-3p; (**D**) sample YLT6L-4p; (**E**) sample YLT6U-r; (**F**) sample YLT6L-f. K, kaolinite; I/S, mixed-layer illite/smectite; C, chamosite; B, berthierine; Q, quartz. Natural-oriented (top trace), glycol-saturated (middle trace) and heated (bottom trace). Numbers represent d-spacings in Ångstrom units.

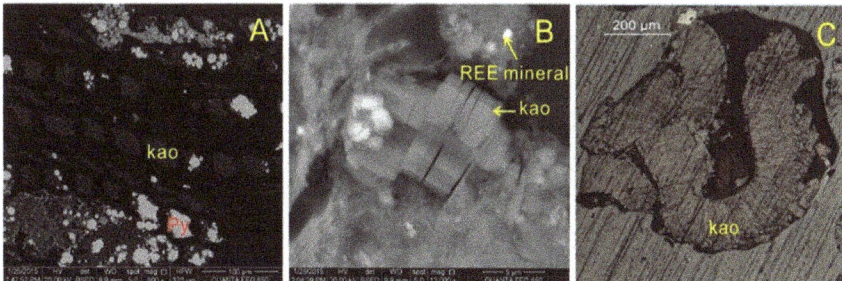

**Figure 10.** *Cont.*

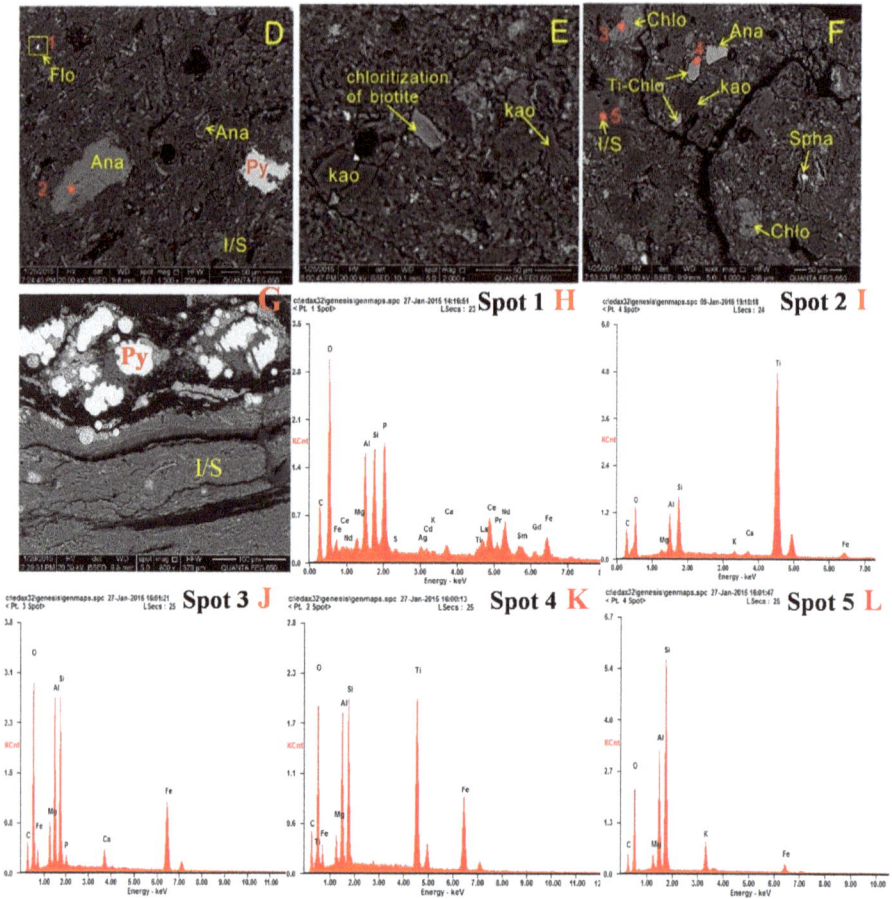

**Figure 10.** Reflected white light optical microscope, SEM back-scattered electron images and selected EDS data in YLT6U and YLT6L partings. (**A**) Fusinite- and semifusinite-cell filling kaolinite and pyrite in sample YLT6L-4p; (**B**) Vermicular kaolinite in sample YLT6U-9p; (**C**) Vermicular kaolinite in sample YLT6U-2p (optical microscope); (**D**) Anatase, I/S, pyrite and monazite in sample YLT6L-4p; (**E**) Kaolinite and chloritized biotite in sample YLT6L-4p; (**F**) I/S, Kaolinite, anatase, sphalerite and chlorite in sample YLT6L-4p; (**G**) Mixed-layer I/S distributed along the bedding planes in sample YLT6U-9p. Sample YLT6L-4p is carbon coated, and sample YLT6U-9p is detected under low vacuum without coating. (**H–L**) EDS spectra of Spots 1–5. Kao, kaolinite; Py, pyrite; Ana, anatase; Chlo, chlorite; Flo, florencite; Spha, sphalerite.

Anatase in the partings occurs as coarse-crystalline (Figure 11A,B) and beaded form (Figures 10D and 11D). They were closely associated with clay minerals and were probably precipitated from Ti-rich fluids with Ti leaching from volcanic ash as described below. The blocky and framboidal pyrite, associated with some fragments of organic matter in sample YLT6U-9p (Figure 10G), suggests that the pyrite is a syngenetic sedimentary material rather than a derivation of volcanic ash. Euhedral pyrite of syngenetic origin also occurs as cubes and octahedrons (Figure 11E). Cell- and cavity-filling pyrite (Figures 10A and 11F) in the YLT6L coal is of authigenic origin.

A number of accessory minerals have also been found in the partings. Epigenetic barite occurs as fracture fillings (Figure 11G), and in some cases, it also coexists with mixed-layer I/S (Figure 11H).

Lathlike sphalerite is embedded in mixed-layer I/S (Figure 11I), suggestive of epigenetic origin. Monazites are present in YLT6L-4p and YLT6U-9p samples (Figure 10B,D). The Energy Dispersive X-ray Spectroscopy (EDS) spectra of minerals in Figure 11 are shown in Figure 12.

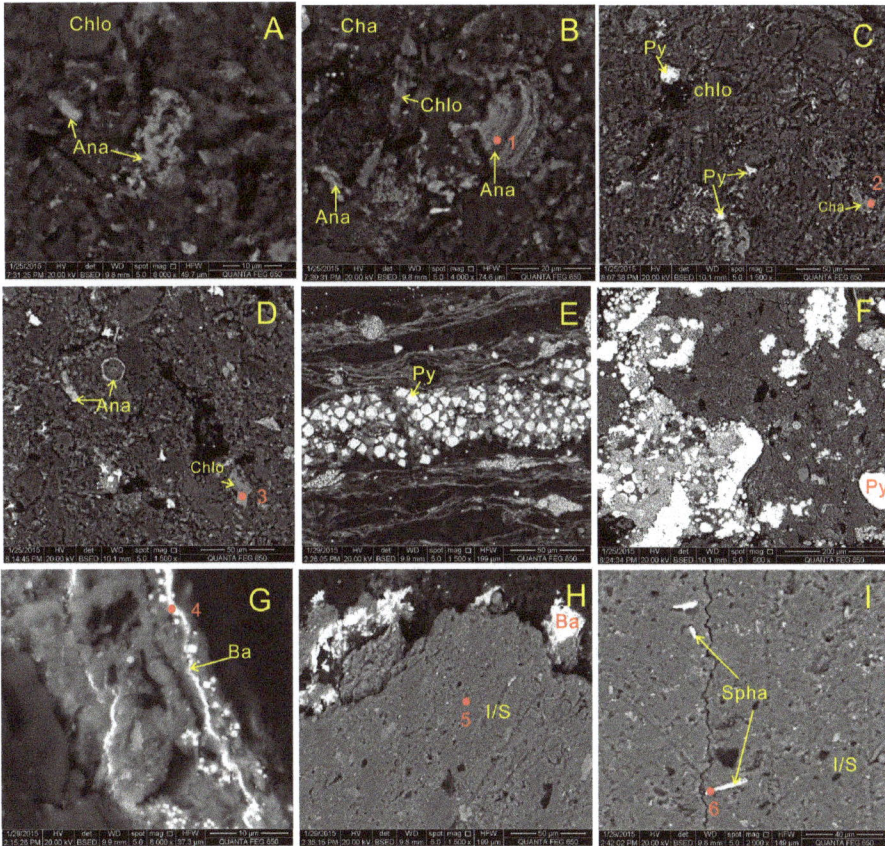

**Figure 11.** SEM back-scattered electron images in YLT6U and YLT6L partings. (**A**) Chlorite in sample YLT6L-4p; (**B**) fracture- and cavity-filling chlorite in sample YLT6L-4p; (**C**) chlorite in sample YLT6L-4p; (**D**) beaded anatase in sample YLT6L-4p; (**E**) euhedral pyrite in sample YLT6U-9p; (**F**) framboidal and cavity-filling pyrite in sample YLT6L-4p; (**G**) fracture-filling barite in sample YLT6U-9p; (**H**) barite coexisting with I/S in sample YLT6U-9p; (**I**) sphalerite in sample YLT6U-9p. Py, pyrite; Ana, anatase; Chlo, chlorite; Cha, chamosite; Ba, barite; Spha, sphalerite.

The roof of the YLT6U coal (YLT6U-r) and the floor of the YLT6L coal (YLT6L-f) are macroscopically and microscopically different. Macroscopically, sample YLT6L-f shows a well-developed bedding plane compared to sample YLT6U-r. Microscopically, samples YLT6U-r and YLT6L-f are mainly composed of mixed-layer illite/smectite and quartz, with a lesser proportion of chlorite, anatase and calcite (Table 2, Figure 4). Siderite and muscovite are also present in sample YLT6U-r, and kaolinite, pyrite and apatite are present in the floor of the YLT6L coal. The clay minerals identified in the host rocks by the clay-fraction studies are kaolinite, chamosite and mixed-layer I/S (Figure 9E,F). The mixed-layer I/S in sample YLT6L-f has a peak of 11.0 Å on the natural-oriented aggregate. The sharp peak of 10.1 Å in sample YLT6L-f indicates the presence of illite on heated aggregate (Figure 9F).

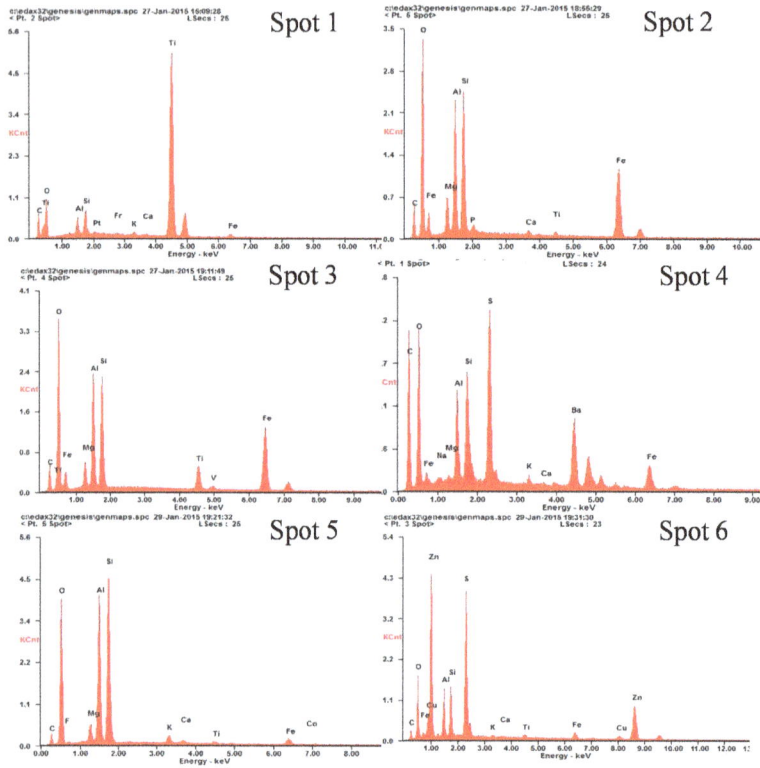

**Figure 12.** EDS spectra of Spots 1–6 in Figure 11.

Quartz in the host rocks occurs mainly as discrete particles with irregular forms (Figure 13A,B). Kaolinite in sample YLT6L-f occurs as discrete particles (Figure 13B,C), and some kaolinite has flocculent shapes (Figure 13D,E). Authigenic anatase in the host rocks is ring shaped (Figure 13A) or occurs as discrete particles (Figure 13B), coarse-crystalline (Figure 13D), platy (Figure 13E), linear (Figure 13F) and fracture-infilling (Figure 13G,H) forms. In addition, some anatases coexist with stripped apatite (Figure 13H). Chamosite coexisting with kaolinite in flocculent forms (Figure 13D,E,H) occurs in the roof strata. Euhedral pyrite is present in the floor (Figure 13I), suggestive of syngenetic origin.

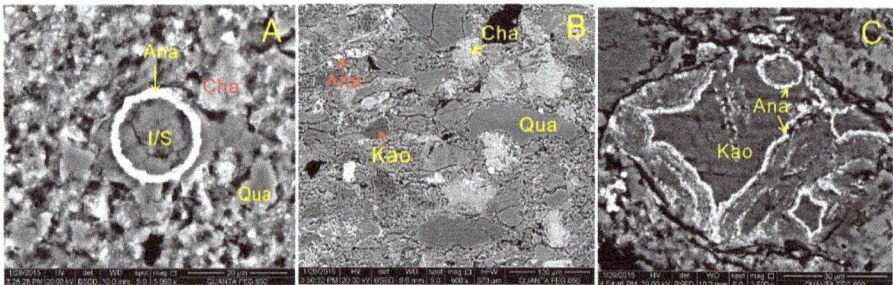

**Figure 13.** *Cont.*

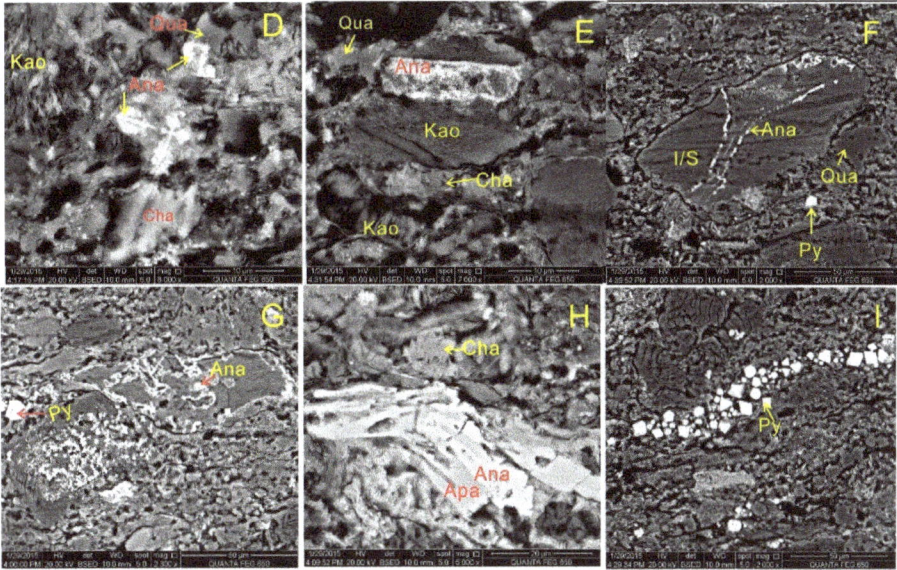

**Figure 13.** SEM back-scattered electron images in the host rocks (under low vacuum). (**A**) Quartz, I/S, chamosite and ring-shaped anatase in sample YLT6U-r; (**B**) Quartz, kaolinite, anatase and chamosite in sample YLT6L-f; (**C**) Kaolinite and anatase in sample YLT6L-f; (**D**) Flocculent kaolinite, anatase and chamosite in sample YLT6L-f; (**E**) Kaolinite and chamosite in sample YLT6L-f; (**F**) Anatase and pyrite in sample YLT6L-f; (**G**) Fracture-filling anatase and pyrite in sample YLT6L-f; (**H**) Anatase, chamosite and apatite in sample YLT6L-f; (**I**) Euhedral pyrite in sample YLT6L-f. Kao, kaolinite; Py, pyrite; Ana, anatase; Cha, chamosite; Qua, quartz; Apa, apatite.

## 5. Discussion

The assemblage of minerals in the No. 6 coal seam of the Yueliangtian coals and associated non-coal samples is attributed to four factors, including sediment-source region, multi-stage injections of hydrothermal fluids, seawater influence and volcanic ash input.

### 5.1. Sediment-Source Region Influence

Similar to other Late Permian coals from southwestern China [36,43–45], the Kangdian Upland (Figure 1) is the sediment-source region for the YLT6 coal [11]. Not only the Emeishan mafic basalts, but also the overlying felsic-intermediate rocks could have been the terrigenous source materials for the Late Permian coals present in this study [46]. Quartz in the YLT6 coal is slightly higher than that in coals from the Dafang (0.8%–11.4% on average) [14] and Zhijin (0.5%–9.4% on average) mines [35] (Figure 1), consistent with Dai *et al.* [15] and indicating that the closer the coals are located to the sediment-source Kangdian Upland region, the higher the quartz in the coal. The abundant authigenic quartz (Figure 5A) is precipitated from the silica-bearing solutions from the weathering of basalt in the Kangdian Upland [15,47,48].

Although the thick lava sequence from the Emeishan large igneous province in southwestern China is basalt, silicic rocks also occur in the uppermost part of the Kangdian upland [47]. There is no quartz and apatite in Emeishan basalt of the Kangdian Upland [49]; the abundant detrital quartz (Figure 5B–D) and authigenic apatite (Figure 13H) probably indicate that the parent rock is also granite or other silicic rocks [47].

*5.2. Multi-Stage Injections of Hydrothermal Fluids*

Hydrothermal fluids play an important role in the enrichment of minerals and trace elements in the coals of Guizhou province [5,12,14,41]. Multi-stage hydrothermal fluids also influenced the mineralogical characteristics of the Yueliangtian coals, partings, roof and floor.

The cell-filling kaolinite (Figure 7A,B) suggests an authigenic origin in coal. Al- and Si-bearing solutions from the sediment-source Kangdian Upland region precipitated in the cells of coal-forming plants during coal formation. Although chlorite (e.g., chamosite) is rare in coals, if present, it is usually observed in high-rank coals or in coals influenced by epigenetic hydrothermal solutions [44,50,51]. Chamosite is present in the YLT6 coal and coexists with kaolinite and quartz (Figure 13D,E) in the roof strata. As suggested by Equation (1), the earlier-precipitated kaolinite was invaded by Fe-Mg-rich hydrothermal fluids and generated kaolinite and chamosite during early diagenesis at temperatures around 165–200 °C [52]. Then, kaolinite and chamosite desilicated and generated quartz, as suggested by Equation (2). Dispersed chamosite (Figure 13B) independent of kaolinite and quartz may have precipitated from Fe-rich hydrothermal solutions, with Fe probably coming from siderite layers in the sedimentary sequence (Figure 2) [53].

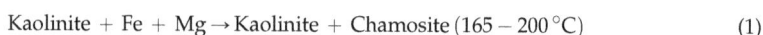

$$\text{Kaolinite } + \text{ Fe } + \text{ Mg} \rightarrow \text{Kaolinite } + \text{ Chamosite} \, (165 - 200 \, ^{\circ}\text{C}) \tag{1}$$

$$\text{Kaolinite } + \text{ Chamosite} \rightarrow \text{Chamosite } + \text{ Quartz} \, (\text{Desilication}) \tag{2}$$

The abundant authigenic quartz (Figure 5A) is probably precipitated from the silicious solution of the weathering product of Emeishan basalt from the Kangdian Upland [47,54]. Epigenetic vein-like calcite (Figure 6D) is probably precipitated by the circulation of Ca-bearing meteoric fluids or by Ca-rich solutions during the coal-formation process [55]. The average formation temperature of vein calcite is about 190 °C [14].

The crystal-shaped cavities commonly occurred in the samples YLT6L-4p (Figures 10D,E and 11C,D) and YLT6L-f (Figure 13E), suggesting that the earlier formed mineral crystals were corroded by hydrothermal solutions [15].

The modes of occurrence of coarse-crystalized anatase (YLT6L-4p; Figure 11A,B), cell-filling calcite (YLT6U-6; Figure 6A), fracture-filling barite (YLT6U-9p; Figure 11G) and cell- or fracture-filling pyrite (YLT6L-4p; Figures 10A and 11F) also suggest solution deposition during different stages of the coal formation [15,41].

*5.3. Seawater Influence*

Although the vertical distance between the floor of the YLT6U and the roof of the YLT6L coals is only 2.94 m, the deposit environments of the two coals are strikingly different.

The total sulfur contents (ranging from 3.61%–13.34%) in the YLT6L coal are higher than those in the YLT6U coal (0.27%–0.75%). Except for the YLT6U-8 and YLT6U-9p samples, no pyrite was present in the YLT6U coal. However, pyrite of syngenetic origin is common in the YLT6L coal, non-coal partings and floor strata (Figure 8B,D,E, Figures 10G, 11E and 13I). It can be inferred that seawater provided a sulfur source both during peat forming and in the diagenetic process of the YLT6L coal. The abundant calcite and mixed-layer I/S in the YLT6L coal also signifies the alkaline, medium environment of seawater (Table 2), which are also beneficial for calcite formation and the transformation from kaolinite to mixed-layer I/S [4]. The paleoenvironment of the Longtan Formation varies from lagoons and tidal flats [11–14]. Thus, seawater had a tremendous influence on YLT6L coal and terminated at the YLT6U-9p sampling interval.

*5.4. Volcanic Ash Input*

As described by Spears [56,57], Dai *et al.* [58,59], Zhou *et al.* [60] and Zhao *et al.* [43], "tonsteins" are partings derived from air-borne material of pyroclastic origin in the peat-forming environment. They have been found in some coal seams of SW China [15,43,58–60].

The three partings (samples YLT6U-2p, YLT6U-9p and YLT6L-4p) have a lateral continuity within the Yueliangtian coal mine. As shown in Table 2, the mineralogy of the three non-coal partings is dominated by kaolinite and mixed-layer I/S, accounting for 80%–89.1% of the mineral compositions of the respective parting. The mixed-layer I/S in the partings mostly occurs as cryptocrystalline matrix (Figures 10D and 11H,I). Kaolinite in samples YLT6U-2p and YLT6U-9p occurs as large crystals with a well-developed vermicular texture (Figure 10B,C). Vermicular kaolinite is thought to indicate air-fall volcanic ash layers altered and deposited in a non-marine, coal-forming environment [61]. Biotite pseudomorphs with chlorite laminae occur in the sample YLT6L-4p (Figure 10E). Such an occurrence of chloritized biotite suggests an *in situ* crystallized origin [15].

High-temperature quartz has been identified in the sample YLT6L-f. The β-quartz (Figure 13B) present in this study shows triangle and irregular forms and is considered to have originated from autochthonous syngenetic felsic to intermediate volcanic ashes. Dai *et al.* [59] and Zhou *et al.* [60] suggested that tonsteins in the lower and upper portions of the Late Permian were mainly of alkali and felsic composition, respectively. The No. 6 coal seam in the present study is located in the upper portion of the Late Permian strata (Figure 2).

The modes of anatase occurrence in the partings (Figure 11D) and floor strata (Figure 13A,F) suggest that it probably either was altered by pyroxene crystals in the original volcanic ash or reprecipitated after labile components was chemically leached in the original volcanic ash [15].

Zircon was not observed by optical microscope or SEM in the tonsteins of the present study, probably either because the relevant volcanic ashes did not contain this mineral phases or because multi-stage hydrothermal fluids as mentioned above altered it.

The mineral assemblages and occurrences of mineral of partings and roof strata in the No. 6 coal seam of the Yueliangtian coal mine suggest that the three partings (samples YLT6U-2p, YLT6U-9p and YLT6L-4p) and roof strata appear to have been derived from felsic volcanic ash.

## 6. Conclusions

(1) The major mineral phases in the YLT6U and YLT6L coals are calcite, quartz, kaolinite and pyrite and, to a lesser extent, chlorite, anatase, illite, mixed-layer illite/smectite, rutile, bassanite and ankerite. The Emeishan basalt and silicic rocks in the Kangdian Upland are the parent rocks of the Yueliangtian coals.

(2) Different modes of occurrence of chamosite are present in the YLT6 coal. This, accompanied with cell-filling quartz, pyrite, and calcite veins, suggests that multi-stage hydrothermal fluids influenced the Yueliangtian coals.

(3) The sedimentary environment is different between the YLT6U and YLT6L coals. Seawater had a tremendous influence on the YLT6L coal and terminated at the YLT6U-9p sampling interval.

(4) Three tonstein (samples YLT6U-2p, YLT6U-9p and YLT6L-4p) layers identified in the coal are probably derived from felsic volcanic ash. These tonsteins are characterized by the occurrence of vermicular kaolinite and chloritized biotite. The roof strata of the YLT6L coal also appear to have been derived from felsic volcanic ash.

**Acknowledgments:** This research was supported by the National Key Basic Research Program of China (No. 2014CB238902), the National Natural Science Foundation of China (Nos. 41420104001 and 41272182), and the Program for Changjiang Scholars and Innovative Research Team in University (IRT13099). The authors are grateful to three anonymous reviewers and editor for their careful comments, which greatly improved the paper quality. We would like to thank Shifeng Dai and Lei Zhao for their constructive suggestions on the earlier version of this paper.

**Author Contributions:** Panpan Xie designed the overall experimental strategy and participated in all of the experiments. Hongjian Song has performed the XRD experiments. Jianpeng Wei and Qingqian Li analyzed the minerals in the samples using SEM-EDX. All authors participated in writing the manuscript.

**Conflicts of Interest:** The authors declare no conflict of interest.

## References

1. Dai, S.F.; Ren, D.Y.; Chou, C.-L.; Finkelman, R.B.; Seredin, V.V.; Zhou, Y.P. Geochemistry of trace elements in Chinese coals: A review of abundances, genetic types, impacts on human health, and industrial utilization. *Int. J. Coal Geol.* **2012**, *94*, 3–21. [CrossRef]

2. Liu, J.J.; Yang, Z.; Yan, X.Y.; Ji, D.P.; Yang, Y.C.; Hu, L.C. Modes of occurrence of highly-elevated trace elements in superhigh-organic-sulfur coals. *Fuel* **2015**, *156*, 190–197. [CrossRef]

3. Dai, S.F.; Ren, D.Y.; Zhou, Y.P.; Chou, C.-L.; Wang, X.B.; Zhao, L.; Zhu, X.W. Mineralogy and geochemistry of a superhigh-organic-sulfur coal, Yanshan Coalfield, Yunnan, China: Evidence for a volcanic ash component and influence by submarine exhalation. *Chem. Geol.* **2008**, *255*, 182–194. [CrossRef]

4. Shao, Y.B.; Guo, Y.H.; Qin, Y.; Shen, Y.L.; Tian, L. Distribution characteristic and geological significance of rare earth elements in Lopingian mudstone of Permian, Panxian county, Guizhou province. *Min. Sci. Technol.* **2011**, *21*, 469–476. [CrossRef]

5. Zhang, J.Y.; Ren, D.Y.; Zheng, C.G.; Zeng, R.S.; Chou, C.-L.; Liu, J. Trace element abundances in major minerals of Late Permian coals from southwestern Guizhou province, China. *Int. J. Coal Geol.* **2002**, *53*, 55–64. [CrossRef]

6. Xiao, K. Analysis of principles for geological structures of Yueliangtian Coal Mine. *Min. Saf. Environ. Prot.* **2003**, *30*, 25–26.

7. Dai, S.F.; Chekryzhov, I.Y.; Seredin, V.V.; Nechaev, V.P.; Graham, I.T.; Hower, J.C.; Ward, C.R.; Ren, D.Y.; Wang, X.B. Metalliferous coal deposits in East Asia (Primorye of Russia and South China): A review of geodynamic controls and styles of mineralization. *Gondwana Res.* **2016**, *29*, 60–82. [CrossRef]

8. Dai, S.F.; Seredin, V.V.; Ward, C.R.; Hower, J.C.; Xing, Y.W.; Zhang, W.G.; Song, W.J.; Wang, P.P. Enrichment of U-Se-Mo-Re-V in coals preserved within marine carbonate successions: Geochemical and mineralogical data from the Late Permian Guiding Coalfield, Guizhou, China. *Miner. Deposita* **2015**, *50*, 159–186. [CrossRef]

9. Ward, C.R. Analysis and significance of mineral matter in coal seams. *Int. J. Coal Geol.* **2002**, *50*, 135–168. [CrossRef]

10. Yuan, M.; Zhang, F. Structure formation and its application to coal exploration in south No. 4 Mining Field, Yueliangtian Coal Mine. *Coal Geol. Explor.* **2004**, *4*, 16–18.

11. Coal Geology Bureau of China. *Sedimentary Environments and Coal Accumulation of Late Permian Coal Formation in Western Guizhou, Southern Sichuan and Eastern Yunnan, China*; Chongqing University Press: Chongqing, China, 1996. (In Chinese)

12. Dai, S.F.; Ren, D.Y.; Tang, Y.G.; Shao, L.Y.; Hao, L.M. Influences of low-temperature hydrothermal fluid on the re-distributions and occurrences of associated elements in coal—A case study from the Late Permian coals in the Zhijin Coalfield, Guizhou Province, southern China. *Acta Geol. Sin.* **2002**, *76*, 437–445.

13. Dai, S.F.; Ren, D.Y.; Tang, Y.G.; Yue, M.; Hao, L.M. Concentration and distribution of elements in Late Permian coals from western Guizhou Province, China. *Int. J. Coal Geol.* **2005**, *61*, 119–137. [CrossRef]

14. Dai, S.F.; Chou, C.-L.; Yue, M.; Luo, K.L.; Ren, D.Y. Mineralogy and geochemistry of a Late Permian coal in the Dafang Coalfield, Guizhou, China: Influence from siliceous and iron-rich calcic hydrothermal fluids. *Int. J. Coal Geol.* **2005**, *61*, 241–258. [CrossRef]

15. Dai, S.F.; Li, T.; Seredin, V.V.; Ward, R.C.; Hower, J.C.; Zhou, Y.P.; Zhang, M.Q.; Song, X.L.; Song, W.J.; Zhao, C.L. Origin of minerals and elements in the Late Permian coals, tonsteins, and host rocks of the Xinde Mine, Xuanwei, eastern Yunnan, China. *Int. J. Coal Geol.* **2014**, *121*, 53–78. [CrossRef]

16. Coal Analysis Laboratory of China Coal Research Institute. *Chinese Standard Method GB/T 482-2008, Sampling of Coal Seams*; National Coal Standardization Technical Committee: Beijing, China, 2008.

17. ASTM International. *Test Method for Moisture in the Analysis Sample of Coal and Coke*; ASTM D3173-11; ASTM International: West Conshohocken, PA, USA, 2011.

18. ASTM International. *Test Method for Volatile Matter in the Analysis Sample of Coal and Coke*; ASTM D3175-11; ASTM International: West Conshohocken, PA, USA, 2011.

19. ASTM International. *Test Method for Ash in the Analysis Sample of Coal and Coke from Coal*; ASTM D3174-11; ASTM International: West Conshohocken, PA, USA, 2011.

20. ASTM International. *Test Methods for Total Sulfur in the Analysis Sample of Coal and Coke*; ASTM D3177-02; ASTM International: West Conshohocken, PA, USA, 2011.

21. ASTM International. *Standard Practice for Preparing Coal Samples for Microscopical Analysis by Reflected Light*; ASTM Standard D2797/D2797M-11a; ASTM International: West Conshohocken, PA, USA, 2011.

22. ASTM International. *Standard Test Method for Microscopical Determination of the Vitrinite Reflectance of Coal*; ASTM Standard D2798-11a; ASTM International: West Conshohocken, PA, USA, 2011.

23. Rietveld, H.M. A profile refinement method for nuclear and magnetic structures. *J. Appl. Crystal.* **1969**, *2*, 65–71. [CrossRef]

24. Taylor, J.C. Computer programs for standardless quantitative analysis of minerals using the full powder diffraction profile. *Powder Diffr.* **1991**, *6*, 2–9. [CrossRef]

25. Ward, C.R.; Spears, D.A.; Booth, C.A.; Staton, I.; Gurba, L.W. Mineral matter and trace elements in coals of the Gunnedah Basin, New South Wales, Australia. *Int. J. Coal Geol.* **1999**, *40*, 281–308. [CrossRef]

26. Ward, C.R.; Matulis, C.E.; Taylor, J.C.; Dale, L.S. Quantification of mineral matter in the Argonne Premium coals using interactive Rietveld-based X-ray diffraction. *Int. J. Coal Geol.* **2001**, *46*, 67–82. [CrossRef]

27. Ruan, C.D.; Ward, C.R. Quantitative X-ray powder diffraction analysis of clay minerals in Australian coals using Rietveld methods. *Appl. Clay Sci.* **2002**, *21*, 227–240. [CrossRef]

28. Dai, S.F.; Yang, J.Y.; Ward, C.R.; Hower, J.C.; Liu, H.D.; Garrison, T.M.; French, D.; O'Keefe, J.M.K. Geochemical and mineralogical evidence for a coal-hosted uranium deposit in the Yili Basin, Xinjiang, northwestern China. *Ore Geol Rev.* **2015**, *70*, 1–30. [CrossRef]

29. Dai, S.F.; Wang, P.P.; Ward, C.R.; Tang, Y.G.; Song, X.L.; Jiang, J.H.; Hower, J.C.; Li, T.; Seredin, V.V.; Wagner, N.J.; *et al.* Elemental and mineralogical anomalies in the coal-hosted Ge ore deposit of Lincang, Yunnan, southwestern China: Key role of $N_2$-$CO_2$-mixed hydrothermal solutions. *Int. J. Coal Geol.* **2014**, *152*, 19–46. [CrossRef]

30. Dai, S.F.; Zhang, W.G.; Ward, C.R.; Seredin, V.V.; Hower, J.C.; Li, X.; Song, W.J.; Wang, X.B.; Kang, H.; Zheng, L.C.; *et al.* Mineralogical and geochemical anomalies of late Permian coals from the Fusui Coalfield, Guangxi province, southern China: Influences of terrigenous materials and hydrothermal fluids. *Int. J. Coal Geol.* **2013**, *105*, 60–84. [CrossRef]

31. China Southwest Oil and Gas Field Company Exploration and Development Institute. *Chinese Petrol and Naturl Gas Industry Standard SY/T 5163-2010, Analysis Method for Clay Minerals and Ordinary Non-Clay Minerals in Sedimentary Rocks by the X-ray Diffraction*; Petroleum Exploration Standardization Technical Committee: Beijing, China, 2010. (In Chinese)

32. ASTM International. *Standard Classification of Coals by Rank*; ASTM D388-12; ASTM International: West Conshohocken, PA, USA, 2012.

33. Coal Analysis Laboratory of China Coal Research Institute. *Chinese Standard Method GB/T 15224.1-2010, Classification for Quality of Coal. Part 1: Ash*; Standardization Administration of China: Beijing, China, 2010. (In Chinese)

34. Dai, S.F.; Ren, D.Y.; Hou, X.Q.; Shao, L.Y. Geochemical and mineralogical anomalies of the late Permian coal in the Zhijin coalfield of southwest China and their volcanic origin. *Int. J. Coal Geol.* **2003**, *55*, 117–138. [CrossRef]

35. Dai, S.F.; Li, D.H.; Ren, D.Y.; Tang, Y.G.; Shao, L.Y.; Song, H.B. Geochemistry of the late Permian No. 30 coal seam, Zhijin Coalfield of Southwest China: Influence of a siliceous low-temperature hydrothermal fluid. *Appl. Geochem.* **2004**, *19*, 1315–1330. [CrossRef]

36. Zhuang, X.G.; Querol, X.; Zeng, R.S.; Xu, W.D.; Alastuy, A.; Lopez-Soler, A.; Plana, F. Mineralogy and geochemistry of coal from the Liupanshui mining district, Guizhou, south China. *Int. J. Coal Geol.* **2000**, *45*, 21–37. [CrossRef]

37. Dai, S.F.; Wang, X.B.; Chen, W.M.; Li, D.H.; Chou, C.-L.; Zhou, Y.P.; Zhu, C.S.; Li, H.; Zhu, X.Y.; Xing, Y.W.; *et al.* A high-pyrite semianthracite of late Permian age in the Songzao Coalfield, southwestern China: Mineralogical and geochemical relations with underlying mafic tuffs. *Int. J. Coal Geol.* **2010**, *83*, 430–445. [CrossRef]

38. Widodo, S.; Oschmann, W.; Bechtel, A.; Sachsenhofer, R.F.; Anggayana, K.; Puettmann, W. Distribution of sulfur and pyrite in coal seams from Kutai Basin (east Kalimantan, Indonesia): Implications for paleoenvironmental conditions. *Int. J. Coal Geol.* **2010**, *81*, 151–162. [CrossRef]

39. Dai, S.F.; Hou, X.Q.; Ren, D.Y.; Tang, Y.G. Surface analysis of pyrite in the No. 9 coal seam, Wuda coalfield, Inner Mongolia, China, using high-resolution time-of-flight secondary ion mass-spectrometry. *Int. J. Coal Geol.* **2003**, *55*, 139–150. [CrossRef]

40. Dai, S.F.; Ren, D.Y.; Tang, Y.G.; Shao, L.Y.; Li, S.S. Distribution, isotopic variation and origin of sulfur in coals in the Wuda coalfield, Inner Mongolia, China. *Int. J. Coal Geol.* **2002**, *52*, 237–250. [CrossRef]

41. Yang, J.Y. Modes of occurrence and origins of noble metals in the Late Permian coals from the Puan Coalfield, Guizhou, southwest China. *Fuel* **2006**, *85*, 1679–1684. [CrossRef]

42. Chou, C.-L. Sulfur in coals: A review of geochemistry and origins. *Int. J. Coal Geol.* **2012**, *100*, 1–13. [CrossRef]

43. Zhao, L.; Ward, C.R.; French, D.; Graham, I.T. Mineralogical composition of Late Permian coal seams in the Songzao Coalfield, southwestern China. *Int. J. Coal Geol.* **2013**, *116–117*, 208–226. [CrossRef]

44. Dai, S.F.; Tian, L.W.; Chou, C.-L.; Zhou, Y.; Zhang, M.Q.; Zhao, L.; Wang, J.M.; Yang, Z.; Cao, H.Z.; Ren, D.Y. Mineralogical and compositional characteristics of Late Permian coals from an area of high lung cancer rate in Xuan Wei, Yunnan, China: Occurrence and origin of quartz and chamosite. *Int. J. Coal Geol.* **2008**, *76*, 318–327. [CrossRef]

45. Dai, S.F.; Zhou, Y.P.; Ren, D.Y.; Wang, X.B.; Li, D.; Zhao, L. Geochemistry and mineralogy of the Late Permian coals from the Songzo Coalfield, Chongqing, southwestern China. *Sci China Ser. D Earth Sci.* **2007**, *50*, 678–688. [CrossRef]

46. Dai, S.F.; Liu, J.J.; Ward, C.R.; Hower, J.C.; French, D.; Jia, S.H.; Hood, M.M.; Garrison, T.M. Mineralogical and geochemical compositions of Late Permian coals and host rocks from the Guxu coalfield, Sichuan Province, China, with emphasis on enrichment of rare metals. *Int. J. Coal Geol.* **2015**. [CrossRef]

47. Wang, X.B.; Wang, R.X.; Wei, Q.; Wang, P.P.; Wei, J.P. Mineralogical and geochemical characteristics of Late Permian coals from the Mahe Mine, Zhaotong Coalfield, Northeastern Yunnan, China. *Mineral* **2015**, *5*, 380–396. [CrossRef]

48. Ren, D.Y. Mineral matter in coal. In *Coal Petrology of China*; Han, D.X., Ed.; Publishing House of China University of Mining and Technology: Xuzhou, China, 1996; pp. 67–78. (In Chinese)

49. Xu, Y.G.; Chung, S.L.; Jahn, B.M.; Wu, G.Y. Petrologic and geochemical constraints on the petrogenesis of Permian-Triassic Emeishan flood basalts in southwestern China. *Lithos* **2001**, *58*, 145–168. [CrossRef]

50. Permana, A.K.; Ward, C.R.; Li, Z.; Gurba, L.W. Distribution and origin of minerals in high-rank coals of the South Walker Creek area, Bowen Basin, Australia. *Int. J. Coal Geol.* **2013**, *116–117*, 185–207. [CrossRef]

51. Dai, S.F.; Chou, C.-L. Occurrence and origin of minerals in a chamosite-bearing coal of Late Permian age, Zhaotong, Yunnan, China. *Am. Mineral.* **2007**, *92*, 1253–1261. [CrossRef]

52. Boles, J.R.; Franks, S.G. Clay diagenesis in Wilcox sandstones of southwest Texas: Implications of smectite diagenesis on sandstone cementation. *J. Sediment. Res.* **1979**, *49*, 55–70.

53. Iijima, A.; Matsumoto, R. Berthierine and chamosite in coal measures of Japan. *Clay Clay Miner.* **1982**, *30*, 264–274. [CrossRef]

54. Wang, X.B.; Dai, S.F.; Chou, C.-L.; Zhang, M.Q.; Wang, J.M.; Song, X.L.; Wang, W.; Jiang, Y.F.; Zhou, Y.P.; Ren, D.Y. Mineralogy and geochemistry of Late Permian coals from the Taoshuping Mine, Yunnan Province, China: Evidences for the sources of minerals. *Int. J. Coal Geol.* **2012**, *96–97*, 49–59. [CrossRef]

55. Kolker, A.; Chou, C.-L. Cleat-filling calcite in Illinois Basin coals: Trace element evidence for meteoric fluid migration in a coal basin. *J. Geol.* **1994**, *102*, 111–116. [CrossRef]

56. Spears, D.A.; Kanaris-Sotiriou, R. A geochemical and mineralogical investigation of some British and other European tonsteins. *Sedimentology* **1979**, *26*, 407–425. [CrossRef]

57. Spears, D.A. The origin of tonsteins, an overview, and links with seatearths, fireclays and fragmental clay rocks. *Int. J. Coal Geol.* **2012**, *94*, 22–31. [CrossRef]

58. Dai, S.F.; Luo, Y.B.; Seredin, V.V.; Ward, C.R.; Hower, J.C.; Zhao, L.; Liu, S.D.; Zhao, C.L.; Tian, H.M.; Zou, J.H. Revisiting the Late Permian coal from the Huayingshan, Sichuan, southwestern China: Enrichment and occurrence modes of minerals and trace elements. *Int. J. Coal Geol.* **2014**, *122*, 110–128. [CrossRef]

59. Dai, S.F.; Wang, X.B.; Zhou, Y.P.; Hower, J.C.; Li, D.H.; Chen, W.M.; Zhu, X.W.; Zou, J.H. Chemical and mineralogical compositions of silicic, mafic, and alkali tonsteins in the Late Permian coals from the Songzao Coalfield, Chongqing, Southwest China. *Chem. Geol.* **2011**, *282*, 29–44. [CrossRef]

*Minerals* **2016**, *6*, 29

60. Zhou, Y.P.; Bohor, B.F.; Ren, Y. Trace element geochemistry of altered volcanic ash layers (tonsteins) in Late Permian coal-bearing formations of eastern Yunnan and western Guizhou Provinces, China. *Int. J. Coal Geol.* **2000**, *44*, 305–324. [CrossRef]

61. Bohor, B.F.; Triplehorn, D.M. Tonsteins: Altered volcanic-ash layers in coal-bearing sequences. *Geol. Soc. Am. Spec. Pap.* **1993**, *285*, 1–44.

*minerals*

MDPI

*Article*

# Mineralogical and Geochemical Characteristics of the Early Permian Upper No. 3 Coal from Southwestern Shandong, China

Xibo Wang [1,*], Lili Zhang [1], Yaofa Jiang [2], Jianpeng Wei [1] and Zijuan Chen [1]

[1] State Key laboratory of Coal Resources and Safety Mining, China University of Mining and Technology, Beijing 100083, China; Zhangliliqingdao@gmail.com (L.Z.); weijianpeng15@gmail.com (J.W.); Z.J.Chen94@gmail.com (Z.C.)

[2] Jiangsu Institute of Architectural Technology, Xuzhou 221116, China; jiangyaofa-xz@163.com

* Correspondence: xibowang@gmail.com; Tel.: +86-10-6234-1868

Academic Editor: Daniel M. Deocampo
Received: 18 April 2016; Accepted: 16 June 2016; Published: 23 June 2016

**Abstract:** The Upper No. 3 coal of the Early Permian age is a major workable seam in the southwestern Shandong coalfield, which is located in the eastern part of North China. From Early Jurassic to Neogene, the coalfield was subjected to intensive tectonic processes, leading to a significant rearrangement in depth of coal seams. In this paper, three Upper No. 3 coals occurring at −228, −670 and −938 m in the Luxi, Liangbaosi, and Tangkou mines, respectively, were collected to investigate their mineralogical and geochemical characteristics, with emphasis on modes of occurrence and origin of epigenetic minerals. The three coal seams are similar in vitrinite reflectance, volatile matter yield, and maceral components, suggesting insignificant influence from the tectonic activities on coal rank. Terrigenous minerals (e.g., kaolinite and quartz) are comparable in both types and distribution patterns in the three coals. The presence of siderite and pyrite of syngenetic or penecontemporaneous origin indicate they were emplaced during peat accumulation. The distribution of epigenetic minerals (e.g., calcite, ankerite, and dolomite) are associated with the underground water activities, which were Ca (Mg, Fe)-bearing.

**Keywords:** coal; tectonic processes; minerals; elements; Early Permian

## 1. Introduction

Coal mainly consists of organic matter (macerals) and mineral matter (including discrete minerals and inorganic elements) [1,2]. The abundance and modes of occurrences of mineral matter are resulted from processes associated with peat accumulation and rank advance, the interaction of the organic matter with basinal fluids, sediment diagenesis, and in some cases, synsedimentary volcanic inputs [3–7]. Thus minerals in coals provide information about the depositional conditions, geologic history of coal-bearing sequences, and regional tectonic evolution [8–10].

The southwestern Shandong coalfield is located in the eastern part of North China (Figure 1). After the Carboniferous-Permian periods, the southwestern Shandong area was subjected to four episodes of tectonic processes: (1) crustal uplift to Jurassic; (2) fold extrusion deformation during Jurassic and Early Cretaceous; (3) extensional deformation during the Early Cretaceous and Oligocene periods, a stage that can further be separated into the North-South rift stage (Jurassic to Paleocene) and West-East rift stage (Eocene to Oligocene); and (4) post-Neogene subsidence [11]. After the third tectonic activity, graben and horst structure developed in the study area, which resulted in rearrangement of the coal-bearing strata at specific depths [11] (e.g., the Upper No. 3 coals in the Luxi, Liangbaosi, and Tangkou mines occur at various depths of −228, −670 and −938 m, respectively, Figure 1).

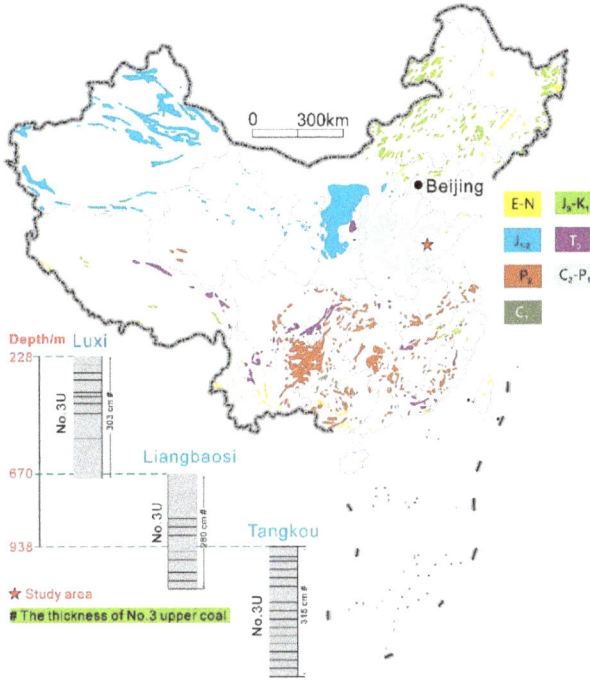

**Figure 1.** Location of the southwestern Shandong coalfield and the three studied Upper No. 3 coals at various depths.

Differences among maceral, mineral and elemental components between the Carboniferous and Permian coals of the southwestern Shandong coalfield were investigated by Zeng et al. [12] and Liu et al. [13,14]. Distributions and modes of occurrences of As, Se, and Hg in the coal from the Xinglongzhuang mine of the coalfield were also discussed in detail [15]. However, few publications have compared the differing mineralogical and geochemical characteristics of the Upper No. 3 coal at various depths. In the current study, new data on the petrology, mineralogy and geochemistry of the Upper No. 3 coal from the Luxi, Liangbaosi, and Tangkou Mines at various depths were investigate to ascertain the coal characteristics, such as minerals, in addition to major and trace elements.

## 2. Geological Setting

The Carboniferous-Permian coal deposits mainly occur in North China as shown in Figure 1. The southwestern Shandong coalfield is located in the eastern part of North China. Deposition of the Late Paleozoic coal in North China began with the Benxi Formation of the Late Carboniferous age, and continued into Late Carboniferous to Early Permian Taiyuan, the Shanxi and lower Shihezi Formations, and the upper Shihezi Formation of early Late Permian age. This sequence terminated with non-coal-bearing red clastic strata of the Late Permian Shiqianfeng Formation. The coal-bearing strata of the southwestern Shandong coalfield contain the Taiyuan and Shanxi Formations. The Taiyuan Formation represents a sequence of shallow marine environments, while the Shanxi Formation is of fluvial plain origin. In the study area, the Taiyuan Formation has a thickness of 120–160 m, and consists of siltstone, mudstone, limestone, and coal. The coal in the Taiyuan Formation occurs in up to 18 seams, in which the No. 16 and 17 coals are mineable. The Shanxi Formation has a thickness of 70–100 m, with 2–3 coal seams. The No. 3 coal was divided into upper and lower coal benches by a stable parting, and both benches are workable.

## 3. Sampling and Methods

The Upper No. 3 coal samples were collected from the Luxi, Liangbaosi and Tangkou mines in the southwestern Shangdong coalfield, China. Due to the rearrangement by tectonic process after coal formation, the burying depth of the No. 3 upper coals in these three mines varies significantly. The coal occurs at −228 m in the Luxi mine, −670 m in the Liangbaosi mine and −938 m in the Tangkou mine, respectively. A total of 43 coal bench samples were collected from the three mines, representing 14 samples from the Luxi mine, 13 from the Liangbaosi mine, and 16 from the Tangkou mine, respectively (Table 1).

Table 1. Moisture, ash yield, volatile matter yield, and total sulfur content (%) in the Upper No. 3 coals from the Luxi, Liangbaosi, and Tangkou mines, southwestern Shandong, China.

| Sample | Thickness (cm) | $M_{ad}$ | $A_d$ | $V_{daf}$ | $S_d$ |
|---|---|---|---|---|---|
| LX3U-1 | 20 | 1.76 | 10.57 | 36.24 | 0.78 |
| LX3U-2 | 20 | 1.8 | 19.43 | 39.60 | 1.22 |
| LX3U-4 | 14 | 1.91 | 16.94 | 35.19 | 0.72 |
| LX3U-5 | 30 | 1.92 | 9.89 | 36.46 | 0.63 |
| LX3U-6 | 10 | 2 | 7.65 | 38.46 | 0.82 |
| LX3U-7 | 17 | 1.89 | 10.3 | 32.24 | 0.55 |
| LX3U-8 | 23 | 1.91 | 12.99 | 37.02 | 0.6 |
| LX3U-10 | 30 | 1.69 | 14.05 | 35.68 | 0.5 |
| LX3U-11 | 15 | 1.78 | 10.61 | 34.77 | 0.68 |
| LX3U-12 | 22 | 1.85 | 12.32 | 36.51 | 0.66 |
| LX3U-13 | 20 | 1.86 | 11.82 | 34.28 | 0.86 |
| LX3U-14 | 40 | 1.68 | 11.48 | 37.00 | 0.73 |
| LX3U-15 | 20 | 1.9 | 18.54 | 38.35 | 0.79 |
| LX3U-16 | 15 | 1.92 | 38.38 | 45.01 | 0.9 |
| Luxi * | - | 1.83 | 14.11 | 36.79 | 0.73 |
| LBS3U-1 | 30 | 1.87 | 41.39 | 37.58 | 1.17 |
| LBS3U-2 | 20 | 2.39 | 8.64 | 37.19 | 0.69 |
| LBS3U-3 | 20 | 2.02 | 19.39 | 37.20 | 1.17 |
| LBS3U-4 | 20 | 2.1 | 9.04 | 38.55 | 0.54 |
| LBS3U-5 | 20 | 1.89 | 6.86 | 40.12 | 0.45 |
| LBS3U-6 | 20 | 2.05 | 11.42 | 35.96 | 0.32 |
| LBS3U-7 | 20 | 2.15 | 8.45 | 35.75 | 0.27 |
| LBS3U-8 | 20 | 2.19 | 8.21 | 38.22 | 0.46 |
| LBS3U-9 | 20 | 2.05 | 16.65 | 38.01 | 0.37 |
| LBS3U-10 | 20 | 2.03 | 17.52 | 37.19 | 0.34 |
| LBS3U-11 | 30 | 1.99 | 10.48 | 36.85 | 0.38 |
| LBS3U-12 | 20 | 2.13 | 9.03 | 33.43 | 0.43 |
| LBS3U-13 | 20 | 1.9 | 26.03 | 35.07 | 0.42 |
| Liangbaosi * | - | 2.05 | 15.65 | 36.90 | 0.56 |
| TK3U-1 | 25 | 2.07 | 9 | 41.63 | 0.34 |
| TK3U-2 | 10 | 2 | 9.51 | 36.62 | 0.51 |
| TK3U-3 | 20 | 1.9 | 7.95 | 40.01 | 0.34 |
| TK3U-4 | 20 | 2.07 | 6.42 | 39.06 | 0.36 |
| TK3U-5 | 20 | 1.97 | 22.05 | 34.97 | 0.26 |
| TK3U-6 | 20 | 1.91 | 14.06 | 36.17 | 0.07 |
| TK3U-7 | 20 | 1.94 | 6.95 | 36.13 | 0.1 |
| TK3U-8 | 20 | 1.87 | 7.93 | 36.99 | 0.25 |
| TK3U-9 | 20 | 1.8 | 9.02 | 33.25 | 0.04 |
| TK3U-10 | 20 | 2.02 | 7.39 | 30.63 | 0.41 |

Table 1. Cont.

| Sample | Thickness (cm) | $M_{ad}$ | $A_d$ | $V_{daf}$ | $S_d$ |
|--------|----------------|----------|-------|-----------|-------|
| TK3U-11 | 20 | 2.06 | 8.76 | 34.40 | 0.18 |
| TK3U-12 | 20 | 1.94 | 18.83 | 34.56 | 0.26 |
| TK3U-13 | 20 | 1.86 | 10.63 | 43.83 | 0.28 |
| TK3U-14 | 20 | 2.02 | 10.03 | 41.63 | 0.14 |
| TK3U-15 | 20 | 2.09 | 8.26 | 36.62 | 0.36 |
| TK3U-16 | 20 | 1.98 | 7.73 | 40.01 | 0.28 |
| Tangkou * | - | 1.97 | 10.29 | 36.88 | 0.25 |

* Average value; M, moisture; A, ash yield; V, volatile matter yield; S, total sulfur content; ad, air dry basis; d, dry basis; daf, air dry and ash free basis.

Proximate analysis for the determination of moisture, volatile matter, and ash yield was performed in accordance with ASTM Standards D3173-11 [16], D3175-11 [17], and D3174-11 [18], respectively. Total sulfur content was determined following the ASTM standard D3177-02 [19]. The reference for vitrinite reflectance determination was an yttrium-aluminum garnet standard (manufacturer Klein and Becker, Idar-Oberstein, Germany) with a certified reflectance of 0.90% for $\lambda = 546$ nm under oil immersion. Macerals were identified using white-light reflectance microscopy under oil immersion and more than 500 particles were counted for each polished pellet. The maceral classification and terminology applied in the current study are based on Taylor et al. [20] and the ICCP System (International Committee for Coal and Organic Petrology), 1994 and 2001 [21,22].

Mineral phases were identified from polished pellets using optical microscopy (Leica DM4500P, Leica Microsystems, Wetzlar, Germany), and by X-ray diffraction (XRD). XRD analysis was performed on a Rigaku D/max 2500pc powder diffractometer (Rigaku, Tokyo, Japan) with Ni-filtered Cu-K$\alpha$ radiation and a scintillation detector. The diffractogram of the powdered sample was recorded over a $2\theta$ interval of $2.6°$–$70°$, with a step size of $0.02°$ and 0.3 mm receiving silt.

Samples were crushed and ground to pass 200 mesh (75 μm) for geochemical analysis. Oxides of major elements such as $SiO_2$, $TiO_2$, $Al_2O_3$, $Fe_2O_3$, $Na_2O$, $K_2O$, MgO, CaO, MnO and $P_2O_5$ in coal ash were determined by X-ray fluorescence (XRF) spectrometry (Thermofisher ARL Advant'XP+, Thermo Fisher Scientific, Waltham, MA, USA). Standard references including ASTM2689, ASTM 2690, and ASTM 2691 were used for calibration of major elements. XRF has an accuracy and precision which deviate by less than 1% from the standard reference values. More details of XRF analysis has been described by Dai et al. [23]. Inductively coupled plasma mass spectrometry (ICP-MS, Thermofisher X series II, Thermo Fisher Scientific), in pulse counting mode (three points per peak), was used to determine trace element concentrations in the samples, except for Hg and F. The ICP-MS analysis and sample microwave digestion program are outlined by Dai et al. [24]. Arsenic and Se were determined by ICP-MS using collision cell technology (CCT) in order to avoid disturbance of polyatomic ions [25]. For ICP-MS analysis, samples were digested using an UltraClave Microwave High Pressure Reactor (Milestone Inc., Shelton, CT, USA). Multi-element standards (Inorganic Ventures: CCS-1, CCS-4, CCS-5, and CCS-6; NIST 2685b and Chinese Standard reference GBW 07114 were used for calibration of trace element concentrations. ICP-MS parameters have an accuracy and precision which deviate by less than 5% from the reference standard values. More details of ICP-MS analysis and its method detection limits for various trace elements were described by Dai et al. [23,24] and Li et al. [25]. Mercury was determined using a Milestone DMA-80 analyzer. Samples are heated to make the evolved Hg selectively captured as an amalgam and then measured by Hg analyzer. The detection limit of Hg is 0.005 μg/g and the relative standard deviation 1.5% [23]. Fluorine was determined by pyro-hydrolysis in conjunction with an ion-selective electrode, following the ASTM method D5987-96 [26]. The detection limit is 10 μg/g. The results of two consecutive determinations carried out in the same laboratory by the same operator using the same apparatus do not differ by more than either 15 μg/g (total fluorine concentration of coal is less than 150 μg/g) or 10% (relative; total fluorine concentration of coal is more than 150 μg/g) [27].

## 4. Results

### 4.1. Coal Chemistry

Table 1 shows the proximate analysis and total sulfur content data for the Upper No. 3 coals from the Luxi, Liangbaosi, and Tangkou mines. Being comparable among the three coals, moisture contents are 1.83%, 2.05% and 1.97% (on an air dry basis), respectively. Similarly, volatile matter yield does not show distinct differences among the coals from the three mines, being 36.79%, 36.90% and 36.88% (on a dry ash free basis), respectively. Likewise, vitrinite random reflectances of the three coals are 0.77%, 0.76% and 0.75%, respectively. The maximum ash yield is 15.65% in the Upper No. 3 coal from the Liangbaosi mine, followed by 14.11% in the Luxi Upper No. 3 coal and 10.29% in the Tangkou Upper No. 3 coal. However, total sulfur contents are quite different among these three coals. Samples from the Luxi, Liangbaosi, and Tangkou mines have a total sulfur content of 0.73%, 0.56%, and 0.25%, respectively. Through the three seam sections, the maximum total sulfur value occurs in the bench closest to the roof and the lowest sulfur content is located in the middle bench (Figure 2).

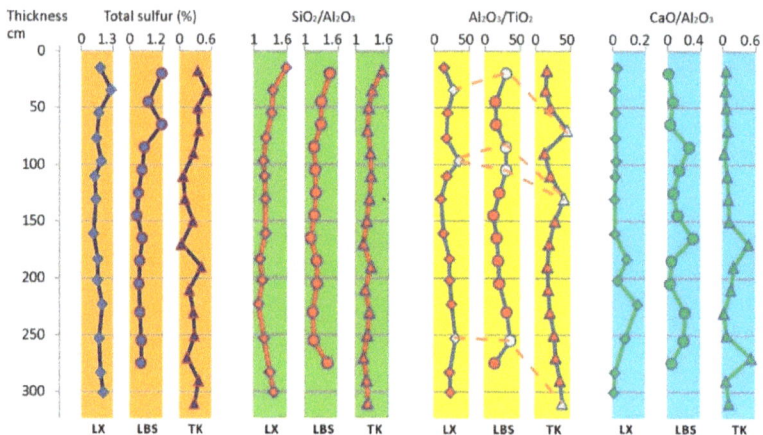

**Figure 2.** Variations of total sulfur content, the $SiO_2/Al_2O_3$ ratio, the $Al_2O_3/TiO_2$ ratio, and the $CaO/Al_2O_3$ ratio in the profiles of the three Upper No. 3 coals from Luxi (LX), Liangbaosi (LBS), and Tangkou (TK) mines.

### 4.2. Maceral Compositions

As listed in Table 2, maceral compositions of these three coals are similar, mainly represented by vitrinite, followed by inertinite and liptinite. Total vitrinite contents are 57.9%, 54.7% and 49.5% respectively, which are mainly dominated by collodetrinite (Figure 3A–C,E) and collotelinite (Figure 3C). Total inertinite contents are 29.4%, 32.9% and 35.7%, respectively, and are represented by semifusinite (Figure 3D,F) and macrinite (Figure 3G,H). The inertinite contents of the three Upper No. 3 coals from southwestern Shandong coalfield are higher than those of other Late Paleozoic coal in northern China (generally less than 25% [28]). Nevertheless, the inertinite content in coal present in this study are lower than the inertinite content of the No.6 coal in Jungar Coalfield (including Guanbanwusu [29], Haerwusu [30] and Heidaigou [31] mines), in the northern Ordos Basin in northern China, in which inertinite contents are 56.7%, 53.7%, 37.4%, respectively. The Carboniferous-Permian coals of the Daqingshan coalfield have an inertinite content of 35.3% [32] (Figure 4). Liptinite contents in the three coals present in this study are 12.8%, 11.4% and 14.9%, respectively, and is dominated by sporinite (Figure 5), minor cutinite (Figure 5C,D) and resinite (Figure 5E,F) (Table 2).

**Table 2.** Maceral compositions of the Upper No. 3 coals from the Luxi, Liangbaosi, and Tangkou mines, southwestern Shandong, China (vol %; on mineral-free basis).

| Sample | T | CT | CP | CD | VD | V | Fus | Sfus | Mac | Mic | Fun | ID | I | Sp | Cut | Res | Sub | L |
|---|---|---|---|---|---|---|---|---|---|---|---|---|---|---|---|---|---|---|
| LX3U-1 | 1.0 | 13.5 | 0.7 | 28.6 | 1.0 | 44.7 | 0.2 | 8.4 | 13.7 | 2.4 | 0.5 | 9.9 | 35.3 | 14.4 | 1.0 | 0.7 | 3.9 | 20.0 |
| LX3U-2 | 1.0 | 11.7 | 1.3 | 39.3 | 2.9 | 56.2 | 0.3 | 8.4 | 7.5 | 0.7 | bdl | 9.1 | 25.7 | 14.0 | 1.0 | bdl | 3.3 | 18.2 |
| LX3U-4 | 4.3 | 10.2 | 0.4 | 23.6 | 2.4 | 40.9 | 0.4 | 10.6 | 13.8 | 2.4 | bdl | 15.8 | 44.1 | 7.5 | 2.8 | 0.4 | 4.3 | 15.0 |
| LX3U-5 | 3.8 | 12.5 | bdl | 40.6 | 0.6 | 57.5 | 0.3 | 5.3 | 12.8 | 3.1 | bdl | 8.1 | 29.4 | 9.1 | 3.4 | 0.3 | 0.3 | 13.1 |
| LX3U-6 | 2.9 | 11.2 | 0.8 | 51.0 | 0.4 | 66.4 | 0.4 | 14.5 | 2.1 | 1.7 | bdl | 6.2 | 24.9 | 7.1 | 0.4 | 0.4 | 0.8 | 8.7 |
| LX3U-7 | 1.6 | 6.2 | bdl | 37.7 | 0.4 | 45.9 | 0.4 | 7.4 | 18.0 | 3.7 | bdl | 4.5 | 34.4 | 17.2 | 0.8 | 0.4 | 1.2 | 19.7 |
| LX3U-8 | 2.4 | 16.2 | bdl | 51.2 | 1.7 | 71.5 | 0.3 | 5.8 | 1.7 | 2.8 | 0.8 | 8.9 | 19.6 | 8.6 | bdl | bdl | 0.3 | 8.9 |
| LX3U-10 | 0.6 | 11.7 | 0.8 | 36.5 | 1.4 | 51.0 | 0.3 | 7.2 | 11.4 | 2.8 | bdl | 12.0 | 34.5 | 10.3 | 1.1 | 0.8 | 2.2 | 14.5 |
| LX3U-11 | 0.7 | 8.1 | bdl | 40.0 | 0.3 | 49.0 | 0.3 | 10.3 | 12.3 | 2.3 | bdl | 10.7 | 36.5 | 10.3 | 0.7 | 1.0 | 2.6 | 14.5 |
| LX3U-12 | 4.6 | 8.1 | 1.6 | 47.6 | 0.8 | 62.7 | 0.3 | 3.8 | 12.4 | 4.9 | 0.3 | 6.2 | 28.1 | 7.0 | 0.8 | 0.8 | 0.5 | 9.2 |
| LX3U-13 | 3.7 | 8.3 | 0.3 | 41.9 | 0.3 | 54.4 | 0.3 | 10.1 | 15.3 | 2.5 | bdl | 7.3 | 36.1 | 7.3 | bdl | 0.6 | 1.5 | 9.5 |
| LX3U-14 | 0.7 | 6.8 | 0.4 | 47.0 | bdl | 54.8 | 0.4 | 10.0 | 11.1 | 0.7 | bdl | 9.0 | 31.2 | 6.8 | 0.7 | bdl | 6.5 | 14.0 |
| LX3U-15 | 3.0 | 14.7 | 3.0 | 39.9 | 9.0 | 69.6 | 0.4 | 4.1 | 6.8 | 1.5 | bdl | 8.7 | 22.2 | 2.6 | 1.1 | bdl | 4.5 | 8.3 |
| LX3U-16 | 7.5 | 18.4 | bdl | 29.7 | 37.9 | 93.5 | 0.3 | bdl | 0.7 | 1.7 | bdl | 2.1 | 4.8 | 1.4 | 0.3 | bdl | bdl | 1.7 |
| Luxi * | 2.5 | 11.0 | 0.7 | 40.3 | 3.4 | 57.9 | 0.3 | 7.4 | 10.4 | 2.3 | 0.2 | 8.6 | 29.4 | 8.8 | 1.1 | 0.4 | 2.5 | 12.8 |
| LBS3U-1 | 1.2 | 28.2 | 3.5 | 29.9 | 5.8 | 68.4 | 0.6 | 4.0 | 5.2 | bdl | bdl | 12.1 | 21.8 | 6.3 | 1.7 | 1.2 | 0.6 | 9.8 |
| LBS3U-2 | 1.0 | 7.5 | 0.5 | 65.0 | bdl | 74.0 | 0.5 | 4.5 | 2.0 | 0.5 | bdl | 4.5 | 12.5 | 12.0 | 0.5 | 1.0 | bdl | 13.5 |
| LBS3U-3 | 0.4 | 13.5 | 3.4 | 50.4 | 0.8 | 68.5 | 0.4 | 2.1 | 10.9 | 0.8 | 0.4 | 5.0 | 20.6 | 7.1 | 1.7 | 0.8 | 1.3 | 10.9 |
| LBS3U-4 | 0.6 | 10.9 | 0.3 | 59.5 | bdl | 71.3 | 0.3 | 6.2 | 6.9 | 1.9 | 0.3 | 4.7 | 20.9 | 5.9 | 0.6 | 0.9 | 0.3 | 7.8 |
| LBS3U-5 | 0.6 | 5.1 | 0.3 | 26.6 | 0.6 | 33.1 | 1.5 | 16.4 | 21.2 | 2.7 | 0.3 | 7.8 | 49.9 | 11.9 | 1.2 | 1.8 | 1.8 | 17.0 |
| LBS3U-6 | 3.1 | 2.8 | 0.7 | 49.5 | 0.4 | 56.4 | 0.7 | 9.3 | 12.1 | 1.0 | 1.0 | 10.7 | 35.0 | 5.2 | 0.4 | 2.4 | 0.7 | 8.7 |
| LBS3U-7 | 1.3 | 1.9 | bdl | 50.3 | bdl | 53.5 | 0.6 | 7.6 | 15.9 | 1.0 | bdl | 10.8 | 36.0 | 6.1 | 2.6 | 0.6 | 1.3 | 10.5 |
| LBS3U-8 | 1.0 | 4.0 | 0.5 | 48.5 | 0.5 | 54.5 | 1.0 | 6.4 | 17.8 | 2.0 | bdl | 5.0 | 32.2 | 6.9 | bdl | 1.0 | 5.5 | 13.4 |
| LBS3U-9 | 1.9 | 2.3 | 0.5 | 38.4 | bdl | 43.1 | bdl | 13.4 | 28.7 | 1.9 | 0.5 | 6.5 | 50.9 | 2.3 | 0.9 | 1.4 | 1.4 | 6.0 |
| LBS3U-10 | bdl | 0.5 | 1.0 | 29.2 | bdl | 30.6 | bdl | 12.4 | 40.2 | 4.8 | bdl | bdl | 57.4 | 4.3 | 1.0 | 1.0 | 0.5 | 6.7 |
| LBS3U-11 | 1.5 | 2.5 | bdl | 35.6 | bdl | 39.6 | 0.5 | 23.3 | 23.3 | 1.0 | bdl | bdl | 48.0 | 4.5 | 2.0 | 0.5 | 3.0 | 9.9 |
| LBS3U-12 | 2.2 | 3.9 | 0.6 | 51.7 | bdl | 58.3 | 0.6 | 9.4 | 17.8 | 2.2 | bdl | bdl | 30.0 | 4.4 | bdl | 1.1 | 2.8 | 8.3 |
| LBS3U-13 | bdl | 1.8 | 0.6 | 58.3 | bdl | 60.7 | bdl | 6.0 | 2.4 | 2.4 | bdl | bdl | 10.7 | 19.1 | 7.1 | bdl | 0.6 | 26.8 |

Table 2. *Cont.*

| Sample | T | CT | CP | CD | VD | V | Fus | Sfus | Mac | Mic | Fun | ID | I | Sp | Cut | Res | Sub | L |
|---|---|---|---|---|---|---|---|---|---|---|---|---|---|---|---|---|---|---|
| Liangbaosi * | 1.2 | 7.1 | 1.0 | 44.7 | 0.8 | 54.7 | 0.5 | 9.6 | 15.6 | 1.6 | 0.2 | 5.2 | 32.9 | 7.3 | 1.5 | 1.0 | 1.5 | 11.4 |
| TK3U-1 | 0.5 | 10.4 | 0.3 | 38.8 | 0.8 | 50.8 | 0.3 | 6.5 | 11.2 | 3.4 | 0.3 | 11.2 | 32.6 | 9.1 | 1.3 | 1.6 | 4.7 | 16.7 |
| TK3U-2 | 0.7 | 9.8 | 0.3 | 41.7 | bdl | 52.5 | 0.3 | 6.8 | 10.5 | 4.8 | 0.3 | 10.2 | 32.9 | 11.2 | 1.0 | 0.3 | 2.0 | 14.6 |
| TK3U-3 | 0.4 | 2.4 | 0.7 | 38.7 | bdl | 42.2 | 0.4 | 11.2 | 16.4 | 0.4 | bdl | 10.1 | 38.3 | 13.9 | 2.1 | 1.7 | 1.7 | 19.5 |
| TK3U-4 | 0.4 | 9.9 | bdl | 51.5 | bdl | 61.8 | 0.4 | 3.9 | 9.0 | 2.6 | bdl | 6.4 | 21.9 | 14.2 | 0.9 | 0.4 | 0.9 | 16.3 |
| TK3U-5 | 0.3 | 14.3 | bdl | 46.1 | 0.3 | 61.1 | 0.3 | 5.8 | 13.0 | 0.7 | bdl | 5.1 | 24.6 | 13.0 | 0.7 | 0.7 | bdl | 14.3 |
| TK3U-6 | 2.3 | 12.3 | 0.3 | 29.1 | 1.1 | 45.0 | 0.3 | 12.3 | 21.1 | 1.4 | 0.3 | 6.0 | 41.6 | 12.3 | bdl | 1.1 | bdl | 13.4 |
| TK3U-7 | 0.8 | 10.6 | bdl | 36.5 | 0.4 | 48.2 | 0.4 | 12.2 | 20.0 | 1.6 | bdl | 5.5 | 40.8 | 9.0 | 0.4 | 0.8 | 0.4 | 11.0 |
| TK3U-8 | 3.1 | 14.0 | bdl | 39.3 | bdl | 56.4 | 0.4 | 9.3 | 14.4 | 2.0 | bdl | 5.5 | 32.3 | 8.2 | 0.4 | 0.4 | 2.3 | 11.3 |
| TK3U-9 | 0.8 | 9.2 | bdl | 35.5 | bdl | 45.4 | 0.4 | 11.6 | 16.3 | bdl | bdl | 6.4 | 34.7 | 12.0 | 4.4 | 1.2 | 2.4 | 19.9 |
| TK3U-10 | 0.4 | 0.8 | bdl | 34.5 | bdl | 35.6 | 0.4 | 5.7 | 24.2 | 0.4 | bdl | 11.7 | 42.1 | 18.2 | 1.1 | 2.7 | 0.4 | 22.4 |
| TK3U-11 | 1.2 | 5.6 | bdl | 40.6 | bdl | 47.4 | 0.4 | 9.6 | 19.5 | 1.6 | bdl | 4.8 | 37.1 | 12.0 | 0.4 | 1.2 | 2.0 | 15.5 |
| TK3U-12 | 0.4 | 4.0 | bdl | 27.8 | bdl | 32.1 | 0.4 | 7.1 | 37.3 | 2.4 | 0.4 | 13.5 | 60.7 | 4.4 | 0.4 | 0.8 | 1.6 | 7.1 |
| TK3U-13 | 1.3 | 2.3 | bdl | 34.8 | bdl | 38.5 | 0.3 | 8.0 | 24.4 | 0.7 | 0.3 | 5.7 | 39.8 | 13.7 | 2.0 | 1.3 | 4.7 | 21.7 |
| TK3U-14 | 4.1 | 7.0 | bdl | 45.6 | bdl | 56.7 | 0.4 | 15.2 | 13.0 | 1.5 | bdl | 1.9 | 34.4 | 5.9 | bdl | 1.5 | 1.5 | 8.9 |
| TK3U-15 | 0.7 | 6.0 | bdl | 44.0 | bdl | 50.7 | 0.3 | 9.3 | 20.0 | 1.0 | 0.3 | 2.3 | 33.7 | 11.3 | 1.7 | 1.3 | 1.3 | 15.7 |
| TK3U-16 | 0.4 | 7.3 | bdl | 60.2 | 0.7 | 68.6 | 0.4 | 9.1 | 8.0 | 0.4 | bdl | 4.4 | 22.6 | 6.9 | 0.7 | 0.4 | 0.7 | 8.8 |
| Tangkou * | 1.1 | 7.9 | 0.1 | 40.2 | 0.2 | 49.5 | 0.4 | 9.0 | 17.5 | 1.5 | 0.1 | 6.9 | 35.7 | 10.9 | 1.1 | 1.1 | 1.7 | 14.9 |

T, telinite; CT, collotelinite; CP, corpogelinite; CD, collodetrinite; VD, vitrodetrinite; V, total vitrinite; Fus, fusinite; Sfus, semifusinite; Mac, macrinite; Mic, micrinite; Fun, funginite; ID, inertodetrinite; I, total inertinites; Sp, sporinite; Cut, cutinite; Res, resinite; Sub, suberinite; L, total liptodetrinite. * Average value; bdl, below detection limit.

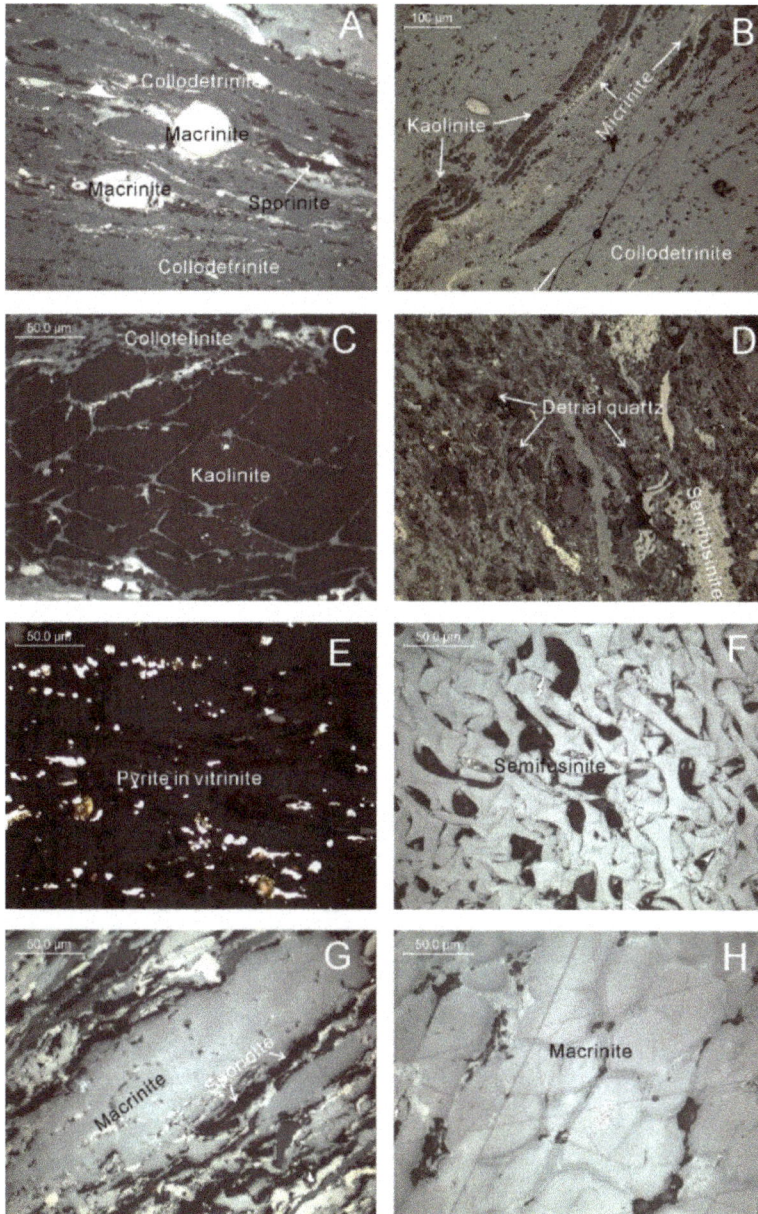

**Figure 3.** Maceral and minerals in the Upper No. 3 coals. (**A**) Collodetrinite, macrinite, and sporinite (LBS3U-13); (**B**) Kaolinite, micrinite in collodetrinite (LBS3U-8); (**C**) Kaolinite in collotelinite (LBS3U-9); (**D**) detrial quartz grains in macerals (LBS3U-4); (**E**) Pyrite in vitrinite (LBS3U-1); (**F**) Semifusinite; (**G**) Macrinite, Sporinite, and colloderitine (LBS3U-3); (**H**) Macrinite (LBS3U-10). (**A,C–H**) is under oil reflectance white light; (**B**) is under white reflectance light.

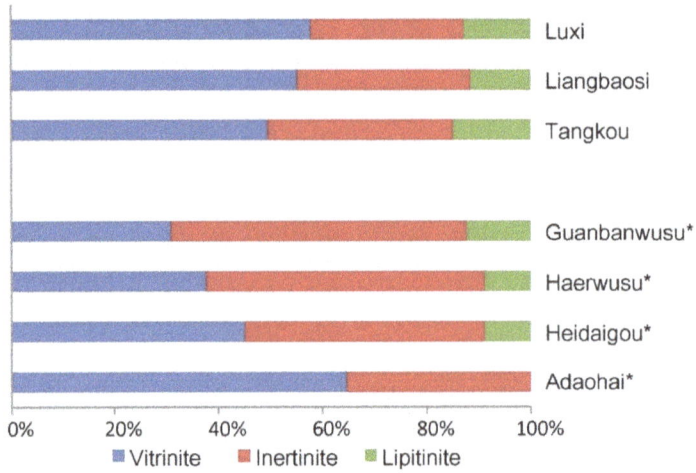

**Figure 4.** Maceral proportions of the three Upper No. 3 coals from the Luxi, Liangbaosi, and Tangkou mines, as well as the coals from the Guanbanwusu [29], Haerwusu [30], Heidaigou [31], and Adaohai [32] mines.

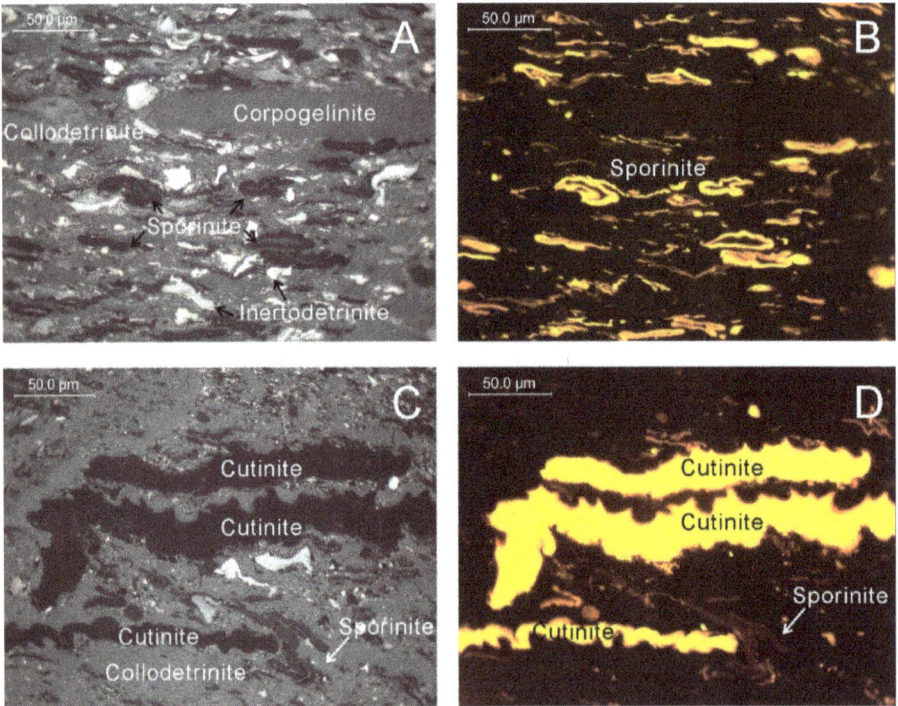

**Figure 5.** *Cont.*

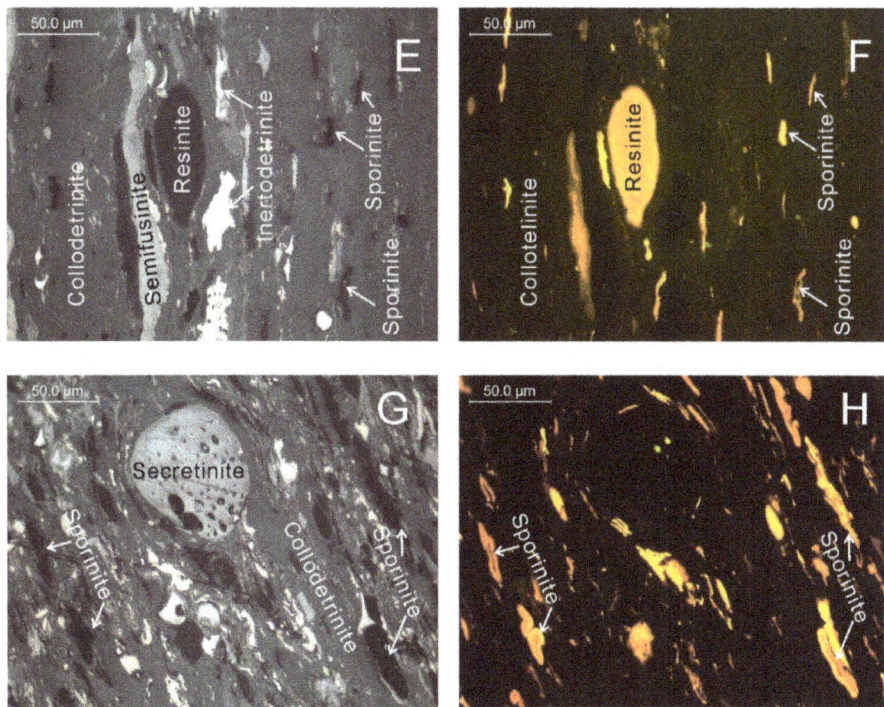

**Figure 5.** Macerals in the Upper No. 3 coals. (**A**) Sporinite, corpogelinite, inertodetrinite, and colledetrinite (LBS3U-4); (**B**) Sporinite under fluorescent light (LBS3U-4); (**C**) Cutinite, sporinite, and colledetrinite (LBS3U-13); (**D**) Cutinite and sporinte under fluorescent light (LBS3U-13); (**E**) Resinite, inertodetrinite, sporinite, semifusinite and colledetrinte (LBS3U-6); (**F**) Resinite and sporinite under fluorescent light (LBS3U-6); (**G**) Secretinite, sporinite, collodetrinite (LBS3U-4); (**H**) Sporinite under fluorescent light (LBS3U-4). **A**, **C**, **E**, and **G** are under oil reflectance white light.

## 4.3. Minerals

Minerals identified by XRD in the Upper No. 3 coals are listed in Figure 6. Kaolinite is the most common mineral, with its occurrence observed in each sample present in this study. Mixed layer I/S is prone to occur in the samples close to roof or floor, with exception of the TK3U-12 in Tangkou mine. Chlorite is only detected at the bottom two benches of the Luxi Upper No. 3 coal. The distribution of quartz is similar to that of the mixed layer I/S and is usually most abundant in the samples close to the roof or floor strata. The distribution patterns of siderite and pyrite show a reverse trend through the coal sections. The presence of siderite is much less in the Luxi Upper No. 3 coal than that in Liangbaosi and Tangkou mines. Comparatively, pyrite is common in the Luxi coal and only occurs in the samples close to the roof strata in the Liangbaosi and Tangkou mines. Carbonate minerals including calcite, ankerite, and dolomite are absent in the Luxi coal, but they are present in the Liangbaosi coal. Calcite and ankerite are present in the Tangkou coal samples. Kaolinite occurs as micro beddings (Figure 3B) or lentoid mixed with maceral (Figure 3C). XRD analysis shows that kaolinite in the benches close to the roof or floor is poorly ordered, while kaolinite in the middle benches is well ordered. Other studies reported a similar phenomenon in some coals from the Sydney Basin [33]. Quartz in the Upper No. 3 coals mainly occurs as irregular particles (Figure 3D) within the organic matrix (macerals). Pyrite in the Upper No. 3 coals is present mainly as disseminated fine particles (Figure 3E) or as framboids. Calcite, ankerite, and dolomite are mainly present as fracture-infillings among various macerals.

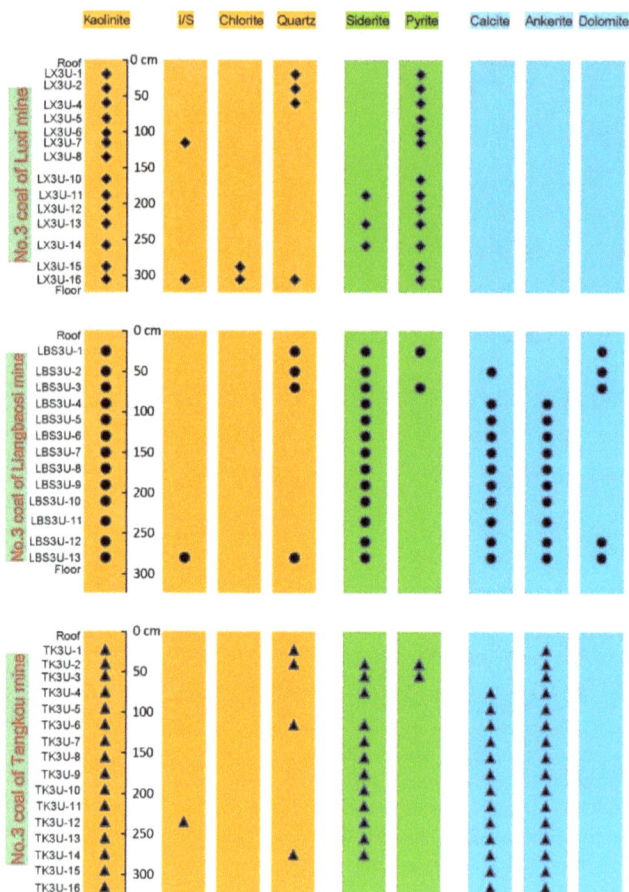

**Figure 6.** Mineral distributions through the three Upper No. 3 coal sections. I/S, mixed layer illite/smectite.

### 4.4. Major and Trace Elements

The major element-oxides in the Upper No. 3 coals are $SiO_2$ and $Al_2O_3$, followed by CaO, $TiO_2$, MgO, $Fe_2O_3$, $Na_2O$ and $K_2O$ (Table 3). Because quartz is absent in most of the Upper No. 3 coal benches, the kaolinite is the major carrier of Si in the coals. In addition to a small proportion of Al in mixed layer I/S, aluminum mainly occurs in the kaolinite. The weight average $SiO_2/Al_2O_3$ ratios for the Upper No. 3 coals from Luxi, Liangbaosi, and Tangkou mines (1.26, 1.30 and 1.25, respectively) are slightly higher than the theoretical value for kaolinite (1.18) but lower than that for common Chinese coal (1.42), as reported by Dai et al. [34]. The $SiO_2/Al_2O_3$ ratio distribution patterns through the coal seam sections in the three mines are similar. Particularly, it is lower in the middle portion than in the top or bottom portions. The content of CaO in the Luxi coal (0.2%) is lower than those in the Liangbaosi (0.49%) and Tangkou (0.5%) coals. The positive correlation coefficient between CaO and $P_2O_5$ (0.98) suggests that CaO is probably associated with apatite in the Luxi coal, although the apatite is too low in content to be detected by XRD technique. The relatively abundant CaO content in the Liangbaosi and Tangkou coals is probably attributed to the epigenetic carbonates (calcite, ankerite, and/or dolomite) deposited from more active underground water.

**Table 3.** Major element contents (%), the ratios of $SiO_2/Al_2O_3$, $CaO/Al_2O_3$, and $Al_2O_3/TiO_2$ in the Upper No. 3 coals from the Luxi, Liangbaosi, and Tangkou mines, southwestern Shandong, China (%).

| Sample | $SiO_2$ | $TiO_2$ | $Al_2O_3$ | $Fe_2O_3$ | MnO | MgO | CaO | $Na_2O$ | $K_2O$ | $P_2O_5$ | $SiO_2/Al_2O_3$ | $Al_2O_3/TiO_2$ | $CaO/Al_2O_3$ |
|---|---|---|---|---|---|---|---|---|---|---|---|---|---|
| LX3U-1 | 5.56 | 0.24 | 3.50 | 0.07 | 0.000 | 0.05 | 0.09 | 0.02 | 0.05 | 0.013 | 1.59 | 14.43 | 0.027 |
| LX3U-2 | 10.37 | 0.28 | 7.71 | 0.25 | 0.001 | 0.11 | 0.09 | 0.03 | 0.18 | 0.017 | 1.34 | 27.88 | 0.012 |
| LX3U-4 | 8.58 | 0.34 | 6.48 | 0.09 | 0.001 | 0.08 | 0.12 | 0.03 | 0.14 | 0.023 | 1.32 | 19.29 | 0.018 |
| LX3U-5 | 5.10 | 0.23 | 4.17 | 0.04 | 0.000 | 0.04 | 0.08 | 0.04 | 0.03 | 0.016 | 1.22 | 17.80 | 0.019 |
| LX3U-6 | 3.68 | 0.09 | 3.12 | 0.04 | 0.000 | 0.04 | 0.07 | 0.07 | 0.02 | 0.011 | 1.18 | 35.90 | 0.023 |
| LX3U-7 | 5.17 | 0.23 | 4.31 | 0.03 | 0.000 | 0.03 | 0.08 | 0.08 | 0.02 | 0.021 | 1.20 | 18.43 | 0.018 |
| LX3U-8 | 6.68 | 0.54 | 5.52 | 0.06 | 0.000 | 0.06 | 0.08 | 0.06 | 0.04 | 0.023 | 1.21 | 10.16 | 0.015 |
| LX3U-10 | 7.10 | 0.42 | 5.83 | 0.07 | 0.000 | 0.08 | 0.09 | 0.08 | 0.06 | 0.037 | 1.22 | 13.79 | 0.016 |
| LX3U-11 | 4.64 | 0.18 | 4.15 | 0.07 | 0.000 | 0.07 | 0.37 | 0.12 | 0.02 | 0.346 | 1.12 | 22.56 | 0.089 |
| LX3U-12 | 5.82 | 0.22 | 5.01 | 0.05 | 0.000 | 0.05 | 0.16 | 0.08 | 0.02 | 0.099 | 1.16 | 22.70 | 0.033 |
| LX3U-13 | 4.90 | 0.18 | 4.49 | 0.07 | 0.000 | 0.08 | 0.69 | 0.09 | 0.01 | 0.514 | 1.09 | 25.26 | 0.154 |
| LX3U-14 | 6.05 | 0.16 | 5.08 | 0.06 | 0.001 | 0.07 | 0.40 | 0.09 | 0.07 | 0.262 | 1.19 | 31.00 | 0.079 |
| LX3U-15 | 9.00 | 0.30 | 6.93 | 0.07 | 0.001 | 0.13 | 0.12 | 0.07 | 0.24 | 0.059 | 1.30 | 22.91 | 0.017 |
| LX3U-16 | 21.18 | 0.65 | 15.53 | 0.45 | 0.003 | 0.38 | 0.17 | 0.06 | 0.72 | 0.024 | 1.36 | 23.96 | 0.011 |
| Luxi * | 7.16 | 0.29 | 5.69 | 0.09 | 0.001 | 0.08 | 0.20 | 0.07 | 0.10 | 0.11 | 1.26 | 19.66 | 0.035 |
| LBS3U-1 | 21.13 | 0.48 | 14.53 | 0.72 | 0.003 | 0.28 | 0.19 | 0.18 | 0.49 | 0.028 | 1.45 | 30.16 | 0.013 |
| LBS3U-2 | 4.23 | 0.22 | 3.24 | 0.05 | 0.002 | 0.08 | 0.22 | 0.03 | 0.03 | 0.006 | 1.30 | 14.93 | 0.068 |
| LBS3U-3 | 9.44 | 0.46 | 7.29 | 0.25 | 0.001 | 0.13 | 0.26 | 0.08 | 0.08 | 0.012 | 1.30 | 15.89 | 0.035 |
| LBS3U-4 | 4.48 | 0.13 | 3.83 | 0.08 | 0.002 | 0.13 | 1.00 | 0.06 | 0.02 | 0.005 | 1.17 | 29.95 | 0.260 |
| LBS3U-5 | 3.15 | 0.09 | 2.63 | 0.03 | 0.001 | 0.08 | 0.38 | 0.12 | 0.03 | 0.004 | 1.19 | 30.28 | 0.143 |
| LBS3U-6 | 5.60 | 0.23 | 4.64 | 0.07 | 0.001 | 0.12 | 0.35 | 0.08 | 0.04 | 0.005 | 1.20 | 20.60 | 0.076 |
| LBS3U-7 | 3.31 | 0.23 | 2.81 | 0.04 | 0.001 | 0.10 | 0.35 | 0.08 | 0.02 | 0.006 | 1.18 | 12.33 | 0.126 |
| LBS3U-8 | 3.09 | 0.16 | 2.79 | 0.05 | 0.001 | 0.13 | 0.86 | 0.08 | 0.01 | 0.009 | 1.11 | 16.99 | 0.310 |
| LBS3U-9 | 8.45 | 0.36 | 6.90 | 0.10 | 0.001 | 0.12 | 0.37 | 0.09 | 0.06 | 0.014 | 1.22 | 19.15 | 0.053 |
| LBS3U-10 | 9.13 | 0.36 | 7.45 | 0.09 | 0.001 | 0.10 | 0.27 | 0.09 | 0.04 | 0.012 | 1.23 | 20.68 | 0.036 |
| LBS3U-11 | 4.43 | 0.12 | 3.85 | 0.06 | 0.002 | 0.13 | 0.83 | 0.12 | 0.03 | 0.020 | 1.15 | 31.07 | 0.217 |
| LBS3U-12 | 3.98 | 0.09 | 3.40 | 0.04 | 0.003 | 0.08 | 0.68 | 0.09 | 0.03 | 0.113 | 1.17 | 36.56 | 0.200 |
| LBS3U-13 | 13.33 | 0.66 | 9.40 | 0.16 | 0.002 | 0.20 | 0.53 | 0.10 | 0.40 | 0.052 | 1.42 | 14.19 | 0.056 |
| Liangbaosi * | 7.61 | 0.28 | 5.85 | 0.15 | 0.002 | 0.13 | 0.49 | 0.10 | 0.11 | 0.02 | 1.30 | 21.05 | 0.083 |

Table 3. Cont.

| Sample | SiO2 | TiO2 | Al2O3 | Fe2O3 | MnO | MgO | CaO | Na2O | K2O | P2O5 | SiO2/Al2O3 | Al2O3/TiO2 | CaO/Al2O3 |
|---|---|---|---|---|---|---|---|---|---|---|---|---|---|
| TK3U-1 | 4.64 | 0.20 | 3.15 | 0.03 | 0.001 | 0.08 | 0.22 | 0.03 | 0.06 | 0.007 | 1.47 | 15.99 | 0.068 |
| TK3U-2 | 4.61 | 0.25 | 3.55 | 0.05 | 0.001 | 0.08 | 0.18 | 0.06 | 0.05 | 0.009 | 1.30 | 14.26 | 0.051 |
| TK3U-3 | 3.78 | 0.15 | 3.05 | 0.04 | 0.001 | 0.08 | 0.22 | 0.07 | 0.02 | 0.007 | 1.24 | 20.65 | 0.072 |
| TK3U-4 | 3.02 | 0.06 | 2.49 | 0.03 | 0.001 | 0.09 | 0.25 | 0.04 | 0.01 | 0.005 | 1.22 | 45.13 | 0.100 |
| TK3U-5 | 11.67 | 0.75 | 9.21 | 0.15 | 0.001 | 0.16 | 0.31 | 0.07 | 0.07 | 0.017 | 1.27 | 12.33 | 0.033 |
| TK3U-6 | 7.13 | 0.26 | 5.57 | 0.08 | 0.002 | 0.17 | 0.53 | 0.12 | 0.05 | 0.012 | 1.28 | 21.07 | 0.094 |
| TK3U-7 | 3.30 | 0.06 | 2.64 | 0.03 | 0.001 | 0.10 | 0.31 | 0.12 | 0.02 | 0.006 | 1.25 | 40.62 | 0.116 |
| TK3U-8 | 3.57 | 0.11 | 2.97 | 0.04 | 0.001 | 0.11 | 0.37 | 0.11 | 0.02 | 0.008 | 1.20 | 27.60 | 0.126 |
| TK3U-9 | 3.56 | 0.16 | 3.11 | 0.05 | 0.002 | 0.13 | 1.49 | 0.08 | 0.01 | 0.012 | 1.14 | 19.14 | 0.481 |
| TK3U-10 | 3.31 | 0.15 | 2.58 | 0.04 | 0.001 | 0.11 | 0.54 | 0.06 | 0.01 | 0.020 | 1.28 | 17.01 | 0.208 |
| TK3U-11 | 3.78 | 0.18 | 3.21 | 0.05 | 0.001 | 0.15 | 0.52 | 0.09 | 0.02 | 0.034 | 1.18 | 18.20 | 0.162 |
| TK3U-12 | 9.96 | 0.39 | 7.92 | 0.14 | 0.001 | 0.21 | 0.24 | 0.09 | 0.07 | 0.015 | 1.26 | 20.56 | 0.030 |
| TK3U-13 | 5.35 | 0.17 | 4.41 | 0.07 | 0.003 | 0.12 | 0.36 | 0.07 | 0.01 | 0.016 | 1.22 | 26.61 | 0.082 |
| TK3U-14 | 3.72 | 0.11 | 3.24 | 0.06 | 0.004 | 0.16 | 1.71 | 0.09 | 0.01 | 0.010 | 1.15 | 28.63 | 0.527 |
| TK3U-15 | 3.91 | 0.09 | 3.23 | 0.03 | 0.001 | 0.06 | 0.24 | 0.08 | 0.01 | 0.008 | 1.21 | 35.07 | 0.073 |
| TK3U-16 | 3.61 | 0.08 | 2.94 | 0.02 | 0.001 | 0.06 | 0.37 | 0.06 | 0.02 | 0.005 | 1.23 | 38.26 | 0.127 |
| Tangkou * | 4.94 | 0.20 | 3.95 | 0.06 | 0.001 | 0.12 | 0.50 | 0.08 | 0.03 | 0.01 | 1.25 | 20.22 | 0.125 |

* Average value.

Table 4. Concentrations of trace elements in the Upper No. 3 coals from the Luxi, Liangbaosi, and Tangkou mines, southwestern Shandong, China (µg/g unless otherwise indicated).

| Sample | Li | Be | F | Sc | V | Cr | Co | Ni | Cu | Zn | Ga | As | Se | Rb | Sr | Zr | Nb | Mo | Cd | Sb | Cs | Ba | REY | Hf | Ta | W | Hg * | Tl | Pb | Bi | Th | U |
|---|---|---|---|---|---|---|---|---|---|---|---|---|---|---|---|---|---|---|---|---|---|---|---|---|---|---|---|---|---|---|---|---|
| LX3U-1 | 29.6 | 4.4 | 83.3 | 7.2 | 198 | 14.2 | 6.2 | 14.7 | 37.0 | 10.8 | 10.2 | 1.7 | 3.8 | 1.8 | 57.7 | 93.4 | 4.3 | 4.1 | 0.5 | 1.6 | 0.2 | 55 | 48 | 3.1 | 0.4 | 0.5 | 149 | 0.2 | 9.7 | 0.2 | 5.0 | 3.0 |
| LX3U-2 | 58.9 | 2.3 | 73.4 | 4.3 | 275 | 23.5 | 9.1 | 19.2 | 96.9 | 14.8 | 11.6 | 3.1 | 8.1 | 7.4 | 38.1 | 137 | 7.7 | 7.7 | 0.8 | 1.2 | 1.1 | 60 | 69 | 4.8 | 0.5 | 0.8 | 184 | 0.5 | 24.8 | 0.5 | 5.1 | 4.8 |
| LX3U-4 | 61.3 | 1.8 | 34.4 | 7.0 | 67.3 | 19.0 | 3.0 | 9.0 | 25.5 | 9.8 | 9.7 | 0.7 | 5.3 | 5.8 | 65.5 | 154.0 | 10.1 | 2.8 | 0.6 | 0.5 | 0.6 | 73 | 107 | 5.6 | 0.9 | 1.1 | 78.6 | 0.2 | 16.3 | 0.4 | 15.5 | 3.5 |
| LX3U-5 | 45.7 | 0.9 | 25.8 | 0.7 | 31.1 | 12.5 | 1.8 | 5.0 | 14.8 | 7.9 | 6.0 | 0.7 | 4.5 | 0.8 | 40.4 | 89.7 | 6.1 | 3.3 | 0.6 | 0.3 | 0.1 | 69 | 17 | 3.2 | 0.6 | 1.0 | 85.9 | 0.1 | 9.2 | 0.2 | 1.9 | 1.4 |
| LX3U-6 | 42.4 | 0.7 | 30.8 | 2.0 | 24.8 | 9.1 | 2.6 | 7.0 | 12.5 | 4.7 | 7.3 | 0.5 | 4.3 | 0.5 | 66.1 | 49.2 | 2.9 | 4.2 | 0.4 | 0.3 | 0.0 | 91 | 83 | 1.6 | 0.2 | 1.5 | 65.5 | 0.1 | 9.9 | 0.2 | 3.1 | 1.1 |
| LX3U-7 | 61.4 | 0.9 | 30.6 | 3.8 | 21.3 | 13.3 | 1.0 | 3.8 | 19.8 | 9.9 | 4.5 | 2.6 | 5.8 | 0.7 | 89.0 | 86.1 | 5.1 | 1.7 | 0.3 | 0.1 | 0.1 | 116 | 107 | 3.1 | 0.5 | 0.6 | 98.2 | 0.2 | 9.3 | 0.2 | 7.3 | 1.5 |
| LX3U-8 | 56.1 | 0.7 | 41.0 | 2.6 | 54.4 | 11.9 | 1.8 | 4.4 | 26.3 | 9.1 | 8.9 | 0.8 | 9.3 | 1.1 | 31.4 | 214 | 10.8 | 2.7 | 0.5 | 0.4 | 0.1 | 82 | 49 | 6.6 | 0.7 | 0.9 | 70.1 | 0.2 | 21.7 | 0.3 | 5.7 | 2.6 |
| LX3U-10 | 88.7 | 1.3 | 189 | 7.2 | 75.6 | 17.2 | 1.7 | 4.5 | 36.3 | 17.9 | 7.5 | 0.8 | 9.1 | 2.1 | 119 | 212 | 8.5 | 2.2 | 0.1 | 0.3 | 0.2 | 112 | 88 | 6.8 | 0.6 | 0.7 | 136 | 0.2 | 18.5 | 0.3 | 13.8 | 3.9 |
| LX3U-11 | 69.4 | 1.1 | 74.0 | 4.1 | 48.1 | 9.3 | 1.9 | 3.5 | 16.9 | 9.1 | 6.5 | 0.9 | 7.1 | 0.7 | 498 | 131 | 5.4 | 2.4 | 0.4 | 0.2 | 0.1 | 228 | 173 | 4.1 | 0.4 | 0.4 | 200 | 0.1 | 10.8 | 0.2 | 9.2 | 2.5 |

**Table 4.** *Cont.*

| Sample | Li | Be | F | Sc | V | Cr | Co | Ni | Cu | Zn | Ga | As | Se | Rb | Sr | Zr | Nb | Mo | Cd | Sb | Cs | Ba | REY | Hf | Ta | W | Hg* | Tl | Pb | Bi | Th | U |
|---|---|---|---|---|---|---|---|---|---|---|---|---|---|---|---|---|---|---|---|---|---|---|---|---|---|---|---|---|---|---|---|---|
| LX3U-12 | 83.3 | 0.9 | 431 | 3.9 | 39.1 | 12.0 | 1.8 | 4.6 | 17.7 | 7.2 | 6.2 | 0.7 | 6.8 | 0.6 | 143 | 101 | 4.6 | 3.3 | 0.3 | 0.2 | 0.1 | 110 | 55 | 3.6 | 0.4 | 0.9 | 54.1 | 0.1 | 10.5 | 0.2 | 5.8 | 1.6 |
| LX3U-13 | 90.4 | 1.3 | 230 | 4.0 | 43.6 | 13.0 | 2.0 | 4.4 | 15.7 | 9.8 | 5.2 | 0.8 | 5.4 | 0.4 | 572 | 77.8 | 4.5 | 2.7 | 0.4 | 0.5 | 0.3 | 297 | 140 | 2.6 | 0.3 | 0.6 | 121 | 0.1 | 8.7 | 0.2 | 5.4 | 1.4 |
| LX3U-14 | 73.5 | 1.2 | 103 | 4.5 | 66.0 | 15.1 | 2.8 | 6.9 | 23.2 | 7.0 | 7.6 | 0.8 | 6.4 | 2.4 | 292 | 59.0 | 3.0 | 3.1 | 0.4 | 0.5 | 0.3 | 161 | 105 | 2.2 | 0.2 | 0.8 | 68.4 | 0.2 | 13.5 | 0.2 | 4.4 | 1.9 |
| LX3U-15 | 93.0 | 1.4 | 188 | 3.9 | 109 | 24.5 | 4.6 | 10.1 | 59.9 | 13.5 | 11.1 | 1.6 | 7.7 | 12.5 | 73.7 | 101 | 5.8 | 3.8 | 0.5 | 0.9 | 1.5 | 101 | 88 | 3.7 | 0.5 | 0.7 | 76.6 | 0.4 | 22.4 | 0.5 | 6.8 | 2.6 |
| LX3U-16 | 159 | 1.4 | 241 | 15.1 | 284 | 54.0 | 6.1 | 17.8 | 115 | 28.1 | 16.2 | 1.7 | 7.1 | 34.0 | 103 | 169 | 9.8 | 3.9 | 0.6 | 0.6 | 4.3 | 150 | 178 | 6.0 | 0.8 | 1.2 | 72.3 | 0.6 | 28.2 | 0.6 | 13.6 | 4.9 |
| Luxi* | 71.9 | 1.4 | 130 | 4.8 | 91.2 | 17.1 | 3.2 | 7.8 | 35.3 | 11.2 | 8.2 | 1.2 | 6.6 | 4.3 | 160 | 119 | 6.2 | 3.4 | 0.4 | 0.5 | 0.5 | 122 | 88 | 4.1 | 0.5 | 0.8 | 102 | 0.2 | 15.1 | 0.2 | 7.0 | 2.6 |
| LBS3U-1 | 64.8 | 4.2 | 53.5 | 12.1 | 382 | 42.9 | 13.3 | 24.5 | 206 | 52.5 | 20.8 | 15.5 | 7.6 | 24.3 | 86.8 | 173 | 9.8 | 5.9 | 1.0 | 2.0 | 5.0 | 85 | 396 | 6.4 | 0.8 | 1.8 | 651 | 2.3 | 38.4 | 0.9 | 14.4 | 8.0 |
| LBS3U-2 | 20.1 | 3.3 | 86.9 | 4.8 | 71.7 | 13.1 | 9.1 | 12.9 | 27.5 | 7.7 | 9.2 | 0.7 | 3.3 | 1.4 | 77.5 | 68.5 | 5.0 | 6.0 | 0.4 | 1.0 | 0.2 | 72 | 146 | 2.4 | 0.4 | 1.3 | 105 | 0.3 | 10.2 | 0.3 | 5.7 | 2.3 |
| LBS3U-3 | 52.9 | 2.7 | 40.7 | 4.4 | 139 | 18.1 | 13.3 | 18.3 | 54.2 | 23.5 | 13.9 | 5.2 | 5.2 | 3.9 | 72.4 | 200 | 16.2 | 6.2 | 0.6 | 1.1 | 0.6 | 89 | 133 | 7.0 | 1.4 | 1.8 | 718 | 0.7 | 19.0 | 0.5 | 20.2 | 4.5 |
| LBS3U-4 | 20.3 | 0.7 | 29.1 | 1.9 | 28.5 | 9.4 | 5.1 | 8.4 | 13.1 | 6.0 | 4.7 | 0.2 | 3.5 | 0.5 | 130 | 56.8 | 2.9 | 3.6 | 0.6 | 0.1 | 0.1 | 126 | 63 | 2.0 | 0.2 | 0.7 | 75.5 | 0.2 | 8.6 | 0.1 | 3.6 | 1.0 |
| LBS3U-5 | 19.1 | 0.7 | 36.6 | 1.5 | 16.3 | 6.5 | 2.3 | 6.3 | 8.4 | 8.8 | 2.2 | 0.2 | 1.9 | 0.7 | 98.6 | 31.7 | 1.4 | 1.5 | 0.4 | 0.1 | 0.1 | 141 | 49 | 1.2 | 0.1 | 0.5 | 43.6 | 0.1 | 4.5 | 0.1 | 2.7 | 0.5 |
| LBS3U-6 | 21.5 | 0.5 | 31.7 | 1.5 | 24.2 | 10.0 | 1.9 | 8.8 | 8.8 | 6.5 | 8.1 | 0.3 | 3.5 | 1.1 | 61.1 | 77.9 | 5.5 | 1.8 | 0.5 | 0.1 | 0.1 | 123 | 65 | 2.7 | 0.6 | 0.5 | 307 | 0.2 | 8.7 | 0.1 | 5.2 | 0.5 |
| LBS3U-7 | 16.6 | 0.5 | 27.4 | 0.6 | 15.8 | 8.8 | 1.6 | 7.9 | 13.1 | 6.3 | 3.7 | 0.3 | 3.9 | 0.4 | 62.1 | 72.3 | 3.0 | 1.4 | 0.2 | 0.1 | 0.0 | 139 | 19 | 2.2 | 0.2 | 0.6 | 130 | 0.2 | 4.5 | 0.1 | 1.4 | 0.2 |
| LBS3U-8 | 28.8 | 0.4 | 38.3 | 1.2 | 28.5 | 10.4 | 1.2 | 7.7 | 14.9 | 11.4 | 4.2 | 0.4 | 3.5 | 0.4 | 138 | 56.5 | 2.9 | 1.5 | 0.2 | 0.2 | 0.2 | 144 | 27 | 1.7 | 0.3 | 0.6 | 116 | 0.2 | 5.7 | 0.2 | 1.1 | 0.2 |
| LBS3U-9 | 44.5 | 0.8 | 48.0 | 3.4 | 40.2 | 14.5 | 1.2 | 6.8 | 15.3 | 26.4 | 15.1 | 0.7 | 8.9 | 1.6 | 81.8 | 158 | 11.8 | 1.4 | 0.5 | 0.3 | 0.2 | 135 | 175 | 5.4 | 1.2 | 1.0 | 175 | 0.2 | 20.3 | 0.5 | 11.8 | 2.3 |
| LBS3U-10 | 41.7 | 0.8 | 37.9 | 3.6 | 43.1 | 15.5 | 0.7 | 5.5 | 19.6 | 24.2 | 13.2 | 0.7 | 7.9 | 0.7 | 56.4 | 146 | 8.9 | 0.9 | 0.5 | 0.2 | 0.2 | 129 | 64 | 4.9 | 0.9 | 0.5 | 207 | 0.2 | 19.5 | 0.5 | 9.9 | 0.9 |
| LBS3U-11 | 50.3 | 0.6 | 119 | 2.4 | 22.6 | 9.6 | 1.4 | 6.9 | 9.1 | 9.7 | 7.9 | 0.3 | 6.2 | 1.1 | 205 | 66.9 | 3.9 | 1.5 | 0.2 | 0.3 | 0.1 | 191 | 49 | 2.2 | 0.3 | 1.0 | 137 | 0.2 | 10.3 | 0.2 | 5.1 | 1.7 |
| LBS3U-12 | 38.9 | 1.1 | 171 | 1.3 | 26.2 | 29.5 | 7.1 | 13.1 | 9.3 | 18.0 | 11.6 | 0.3 | 2.8 | 1.2 | 211 | 42.6 | 2.1 | 3.8 | 0.3 | 0.3 | 0.1 | 186 | 70 | 1.4 | 0.3 | 1.0 | 117 | 0.2 | 8.9 | 0.1 | 2.6 | 0.6 |
| LBS3U-13 | 41.0 | 1.5 | 92.6 | 4.7 | 87.2 | 30.6 | 9.3 | 17.1 | 16.4 | 24.2 | 20.1 | 0.4 | 8.3 | 8.0 | 62.1 | 269 | 14.1 | 2.4 | 0.4 | 0.5 | 1.3 | 106 | 64 | 8.0 | 1.2 | 3.2 | 186 | 0.3 | 44.6 | 0.4 | 6.6 | 2.9 |
| Liangbaosi* | 37.0 | 1.4 | 64.2 | 3.6 | 80.5 | 17.5 | 5.3 | 11.4 | 37.4 | 17.3 | 10.6 | 2.4 | 5.3 | 4.1 | 106 | 110 | 6.7 | 3.0 | 0.4 | 0.5 | 0.7 | 129 | 102 | 3.7 | 0.6 | 1.1 | 240 | 0.5 | 16.2 | 0.3 | 7.1 | 2.2 |
| TK3U-1 | 11.1 | 3.2 | 58.0 | 5.6 | 121 | 21.2 | 5.9 | 11.3 | 44.8 | 10.7 | 8.4 | 0.5 | 5.0 | 1.4 | 73.2 | 95.3 | 4.8 | 3.1 | 0.4 | 0.7 | 0.1 | 100 | 91 | 3.0 | 0.4 | 1.3 | 60.9 | 0.1 | 13.3 | 0.3 | 6.3 | 2.9 |
| TK3U-2 | 16.9 | 1.6 | 41.2 | 6.0 | 179 | 18.6 | 4.9 | 9.8 | 40.0 | 7.7 | 6.8 | 1.1 | 5.3 | 1.5 | 75.7 | 75.0 | 4.7 | 3.4 | 0.4 | 0.5 | 0.1 | 117 | 111 | 2.4 | 0.4 | 0.9 | 99.1 | 0.1 | 12.7 | 0.3 | 5.9 | 2.7 |
| TK3U-3 | 15.6 | 1.1 | 29.8 | 1.2 | 85.4 | 13.5 | 3.4 | 7.4 | 19.0 | 5.7 | 4.1 | 0.8 | 3.3 | 0.6 | 30.3 | 49.1 | 3.1 | 2.7 | 0.4 | 0.3 | 0.0 | 119 | 33 | 1.6 | 0.3 | 0.5 | 68.6 | 0.1 | 8.7 | 0.1 | 1.9 | 0.4 |
| TK3U-4 | 11.6 | 0.5 | 65.1 | 0.7 | 35.1 | 9.0 | 3.4 | 9.1 | 13.8 | 5.1 | 4.0 | 0.3 | 2.6 | 1.6 | 28.1 | 28.9 | 1.3 | 3.1 | 0.4 | 0.4 | 0.1 | 102 | 39 | 1.0 | 0.3 | 0.9 | 37.3 | 0.2 | 5.2 | 0.1 | 1.0 | 0.2 |
| TK3U-5 | 29.0 | 0.5 | 47.5 | 4.9 | 61.3 | 17.5 | 2.2 | 9.7 | 34.5 | 14.0 | 12.6 | 0.4 | 7.3 | 1.6 | 52.2 | 83.2 | 16.8 | 2.7 | 0.4 | 0.4 | 0.1 | 118 | 111 | 7.7 | 1.5 | 1.4 | 511 | 0.2 | 19.1 | 0.2 | 16.7 | 4.4 |
| TK3U-6 | 25.0 | 0.5 | 31.5 | 1.1 | 29.6 | 14.4 | 1.7 | 5.6 | 14.6 | 5.9 | 4.8 | 0.4 | 4.1 | 1.4 | 22.9 | 27.4 | 4.9 | 1.5 | 0.4 | 0.2 | 0.1 | 120 | 22 | 2.8 | 0.4 | 0.7 | 149 | 0.1 | 10.3 | 0.2 | 1.6 | 0.5 |
| TK3U-7 | 20.7 | 0.5 | 43.3 | 1.3 | 15.8 | 7.0 | 1.5 | 6.2 | 7.7 | 12.1 | 2.8 | 0.4 | 2.3 | 1.1 | 96.1 | 43.6 | 1.4 | 1.5 | 0.5 | 0.1 | 0.1 | 152 | 53 | 0.9 | 0.1 | 0.3 | 25.8 | 0.1 | 4.9 | 0.1 | 2.0 | 0.5 |
| TK3U-8 | 20.0 | 0.6 | 47.5 | 1.3 | 18.4 | 10.5 | 1.4 | 6.7 | 6.4 | 6.1 | 5.3 | 0.3 | 4.4 | 0.7 | 99.4 | 47.2 | 3.9 | 1.6 | 0.5 | 0.1 | 0.1 | 146 | 57 | 1.5 | 0.1 | 0.5 | 46.8 | 0.1 | 8.4 | 0.1 | 3.5 | 0.9 |
| TK3U-9 | 25.3 | 0.3 | 48.6 | 1.2 | 11.5 | 6.2 | 1.5 | 6.8 | 9.9 | 9.9 | 3.4 | 0.4 | 3.7 | 0.4 | 145 | 44.8 | 2.1 | 1.3 | 0.4 | 0.1 | 0.0 | 173 | 38 | 1.6 | 0.2 | 0.8 | 69.2 | 0.1 | 4.9 | 0.1 | 2.2 | 0.4 |
| TK3U-10 | 28.2 | 0.6 | 49.7 | 2.5 | 13.6 | 8.6 | 1.3 | 6.8 | 14.4 | 10.2 | 2.4 | 0.5 | 3.7 | 0.5 | 140 | 52.4 | 2.2 | 0.9 | 0.3 | 0.2 | 0.1 | 183 | 58 | 1.5 | 0.2 | 0.4 | 98.3 | 0.1 | 3.0 | 0.1 | 2.8 | 0.6 |
| TK3U-11 | 28.0 | 0.5 | 62.3 | 0.5 | 17.3 | 9.7 | 1.6 | 7.1 | 12.0 | 10.6 | 3.9 | 0.3 | 3.9 | 0.5 | 86.1 | 52.4 | 2.9 | 1.0 | 0.6 | 0.2 | 0.0 | 174 | 52 | 1.8 | 0.2 | 0.7 | 238 | 0.1 | 6.6 | 0.2 | 2.5 | 0.4 |
| TK3U-12 | 46.2 | 1.0 | 44.9 | 2.0 | 27.6 | 17.2 | 1.3 | 8.8 | 15.1 | 8.2 | 9.1 | 0.0 | 10.0 | 1.7 | 21.6 | 167 | 11.7 | 0.8 | 0.5 | 0.2 | 0.2 | 109 | 53 | 5.8 | 1.1 | 0.7 | 231 | 0.1 | 21.2 | 0.5 | 10.6 | 2.5 |
| TK3U-13 | 52.9 | 0.9 | 44.2 | 3.4 | 16.4 | 9.0 | 2.8 | 10.9 | 13.0 | 9.3 | 4.5 | 1.0 | 4.7 | 0.6 | 117 | 63.6 | 2.9 | 0.9 | 0.3 | 0.2 | 0.1 | 183 | 90 | 2.3 | 0.3 | 0.5 | 73.4 | 0.2 | 6.5 | 0.2 | 6.0 | 1.3 |
| TK3U-14 | 34.5 | 0.7 | 35.9 | 4.2 | 22.9 | 9.4 | 4.9 | 11.8 | 10.6 | 9.9 | 4.1 | 0.3 | 3.6 | 0.7 | 165 | 52.2 | 3.0 | 1.6 | 0.4 | 0.2 | 0.1 | 156 | 70 | 1.7 | 0.3 | 0.6 | 129 | 0.1 | 6.1 | 0.1 | 3.8 | 0.9 |
| TK3U-15 | 24.2 | 0.9 | 44.4 | 1.6 | 20.2 | 11.9 | 8.2 | 11.8 | 10.4 | 9.8 | 3.9 | 0.9 | 3.4 | 0.5 | 47.8 | 48.2 | 2.3 | 1.9 | 0.5 | 0.1 | 0.0 | 121 | 58 | 1.6 | 0.2 | 0.7 | 79.1 | 0.1 | 4.0 | 0.1 | 3.0 | 0.4 |
| TK3U-16 | 18.8 | 1.7 | 54.5 | 4.0 | 25.8 | 9.8 | 8.5 | 12.3 | 10.2 | 13.5 | 10.1 | 0.6 | 3.9 | 1.0 | 109 | 64.8 | 2.1 | 2.3 | 0.7 | 0.4 | 0.1 | 133 | 64 | 2.2 | 0.4 | 0.8 | 67.3 | 0.1 | 10.4 | 0.1 | 3.8 | 1.3 |
| Tangkou* | 25.5 | 1.0 | 47.1 | 2.4 | 40.7 | 10.5 | 8.9 | 8.9 | 17.0 | 9.0 | 5.6 | 0.5 | 4.4 | 0.9 | 81.9 | 74.0 | 4.4 | 1.9 | 0.4 | 0.4 | 0.1 | 138 | 61 | 2.5 | 0.4 | 0.7 | 124 | 0.1 | 9.0 | 0.2 | 4.6 | 1.2 |
| Word value* | 14.0 | 2.0 | 82.0 | 3.7 | 28.0 | 17.0 | 6.0 | 17.0 | 16.0 | 28.0 | 6.0 | 8.3 | 1.3 | 18.0 | 100 | 36.0 | 4.0 | 2.1 | 0.2 | 1.0 | 1.1 | 150 | 69 | 1.2 | 0.3 | 1.0 | 100 | 0.6 | 9.0 | 1.1 | 3.2 | 1.9 |

* Average value; Word value from Ketris and Yudovich [35]; Hg as ng/g.

Compared to average values for world hard coals reported by Ketris and Yodovich [35], most trace elements are lower in the Upper No. 3 coal, with the exception of Se, Zr, and Hf which are slightly higher (Table 4, Figure 7). The elevated concentrations of Se, Zr, and Hf have a positive correlation with ash yields, suggesting they were probably associated with the clay mineral (e.g., kaolinite). The enriched Li in the Luxi and Liangbaosi coals may also have similar modes of occurrences. Although the potential hazardous elements As and Hg are overall low in content, they are enriched in some benches. For example, samples LBS3U-1 and LBS3U-3 have high contents of As and Hg, 15.54 µg/g vs. 5.20 µg/g and 651 ng/g vs. 718 ng/g, respectively (Table 4). This is probably a result of their higher total sulfur content of 1.17% and 1.17% (Table 1). The weighted average REY (rare earth elements plus ytrrium [7]) content of the Upper No. 3 coal from the Luxi, Liangbaosi, and Tangkou mines is 88, 102 and 61 µg/g, respectively, close to and lower than the averages for world hard coals (69 µg/g) [35] and Chinese coals (136 µg/g), respectively [34].

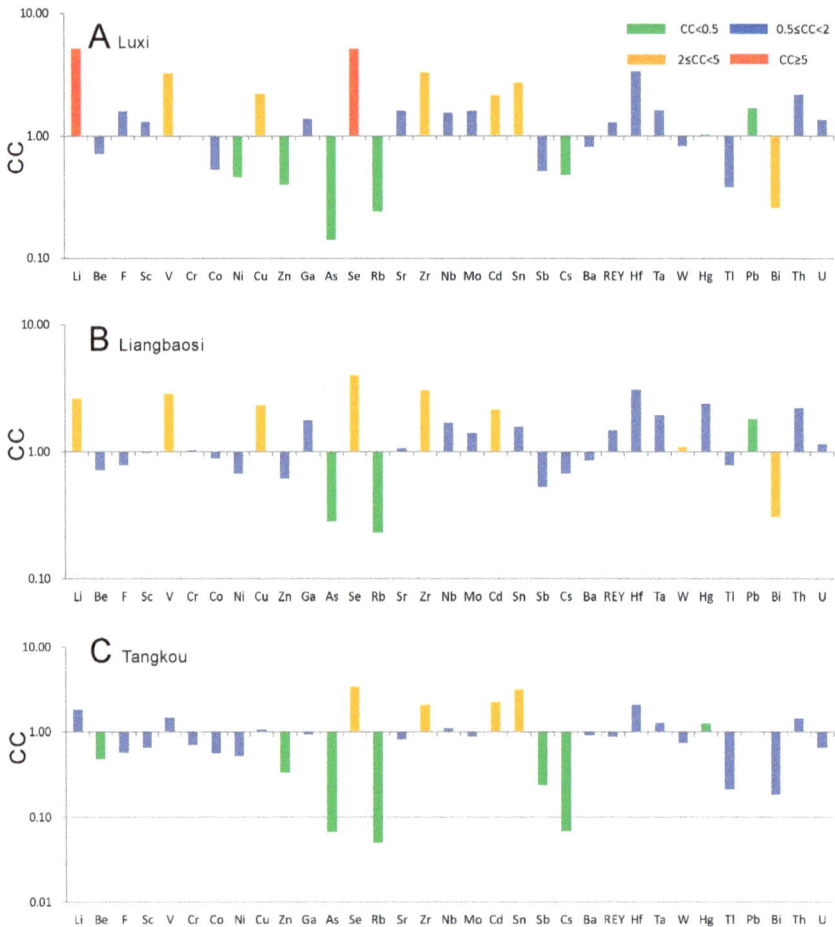

**Figure 7.** Concentration coefficients (CC) of the trace elements in the three Upper No. 3 coals vs. the world hard coal. Data of world hard coals are from Ketris and Yodovich [35]. (**A**) the Luxi Upper No. 3 coal; (**B**) the Luxi Upper No. 3 coal; (**C**) the Tangkou Upper No. 3 coal.

## 5. Discussion

### 5.1. Correlation of the Three Upper No. 3 Coals

The tectonic processes rearranged the Upper No. 3 coal at shallow, middle, and deep depths in the Luxi (−228 m), Liangbaosi (−670 m), and Tangkou (−938 m) mines, respectively. Although the studied areas have been subjected to drastic multi-stage tectonic activities, the processes have no or little significant impact on rank, or maceral compositions as described above. Firstly, the three Upper No. 3 coals are comparable in thickness, e.g., 3.03, 2.8 and 3.15 m, respectively. Secondly, the three coals are similar in rank, which is supported by their volatile matter yield (36.79%, 36.90% and 36.88%, respectively) (Table 1) and vitrinite random reflectance (0.77%, 0.76% and 0.75%, respectively), indicating a coal rank of high volatile bituminous (ASTM D388 [36]). Additionally, the maceral components and proportions have no significant differences (Figures 3 and 5).

### 5.2. Differentiation of Minerals

Minerals in the three Upper No. 3 coals are of terrigenous, authigenic, and epigenetic origins. Terrigenous minerals (e.g., kaolinite and quartz) were attributed to the detrital source during peat accumulating [37]. Authigenic minerals (e.g., pyrite and siderite) were associated with their sedimentary environment during the syngenetic or penecontemporaneous stage of coal formation [38,39]. However, epigenetic minerals (e.g., calcite, ankerite, and dolomite) are related to fluid activities after coalification [37,40].

#### 5.2.1. Terrigenous Minerals

Terrigenous minerals, mainly kaolinite and quartz, share the same distribution patterns through the three coal seam sections, i.e., kaolinite occurs in all the benches and quartz is only present in the benches close to the roof and floor strata. During peat accumulation, the vegetation in and around the mire acted as a filter, preventing quartz particles from penetrating into the mire, while kaolinite is fine enough to be carried into and preserved in the peat mire [37,41]. In the beginning and ending of peat accumulation of the No. 3 coal, the vegetation prevention was weak, thereby allowing quartz particles to be moved into and deposited within the benches close to the floor or roof strata. This explains why the $SiO_2/Al_2O_3$ ratio is comparable to the theoretical ratio of kaolinite (1.18) in the middle benches but is higher in the upper or bottom benches of the three seam sections (Table 3, Figure 2).

The $Al_2O_3/TiO_2$ ratio is a valuable provenance indicator of sedimentary rocks [28,29], because the ratio of $Al_2O_3$ versus $TiO_2$ in mudstones/sandstones to the same ratio in their parent rocks [42]. The $Al_2O_3/TiO_2$ ratios is also applied as a provenance indicator for coal and coal-bearing strata [2,40,43–46]. The $Al_2O_3/TiO_2$ ratios of the Upper No. 3 coals from the Luxi, Liangbaosi, and Tangkou mines are similar in average value (19.66, 21.05, and 20.22, respectively). This is close to the lower end of felsic igneous rocks (21–70) [42]. During Late Carboniferous and Early Permian, the terrigenous source of the coal-bearing basin in North China was dominantly from the northern Yinshan Upland, which is mainly made up of alkaline granite [28]. This is also the case for the Upper No. 3 coal in the southwestern Shandong coalfield, which is supported by the $Al_2O_3/TiO_2$ ratio.

#### 5.2.2. Authigenic Minerals

Pyrite and siderite are of authigenic origin and they formed syngenetically during peat accumulation [37]. The presence of siderite and pyrite was controlled by the sedimentary environment, where sulfur supply is important because iron in solution would otherwise combine with bacterially produced $H_2S$ instead of reacting with dissolved $CO_2$, released by fermentation of organic matter [37,47]. The elevated sulfur content (0.73%) is attributed to the common pyrite in the Luxi Upper No. 3 coals, while the lower sulfur content (0.56% and 0.25%) is consistent with a lack of pyrite and frequent presence of siderite in the Liangbaosi and Tangkou coals (Figure 4).

For low sulfur coal (<1% S), sulfur is derived primarily from parent plant material [48]. In medium (1% to <3% S) to high sulfur (⩾3% S) coals, the sulfur is partly inherited from plant and largely from sulfate in seawater that flooded into peat [38,48]. Although the average sulfur content is lower than 1%, the elevated sulfur in the upper benches of the three profiles suggests a seawater influence during the end of the peat accumulation of the Upper No. 3 coal. Chen has reported that the bottom section of Shanxi Formation was subjected to seawater influence [49]. In addition, the variations of sulfur content among the three coals suggest that the seawater influence on the coal-forming peat mire decreases by location, from the Luxi, to the Liangbaosi, and to the Tangkou mine. This is supported by the Sr/Ba ratio in the three coals, 1.31, 0.83 and 0.59, respectively. Because the solubility of Ba compounds is lower than that of Sr, once Ba is precipitated as $BaSO_4$, this compound is difficult to dissolve when sulfate exists in the water [50]. Thus, Sr can move farther seaward than Ba. Therefore, the Sr/Ba ratio is a useful indicator of marine and terrigenous environments, where the ratio increases from terrigenous, paralic, and marine lithofacies [50–52]. The three coals in the present work are different from the Upper No. 3 coal at the Xinglongzhuang mine, southwestern Shandong coalfield. The Xinglongzhuang Upper No. 3 coal has an average sulfur content of 1.49%, with the maximum value occurring in the middle section (3.75%). Pyrite in the coal from the Xinglongzhuang mine occurs mainly as cell or fracture-infillings, with a small proportion of pyrite as disseminated fine particles or framboidal crystals; the sulfur content in the coal from this coalfield is thus attributed to epigenetic invasion as opposed to seawater influence of early diagenesis [15].

### 5.2.3. Epigenetic Minerals

Calcite in coal is mainly epigenetic origin and occurs as fracture- or cleat-fillings [37]. Detrital calcite is rare in coal because calcite can easily be decomposed under the acidic conditions [53,54]. Occurring as fracture-infillings, carbonate minerals including calcite, ankerite, and/or dolomite in the Upper No. 3 coal are therefore suggestive of an epigenetic origin. They were probably precipitated by circulation of Ca (Mg, Fe)-bearing underground water precipitation [55]. No carbonate minerals are detected in the Luxi coal. However, calcite and ankerite are present in the Liangbaosi and Tangkou coals; dolomite is only present in the Liangbaosi coal (Figure 6). The increase in epigenetic fluids with depth invasion significantly increased the calcium input in the coal. The Luxi coal has a $CaO/Al_2O_3$ ratio of 0.035, while the same ratio in the Liangbaosi and Tangkou coals is more than double, e.g., 0.083 in the Liangbaosi coal and 0.125 in the Tangkou coal (Table 3, Figure 2). The epigenetic carbonate minerals in coal can be formed in different stages [55–58]. The distribution differences between dolomite and calcite/ankerite suggest that they were not precipitated simultaneously and were probably derived from various fluids with different compositions. Calcite and ankerite were also not from in the same period. For example, the TK3U-1, 2, 3 bench samples in the Tangkou mine have ankerite and no calcite, whereas the sample LBS3U-2 in the Liangbaosi mine has calcite but no ankerite. The current data support that calcite, ankerite, and dolomite in the Liangbaosi coal were derived from various fluids with different compositions and there were at least three stages for epigenetic carbonate mineral precipitation.

## 6. Conclusions

Due to the tectonic activities after Jurassic, the positions of the Early Permian Upper No. 3 coals were significantly rearranged in depth in the Luxi, Liangbaosi, and Tangkou mines. The three Upper No. 3 coals are similar in rank and maceral compositions, suggesting that there were no significant influences from the tectonic processes. Although the Upper No. 3 coals are low in sulfur, they may still have derived from marine influence. Terrigenous minerals are comparable in both types and distribution patterns among the three coal seam sections; siderite and pyrite signify minerals of syngenetic or penecontemporaneous precipitation origin rather than an epigenetic origin. Epigenetic minerals (e.g., calcite, ankerite, and dolomite) were attributed to invasion by Ca, Mg or Fe-bearing fluids.

**Acknowledgments:** This research was supported by the National Key Basic Research Program of China (No. 2014CB238900), the National Natural Science Foundation of China (No. 41202121), the Program for Changjiang Scholars and Innovative Research Team in University (IRT13099).

**Author Contributions:** Xibo Wang conceived the overall experimental strategy and performed major and trace elements measurement. Yaofa Jiang did microscopic experiments. Lili Zhang, Jianpeng Wei and Zijuan Chen did the XRD and coal chemistry analysis. All authors participated in writing the manuscript.

**Conflicts of Interest:** The authors declare no conflict of interest.

## References

1. Ward, C.R. Mineral matter in low-rank coals and associated strata of the Mae Moh basin, northern Thailand. *Int. J. Coal Geol.* **1991**, *17*, 69–93. [CrossRef]
2. Dai, S.; Seredin, V.V.; Ward, C.R.; Hower, J.C.; Xing, Y.; Zhang, W.; Song, W.; Wang, P. Enrichment of U–Se–Mo–Re–V in coals preserved within marine carbonate successions: Geochemical and mineralogical data from the Late Permian Guiding Coalfield, Guizhou, China. *Miner. Depos.* **2015**, *50*, 159–186. [CrossRef]
3. Hower, J.C.; Eble, C.F.; Dai, S.; Belkin, H.E. Distribution of rare earth elements in eastern Kentucky coals: Indicators of multiple modes of enrichment? *Int. J. Coal Geol.* **2016**. [CrossRef]
4. Ward, C.R. Mineralogical analysis in hazard assessment. In *Geological Hazards-the Impact to Mining*; Doyle, R., Moloney, J., Eds.; Coalfield Geology Council of New South Wales: Newcastle, Australia, 2001; pp. 81–88.
5. Ward, C.R.; Corcoran, J.F.; Saxby, J.D.; Read, H.W. Occurrence of phosphorus minerals in Australian coal seams. *Int. J. Coal Geol.* **1996**, *31*, 185–210. [CrossRef]
6. Ward, C.R.; Spears, D.A.; Booth, C.A.; Staton, I.; Gurba, L.W. Mineral matter and trace elements in coals of the Gunnedah Basin, New South Wales, Australia. *Int. J. Coal Geol.* **1999**, *40*, 281–308. [CrossRef]
7. Dai, S.; Graham, I.T.; Ward, C.R. A review of anomalous rare earth elements and yttrium in coal. *Int. J. Coal Geol.* **2016**, *159*, 82–95. [CrossRef]
8. Dai, S.; Ren, D.; Tang, Y.; Yue, M.; Hao, L. Concentration and distribution of elements in Late Permian coals from western Guizhou Province, China. *Int. J. Coal Geol.* **2005**, *61*, 119–137. [CrossRef]
9. Liu, J.; Yang, Z.; Yan, X.; Ji, D.; Yang, Y.; Hu, L. Modes of occurrence of highly-elevated trace elements in superhigh-organic-sulfur coals. *Fuel* **2015**, *156*, 190–197. [CrossRef]
10. Dai, S.; Zeng, R.; Sun, Y. Enrichment of arsenic, antimony, mercury, and thallium in a Late Permian anthracite from Xingren, Guizhou, Southwest China. *Int. J. Coal Geol.* **2006**, *66*, 217–226. [CrossRef]
11. Li, W.C. Workflows of identification of the coalfield tectonic types. *Coal Geol. China* **1998**, *10*, 4–9.
12. Zeng, R.S.; Zhuang, X.G.; Yang, S.K. Quality of the coals from middle area of coal-bearing district of western Shandong. *Coal Geol. China* **2000**, *12*, 10–15.
13. Liu, G.; Yang, P.; Peng, Z.; Chou, C.-L. Petrographic and geochemical contrasts and environmentally significant trace elements in marine-influenced coal seams, Yanzhou mining area, China. *Int. J. Coal Geol.* **2004**, *23*, 491–506. [CrossRef]
14. Liu, G.J.; Zheng, L.G.; Gao, L.F.; Zhang, H.Y.; Peng, Z.C. The characterization of coal quality from the Jining coalfield. *Energy* **2005**, *30*, 1903–1914. [CrossRef]
15. Liu, G.J.; Zheng, L.G.; Zhang, Y.; Qi, C.C.; Chen, Y.W.; Peng, Z.C. Distribution and mode of occurrence of As, Hg and Se and sulfur in coal Seam 3 of the Shanxi Formation, Yanzhou coalfield, China. *Int. J. Coal Geol.* **2007**, *71*, 371–385. [CrossRef]
16. ASTM International. *Test Method for Moisture in the Analysis Sample of Coal and Coke*; ASTM Standard D3173-11; ASTM International: West Conshohocken, PA, USA, 2011.
17. ASTM International. *Test Method for Volatile Matter in the Analysis Sample of Coal and Coke*; ASTM Standard D3175-11; ASTM International: West Conshohocken, PA, USA, 2011.
18. ASTM International. *Annual Book of ASTM Standards. Test Method for Ash in the Analysis Sample of Coal and Coke*; ASTM Standard D3174-11; ASTM International: West Conshohocken, PA, USA, 2011.
19. ASTM International. *Test Methods for Total Sulfur in the Analysis Sample of Coal and Coke*; ASTM Standard D3177-02; ASTM International: West Conshohocken, PA, USA, 2011.
20. Taylor, G.H.; Teichmüller, M.; Davis, A.; Diessel, C.F.K.; Littke, R.; Robert, P. *Organic Petrology*; Gebrüder Borntraeger: Berlin, Germany, 1998; pp. 162–174.

21. International Committee for Coal Petrology (ICCP). The new vitrinite classification (ICCP System 1994). *Fuel* **1998**, *77*, 349–358.

22. International Committee for Coal Petrology (ICCP). The new inertinite classification (ICCP System 1994). *Fuel* **2001**, *80*, 459–471.

23. Dai, S.; Hower, J.C.; Ward, C.R.; Guo, W.; Song, H.; O'Keefe, J.M.K.; Xie, P.; Hood, M.M.; Yan, X. Elements and phosphorus minerals in the middle Jurassic inertinite-rich coals of the Muli Coalfield on the Tibetan Plateau. *Int. J. Coal Geol.* **2015**, *144–145*, 23–47. [CrossRef]

24. Dai, S.; Wang, X.; Zhou, Y.; Hower, J.C.; Li, D.; Chen, W.; Zhu, X. Chemical and mineralogical compositions of silicic, mafic, and alkali tonsteins in the late Permian coals from the Songzao Coalfield, Chongqing. Southwest China. *Chem. Geol.* **2011**, *282*, 29–44. [CrossRef]

25. Li, X.; Dai, S.; Zhang, W.; Li, T.; Zheng, X.; Chen, W. Determination of As and Se in coal and coal combustion products using closed vessel microwave digestion and collision/reaction cell technology (CCT) of inductively coupled plasma mass spectrometry (ICP-MS). *Int. J. Coal Geol.* **2014**, *124*, 1–4. [CrossRef]

26. ASTM International. *Standard Test Method for Total Fluorine in Coal and Coke by Pyrohydrolytic Extraction and Ion Selective Electrode or Ion Chromatograph Methods*; ASTM Standard D5987–96; ASTM International: West Conshohocken, PA, USA, 2011.

27. Dai, S.; Ren, D. Fluorine concentration of coals in China—An estimation considering coal reserves. *Fuel* **2006**, *85*, 929–935. [CrossRef]

28. Han, D.; Ren, D.; Wang, Y.; Jin, K.; Mao, H.; Qin, Y. *Coal Petrology in China*; China University of Mining & Technology Press: Xuzhou, China, 1996. (In Chinese)

29. Dai, S.; Jiang, Y.; Ward, C.; Gu, L.; Seredin, V.; Liu, H.; Zhou, D.; Wang, X.; Sun, Y.; Zou, J.; et al. Mineralogical and geochemical compositions of the coal in the Guanbanwusu Mine, Inner Mongolia, China: Further evidence for the existence of an Al (Ga and REE) ore deposit in the Jungar Coalfield. *Int. J. Coal Geol.* **2012**, *98*, 10–40. [CrossRef]

30. Dai, S.; Li, D.; Chou, C.-L.; Zhao, L.; Zhang, Y.; Ren, D.; Ma, Y.; Sun, Y. Mineralogy and geochemistry of boehmite-rich coals: New insights from the Haerwusu Surface Mine, Jungar Coalfield, Inner Mongolia, China. *Int. J. Coal Geol.* **2008**, *74*, 185–202. [CrossRef]

31. Dai, S.; Ren, D.; Chou, C.-L.; Li, S.; Jiang, Y. Mineralogy and geochemistry of the No. 6 coal (Pennsylvanian) in the Junger Coalfield, Ordos Basin, China. *Int. J. Coal Geol.* **2006**, *66*, 253–270. [CrossRef]

32. Dai, S.; Zou, J.; Jiang, Y.; Ward, C.R.; Wang, X.; Li, T.; Xue, W.; Liu, S.; Tian, H.; Sun, X.; et al. Mineralogical and geochemical compositions of the Pennsylvanian coal in the Adaohai Mine, Daqingshan Coalfield, Inner Mongolia, China: Modes of occurrence and origin of diaspore, gorceixite, and ammonian illite. *Int. J. Coal Geol.* **2012**, *94*, 250–270. [CrossRef]

33. Ward, C.R.; Warbrooke, P.R.; Roberts, I. Geochemical and mineralogical changes in a coal seam due to contact metamorphism, Sydney Basin, New South Wales, Australia. *Int. J. Coal Geol.* **1989**, *11*, 105–125. [CrossRef]

34. Dai, S.; Ren, D.; Chou, C.-L.; Finkelman, R.B.; Seredin, V.V.; Zhou, Y. Geochemistry of trace elements in Chinese coals: A review of abundances, genetic types, impacts on human health, and industrial utilization. *Int. J. Coal Geol.* **2012**, *94*, 3–21. [CrossRef]

35. Ketris, M.P.; Yudovich, Y.E. Estimations of Clarkes for carbonaceous biolithes: World average for trace element contents in black shales and coals. *Int. J. Coal Geol.* **2009**, *78*, 135–148. [CrossRef]

36. ASTM International. *Standard Classification of Coals by Rank*; ASTM Standard D388-07; ASTM International: West Conshohocken, PA, USA, 2011.

37. Ward, C.R. Analysis and significance of mineral matter in coal seams. *Int. J. Coal Geol.* **2002**, *50*, 135–168. [CrossRef]

38. Chou, C-L. Sulfur in coals: A review of geochemistry and origins. *Int. J. Coal Geol.* **2012**, *100*, 1–13. [CrossRef]

39. Dai, S.; Hou, X.; Ren, D.; Tang, Y. Surface analysis of pyrite in the No. 9 coal seam, Wuda Coalfield, Inner Mongolia, China, using high-resolution time-of-flight secondary ion mass-spectrometry. *Int. J Coal Geol.* **2003**, *55*, 139–150. [CrossRef]

40. Johnston, M.N.; Hower, J.C.; Dai, S.; Wang, P.; Xie, P.; Liu, J. Petrology and geochemistry of the Harlan, Kellioka, and Darby coals from the Louellen 7.5-Minute quadrangle, Harlan County, Kentucky. *Minerals* **2015**, *5*, 894–918. [CrossRef]

41. Kravits, C.M.; Crelling, J.C. Effects of overbank deposition on the quality and maceral composition of the Herrin (No.6) coal (Pennsylvanian) of Southern Illinois. *Int. J. Coal Geol.* **1981**, *1*, 195–212. [CrossRef]

42. Hayashi, K.I.; Fujisawa, H.; Holland, H.D.; Ohmoto, H. Geochemistry of 1.9 Ga sedimentary rocks from northeastern Labrador, Canada. *Geochim. Cosmochim. Acta* **1997**, *61*, 4115–4137. [CrossRef]

43. He, B.; Xu, Y.G.; Zhong, Y.T.; Guan, J.P. The Guadalupian–Lopingian boundary mudstones at Chaotian (SW China) are clastic rocks rather than acidic tuffs: Implication for a temporal coincidence between the end-Guadalupian mass extinction and the Emeishan volcanism. *Lithos* **2010**, *119*, 10–19. [CrossRef]

44. Dai, S.; Yang, J.; Ward, C.R.; Hower, J.C.; Liu, H.; Garrison, T.M.; French, D.; O'Keefe, J.M.K. Geochemical and mineralogical evidence for a coal-hosted uranium deposit in the Yili Basin, Xinjiang, northwestern China. *Ore Geol. Rev.* **2015**, *70*, 1–30. [CrossRef]

45. Hower, J.C.; Eble, C.F.; O'Keefe, J.M.K.; Dai, S.; Wang, P.; Xie, P.; Liu, J.; Ward, C.R.; French, D. Petrology, palynology, and geochemistry of Gray Hawk Coal (Early Pennsylvanian, Langsettian) in Eastern Kentucky, USA. *Minerals* **2015**, *5*, 592–622. [CrossRef]

46. Zhao, L.; Ward, C.R.; French, D.; Graham, I.T. Major and trace element geochemistry of coals and intra-seam claystones from the Songzao Coalfield, SW China. *Minerals* **2015**, *5*, 870–893. [CrossRef]

47. Gould, K.W.; Smith, J.W. The genesis and isotopic composition of carbonates associated with some Permian Australian coals. *Chem. Geol.* **1979**, *24*, 137–150. [CrossRef]

48. Dai, S.; Ren, D.; Tang, Y.; Shao, L.; Li, S. Distribution, isotopic variation and origin of sulfur in coals in the Wuda coalfield, Inner Mongolia, China. *Int. J. Coal Geol.* **2015**, *51*, 237–225. [CrossRef]

49. Chen, Z.H.; Wu, F.D.; Zhang, S.L.; Zhang, N.M.; Ma, J.X.; Ge, L.G. *Sedimentary Environments and Coal Accumulation of Late Paleozoic Coal Formation in Northern China*; Press of China University of Geosciences: Beijing, China, 1993; p. 190. (In Chinese)

50. Lan, X.H.; Ma, D.X.; Xu, M.G. Some geochemical indicators of the Pearl River Delta and their facies significance. *Mar. Geol. Quat. Geol.* **1987**, *7*, 39–49. (In Chinese)

51. Chen, Z.; Chen, Z.; Zhang, W. Quaternary stratigraphy and trace-element indices of the Yangtze Delta, Eastern China, with special reference to marine transgressions. *Quat. Res.* **1997**, *47*, 181–191. [CrossRef]

52. Chagué-Goff, C. Chemical signatures of palaeotsunamis: A forgotten proxy? *Mar. Geol.* **2010**, *271*, 67–71. [CrossRef]

53. Bouška, V.; Pešek, J.; Sykorova, I. Probable modes of occurrence of chemical elements in coal. *Acta Montana Ser. B Fuel Carbon Mineral Process. Praha* **2000**, *10*, 53–90.

54. Dai, S.; Chou, C.-L. Occurrence and origin of minerals in a chamosite-bearing coal of Late Permian age, Zhaotong, Yunnan, China. *Am. Mineral.* **2007**, *92*, 1253–1261. [CrossRef]

55. Kolker, A.; Chou, C.-L. Cleat-filling calcite in Illinois Basin coals: Trace element evidence for meteoric fluid migration in a coal basin. *J. Geol.* **1994**, *102*, 111–116. [CrossRef]

56. Shields, A.J. Secondary Mineralization in Coal: Case Study of the Bunnerong PHKB 1 Borehole, Sydney Basin, Australia. Unpublished. Bachelor's (Honours) Thesis, University of Sydney, Sydney, Australia, 1994. p. 101.

57. Dai, S.; Chou, C.-L.; Yue, M.; Luo, K.; Ren, D. Mineralogy and geochemistry of a Late Permian coal in the Dafang Coalfield, Guizhou, China: Influence from siliceous and iron-rich calcic hydrothermal fluids. *Int. J. Coal Geol.* **2005**, *61*, 241–258. [CrossRef]

58. Dai, S.; Liu, J.; Ward, C.R.; Hower, J.C.; French, D.; Jia, S.; Hood, M.M.; Garrison, T.M. Mineralogical and geochemical compositions of Late Permian coals and host rocks from the Guxu Coalfield, Sichuan Province, China, with emphasis on enrichment of rare metals. *Int. J. Coal Geol.* **2016**. [CrossRef]

*Article*

# Geological Controls on Mineralogy and Geochemistry of an Early Permian Coal from the Songshao Mine, Yunnan Province, Southwestern China

**Ruixue Wang [1,2]**

[1] College of Geoscience and Surveying Engineering, China University of Mining and Technology (Beijing), Beijing 100083, China; wangruixue504@gmail.com; Tel.: +86-10-8232-0638
[2] State Key Laboratory of Coal Resources and Safe Mining, China University of Mining and Technology (Beijing), Beijing 100083, China

Academic Editor: Thomas N. Kerestedjian
Received: 13 April 2016; Accepted: 29 June 2016; Published: 5 July 2016

**Abstract:** This paper discusses the content, distribution, modes of occurrence, and enrichment mechanism of mineral matter and trace elements of an Early Permian coal from Songshao (Yunnan Province, China) by means of coal-petrological, mineralogical, and geochemical techniques. The results show that the Songshao coal is characterized by high total and organic sulfur contents (3.61% and 3.87%, respectively). Lithium (170.39 µg/g) and Zr (184.55 µg/g) are significantly enriched in the Songshao coal, and, to a lesser extent, elements such as Hg, La, Ce, Nd, Th, Sr, Nb, Sn, Hf, V, and Cr are also enriched. In addition to Hg and Se that are enriched in the roof and floor strata of the coal seam, Li, La, Ce, Pr, Nd, Sm, Gd, Y, Cd, and Sb are slightly enriched in these host rocks. Compared to the upper continental crust, rare earth elements and yttrium in the host rocks and coal samples are characterized by a light-REE enrichment type and have negative Eu, positive Ce and Gd anomalies. Major minerals in the samples of coal, roof, and floor are boehmite, clay minerals (kaolinite, illite, and mixed layer illite-smectite), pyrite, and anatase. Geochemical and mineralogical anomalies of the Songshao coal are attributed to hydrothermal fluids, seawater, and sediment-source rocks.

**Keywords:** coal; mineral; elements; Early Permian; genetic types

## 1. Introduction

Yunnan is one of the most coal-rich provinces in Southern China. Coal resources are mainly concentrated in the east and south of Yunnan. The Songshao Mine is located in the eastern part of Yunnan Province (Figure 1). Previous studies have focused on geologic structure, coal-bearing sequences, and coal quality of the Songshao Mine [1]. This paper aims to discuss geological controls on mineralogy and geochemistry of Early Permian coal from the Songshao Mine. Studies on mineral matter in coal are important because the process of coal formation, including peat accumulation, the interaction of the organic matter with basinal fluids, sediment diagenesis, and sometimes synsedimentary volcanic inputs, may result in enrichment of mineral matter in coal. Therefore, investigations on the mineral matter in coal could help to better understand the process of coal formation [2–4]. From an economic perspective, mineral matter in coal can serve as carriers for some valuable elements (e.g., Ga, Al, rare earth elements and yttrium) that have been recovered, or have such potential, from coal combustion wastes [5–8]. Geochemical anomalies in coals from Eastern Yunnan have previously been reported; for instance, Zhou et al. [9] reported trace-element geochemistry of altered volcanic ash layers (tonsteins) in Late Permian coal-bearing sequences in Eastern Yunnan and Western Guizhou Provinces; Dai et al. [10] described modes of occurrence and origin of quartz and chamosite in Xuanwei coals. Dai et al. [11] found a new type of Nb(Ta)–Zr(Hf)–REE–Ga polymetallic deposit of volcanic origin in the Late

Permian coal-bearing strata of Eastern Yunnan; Dai et al. [12] reported on the mineralogical and geochemical compositions of Late Permian C2 and C3 coals (both medium-volatile bituminous) from the Xinde Mine in Xuanwei, Eastern Yunnan. Geological factors controlling these geochemical and mineralogical anomalies have previously been analyzed [3,13–16]. For example, Dai et al. [13] documented mineralogical and geochemical anomalies of Late Permian coals from the Fusui coalfield, which were caused by influences from terrigenous materials and hydrothermal fluids.

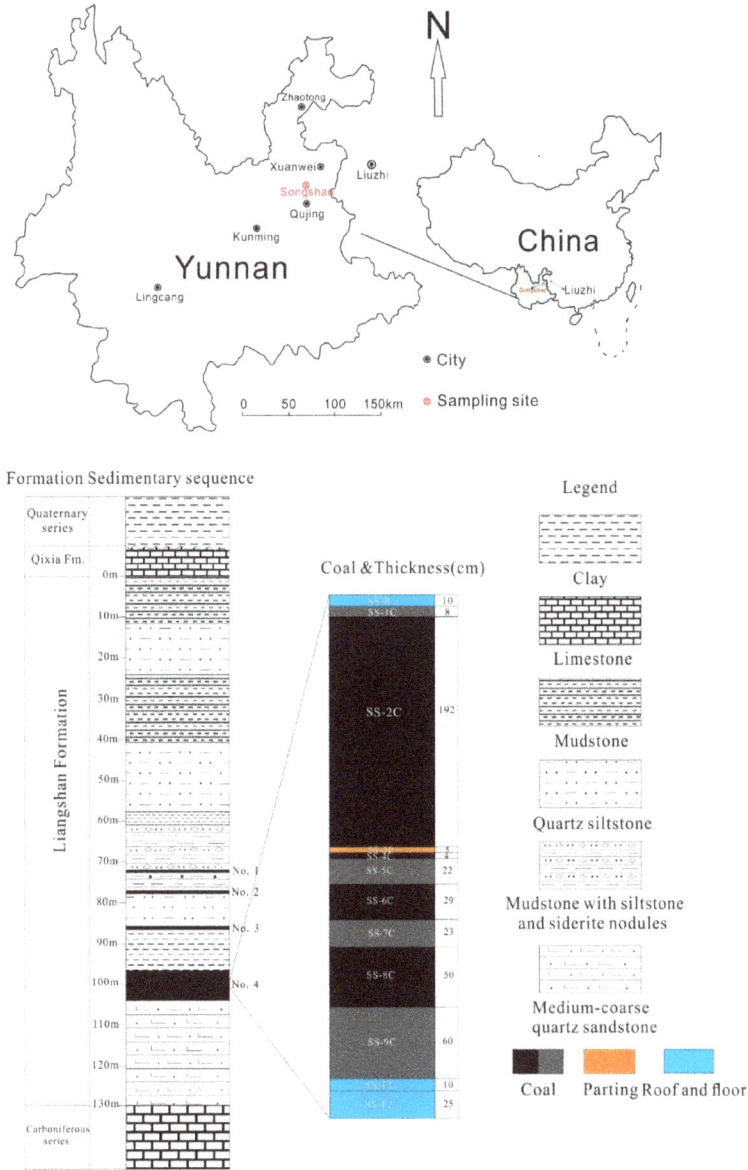

**Figure 1.** Location and sedimentary sequence of the Songshao Mine, Eastern Yunnan Province, China.

In the present study, new data on the Early Permian Coal from Songshao are reported with the aim to: (1) investigate geochemical and mineralogical compositions and modes of occurrence of elements and minerals; and (2) discuss the geological factors influencing mineralogical and geochemical anomalies.

## 2. Geological Setting

The sedimentary sequences in the Songshao Mine include Upper Carboniferous strata, Lower Permian Liangshan Formation, Lower Permian Qixia Formation, and the Quaternary system (Figure 1).

The Liangshan Formation comprises the coal-bearing seam, which underlies limestone of the Qixia Formation, and conformably overlies Carboniferous limestone. The lower part of the Liangshan Formation consists of coarse quartzose sandstones; the middle part is mainly composed of mudstones, interlayered with coal seams; and the upper part of the formation consists mainly of mudstones interbedded with quartzose sandstones. The thickness of the Liangshan Formation is about 131 m, and the coal seams are numbered 1 to 4. Only the No. 4 coal seam is minable and contains several gritty mudstone partings. The thickness of the No. 4 coal varies from 2 to 6 m. Samples, taken from No. 4 coal bed were numbered as SS-R, SS-1C, SS-2C, SS-3P, SS-4C, SS-5C, SS-6C, SS-7C, SS-8C, SS-9C, SS-F1 and SS-F2 (SS-R is roof; SS-3P is parting, SS-F1 and SS-F2 are floors and others are coals).

The roof of No. 4 coal is made up of grey black carbonaceous mudstone, and the floor consists of quartzose sandstones. The Upper Carboniferous Formation is made up of dolomite limestone. The Qixia Formation consists of medium-thick and thick limestone. The Quaternary system is composed of colluvial deposits.

## 3. Samples and Analytical Procedures

The samples selected for this study were collected from the faces of the mined coal seams in the Songshao Mine (No. 4 coal; Figure 1). All collected samples were immediately stored in plastic bags to minimize contamination and oxidation.

All the analyses were conducted at the State Key Laboratory of Coal Resources and Safe Mining (China University of Mining and Technology, Beijing, China). Proximate analysis was performed according to Chinese Standard GB/T 212-2008 [17]. The total sulfur and forms of sulfur were conducted following Chinese Standard GB/T 214-2007 [18] and GB/T 215-2003 [19], respectively. Mean random vitrinite reflectance ($R_r$) and maceral composition followed ISO 7404-5: 2009 [20]. Percentages of major-element oxides were determined by X-ray fluorescence (XRF) spectrometry (Thermofisher ARL Advant'XP+, ThermoFisher Scientific, Waltham, MA, USA). Trace-element concentrations were determined by inductively coupled plasma mass spectrometry (ICP-MS, Thermofisher X series II, Thermo Fisher Scientific), with exception of Hg and F. Mercury was determined using a Milestone DMA-80 Hg analyzer (Milestone, Sorisole, Italy), with a detection limit of 0.005-ng Hg, 1.5% relative standard deviation (RSD), and 0–1000 ng linearity for the calibration [21]. Fluorine was determined by pyrohydrolysis in conjunction with an ion-selective electrode, following the method described in Chinese National Standard GB/T 4633-1997 (1997) [22]. The mineralogy was determined by X-ray powder diffraction (XRD, Rigaku, Tokyo, Japan) plus Siroquant™ (Sietronics Pty Ltd, Canberra, Australia), optical microscopic observation, and a Field Emission-Scanning Electron Microscope (FE-SEM, FEI Quanta™ 650 FEG, Hillsboro, OR, USA), in conjunction with an energy-dispersive X-Ray spectrometer(EDS, Genesis Apex 4, EDAX Inc., Mahwah, NJ, USA), and these analytical procedures were described by Dai et al. [13].

## 4. Results

### 4.1. Ultimate and Proximate Analyses and Coal Rank

The vitrinite reflectance (1.24%) and volatile matter (30.91 wt %, daf) of the coal bench samples (Table 1) indicate a middle- to high-volatile bituminous rank, according to the ASTM classification

(ASTM D388-12, 2012) [23]. The coal is a medium-ash and high-sulfur coal according to Chinese Standards GB/T 15224.1-2004 [24] (coals with ash yield 20.01%–30.00% are high-ash coal) and GB/T 15224.2-2004 [25] (coals with total sulfur content >3% are high-sulfur coal). Organic sulfur accounts for most of the otal sulfur (Table 1). However, sample SS-1C has a higher content of pyrite sulfur than other coal benches.

Table 1. Proximate and ultimate analyses (wt %), and forms of sulfur (wt %) in Songshao coal.

| Sample | $R_r$ | $M_{ad}$ | $A_d$ | $VM_{daf}$ | $S_{t,d}$ | $S_{s,d}$ | $S_{p,d}$ | $S_{o,daf}$ | $N_{daf}$ | $C_{daf}$ | $H_{daf}$ |
|---|---|---|---|---|---|---|---|---|---|---|---|
| SS-R | nd | nd | 79.88 | nd | 5.35 | nd | nd | nd | nd | nd | nd |
| SS-1C | 1.35 | 1.40 | 33.35 | 29.84 | 13.03 | 0.31 | 8.06 | 6.99 | 0.70 | 75.50 | 4.04 |
| SS-2C | 1.21 | 0.64 | 19.22 | 30.68 | 3.85 | 0.05 | 0.71 | 3.83 | 1.10 | 85.25 | 4.84 |
| SS-3P | 1.17 | 0.47 | 47.19 | 77.10 | 1.68 | 0.20 | 0.04 | 2.72 | 0.62 | 64.71 | 3.16 |
| SS-4C | 1.26 | 0.63 | 9.67 | 27.66 | 3.78 | bdl | 0.11 | 4.09 | 1.16 | 87.57 | 4.73 |
| SS-5C | 1.29 | 0.59 | 26.35 | 28.84 | 2.86 | bdl | 0.05 | 3.84 | 1.21 | 84.75 | 4.98 |
| SS-6C | 1.23 | 0.52 | 24.20 | 29.81 | 2.83 | bdl | 0.07 | 3.68 | 1.11 | 84.97 | 4.93 |
| SS-7C | 1.41 | 0.44 | 17.84 | 28.31 | 3.26 | 0.01 | 0.10 | 3.83 | 0.98 | 86.05 | 4.82 |
| SS-8C | 1.22 | 0.58 | 32.77 | 31.62 | 2.81 | bdl | 0.47 | 3.55 | 1.11 | 83.36 | 5.18 |
| SS-9C | 1.24 | 0.52 | 29.85 | 29.87 | 3.16 | 0.07 | 0.25 | 4.04 | 1.01 | 83.63 | 4.97 |
| SS-F1 | nd | nd | 50.33 | nd | 1.53 | nd | nd | nd | nd | nd | nd |
| SS-F2 | nd | nd | 80.78 | nd | 0.45 | nd | nd | nd | nd | nd | nd |
| WA | 1.24 | 0.60 | 23.80 | 30.91 | 3.61 | 0.04 | 0.62 | 3.87 | 1.07 | 84.32 | 4.88 |

M, moisture; A, ash yield; VM, volatile matter; C, carbon; H, hydrogen; N, nitrogen; $S_t$, total sulfur; $S_s$, sulfate sulfur; $S_p$, pyrite sulfur; $S_o$, organic sulfur; ad, air-dry basis; d, dry basis; daf, dry and ash-free basis; WA, weighted average for coals (weighted by thickness of sample interval); bdl, below detection limit; nd, not detected.

*4.2. Geochemical Composition*

4.2.1. Major Element Oxides

The major-element oxides are mainly represented by $SiO_2$ and $Al_2O_3$ (Table 2). When considered on a whole-coal basis, Songshao coal contains higher proportions of $Al_2O_3$, $SiO_2$, and, to a lesser extent, MgO, $TiO_2$, and $K_2O$, than the average values for Chinese coals reported by Dai et al. [26]; other major element oxides, however, are either lower than, or close to, corresponding average values for Chinese coals. The $SiO_2/Al_2O_3$ ratios for the coals are much lower than those of common Chinese coals (1.42) [26] and the theoretical $SiO_2/Al_2O_3$ ratio of kaolinite (1.18).

Table 2. Percentages of major-element oxides in the Songshao coal, parting, roof and floor rocks.

| Sample | Th (cm) | LOI | $SiO_2$ | $TiO_2$ | $Al_2O_3$ | $Fe_2O_3$ | MnO | MgO | CaO | $Na_2O$ | $K_2O$ | $P_2O_5$ | $SiO_2/Al_2O_3$ |
|---|---|---|---|---|---|---|---|---|---|---|---|---|---|
| SS-R | 10 | 20.12 | 40.27 | 0.86 | 28.27 | 6.41 | 0.01 | 1.03 | 0.35 | 0.05 | 2.15 | 0.06 | 1.42 |
| SS-1C | 8 | 66.65 | 9.08 | 0.2 | 8.73 | 13.83 | 0.01 | 0.13 | 0.14 | 0.04 | 0.11 | 0.2 | 1.04 |
| SS-2C | 192 | 80.78 | 8.35 | 0.46 | 8.07 | 0.9 | 0 | 0.14 | 0.37 | 0.03 | 0.1 | 0.12 | 1.03 |
| SS-3P | 5 | 52.81 | 0.76 | 0.02 | 1.6 | 0.59 | 0.08 | 0.26 | 36.97 | 0.02 | 0 | 0.01 | 0.48 |
| SS-4C | 4 | 90.33 | 3.07 | 0.13 | 4.8 | 0.2 | 0 | 0.18 | 0.41 | 0.19 | 0.03 | 0.03 | 0.64 |
| SS-5C | 22 | 73.65 | 11.55 | 0.44 | 13.08 | 0.19 | 0 | 0.36 | 0.1 | 0.04 | 0.19 | 0.03 | 0.88 |
| SS-6C | 29 | 75.8 | 10.36 | 0.42 | 12.28 | 0.22 | 0 | 0.3 | 0.07 | 0.02 | 0.19 | 0.02 | 0.84 |
| SS-7C | 23 | 82.16 | 7.72 | 0.29 | 8.86 | 0.13 | 0 | 0.27 | 0.1 | 0.02 | 0.13 | 0.03 | 0.87 |
| SS-8C | 50 | 67.23 | 14.57 | 0.57 | 15.75 | 0.61 | 0 | 0.4 | 0.11 | 0.03 | 0.33 | 0.06 | 0.93 |
| SS-9C | 60 | 70.15 | 13.72 | 0.54 | 13.74 | 0.66 | 0 | 0.31 | 0.08 | 0.02 | 0.42 | 0.03 | 1 |
| SS-F1 | 10 | 49.67 | 25.24 | 1.12 | 22.83 | 0.19 | 0 | 0.21 | 0.07 | 0.01 | 0.39 | 0.03 | 1.11 |
| SS-F2 | 25 | 19.22 | 39.79 | 1.65 | 37.45 | 0.46 | 0 | 0.31 | 0.05 | 0.01 | 0.64 | 0.06 | 1.06 |
| SS Coal | - | 76.5 | 10.24 | 0.46 | 10.56 | 0.95 | 0 | 0.23 | 0.23 | 0.03 | 0.19 | 0.08 | 0.97 |
| Chinese Coal [a] | n.d. | n.d. | 8.47 | 0.33 | 5.98 | 4.85 | 0.02 | 0.22 | 1.23 | 0.16 | 0.19 | 0.09 | 1.42 |
| CC | n.d. | n.d. | 1.21 | 1.4 | 1.77 | 0.2 | 0.15 | 1.06 | 0.19 | 0.18 | 1.01 | 0.88 | 0.68 |

Th, thickness; LOI, loss on ignition; CC, weighted average for coal samples; [a] From Reference [26]; n.d., no data.

The content of $Fe_2O_3$ is significantly higher in the roof and sample SS-1C than in other samples. The content of CaO in the parting SS-3P is particularly high. Based on the variation in concentrations through the seam section, two groups of major element oxides can be classified:

Group 1 includes $SiO_2$, $TiO_2$, $Al_2O_3$, and $K_2O$, all of which show a saw-like distribution, similar to the ash yield variation through the seam section. These oxides are enriched in the host rocks (roof and floor), but are at a low level in the coal benches and parting (Figure 2).

Group 2 consists of $Na_2O$, MgO, $P_2O_5$, MnO, and $Fe_2O_3$. The concentrations of these oxides do not show distinct variation through the seam section.

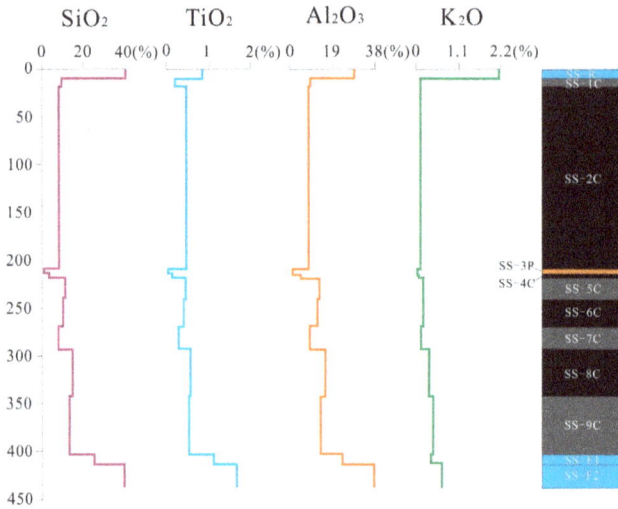

**Figure 2.** Variations of $SiO_2$, $TiO_2$, $Al_2O_3$ and $K_2O$ through the seam section.

### 4.2.2. Trace Elements

The trace-element contents of Songshao coal are listed in Table 3. Compared to the averages for world hard coals [27], and based on the enrichment classification of trace elements in coal by Dai et al. [28], a number of trace elements are enriched in Songshao coal. Lithium has a concentration coefficient (CC: ratio of element concentration in Songshao coals vs. world hard coals) >10. Zirconium displays CC values between 5 and 10. Elements, including Hg, Sr, Nb, Sn, Hf, V, Cr, Se, Th, and some light rare earth elements (La, Ce, and Nd), are slightly enriched in the coal (CC = 2–5). However, Rb, Ba, Bi, Mo, and Sb, are depleted in the coal (CC < 0.5). The concentrations of the remaining elements (0.5 < CC < 2) are close to the corresponding averages for world hard coals [27].

Particularly, a number of trace elements are enriched in sample SS-1C. Trace elements with a concentration coefficient >10 include Hg (CC = 19.73) and V (CC = 13.69); trace elements with a CC of 5–10 include Sc, La, Ce, Pr, Sm, Eu, Th, Li, Sr, Se, Zr, Cr, and Tl.

Trace elements in the roof and floor rocks with a concentration coefficient (CC = ratio of element concentration in Songshao coal vs. UCC) higher than 10 include Hg and Se. Lithium shows a CC of 5–10. Lanthanum, Ce, Pr, Nd, Sm, Gd, Y, Cd, and Sb are slightly enriched in the roof and floor samples (CC = 2–5).

The anomalous geochemical composition of Songshao coal is represented by ratios of some trace elements. On average, Songshao coal has higher Nb/Ta, Zr/Hf, and especially Li/Rb values, as well as lower values of Ba/Sr and Rb/Cs than the averages for the world hard coals. From top (roof) to bottom (floor), the ratios of Ba/Sr and Nb/Ta show similar distribution patterns, which gradually increase; the ratio of Rb/Cs, however, shows an adverse trend.

**Table 3.** Trace-element concentrations (µg/g) in the Songshao coal, parting, and roof and floor rocks.

| Sample | SS-R | SS-1C | SS-2C | SS-3P | SS-4C | SS-5C | SS-6C | SS-7C | SS-8C | SS-9C | SS-F1 | SS-F2 | World | SC | C1 | UCC | PC | C2 |
|---|---|---|---|---|---|---|---|---|---|---|---|---|---|---|---|---|---|---|
| Li | 160.70 | 84.29 | 140.24 | 17.10 | 72.32 | 170.02 | 184.16 | 124.26 | 231.48 | 257.89 | 844.97 | 944.01 | 14.00 | 170.39 | 12.17 | 20.00 | 116.01 | 5.80 |
| Be | 3.45 | 3.29 | 0.98 | 0.37 | 0.78 | 1.88 | 1.87 | 1.30 | 1.89 | 1.92 | 3.24 | 3.19 | 2.00 | 1.41 | 0.71 | 3.00 | 1.62 | 0.54 |
| F | 670.30 | 138.20 | 131.09 | 49.00 | 154.90 | 176.81 | 175.67 | 158.68 | 204.91 | 174.03 | 165.51 | 293.06 | 82.00 | 153.84 | 1.88 | - | 348.55 | - |
| Sc | 25.53 | 25.02 | 3.81 | 1.61 | 7.10 | 5.07 | 6.41 | 4.13 | 4.06 | 6.87 | 10.47 | 30.63 | 3.70 | 5.03 | 1.36 | 11.00 | 4.57 | 0.42 |
| V | 231.22 | 383.34 | 57.25 | 9.74 | 34.93 | 76.15 | 72.88 | 55.08 | 88.81 | 99.12 | 231.22 | 120.60 | 28.00 | 75.55 | 2.70 | 60.00 | 59.02 | 0.98 |
| Cr | 149.02 | 87.56 | 35.44 | 6.44 | 19.58 | 41.68 | 36.16 | 34.45 | 43.83 | 54.30 | 149.02 | 111.25 | 17.00 | 40.26 | 2.37 | 35.00 | 32.18 | 0.92 |
| Co | 20.24 | 19.44 | 5.76 | 6.38 | 3.70 | 3.28 | 2.23 | 4.35 | 2.78 | 2.92 | 20.24 | 2.46 | 6.00 | 4.73 | 0.79 | 10.00 | 5.17 | 0.52 |
| Ni | 55.07 | 34.39 | 14.59 | 11.27 | 6.26 | 20.50 | 14.34 | 15.10 | 23.40 | 27.39 | 55.07 | 38.61 | 17.00 | 18.28 | 1.08 | 20.00 | 17.84 | 0.89 |
| Cu | 60.63 | 20.57 | 23.98 | 3.82 | 23.53 | 34.69 | 32.71 | 34.61 | 47.51 | 32.42 | 60.63 | 17.88 | 16.00 | 29.80 | 1.86 | 25.00 | 25.00 | 1.00 |
| Zn | 150.53 | 9.04 | 2.42 | 11.40 | 0.60 | 2.88 | 2.93 | 3.77 | 9.96 | 7.09 | 150.53 | 8.69 | 28.00 | 4.46 | 0.16 | 71.00 | 12.65 | 0.18 |
| Ga | 50.30 | 15.91 | 7.43 | 1.10 | 3.98 | 10.80 | 9.00 | 10.02 | 11.37 | 14.08 | 36.39 | 24.71 | 6.00 | 9.46 | 1.58 | 17.00 | 8.26 | 0.49 |
| Ge | 1.46 | 0.84 | 1.64 | 0.15 | 0.24 | 0.65 | 0.88 | 1.32 | 1.11 | 0.98 | 0.86 | 1.28 | 2.40 | 1.29 | 0.54 | 1.60 | 1.68 | 1.05 |
| As | 9.34 | 12.31 | 2.87 | 1.06 | 2.11 | 1.53 | 1.40 | 1.51 | 2.79 | 4.02 | 2.18 | 2.25 | 9.00 | 2.96 | 0.33 | 2.00 | 3.81 | 1.91 |
| Se | 7.22 | 9.17 | 4.16 | 0.94 | 4.60 | 5.53 | 6.18 | 4.08 | 7.01 | 6.22 | 3.90 | 2.61 | 1.60 | 5.18 | 3.24 | 0.08 | 3.92 | 47.23 |
| Rb | 96.10 | 3.71 | 1.97 | 0.09 | 0.55 | 2.77 | 2.71 | 1.97 | 4.43 | 4.02 | 2.94 | 4.34 | 18.00 | 2.69 | 0.15 | 112.00 | 8.02 | 0.07 |
| Sr | 104.84 | 588.93 | 493.21 | 180.81 | 139.62 | 22.06 | 22.96 | 16.11 | 31.17 | 17.25 | 19.01 | 10.42 | 100.00 | 267.14 | 2.67 | 350.00 | 209.03 | 0.60 |
| Zr | 209.44 | 227.44 | 219.24 | 7.06 | 40.47 | 151.48 | 136.14 | 93.08 | 178.18 | 168.14 | 422.17 | 589.72 | 36.00 | 184.55 | 5.13 | 190.00 | 132.90 | 0.70 |
| Nb | 18.43 | 4.15 | 8.89 | 0.94 | 2.56 | 8.21 | 8.15 | 5.82 | 8.51 | 7.56 | 21.85 | 32.61 | 4.00 | 8.10 | 2.03 | 25.00 | 6.68 | 0.27 |
| Mo | 0.69 | 2.08 | 0.39 | 0.14 | 0.22 | 1.77 | 0.50 | 0.57 | 0.80 | 0.73 | 0.88 | 1.23 | 2.10 | 0.62 | 0.29 | 1.50 | 1.13 | 0.76 |
| Ag | 0.97 | 1.29 | 1.18 | bdl | 0.17 | 0.83 | 0.76 | 0.44 | 0.96 | 0.97 | 0.97 | 2.54 | bdl | 1.00 | bdl | 0.05 | 0.65 | 13.07 |
| Cd | 0.53 | 0.56 | 0.29 | 0.05 | 0.19 | 0.30 | 0.28 | 0.23 | 0.63 | 0.50 | 0.53 | 0.67 | 0.20 | 0.36 | 1.81 | 0.10 | 0.31 | 3.05 |
| In | 0.19 | 0.10 | 0.04 | 0.13 | 0.02 | 0.05 | 0.04 | 0.05 | 0.06 | 0.09 | 0.25 | 0.13 | 0.04 | 0.05 | 1.30 | - | 0.05 | - |
| Sn | 4.97 | 5.12 | 2.54 | 38.34 | 0.68 | 2.02 | 1.51 | 2.50 | 2.16 | 6.26 | 8.68 | 5.73 | 1.40 | 3.45 | 2.46 | 5.50 | 2.73 | 0.50 |
| Sb | 0.97 | 0.62 | 0.26 | 0.12 | 0.23 | 0.37 | 0.30 | 0.18 | 0.46 | 0.51 | 0.47 | 0.66 | 1.00 | 0.33 | 0.33 | 0.20 | 0.57 | 2.83 |
| Cs | 15.02 | 0.69 | 0.36 | 0.14 | 0.12 | 1.51 | 1.14 | 0.77 | 2.08 | 2.72 | 3.13 | 3.44 | 1.10 | 1.09 | 0.99 | 3.70 | 1.09 | 0.29 |
| Ba | 131.56 | 77.11 | 24.02 | 6.63 | 5.59 | 10.54 | 7.81 | 6.70 | 15.72 | 16.17 | 10.38 | 5.04 | 150.00 | 19.47 | 0.13 | 550.00 | 64.86 | 0.12 |
| La | 100.65 | 29.82 | 5.99 | 25.27 | 34.61 | 36.14 | 23.61 | 45.01 | 36.70 | 60.67 | 102.44 | 57.28 | 11.00 | 34.27 | 3.12 | 30.00 | 83.12 | 2.77 |
| Ce | 185.72 | 58.81 | 12.06 | 41.35 | 77.86 | 80.91 | 54.56 | 95.52 | 88.70 | 120.63 | 240.78 | 110.15 | 23.00 | 72.30 | 3.14 | 64.00 | 185.05 | 2.89 |
| Pr | 18.18 | 5.23 | 1.41 | 4.04 | 6.98 | 7.77 | 5.11 | 8.88 | 8.64 | 14.30 | 17.93 | 12.74 | 3.40 | 6.70 | 1.97 | 7.10 | 15.97 | 2.25 |
| Nd | 68.89 | 18.92 | 5.82 | 14.89 | 25.38 | 29.32 | 19.84 | 32.67 | 33.62 | 54.74 | 68.81 | 47.63 | 12.00 | 24.91 | 2.08 | 26.00 | 60.98 | 2.35 |
| Sm | 11.82 | 3.02 | 1.10 | 2.61 | 3.95 | 4.93 | 3.44 | 5.13 | 5.92 | 9.47 | 11.40 | 7.43 | 2.20 | 4.10 | 1.86 | 4.50 | 10.09 | 2.24 |
| Eu | 2.44 | 0.53 | 0.21 | 0.46 | 0.71 | 0.88 | 0.59 | 0.87 | 1.07 | 1.59 | 1.73 | 1.25 | 0.43 | 0.73 | 1.70 | 0.90 | 1.59 | 1.77 |
| Gd | 11.63 | 3.42 | 1.30 | 2.90 | 4.55 | 5.37 | 3.51 | 5.37 | 6.23 | 9.76 | 10.35 | 7.14 | 2.70 | 4.44 | 1.65 | 3.80 | 9.50 | 2.50 |
| Tb | 1.10 | 0.38 | 0.16 | 0.36 | 0.48 | 0.59 | 0.38 | 0.55 | 0.72 | 1.20 | 1.26 | 0.80 | 0.31 | 0.49 | 1.57 | 0.60 | 1.14 | 1.90 |
| Dy | 5.30 | 2.14 | 0.87 | 1.97 | 2.61 | 3.10 | 1.98 | 2.68 | 3.91 | 6.74 | 6.77 | 4.69 | 2.10 | 2.61 | 1.24 | 3.50 | 6.30 | 1.80 |
| Y | 25.96 | 10.69 | 5.05 | 10.31 | 13.23 | 15.34 | 9.04 | 12.34 | 18.96 | 37.75 | 56.89 | 25.90 | 8.40 | 12.79 | 1.52 | 22.00 | 45.75 | 2.08 |

Table 3. *Cont.*

| Sample | SS-R | SS-1C | SS-2C | SS-3P | SS-4C | SS-5C | SS-6C | SS-7C | SS-8C | SS-9C | SS-F1 | SS-F2 | World | SC | C1 | UCC | PC | C2 |
|---|---|---|---|---|---|---|---|---|---|---|---|---|---|---|---|---|---|---|
| Ho | 1.02 | 0.41 | 0.16 | 0.37 | 0.47 | 0.56 | 0.35 | 0.47 | 0.71 | 1.23 | 1.19 | 0.91 | 0.57 | 0.48 | 0.85 | 0.80 | 1.14 | 1.42 |
| Er | 3.20 | 1.26 | 0.48 | 1.12 | 1.40 | 1.67 | 1.00 | 1.34 | 2.08 | 3.62 | 3.49 | 3.02 | 1.00 | 1.45 | 1.45 | 2.30 | 3.42 | 1.49 |
| Tm | 0.44 | 0.18 | 0.06 | 0.15 | 0.18 | 0.22 | 0.13 | 0.17 | 0.27 | 0.45 | 0.44 | 0.43 | 0.30 | 0.20 | 0.65 | 0.30 | 0.44 | 1.46 |
| Yb | 3.26 | 1.28 | 0.40 | 1.08 | 1.27 | 1.56 | 0.88 | 1.16 | 1.88 | 3.06 | 2.96 | 3.19 | 1.00 | 1.38 | 1.38 | 2.20 | 3.03 | 1.38 |
| Lu | 0.46 | 0.17 | 0.05 | 0.15 | 0.16 | 0.20 | 0.12 | 0.15 | 0.25 | 0.40 | 0.39 | 0.45 | 0.20 | 0.18 | 0.91 | 0.30 | 0.40 | 1.34 |
| Hf | 5.34 | 3.83 | 5.56 | 0.21 | 1.12 | 4.25 | 3.83 | 2.48 | 4.76 | 4.75 | 10.00 | 13.78 | 1.20 | 4.81 | 4.01 | 5.80 | 3.55 | 0.61 |
| Ta | 1.28 | 0.36 | 0.70 | 0.29 | 0.17 | 0.59 | 0.59 | 0.38 | 0.42 | 0.27 | 1.41 | 2.15 | 0.30 | 0.55 | 1.83 | 2.20 | 0.46 | 0.21 |
| W | 5.17 | 0.24 | 1.20 | 0.14 | 0.16 | 0.56 | 0.50 | 0.62 | 0.58 | 0.66 | 2.65 | 11.90 | 0.99 | 0.87 | 0.88 | 2.00 | 0.91 | 0.46 |
| Hg | 0.26 | 1.97 | 0.17 | 0.02 | 0.12 | 0.14 | 0.13 | 0.10 | 0.21 | 0.24 | 0.17 | 0.24 | 0.10 | 0.21 | 2.11 | 0.02 | 0.17 | 8.63 |
| Tl | 0.79 | 3.45 | bdl | bdl | bdl | bdl | bdl | bdl | 0.02 | 0.03 | 0.79 | 0.02 | 0.58 | 0.07 | 0.12 | 0.80 | 0.25 | 0.31 |
| Pb | 34.89 | 11.87 | 10.11 | 0.97 | 6.79 | 15.73 | 17.38 | 9.81 | 18.75 | 21.37 | 34.89 | 7.12 | 9.00 | 13.65 | 1.52 | 20.00 | 12.03 | 0.60 |
| Bi | 1.80 | 0.04 | 0.21 | 0.60 | bdl | 0.23 | 0.19 | bdl | 0.13 | 1.06 | 1.03 | 1.84 | 1.10 | 0.32 | 0.29 | - | 0.59 | - |
| Th | 25.72 | 16.97 | 10.34 | 0.99 | 5.22 | 7.90 | 9.60 | 7.37 | 10.36 | 11.75 | 20.28 | 38.40 | 3.20 | 10.16 | 3.17 | 10.70 | 7.74 | 0.72 |
| U | 5.62 | 3.91 | 2.25 | 0.48 | 1.18 | 3.70 | 3.24 | 2.92 | 3.96 | 4.09 | 8.27 | 8.69 | 1.90 | 2.94 | 1.55 | 2.00 | 2.58 | 1.29 |

World, world hard coals, from Ketris and Yudovich [27]; SC, weighted average for coal samples; C1, concentration coefficient of coal samples; UCC, upper continental crust; PC, weighted average for host rock; C2, concentration coefficient of host rock; -, without the data item.

### 4.2.3. Rare Earth Elements

Rare earth elements (REE) in coal are generally associated with minerals, especially clay minerals and phosphate, which are generally associated with ash yield [29]; in some cases, heavy rare earth elements are associated with the organic matter in coal [30]. A threefold classification of REE was used for this study: Light (LREE: La, Ce, Pr, Nd, and Sm), medium (MREE: Eu, Gd, Tb, Dy, and Y), and heavy (HREE: Ho, Er, Tm, Yb, and Lu) REE [31]. Accordingly, in comparison with the upper continental crust [32], three enrichment types are identified [31]: L-type (light-REE; $La_N/Lu_N > 1$), M-type (medium-REE; $La_N/Sm_N < 1$ and $Gd_N/Lu_N > 1$), and H-type (heavy REE; $La_N/Lu_N < 1$).

With the exception of sample SS-9C, the REE patterns in the coal benches are characterized by L-REE enrichment, Eu negative anomalies, and Gd-maximum (Gd reaches to the peak of the patterns) (Figure 3). In particular, sample SS-1C has the highest REE concentration relative to other coal samples and does not show an Eu anomaly.

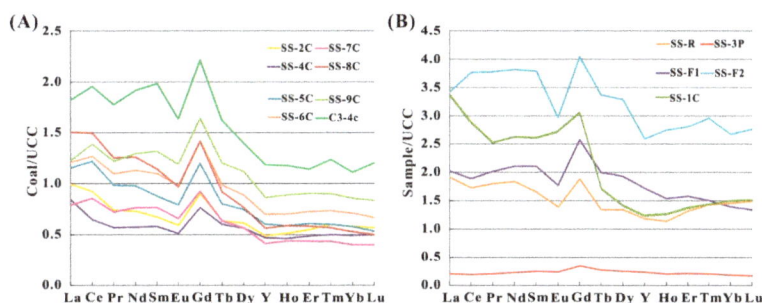

**Figure 3.** Distribution patterns of rare earth elements (REE) of samples in the Songshao coal mine. (**A**) Plots for the Songshao coal in comparison with C3-4c from the Xinde Mine. (**B**) Plots of SS-1C, roof and floor, and parting from the Songshao Coal Mine. REE are normalized by upper continental crust (Taylor and McLennan [32]).

The lower REE abundance of the parting is attributed to its lithological composition, which is represented by 97% calcite (Table 4). According to previous studies, the content of REE in limestone is lower in comparison with other lithology, such as clay and oil shale [33,34]. The REE distribution pattern for the parting is characterized by an M-REE enrichment type and almost no Eu anomalies.

**Table 4.** Low temperature ash (LTA) yields of coal samples and mineral compositions (%) of LTAs, parting, roof and floor determined by XRD and Siroquant. "-"means below the detection limit of Siroquant analysis.

| Sample | LTA | Kaolinite | Illite | I/S | Marcasite | Pyrite | Calcite | Anatase | Boehmite | Diaspore | Brucite |
|--------|-----|-----------|--------|-----|-----------|--------|---------|---------|----------|----------|---------|
| SS-R   | -    | 54.5 | 39.2 | -    | -    | 5.1  | -    | 1.2 | -    | -    | -   |
| SS-1C  | 47.2 | 41.3 | -    | -    | 12.7 | 44.6 | -    | 0.9 | -    | -    | -   |
| SS-2C  | 22.9 | 55.2 | 17.9 | -    | -    | -    | 6.1  | 1.3 | 3.4  | 14.9 | 1.1 |
| SS-3P  | 73.8 | -    | -    | -    | -    | -    | 97.7 | -   | 2.1  | 0.1  | 0   |
| SS-4C  | 11.0 | 22.6 | 17.3 | 14.7 | -    | -    | 4.7  | 1.7 | 34.3 | 4.3  | 0.4 |
| SS-5C  | 31.9 | 50.8 | 9.1  | 14.6 | -    | -    | -    | 0.6 | 24.9 | -    | -   |
| SS-6C  | 23.7 | 60.5 | 6.7  | 1.7  | -    | -    | -    | 0.9 | 30.3 | -    | -   |
| SS-7C  | 20.7 | 51.8 | 6.1  | -    | -    | -    | -    | 0.1 | 34.2 | 7.6  | 0.3 |
| SS-8C  | 39.0 | 53.7 | 14.8 | -    | -    | -    | -    | 0.4 | 25.3 | 5.1  | 0.7 |
| SS-9C  | 35.1 | 63.6 | 14.7 | -    | -    | -    | -    | -   | 20.9 | -    | 0.4 |
| SS-F1  | -    | 83.6 | 6.3  | -    | -    | -    | -    | 0.7 | 9.4  | -    | -   |
| SS-F2  | -    | 78.5 | 7.7  | -    | -    | -    | -    | 1.0 | 12.9 | -    | -   |

In addition, the floor and roof samples have similar REE distribution patterns, with weak negative Eu anomalies, and L- and M-REE enrichment types.

REE in all the samples are characterized by a Gd-maximum. The distribution patterns of REE in Songshao coal are similar to that in sample C3-4c from the Xinde Mine in Eastern Yunnan, China [12]. Sample C3-4c is characterized by M-type REE spectra with a Gd-maximum (Figure 3); such patterns in coal are typical of a great deal of acid water, including high $pCO_2$-waters in coal basins [35,36].

### 4.3. Mineralogical Composition

#### 4.3.1. Mineral Phases

The proportions of each crystalline phase, identified from the X-ray diffractograms of the coal LTA (low temperature ash), parting, roof, and floor samples, are given in Table 4. The phases identified in the coal LTAs include kaolinite, illite, I/S (mixed layer of illite and smectite), marcasite, pyrite, calcite, anatase, boehmite, diaspore, and a trace amount of brucite. Additionally, stannite, fluorapatite, apatite, zircon, halotrichite, and some REE-bearing minerals, including xenotime, florencite, and silicorhabdophane, are also identified in sample SS-7C by SEM-EDS analysis.

Calcite is the dominant mineral in parting. Other minerals including boehmite and diaspore are also identified from parting LTA.

Minerals in the roof sample (SS-R) include kaolinite, illite, pyrite, and small proportions of anatase, quartz, and gypsum. The floors samples (SS-F1 and SS-F2) are composed of kaolinite, illite, anatase, and boehmite.

#### 4.3.2. Comparison between Mineralogical and Chemical Compositions

The chemical composition of the (high-temperature) coal ash calculated from the XRD and Siroquant analyses of each LTA or roof and floor samples are listed in Table 4. The two data sets, derived respectively from the XRD and the XRF data, have been compared and are presented as X–Y plots (Figure 4), with a diagonal line on each plot indicating where the points would fall if the estimates from the two different techniques were equal. The points for $SiO_2$, and $Al_2O_3$ in Figure 4 plot close to the equality line, suggesting that the XRD results are generally compatible with the chemical analysis data.

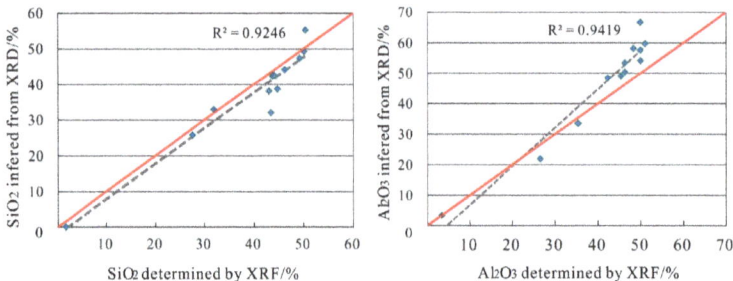

**Figure 4.** Comparison of observed normalized Si and Al oxide percentages from chemical analysis (*X*-axis) to oxide percentages for sample ash inferred from XRD analysis data (*Y*-axis). The diagonal line in each plot indicates equality.

#### 4.3.3. Modes of Mineral Occurrence

Clay Minerals

Clay minerals, including kaolinite, I/S (mixed-layer illite-smectite), illite, and smectite, are identified using XRD and SEM-EDS. They normally occur in the matrix of organic matter of the coal (Figure 5A). Kaolinite is the dominant mineral both in the coals and the roof and floor strata. In some cases, kaolinite distributes along the bedding (Figure 5B), and sometimes it occurs in detrital forms.

Other modes of occurrence of kaolinite include fracture- and cell-fillings (Figure 5C) of authigenic origin [37]. I/S, illite, and smectite distribute along the bedding planes (Figure 5B); occur as cell-fillings (Figure 5D,E); or distribute around boehmite (Figure 5F).

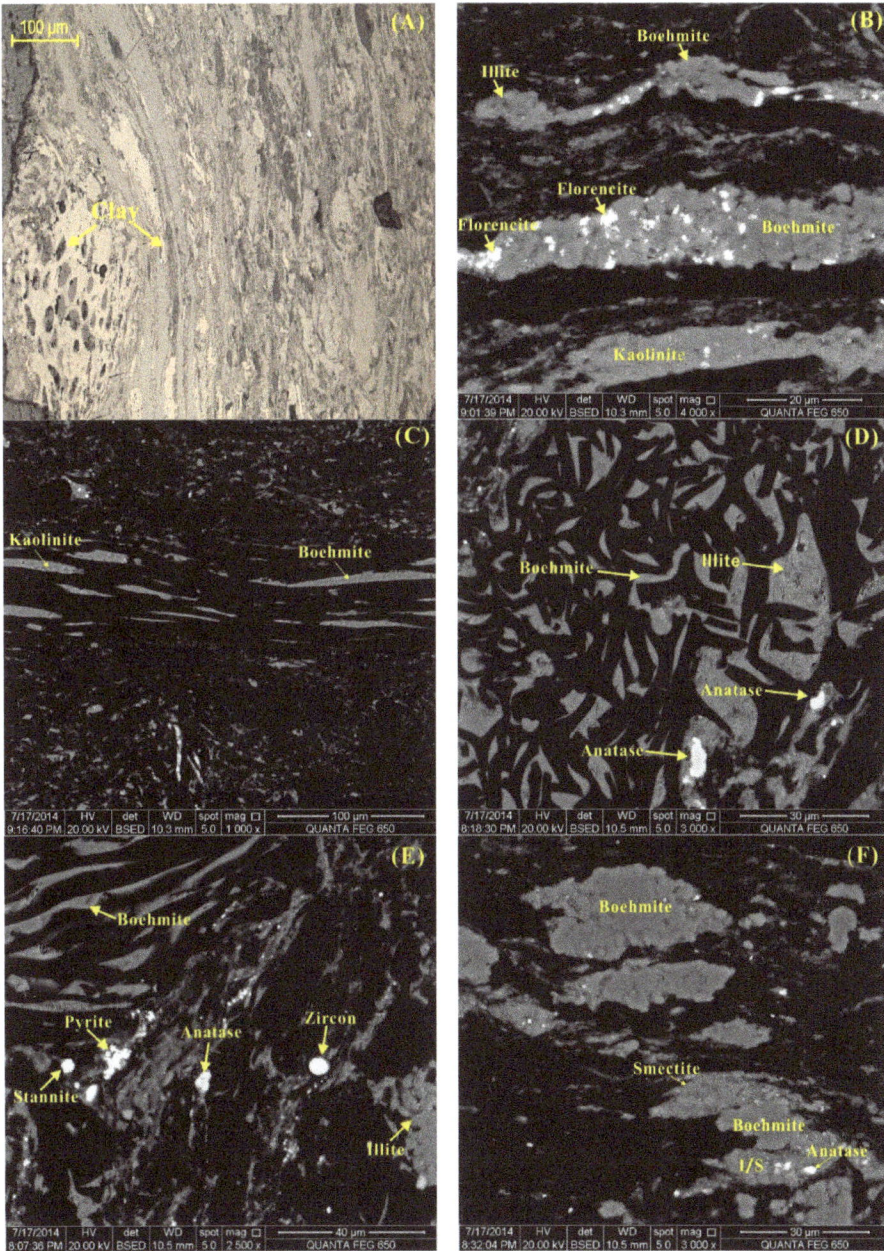

**Figure 5.** *Cont.*

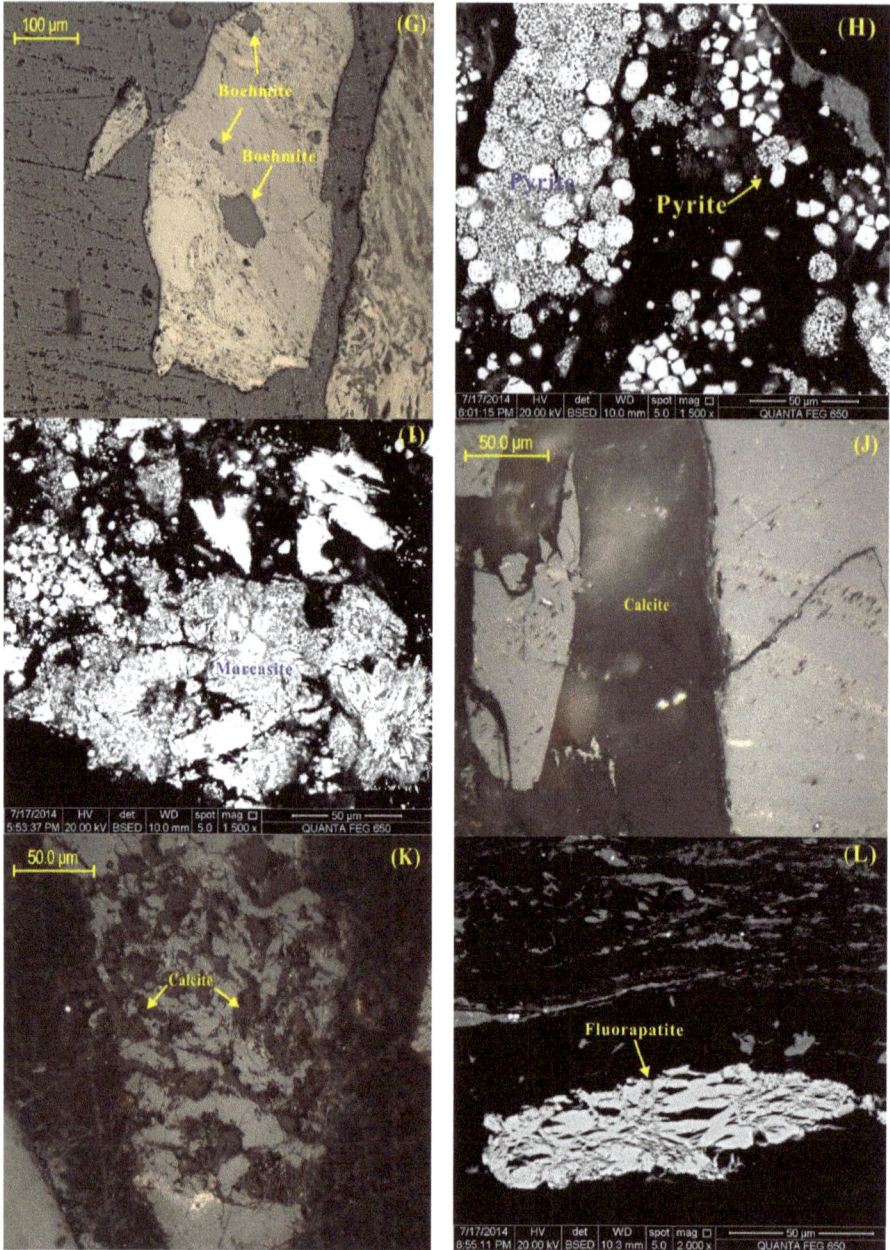

**Figure 5.** Minerals in the Songshao coal. (**A**) clay minerals in SS-9C; (**B**) kaolinite, illite, boehmite and florencite in SS-7C; (**C**) cell-filling kaolinite and boehmite in SS-7C; (**D**) cell-filling boehmite, illite, and anatase in SS-7C; (**E**) cell-filling boehmite, ditrital zircon, anatase, pyrite and stannite in SS-7C; (**F**) boehmite occurs small lumps and bead-like block embedded in the clay minerals and ditrital anatase in SS-7C; (**G**) boehmite in SS-9C; (**H**) framboidial pyrite in SS-1C; (**I**) plate, block and radial form marcasite in SS-1C; (**J**) vein-filling calcite in SS-3P; (**K**) cell-filling calcite in SS-3P; (**L**) fracture-filling florencite in SS-7C. B-F, H, I, L, SEM and back-scattered electron images; (**A**), (**G**), (**J**), (**K**), reflected light.

Boehmite, Disapore, Brucite

Oxyhydroxide minerals, such as boehmite, disapore, and brucite, were identified in Songshao coal. High boehmite content (mean 4.33%) was identified in Songshao coal, occurring as fracture-fillings (Figure 5B), small lumps (Figure 5F), cell-fillings (Figure 5D,E), and bead-like block embedded in the clay minerals (Figure 5F). The lumps show different shapes and variable sizes, from a few to one hundred micrometers. The surface of the boehmite in the coal is much smoother than that of the clay minerals (Figure 5F) and the relief of boehmite is higher under the polarizing microscope (Figure 5G). Minerals associated with the boehmite in the coal include goyazite, rutile, zircon, and Pb-bearing minerals (galena, clausthalite, and selenio-galena) [38]. Additionally, zircon is also identified in Songshao coal.

Pyrite and Marcasite

Pyrite is only detected in samples SS-R and SS-1C using the XRD technique. A trace amount of pyrite is also detected in sample SS-7C using SEM-EDS. Pyrite in sample SS-1C occurs as framboidal massive (Figure 5H) and euhedral crystal forms (Figure 5H), or distributes along the bedding (Figure 5E). Marcasite is only detected in sample SS-1C and coexists with pyrite, occurring as plate, block and radial forms (Figure 5I). This kind of marcasite may be of a syngenetic origin.

Calcite

Calcite is detected in samples SS-2C, SS-3P, and SS-4C using XRD and it accounts for 72.1% in the parting (SS-3P). Calcite occurs as vein- (Figure 5J), fracture- or cells-fillings (Figure 5K), indicating an epigenetic origin.

Antase and Zircon

Antase occurs as discrete particles in clay minerals (Figure 5E,F) or as cell-fillings (Figure 5D). Zircon occurs as detrital particles of terrigenous origin (Figure 5E).

The mode of occurrence of zircon indicates a detrital material of terrigenous origin. Zircon may be a pyroclastic mineral in some tonsteins [4,39] or it may occur as detrital material derived from the sediment source region.

Apatite, Halotrichite, and Stannite

Apatite in coal is classified as fluorapatite and zwiesellite. Fluorapatite is a common mineral in coal. It looks like bamboo leaves filling in the fracture (Figure 5L), indicating an epigenetic chemical deposition. A spot of halotrichite was also observed in sample SS-7C. The copiapite-group of minerals is one of the most common Fe-sulfates reported from many coals, such as coals from the Jaintia Hills coalfield in Meghalaya, India [40]. A trace of stannite was detected using SEM-EDS (Figure 5E).

REE-Bearing Minerals

REE-bearing minerals are at a low concentrations, below the detection limit of the XRD and Siroquant analysis, but were observed under SEM-EDS in sample SS-7C. These minerals are Xenotime, florencite, rhabdophane, and silicorhabdophane.

Phosphate minerals, such as apatite, monazite, and, in some cases, xenotime, zircon, and some clay minerals, are usually the carriers of thorium in coal [41]. In this study, thorium was not detected, but Dy was detected in xenotime.

Florencite occurs as fracture-fillings, coexisting with kaolinite (Figure 5B). However, the small size of these particles makes it difficult to obtain a representative EDS spectrum.

## 5. Factors Controlling Enrichment of Trace Elements and Minerals in Songshao Coal

It has been reported that seven factors control the enrichment of elements and minerals in Chinese coals: The sediment-source rocks, low-temperature hydrothermal fluids, marine environments, volcanic ash, magmatic fluids, submarine exhalation, and groundwater [3]. The first three factors are likely to be responsible for the geochemical and mineralogical anomalies of Songshao coal.

### 5.1. Input from Sediment-Source Region

As described above, Songshao coal is enriched in Al, Si, Ti, Li, Zr, Hf, Th, Sr, and V. Aluminum and Si are largely contained in clay minerals. Two modes of kaolinite occurrence in Songshao coal were identified: Detrital kaolinite of terrigenous origin and cell-filling kaolinite of authigenic origin. Anatase, which occurs as discrete particles in the clay mineral matrix, is the host of Ti and was probably derived from sediment-source sources. Lithium generally occurs in minerals, or can be absorbed by bauxite minerals, which can then be named as a Li deposit [42] if Li is highly concentrated. Moreover, Li may occur in silicate minerals, because it tends to be absorbed by clay minerals, which are formed during a weathering process [43]. It was reported that the Langdaisa Li-bearing claystone of the Permian Liangshan Formation contains 0.12%–0.74% Li [44,45]. Langdaisa is located in the Liuzhi District of Guizhou and Liuzhi is situated to the northeast of the Songshao Mine, indicating the same sediment source for the Langdaisa Li-bearing claystone and Songshao coal. Thorium is not easily altered by the processes of weathering and transportation, and, thus, it can serve as an effective indicator for deducing the source region. Thorium is generally absorbed by clay and more likely occurs in bauxite [43]. In addition, thorium can occur in apatite, xenotime, or zircon in coal [41]. In this study, minerals including apatite, xenotime, and zircon have been identified in sample SS-7C by SEM-EDX; however, thorium was not found to occur in these minerals. The correlation coefficients of Zr-Th and Zr-Hf are 0.814 and 0.974, respectively (Figure 6). Zircon is the carrier of Zr and also contains Th and Hf. The mode of zircon occurrence indicates a detrital material of terrigenous origin. Elements Hf, Zr, and Li show a similar variation tendency through the seam section, suggesting a same source (Figure 6).

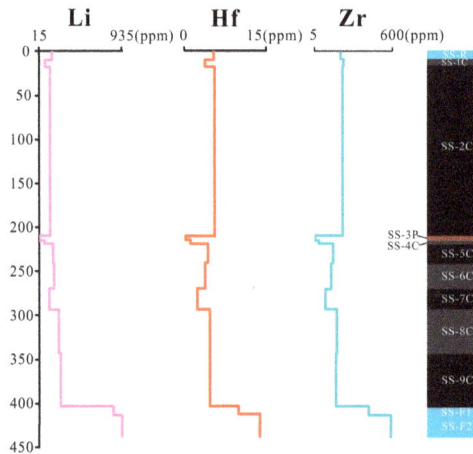

**Figure 6.** Variations of Li, Hf, and Zr through the seam section.

Seredin and Dai [31] suggested that the L-type distribution of REE in coal are attributed to terrigenous or tuffaceous origins during peat accumulation stage. In the first case, the REE may be transported as colloidal and ionic forms from uplift containing magmatic rocks enriched in light REE, such as granite and carbonatite [31]. The light REE enrichment type in Songshao coal indicates a terrigenous origin. A similar instance occurred in the No. 6 Coal seam in the Jungar deposit in Inner Mongolia, China [46]. REE are present in Songshao coal as REE-rich minerals, such as xenotime and florencite.

*5.2. Hydrothermal Fluid Influences*

Hydrothermal fluid is another critical factor responsible for the trace-element and mineral anomalies in the Songshao coal, parting, and roof and floor rocks. The hydrothermal fluid includes syngenetic hydrothermal solutions injected into the peat mire during peat accumulation. Mineralogical evidence of hydrothermal influence on Songshao coal includes:

(1) Pyrite and calcite occur as micro-veinlets (Figure 5H,J). Multi-generation pyrite occurs in sample SS-1C. Framboidal pyrite is of an early diagenetic origin [47]. It occurs as lenses and spheres and the aggregates are composed of microcrystalline paticles. The framboids are commonly coated by late-formed sulfides. Meanwhile, sample SS-1C is enriched in As, Se, and Hg. According to previous research [48–51], some trace elements, for instance, arsenic and mercury, are concentrated in the late stage of hydrothermal fluids. In some cases, arsenic was syngenetically derived from hydrothermal solutions [6]. Particularly, the enrichment of arsenic was associated with late-stage pyrite that coats early framboids and with microscale faulting. This phenomenon also occurred in the West Virginia coals from the Appalachian region, USA [52]. Therefore, hydrothermal fluid may be responsible for the enrichment of trace elements in sample SS-1C.

(2) Clay minerals, boehmite and calcite occur as cell-fillings (Figure 5B,C,K). Under favorable conditions, solutions that contained soluble Al and Si entered the coal basin, filled and then were deposited in the cells during peat accumulation.

(3) High boehmite content (mean 4.33%) in Songshao coal is not commonly observed; the content of boehmite in the coal present in this study is close to that (mean 6.1%) in the No. 6 Coal from the Jungar Coalfield [46] and from some mines of the Daqingshan Coalfield [53]. In addition, boehmite may be the result of low-temperature hydrothermal fluids, generally associated with zeolites [54]. However, zeolites have not been observed in the present study. The occurrence mode of boehmite in Songshao coal is similar to that in the No. 6 Coal in the Jungar Coalfield [46] and the Daqingshan Coalfield [53]. The boehmite in Jungar coals is the major carrier of Ga and Th, which are significantly enriched in the coal (44.8 µg/g Ga; 17.8 µg/g Th; on a whole coal basis). However, trace elements were not enriched with boehmite in Songshao coal. The difference in element abundance between boehmites in the coals from the different coalfields was probably due to different origins of boehmite. An explanation for high boehmite content is that in the humid–warm climatic conditions, rock in sediment source suffered from weathering and erosion, which can cause Al to be concentrated in weathering crust. Colloidal solution containing Al removed from the crust flowed into the peat mire, and boehmite was formed by compaction and dehydration of Al colloidal solution during peat accumulation and early diagenesis.

(4) The presence and mode of occurrence of xenotime, florencite, rhabdophane and silicorhabdophane in the coal present in this study indicate that these minerals were deposited from hydrothermal fluids at syngenetic or early diagenetic stages. Rhabdophane, florencite, and silicorhabdophane are the major carriers of light rare earth elements in the samples of the present study. These minerals in coal are generally derived from hydrothermal fluids, and they are also the major REE carriers in some other REE-rich coals as well [31,54–60].

(5) S positively correlates with Hg and Fe, with correlation coefficients $r_{S-Hg}$ = 0.85 and $r_{S-Fe}$ = 0.88 (Figure 7), respectively. Fluorine is slightly enriched in Songshao coal with CC = 1.88. Syngenetic hydrothermal solutions may be rich in Hg and F, and this could lead to enrichment of Hg in upper and lower portions of the coal seam and high F concentrations throughout the whole seam section. Hydrothermal fluids leading to enrichment of Hg and F has also been reported in some coals from other coalfields [61–64]. Fluorine-bearing fluorapatite was found to occur as fracture-fillings in sample SS-7C. Epigenetic hydrothermal fluids led to the enrichment of F, Fe, Se and S in the roof rock. Owing to the injection of epigenetic hydrothermal fluids into the coal seam, the concentrations of epithermal-associated elements (e.g., F, Se, S, and Hg) are highly elevated near the contact between the coal bench and roof strata.

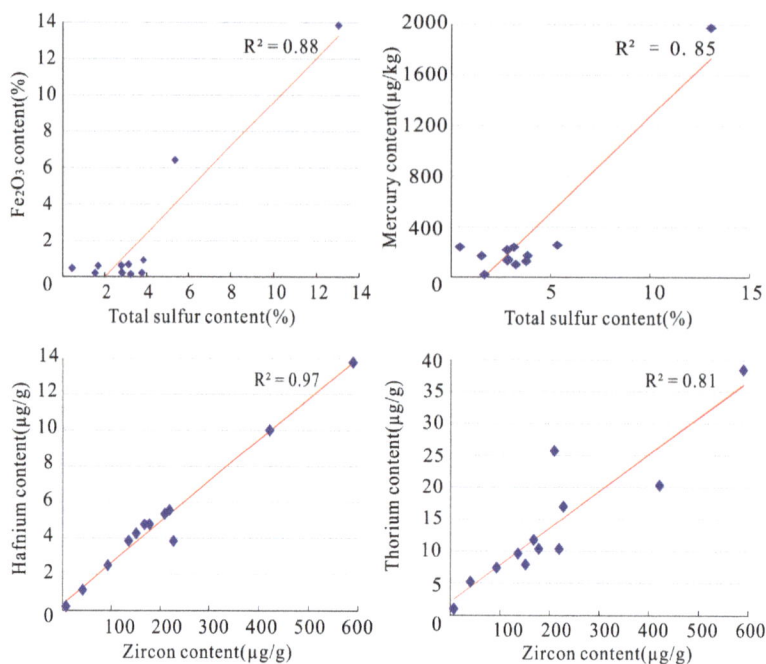

**Figure 7.** Correlation coefficients between two elements.

### 5.3. Marine Environments

Because Songshao is located in the area of arenaceous and argillaceous facies in litoral zone (Figure 8), the peat swamp of the coal had been subjected to a marine environment, as indicated by the following evidence:

(1) A number of researches showed that coals formed in marine-influenced environment are generally enriched in some elements, including S, B, V, Br, Rb, Sr, Mo, and U [65–69], because these elements are enriched in seawater in comparison with fresh water. Songshao coal, as described above, is enriched in V and Sr.

(2) Songshao coal is a high-sulfur coal with average total sulfur and organic sulfur contents 3.61% and 3.87%, respectively. Previous work showed that sulfate in seawater that flooded peat swamps is the major sources of sulfur for medium- and high-sulfur coals [70]. Some studies also indicated that submarine exhalation can also lead to sulfur enrichment in coal [71]. The elevated content of sulfur in Songshao coal is attributed to seawater that injected into the peat swamp during peat accumulation. Sulfates in the seawater were reduced by anaerobic sulfate-reducing bacteria. Various forms of sulfur, which decrease in interstitial water, react with iron ions and organic matters. Then, sulfur-bearing minerals (pyrite mostly) and organic sulfide formed [72].

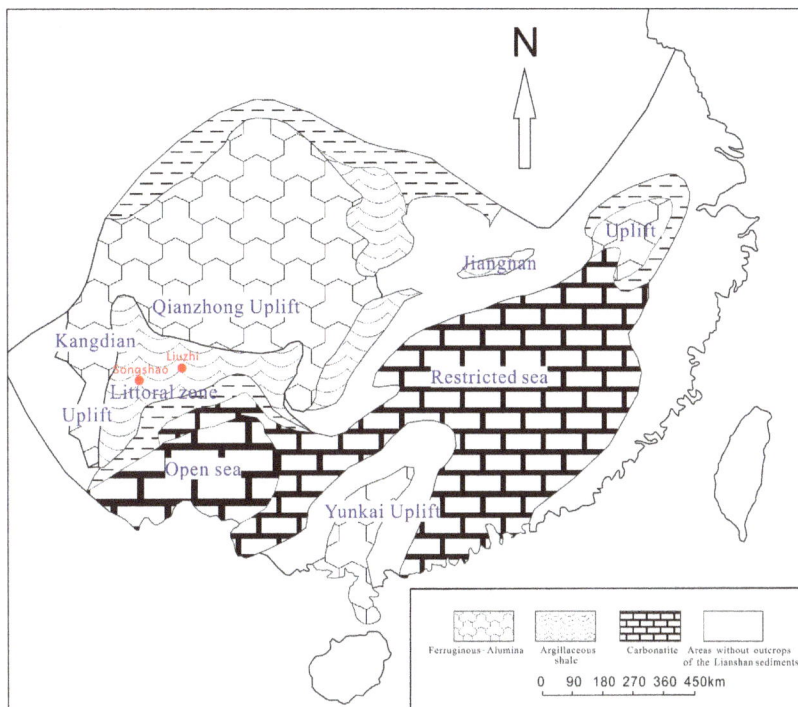

**Figure 8.** Sketch map of lithological facies and paleogeography in the study area during the Liangshan Stage. Modified from [73,74].

## 6. Conclusions

Songshao coal is medium- to high-volatile bituminous in rank, with high total and organic sulfur contents. Major-element oxides are mainly represented by $SiO_2$ and $Al_2O_3$. Trace elements, including Li, Zr, Hg, Sr, Nb, Sn, Hf, V, Cr, Se, Th, and some light REE, are enriched. The REE enrichment patterns in the coal seam are L-types (light REE enrich) and are characterized by negative Eu anomalies and a Gd-maximum. The mineral phases identified in the coal LTAs include kaolinite, illite, I/S (mixed layer illite-smectite), pyrite, calcite, anatase, boehmite, diaspore, and some REE-bearing minerals. Clay minerals occur in the matrix of organic matter. Boehmite in the coal occurs as fracture-fillings, small lumps, as cell-fillings, or as bead-like blocks embedded in the clay minerals. Pyrite occurs as framboidal forms; calcite occurs as fracture-fillings or cell-fillings. Other minerals occur as debris in the samples.

High total and organic sulfur contents and the enrichment Sr and V are attributed to a marine influence. Hydrothermal fluids are responsible for the high boehmite content (mean 4.33%) and its modes of occurrence; the presence and modes of REE-bearing minerals, pyrite and calcite; the enrichment of Hg and F in the coal, roof and floor strata rocks. Songshao coal and the roof and floor strata rocks are also enriched in $SiO_2$, $Al_2O_3$, Li, Zr, and light rare earth elements, which were derived from the sediment source.

**Acknowledgments:** This research was supported by the National Key Basic Research Development Program (973 Program, No. 2014CB238902). Special thanks are given to Shifeng Dai and Xiaolin Song for providing samples and constructive suggestion for this paper. Shifeng Dai also helped with SEM-EDX experiments. Peipei Wang, Qin Zhu, Shaohui, Jia and Jihua Sun helped the determination of major and trace elements.

**Conflicts of Interest:** The author declares no conflict of interest.

## References

1. Yunnan Coal Geology Prospecting Team No. 143. *Report on Coal Geological Exploration of Songshao Mine*; Yunnan Coal Geology Prospecting Team No. 143: Qujing, China, 1996.
2. Dai, S.; Wang, X.; Zhou, Y.; Hower, J.C.; Li, D.; Chen, W.; Zhu, X.; Zou, J. Chemical and mineralogical compositions of silicic, mafic, and alkali tonsteins in the late Permian coals from the Songzao Coalfield, Chongqing, Southwest China. *Chem. Geol.* **2011**, *282*, 29–44. [CrossRef]
3. Ren, D.; Zhao, F.; Dai, S.; Zhang, J.; Luo, K. *Geochemistry of Trace Elements in Coal*; Science Press: Beijing, China, 2006; pp. 40–59. (In Chinese)
4. Ward, C.R. Analysis and significance of mineral matter in coal seams. *Int. J. Coal Geol.* **2002**, *50*, 135–168. [CrossRef]
5. Wang, X.; Dai, S.; Ren, D.; Yang, J. Mineralogy and geochemistry of Al-hydroxide/oxyhydroxide mineral-bearing coals of Late Paleozoic age from the Weibei coalfield, southeastern Ordos Basin, North China. *Appl. Geochem.* **2011**, *26*, 1086–1096. [CrossRef]
6. Dai, S.; Wang, X.; Seredin, V.V.; Hower, J.C.; Ward, C.R.; O'Keefe, J.M.K.; Huang, W.; Li, T.; Li, X.; Liu, H.; et al. Petrology, mineralogy, and geochemistry of the Ge-rich coal from the Wulantuga Ge ore deposit, Inner Mongolia, China: New data and genetic implications. *Int. J. Coal Geol.* **2012**, *90*, 72–99. [CrossRef]
7. Dai, S.; Jiang, Y.; Ward, C.R.; Gu, L.; Seredin, V.V.; Liu, H.; Zhou, D.; Wang, X.; Sun, Y.; Zou, J.; et al. Mineralogical and geochemical compositions of the coal in the Guanbanwusu Mine, Inner Mongolia, China: further evidence for the existence of an Al (Ga and REE) ore deposit in the Jungar Coalfield. *Int. J. Coal Geol.* **2012**, *98*, 10–40. [CrossRef]
8. Dai, S.; Seredin, V.V.; Ward, C.R.; Jiang, J.; Hower, J.C.; Song, X.; Jiang, Y.; Wang, X.; Gornostaeva, T.; Li, X.; et al. Composition and modes of occurrence of minerals and elements in coal combustion products derived from high-Ge coals. *Int. J. Coal Geol.* **2014**, *121*, 79–97. [CrossRef]
9. Zhou, Y.; Bohor, B.F.; Ren, Y. Trace element geochemistry of altered volcanic ash layers (tonsteins) in Late Permian coal-bearing formations of eastern Yunnan and western Guizhou Provinces, China. *Int. J. Coal Geol.* **2000**, *44*, 305–324. [CrossRef]
10. Dai, S.; Tian, L.; Chou, C.-L.; Zhou, Y.; Zhang, M.; Zhao, L.; Wang, J.; Yang, Z.; Cao, H.; Ren, D. Mineralogical and compositional characteristics of Late Permian coals from an area of high lung cancer rate in Xuan Wei, Yunnan, China: Occurrence and origin of quartz and chamosite. *Int. J. Coal Geol.* **2008**, *76*, 318–327. [CrossRef]
11. Dai, S.; Zhou, Y.; Zhang, M.; Wang, X.; Wang, J.; Song, X.; Jiang, Y.; Luo, Y.; Song, Z.; Yang, Z.; et al. A new type of Nb (Ta)–Zr (Hf)–REE–Ga polymetallic deposit in the late Permian coal-bearing strata, eastern Yunnan, southwestern China: possible economic significance and genetic implications. *Int. J. Coal Geol.* **2010**, *83*, 55–63. [CrossRef]
12. Dai, S.; Li, T.; Seredin, V.V.; Ward, C.R.; Hower, J.C.; Zhou, Y.; Zhang, M.; Song, X.; Song, W.; Zhao, C. Origin of minerals and elements in the Late Permian coals, tonsteins, and host rocks of the Xinde Mine, Xuanwei, eastern Yunnan, China. *Int. J. Coal Geol.* **2014**, *121*, 53–78. [CrossRef]
13. Dai, S.; Zhang, W.; Ward, C.R.; Seredin, V.V.; Hower, J.C.; Li, X.; Song, W.; Wang, X.; Kang, H.; Zheng, L.; et al. Mineralogical and geochemical anomalies of late Permian coals from the Fusui Coalfield, Guangxi Province, southern China: Influences of terrigenous materials and hydrothermal fluids. *Int. J. Coal Geol.* **2013**, *105*, 60–84. [CrossRef]
14. Dai, S.; Zhang, W.; Seredin, V.V.; Ward, C.R.; Hower, J.C.; Song, W.; Wang, X.; Li, X.; Zhao, L.; Kang, H.; et al. Factors controlling geochemical and mineralogical compositions of coals preserved within marine carbonate successions: A case study from the Heshan Coalfield, southern China. *Int. J. Coal Geol.* **2013**, *109*, 77–100. [CrossRef]
15. Dai, S.; Luo, Y.; Seredin, V.V.; Ward, C.R.; Hower, J.C.; Zhao, L.; Liu, S.; Zhao, C.; Tian, H.; Zou, J. Revisiting the late Permian coal from the Huayingshan, Sichuan, southwestern China: Enrichment and occurrence modes of minerals and trace elements. *Int. J. Coal Geol.* **2014**, *122*, 110–128. [CrossRef]
16. Li, B.; Zhuang, X.; Li, J.; Querol, X.; Font, O.; Moreno, N. Geological controls on mineralogy and geochemistry of the Late Permian coals in the Liulong Mine of the Liuzhi Coalfield, Guizhou Province, Southwest China. *Int. J. Coal Geol.* **2016**, *154*, 1–15. [CrossRef]

17. Standardization Administration of China; General Administration of Quality Supervision, Inspection and Quarantine of the China. *Chinese National Standard GB/T 212-2008; Proximate Analysis of Coal*; Standand Press of China: Beijing, China, 2008; pp. 3–5. (In Chinese)

18. Standardization Administration of China; General Administration of Quality Supervision, Inspection and Quarantine of the China. *Chinese National Standard GB/T 214-2007; The Determination Methods of Total Sulfur in Coal*; Standand Press of China: Beijing, China, 2007; pp. 1–5. (In Chinese)

19. Standardization Administration of China; General Administration of Quality Supervision, Inspection and Quarantine of the China. *Chinese National Standard GB/T 215-2003; The Determination Methods of Form Sulfur in Coal*; Standand Press of China: Beijing, China, 2003; pp. 1–5. (In Chinese)

20. International Organization for Standardization. *Methods for the Petrographic Analysis of Bituminous Coal and Anthracite-Part 5: Method of Determining Microscopically the Reflectance of Vitrinite (ISO 7404-5)*; International Organization for Standardization (ISO): Geneva, Switzerland, 2009.

21. Dai, S.; Hower, J.C.; Ward, C.R.; Guo, W.; Song, H.; O'Keefe, J.M.K.; Xie, P.; Hood, M.M.; Yan, X. Elements and phosphorus minerals in the middle Jurassic inertinite-rich coals of the Muli Coalfield on the Tibetan Plateau. *Int. J. Coal Geol.* **2015**, *144*, 23–47. [CrossRef]

22. Standardization Administration of China; General Administration of Quality Supervision, Inspection and Quarantine of the China. *Chinese National Standard GB/T 4633-1997; Determination of Fluorine in Coal*; Standard Press of China: Beijing, China, 1997; pp. 1–5. (In Chinese)

23. ASTM International. *Standard Classification of Coals by Rank*; ASTM D388-12; ASTM International: West Conshohocken, PA, USA, 2012.

24. Standardization Administration of China; General Administration of Quality Supervision, Inspection and Quarantine of the China. *Chinese National Standard GB/T 15224.1-2004; Classification for Quality of Coal-Part 1: Ash*; Standard Press of China: Beijing, China, 2004; pp. 1–2. (In Chinese)

25. Standardization Administration of China; General Administration of Quality Supervision, Inspection and Quarantine of the China. *Chinese National Standard GB/T 15224.2-2004; Classification for Quality of Coal-Part 2: Sulfur Content*; Standard Press of China: Beijing, China, 2004; pp. 1–2. (In Chinese)

26. Dai, S.; Ren, D.; Chou, C.-L.; Finkelman, R.B.; Seredin, V.V.; Zhou, Y. Geochemistry of trace elements in Chinese coals: A review of abundances, genetic types, impacts on human health, and industrial utilization. *Int. J. Coal Geol.* **2012**, *94*, 3–21. [CrossRef]

27. Ketris, M.P.; Yudovich, Y.E. Estimations of Clarkes for Carbonaceous biolithes: World averages for trace element contents in black shales and coals. *Int. J. Coal Geol.* **2009**, *78*, 135–148. [CrossRef]

28. Dai, S.; Seredin, V.V.; Ward, C.R.; Hower, J.C.; Xing, Y.; Zhang, W.; Song, W.; Wang, P. Enrichment of U–Se–Mo–Re–V in coals preserved within marine carbonate successions: Geochemical and mineralogical data from the Late Permian Guiding Coalfield, Guizhou, China. *Miner. Deposita* **2015**, *50*, 159–186. [CrossRef]

29. Dai, S.; Li, D.; Chou, C.-L.; Zhao, L.; Zhang, Y.; Ren, D.; Ma, Y.; Sun, Y. Mineralogy and geochemistry of boehmite-rich coals: new insights from the Haerwusu Surface Mine, Jungar Coalfield, Inner Mongolia, China. *Int. J. Coal Geol.* **2008**, *74*, 185–202. [CrossRef]

30. Dai, S.; Chekryzhov, I.Y.; Seredin, V.V.; Nechaev, V.P.; Graham, I.T.; Hower, J.C.; Ward, C.R.; Ren, D.; Wang, W. Metalliferous coal deposits in East Asia (Primorye of Russia and South China): A review of geodynamic controls and styles of mineralization. *Gondwana Res.* **2016**, *29*, 60–82. [CrossRef]

31. Seredin, V.V.; Dai, S. Coal deposits as potential alternative sources for lanthanides and yttrium. *Int. J. Coal Geol.* **2012**, *94*, 67–93. [CrossRef]

32. Taylor, S.R.; McLennan, S.M. *The Continental Crust: Its Composition and Evolution*; Blackwell Scientific Publications: Palo Alto, CA, USA, 1985; p. 312.

33. Zamanian, H.; Ahmadnejad, F.; Zarasvandi, A. Mineralogical and geochemical investigations of the Mombi bauxite deposit, Zagros Mountains, Iran. *Chem. Erde Geochem.* **2016**, *76*, 13–37. [CrossRef]

34. Fu, X.; Wang, J.; Zeng, Y.; Tan, F.; Feng, X. REE geochemistry of marine oil shale from the Changshe Mountain area, northern Tibet, China. *Int. J. Coal Geol.* **2010**, *81*, 191–199. [CrossRef]

35. Shand, P.; Johannesson, K.H.; Chudaev, O.; Chudaeva, V.; Edmunds, W.M. Rare Earth Element Contents of High pCO$_2$ Groundwaters of Primorye, Russia: Mineral Stability and Complexation Controls. In *Rare Earth Elements in Groundwater Flow Systems*; Johannesson, K.H., Ed.; Springer: Dordrecht, The Netherlands, 2005; Volume 7, pp. 161–186.

36.  Dai, S.; Graham, I.T.; Ward, C.R. A review of anomalous rare earth elements and yttrium in coal. *Int. J. Coal Geol.* **2016**, *159*, 82–95. [CrossRef]

37.  Ward, C.R. Minerals in bituminous coals of the Sydney Basin (Australia) and the Illinois Basin (USA). *Int. J. Coal Geol.* **1989**, *13*, 455–479. [CrossRef]

38.  Hrinko, V. Technological, chemical, and mineralogical characteristics of bauxites and country rocks near Drienovec. *Miner. Slovaca* **1986**, *18*, 551–555.

39.  Spears, D.A. The origin of tonsteins, an overview, and links with seatearths, fireclays and fragmental clay rocks. *Int. J. Coal Geol.* **2012**, *94*, 22–31. [CrossRef]

40.  Sahoo, P.K.; Tripathy, S.; Panigrahi, M.K.; Equeenuddin, S.M. Geochemical characterization of coal and waste rocks from a high sulfur bearing coalfield, India: Implication for acid and metal generation. *J. Geochem. Explor.* **2014**, *145*, 135–147. [CrossRef]

41.  Finkelman, R.B. Environmental Aspects of Trace Elements in Coal.   In *Modes of Occurrence of Environmentally-Sensitive Trace Elements in Coal*; Swaine, D.J., Goodarzi, F., Eds.; Springer: Dordrecht, The Netherlands, 1995; Volume 2, pp. 24–50.

42.  Chen, P.; Chai, D. *Sedimentary Geochemistry of Carboniferous Bauxite Deposits in Shanxi Massif*; Shanxi Science and Technology Press: Shanxi, China, 1997; pp. 1–194. (In Chinese)

43.  Liu, Y.; Cao, L.; Li, Z.; Wang, H.; Chu, T.; Zhang, J. *Element Geochemistry*; Geological publishing press of Beijing: Beijing, China, 1984; pp. 125–136. (In Chinese)

44.  Qian, D.D. *History of China's Deposit Discovery, Guizhou Volume*; Geological Publishing House: Beijing, China, 1996; pp. 177–182. (In Chinese)

45.  Wang, D.; Li, P.; Qu, W.; Yin, L.; Zhao, Z.; Lei, Z.; Wen, S. Discovery and preliminary study of the high tungsten and lithium contents in the Dazhuyuan bauxite deposit, Guizhou, China. *Sci. China Earth Sci.* **2013**, *56*, 145–152. [CrossRef]

46.  Dai, S.; Ren, D.; Chou, C.-L.; Li, S.; Jiang, Y. Mineralogy and geochemistry of the No. 6 coal (Pennsylvanian) in the Junger coalfield, Ordos basin, China. *Int. J. Coal Geol.* **2006**, *66*, 253–270. [CrossRef]

47.  Wilkin, R.T.; Barnes, H.L. Formation processes of framboidal pyrite. *Geochim. Cosmochim. Acta* **1997**, *61*, 323–339. [CrossRef]

48.  Dai, S.; Hou, X.; Ren, D.; Tang, Y. Surface analysis of pyrite in the No. 9 coal seam, Wuda Coalfield, Inner Mongolia, China, using high-resolution time-of-flight secondary ion mass-spectrometry. *Int. J. Coal Geol.* **2003**, *55*, 139–150. [CrossRef]

49.  Diehl, S.F.; Goldhaber, M.B.; Hatch, J.R. Modes of occurrence of mercury and other trace elements in coals from the warrior field, Black Warrior Basin, Northwestern Alabama. *Int. J. Coal Geol.* **2004**, *59*, 193–208. [CrossRef]

50.  Zhang, J.; Ren, D.; Zheng, C.; Zeng, R.; Chou, C.-L.; Liu, J. Trace element abundances in major minerals of Late Permian coals from southwestern Guizhou province, China. *Int. J. Coal Geol.* **2002**, *53*, 55–64. [CrossRef]

51.  Dai, S.; Yang, J.; Ward, C.R.; Hower, J.C.; Liu, H.; Garrison, T.M.; French, D.; O'Keefe, J.M.K. Geochemical and mineralogical evidence for a coal-hosted uranium deposit in the Yili Basin, Xinjiang, northwestern China. *Ore Geol. Rev.* **2015**, *70*, 1–30. [CrossRef]

52.  Diehl, S.F.; Goldhaber, M.B.; Koenig, A.E.; Lowers, H.A.; Ruppert, L.F. Distribution of arsenic, selenium, and other trace elements in high pyrite Appalachian coals: Evidence for multiple episodes of pyrite formation. *Int. J. Coal Geol.* **2012**, *94*, 238–249. [CrossRef]

53.  Dai, S.; Zou, J.; Jiang, Y.; Ward, C.R.; Wang, X.; Li, T.; Xue, W.; Liu, S.; Tian, H.; Sun, X.; et al. Mineralogical and geochemical compositions of the Pennsylvanian coal in the Adaohai Mine, Daqingshan Coalfield, Inner Mongolia, China: Modes of occurrence and origin of diaspore, gorceixite, and ammonian illite. *Int. J. Coal Geol.* **2012**, *94*, 250–270. [CrossRef]

54.  Dai, S.; Ren, D.; Li, S.; Chou, C.-L. A discovery of extremely-enriched boehmite from coal in the Junger coalfield, the Northeastern Ordos Basin. *Acta Geol. Sin.* **2006**, *80*, 294–300. (In Chinese)

55.  Dai, S.; Yan, X.; Ward, C.R.; Hower, J.C.; Zhao, L.; Wang, X.; Zhao, L.; Ren, D.; Finkelman, R.B. Valuable elements in Chinese coals: A review. *Int. Geol. Rev.* **2016**. [CrossRef]

56.  Hower, J.C.; Eble, C.F.; Dai, S.; Belkin, H.E. Distribution of rare earth elements in eastern Kentucky coals: Indicators of multiple modes of enrichment? *Int. J. Coal Geol.* **2016**, *160–161*, 73–81. [CrossRef]

57. Zhao, L.; Dai, S.; Graham, I.; Wang, P. Clay mineralogy of coal-hosted Nb-Zr-REE-Ga mineralized beds from Late Permian strata, eastern Yunnan, SW China: Implications for palaeotemperature and origin of the micro-quartz. *Minerals* **2016**, *6*, 45. [CrossRef]

58. Johnston, M.N.; Hower, J.C.; Dai, S.; Wang, P.; Xie, P.; Liu, J. Petrology and Geochemistry of the Harlan, Kellioka, and Darby Coals from the Louellen 7.5-Minute Quadrangle, Harlan County, Kentucky. *Minerals* **2015**, *5*, 894–918. [CrossRef]

59. Dai, S.; Liu, J.; Ward, C.R.; Hower, J.C.; French, D.; Jia, S.; Hood, M.M.; Garrison, T.M. Mineralogical and geochemical compositions of Late Permian coals and host rocks from the Guxu Coalfield, Sichuan Province, China, with emphasis on enrichment of rare metals. *Int. J. Coal Geol.* **2015**. [CrossRef]

60. Liu, J.; Yang, Z.; Yan, X.; Ji, D.; Yang, Y.; Hu, L. Modes of occurrence of highly-elevated trace elements in superhigh-organic-sulfur coals. *Fuel* **2015**, *156*, 190–197. [CrossRef]

61. Dai, S.; Wang, P.; Ward, C.R.; Tang, Y.; Song, X.; Jiang, J.; Hower, J.C.; Li, T.; Seredin, V.V.; Wagner, N.J.; et al. Elemental and mineralogical anomalies in the coal-hosted Ge ore deposit of Lincang, Yunnan, southwestern China: Key role of $N_2$-$CO_2$-mixed hydrothermal solutions. *Int. J. Coal Geol.* **2015**, *152*, 19–46. [CrossRef]

62. Dai, S.; Liu, J.; Ward, C.R.; Hower, J.C.; Xie, P.; Jiang, Y.; Hood, M.M.; O'Keefe, J.M.K.; Song, H. Petrological, geochemical, and mineralogical compositions of the low-Ge coals from the Shengli Coalfield, China: A comparative study with Ge-rich coals and a formation model for coal-hosted Ge ore deposit. *Ore Geol. Rev.* **2015**, *71*, 318–349. [CrossRef]

63. Dai, S.; Li, T.; Jiang, Y.; Ward, C.R.; Hower, J.C.; Sun, J.; Liu, J.; Song, H.; Wei, P.; Li, Q.; et al. Mineralogical and geochemical compositions of the Pennsylvanian coal in the Hailiushu Mine, Daqingshan Coalfield, Inner Mongolia, China: Implications of sediment-source region and acid hydrothermal solutions. *Int. J. Coal Geol.* **2015**, *137*, 92–110. [CrossRef]

64. Eskenazy, G.; Dai, S.; Li, X. Fluorine in Bulgarian coals. *Int. J. Coal Geol.* **2013**, *105*, 16–23.

65. Hower, J.C.; Eble, C.F.; O'Keefe, J.M.K.; Dai, S.; Wang, P.; Xie, P.; Liu, J.; Ward, C.R.; French, D. Petrology, Palynology, and Geochemistry of Gray Hawk Coal (Early Pennsylvanian, Langsettian) in Eastern Kentucky, USA. *Minerals* **2015**, *5*, 592–622. [CrossRef]

66. Dai, S.; Ren, D.; Tang, Y.; Shao, L.; Li, S. Distribution, isotopic variation and origin of sulfur in coals in the Wuda coalfield, Inner Mongolia, China. *Int. J. Coal Geol.* **2002**, *51*, 237–250. [CrossRef]

67. O'Keefe, J.M.K.; Bechtel, A.; Christanis, K.; Dai, S.; DiMichele, W.A.; Eble, C.F.; Esterle, J.S.; Mastalerz, M.; Raymond, A.L.; Valentim, B.V.; et al. On the fundamental difference between coal rank and coal type. *Int. J. Coal Geol.* **2013**, *118*, 58–87. [CrossRef]

68. Goodarzi, F.; Swaine, D.J. *Paleoenvironmental and Environmental Implications of the Boron Content of Coals*; Geological Survey of Canada: Calgary, AB, Canada, 1994; Volume 471, p. 82.

69. Shao, L.; Jones, T.; Gayer, R.; Dai, S.; Li, S.; Jiang, Y.; Zhang, P. Petrology and geochemistry of the high-sulphur coals from the Upper Permian carbonate coal measures in the Heshan Coalfield, southern China. *Int. J. Coal Geol.* **2003**, *55*, 1–26. [CrossRef]

70. Dai, S.; Xie, P.; Jia, S.; Ward, C.R.; Hower, J.C.; Yan, X.; French, D. Enrichment of U-Re-V-Cr-Se and rare earth elements in the Late Permian coals of the Moxinpo Coalfield, Chongqing, China: Genetic implications from geochemical and mineralogical data. *Ore Geol. Rev.* **2016**. [CrossRef]

71. Dai, S.; Ren, D.; Zhou, Y.; Chou, C.-L.; Wang, X.; Zhao, L.; Zhu, X. Mineralogy and geochemistry of a superhigh-organic-sulfur coal, Yanshan Coalfield, Yunnan, China: Evidence for a volcanic ash component and influence by submarine exhalation. *Chem. Geol.* **2008**, *255*, 182–194. [CrossRef]

72. Chou, C.-L. Sulfur in coals: A review of geochemistry and origins. *Int. J. Coal Geol.* **2012**, *100*, 1–13. [CrossRef]

73. Jin, Y.; Fang, R. Early Permian brachiopods from the Kuangshan Formation in Luliang County, Yunnan, with notes on paleogeography of South China during the Liangshan Stage. *Acta Palaeontol. Sin.* **1985**, *24*, 216–228.

74. Zhao, L.; Dai, S.; Graham, I.T.; Li, X.; Liu, H.; Song, X.; Hower, J.C.; Zhou, Y. Cryptic sediment-hosted critical element mineralization from eastern Yunnan Province, southwestern China: Mineralogy, geochemistry, relationship to Emeishan alkaline magmatism and possible origin. *Ore Geol. Rev.* **2016**. [CrossRef]

*minerals*

MDPI

*Article*

# Origin of Minerals and Elements in the Late Permian Coal Seams of the Shiping Mine, Sichuan, Southwestern China

Yangbing Luo * and Mianping Zheng

MLR Key Laboratory of Saline Lake Resources and Environments, Institute of Mineral Resources, Chinese Academy of Geological Sciences (CAGS), Beijing 100037, China; mpzheng@126.com
* Correspondence: luoyangbing0319@gmail.com; Tel.: +86-10-6899-9837

Academic Editor: Thomas N. Kerestedjian
Received: 27 January 2016; Accepted: 13 July 2016; Published: 19 July 2016

**Abstract:** Volcanic layers in coal seams in southwestern China coalfields have received much attention given their significance in coal geology studies and their potential economic value. In this study, the mineralogical and geochemical compositions of C19 and C25 coal seams were examined, and the following findings were obtained. (1) Clay minerals in sample C19-r are argillized, and sedimentary layering is not observed. The acicular idiomorphic crystals of apatite and the phenocrysts of Ti-augite coexisting with magnetite in roof sample C19-r are common minerals in basaltic rock. The rare earth elements (REE) distribution pattern of C19-r, which is characterized by positive Eu anomalies and M-REE enrichment, is the same as that of high-Ti basalt. The concentrations of Ti, V, Co, Cr, Ni, Cu, Zn, Nb, Ta, Zr, and Hf in C19-r are closer to those of high-Ti basalt. In conclusion, roof sample C19-r consists of tuffaceous clay, probably with a high-Ti mafic magma source. (2) The geochemical characteristics of the C25 coals are same as those reported for coal affected by alkali volcanic ash, enrichment in Nb, Ta, Zr, Hf, and REE, causing the C25 minable coal seams to have higher potential value. Such a vertical study of coals and host rocks could provide more information for coal-forming depositional environment analysis, for identification of volcanic eruption time and magma intrusion, and for facilitating stratigraphic subdivision and correlation.

**Keywords:** southeastern Sichuan; Late Permian; Shiping mine; volcanic layers in coal seams; high-Ti basalt

---

## 1. Introduction

Volcanic ash ejected from volcanoes falls into peat swamps and subsequently can form a thin and stable stratum in coal seams, which is usually "tonstein" and in some cases, occurs as roof and floor strata of coal seams [1,2]. These volcanic layers in coal seams get much attention as they can help to identify times of volcanic eruption and magma intrusion in this area, assess the coal-forming environments, and facilitate stratigraphic subdivision and correlation [1–8]. Moreover, some volcanic layers in the coal seams may contain valuable trace elements (such as rare earth elements, Nb, Ta, Zr, Hf, and Ga) that could increase the potential value of the coal [4,5]. Consequently, host rocks (tonsteins, floor and roof) in coal seams related to volcanic ash are of great importance in coal geology [2,3,6,7].

Volcanic activity frequently occurred in the Late Permian age in southwestern China [1,2,9–13], and volcanic ashes deposited in the peat-forming environment have been found in some coal seams in this area [1–5]. These volcanic layers found in southwest China can be divided into four types: felsic [2,14], alkali [2,4,5,15,16], mafic [1,3], and dacitic [3]. It was reported that felsic and dacitic volcanic layers are common in Late Permian coal seams in southwestern China [1,2,6,7]. Alkali volcanic layer in the coal seams in southwestern China have attracted much attention given that coal seams affected by alkali volcanic ash are enriched in Nb, Ta, Zr, Hf, and REE [2,4,5,15,16]. However, because mafic eruptions generally do not form tuffs, mafic volcanic layers occur rarely in coal-bearing strata around the world and only a few have been found in the Late Permian coals from southwestern China [1,3].

More attention has been paid to tonsteins in coal seams [5,16]; however, research on the roof and floor strata with a volcanic source was scarcely reported [3]. In this paper, geochemical and mineralogical characteristics of the roof sample in the C19 coal seam with a high-Ti mafic magma source were investigated. Variation in the element geochemistry and mineralogy of the C19 and C25 coals and floor sample in Late Permian from the Shiping mine, Sichuan, southwestern China was also described, with an emphasis on elements and minerals of volcanic origin in these coals and hosts rocks. Such a vertical study of coals and host rocks could provide more information for coal-forming depositional environments analysis, for identification of volcanic eruption time and magma intrusion, and for facilitating stratigraphic subdivision and correlation.

## 2. Materials and Methods

The Shiping mine is located in the southeastern part of Sichuan Province, southwestern China (Figure 1).

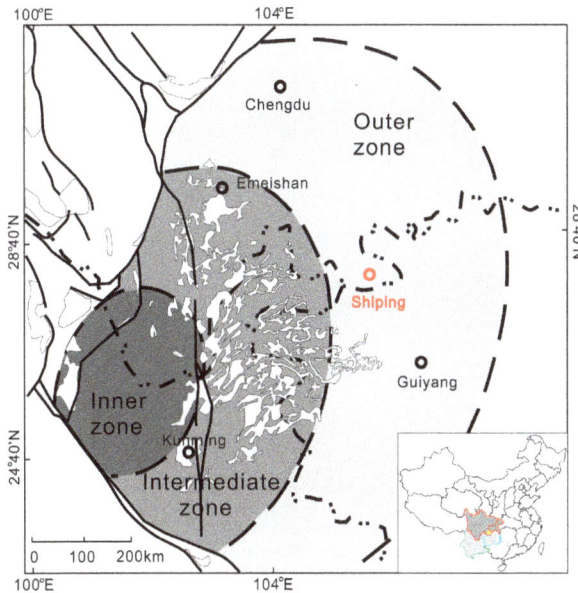

**Figure 1.** Distribution of Late Permian Emeishan basalts in southwestern China [13].

The sedimentary sequences in the Shiping mine include the Quaternary, Upper Triassic Xujiahe Formation, Middle Triassic Leikoupo Formation, Lower Triassic Jialingjiang and Feixianguan Formations, Upper Permian Changxing and Longtan Formations, Middle Permian Maokou and Xixia Formation, Lower Permian Liangshan Formation, and Silurian.

The coal-bearing sequence of the Shiping mine is the Late Permian Longtan Formation, which is composed mainly of mudstone, siltstone, sandstone, flint-bearing limestone, muddy sandstone, claystone, and six coal seams (Figure 2).

The Longtan Formation contains two major minable coal seams, namely the C19 and the C25. The C25 coal is the lowermost coal seam in the Late Permian strata of southwestern China.

The thickness of the C25 coal seam is 0.80–1.30 m. The C19 coal, with a thickness of 1.50–2.00 m, is the lower-middle coal seam in the Late Permian strata of southwestern China.

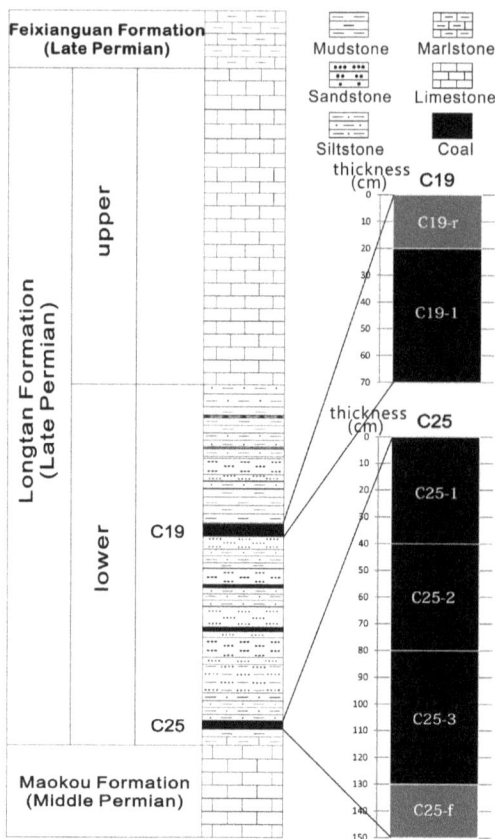

**Figure 2.** The sedimentary sequences of the Shiping mine.

## 3. Samples and Analytical Procedures

Six samples, including four coal bench samples (C19-1, C25-1, C25-2, and C25-3), one roof sample (C19-r), and one floor sample (C25-f), were taken from C19 and C25 coal seams mined at the coal working face of the Shiping mine (Figure 2). The identifications of all samples were shown in Figure 2. Each sample was cut over an area 10 cm wide and 10 cm deep. All samples were stored immediately in plastic bags to minimize contamination and oxidation.

Proximate analysis was conducted with ASTM Standards D3173-11 [17], D3174-11 [18], and D3175-11 [19]. Total sulfur and forms of sulfur were determined under ASTM Standards D3177-02 [20] and D2492-02 [21], respectively. Petrographic examination of the coals was performed under optical microscope following ASTM Standard D2797/D2797M-11a [22]. Mean random reflectance of vitrinite (percent $R_{o,ran}$) was determined by using a Leica DM-4500P microscope (Leica Camera AG, Wetzlar, Germany). Maceral constituents were identified under white-light reflectance oil immersion microscopy.

A field emission-scanning electron microscope (FE-SEM, FEI Quanta™ 650 FEG, FEI, Hillsboro, OR, USA), in conjunction with an EDAX energy-dispersive X-ray spectrometer (Genesis Apex 4, EDAX Inc., Mahwah, NJ, USA), was used to study morphology and microstructure and also to determine the distribution of some elements in the coal and rock samples. Low-temperature (oxygen-plasma) ashing (LTA) was performed, using an Emitech K1050X plasma asher (Quorum Inc., Lewes, UK), to remove organic matter in the coal prior to XRD analysis. The residues of this process were then analyzed by X-ray diffraction (XRD) using a D/max-2500/PC powder diffractometer (Rigaku, Tokyo, Japan) with Ni-filtered Cu-Kα radiation and a scintillation detector. The XRD pattern was recorded over a 2θ interval of 2.6°–70°, with a step size of 0.01°.

Concentrations of major element oxides in the samples (on ash basis; ashing temperature of 815 °C) were obtained by X-ray fluorescence (XRF) spectrometry. Mercury was determined by a Milestone DMA-80 Hg analyzer (Milestone, Sorisole, Italy). Fluorine analysis was conducted using ASTM Standard D5987-96 [23]. Inductively coupled plasma mass spectrometry (Thermo Fisher, Edmonton, AB, Canada, X series II ICP-MS) was used to determine the trace elements in the samples. All samples were digested using an UltraClave Microwave High Pressure Reactor (Milestone Inc., Shelton, CT, USA). Details for these coal-related sample digestion and ICP-MS analysis techniques are given by Dai et al. [1]. Arsenic and selenium were determined by ICP-MS, using collision cell technology (CCT), in order to avoid disturbance of polyatomic ions [24].

## 4. Results

### 4.1. Coal Chemistry and Coal Petrology

The vitrinite random reflectance ($R_{o,ran}$ average 2.21%) and the weighted average volatile matter ($V_{daf}$ average 15.98%) of the coal bench samples (Table 1) indicate a bituminous coal according to the ASTM classification D388-12, 2012 [25]. The total sulfur content varies considerably between the C19 (0.59%) and C25 (average 3.52%) coals, and the coals can be classified as low-sulfur coal and high-sulfur coal, respectively, according to Chinese standard GB 15224.1-2004 [26].

**Table 1.** Bench thickness (cm), proximate analysis (%), vitrinite random reflectance (%), and gross calorific values (MJ/kg) of coal benches from the Shiping mine.

| Samples | Thickness | $M_{ad}$ | $A_d$ | $V_{daf}$ | $S_{t,d}$ | $S_{s,d}$ | $S_{p,d}$ | $S_{o,d}$ | $R_{o,ran}$ | $Q_{gr,d}$ |
|---|---|---|---|---|---|---|---|---|---|---|
| C19-1 | 50 | 1.75 | 17.42 | 9.14 | 0.59 | nd | nd | nd | 2.30 | 29.35 |
| C25-1 | 40 | 2.70 | 16.52 | 13.90 | 6.53 | 2.11 | 3.58 | 0.84 | 2.21 | 28.13 |
| C25-2 | 40 | 1.32 | 16.66 | 13.04 | 1.45 | 0.33 | 1.08 | 0.05 | 2.23 | 28.99 |
| C25-3 | 50 | 1.58 | 44.54 | 27.84 | 2.59 | 0.81 | 1.40 | 0.38 | 2.10 | 15.45 |

M, moisture; A, ash yield; V, volatile matter; $S_t$, total sulfur; ad, air-dry basis; d, dry basis; daf, dry and ash-free basis; $R_{o,ran}$, random reflectance; $Q_{gr,d}$, gross calorific value, on a dry basis; nd, not detected.

The C19 and C25 coals contain abundant vitrinite (Table 2), with collodetrinite (Figure 3A) being the most abundant maceral, followed by collotelinite (Figure 3A), along with small proportions of telinite (Figure 3B) and vitrodetrinite (Figure 3C).

Inertinite macerals occur in lesser proportions (Table 2) and are dominated by semifusinite (Figure 3A), followed by macrinite (Figure 3D) and inertodetrinite (Figure 3A), with trace amounts of micrinite (Figure 3E) and fusinite (Figure 3F). The cell structures of the semifusinite and fusinite are better preserved and have swelled and deformed form (Figure 3A,F).

**Table 2.** Maceral composition (vol. %; on mineral-free basis) of the Shiping coals.

| Samples | Cd | Ct | T | Cg | Vd | T-V | F | Sf | Ma | Mi | Sc | Id | T-I |
|---|---|---|---|---|---|---|---|---|---|---|---|---|---|
| C19-1 | 47.9 | 32.7 | 0.4 | bdl | 2.7 | 83.7 | 0.4 | 8.6 | 1.6 | 0.4 | bdl | 5.4 | 16.3 |
| C25-1 | 54.5 | 21.8 | bdl | bdl | 2.3 | 78.6 | bdl | 14.4 | 4.3 | bdl | bdl | 2.7 | 21.4 |
| C25-2 | 60.8 | 25.2 | 0.4 | bdl | 0.7 | 87.1 | bdl | 7.2 | 2.9 | bdl | bdl | 2.9 | 12.9 |
| C25-3 | 51.1 | 19.6 | 0.9 | bdl | 2.3 | 74.0 | bdl | 10.0 | 8.7 | 0.5 | bdl | 6.8 | 26.0 |

Cd, collodetrinite; Ct, collotelinite; T, telinite; Cg, corpogelinite; Vd, vitrodetrinite; T-V, total vitrinite; F, fusinite; Sf, semifusinite; Ma, macrinite; Mi, micrinite; Id, inertodetrinite; T-I, total inertinite. bdl, below detection limit.

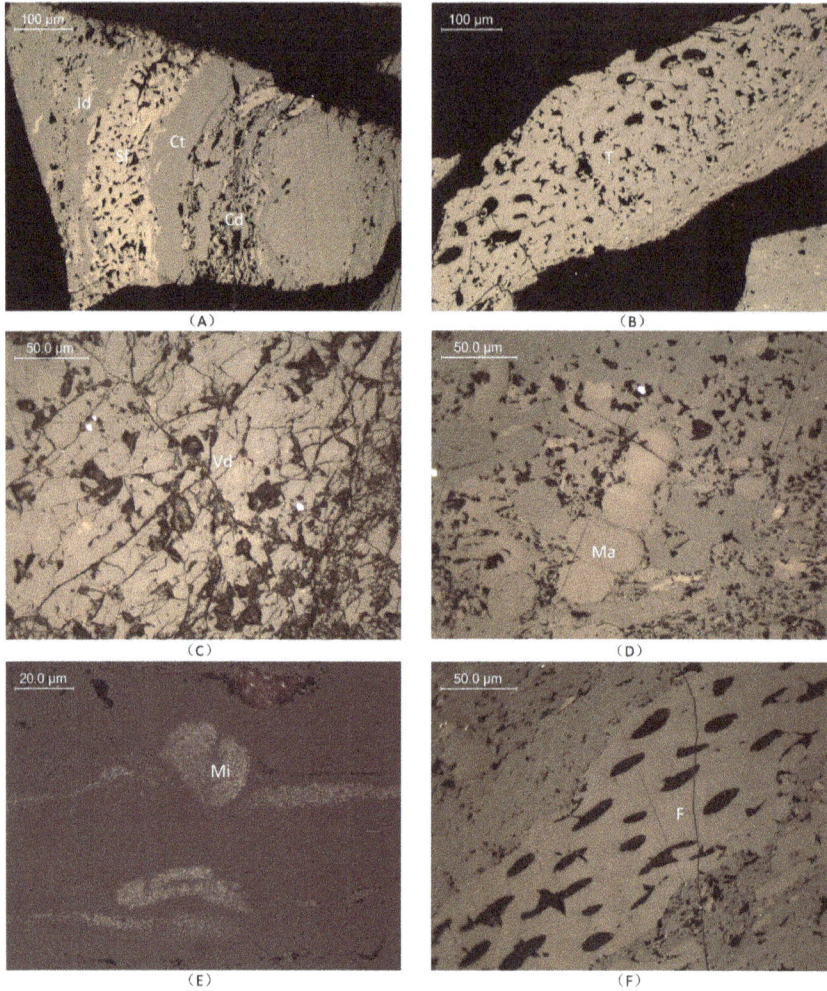

**Figure 3.** Macerals in the coal samples, reflected light, and oil immersion. (**A**) collodetrinite, collotelinite, semifusinite, and inertodetrinite in sample C25-1; (**B**) telinite in sample C25-2; (**C**) vitrodetrinite in sample C25-1; (**D**) fusinite with swelling cells in sample C25-3; (**E**) micrinite in sample C25-3; (**F**) macrinite in sample C25-3.

### 4.2. Modes of Occurrence of Minerals

The mineral compositions of the C19 and C25 coal low-temperature ashes (LTA), roof, and floor samples, as determined by XRD and Siroquant software, are listed in Table 3.

All samples contain clay minerals and pyrite (Table 3). The clay minerals in the C19 coal seams are mainly kaolinite, followed by illite and illite-smectite mixed layer clays. Kaolinite is the only clay mineral in the C25 coal seams. Kaolinite in the coals occurs as discrete particles (Figure 4A), cell-fillings (Figure 4B), and fracture-fillings (Figure 4A). Pyrite occurs as discrete particle aggregates (Figure 5A) and as cell fillings (Figure 5B), but it is present mainly as framboidal (Figure 5C) and needle-like forms together with marcasite (Figure 5D).

**Table 3.** LTA yields of coal samples and mineral compositions (%) of coal LTAs, partings, roofs, and floors determined by XRD and Siroquant.

| Minerals | C19-r | C19-1 | C25-1 | C25-2 | C25-3 | C25-f |
|---|---|---|---|---|---|---|
| LTAs | - | 17.58 | 24.94 | 20.21 | 49.35 | - |
| Kaolinite | 34.5 | 34.8 | 51.2 | 27.6 | 65.5 | 80.3 |
| Illite | 9.8 | 15.5 | - | - | - | - |
| I/S mixed-layer | 26.6 | 1.6 | - | - | - | - |
| Quartz | 19.8 | 26.6 | 1.4 | - | - | - |
| Pyrite | 0.4 | 7.4 | 27.9 | 3 | 3.4 | 5.4 |
| Marcasite | - | 10.3 | - | - | - | - |
| Anatase | 4.9 | - | 3.1 | - | - | - |
| Rutile | - | - | 6.8 | - | - | - |
| Calcite | 0.4 | - | 7.3 | 66 | 28.4 | 4.1 |
| Siderite | 3 | - | - | - | - | - |
| Jarosite | 0.6 | - | - | - | - | 2 |
| Bassanite | - | 3.7 | 2.2 | 3.4 | 2.7 | - |
| Gypsum | - | - | - | - | - | 8.2 |

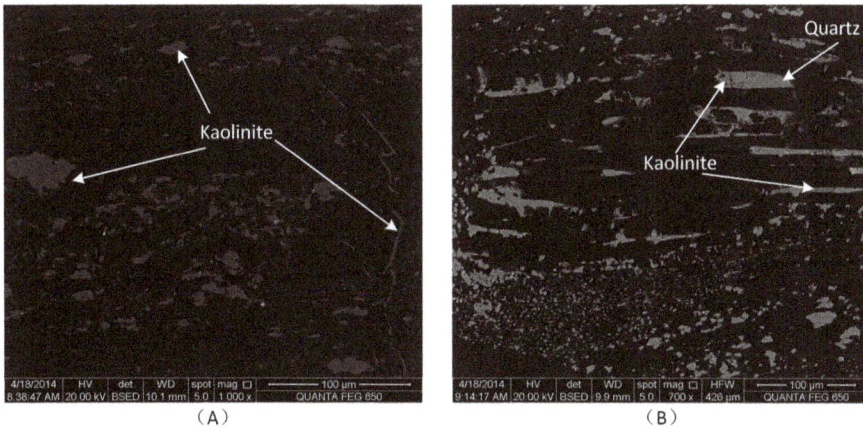

(A)    (B)

**Figure 4.** SEM back-scattered electron images of discrete particles and fracture-filling kaolinite in sample C19-1 (**A**); and cell-filling kaolinite and quartz in sample C25-2 (**B**).

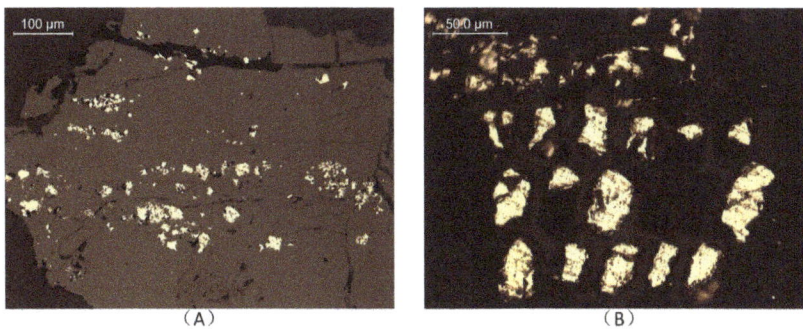

(A)    (B)

**Figure 5.** *Cont.*

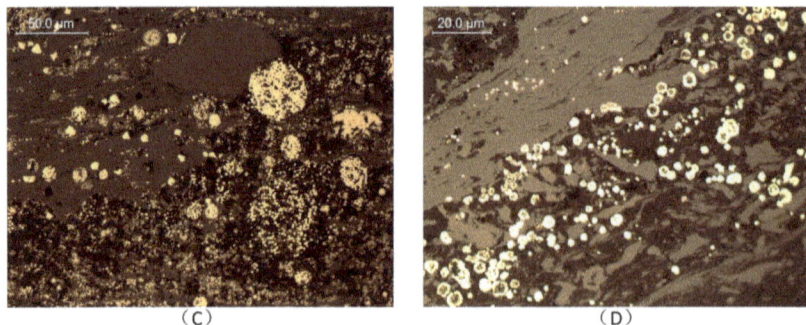

**Figure 5.** Pyrite in the C25 coal. (**A**) Particles of pyrite in sample C25-2; (**B**) cell-filling pyrite in sample C25-3; (**C**) framboidal pyrite in sample C25-3; (**D**) needle-like forms combined with marcasite in sample C25-3. Optical microscope, reflected light.

Quartz and marcasite are mainly contained in the C19 coal (Table 3). Quartz in coals occurs as discrete particles (Figure 6A), cell-fillings (Figure 4B), and fracture-fillings (Figure 6B), and marcasite occurs as needle-like forms together with pyrite (Figure 5D).

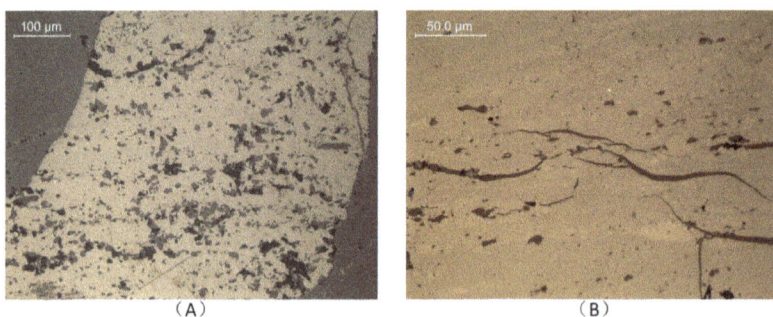

**Figure 6.** Quartz in sample C19-1. (**A**) Particles of quartz; (**B**) fracture-filling quartz. Optical microscope, reflected light.

Calcite, Ti-oxide minerals, and fluocerite are mainly contained in the C25 coals. The calcite in the C25 coals occurs as fracture fillings (Figure 7). As observed by SEM-EDS analysis of C25 coals, anatase, rutile (Figure 8A), and fluocerite (Figure 8B) are distributed in the kaolinite matrix.

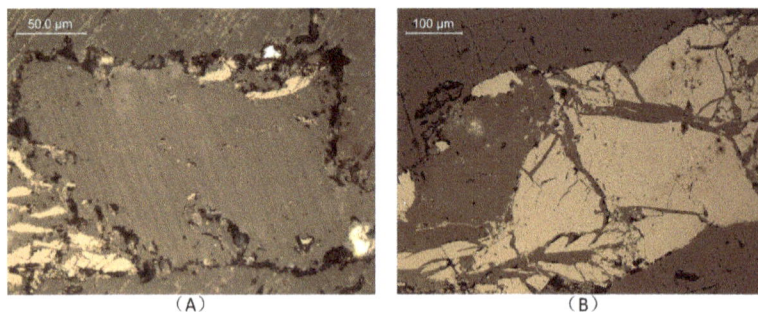

**Figure 7.** Calcite in C25-2. Optical microscope, reflected light. (**A**) Fracture-filling calcite and; (**B**) Fracture-filling calcite.

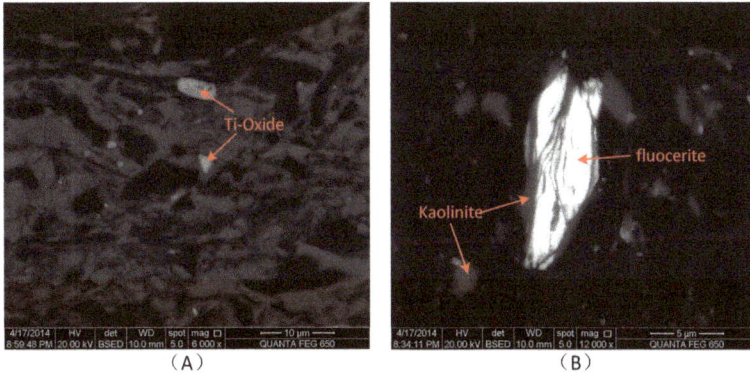

**Figure 8.** Back-scattered electron images of (**A**) Ti-oxide and (**B**) fluocerite in C25-3.

Chalcopyrite, titaniferous magnetite, apatite, and Ti-augite are present only in roof sample C19-r. Chalcopyrite occurs as fracture fillings (Figure 9A). Barite occurs in the form of discrete particles (Figure 9B). Titaniferous magnetite occurs as irregular granular. Apatite occurs as acicular idiomorphic crystals (Figure 9C). Ti-augite and rare magnetite occur as phenocrysts (Figure 9D). Marcasite was observed by SEM-EDS analysis in sample C25-f (Figure 10).

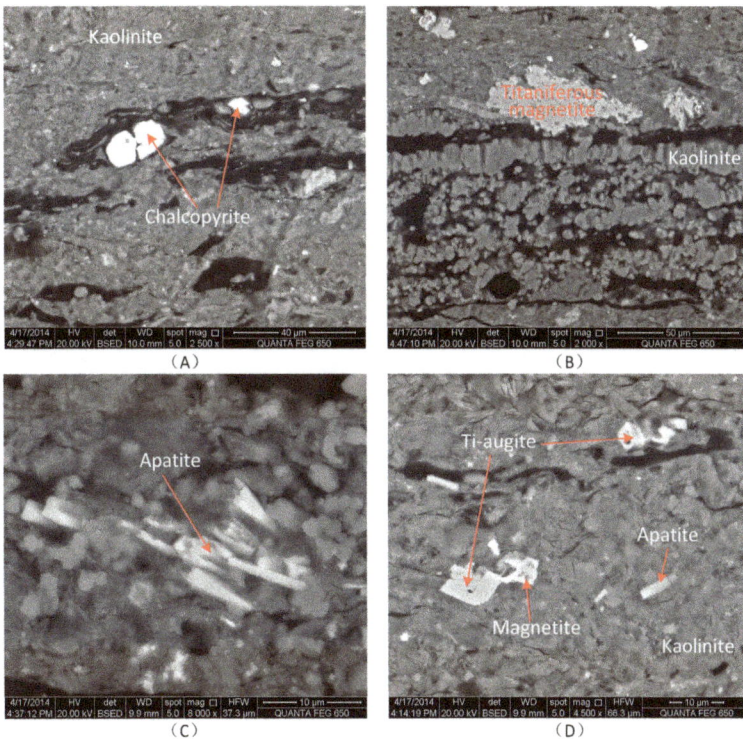

**Figure 9.** SEM back-scattered electron images of (**A**) fracture-filling chalcopyrite; (**B**) irregular granular of titaniferous magnetite and a vermicular texture in the kaolinite; (**C**) acicular idiomorphic crystals of apatite; and (**D**) phenocrysts of Ti-augite in C19-r.

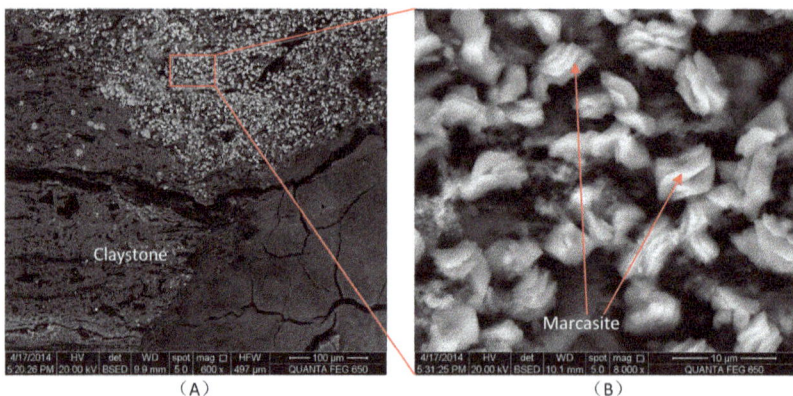

**Figure 10.** SEM back-scattered electron images of marcasite in C25-f. (**A**) Marcasite and claystone; (**B**) Marcasite.

### 4.3. Concentration and Distribution of Major and Trace Elements

#### 4.3.1. Major Element Oxides

The element compositions of the C19 and C25 coals, roof, and floor are listed in Table 4.

**Table 4.** Bench thickness (cm), major element oxides (%), loss on ignition (%), and trace elements (µg/g) in the Shiping coals and host rocks.

| Elements | C19-r | C19-1 | C25-1 | C25-2 | C25-3 | C25-f |
|---|---|---|---|---|---|---|
| Thickness | - | 50 | 40 | 40 | 50 | - |
| $SiO_2$ | 43.4 | 13.5 | 4.75 | 4.20 | 16.1 | 25.5 |
| $TiO_2$ | 3.88 | 0.279 | 0.291 | 0.229 | 0.482 | 0.934 |
| $Al_2O_3$ | 22.5 | 1.88 | 4.51 | 4.12 | 14.5 | 22.4 |
| $Fe_2O_3$ | 1.90 | 0.337 | 5.50 | 1.40 | 2.70 | 4.49 |
| MnO | 0.007 | 0.003 | 0.004 | 0.037 | 0.017 | 0.004 |
| MgO | 0.363 | 0.096 | 0.066 | 0.064 | 0.176 | 0.192 |
| CaO | 0.708 | 0.426 | 0.389 | 3.78 | 5.75 | 2.63 |
| $Na_2O$ | 0.480 | 0.053 | 0.050 | 0.022 | 0.055 | 0.057 |
| $K_2O$ | 1.15 | 0.017 | 0.066 | 0.031 | 0.122 | 0.228 |
| $P_2O_5$ | 0.403 | 0.007 | 0.009 | 0.006 | 0.016 | 0.026 |
| LOI | 24.6 | 82.9 | 83.9 | 83.6 | 56.2 | 39.8 |
| $SiO_2/Al_2O_3$ | 1.93 | 7.19 | 1.05 | 1.02 | 1.11 | 1.14 |
| Li | 49.1 | 9.97 | 21.7 | 33.2 | 178 | 343 |
| Be | 3.38 | 2.89 | 7.73 | 16.2 | 9.47 | 9.29 |
| F | 787 | 36.8 | 77.2 | 52.4 | 342 | 480 |
| Sc | 22.6 | 5.00 | 6.79 | 4.07 | 8.96 | 15.3 |
| V | 331 | 38.5 | 61.0 | 26.3 | 646 | 2904 |
| Cr | 118 | 12.7 | 24.1 | 13.8 | 282 | 1837 |
| Co | 16.2 | 17.9 | 2.61 | 2.35 | 3.83 | 20.0 |
| Ni | 36.9 | 34.6 | 14.8 | 9.65 | 46.2 | 166 |
| Cu | 278 | 20.6 | 14.8 | 9.42 | 29.2 | 95.8 |
| Zn | 300 | 13.9 | 20.4 | 14.7 | 80.6 | 71.5 |
| Ga | 38.0 | 4.06 | 12.8 | 9.01 | 36.5 | 75.3 |
| Ge | 2.47 | 0.93 | 3.79 | 8.62 | 10.3 | 4.80 |
| As | 11.4 | 1.66 | 3.29 | 2.67 | 7.98 | 25.2 |
| Se | 22.2 | 4.02 | 5.09 | 4.09 | 22.8 | 63.2 |
| Rb | 33.7 | 0.496 | 1.53 | 0.554 | 3.43 | 6.02 |
| Sr | 759 | 77.9 | 164 | 325 | 301 | 177 |

Table 4. *Cont.*

| Elements | C19-r | C19-1 | C25-1 | C25-2 | C25-3 | C25-f |
|----------|-------|-------|-------|-------|-------|-------|
| Y | 69.2 | 10.6 | 20.2 | 17.1 | 197 | 137 |
| Zr | 613 | 82.7 | 228 | 125 | 1744 | 1648 |
| Nb | 91.3 | 11.4 | 17.0 | 19.1 | 192 | 131 |
| Mo | 20.0 | 0.868 | 1.31 | 0.707 | 3.82 | 24.2 |
| Cd | 1.63 | 0.210 | 0.666 | 0.432 | 14.6 | 17.9 |
| In | 0.183 | 0.046 | 0.020 | 0.045 | 0.226 | 0.563 |
| Sn | 4.88 | 0.831 | 1.12 | 1.18 | 7.89 | 12.6 |
| Sb | 1.26 | 0.097 | 0.251 | 0.177 | 0.966 | 7.63 |
| Cs | 4.07 | 0.071 | 0.206 | 0.069 | 1.06 | 1.82 |
| Ba | 223 | 13.5 | 18.3 | 9.47 | 46.3 | 78.0 |
| REE | 703 | 40.7 | 73.8 | 124 | 1037 | 1323 |
| La | 127 | 5.48 | 11.4 | 25.4 | 173 | 221 |
| Ce | 274 | 11.7 | 19.5 | 44.3 | 322 | 533 |
| Pr | 29.8 | 1.29 | 2.13 | 5.21 | 42.5 | 54.6 |
| Nd | 117 | 4.72 | 7.72 | 18.1 | 159 | 224 |
| Sm | 26.0 | 1.08 | 1.77 | 3.15 | 30.6 | 39.8 |
| Eu | 6.45 | 0.254 | 0.416 | 0.436 | 2.82 | 3.26 |
| Gd | 22.5 | 1.34 | 2.32 | 3.44 | 32.2 | 36.5 |
| Tb | 2.76 | 0.222 | 0.438 | 0.493 | 4.87 | 5.07 |
| Dy | 13.1 | 1.39 | 2.97 | 2.67 | 28.6 | 27.3 |
| Ho | 2.46 | 0.313 | 0.636 | 0.535 | 6.01 | 5.58 |
| Er | 6.35 | 0.969 | 1.92 | 1.46 | 17.5 | 15.6 |
| Tm | 0.878 | 0.149 | 0.282 | 0.203 | 2.50 | 2.37 |
| Yb | 5.48 | 0.994 | 1.85 | 1.26 | 16.1 | 15.6 |
| Lu | 0.804 | 0.159 | 0.280 | 0.189 | 2.35 | 2.39 |
| Hf | 13.7 | 1.78 | 3.51 | 3.01 | 38.2 | 28.0 |
| Ta | 5.43 | 0.619 | 0.835 | 0.930 | 5.16 | 5.93 |
| Hg (ppb) | 59.8 | 19.5 | 396 | 197 | 372 | 621 |
| Tl | 0.077 | 0.004 | 0.188 | 0.045 | 0.083 | 0.115 |
| Pb | 66.6 | 6.18 | 12.5 | 2.19 | 12.6 | 29.1 |
| Bi | 0.164 | 0.212 | 0.146 | 0.205 | 0.427 | 0.598 |
| Th | 17.8 | 3.20 | 4.19 | 4.61 | 16.9 | 28.0 |
| U | 4.34 | 0.753 | 2.33 | 1.30 | 155 | 505 |

The percentages of $Al_2O_3$ and CaO in the C19 coal (C19-1) are lower than those in the more normal Chinese coals reported by Dai et al. [27]. The $SiO_2/Al_2O_3$ ratio of sample C19-1 (7.19 on average) is much higher than that of other Chinese coals (1.42) [27] and also than the theoretical ratio for kaolinite (1.18), indicating free $SiO_2$ in the coal. The percentages of other major element oxides, however, are either lower than or close to those in common Chinese coals.

The percentages of $Al_2O_3$ and CaO in the C25 coals are higher than those in the more normal Chinese coals reported by Dai et al. [27]. The percentages of other major element oxides, however, are either lower than or close to those in common Chinese coals. The $SiO_2/Al_2O_3$ ratio of the C25 coals (1.06 on average) is lower than that of other Chinese coals (1.42) [27], and also than the theoretical ratio for kaolinite (1.18), indicating no free $SiO_2$ in the coal.

It should be noted that the percentages of $TiO_2$ are high in roof sample C19-r (3.88%) but low in the C19 coal, C25 coals, and the floor sample (sample C25-f) (0.23–0.93%). Oxides $SiO_2$, $TiO_2$, $Al_2O_3$, MgO, $Na_2O$, $K_2O$, and $P_2O_5$ show a pattern of vertical variation the same as that of the ash yield through the seam section (Figure 11).

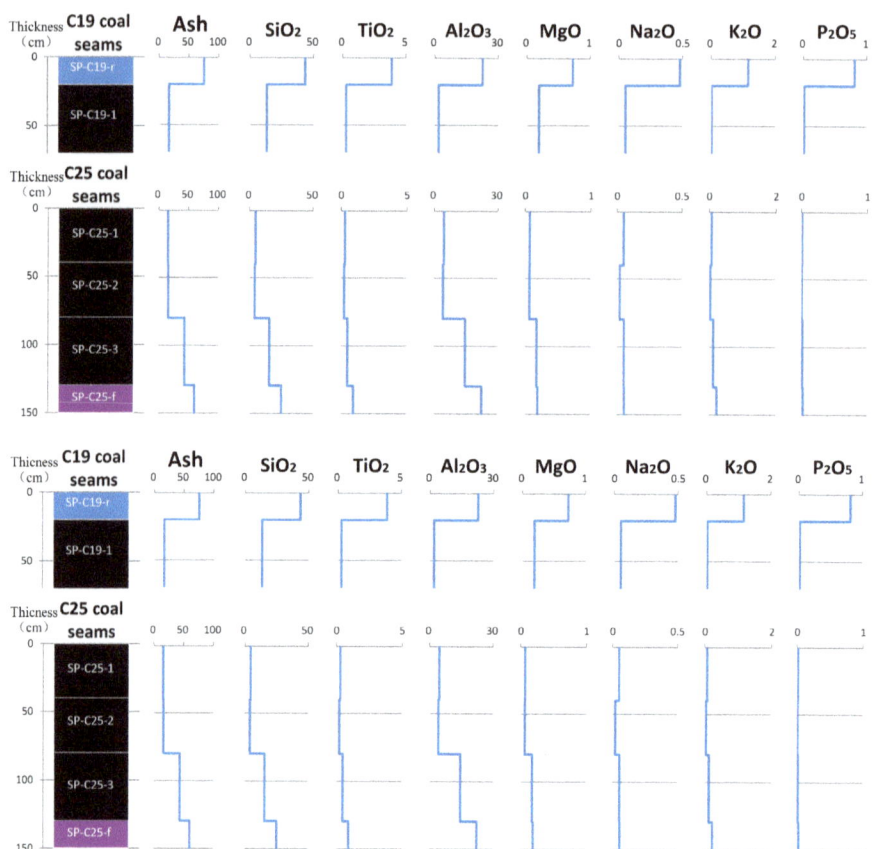

**Figure 11.** Variations of ash yield and selected major elements (%) through the roof, coal seam, and floor section of the Shiping C19 and C25 coal.

### 4.3.2. Trace Elements

Trace elements in most of the coal samples of this study are lower than or close to those of hard coals of the world [28]. However, a large number of trace elements are enriched in sample C25-3 (Figure 12).

Compared to the average for hard coals of the world [28] and based on the trace-element enrichment classification [29], a large number of trace elements are depleted in samples C19-1, C25-1, and C25-2. The trace elements with a concentration coefficient (CC = ratio of element concentration in Shiping coals/concentration hard coals of the world [28]) <0.5 include As, Rb, Sb, Cs, Ba, Tl, and Bi. Only Zr in sample C25-1 (5 < CC < 10) and Be in sample C25-2 (5 < CC < 10) are enriched.

Compared to the average for hard coals of the world [28], a large number of trace elements, including Li, V, Cr, Se, Zr, Nb, Cd, REE, Hf, Ta, and U, are enriched in C25-3 (CC > 10). Trace elements with a CC of 5–10 include Ga, In, Sn, and Th. Many other elements, including Be, F, Sc, Ni, Zn, Ge, Sr, and Hg, are slightly enriched (CC = 2–5) in the coal. The concentrations of the remaining elements are depleted or close to the average for hard coals of the world [28].

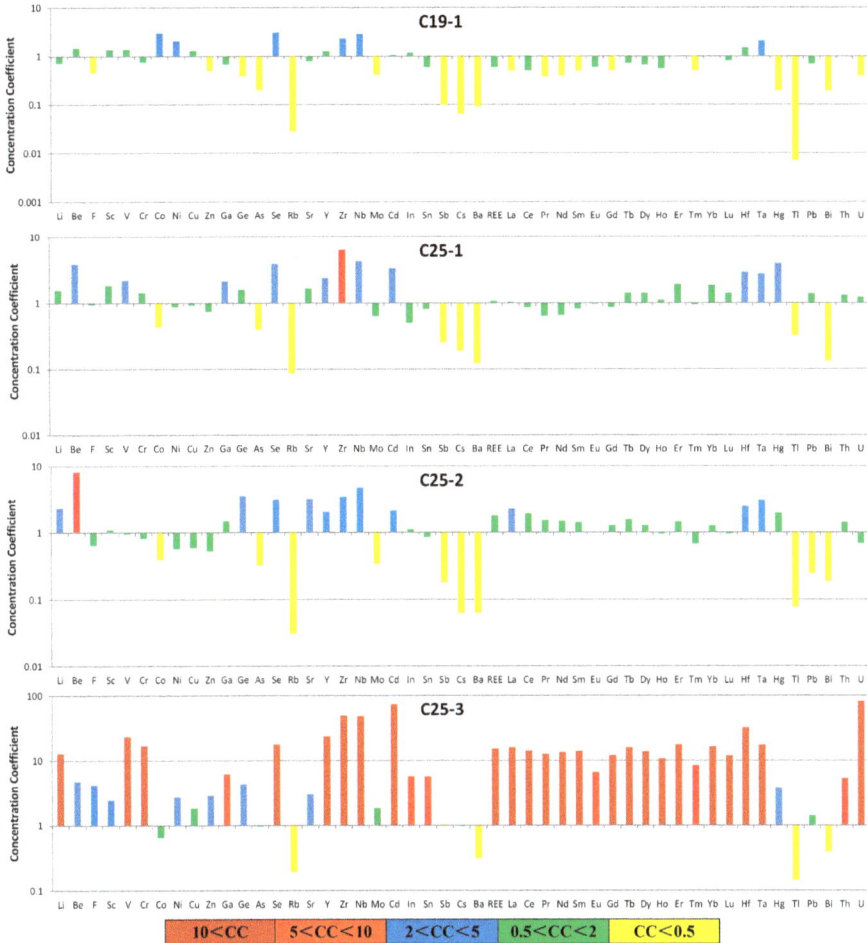

**Figure 12.** Concentration coefficients (CCs) of trace elements in the Shiping coals, normalized by average trace element concentrations in hard coals of the world [28] and based on the trace-element enrichment classification [29].

Compared to the average for the world clay [30] (Figure 13), trace elements Se and Mo are enriched in roof sample C19-r (CC > 10). Trace elements with a CC of 5–10 include Cu, Nb, and Eu. Many other elements, including Rb, Cs, Ba, Tl, and Bi are depleted (CC < 0.5) in C19-r. The concentrations of the remaining elements are slightly enriched (CC = 2–5) or close to (CC= 0.5–2) the average for world clay [30].

Compared to the average for the world clay [30] (Figure 13), a large number of trace elements, including V, Cr, Se, Nb, Mo, Cd, and U, are enriched in floor sample C25-f (CC > 10). Trace elements with a CC of 5–10 include Li, Zr, In, Sb, REE, Hf, Ta, and Hg. Some elements, including Rb, Cs, Ba, and Tl, are depleted (CC < 0.5) in sample C25-f. The concentrations of the remaining elements are slightly enriched (CC = 2–5) or close to (CC = 0.5–2) the average for world clay [30].

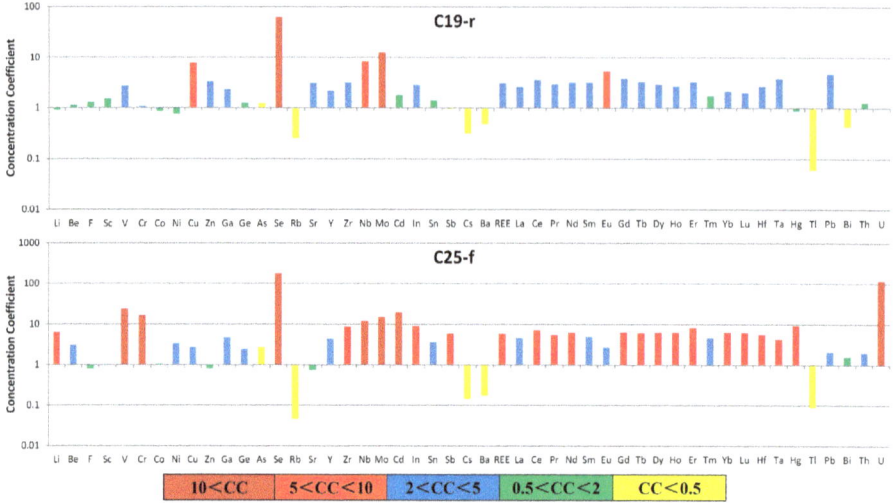

**Figure 13.** Concentration coefficients (CCs) of trace elements in the Shiping roof and floor samples, normalized by average trace element concentrations in clay of the world [30].

### 4.3.3. Rare Earth Elements

A threefold classification of rare earth elements [31] was used in this study. By comparison with the upper continental crust (UCC), three enrichment types were identified: L-REE (light REE; $La_N/Lu_N > 1$), M-REE (medium REE; $La_N/Sm_N < 1$, $Gd_N/Lu_N > 1$), and H-REE (heavy REE; $La_N/Lu_N < 1$) [31].

The REE enrichment patterns in the C19-1 and C25-1 coal benches are characterized by weak Eu anomalies and H- and M-REE enrichment (Figure 14A). The REE enrichment patterns in sample C25-2 and C25-3 coal benches are characterized by strong negative Eu anomalies and M-REE enrichment (Figure 14B,C).

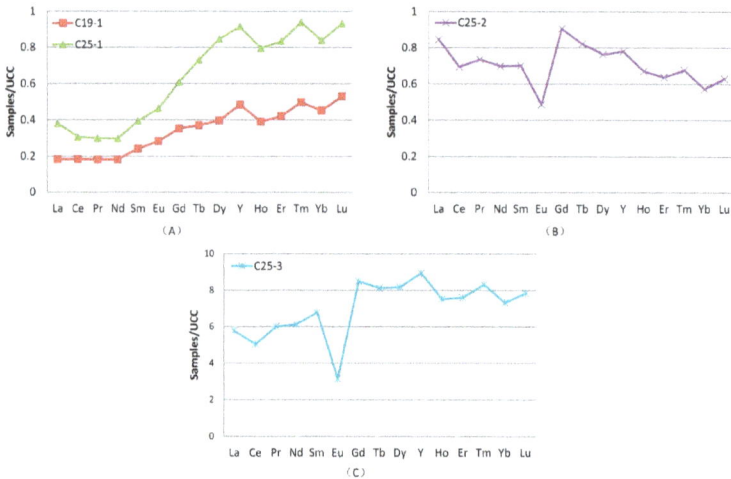

**Figure 14.** Distribution patterns of REE in the coal samples from the Shiping mine. REE are normalized by Upper Continental Crust (UCC) [32]. (**A**) Distribution patterns of REE in sample C19-1 and C25-1; (**B**) Distribution patterns of REE in sample C25-2; (**C**) Distribution patterns of REE in sample C25-3.

The REE enrichment patterns in the roof sample C19-r (Figure 15A) are characterized by positive Eu anomalies and M-REE enrichment. Floor sample C25-f has the same REE enrichment patterns as the C25 coals (Figure 15B), which are characterized by strong negative Eu anomalies and M-REE enrichment.

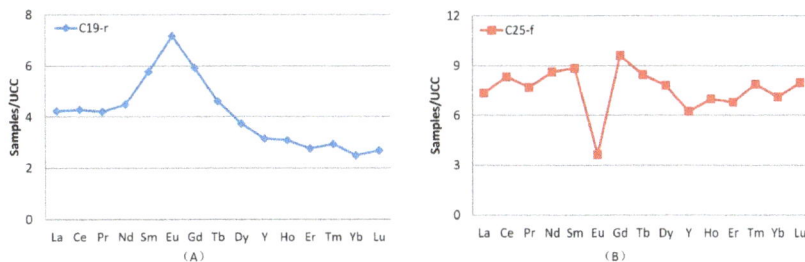

**Figure 15.** Distribution patterns of REE in the roof and floor samples from the Shiping mine. REE are normalized by Upper Continental Crust (UCC) [32]. (**A**) Distribution patterns of REE in sample C19-r; (**B**) Distribution patterns of REE in sample C25-f.

## 5. Discussion

A number of volcanic layers [33–35] and dispersed volcanic ashes in organic matter of coal seams [36–40] have been observed in the southwestern China coalfield, and these volcanic layers can be divided into four types: felsic, alkali, mafic, and dacitic [1,2,10]. In this paper, we have discussed the characteristics of the elements and minerals in the C19 and C25 coal seams, which are affected by different types of volcanic activity.

### 5.1. C19 Coal Seam and Basalt

The petrological characteristics of roof sample C19-r are different from those of other normal roof deposits. The clay minerals in C19-r are argillized (Figure 9A,D), and sedimentary layering is not observed (Figure 16), indicating that roof sample C19-r probably is a tuffaceous clay.

The contents of kaolinite, Illite, I/S, etc. are the same as in the floors of the C2 and C3 Coals in Xinde Mine, Xuanwei, eastern Yunnan, China [3]. These floor strata are identified as fully argillized, fine-grained, tuffaceous clays with high-Ti mafic magma source [3].

A vermicular texture in the kaolinite of roof sample C19-r (Figure 9B) is often used as an indicator of a volcanic origin [2,3,41–43]. The acicular idiomorphic crystals of apatite and the phenocrysts of Ti-augite coexisting with magnetite in roof sample C19-r (Figure 9C,D) are common minerals in basaltic rock [2,12,44–46]. Irregular granular titaniferous magnetite in C19-r (Figure 9B) is a typical mineral in basalt [47–50]. All the crystal modes of minerals indicate that roof sample C19-r has a high-Ti basalt origin.

The major elements $TiO_2/Al_2O_3$ ratio can indicate the acidic/basic/intermediate property of deposits, including normal sedimentary rocks, coal seam, and tonsteins [5,6,14,51–60]. The $TiO_2/Al_2O_3$ ratios are >0.08 for mafic, 0.08–0.02 for intermediate, and <0.02 for silicic rocks [1–3]. The $TiO_2/Al_2O_3$ ratio for C19-r is 0.17, indicating a mafic origin. Trace elements, including V, Co, Cr, Ni, Cu, Zn, Nb, Ta, Zr, Hf, and $TiO_2$ in C19-r, are closer to those of mafic tuff rather than alkalic and silicic tuff [1].

The concentration of $TiO_2$ in C19-r is high (3.88%). Concentrations of $TiO_2$ in basalt higher than 2.8% are indicative of high-Ti basalt [10,11]. The concentrations of trace elements, including V, Co, Cr, Ni, Cu, Zn, Nb, Ta, Zr, and Hf, in sample C19-r are closer to those of high-Ti basalt than to those of low-Ti basalt (Table 5). The elemental compositions indicate that roof sample C19-r has a high-Ti basaltic volcanic source.

The REE distribution patterns of C19-r are characterized by positive Eu anomalies and M-REE-type enrichment. This is different from those of normal deposits of Late Permian clay rocks in the Emeishan large igneous province, which are generally characterized by weakly negative or no Eu anomalies [1,51], and are the same as those of high-Ti [10] (Figure 17A). To summarize, roof sample C19-r consists of tuffaceous clays and probably has a high-Ti mafic magma source.

**Figure 16.** Images of C19-r and C25-f samples collected from the Shiping mine.

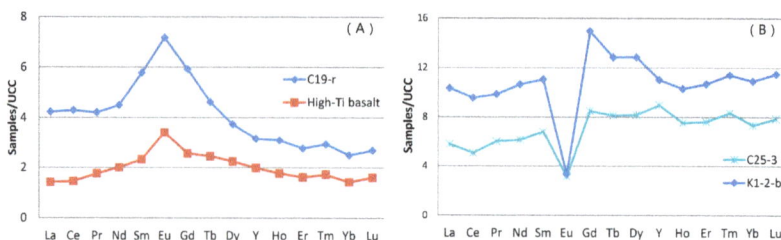

**Figure 17.** Distribution patterns of REE in the (**A**) high-Ti basalts [10] and C19-r; and (**B**) the coal in Lvshuidong [5] and C25-3. REE are normalized by Upper Continental Crust (UCC) [32].

**Table 5.** The trace element ($\mu g/g$) in sample C19-r compared with high/low Ti basalts [10].

| Samples | V | Cr | Co | Ni | Cu | Zn | Ga | Zr | Nb | Hf | Ta |
|---------|-----|-----|------|------|-----|-----|------|-----|------|------|-------|
| C19-r | 331 | 118 | 16.2 | 36.9 | 278 | 300 | 38.0 | 613 | 91.3 | 13.7 | 5.43 |
| High-Ti Basalt | 374 | 71.0 | 40.3 | 63.5 | 248 | 131 | 23.9 | 391 | 48.5 | 8.51 | 2.97 |
| Low-Ti Basalt | 289 | 230 | 44.5 | 124 | 109 | 90.7 | 19.9 | 129 | 13.4 | 3.05 | 0.773 |

Coal sample C19-1 has trace element and REE distribution patterns similar to those of normal coal deposits derived from the Kangdian Upland source region [6,7], indicating that sample C19-r is a normal sedimentary rock.

*5.2. C25 Coals and Volcanic Ashes of Alkali Rhyolites*

In the early part of the Late Permian, the volcanic ash had mainly alkalic composition [1,2]. The C25 coals examined in this study are the lowermost coal seam of the Late Permian strata, and these coals are in the same layer as other coals that were affected by alkali volcanics in southwestern China [4,5,33]. Moreover, the Shiping mine is within the area of alkali tonstein distribution [16].

High field strength elements, including Nb, Ta, Zr, Hf, and REE, are significantly enriched in C25 coals, and other elements such as Sc, Ti, V, Cr, Co, Ni, Cu, and Zn are depleted in C25 coals [4,13]. This characteristic is the same as that of alkali tonsteins and other coals affected by alkali volcanics in southwestern China [4,5,61]. This indicates that the C25 coal in the Shiping mine had been subjected to alkalic volcanic ash.

The REE distribution patterns of the C25 coals are characterized by strongly negative Eu anomalies and M-REE-type enrichment, the same as some Chinese alkali granites [62,63], some alkali tonsteins [1] reported in the south of China, and some coals affected by alkali volcanics [5] (Figure 17B). Overall, the C25 coals have high concentrations of Nb, Ta, Zr, Hf, and REE, which came mainly from alkali volcanics. The origin of these rare metals in the C25 coal present in this study is similar to those in the coals or coal-bearing sedimentary sequences in the surrounding areas [63–67].

The floor sample (C25-f) has high concentrations of Nb, Ta, Zr, Hf, and REE and has an REE distribution pattern similar to that of the C25 coal. It is the same as some alkali tonsteins reported in the south of China [1]. These results indicate that floor sample C25-f probably has alkali volcanic origin.

## 6. Conclusions

Sample C19-r does not have distinct stratification, indicating that the roof stratum is not a normal rock. The acicular idiomorphic crystals of apatite and the phenocrysts of Ti-augite coexisting with magnetite in roof sample C19-r are common minerals in basaltic volcanic rock. The REE distribution patterns of C19-r are characterized by positive Eu anomalies and M-REE-type enrichment. All of the above indicates that the roof sample C19-r probably has a basaltic origin.

The concentration of $TiO_2$ in C19-r is high (3.88%). Concentrations of $TiO_2$ in basalt higher than 2.8% are indicative of high-Ti basalt. The concentrations of trace elements, including V, Co, Cr, Ni, Cu, Zn, Nb, Ta, Zr, and Hf, in C19-r are closer to those of high-Ti basalt instead of those of low-Ti basalt. To summarize, roof sample C19-r probably has a high-Ti basaltic origin.

Nb, Ta, Zr, Hf, and REE are significantly enriched in C25 coals, and other elements such as Sc, Ti, V, Cr, Co, Ni, Cu, and Zn are depleted in C25 coals. The REE distribution patterns of the C25 coals are characterized by strongly negative Eu anomalies and M-REE-type enrichment. This indicates that the C25 coals in the Shiping mine were probably affected by alkalic volcanic ashes.

**Acknowledgments:** This research was supported by the National Key Basic Research Program of China (No. 2014CB238902), the National Natural Science Foundation of China (Nos. 41420104001, 41272182, and U1407207), and the Program for Changjiang Scholars and Innovative Research Team in University (IRT13099). Many thanks are given to Shifeng Dai for his constructive comments, which greatly improved the manuscript quality. We are grateful to the three anonymous reviewers for their careful comments.

**Author Contributions:** All co-authors participated in the work of this study. Yangbing Luo carried out the mineralogy and geochemistry analyses data. Mianping Zheng helped to design the research and structure of the manuscript.

**Conflicts of Interest:** The authors declare no conflict of interest.

## References

1. Dai, S.; Wang, X.; Zhou, Y.; Hower, J.C.; Li, D.; Chen, W.; Zhu, X.; Zou, J. Chemical and mineralogical compositions of silicic, mafic, and alkali tonsteins in the late Permian coals from the Songzao Coalfield, Chongqing, Southwest China. *Chem. Geol.* **2011**, *282*, 29–44. [CrossRef]

2. Zhou, Y.; Bohor, B.F.; Ren, Y. Trace element geochemistry of altered volcanic ash layers (tonsteins) in Late Permian coal-bearing formations of eastern Yunnan and western Guizhou Provinces, China. *Int. J. Coal Geol.* **2000**, *44*, 305–324. [CrossRef]

3. Dai, S.; Li, T.; Seredin, V.V.; Ward, C.R.; Hower, J.C.; Zhou, Y.; Zhang, M.; Song, X.; Song, W.; Zhao, C. Origin of minerals and elements in the Late Permian coals, tonsteins, and host rocks of the Xinde Mine, Xuanwei, eastern Yunnan, China. *Int. J. Coal Geol.* **2014**, *121*, 53–78. [CrossRef]

4. Dai, S.; Liu, J.; Ward, C.R.; Hower, J.C.; French, D.; Jia, S.; Hood, M.M.; Garrison, T.M. Mineralogical and geochemical compositions of Late Permian coals and host rocks from the Guxu Coalfield, Sichuan Province, China, with emphasis on enrichment of rare metals. *Int. J. Coal Geol.* **2015**. [CrossRef]

5. Dai, S.; Luo, Y.; Seredin, V.V.; Ward, C.R.; Hower, J.C.; Zhao, L.; Liu, S.; Zhao, C.; Tian, H.; Zou, J. Revisiting the late Permian coal from the Huayingshan, Sichuan, southwestern China: Enrichment and occurrence modes of minerals and trace elements. *Int. J. Coal Geol.* **2014**, *122*, 110–128. [CrossRef]

6. Wang, X.; Zhang, Y.; Pan, Y.; Liu, C. *The Sedimentary Environment and the Regularity of Coal Accumulation of Late Permian Coals in Southern China*; Chongqing University Press: Chongqing, China, 1996; pp. 74–92. (In Chinese)

7. Zhang, Y. *The Sedimentary Environment and Coal Accumulation of the Late Permian Coals in Southern Sichuan, China*; Guizhou Science and Technology Press: Guiyang, China, 1993; pp. 44–65. (In Chinese)

8. Arbuzov, S.I.; Mezhibor, A.M.; Spears, D.A.; Ilenok, S.S.; Shaldybin, M.V.; Belaya, E.V. Nature of tonsteins in the Azeisk deposit of the Irkutsk Coal Basin (Siberia, Russia). *Int. J. Coal Geol.* **2016**, *153*, 99–111. [CrossRef]

9. Hou, T.; Zhang, Z.; Kusky, T.; Du, Y.; Liu, J.; Zhao, Z. A reappraisal of the high-Ti and low-Ti classification of basalts and petrogenetic linkage between basalts and mafic-ultramafic intrusions in the Emeishan Large Igneous Province, SW China. *Ore Geol. Rev.* **2011**, *41*, 133–143. [CrossRef]

10. Xiao, L.; Xu, Y.; Mei, H.; Zheng, Y.; He, B.; Pirajno, F. Distinct mantle sources of low-Ti and high-Ti basalts from the western Emeishan large igneous province, SW China: Implications for plume-lithosphere interaction. *Earth Planet. Sci. Lett.* **2004**, *228*, 525–546. [CrossRef]

11. Xu, Y.; Chung, S. The Emeishan large igneous province: Evidence for mantle plume activity and melting conditions. *Geochimica* **2001**, *30*, 1–9.

12. Xu, Y.; Chung, S.; Shao, H.; He, B. Silicic magmas from the Emeishan large igneous province, Southwest China: Petrogenesis and their link with the end-Guadalupian biological crisis. *Lithos* **2010**, *119*, 47–60. [CrossRef]

13. Xu, Y.; He, B.; Chung, S.; Menzies, M.A.; Frey, F. A geologic, geochemical, and geophysical consequences of plume involvement in the Emeishan flood-basalt province. *Geology* **2004**, *32*, 917–920. [CrossRef]

14. Wang, P.; Ji, D.; Yang, Y.; Zhao, L. Mineralogical compositions of Late Permian coals from the Yueliangtian mine, western Guizhou, China: Comparison to coals from eastern Yunnan, with an emphasis on the origin of the minerals. *Fuel* **2016**, *181*, 859–869. [CrossRef]

15. Zhao, L.; Ward, C.R.; French, D.; Graham, I.T. Mineralogical composition of Late Permian coal seams in the Songzao Coalfield, southwestern China. *Int. J. Coal Geol.* **2013**, *116*, 208–226. [CrossRef]

16. Zhou, Y. The alkali pyroclastic tonsteins of early stage of late Permian age in southwestern China. *Coal Geol Explor.* **1999**, *27*, 5–9. (In Chinese)

17. ASTM International. *Test Method for Moisture in the Analysis Sample of Coal and Coke*; ASTM Standard D3173-11; ASTM International: West Conshohocken, PA, USA, 2011.

18. ASTM International. *Annual Book of ASTM Standards. Test Method for Ash in the Analysis Sample of Coal and Coke*; ASTM Standard D3174-11; ASTM International: West Conshohocken, PA, USA, 2011.

19. ASTM International. *Test Method for Volatile Matter in the Analysis Sample of Coal and Coke*; ASTM Standard D3175-11; ASTM International: West Conshohocken, PA, USA, USA, 2011.

20. ASTM International. *Test Methods for Total Sulfur in the Analysis Sample of Coal and Coke*; ASTM Standard D3177-02; ASTM International: West Conshohocken, PA, USA, USA, 2002.

21. ASTM International. *Standard Test Method for Forms of Sulfur in Coal*; ASTM Standard D2492-02; ASTM International: West Conshohocken, PA, USA, 2002.

22. ASTM International. *Standard Practice for Preparing Coal Samples for Microscopical Analysis by Reflected Light*; ASTM Standard D2797/D2797M-11a; ASTM International: West Conshohocken, PA, USA, 2011.

23. ASTM International. *Standard Test Method for Total Fluorine in Coal and Coke by Pyrohydrolytic Extraction and Ion Selective Electrode or Ion Chromatograph Methods*; ASTM Standard D5987-96; ASTM International: West Conshohocken, PA, USA, 2002.

24. Li, X.; Dai, S.; Zhang, W.; Li, T.; Zheng, X.; Chen, W. Determination of As and Se in coal and coal combustion products using closed vessel microwave digestion and collision/reaction cell technology (CCT) of inductively coupled plasma mass spectrometry (ICP-MS). *Int. J. Coal Geol.* **2014**, *124*, 1–4. [CrossRef]

25. ASTM International. *Standard Classification of Coals by Rank*; ASTM Standard D388-12; ASTM International: West Conshohocken, PA, USA, 2012.

26. Standardization Administration of China; General Administration of Quality Supervision, Inspection and Quarantine of the China. *Classification for Quality of Coal-Part 1: Ash*; Chinese National Standard GB/T 15224.1-2004; Standard Press of China: Beijing, China, 2004. (In Chinese)

27. Dai, S.; Ren, D.; Chou, C.-L.; Finkelman, R.B.; Seredin, V.V.; Zhou, Y. Geochemistry of trace elements in Chinese coals: A review of abundances, genetic types, impacts on human health, and industrial utilization. *Int. J. Coal Geol.* **2012**, *94*, 3–21. [CrossRef]

28. Ketris, M.; Yudovich, Y.E. Estimations of clarkes for carbonaceous biolithes: World averages for trace element contents in black shales and coals. *Int. J. Coal Geol.* **2009**, *78*, 135–148. [CrossRef]

29. Dai, S.; Seredin, V.V.; Ward, C.R.; Hower, J.C.; Xing, Y.; Zhang, W.; Song, W.; Wang, P. Enrichment of U-Se-Mo-Re-V in coals preserved within marine carbonate successions: Geochemical and mineralogical data from the Late Permian Guiding Coalfield, Guizhou, China. *Miner. Deposita* **2015**, *50*, 159–186. [CrossRef]

30. Grigoriev, N. *Chemical Element Distribution in the Upper Continental Crust*; UB RAS: Ekaterinburg, Russia, 2009; p. 382. (In Russian)

31. Seredin, V.V.; Dai, S. Coal deposits as potential alternative sources for lanthanides and yttrium. *Int. J. Coal Geol.* **2012**, *94*, 67–93. [CrossRef]

32. Taylor, S.R.; McLennan, S.M. *The Continental Crust: Its Composition and Evolution*; Blackwell: Oxford, UK, 1985; p. 312.

33. Dai, S.F.; Chekryzhov, I.Y.; Seredin, V.V.; Nechaev, V.P.; Graham, I.T.; Hower, J.C.; Ward, C.R.; Ren, D.Y.; Wang, X.B. Metalliferous coal deposits in East Asia (Primorye of Russia and South China): A review of geodynamic controls and styles of mineralization. *Gondwana Res.* **2016**, *29*, 60–82. [CrossRef]

34. Dai, S.; Graham, I.T.; Ward, C.R. A review of anomalous rare earth elements and yttrium in coal. *Int. J. Coal Geol.* **2016**, *159*, 82–95. [CrossRef]

35. Dai, S.; Zhou, Y.; Zhang, M.; Wang, X.; Wang, J.; Song, X.; Jiang, Y.; Luo, Y.; Song, Z.; Yang, Z.; et al. A new type of Nb(Ta)–Zr(Hf)–REE–Ga polymetallic deposit in the late Permian coal-bearing strata, eastern Yunnan, southwestern China: Possible economic significance and genetic implications. *Int. J. Coal Geol.* **2010**, *83*, 55–63. [CrossRef]

36. Liu, J.; Yang, Z.; Yan, X.; Ji, D.; Yang, Y.; Hu, L. Modes of occurrence of highly-elevated trace elements in superhigh-organic-sulfur coals. *Fuel* **2015**, *156*, 190–197. [CrossRef]

37. Dai, S.; Ren, D.; Zhou, Y.; Chou, C.-L.; Wang, X.; Zhao, L.; Zhu, X. Mineralogy and geochemistry of a superhigh-organic-sulfur coal, Yanshan Coalfield, Yunnan, China: Evidence for a volcanic ash component and influence by submarine exhalation. *Chem. Geol.* **2008**, *255*, 182–194. [CrossRef]

38. Dai, S.; Wang, X.; Chen, W.; Li, D.; Chou, C.-L.; Zhou, Y.; Zhu, C.; Li, H.; Zhua, X.; Xing, Y.; et al. A high-pyrite semianthracite of Late Permian age in the Songzao Coalfield, southwestern China: Mineralogical and geochemical relations with underlying mafic tuffs. *Int. J. Coal Geol.* **2010**, *83*, 430–445. [CrossRef]

39. Dai, S.; Ren, D.; Hou, X.; Shao, L. Geochemical and mineralogical anomalies of the late Permian coal in the Zhijin coalfield of southwest China and their volcanic origin. *Int. J. Coal Geol.* **2003**, *55*, 117–138. [CrossRef]

40. Dai, S.; Chou, C.-L.; Yue, M.; Luo, K.; Ren, D. Mineralogy and geochemistry of a Late Permian coal in the Dafang Coalfield, Guizhou, China: Influence from siliceous and iron-rich calcic hydrothermal fluids. *Int. J. Coal Geol.* **2005**, *61*, 241–258. [CrossRef]

41. Zhao, L.; Ward, C.R.; French, D.; Graham, I.T. Mineralogy of the volcanic-influenced Great Northern coal seam in the Sydney Basin, Australia. *Int. J. Coal Geol.* **2012**, *94*, 94–110. [CrossRef]

42. Ruppert, L.F.; Moore, T.A. Differentiation of volcanic ash-fall and water-borne detrital layers in the Eocene Senakin coal bed, Tanjung Formation, Indonesia. *Org. Geochem.* **1993**, *20*, 233–247. [CrossRef]

43. Spears, D.A. The origin of tonsteins, an overview, and links with seatearths, fireclays and fragmental clay rocks. *Int. J. Coal Geol.* **2012**, *94*, 22–31. [CrossRef]

44. Dan, W.; Wang, Q.; Wang, X.; Liu, Y.; Wyman, D.A.; Liu, Y. Overlapping Sr–Nd–Hf–O isotopic compositions in Permian mafic enclaves and host granitoids in Alxa Block, NW China: Evidence for crust–mantle interaction and implications for the generation of silicic igneous provinces. *Lithos* **2015**, *230*, 133–145. [CrossRef]

45. Liu, W.; Zhang, J.; Sun, T.; Zhou, L.; Liu, A. Low-Ti iron oxide deposits in the Emeishan large igneous province related to low-Ti basalts and gabbroic intrusions. *Ore Geol. Rev.* **2015**, *65*, 180–197. [CrossRef]

46. Shellnutt, J.G.; Lee, T.; Yang, C.; Hu, S.; Wu, J.; Iizuka, Y. A mineralogical investigation of the Late Permian Doba gabbro, southern Chad: Constraints on the parental magma conditions and composition. *J. Afr. Earth Sci.* **2016**, *114*, 13–20. [CrossRef]

47. Mathison, C.I. Magnetites and ilmenites in the Somerset dam layered basic intrusion, southeastern Queensland. *Lithos* **1975**, *8*, 93–111. [CrossRef]

48. Hou, M.; Deng, M.; Zhang, B.; Wang, W.; Li, X.; Wang, W.; Pei, S.; Yang, Y. A major Ti-bearing mineral in Emeishan basalts: The occurrence, characters and genesis of sphene. *Acta Petrol. Sin.* **2011**, *27*, 2487–2499. (In Chinese)

49. Liu, X.; Cai, Y.; Lu, Q.; Tao, Z.; Zhao, F.; Cai, F.; Li, C.; Song, X. Actual traces of mantle fluid from alkali-rich porphyries in western Yunnan, and associated implications to metallogenesis. *Earth Sci. Front.* **2010**, *17*, 114–136. (In Chinese)

50. Du, W.; Han, B.; Zhang, W.; Liu, Z. The discovery of peridotite xenoliths and megacrysts in Jining, Inner Mongolia. *Acta Petrol. Mineral.* **2016**, *25*, 13–24. (In Chinese) [CrossRef] [PubMed]

51. He, B.; Xu, Y.; Zhong, Y.; Guan, J. The Guadalupian–Lopingian boundary mudstones at Chaotian (SW China) are clastic rocks rather than acidic tuffs: Implication for a temporal coincidence between the end-Guadalupian mass extinction and the Emeishan volcanism. *Lithos* **2010**, *119*, 10–19. [CrossRef]

52. Dai, S.; Hower, J.C.; Ward, C.R.; Guo, W.; Song, H.; O'Keefe, J.M.K.; Xie, P.; Hood, M.M.; Yan, X. Elements and phosphorus minerals in the middle Jurassic inertinite-rich coals of the Muli Coalfield on the Tibetan Plateau. *Int. J. Coal Geol.* **2015**, *144–145*, 23–47. [CrossRef]

53. Zhao, L.; Graham, I. Origin of the alkali tonsteins from southwest China: Implications for alkaline magmatism associated with the waning stages of the Emeishan Large Igneous Province. *Aust. J. Earth Sci.* **2016**, *63*, 123–128. [CrossRef]

54. Dai, S.; Yang, J.; Ward, C.R.; Hower, J.C.; Liu, H.; Garrison, T.M.; French, D.; O'Keefe, J.M.K. Geochemical and mineralogical evidence for a coal-hosted uranium deposit in the Yili Basin, Xinjiang, northwestern China. *Ore Geol. Rev.* **2015**, *70*, 1–30. [CrossRef]

55. Dai, S.; Li, T.; Jiang, Y.; Ward, C.R.; Hower, J.C.; Sun, J.; Liu, J.; Song, H.; Wei, J.; Li, Q.; et al. Mineralogical and geochemical compositions of the Pennsylvanian coal in the Hailiushu Mine, Daqingshan Coalfield, Inner Mongolia, China: Implications of sediment-source region and acid hydrothermal solutions. *Int. J. Coal Geol.* **2015**, *137*, 92–110. [CrossRef]

56. Johnston, M.N.; Hower, J.C.; Dai, S.; Wang, P.; Xie, P.; Liu, J. Petrology and Geochemistry of the Harlan, Kellioka, and Darby Coals from the Louellen 7.5-Minute Quadrangle, Harlan County, Kentucky. *Minerals* **2015**, *5*, 894–918. [CrossRef]

57. Hower, J.C.; Eble, C.F.; O'Keefe, J.M.K.; Dai, S.; Wang, P.; Xie, P.; Liu, J.; Ward, C.R.; French, D. Petrology, Palynology, and Geochemistry of Gray Hawk Coal (Early Pennsylvanian, Langsettian) in Eastern Kentucky, USA. *Minerals* **2015**, *5*, 592–622. [CrossRef]

58. Dai, S.; Zhang, W.; Ward, C.R.; Seredin, V.V.; Hower, J.C.; Li, X.; Song, W.; Wang, X.; Kang, H.; Zheng, L.; et al. Mineralogical and geochemical anomalies of Late Permian coals from the Fusui Coalfield, Guangxi Province, southern China: Influences of terrigenous materials and hydrothermal fluids. *Int. J. Coal Geol.* **2013**, *105*, 60–84. [CrossRef]

59. Dai, S.; Wang, P.; Ward, C.R.; Tang, Y.; Song, X.; Jiang, J.; Hower, J.C.; Li, T.; Seredin, V.V.; Wagner, N.J.; et al. Elemental and mineralogical anomalies in the coal-hosted Ge ore deposit of Lincang, Yunnan, southwestern China: Key role of $N_2$-$CO_2$-mixed hydrothermal solutions. *Int. J. Coal Geol.* **2015**, *152*, 19–46. [CrossRef]

60. Dai, S.; Liu, J.; Ward, C.R.; Hower, J.C.; Xie, P.; Jiang, Y.; Hood, M.M.; O'Keefe, J.M.K.; Song, H. Petrological, geochemical, and mineralogical compositions of the low-Ge coals from the Shengli Coalfield, China: A comparative study with Ge-rich coals and a formation model for coal-hosted Ge ore deposit. *Ore Geol. Rev.* **2015**, *71*, 318–349. [CrossRef]

61. Zhuang, X.; Su, S.; Xiao, M.; Li, J.; Alastuey, A.; Querol, X. Mineralogy and geochemistry of the Late Permian coals in the Huayingshan coal-bearing area, Sichuan Province, China. *Int. J. Coal Geol.* **2012**, *94*, 271–282. [CrossRef]

62. Zhao, Z.; Zhou, L. Geochemistry of rare earth elements in some Chinese alkali-rich intrusive rocks. *Sci. China Ser. B* **1994**, *24*, 1109–1120. (In Chinese)

63. Dai, S.; Xie, P.; Jia, S.; Ward, C.R.; Hower, H.C.; Yan, X.; French, D. Enrichment of U-Re-V-Cr-Se and rare earth elements in the Late Permian coals of the Moxinpo Coalfield, Chongqing, China: Genetic implications from geochemical and mineralogical data. *Ore Geol. Rev.* **2017**, *80*, 1–17. [CrossRef]

64. Zhao, L.; Dai, S.; Graham, I.T.; Li, X.; Liu, H.; Song, X.; Hower, J.C.; Zhou, Y. Cryptic sediment-hosted critical element mineralization from eastern Yunnan Province, southwestern China: Mineralogy, geochemistry, relationship to Emeishan alkaline magmatism and possible origin. *Ore Geol. Rev.* **2017**, *80*, 116–140. [CrossRef]

65. Dai, S.; Yan, X.; Ward, C.R.; Hower, J.C.; Zhao, L.; Wang, X.; Zhao, L.; Ren, D.; Finkelman, R.B. Valuable elements in Chinese coals: A review. *Int. Geol. Rev.* **2016**. [CrossRef]

66. Hower, J.C.; Eble, C.F.; Dai, S.; Belkin, H.E. Distribution of rare earth elements in eastern Kentucky coals: Indicators of multiple modes of enrichment? *Int. J. Coal Geol.* **2016**, *160–161*, 73–81. [CrossRef]

67. Zhao, L.; Dai, S.; Graham, I.; Wang, P. Clay mineralogy of coal-hosted Nb-Zr-REE-Ga mineralized beds from Late Permian strata, eastern Yunnan, SW China: Implications for palaeotemperature and origin of the micro-quartz. *Minerals* **2016**, *6*, 45. [CrossRef]

*minerals*

MDPI

*Article*

# Geochemistry and Mineralogy of Tuff in Zhongliangshan Mine, Chongqing, Southwestern China

Jianhua Zou [1,2,*], Heming Tian [2] and Tian Li [2]

[1]   College of Geosciences and Survey Engineering, China University of Mining and Technology (Beijing), Beijing 100083, China

[2]   Chongqing Key Laboratory of Exogenic Mineralization and Mine Environment, Chongqing Institute of Geology and Mineral Resources, Chongqing 400042, China; heming1986824@126.com (H.T.); lt_litian@163.com (T.L.)

*   Correspondence: zoujianhua1200@gmail.com; Tel.: +86-23-8831-6044

Academic Editors: Shifeng Dai and Thomas N. Kerestedjian
Received: 30 January 2016; Accepted: 11 May 2016; Published: 20 May 2016

**Abstract:** Coal-bearing strata that host rare metal deposits are currently a hot issue in the field of coal geology. The purpose of this paper is to illustrate the mineralogy, geochemistry, and potential economic significance of rare metals in the late Permian tuff in Zhongliangshan mine, Chongqing, southwestern China. The methods applied in this study are X-ray fluorescence spectrometry (XRF), inductively coupled mass spectrometry (ICP-MS), X-ray diffraction analysis (XRD) plus Siroquant, and scanning electron microscopy in conjunction with an energy-dispersive X-ray spectrometry (SEM-EDX). The results indicate that some trace elements including Li, Be, Sc, V, Cr, Co, Ni, Cu, Zn, Ga, Zr, Nb, Cd, Sb, REE, Hf, Ta, Re, Th, and U are enriched in the tuff from Zhongliangshan mine. The minerals in the tuff mainly include kaolinite, illite, pyrite, anatase, calcite, gypsum, quartz, and traces of minerals such as zircon, florencite, jarosite, and barite. The tuff is of mafic volcanic origin with features of alkali basalt. Some minerals including florencite, gypsum, barite and a portion of anatase and zircon have been derived from hydrothermal solutions. It is suggested that Zhongliangshan tuff is a potential polymetallic ore and the recovery of these valuable elements needs to be further investigated.

**Keywords:** geochemistry; mineralogy; origin; tuff; Chongqing

## 1. Introduction

With the depletion of traditional rare metal deposits, coal deposits as promising alternative sources for rare metals have attracted much attention in recent years [1–9]. At present, germanium is the most successful rare metal element that has been extracted from coal ash [4,8,9]. The three well-known coal-bearing strata hosted Ge deposits include Lincang (Yunnan Province) and Wulantuga (Inner Mongolia) of China, and Spetzugli of Russia, are the main sources for the industrial Ge at present and for the foreseeable future [1,4,8,9]. The super-large coal-bearing strata hosted gallium deposit in the Jungar Coalfield (Inner Mongolia), China, is another typical example discovered in 2006 [10], which was considered as the third and the most outstanding discovery after the coal-bearing strata hosted uranium and germanium deposits [2,10]. Moreover, aluminum is also enriched in Jungar coalfield [1,10]. In 2010, another new type of coal-bearing strata hosted Nb (Ta)-Zr (Hf)-REE-Ga polymetallic deposit of volcanic origin was discovered in the late Permian coal-bearing strata of eastern Yunnan, southwestern China [11]. Similar polymetallic deposits have since been discovered in some coalfields from southern China [1,3]. Similar to most typical areas enriched in rare metals in coal-bearing strata, the tectonic controls on the localization of the metalliferous coal deposits and the mechanisms of rare-metal mineralization in south China and south Primorye of Russia have been studied comparatively in detail [3]. The possible recovery of rare earth elements from coal and its combustion products such as fly ash is an exciting new research area [2,12–16], because coal and its combustion derivation (fly ash) may have elevated concentrations of these rare metals.

The purpose of this paper is to discuss the mineralogical and geochemical compositions of tuff layer in late Permian coal-bearing strata of Zhongliangshan mine, Chongqing, southwestern China. It also contributes to the discussion on the origin and potential prospects of rare metals mineralization of the tuff.

## 2. Geological Setting

The Zhongliangshan mine is located in the urban area of Chongqing, southwestern China (Figure 1). The coal-bearing sequence is the late Permian Longtan Formation ($P_3l$), which is composed of the light gray, gray, dark gray mudstone, sandy mudstone, siltstone, sandstone and coal seams. This formation is enriched in brachiopods, fern, cephalopods, bivalves, trilobite and other fossils. The Longtan Formation was deposited in a continental–marine transitional environment and has a thickness varying from 26.5 to 105.02 m, with an average of 71.08 m. It contains 10 coal seams, which are identified as K1 to K10 from top to bottom. The Changxin Formation conformably overlies the Longtan Formation and is mainly composed of thick layers of brown-gray, dark gray limestone intercalated with thin layers of mudstone and flint nodules. Some fossils including brachiopods, spindle dragonflies, sponges, corals, and trilobites are enriched in the Changxin Formation. The Maokou Formation disconformably underlies the Longtan Formation, which consists of thick layers of light gray to dark gray bioclastic limestone.

The tuff layer, with a thickness mostly of 2–5 m, light-gray or light-gray–white in color, and a conchoidal fracture and a soapy feel, is located at the lowermost Longtan Formation. The K10 coal seam conformably overlies the tuff layer, which has a disconformable contact with the underlying Maokou Formation (middle Permian) (Figure 2). The tuff is enriched in pyrite and shows massive bedding structure. The tuff was derived from the basalt eruption and deposited directly on the weathered surface of the Maokou Formation limestone, and then was subjected to weathering, leaching, and eluviation [17,18]. It is usually described as bauxite or bauxitic mudstone during core sample identification or field lithological description [17,18].

## 3. Samples and Analytical Procedures

A total of 21 bench samples were taken from the tuff layer in the Zhongliangshan mine, following the Chinese Standard GB/T 482-2008 [19]. Each tuff bench sample was cut over an area 10-cm wide, 10-cm deep and 10-cm thick. All collected samples were immediately stored in plastic bags to minimize contamination and oxidation. Large chips were selected at random from each sample for preparation of polished sections and also kept for later reference if required. The remainder of each sample was crushed and ground to pass through the 200-mesh sieve for analysis.

The loss of ignition (LOI) of each sample was determined according to ASTM standard D3174 [20]. All samples were analyzed by X-ray diffraction (XRD) using a D8 advance powder diffractmeter with Ni-filtered Cu-Kα radiation and a scintillation detector. The XRD pattern was recorded over a 2θ interval of 2.6°–70°, with a step size of 0.02°. X-ray diffractograms of the tuff samples were subjected to quantitative mineralogical analysis using Siroquant™ of China University of Mining and Technology (Beijing), a commercial interpretation software developed by Taylor [21] based on the principles for diffractogram profiling set out by Rietveld [22]. Further details indicating the use of this technique for coal-related materials are given by Ward *et al.* [23,24] and Ruan and Ward [25]. A Scanning Electron Microscope in conjunction with an energy-dispersive X-ray spectrometer (SEM-EDX, JEOL JSM-6610LV+OXFORD X-max, Tokyo, Japan), with an accelerating voltage of 20 kV, was used to study morphology and microstructure of minerals, and also to determine the distribution of some elements in tuff samples under a high vacuum mode in Chongqing Institute of Geology and Mineral Resources.

*Minerals* **2016**, *6*, 47

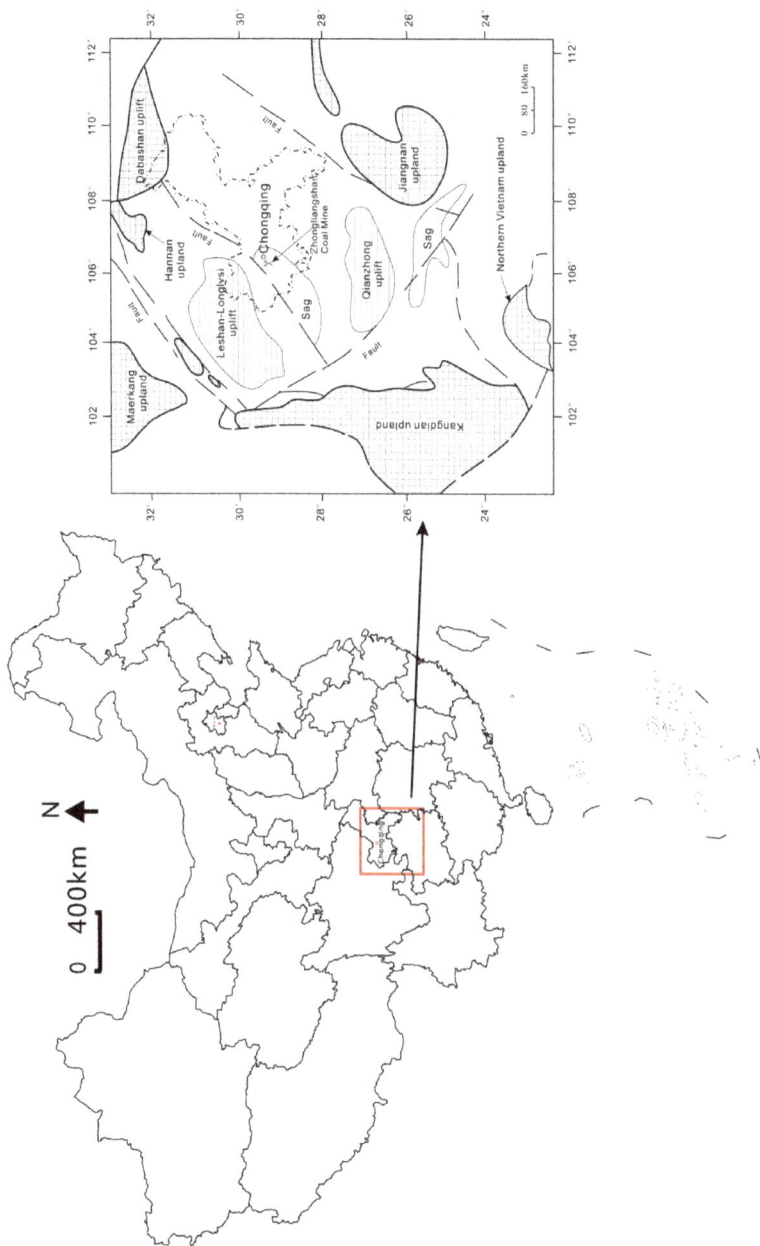

**Figure 1.** Location of the Zhongliangshan Mine, Chongqing, southwestern China.

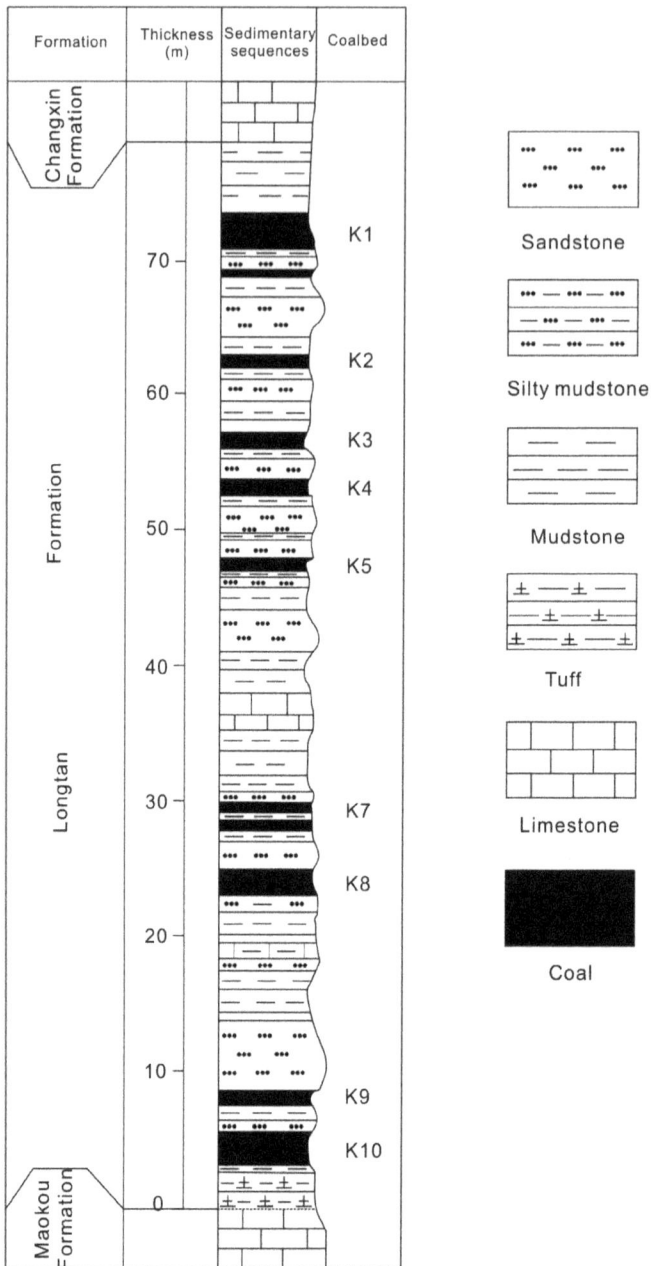

**Figure 2.** Generalized sedimentary sequence at the Zhongliangshan Mine, Chongqing.

Percentages of major element oxides including $SiO_2$, $TiO_2$, $Al_2O_3$, $Fe_2O_3$, MnO, MgO, CaO, $Na_2O$, $K_2O$, and $P_2O_5$ in the tuff samples were determined by X-ray fluorescence spectrometry (XRF) in Chongqing Institute of Geology and Mineral Resources. The contents of trace elements were determined by inductively coupled mass spectrometry (Thermo X series II ICP-MS, Thermo

Fisher Scientific, Waltham, MA, USA) in Chongqing Institute of Geology and Mineral Resources. The procedures of ICP-MS were: weigh 0.25 g sample in a 50 mL Teflon beaker; add 20 mL $HNO_3$-$HClO_4$-HF (volume ratio of 4:1:5) and 2 mL $H_2SO_4$; place on a temperature controlled heating plate and heat to 230 °C until like wet salt; then heat to 280 °C and evaporate to dryness; turn off the heating plate to cool the sample for 3 min; add 8 mL concentrated aqua regia; incubate for 10 min; transfer the solution to a 25 mL plastic flask; mix and volume; take 5 mL of solution to a 25 mL volumetric flask and dilute to the mark; and study using high resolution inductively coupled plasma mass spectrometry.

## 4. Results

### 4.1. Minerals

The proportion of each crystalline phase of the tuff identified by X-ray diffractometry plus Siroquant is given in Table 1. The minerals in Zhongliangshan tuff mainly include kaolinite, illite, pyrite, anatase, calcite, gypsum, and quartz. Some trace minerals such as zircon, florencite, jarosite, and barite, are observed under SEM-EDX.

**Table 1.** Mineral compositions of Tuff by XRD analysis and Siroquant (%).

| Sample | Kaolinite | Illite | Pyrite | Anatase | Calcite | Quartz | Gypsum |
|--------|-----------|--------|--------|---------|---------|--------|--------|
| S140SE7-1 | 83.2 | - | 13.6 | 2.7 | - | 0.5 | - |
| S140SE7-2 | 85.1 | - | 9.3 | 3.5 | - | 2.1 | - |
| S140SE7-3 | 65.4 | - | 11.6 | 8.2 | - | 14.8 | - |
| S140SE7-4 | 69.9 | - | 20.2 | 6.7 | - | 3.3 | - |
| S140SE7-5 | 53.7 | - | 38.2 | 2.8 | - | - | 5.3 |
| S140SE7-6 | 61.4 | - | 30.8 | 3.2 | - | - | 4.6 |
| S140SE7-7 | 61.9 | - | 19.8 | 4.1 | - | - | 14.2 |
| S140SE7-8 | 78.7 | - | 14.8 | 3.3 | - | 0.1 | 3.1 |
| S140SE7-9 | 80 | - | 10.6 | 7.5 | - | 0.3 | 1.7 |
| S140SE7-10 | 65.8 | - | 26.5 | 4.7 | - | 0.3 | 2.7 |
| S140SE7-11 | 72.4 | - | 16.1 | 8.8 | - | 0.3 | 2.5 |
| S140SE7-12 | 73.4 | 6.7 | 2.8 | 14.3 | - | 0.4 | 2.4 |
| S140SE7-13 | 49.9 | 18.6 | 10.6 | 9.7 | 6.1 | 0.3 | 4.8 |
| S140SE7-14 | 11.4 | 3.8 | 28.8 | 0.8 | 45 | - | 10.3 |
| S140SE7-15 | 52.9 | 21 | 3.1 | 8.4 | 10.7 | 0.5 | 3.5 |
| S140SE7-16 | 81.5 | 4.4 | 1.4 | 6.9 | 3.8 | 0.1 | 1.8 |
| S140SE7-17 | 86.4 | 4.4 | 1.1 | 6.9 | - | - | 1.2 |
| S140SE7-18 | 80.7 | 6 | 2.4 | 7.2 | 0.4 | 0.4 | 2.8 |
| S140SE7-19 | 34.5 | 31.3 | 1.9 | 3.6 | 24.7 | 0.3 | 3.7 |
| S140SE7-20 | 67.2 | 18.6 | 1 | 10.5 | 0.1 | 0.5 | 2.2 |
| S140SE7-21 | 56 | 35.3 | 0.4 | 7.2 | 0.3 | - | 0.7 |
| Average | 65.30 | 15.01 | 12.62 | 6.24 | 11.39 | 1.61 | 3.97 |

#### 4.1.1. Kaolinite and Illite

Kaolinite is the dominant mineral of the tuff in Zhongliangshan (Figure 3). The average content of kaolinite is up to 65.3%, and all the samples are richer than 50% except for samples S140SE7-13, S140SE7-14, and S140SE7-19. Kaolinite occurs mainly as matrix material (Figure 4A), and to a lesser extent, as vermicular (Figure 4B) and individual massive (Figure 4C). Illite occurs at the lower part of the profile (Figure 3).

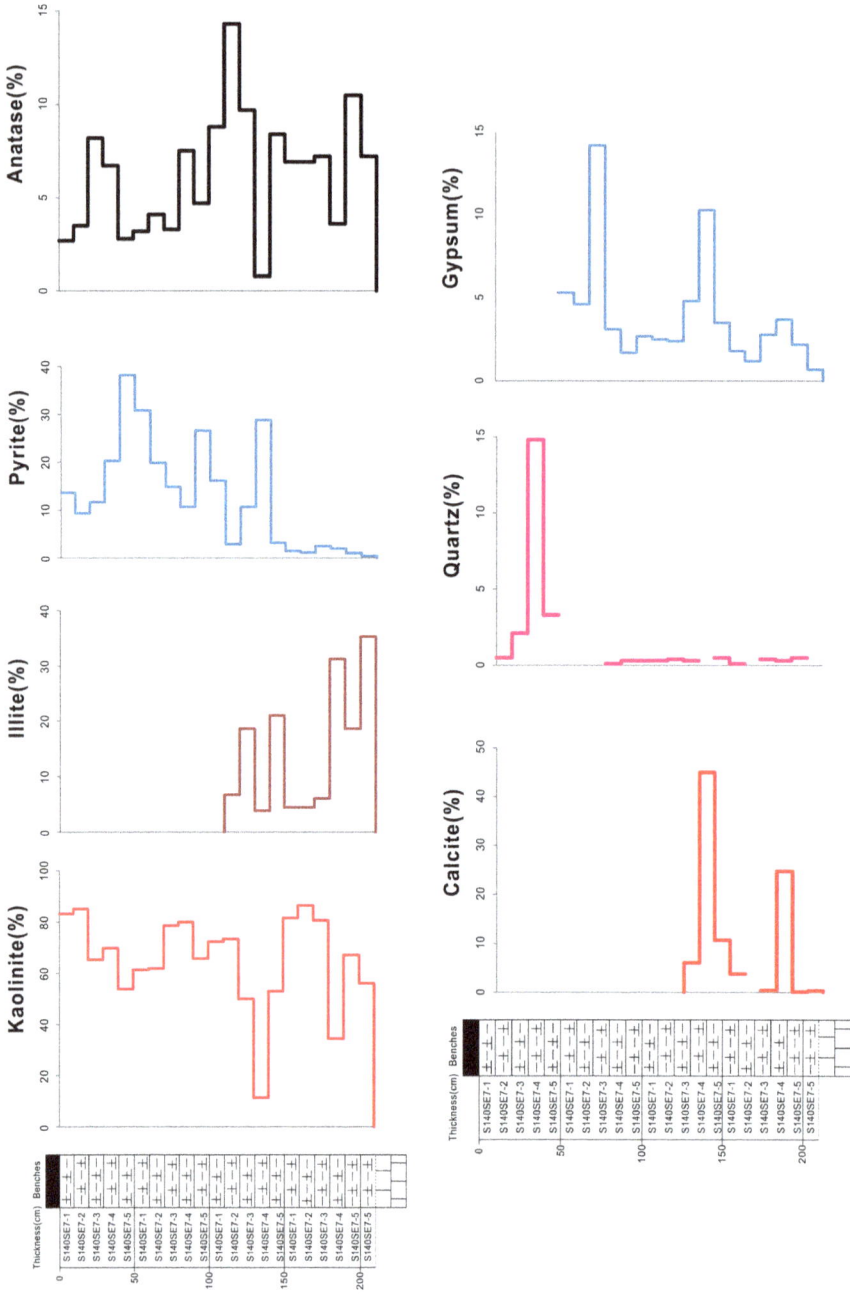

**Figure 3.** Vertical variations of minerals from the tuff in the Zhongliangshan mine.

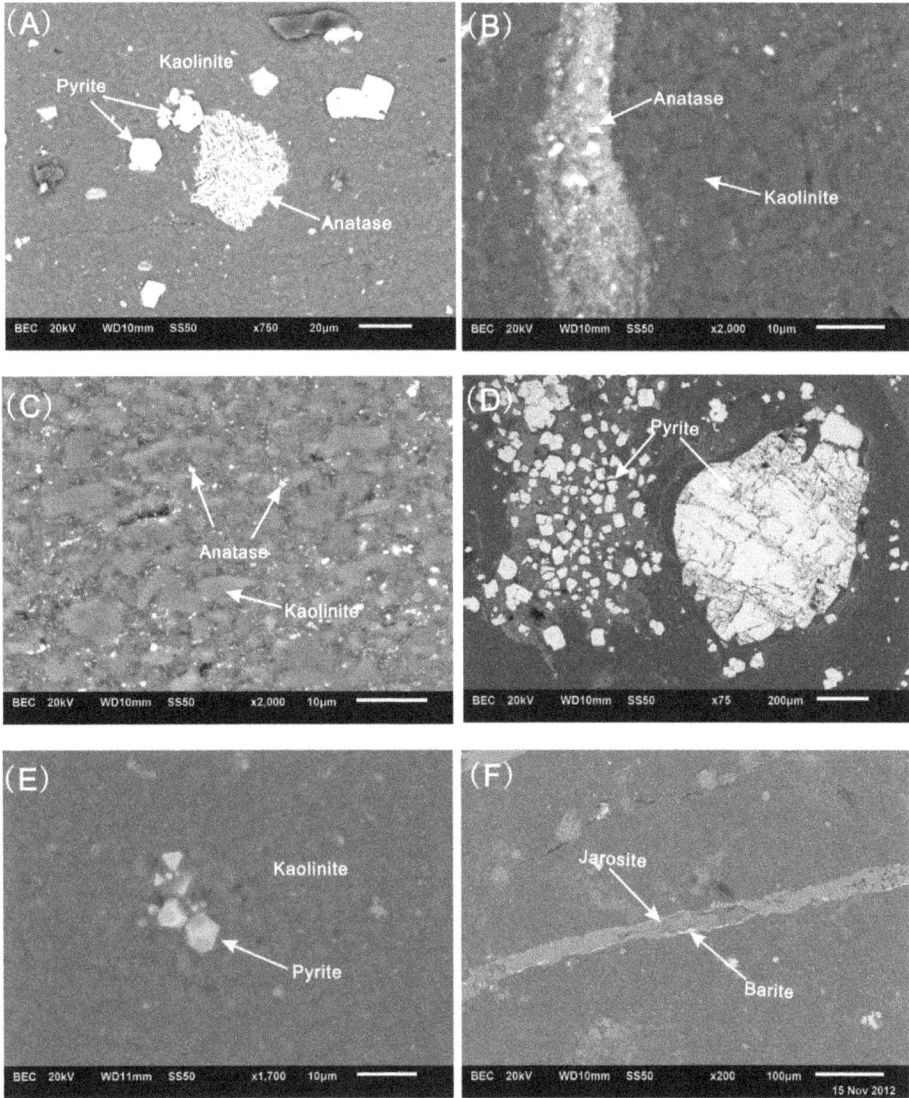

**Figure 4.** Back scattered images of minerals in the Zhongliangshan tuff: (**A**) kaolinite, pyrite and anatase in sample S140SE7-1; (**B**) kaolinite and anatase in sample S140SE7-8; (**C**) kaolinite and anatase in sample S140SE7-1; (**D**) pyrite in sample S140SE7-4; (**E**) pyrite and kaolinite in sample S140SE7-6; and (**F**) jarosite and barite in sample S140SE7-18.

### 4.1.2. Pyrite, Jarosite and Barite

Pyrite distributes widely in the tuff samples and its content varies from 0.4% to 38.2% (12.6% on average). Its content is gradually decreasing from top to bottom (Figure 3), suggesting that the upper portion has been more subjected to seawater. Pyrite mainly occurs as discrete particles (Figure 4A,D), lumps (Figure 4D), and in some cases, as cubic crystal and pentagonal dodecahedron (Figure 4E). Jarosite occurs as fracture-fillings (Figure 4F), indicating a weathering product of pyrite. Barite is located on the edge of jarosite (Figure 4F), which may be formed by the reaction of jarosite with the hydrothermal solution containing Ba.

### 4.1.3. Anatase

Anatase is present evenly in the tuff samples and varies from 0.8% to 14.3% with an average of 6.2%. The content of Nb in the anatase is up to 0.18% determined by SEM-EDX. Anatase occurs mainly as irregular fine particles (Figure 4B,C) or as colloidal (Figures 4A and 5A) in the kaolinite matrix.

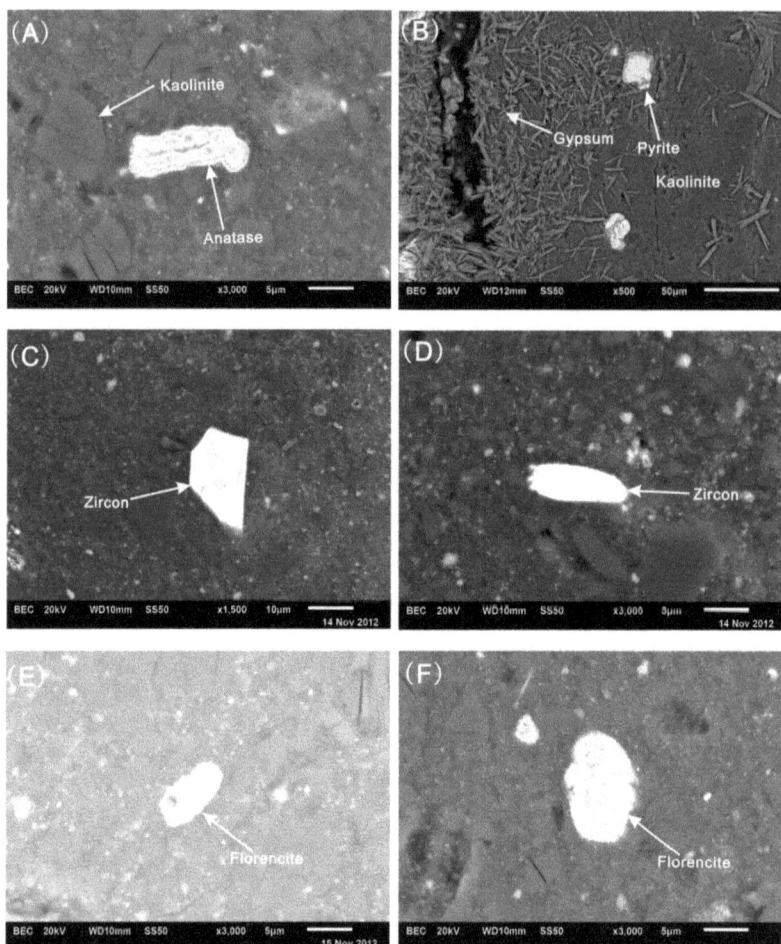

**Figure 5.** Back scattered images of minerals in the Zhongliangshan tuff: (**A**) kaolinite and anatase in sample S140SE7-4; (**B**) gypsum, pyrite, and kaolinite in sample S140SE7-10; (**C**) zircon in sample S140SE7-15; (**D**) zircon in the sample S140SE7-15; (**E**) florencite in sample S140SE7-21; and (**F**) florencite in sample S140SE7-4.

### 4.1.4. Calcite and Gypsum

Calcite distributes at the lower portion of the profile (Table 1, Figure 3), similar to that of illite. Gypsum occurs as radiating forms in the tuff and is present on the edge of fractures (Figure 5B), indicating an epigenetic origin.

### 4.1.5. Zircon and Florencite

Although zircon and florencite are at concentration below the detection limit of the XRD and Siroquant analysis, they have been observed under SEM-EDX in the tuff samples of the present study. Zircon occurs as subhedral (Figure 5C) and long axis (Figure 5D) in the kaolinite matrix. Florencite occurs as ellipsoidal form in kaolinite; however, minerals containing medium (M-REE) and heavy-rare earth elements (H-REE) have not been observed (M-REE include Eu, Gd, Tb, Dy, and Y; and H-REE include Ho, Er, Tm, Yb, and Lu [7]).

### 4.2. Major Elements

The loss of ignition of the tuff samples varies from 13.94% to 23.56%, with an average of 17.7%. The major element oxides are mainly represented by $SiO_2$ (35.3% on average) and $Al_2O_3$ (29.23%), followed by $Fe_2O_3$ (10.95%) and $TiO_2$ (3.82%) (Table 2). The ratio of $SiO_2/Al_2O_3$ is from 1.16 to 1.26 and averages 1.21, higher than the theoretical value of kaolinite (1.18). The ratio of $TiO_2/Al_2O_3$ is from 0.09 to 0.15, with an average of 0.13.

### 4.3. Trace Elements

Compared with the average concentration of the Upper Continental Crust (UCC) [26], some trace elements are enriched in the tuff samples from Zhongliangshan mine (Table 2). The concentration coefficients (CC, the ratio of the trace-element concentrations in investigated samples *vs.* UCC) of trace elements higher than 10 include Li, Cr, Cu, Cd, Sb and Re; whereas the elements with CC between 5 and 10 include V, Ni, Zr, Hf, and U. Elements Be, Sc, Co, Zn, Ga, Nb, REE, Ta, and Th, have a CC between 2 and 5. Elements Rb, Sr, Ba, and Tl are depleted, with a CC < 0.5. Other trace elements have concentrations close to the UCC, with CC between 0.5 and 2.

#### 4.3.1. Scandium

The average content of Sc in tuff samples is 30.1 µg/g, which is close to these of the tuffs from Songzao (29.8 µg/g), Nanchuan (26.3 µg/g) and the mafic rocks (29 µg/g, 1060 samples) [27]. Scandium is immobile during weathering and alteration and thus can be used as a reliable indicator for the source of tonsteins in coal-bearing strata system [28,29].

#### 4.3.2. Vanadium, Cr, Co and Ni

The average contents of V, Cr, Co and Ni in the investigated samples are 576, 360, 39.8, and 114 µg/g, respectively, close to the tuff from Songzao (V, Cr, Co, and Ni being 576, 549, 37.9, and 164 µg/g, respectively) [1,3] and the normal detrital sediments (888 samples) in the south of Sichuan Province surrounding Chongqing (V, Cr, Co, and Ni being 442, 206, 31, and 61 µg/g, respectively) [30]. The contents of V and Cr have the same variations through the seam section, gradually increasing from top to bottom (Figure 6). However, the contents of Co and Ni are higher in the middle relative to the upper and lower portions (Figure 6). The terrigenous source of the inorganic matter in the late Permian coals and normal sediments in southwestern China is the Emeishan Basalt of the Kangdian Upland, which is enriched in V, Cr, Co, and Ni [31,32]. The values of tuff samples in the Zhongliangshan mine are close to those in normal sediments, indicating the normal sediments in southwestern China and tuff in Zhongliangshan have the same magmatic sources (the Emeishan basalt magma enriched in V, Cr, Co, and Ni). Dai *et al.* [1,3,18] suggested that some dark minerals such as basic plagioclase and pyroxene in the basalt rocks could be easily decomposed under weathering conditions and then transported into coal-bearing basin as complex anions.

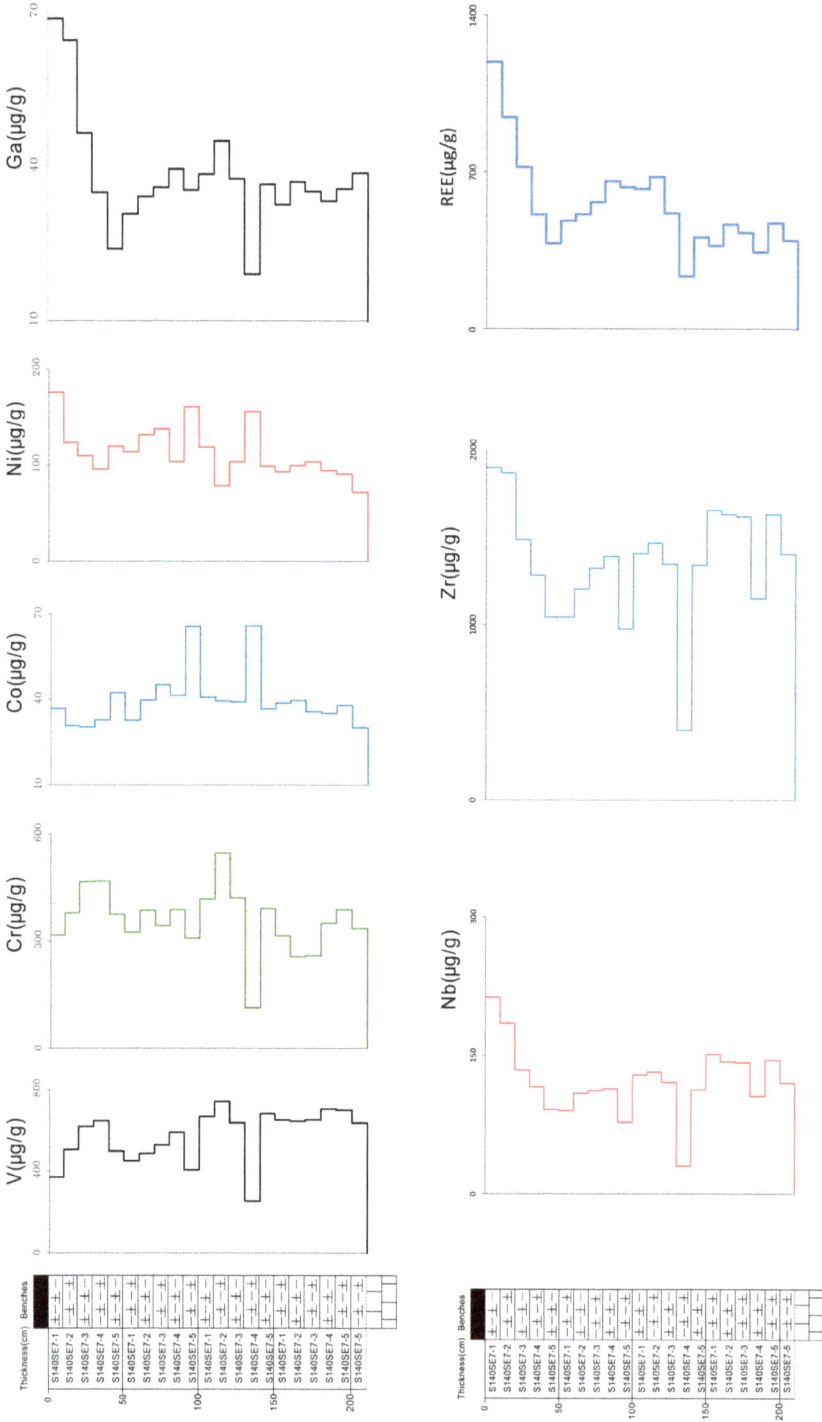

**Figure 6.** Vertical variations of selected trace elements of the tuff in Zhongliangshan mine, Chongqing.

**Table 2.** Elemental concentrations in Tuff samples from the Zhongliangshan Mine (elements in µg/g, Oxides in %).

| Sample | LOI | SiO₂ | TiO₂ | Al₂O₃ | Fe₂O₃ | MnO | MgO | CaO | Na₂O | K₂O | P₂O₅ | FeO | SiO₂/Al₂O₃ | TiO₂/Al₂O₃ | Li | Be | Sc | V | Cr | Co | Ni | Cu |
|---|---|---|---|---|---|---|---|---|---|---|---|---|---|---|---|---|---|---|---|---|---|---|
| S140SE7-1 | 19.87 | 36.25 | 2.77 | 31.23 | 9.1 | 0.012 | 0.26 | 0.12 | 0.14 | 0.1 | 0.075 | 0.6 | 1.16 | 0.09 | 454 | 9.46 | 33.4 | 371 | 317 | 36.8 | 176 | 247 |
| S140SE7-2 | 17.87 | 38.9 | 3.21 | 33.6 | 5.52 | 0.009 | 0.26 | 0.1 | 0.31 | 0.1 | 0.04 | 0.55 | 1.16 | 0.1 | 445 | 9.91 | 38 | 507 | 379 | 30.8 | 124 | 243 |
| S140SE7-3 | 18.15 | 36.84 | 3.82 | 30.98 | 9.52 | 0.02 | 0.2 | 0.11 | 0.22 | 0.1 | 0.051 | 0.5 | 1.19 | 0.12 | 349 | 7.55 | 31.8 | 621 | 467 | 30.3 | 110 | 245 |
| S140SE7-4 | 19.66 | 32.69 | 3.63 | 27.44 | 15.46 | 0.21 | 0.19 | 0.14 | 0.17 | 0.1 | 0.051 | 0.8 | 1.19 | 0.13 | 367 | 6.33 | 29.1 | 648 | 469 | 32.8 | 96.2 | 259 |
| S140SE7-5 | 23.56 | 24.78 | 2.57 | 20.01 | 28.09 | 0.015 | 0.18 | 0.36 | 0.13 | 0.085 | 0.037 | 0.75 | 1.24 | 0.13 | 337 | 5.66 | 25.6 | 500 | 375 | 42.3 | 120 | 258 |
| S140SE7-6 | 21.65 | 28.05 | 2.92 | 23.17 | 23 | 0.015 | 0.22 | 0.31 | 0.15 | 0.094 | 0.046 | 0.6 | 1.21 | 0.13 | 299 | 5.42 | 24 | 453 | 325 | 32.7 | 114 | 235 |
| S140SE7-7 | 19.52 | 32.7 | 3.28 | 27.62 | 16.08 | 0.012 | 0.27 | 0.12 | 0.15 | 0.11 | 0.043 | 0.4 | 1.18 | 0.12 | 421 | 6.95 | 28.2 | 489 | 386 | 39.7 | 132 | 286 |
| S140SE7-8 | 19.06 | 33.86 | 3.35 | 28.75 | 14.17 | 0.013 | 0.31 | 0.12 | 0.17 | 0.13 | 0.048 | 0.4 | 1.18 | 0.12 | 401 | 6.61 | 32.4 | 530 | 344 | 45 | 138 | 278 |
| S140SE7-9 | 16.45 | 38.23 | 3.87 | 32.89 | 7.81 | 0.008 | 0.25 | 0.11 | 0.16 | 0.11 | 0.049 | 0.35 | 1.16 | 0.12 | 442 | 7.11 | 33.3 | 592 | 388 | 41.4 | 104 | 251 |
| S140SE7-10 | 20.62 | 29.97 | 2.89 | 24.88 | 20.9 | 0.011 | 0.23 | 0.17 | 0.11 | 0.13 | 0.03 | 0.55 | 1.2 | 0.12 | 311 | 4.87 | 24.7 | 408 | 308 | 65.4 | 161 | 282 |
| S140SE7-11 | 17.82 | 35.72 | 4.38 | 29.64 | 11.54 | 0.004 | 0.15 | 0.26 | 0.21 | 0.17 | 0.066 | 0.5 | 1.21 | 0.15 | 321 | 7.34 | 33.2 | 669 | 419 | 40.7 | 119 | 316 |
| S140SE7-12 | 15.42 | 40.12 | 4.65 | 33.07 | 5.18 | 0.007 | 0.2 | 0.26 | 0.49 | 0.43 | 0.086 | 0.25 | 1.21 | 0.14 | 248 | 7.59 | 36.5 | 743 | 547 | 39.4 | 78.8 | 330 |
| S140SE7-13 | 17.93 | 35.93 | 4.39 | 29.44 | 9.57 | 0.008 | 0.2 | 1.56 | 0.3 | 0.43 | 0.067 | 0.6 | 1.22 | 0.15 | 219 | 6.48 | 30.7 | 639 | 422 | 39.1 | 104 | 266 |
| S140SE7-14 | 18.09 | 15.54 | 1.55 | 11.23 | 31.32 | 0.039 | 0.43 | 20.38 | 0.16 | 0.65 | 0.017 | 1.75 | 1.38 | 0.14 | 31.5 | 2.16 | 12.6 | 253 | 114 | 65.6 | 156 | 146 |
| S140SE7-15 | 15.61 | 38.45 | 4.33 | 30.94 | 5.28 | 0.01 | 0.33 | 3.4 | 0.53 | 0.82 | 0.05 | 0.85 | 1.24 | 0.14 | 170 | 6.3 | 31.3 | 683 | 391 | 36.6 | 99.4 | 188 |
| S140SE7-16 | 14.89 | 40.64 | 5.28 | 34.4 | 1.94 | 0.006 | 0.16 | 1.52 | 0.5 | 0.25 | 0.045 | 0.4 | 1.18 | 0.15 | 254 | 7.2 | 35.7 | 652 | 315 | 38.7 | 93.4 | 265 |
| S140SE7-17 | 14.91 | 41.58 | 5.03 | 35.15 | 1.99 | 0.004 | 0.21 | 0.13 | 0.49 | 0.37 | 0.056 | 0.35 | 1.18 | 0.14 | 299 | 6.89 | 34.8 | 646 | 257 | 39.6 | 100 | 255 |
| S140SE7-18 | 15.35 | 40.55 | 4.89 | 34.01 | 3.47 | 0.004 | 0.24 | 0.4 | 0.49 | 0.49 | 0.059 | 0.85 | 1.19 | 0.14 | 258 | 6.31 | 33.1 | 653 | 260 | 35.7 | 104 | 265 |
| S140SE7-19 | 15.5 | 37.44 | 4.08 | 29.19 | 4.03 | 0.013 | 0.59 | 6.57 | 0.72 | 1.47 | 0.045 | 0.9 | 1.28 | 0.14 | 153 | 5.66 | 25 | 707 | 351 | 35.1 | 94.7 | 199 |
| S140SE7-20 | 13.94 | 42.07 | 5.01 | 34.88 | 1.83 | 0.003 | 0.32 | 0.47 | 0.47 | 0.76 | 0.06 | 0.55 | 1.21 | 0.14 | 246 | 6.79 | 32.1 | 701 | 388 | 37.9 | 91.2 | 281 |
| S140SE7-21 | 15.82 | 40.92 | 4.24 | 32.45 | 3.12 | 0.004 | 0.46 | 1.12 | 0.53 | 1.26 | 0.055 | 1.3 | 1.26 | 0.13 | 168 | 5.58 | 26.8 | 638 | 336 | 30.1 | 72 | 193 |
| Average | 17.7 | 35.3 | 3.82 | 29.28 | 10.9 | 0.02 | 0.27 | 1.8 | 0.31 | 0.39 | 0.05 | 0.66 | 1.21 | 0.13 | 295 | 6.58 | 30.1 | 576 | 360 | 39.8 | 114 | 252 |
| UCC | nd | nd | nd | nd | nd | nd | nd | nd | nd | nd | nd | nd | nd | nd | 20 | 3 | 11 | 60 | 35 | 10 | 20 | 25 |
| CC | nd | nd | nd | nd | nd | nd | nd | nd | nd | nd | nd | nd | nd | nd | 14.7 | 2.2 | 2.7 | 9.6 | 10.3 | 4 | 5.7 | 10.1 |

Table 2. Cont.

| Sample | Zn | Ga | Rb | Sr | Zr | Nb | Mo | Cd | In | Sb | Cs | Ba | Hf | Ta | W | Re | Tl | Pb | Bi | Th | U |
|---|---|---|---|---|---|---|---|---|---|---|---|---|---|---|---|---|---|---|---|---|---|
| S140SE7-1 | 237 | 68.7 | 3.98 | 292 | 1898 | 213 | 3.4 | 3.13 | 0.615 | 4.22 | 1.17 | 146 | 57.6 | 15.9 | 4.02 | 0.042 | 0.11 | 58.1 | 1.41 | 50.3 | 25.7 |
| S140SE7-2 | 190 | 64.5 | 2.86 | 190 | 1868 | 185 | 2.12 | 2.04 | 0.654 | 2.82 | 1.15 | 113 | 51.4 | 14.6 | 3.52 | 0.02 | 0.058 | 40.4 | 1.21 | 49.7 | 24.5 |
| S140SE7-3 | 143 | 46.4 | 2.96 | 282 | 1487 | 134 | 1.73 | 0.952 | 0.587 | 3.51 | 1.48 | 141 | 37.7 | 8.88 | 4.44 | 0.018 | 0.085 | 39 | 1.62 | 33.4 | 9.83 |
| S140SE7-4 | 123 | 34.9 | 3.52 | 256 | 1284 | 116 | 1.8 | 1.7 | 0.719 | 3.7 | 1.4 | 122 | 33.4 | 8.29 | 3.44 | 0.013 | 0.087 | 47.4 | 1.63 | 28.2 | 9.71 |
| S140SE7-5 | 107 | 23.9 | 3.2 | 166 | 1046 | 91.6 | 1.56 | 1.46 | 0.457 | 4.13 | 1.07 | 92 | 26 | 6.91 | 3.16 | 0.014 | 0.207 | 61.3 | 1.22 | 23.8 | 8.11 |
| S140SE7-6 | 100 | 30.7 | 2.03 | 191 | 1046 | 90.2 | 1.22 | 1.31 | 0.437 | 4.23 | 0.892 | 89.6 | 27.1 | 5.96 | 2.75 | 0.007 | 0.24 | 53.5 | 1.15 | 23.4 | 8.97 |
| S140SE7-7 | 109 | 34.1 | 3.69 | 188 | 1206 | 109 | 0.985 | 1.22 | 0.494 | 3.41 | 1.1 | 101 | 30.7 | 7.6 | 2.83 | 0.006 | 0.103 | 49.3 | 1.35 | 27.9 | 11.4 |
| S140SE7-8 | 124 | 35.9 | 3.1 | 232 | 1324 | 112 | 1.04 | 1.31 | 0.541 | 3.91 | 1.1 | 115 | 33.8 | 7.82 | 3.66 | 0.009 | 0.189 | 55.2 | 1.51 | 29 | 13.2 |
| S140SE7-9 | 115 | 39.5 | 2.66 | 194 | 1392 | 114 | 1.03 | 1.5 | 0.519 | 2.23 | 1.05 | 99.3 | 36.9 | 8.57 | 3.67 | 0.006 | 0.058 | 34 | 1.48 | 30.4 | 16.6 |
| S140SE7-10 | 103 | 35.4 | 2.46 | 154 | 978 | 77.7 | 1.45 | 1.83 | 0.516 | 6.79 | 0.927 | 79.1 | 25.4 | 5.68 | 2.8 | 0.01 | 0.09 | 75.2 | 1.14 | 21.7 | 12.6 |
| S140SE7-11 | 192 | 38.5 | 5.34 | 287 | 1408 | 129 | 1.54 | 2.77 | 0.6 | 2.73 | 1.94 | 127 | 36.1 | 8.79 | 3.88 | 0.006 | 0.093 | 44 | 1.73 | 30 | 16 |
| S140SE7-12 | 178 | 44.9 | 14.9 | 325 | 1467 | 132 | 1.68 | 4.45 | 0.572 | 1.71 | 5.03 | 141 | 37.9 | 8.98 | 4.05 | 0.013 | 0.118 | 23.6 | 2.06 | 31.8 | 17.7 |
| S140SE7-13 | 161 | 37.6 | 12.5 | 341 | 1348 | 121 | 1.67 | 5.14 | 0.627 | 2.21 | 3.52 | 137 | 33.2 | 8.27 | 3.41 | 0.006 | 0.134 | 33.8 | 1.69 | 28.5 | 14 |
| S140SE7-14 | 108 | 19 | 14.9 | 503 | 400 | 30.7 | 0.712 | 18.4 | 0.211 | 3.32 | 1.66 | 60.3 | 9.45 | 2.28 | 1.7 | 0.008 | 0.292 | 34.9 | 0.595 | 8.57 | 5.46 |
| S140SE7-15 | 148 | 36.5 | 23.7 | 404 | 1342 | 113 | 1.37 | 6.96 | 0.501 | 2.08 | 5.27 | 143 | 34.9 | 7.88 | 3.69 | 0.006 | 0.156 | 26.9 | 1.52 | 28.7 | 14.2 |
| S140SE7-16 | 170 | 32.6 | 6.83 | 279 | 1654 | 151 | 1.78 | 3.52 | 0.628 | 1.27 | 1.96 | 131 | 41.3 | 10.4 | 4.61 | 0.018 | 0.071 | 19.6 | 1.34 | 34 | 17.4 |
| S140SE7-17 | 166 | 37 | 9.09 | 261 | 1631 | 143 | 0.996 | 4 | 0.567 | 1.68 | 2.37 | 131 | 41.4 | 10.1 | 3.76 | 0.025 | 0.08 | 23.8 | 1.36 | 32.2 | 18.5 |
| S140SE7-18 | 147 | 35.1 | 12.1 | 286 | 1619 | 142 | 1.4 | 3.55 | 0.531 | 1.99 | 2.25 | 137 | 39.8 | 9.76 | 3.74 | 0.013 | 0.08 | 26.1 | 1.39 | 32 | 19.5 |
| S140SE7-19 | 139 | 33.3 | 39.5 | 387 | 1153 | 106 | 1.49 | 6.92 | 0.505 | 2.42 | 5.33 | 160 | 28.8 | 7.07 | 3.48 | 0.036 | 0.287 | 21.8 | 1.39 | 27 | 20.1 |
| S140SE7-20 | 174 | 35.6 | 19.8 | 322 | 1631 | 145 | 1.6 | 4.14 | 0.578 | 1.76 | 3.18 | 158 | 41.4 | 9.92 | 4.05 | 0.009 | 0.164 | 23 | 1.55 | 34 | 26.4 |
| S140SE7-21 | 153 | 38.7 | 31.1 | 323 | 1405 | 120 | 1.48 | 3.47 | 0.526 | 1.5 | 4.24 | 165 | 35 | 8.39 | 3.43 | 0.015 | 0.236 | 25.2 | 1.27 | 29 | 23.8 |
| Average | 147 | 38.2 | 10.5 | 279 | 1361 | 123 | 1.53 | 3.8 | 0.54 | 2.93 | 2.29 | 123 | 35.2 | 8.67 | 3.53 | 0.01 | 0.14 | 38.9 | 1.41 | 30.2 | 15.9 |
| UCC | 71 | 17 | 112 | 350 | 190 | 25 | 1.5 | 0.1 | nd | 0.2 | 3.7 | 550 | 5.8 | 2.2 | 2 | 0.0004 | 0.8 | 20 | nd | 10.7 | 2 |
| CC | 2.1 | 2.2 | 0.1 | 0.8 | 7.2 | 4.9 | 1 | 38 | nd | 14.7 | 0.6 | 0.2 | 6.1 | 3.9 | 1.8 | 35.7 | 0.2 | 1.9 | nd | 2.8 | 7.9 |

UCC, the Upper Continental Crust; CC, concentration coefficient of trace elements in the tuff, normalized by average trace element concentrations in UCC [26]; nd, no data.

4.3.3. Niobium, Ta, Zr and Hf

The average contents of Nb, Ta, Zr, and Hf of tuff in Zhongliangshan mine are 123, 8.67, 1361, and 35.2 µg/g, respectively, being close to those of the tuff from Songzao (Nb, Ta, Zr and Hf being 118, 9.46, 1377, and 41.5 µg/g, respectively). The Nb and Zr display a similar trend, both gradually decreasing from top to bottom (Figure 6).

The concentration of (Nb, Ta)$_2$O$_5$ of tuff in Zhongliangshan mine varies from 47 to 324 µg/g and averages 186 µg/g, lower than the concentration of the late Permian "Nb (Ta)-Zr (Hf)-Ga-REE" polymetallic deposit discovered in eastern Yunnan, southwestern China [11]. (Zr, Hf)O$_2$ varies from 551 to 2632 µg/g and averages 1880 µg/g, which does not meet the minimum industrial grade of the weathering crust type deposit (8000 µg/g) [33].

The common Nb-, Zr-, REE-, and Ga-bearing minerals have rarely been observed in the tuff, and thus it is suggested that these rare metals probably occur as absorbed ions [11,29]. However, Nb may occur as isomorph in the Ti-bearing minerals (Figure 4B,C) and Zr occurs as zircon (Figure 5C,D) in studied samples.

4.3.4. Gallium

The concentration of Ga in Zhongliangshan tuff varies from 19 to 68.7 µg/g and averages 38.2 µg/g, higher than the minimum industrial grade in bauxite (20 µg/g) and coal (30 µg/g) [34], but lower than the concentration of the late Permian "Nb(Ta)-Zr(Hf)-Ga-REE" polymetallic deposit in eastern Yunnan, southwestern China [11]. From top to bottom, the concentration of Ga gradually decreases, consistent with those of the Nb and Zr. Because the geochemical nature of Ga is similar to Al [1,8], it may occur as isomorph in Al-bearing minerals (e.g., kaolinite).

*4.4. Rare Earth Elements (REE)*

In this study, REE is used to specifically represent the elemental suite La, Ce, Pr, Nd, Sm, Eu, Gd, Tb, Dy, Y, Ho, Er, Tm, Yb, and Lu [35]. The abundances and geochemical parameters of REE in the tuff samples are listed in Tables 3 and 4 respectively. The concentration of REE varies from 234 to 1189 µg/g, with an average of 548 µg/g. The concentration of REE gradually decreases from top to bottom, similar to that of Nb and Zr. Yttrium is closely associated with lanthanides in nature, because its ionic radius is very similar and its ionic charge is equal to that of Ho [7]. For this reason, yttrium is generally placed between Dy and Ho in normalized REE patterns [36]. Based on Seredin-Dai's classification [7], a three-fold geochemical classification of REE was used in the present study, including light (L-REE: La, Ce, Pr, Nd, and Sm), medium (M-REE: Eu, Gd, Tb, Dy, and Y), and heavy (H-REE: Ho, Er, Tm, Yb, and Lu) REE [7]. Accordingly, three enrichment types are identified, L-type (light REE; La$_N$/Lu$_N$ > 1), M-type (medium REE; La$_N$/Sm$_N$ < 1, Gd$_N$/Lu$_N$ > 1), and H-type (heavy REE: La$_N$/Lu$_N$ < 1), in comparison with the upper continental crust [7]. This classification has been widely adopted and used in recent years [2].

Table 4 and Figure 7 illustrate that the tuff in the Zhongliangshan mine is mainly enriched in heavy REE. Only samples S140SE7-1, S140SE7-3, S140SE7-4, S140SE7-6, and S140SE7-8 are enriched in light REE; and samples S140SE7-2, S140SE7-10, and 140SE7-14 are enriched in medium REE. From top to bottom, the light REE enrichment only occurs in the upper portion of the profile, while the lower portion is enriched in heavy REE and the medium REE enrichment occasionally occurs in the middle portion.

The Ce-anomaly (expressed as δCe) values vary from 0.70 to 1.77, with an average of 1.41, indicating a well-pronounced Ce positive anomaly. The REE distribution patterns of the tuff display positive Ce anomalies, owing to the in-situ precipitation of Ce$^{4+}$ in the process of weathering, leaching, and eluviation [35]. The Eu-anomaly (δEu) values varying from 0.76 to 1.51, with an average of 1.06, show a slight Eu positive anomaly, indicating the tuff and the Emeishan basalt have the same origin [35]. From top to bottom, δCe and δEu markedly increase. The distribution of REE of the tuff in the Zhongliangshan mine appears as a sawtooth shape, the portions of La-Sm and Gd-Lu occurring gentle and small slope, which indicates that the fractionation of REE is low.

**Table 3.** Rare earth elements in the tuff samples collected from the Zhongliangshan Mine (µg/g).

| Sample | La | Ce | Pr | Nd | Sm | Eu | Gd | Tb | Dy | Y | Ho | Er | Tm | Yb | Lu |
|---|---|---|---|---|---|---|---|---|---|---|---|---|---|---|---|
| S140SE7-1 | 266 | 372 | 55.7 | 212 | 34.3 | 5.38 | 30 | 4.92 | 28.3 | 142 | 5.19 | 15.7 | 2.38 | 13.2 | 1.88 |
| S140SE7-2 | 180 | 314 | 41.8 | 174 | 32.2 | 4.79 | 25.9 | 4.22 | 22.9 | 111 | 4.79 | 12 | 2.08 | 10.8 | 1.67 |
| S140SE7-3 | 134 | 313 | 26.8 | 93.9 | 16.9 | 2.99 | 15.4 | 2.94 | 16.2 | 76.8 | 2.99 | 8.08 | 1.23 | 8.02 | 1.25 |
| S140SE7-4 | 103 | 224 | 17.8 | 61.1 | 8.7 | 1.83 | 10.4 | 1.87 | 10.9 | 52.7 | 2.11 | 5.66 | 0.929 | 5.45 | 0.897 |
| S140SE7-5 | 74.6 | 152 | 12.8 | 48.1 | 8.55 | 1.79 | 8.32 | 1.76 | 9.73 | 46.7 | 1.81 | 5.28 | 0.843 | 5.4 | 0.837 |
| S140SE7-6 | 91.5 | 201 | 18.2 | 66.4 | 13.7 | 2.58 | 11.2 | 1.97 | 9.95 | 49.5 | 1.73 | 5.21 | 0.796 | 5.23 | 0.763 |
| S140SE7-7 | 95.7 | 237 | 16.5 | 52.9 | 9.71 | 2.09 | 9.78 | 1.93 | 9.99 | 56.6 | 2.02 | 5.97 | 1.03 | 6.63 | 0.996 |
| S140SE7-8 | 108 | 265 | 17.6 | 59.6 | 10.7 | 2.05 | 10.2 | 2.03 | 11.1 | 59.1 | 2.15 | 6.5 | 1.21 | 7.37 | 0.986 |
| S140SE7-9 | 111 | 338 | 19.8 | 67.4 | 11.9 | 2.5 | 11.7 | 2.24 | 11.8 | 62.2 | 2.27 | 7.23 | 1.22 | 7.69 | 1.19 |
| S140SE7-10 | 77.7 | 293 | 22.9 | 107 | 29 | 5.19 | 18.7 | 2.42 | 10.9 | 50 | 1.87 | 5.5 | 0.895 | 5.65 | 0.866 |
| S140SE7-11 | 103 | 297 | 22.3 | 79.7 | 14.7 | 2.47 | 11.4 | 2.23 | 12.4 | 58.3 | 2.42 | 7.58 | 1.28 | 8.05 | 1.23 |
| S140SE7-12 | 106 | 326 | 24.4 | 87.8 | 15.7 | 2.91 | 12.5 | 2.33 | 13.6 | 62.5 | 2.65 | 8.38 | 1.42 | 9.21 | 1.43 |
| S140SE7-13 | 76.5 | 251 | 17.7 | 60.4 | 10.4 | 2.46 | 9.64 | 2.04 | 12 | 52.7 | 2.26 | 6.79 | 1.18 | 7.91 | 1.12 |
| S140SE7-14 | 18.9 | 101 | 9.12 | 45.4 | 12.7 | 2.56 | 7.86 | 1.23 | 5.44 | 22.9 | 0.907 | 2.45 | 0.367 | 2.45 | 0.326 |
| S140SE7-15 | 57.3 | 184 | 13 | 47.7 | 9 | 2.22 | 8.96 | 1.98 | 10.7 | 53.5 | 2.16 | 6.45 | 1.13 | 7.54 | 1.04 |
| S140SE7-16 | 46.2 | 163 | 10.7 | 40.3 | 9.95 | 2.44 | 9.47 | 2.13 | 12.2 | 54.7 | 2.17 | 6.58 | 1.12 | 7.07 | 1.06 |
| S140SE7-17 | 67.2 | 222 | 15.4 | 55.7 | 9.3 | 2.59 | 9.83 | 2.01 | 11.9 | 50.8 | 2.2 | 6.38 | 1.13 | 6.99 | 1.08 |
| S140SE7-18 | 61.8 | 201 | 13.7 | 50.2 | 9.63 | 2.64 | 9.67 | 1.96 | 10.5 | 48.9 | 2.09 | 6 | 1.01 | 6.68 | 1.04 |
| S140SE7-19 | 50.7 | 158 | 11.1 | 37.9 | 8.35 | 2.43 | 7.86 | 1.45 | 8.4 | 41.1 | 1.56 | 4.65 | 0.799 | 5.21 | 0.714 |
| S140SE7-20 | 67.3 | 241 | 14.3 | 49.6 | 10.3 | 3.03 | 10.7 | 1.73 | 10.1 | 46.8 | 1.86 | 5.59 | 0.939 | 6.37 | 0.913 |
| S140SE7-21 | 56.1 | 197 | 12.2 | 43.3 | 8.68 | 2.79 | 8.27 | 1.46 | 8.61 | 40 | 1.67 | 4.98 | 0.817 | 5.35 | 0.827 |
| Average | 93.0 | 240 | 19.7 | 73.4 | 14.0 | 2.84 | 12.3 | 2.23 | 12.3 | 58.99 | 2.33 | 6.81 | 1.13 | 7.06 | 1.05 |

**Table 4.** Rare earth elements geochemical parameters of Zhongliangshan tuff.

| Sample | REE (µg/g) | LREE (µg/g) | MREE (µg/g) | HREE (µg/g) | L/M | L/H | M/H | (La/Lu)$_N$ | (La/Sm)$_N$ | (Gd/Lu)$_N$ | δCe | δEu |
|---|---|---|---|---|---|---|---|---|---|---|---|---|
| S140SE7-1 | 1189 | 940 | 211 | 38.4 | 4.46 | 24.51 | 5.49 | 1.41 | 1.16 | 1.26 | 0.70 | 0.77 |
| S140SE7-2 | 942 | 742 | 169 | 31.3 | 4.40 | 23.68 | 5.39 | 1.08 | 0.84 | 1.22 | 0.83 | 0.76 |
| S140SE7-3 | 721 | 585 | 114 | 21.6 | 5.11 | 27.10 | 5.30 | 1.07 | 1.19 | 0.97 | 1.19 | 0.85 |
| S140SE7-4 | 507 | 415 | 77.7 | 15.0 | 5.34 | 27.56 | 5.16 | 1.15 | 1.78 | 0.92 | 1.19 | 0.88 |
| S140SE7-5 | 379 | 296 | 68.3 | 14.2 | 4.33 | 20.89 | 4.82 | 0.89 | 1.31 | 0.78 | 1.12 | 0.98 |
| S140SE7-6 | 480 | 391 | 75.2 | 13.7 | 5.20 | 28.47 | 5.48 | 1.20 | 1.00 | 1.16 | 1.12 | 0.96 |
| S140SE7-7 | 509 | 412 | 80.4 | 16.6 | 5.12 | 24.74 | 4.83 | 0.96 | 1.48 | 0.78 | 1.36 | 0.99 |
| S140SE7-8 | 564 | 461 | 84.5 | 18.2 | 5.46 | 25.30 | 4.64 | 1.10 | 1.51 | 0.82 | 1.39 | 0.90 |
| S140SE7-9 | 658 | 548 | 90.4 | 19.6 | 6.06 | 27.96 | 4.61 | 0.93 | 1.40 | 0.78 | 1.64 | 0.97 |
| S140SE7-10 | 632 | 530 | 87.2 | 14.8 | 6.07 | 35.83 | 5.90 | 0.90 | 0.40 | 1.70 | 1.58 | 1.02 |
| S140SE7-11 | 624 | 517 | 86.8 | 20.6 | 5.95 | 25.13 | 4.22 | 0.84 | 1.05 | 0.73 | 1.41 | 0.88 |
| S140SE7-12 | 677 | 560 | 93.8 | 23.1 | 5.97 | 24.25 | 4.06 | 0.74 | 1.01 | 0.69 | 1.46 | 0.95 |
| S140SE7-13 | 514 | 416 | 78.8 | 19.3 | 5.28 | 21.60 | 4.09 | 0.68 | 1.10 | 0.68 | 1.56 | 1.13 |
| S140SE7-14 | 234 | 187 | 40.0 | 6.5 | 4.68 | 28.79 | 6.15 | 0.58 | 0.22 | 1.90 | 1.75 | 1.18 |
| S140SE7-15 | 407 | 311 | 77.4 | 18.3 | 4.02 | 16.98 | 4.22 | 0.55 | 0.96 | 0.68 | 1.54 | 1.14 |
| S140SE7-16 | 369 | 270 | 80.9 | 18.0 | 3.34 | 15.01 | 4.50 | 0.44 | 0.70 | 0.71 | 1.67 | 1.15 |
| S140SE7-17 | 465 | 370 | 77.1 | 17.8 | 4.79 | 20.79 | 4.34 | 0.62 | 1.08 | 0.72 | 1.57 | 1.24 |
| S140SE7-18 | 427 | 336 | 73.7 | 16.8 | 4.57 | 20.00 | 4.38 | 0.59 | 0.96 | 0.73 | 1.58 | 1.26 |
| S140SE7-19 | 340 | 266 | 61.2 | 12.9 | 4.34 | 20.57 | 4.74 | 0.71 | 0.91 | 0.87 | 1.52 | 1.38 |
| S140SE7-20 | 471 | 383 | 72.4 | 15.7 | 5.29 | 24.41 | 4.62 | 0.74 | 0.98 | 0.93 | 1.77 | 1.33 |
| S140SE7-21 | 392 | 317 | 61.1 | 13.6 | 5.19 | 23.25 | 4.48 | 0.68 | 0.97 | 0.79 | 1.72 | 1.51 |
| Average | 548 | 441 | 88.6 | 18.4 | 5.00 | 24.13 | 4.83 | 0.85 | 1.05 | 0.94 | 1.41 | 1.06 |

ΣREE, sum of La, Ce, Pr, Nd, Sm, Eu, Gd, Tb, Dy, Y, Ho, Er, Tm, Yb, and Lu; LREE, sum of La, Ce, Pr, Nd, and Sm; MREE, sum of Eu, Gd, Tb, Dy and Y; HREE, sum of Ho, Er, Tm, Yb, and Lu; L/M, ratio of LREE and MREE; L/H, ratio of LREE and HREE; M/H, ratio of MREE and HREE; (La/Lu)$_N$, ratio of (La)$_N$ and (Lu)$_N$; (La/Sm)$_N$, ratio of (La)$_N$ and (Sm)$_N$; (Gd/Lu)$_N$, ratio of (Gd)$_N$ and (Lu)$_N$; δCe = Ce$_N$/(La$_N$ × Pr$_N$)$^{1/2}$; δEu = Eu$_N$/(Sm$_N$ × Gd$_N$)$^{1/2}$; N, REE are normalized by Upper Continental Crust (UCC) [26].

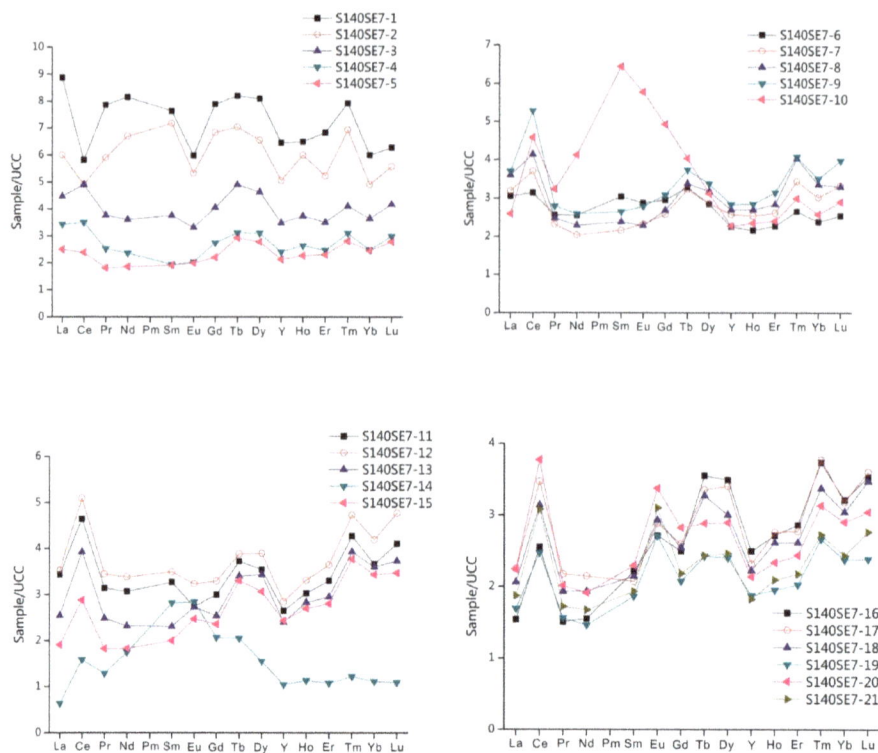

**Figure 7.** Distribution patterns of REE in the tuff samples from Zhongliangshan mine. REE are normalized by Upper Continental Crust [26].

Two reasons may be responsible for the H-REE enrichment of the tuff samples in the Zhongliangshan mine. First, L-REE can be easily leached by groundwater than H-REE; Second, L-REE can be easily adsorbed on the organic matter than the H-REE [37], which may be adsorbed by the coal seam overlying the tuff. The REE enrichment mode in the Zhongliangshan tuff is similar to that of Songzao Coalfield. Some studies have shown that L-REE are more easily to be leached by groundwater and are more apt to be adsorbed by organic matter [38–42].

## 5. Discussion

### 5.1. Origin of Tuff

In the late Permian Age, the Dongwu movement, one of the most important tectonic events in southern China, caused the upper Yangtze basin uplifting and the subsequent sea regression, which led to an extensive erosion in the area. The upper part of Maokou limestone of Sichuan Basin had been subjected to a serious erosion, resulting in the formation of a vast weathering residual plain, where peat subsequently accumulated. Meanwhile, the Emeishan basalt volcano began erupting and reached a climax in the early stage of late Permian, leading to a tuff layer overlying the Maokou limestone [17].

$Al_2O_3$ and $TiO_2$ are both stable components in the rock and would be little altered during alteration, so the ratio of $TiO_2/Al_2O_3$ (KAT) would be kept constant and can frequently be used to study the origin of volcanic ash [29,42]. It is suggested that KAT values for silicic volcanic ash are <0.02, and those for mafic and alkali volcanic ashes are >0.08 and between 0.02 and 0.08, respectively [43,44]. The KAT ratios of the tuff in the Zhongliangshan mine are >0.08 (Figure 8), suggesting a mafic volcanic origin. In addition, the tuff samples fall in the area basalt to alkali basalt from the La/Yb-REE diagram (Figure 9), indicating a feature of alkali basalt.

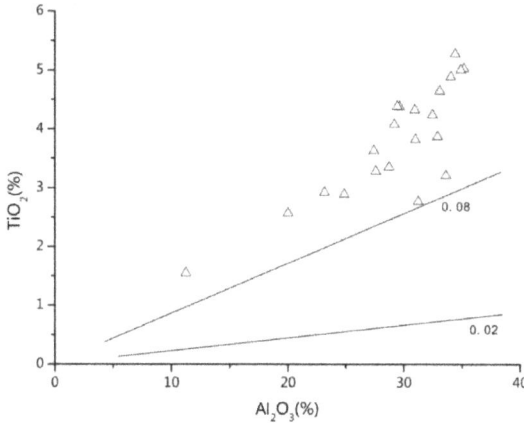

**Figure 8.** Plot for $TiO_2$ *vs.* $Al_2O_3$ of tuff samples in the Zhongliangshan mine, Chongqing.

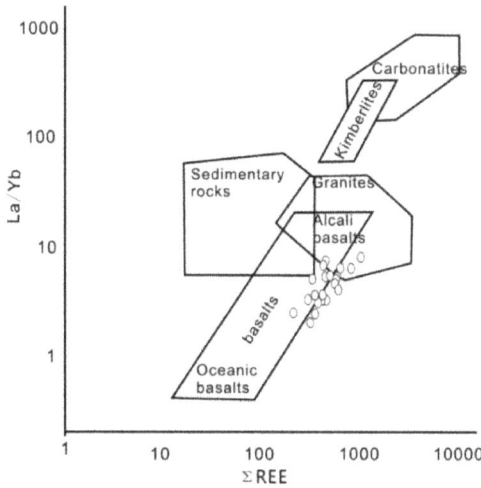

**Figure 9.** Relation between REE and La/Yb of tuff samples in the Zhongliangshan mine, Chongqing.

*5.2. Hydrothermal Solution*

Some researchers have shown that there have been activities of low-temperature hydrothermal solutions in the late Permian Age in southwestern China, which resulted in enrichment of trace elements and minerals in some coal [1,45–51]. Similarly, some minerals of tuff in Zhongliangshan are formed owing to the influence of hydrothermal solution.

In addition to the derivation from volcanic ash, anatase and zircon might have been derived from hydrothermal alteration in the Zhongliangshan tuffs. Anatase of various particle sizes is distributed in the kaolinite matrix (Figure 4B,C). Figures 4A and 5A illustrate that part of anatase could be formed by hydrothermal alteration. Zircon from Figure 5D displays long axis and could be formed by the effect of hydrothermal alteration. Zircon in Figure 5D exclusively contains Zr, Si and O determined by the SEM-EDX. Finkelman [52] has demonstrated that Hf, Th, U, Y and HREE occur in the volcanogenic zircon, but were not identified in authigenic ziron, in accordance with the results of the Zhongliangshan tuff samples.

Florencite, the main carrier of REE in the Zhongliangshan tuff samples, occurs as ellipsoidal in the kaolinite matrix (Figure 5E,F), indicating a syngenetic or early diagenetic hydrothermal origin. Dai *et al.* [14] have also demonstrated that florencite is one of the important carriers of REE in the late Permian coals in southwestern China [3].

Gypsum (Figure 5B) and barite (Figure 4F) occur as crack-fillings, the former occurring as radiating and the latter on the edge of jarosite, indicating an epigenetic hydrothermal origin.

### 5.3. Preliminary Evaluation of Rare Metals

Coal and coal-bearing strata have recently become alternative sources for recovery of rare metals [2,3,7,8]. The U.S. Department of Energy's National Energy Technology Laboratory has selected 10 projects to receive funding for research in support of the lab's program on recovery of rare earth elements from coal and coal byproducts since 2015 [2,53].

Based on the Chinese industry standards [33], the required $(Nb,Ta)_2O_5$ concentrations for marginal and industrial grade Nb(Ta) ore deposits of weathered crust type are 80–100 and 160–200 μg/g, respectively; equivalent concentrations are 40–60 and 100–120 μg/g for Nb(Ta) ore deposits of river placer type. The concentration of $(Nb,Ta)_2O_5$ varies from 47 to 324 μg/g, with an average of 186 μg/g, higher than the marginal and industrial grade for weathered crust and placer deposit types. Concentration of $TiO_2$ varies from 1.55% to 5.28% and averages 3.82%, higher than the industrial grade of Chinese industry standard [54]. The average concentration of Ga (38.2 μg/g) is also up to the standards for industrial utilization in bauxite (20 μg/g) and coal mining (30 μg/g) [34]. In addition, the concentrations of REE vary from 234 to 1189 μg/g and averages 548 μg/g, higher than the cut-off grade of Chinese weathering crust ion adsorption type rare earth elements deposits (500 μg/g) [55].

The Nb, Ti, Ga, and REE all exceed their respective industrial grade of China in the tuff of the Zhongliangshan mine. It is considered that the Zhongliangshan tuff is a potential polymetallic ore worth in-depth study.

## 6. Conclusions

Compared with the Upper Continental Crust, some trace elements including Li, Be, Sc, V, Cr, Co, Ni, Cu, Zn, Ga, Zr, Nb, Cd, Sb, REE, Hf, Ta, Re, Th, and U are enriched in tuff from Zhongliangshan mine, Chongqing, southwestern China. The minerals mainly include kaolinite, illite, pyrite, anatase, calcite, gypsum, quartz, and traces of minerals such as zircon, florencite, jarosite, and barite. The tuff is of mafic volcanic origin with features of alkali basalt. The H-REE enriched in the tuff due to L-REE being leached easier by groundwater and adsorbed in the organic matter of the coal seam overlying the tuff. Some minerals including florencite, gypsum, barite, and a portion of anatase and zircon are precipitated from hydrothermal solution. It is suggested that Zhongliangshan tuff is a potential polymetallic ore and the opportunity for recovery of these valuable elements needs to be studied in depth.

**Acknowledgments:** The authors wish to express their appreciation to Shifeng Dai for revision suggestions and English polishing. We thank Lei Zhao and Lixin Zhao, who helped to identify the minerals under SEM-EDX. We also thank Peipei Wang for the mineral quantitative analysis using Siroquant software. The authors are indebted to three anonymous reviewers for their careful reviews and constructive comments, which greatly improved the manuscript. This research was supported by the National Key Basic Research and Development Program (No.2014CB238902) and National Natural Science Foundation of China (No. 41502162).

**Author Contributions:** Jianhua Zou and Heming Tian collected tuff samples in Zhongliangshan mine. Jianhua Zou conducted the determinations of major-element contents. Tian Li and Heming Tian were responsible for the analysis of trace-element concentrations. Jianhua Zou and Tian Li were responsible for the mineralogy investigation using XRD and SEM-EDX.

**Conflicts of Interest:** The authors declare no conflict of interest.

## References

1. Dai, S.F.; Ren, D.Y.; Chou, C.-L.; Finkelman, R.B.; Seredin, V.V.; Zhou, Y.P. Geochemistry of trace elements in Chinese coals: A review of abundances, genetic types, impacts on human health, and industrial utilization. *Int. J. Coal Geol.* **2012**, *94*, 3–21. [CrossRef]
2. Hower, J.C.; Granite, E.J.; Mayfield, D.B.; Lewis, A.S.; Finkelman, R.B. Notes on Contributions to the Science of Rare Earth Element Enrichment in Coal and Coal Combustion Byproducts. *Minerals* **2016**, *6*, 32. [CrossRef]
3. Dai, S.F.; Chekryzhov, I.Y.; Seredin, V.V.; Nechaev, V.P.; Graham, I.T.; Hower, J.C.; Ward, C.R.; Ren, D.Y.; Wang, X.B. Metalliferous coal deposits in East Asia (Primorye of Russia and South China): A review of geodynamic controls and styles of mineralization. *Gondwana Res.* **2016**, *29*, 60–82. [CrossRef]
4. Dai, S.F.; Wang, P.P.; Ward, C.R.; Tang, Y.G.; Song, X.L.; Jiang, J.H.; Hower, J.C.; Li, T.; Seredin, V.V.; Wagner, N.J.; *et al.* Elemental and mineralogical anomalies in the coal-hosted Ge ore deposit of Lincang, Yunnan, southwestern China: Key role of $N_2$-$CO_2$-mixed hydrothermal solutions. *Int. J. Coal Geol.* **2015**, *152*, 19–46. [CrossRef]
5. Johnston, M.N.; Hower, J.C.; Dai, S.; Wang, P.; Xie, P.; Liu, J. Petrology and Geochemistry of the Harlan, Kellioka, and Darby Coals from the Louellen 7.5-Minute Quadrangle, Harlan County, Kentucky. *Minerals* **2015**, *5*, 894–918. [CrossRef]
6. Seredin, V.V.; Finkelman, R.B. Metalliferous coals: A review of the main genetic and geochemical types. *Int. J. Coal Geol.* **2008**, *76*, 253–289. [CrossRef]
7. Seredin, V.V.; Dai, S.F. Coal deposits as potential alternative sources for lanthanides and yttrium. *Int. J. Coal Geol.* **2012**, *94*, 67–93. [CrossRef]
8. Seredin, V.V.; Dai, S.F.; Sun, Y.Z.; Chekryzhov, I.Y. Coal deposits as promising sources of rare metals for alternative power and energy-efficient technologies. *Appl. Geochem.* **2013**, *31*, 1–11. [CrossRef]
9. Dai, S.F.; Seredin, V.V.; Ward, C.R.; Jiang, J.H.; Hower, J.C.; Song, X.L.; Jiang, Y.F.; Wang, X.B.; Gornostaeva, T.; Li, X.; *et al.* Composition and modes of occurrence of minerals and elements in coal combustion products derived from high-Ge coals. *Int. J. Coal Geol.* **2014**, *121*, 79–97. [CrossRef]
10. Seredin, V.V. From coal science to metal production and environmental protection: A new story of success. *Int. J. Coal Geol.* **2012**, *90–91*, 1–3. [CrossRef]
11. Dai, S.F.; Zhou, Y.P.; Zhang, M.Q.; Wang, X.B.; Wang, J.M.; Song, X.L.; Jiang, Y.F.; Luo, Y.B.; Song, Z.T.; Yang, Z.; *et al.* A new type of Nb (Ta)–Zr(Hf)–REE–Ga polymetallic deposit in the late Permian coal-bearing strata, eastern Yunnan, southwestern China: Possible economic significance and genetic implications. *Int. J. Coal Geol.* **2010**, *83*, 55–63. [CrossRef]
12. Dai, S.F.; Seredin, V.V.; Ward, C.R.; Hower, J.C.; Xing, Y.W.; Zhang, W.G.; Song, W.J.; Wang, P.P. Enrichment of U-Se-Mo-Re-V in coals preserved within marine carbonate successions: Geochemical and mineralogical data from the Late Permian Guiding Coalfield, Guizhou, China. *Miner. Deposita* **2015**, *50*, 159–186. [CrossRef]
13. Hower, J.C.; Eble, C.F.; O'Keefe, J.M.K.; Dai, S.F.; Wang, P.P.; Xie, P.P.; Liu, J.J.; Ward, C.R.; French, D. Petrology, Palynology, and Geochemistry of Gray Hawk Coal (Early Pennsylvanian, Langsettian) in Eastern Kentucky, USA. *Minerals* **2015**, *5*, 592–622. [CrossRef]
14. Dai, S.F.; Zhao, L.; Hower, J.C.; Johnston, M.N.; Song, W.J.; Wang, P.P.; Zhang, S.F. Petrology, mineralogy, and chemistry of size-fractioned fly ash from the Jungar power plant, Inner Mongolia, China, with emphasis on the distribution of rare earth elements. *Energy Fuels* **2014**, *28*, 1502–1514. [CrossRef]
15. Zhuang, X.G.; Su, S.C.; Xiao, M.G.; Li, J.; Alastuey, A.; Querol, X. Mineralogy and geochemistry of the Late Permian coals in the Huayingshan coal-bearing area, Sichuan Province, China. *Int. J. Coal Geol.* **2012**, *94*, 271–282. [CrossRef]
16. Dai, S.F.; Yang, J.Y.; Ward, C.R.; Hower, J.C.; Liu, H.D.; Garrison, T.M.; French, D.; O'Keefe, J.M.K. Geochemical and mineralogical evidence for a coal-hosted uranium deposit in the Yili Basin, Xinjiang, northwestern China. *Ore Geol. Rev.* **2015**, *70*, 1–30. [CrossRef]
17. China Coal Geology Bureau. *Sedimentary Environments and Coal Accumulation of Late Permian Coal Formation in Western Guizhou, Southern Sichuan and Eastern Yunnan, China*; Chongqing University Press: Chongqing, China, 1996. (In Chinese).
18. Dai, S.F.; Liu, J.J.; Ward, C.R.; Hower, J.C.; French, D.; Jia, S.H.; Hood, M.M.; Garrison, T.M. Mineralogical and geochemical compositions of Late Permian coals and host rocks from the Guxu Coalfield, Sichuan Province, China, with emphasis on enrichment of rare metals. *Int. J. Coal Geol.* **2016**. [CrossRef]

19. Standardization Administration of the People's Republic of China. *Sampling of Coal Seams*; Chinese Standard GB/T 482–2008. Standardization Administration of the People's Republic of China: Beijing, China, 2008. (In Chinese)

20. ASTM International. *Test Method for Ash in the Analysis Sample of Coal and Coke from Coal*; ASTM D3174–11; ASTM International: West Conshohocken, PA, USA, 2011.

21. Taylor, J.C. Computer programs for standardless quantitative analysis of minerals using the full powder diffraction profile. *Powder Diffr.* **1991**, *6*, 2–9. [CrossRef]

22. Rietveld, H.M. A profile refinement method for nuclear and magnetic structures. *Appl. Crystallogr.* **1969**, *2*, 65–71. [CrossRef]

23. Ward, C.R.; Spears, D.A.; Booth, C.A.; Staton, I.; Gurba, L.W. Mineral matter and trace elements in coals of the Gunnedah Basin, New South Wales, Australia. *Int. J. Coal Geol.* **1999**, *40*, 281–308.

24. Ward, C.R.; Matulis, C.E.; Taylor, J.C.; Dale, L.S. Quantification of mineral matter in the Argonne Premium coals using interactive Rietveld-based X-ray diffraction. *Int. J. Coal Geol.* **2001**, *46*, 67–82. [CrossRef]

25. Ruan, C.D.; Ward, C.R. Quantitative X-ray powder diffraction analysis of clay minerals in Australian coals using Rietveldmethods. *Appl. Clay Sci.* **2002**, *21*, 227–240. [CrossRef]

26. Taylor, S.R.; McLennan, S.M. *The Continental Crust: Its Composition and Evolution*; Blackwell: Oxford, UK, 1985; p. 312.

27. Chi, Q.H.; Yan, M.C. *Handbook of Elemental Abundance for Applied Geochemistry*; Geological Publishing House: Beijing, China, 2007; pp. 1–148. (In Chinese)

28. Zhou, Y.P.; Ren, Y.L. Element gochemistry of volcanic ash derived tonsteins in late Permian coal-bearing formation of eastern Yunnan and western Guizhou, China. *Acta Sedimentol Sin.* **1994**, *12*, 123–132. (In Chinese)

29. Zhou, Y.P.; Bohor, B.F.; Ren, Y.L. Trace element geochemistry of altered volcanic ash layers (tonsteins) in late Permian coal-bearing formations of eastern Yunnan and western Guizhou provinces, China. *Int. J. Coal Geol.* **2000**, *44*, 305–324. [CrossRef]

30. Sichuan Bureau of Coal Geology; Sichuan Institute of Coal Geology. *Sedimentary Environment and Coal Accumulating Regulations of Late Permian Coal-Bearing Formation in Southern Sichuan*; Guizhou Sceinece and Technology Press: Guiyang, China, 1994. (In Chinese)

31. Dai, S.F.; Chou, C.-L.; Yue, M.; Luo, K.L.; Ren, D.Y. Mineralogy and geochemistry of a Late Permian coal in the Dafang Coalfield, Guizhou, China: Influence from siliceous and iron-rich calcic hydrothermal fluids. *Int. J. Coal Geol.* **2005**, *61*, 241–258. [CrossRef]

32. Dai, S.F.; Ren, D.Y.; Zhou, Y.P.; Chou, C.-L.; Wang, X.B.; Zhao, L.; Zhu, X.W. Mineralogy and geochemistry of a superhigh-organic-sulfur coal, Yanshan Coalfield, Yunnan, China: Evidence for a volcanic ash component and influence by submarine exhalation. *Chem. Geol.* **2008**, *255*, 182–194. [CrossRef]

33. DZ/T 0203–2002. *Geology Mineral Industry Standard of P.R. China: Specifications for Rare Metal Mineral Exploration*; Geological Press: Beijing, China, 2002. (In Chinese)

34. Mineral Resources Industry Requirements Manual Editorial Board. *Mineral Resources Industry Requirements Manual*; Geological Press: Beijing, China, 2010. (In Chinese)

35. Dai, S.F.; Graham, I.T.; Chou, C.-L.; Ward, C.R. A review of anomalous rare earth elements and yttrium in coal. *Int. J. Coal Geol.* **2016**, *159*, 82–95. [CrossRef]

36. Bao, Z.W.; Zhao, Z.H. Geochemistry of mineralization with exchangeable REY in the weathering crusts of granitic rocks in South China. *Ore Geol. Rev.* **2008**, *33*, 519–535. [CrossRef]

37. Dai, S.F.; Li, D.; Chou, C.-L.; Zhao, L.; Zhang, Y.; Ren, D.Y.; Ma, Y.W.; Sun, Y.Y. Mineralogy and geochemistry of boehmite-rich coals: New insights from the Haerwusu Surface Mine, Jungar Coalfield, Inner Mongolia, China. *Int. J. Coal Geol.* **2008**, *74*, 185–202. [CrossRef]

38. Eskenazy, G.M. Rare earth elements in a sampled coal from the Pirin Deposit, Bulgaria. *Int. J. Coal Geol.* **1987**, *7*, 301–314. [CrossRef]

39. Crowley, S.S.; Stanton, R.W.; Ryer, T.A. The effects of volcanic ash on the maceral and chemical composition of the C coal bed, Emery Coal Field, Utah. *Org. Geochem.* **1989**, *14*, 315–331. [CrossRef]

40. Hower, J.C.; Ruppert, L.F.; Eble, C.F. Lanthanide, yttrium, and zirconium anomalies in the fire Clay coal bed, Eastern Kentucky. *Int. J. Coal Geol.* **1999**, *39*, 141–153. [CrossRef]

41. Dai, S.F.; Ren, D.Y.; Chou, C.-L.; Li, S.S.; Jiang, Y.F. Mineralogy and geochemistry of the No. 6 coal (Pennsylvanian) in the Jungar Coalfield, Ordos Basin, China. *Int. J. Coal Geol.* **2006**, *66*, 253–270. [CrossRef]

42. Dai, S.F.; Li, T.J.; Jiang, Y.F.; Ward, C.R.; Hower, J.C.; Sun, J.H.; Liu, J.J.; Song, H.J.; Wei, J.P.; Li, Q.Q.; *et al.* Mineralogical and geochemical compositions of the Pennsylvanian coal in the Hailiushu Mine, Daqingshan Coalfield, Inner Mongolia, China: Implications of sediment-source region and acid hydrothermal solutions. *Int. J. Coal Geol.* **2015**, *137*, 92–110. [CrossRef]

43. Addison, R.; Harrison, R.K.; Land, D.H.; Young, B.R.; Davis, A.E.; Smith, T.K. Volcanogenic tonsteins from tertiary coal measures, East Kalimantan, Indonesia. *Int. J. Coal Geol.* **1983**, *3*, 1–30. [CrossRef]

44. Burger, K.; Zhou, Y.P.; Ren, D.Y. Petrography and geochemistry of tonsteins from the 4th Member of the Upper Triassic Xujiahe Formation in southern Sichuan Province, China. *Int. J. Coal Geol.* **2002**, *49*, 1–17. [CrossRef]

45. Dai, S.F.; Zhang, W.G.; Seredin, V.V.; Ward, C.R.; Hower, J.C.; Song, W.J.; Wang, X.B.; Li, X.; Zhao, L.; Kang, H.; *et al.* Factors controlling geochemical and mineralogical compositions of coals preserved within marine carbonate successions: A case study from the Heshan Coalfield, southern China. *Int. J. Coal Geol.* **2013**, *109–110*, 77–100. [CrossRef]

46. Dai, S.F.; Zhang, W.G.; Ward, C.R.; Seredin, V.V.; Hower, J.C.; Li, X.; Song, W.J.; Wang, X.B.; Kang, H.; Zheng, L.C.; *et al.* Mineralogical and geochemical anomalies of Late Permian coals from the Fusui Coalfield, Guangxi Province, southern China: Influences of terrigenous materials and hydrothermal fluids. *Int. J. Coal Geol.* **2013**, *105*, 60–84. [CrossRef]

47. Ren, D.Y.; Zhao, F.H.; Dai, S.F.; Zhang, J.Y.; Luo, K.L. *Geochemistry of Trace Elements in Coal*; Science Press: Beijing, China, 2006. (In Chinese)

48. Zhou, Y.P.; Ren, Y.L. Distribution of arsenic in coals of Yunnan Province, China, and its controlling factors. *Int. J. Coal Geol.* **1992**, *20*, 85–98. [CrossRef]

49. Dai, S.F.; Chou, C.-L. Occurrence and origin of minerals in a chamosite-bearing coal of Late Permian age, Zhaotong, Yunnan, China. *Am. Mineral.* **2007**, *92*, 1253–1261. [CrossRef]

50. Wang, X.B.; Dai, S.F.; Chou, C.-L.; Zhang, M.Q.; Wang, J.M.; Song, X.L.; Wang, W.; Jiang, Y.F.; Zhou, Y.P.; Ren, D.Y. Mineralogy and geochemistry of Late Permian coals from the Taoshuping Mine, Yunnan Province, China: Evidences for the sources of minerals. *Int. J. Coal Geol.* **2012**, *96–97*, 49–59. [CrossRef]

51. Dai, S.F.; Tian, L.W.; Chou, C.-L.; Zhou, Y.P.; Zhang, M.Q.; Zhao, L.; Wang, J.M.; Yang, Z.; Cao, H.Z.; Ren, D.Y. Mineralogical and compositional characteristics of Late Permian coals from an area of high lung cancer rate in Xuan Wei, Yunnan, China: Occurrence and origin of quartz and chamosite. *Int. J. Coal Geol.* **2008**, *76*, 318–327. [CrossRef]

52. Finkelman, R.B. *Modes of Occurrence of Trace Elements in Coal*; US Geological Survey Open-File Report; United States Geological Survey: Reston, VA, USA, 1981; No. 81–99; p. 322.

53. The U.S. Department of Energy. Available online: http://www.energy.gov/fe/articles/doe-selects-projects-enhance-its\T1\textquoterights-research-recovery-rare-earth-elements-coal-and-coal.2015 (accessed on 2 December 2015).

54. DZ/T 0208–2002. *Geology Mineral Industry Standard of P.R. China: Specifications for Placer (Metallic Mineral) Exploration*; Geological Press: Beijing, China, 2002. (In Chinese)

55. DZ/T 0204–2002. *Geology Mineral Industry Standard of P.R. China: Specifications for Rare Earth Mineral Exploration*; Geological Press: Beijing, China, 2002. (In Chinese)

*minerals*

MDPI

Article

# Petrology, Palynology, and Geochemistry of Gray Hawk Coal (Early Pennsylvanian, Langsettian) in Eastern Kentucky, USA

James C. Hower [1,*], Cortland F. Eble [2], Jennifer M. K. O'Keefe [3], Shifeng Dai [4], Peipei Wang [4], Panpan Xie [4], Jingjing Liu [4], Colin R. Ward [5] and David French [5]

[1]  Center for Applied Energy Research, University of Kentucky, 2540 Research Park Drive, Lexington, KY 40511, USA
[2]  Kentucky Geological Survey, Lexington, KY 40506, USA; eble@uky.edu
[3]  Department of Earth & Space Science, Morehead State University, Morehead, KY 40351, USA; j.okeefe@moreheadstate.edu
[4]  State Key Laboratory of Coal Resources and Safe Mining, China University of Mining and Technology (Beijing), Beijing 100083, China; daishifeng@gmail.com (S.D.); wangpeipei1100@gmail.com (P.W.); xiepanpan90@163.com (P.X.); liujj.cumtb@gmail.com (J.L.)
[5]  School of Biological, Earth and Environmental Sciences, University of New South Wales, Sydney, NSW 2052, Australia; c.ward@unsw.edu.au (C.R.W.); d.french@unsw.edu.au (D.F.)
*  Author to whom correspondence should be addressed; james.hower@uky.edu; Tel.: +1-859-257-0261; Fax: +1-859-257-0360.

Academic Editor: Karen Hudson-Edwards
Received: 13 July 2015; Accepted: 31 August 2015; Published: 11 September 2015

**Abstract:** This study presents recently collected data examining the organic petrology, palynology, mineralogy and geochemistry of the Gray Hawk coal bed. From the Early Pennsylvanian, Langsettian substage, Gray Hawk coal has been mined near the western edge of the eastern Kentucky portion of the Central Appalachian coalfield. While the coal is thin, rarely more than 0.5-m thick, it has a low-ash yield and a low-S content, making it an important local resource. The Gray Hawk coal palynology is dominated by *Lycospora* spp., and contains a diverse spectrum of small lycopods, tree ferns, small ferns, calamites, and gymnosperms. The maceral assemblages show an abundance of collotelinite, telinite, vitrodetrinite, fusinite, and semifusinite. Fecal pellet-derived macrinite, albeit with more compaction than is typically seen in younger coals, was observed in the Gray Hawk coal. The minerals in the coal are dominated by clay minerals (e.g., kaolinite, mixed-layer illite/smectite, illite), and to a lesser extent, pyrite, quartz, and iron III hydroxyl-sulfate, along with traces of chlorite, and in some cases, jarosite, szomolnokite, anatase, and calcite. The clay minerals are of authigenic and detrital origins. The occurrence of anatase as cell-fillings also indicates an authigenic origin. With the exception of Ge and As, which are slightly enriched in the coals, the concentrations of other trace elements are either close to or much lower than the averages for world hard coals. Arsenic and Hg are also enriched in the top bench of the coal and probably occur in pyrite. The elemental associations (e.g., $Al_2O_3/TiO_2$, Cr/Th-Sc/Th) indicate a sediment-source region with intermediate and felsic compositions. Rare metals, including Ga, rare earth elements and Ge, are highly enriched in the coal ashes, and the Gray Hawk coals have a great potential for industrial use of these metals. The rare earth elements in the samples are weakly fractionated or are characterized by heavy-REE enrichment, indicating an input of natural waters or probably epithermal solutions.

**Keywords:** trace elements in coal; minerals in coal; epiphyllous fungi; macrinite; Gray Hawk coal

## 1. Introduction

The Lower Pennsylvanian Langsettian-substage Gray Hawk coal bed (Figure 1) has been mined in a small area in the western part of the Kentucky portion of the Central Appalachian coalfield; the samples considered here are from five $7\frac{1}{2}$-minute quadrangles in three counties (Figure 2). The Gray Hawk coal is one of the uppermost Langsettian (late Westphalian A) coals in the region, with the overlying Lily/Manchester/River Gem (also known by at least 18 other names) coal being the oldest truly widespread coal in eastern Kentucky [1]. Despite its restricted geographic range, the thin coal (with most sites having coal <45.7 cm (<18 inches) thick) has been a valuable local resource owing to its relatively low-ash yield and low-S content, allowing it to be shipped as a run-of-mine coal. (This observation is based on observations by Hower and Eble in their several decades of experience with the Kentucky coal mining industry; for example, one mine, bidding for a university contract requiring "washed" (implying beneficiated) coal, sprayed the mined coal with water (represented here as samples 5048–5050 in Table 1)).

In this study, we briefly summarize historical data from Kentucky Geological Survey, U.S. Geological Survey, and University of Kentucky Center for Applied Energy Research (CAER) sampling, with the emphasis on the addition of more detailed results from a 2007 sampling of the coal. Geochemistry and mineralogy of the latter samples are discussed in this work. In addition, examples of the inertinite-maceral macrinite in the Gray Hawk were discussed by Hower *et al.* [2], and are further discussed here in light of their implications for our understanding of the role of insects and other arthropods in the development of coal macerals [2,3].

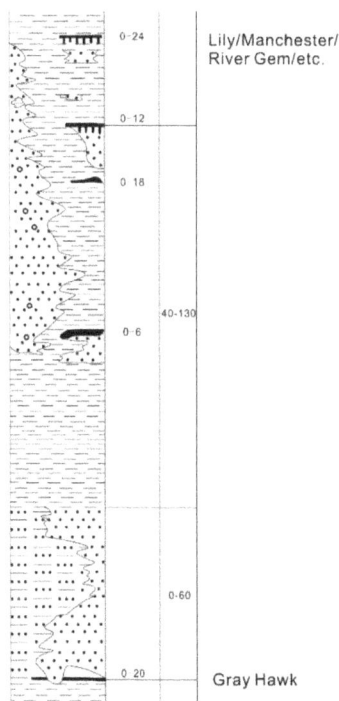

**Figure 1.** A portion of the Langsettian (Pennsylvanian) geologic section from the Parrot 7.5′ quadrangle [4]. The position of the Gray Hawk coal relative to the Lily and correlative coals is based on the section from the map of the McKee 7.5′ quadrangle [5]. Coal thicknesses are in inches (1 inch = 2.54 cm) and the interval thicknesses are in feet (1 foot = 0.3048 m).

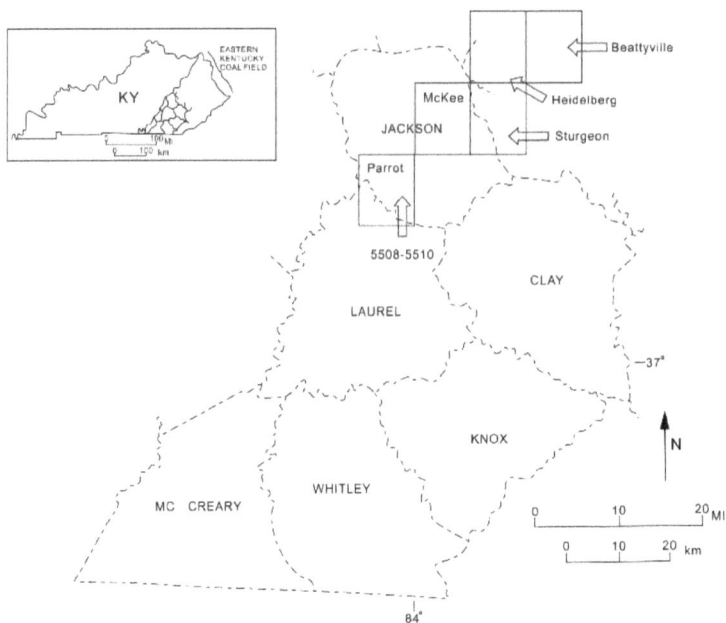

**Figure 2.** Map of a portion of southeastern Kentucky showing the location of the 7.5′ quadrangles noted on Table 1 and the location of the site of samples 5508–5510 in the Parrot 7.5′ quadrangle.

## 2. Geological Setting

While broad, basin-wide studies have been made of eastern Kentucky coals [6], no specific studies of the Gray Hawk coal precede this effort. Considering the low-S, and low-ash nature of the coal, perhaps we can draw the low-ash and low-sulfur analogies to the Blue Gem coal [7] and the low-S, low-ash portions of the Blue Gem-correlative Pond Creek coal [8–10]. The implications for the Gray Hawk coal are discussed in Section 4.1, below. Having said that, we observe that Crowder [4] noted occasional black shale above what is probably the Gray Hawk coal.

## 3. Methods

As noted above, 1980s vintage collections (Table 1) were supplemented by fresh sampling of Gray Hawk coal in a surface mine in the Parrot $7\frac{1}{2}$′ quadrangle (samples 5508, 5509 and 5510; 37.319369° N latitude/84.030483° W longitude, Jackson County, Kentucky, USA). The three samples were collected as benches/lithotypes from the full thickness of the mined coal.

Proximate analysis was performed following ASTM (American Society for Testing and Materials) Standards D3173-11 [11], D3175-11 [12] and D3174-11 [13]. Total sulfur and forms of sulfur were determined following ASTM Standards D3177-02 [14] and D2492-02 [15], respectively. Ultimate analysis and heating value determinations were performed based on ASTM Standards D3176-15 [16] and D5865-13 [17], respectively. Oxides of major elements were determined on a Philips PW2404 X-ray Spectrometer (X-ray fluorescence, Philips, Amsterdam, The Netherlands) following procedures outlined by Hower and Bland [8].

**Table 1.** Location and thickness of 1980s vintage collections.

| Sample | Quadrangle | County | N Latitude | W Longitude | Thickness (cm) |
|--------|-----------|--------|-----------|-------------|----------------|
| 5139 | McKee | Jackson | 37.39278 | 83.95139 | 32.00 |
| 5028 | McKee | Jackson | 37.48361 | 83.89389 | 40.13 |
| 5091 | Parrot | Jackson | 37.28833 | 84.07111 | 32.00 |
| 5023 | Beattyville | Lee | 37.58444 | 83.70028 | 35.05 |
| 5029 | Heidelberg | Lee | 37.50972 | 83.79500 | 40.64 |
| 5048 | Sturgeon | Lee | 37.49972 | 83.82389 | 37.49 |
| 5049 | Sturgeon | Lee | 37.49917 | 83.82417 | 42.98 |
| 5050 | Sturgeon | Lee | 37.49889 | 83.82361 | 42.67 |
| 5185 | Sturgeon | Owsley | 37.44417 | 83.78111 | 45.72 |
| - | Sturgeon | Owsley | 37.44432 | 83.78113 | 45.01 |

Inductively coupled plasma mass spectrometry (X series II ICP-MS) (ThermoFisher, Waltham, MA, USA), in pulse counting mode (three points per peak), was used to determine trace elements in the coal samples. For ICP-MS analysis, samples were digested using an UltraClave Microwave High Pressure Reactor (Milestone, Milano, Italy) [18]. Arsenic and Se were determined by ICP-MS using collision cell technology (CCT) in order to avoid disturbance of polyatomic ions [19]. Multi-element standards (Inorganic Ventures: CCS-1, CCS-4, CCS-5 and CCS-6; NIST 2685b and Chinese standard reference GBW 07114) were used for calibration of trace element concentrations. The method detection limit (MDL) for each of the trace elements, calculated as three times the standard deviation of the average from the blank samples ($n = 10$), is listed in Table 2.

**Table 2.** Method detection limit (MDL; µg/L) of trace elements for ICP-MS analysis.

| Trace Elements | MDL | Trace Elements | MDL | Trace Elements | MDL | Trace Elements | MDL |
|----------------|-----|----------------|-----|----------------|-----|----------------|-----|
| Li | 0.008 | Se | 0.151 | Ba | 0.0207 | Tm | 0.0058 |
| Be | 0.0034 | Rb | 0.3376 | La | 0.0063 | Yb | 0.0055 |
| Sc | 0.0406 | Sr | 0.0122 | Ce | 0.0067 | Lu | 0.0055 |
| V | 0.0131 | Y | 0.008 | Pr | 0.007 | Hf | 0.0121 |
| Cr | 0.0215 | Zr | 0.024 | Nd | 0.011 | Ta | 0.0347 |
| Co | 0.0067 | Nb | 0.0455 | Sm | 0.0067 | W | 0.1349 |
| Ni | 0.1031 | Mo | 0.1124 | Eu | 0.0077 | Tl | 0.0843 |
| Cu | 0.0104 | Cd | 0.0048 | Gd | 0.003 | Pb | 0.0143 |
| Zn | 0.1542 | Sn | 0.0486 | Tb | 0.0064 | Bi | 0.0362 |
| Ga | 0.0015 | Sb | 0.0152 | Dy | 0.0091 | Th | 0.0713 |
| Ge | 0.0168 | In | 0.0024 | Ho | 0.0052 | U | 0.001 |
| As | 0.0955 | Cs | 0.0084 | Er | 0.0009 | | |

Mercury was determined using a Milestone DMA-80 analyzer (Milestone, Milan, Italy). Solid coal samples are directly (without digestion) heated and the evolved Hg is selectively captured as an amalgam and measured by atomic absorption spectrophotometry. The detection limit of Hg is 0.005 ng, the relative standard deviation (RSD) from 11 runs on Hg standard reference is 1.5%, and the linearity of the calibration is in the range 0–1000 ng.

A field emission-SEM (FEI Quanta™ 650 FEG (FEI, Hillsboro, OR, USA), in conjunction with an EDAX energy-dispersive X-ray spectrometer (Genesis Apex 4) (EDAX Inc., Mahwah, NJ, USA), was used to study the modes of occurrence of the minerals, and also to determine the occurrence of selected elements. Samples were carbon-coated using a Quorum Q150T ES sputtering coater (Quorum Technologies Ltd., Lewes, UK), and were then mounted on standard aluminum SEM stubs using sticky conductive carbon tabs. The working distance of the FE-SEM-EDS was 10 mm, beam voltage 20.0 kV, aperture 6, and spot size 5.0. The images were captured via a retractable solid state back-scattered electron detector.

The mineral compositions were determined by X-ray powder diffraction (XRD) (Rigaku, Tokyo, Japan) of low-temperature (oxygen-plasma) ashes of the powdered samples, supplemented by SEM-EDS (FEI, Hillsboro, OR, USA and EDAX Inc., Mahwah, NJ, USA) analysis of the coals in polished section (see below). Low-temperature ashing was carried out using an EMITECH K1050X plasma asher prior to XRD analysis. XRD analysis of the low-temperature ashes was performed on a D/max-2500/PC powder diffractometer with Ni-filtered Cu-Kα radiation and a scintillation detector. Each XRD pattern was recorded over a 2θ interval of 2.6°–70°, with a step size of 0.01°. X-ray diffractograms of the LTAs were subjected to quantitative mineralogical analysis using Siroquant™, commercial interpretation software developed by Taylor [20] based on the principles for diffractogram profiling set out by Rietveld [21]. Further details indicating the use of this technique for coal-related materials are given by Ward *et al.* [22,23] and Ruan and Ward [24].

Petrology was determined on 2.54-cm-diameter epoxy-bound particulate pellets prepared to a final 0.05-μm alumina polish on Leitz Orthoplan microscopes equipped with a 50× reflected-light, oil-immersion objective. Vitrinite reflectance was measured with the incoming light polarized at 45° and the reflected light passing through a 546-nm bandwidth filter on the path to the photomultiplier. The photomultiplier was standardized using glass standards of known reflectance. Maceral identification was based on nomenclature from the International Committee for Coal and Organic Petrology [25,26]. One pellet each of samples 5508, 5509 and 5510 was etched in an acidified saturated solution of potassium permanganate ($KMnO_4$) in order to show details of vitrinite macerals (procedures after Eble *et al.*) [27].

Palynomorphs were liberated by first oxidizing 2–3 g of < 20 mesh coal with Schulze's Solution (concentrated nitric acid saturated with potassium chlorate). Following oxidation, the samples were digested with 5% potassium hydroxide, repeatedly washed with distilled water, and concentrated with zinc chloride (specific gravity 1.9). Amorphous organic matter (AOM) was removed from the residues using ethylene glycol monoethyl ether (2-ethoxyethanol), ultrasonic vibration, and a short period of centrifugation. Samples were strew-mounted onto 25-mm square cover glasses with polyvinyl alcohol, and fixed to 75 × 25-mm microscope slides with acrylic resin. Spore and pollen abundances are based on a count of 250 palynomorphs for each sample.

## 4. Results and Discussion

### 4.1. Chemistry

The basic coal chemistry and coal petrology data for the previously collected Gray Hawk coal samples are shown in Table 3. With one exception, sample 5028 in the McKee $7\frac{1}{2}$' quadrangle ($S_t$ = 3.98%), the coal has a low-ash yield and low-S content, mostly varying from 1.68%–5.79% and 0.55%–0.65%, respectively.

Table 4 lists the basic coal chemistry data for the newly-collected Gray Hawk coal samples (samples 5508, 5509 and 5510). These samples exhibit low-ash yields and low-sulfur contents, a characteristic of ombrotrophic peat-forming systems. Such settings have been noted for many coals in the region, including the low-ash, low-S Blue Gem coal [7] and portions of the correlative Pond Creek coal [9,10]. As a consequence of the collapse of the peat dome, leading to a rheotrophic ecosystem, and a brackish or marine incursion, the uppermost coal bench has a medium-sulfur content (1.55%) dominated by pyritic sulfur (0.99%). High-S content has been observed at the top of the coal bed in other eastern Kentucky coals [28–30] as well as coals from other areas (e.g., Chou [31]).

### 4.2. Petrology

The coal at the Parrott $7\frac{1}{2}$'-quadrangle site is high volatile A bituminous, with a maximum vitrinite reflectance in the 0.8%–0.9% $R_{max}$ range (Table 5). This is generally higher than the rank of the previously studied samples (Table 3). The current study location is further to the southeast and closer to the locus of the high-rank region on the northwest side of the Pine Mountain thrust fault [29,32]. Maceral assemblages are dominated by collotelinite, telinite and vitrodetrinite; fusinite and semifusinite with lesser amounts of micrinite and macrinite; and sporinite with lesser amounts of cutinite (Table 5).

**Table 3.** As-determined moisture, ash and forms of sulfur; mineral-free-basis maceral groups, and maximum vitrinite reflectance of Gray Hawk coal samples from early 1980s Kentucky Geological Survey and Center for Applied Energy Research collections.

| Sample | Proximate Analysis (%) | | Sulfur and Forms of Sulfur (%) | | | | HV (MJ/kg) | Maceral (vol %) | | | $R_{o,max}$ |
|---|---|---|---|---|---|---|---|---|---|---|---|
| | M | Ash | $S_t$ | $S_{py}$ | $S_{sulf}$ | $S_{org}$ | | V | I | L | |
| 5139 | 6.00 | 2.20 | 0.60 | 0.14 | 0.01 | 0.45 | 32.05 | 82.6 | 11.0 | 6.4 | 0.73 |
| 5028 | 4.69 | 5.79 | 3.98 | 2.07 | 0.01 | 1.90 | 30.77 | 73.1 | 17.9 | 9.0 | 0.73 |
| 5091 | 5.28 | 1.88 | 0.65 | 0.10 | 0.01 | 0.54 | 31.89 | 79.8 | 11.5 | 8.7 | 0.77 |
| 5023 | 4.11 | 3.44 | 0.60 | 0.08 | 0.00 | 0.52 | 30.87 | 81.1 | 11.9 | 7.0 | 0.72 |
| 5029 | 5.22 | 1.68 | 0.63 | 0.35 | 0.01 | 0.27 | 31.87 | 77.3 | 12.2 | 10.5 | 0.81 |
| 5048 | 6.15 | 1.99 | 0.55 | 0.23 | 0.00 | 0.32 | 31.66 | 79.5 | 13.2 | 7.3 | 0.78 |
| 5049 | 5.47 | 2.28 | 0.58 | 0.30 | 0.00 | 0.28 | 31.70 | 77.1 | 11.9 | 11.0 | 0.78 |
| 5050 | 5.61 | 2.74 | 1.09 | 0.30 | 0.01 | 0.78 | 31.35 | 70.4 | 21.7 | 7.9 | 0.75 |
| 5185 | 5.90 | 1.74 | 0.64 | 0.01 | 0.02 | 0.61 | 32.15 | 78.3 | 14.9 | 6.8 | 0.83 |

M, Moisture. Ash, ash yield. $S_t$, total sulfur. $S_{py}$, pyrite sulfur. $S_{sulf}$, sulfate sulfur. $S_{org}$, organic sulfur. HV, heat value. V, vitrinite. I, inertinite. L, liptinite. $R_{o,max}$, maximum vitrinite reflectance.

**Table 4.** As-determined proximate and ultimate analyses; forms of sulfur, as-determined heating value of Gray Hawk benches. Unit for moisture, ash yield, volatile matter, fixed carbon, C, H, N, O, and sulfur content, is wt %. Unit for heating value is MJ/kg.

| Sample | Bench | Thick (cm) | M | Ash | VM | FC | C | H | N | O | $S_t$ | $S_{py}$ | $S_{sulf}$ | $S_{org}$ | HV |
|---|---|---|---|---|---|---|---|---|---|---|---|---|---|---|---|
| 5508 | 1 of 3 (top) | 12.2 | 2.77 | 3.04 | 38.49 | 55.69 | 79.63 | 5.71 | 1.64 | 8.43 | 1.55 | 0.99 | 0.02 | 0.54 | 32.93 |
| 5509 | 2 of 3 | 11.8 | 3.25 | 1.42 | 39.03 | 56.29 | 80.46 | 5.76 | 1.66 | 9.98 | 0.72 | 0.10 | 0.01 | 0.61 | 33.41 |
| 5510 | 3 of 3 | 8.5 | 2.76 | 4.07 | 40.44 | 52.73 | 78.77 | 5.78 | 1.61 | 9.08 | 0.69 | 0.08 | 0.01 | 0.60 | 32.84 |

M, Moisture. Ash, ash yield. VM, volatile matter. St, total sulfur. Spy, pyrite sulfur. Ssulf, sulfate sulfur. Sorg, organic sulfur. HV, heating value.

**Table 5.** Petrological compositions and vitrinite maximum reflectance of samples 5508, 5509 and 5510 (%).

| Sample | T | CT | VD | CG | T-V | F | SF | Mic | Mac | ID | T-I | Sp | Cut | Res | LipD | T-L | $R_{o,max}$ |
|---|---|---|---|---|---|---|---|---|---|---|---|---|---|---|---|---|---|
| 5508 | 12.6 | 42.9 | 6.7 | 0.8 | 63.0 | 14.0 | 8.1 | 2.6 | 1.4 | 0.4 | 26.5 | 7.7 | 2.6 | 0.0 | 0.2 | 10.5 | 0.89 |
| 5509 | 14.0 | 56.5 | 6.6 | 0.6 | 77.8 | 2.2 | 4.8 | 1.6 | 1.6 | 0.2 | 10.4 | 10.6 | 1.2 | 0.0 | 0.0 | 11.8 | 0.86 |
| 5510 | 12.8 | 38.5 | 12.2 | 3.2 | 66.7 | 9.8 | 5.0 | 1.2 | 0.8 | 0.2 | 17.0 | 14.2 | 1.0 | 0.4 | 0.6 | 16.2 | 0.82 |

T, telinite; CT, collotelinite; VD, vitrodetrinite; CG, corpogelinite; T-V, total vitrinite; F, fusinite; SF, semifusinite; Mic, micrinite; Mac, macrinite; ID, inertodetrinite; T-I, total inertinite; Sp, sporinite; Cut, cutinite; Res, resinite; LipD, liptodetrinite; T-L, total liptinite; $R_{o,max}$, random vitrinite maximum reflectance.

Cutinite was noted to occur in association with possible epiphyllous fungi (funginite) (Figure 3). While such forms are not well known before the Cretaceous, Taylor *et al.* [33] and Burlington [34] did note epiphyllous fungi in association with Stephanian pteridosperms. *Verrucosporites*, not observed in the palynomorph counts, was noted in the bottom bench during petrography studies (sample 5510) (Figure 4A,B). The megaspore *Triletes globosus* (Figure 4C,D) would have been excluded from the miospore slides due to the screening step during processing. Thin vitrinertoliptite cannel bands were noted in the bottom bench (sample 5510) (Figure 5). Without oriented block samples, we cannot establish the location of the cannel within the bench.

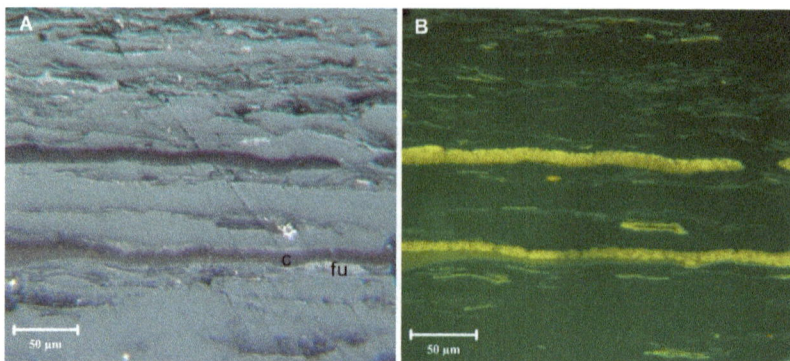

**Figure 3.** Cutinite in sample 5510. Epiphyllous fungi (fu) growing on cutinite (c). (**A**) Reflected light, oil immersion; (**B**) Blue-light.

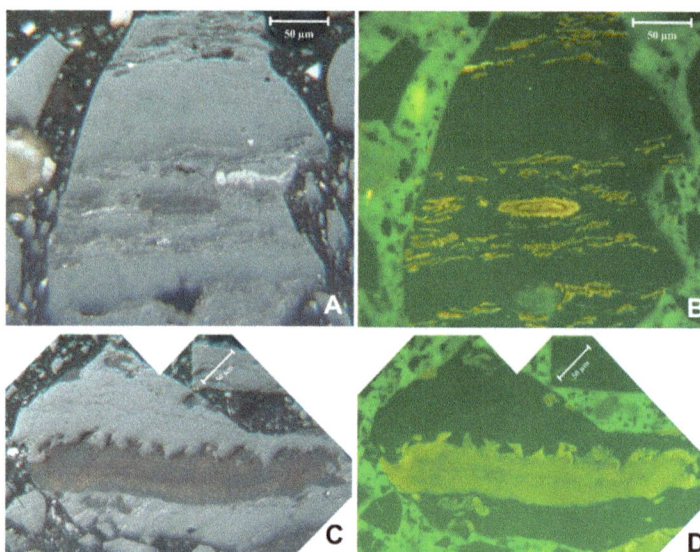

**Figure 4.** Sporinite in sample 5510. (**A,B**) *Verrucosporites* (larger ornamented spore in lower center) with smaller *Lycospora* in mixed maceral matrix. (**C,D**) megaspore, probably *Triletes globosus*. (**A,C**) reflected light, oil immersion. (B) and (D) blue-light.

Fusinite and semifusinite occur as both primary, direct fire-derived varieties and forms which were degraded and then charred. The primary, fire-derived origin of fusinite has long been established [35,36], with more recent advocacy on the part of Scott and his co-workers, among others [37–48]. Certain inertinite macerals with a fusinite and semifusinite reflectance took a more convoluted path, undergoing varying degrees of degradation prior to charring [49,50], and likely charring at a range of temperatures [47,48,51].

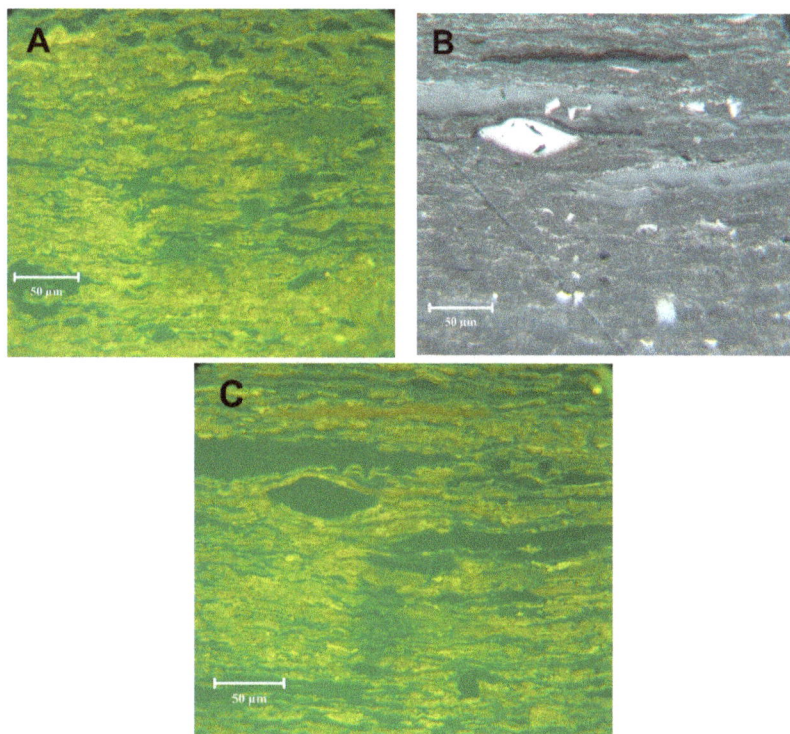

**Figure 5.** Cannel in sample 5510. (**A**) cannel; (**B,C**) cannel with vitrinite bands and inertinite fragments; (**A,C**) blue-light; (**B**) reflected light, oil immersion.

A variety of primary fusinite (such as in Figure 6A,B) and fusinite- and semifusinite-levels of reflectances occur in inertinites subjected to charring after degradation (Figure 6A,C,D). Both the fusinite- and semifusinite-reflectance macerals in Figure 6C and, in particular, the semifusinite-reflectance maceral in Figure 6D have swollen cell walls typically associated with fungal or bacterial degradation [52,53].

Macrinite, at least in part, is a product of the ingestion and excretion of woody material with possible (or likely) successive cycles of fungal and bacterial colonization and/or coprophagous recycling [49,50,54–57]. In part, the complexity of macrinite structure, particularly the inclusion of fusinite and liptinite within the macrinite, underwent significant changes from the Devonian [58–60] to the Pennsylvanian [4,61,62], the Cretaceous and later [5,49,54,63]. Much of this was due to the expansion and evolution of insects and other arthropods. Raymond *et al.* [64] note that, while the three extant digestive schemes employed by modern detritivores (external rumen, facultative mutualism and symbiosis) existed in the Pennsylvanian, the faunal participants were different. This difference is exemplified in the difficulty of assigning fossil arthropod coprolites to their producers [58,59,65–68]. Further, the rate of fungal, bacterial, and detritivore decomposition was slower in the Pennsylvanian compared to younger settings [64]. In the latter context, the inclusions in the macrinite on the right side of Figure 7A demonstrate that non-discriminating feeding activity was part of the Langsettian ecosystem. Much more common in Pennsylvanian coals is relatively inclusion-free macrinite (Figure 7A,B,F,G and the macrinite across the center of Figure 7H). Fecal pellet-derived macrinite (Figure 7C–E,H), common in younger coals [49], was previously noted in a survey of Pennsylvanian macrinite [4]. This should not be a surprise; if something was ingesting the wood, excretion would be a necessary function. The

Cretaceous forms are dominated by individual pellets while the Gray Hawk forms appear to be merged into larger masses, consistent with Type 1 coprolites as defined by Taylor and Scott [62], albeit with some indication that the origin was from individual pellets. The compaction and merger may be a function of the moisture content of the original pellets, the vegetal composition of the diet and the subsequent fecal pellet, the wet *vs.* dry nature of the depositional (and subsequent) environment, or a combination of all of the above plus coalification and other unaccounted for influences. Given the small size of the individual pellets forming the coprolite masses, collembola or perhaps an extinct group of herbivores or detritovores, may have produced them [49].

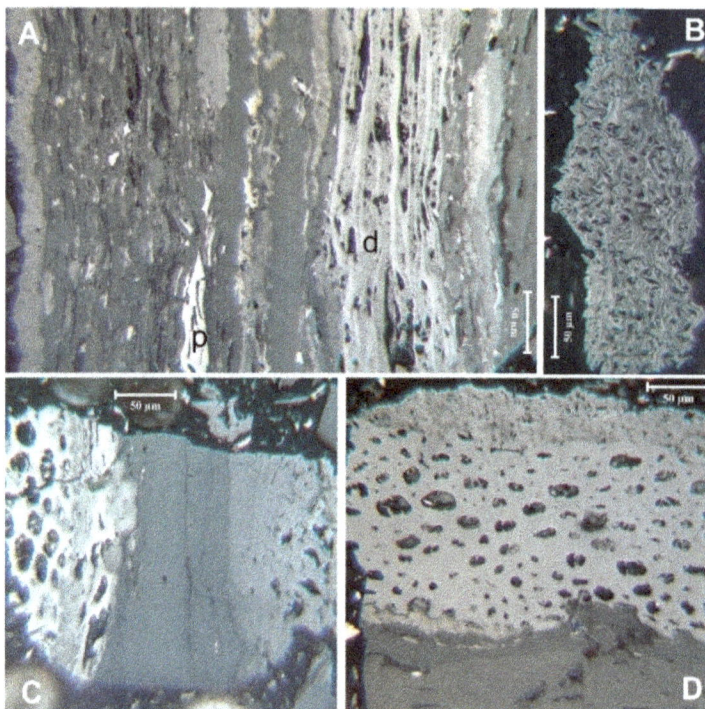

**Figure 6.** Fusinite and semifusinite in the coal. (**A**) mix of primary (p) and degraded (d) fusinite and semifusinite; (**B**) primary fusinite; (**C**) degraded fusinite and semifusinite; (**D**) degraded semifusinite. Reflected light, oil immersion.

Vitrinite counts were made on the unetched pellets; the observations here are based on the details revealed in the etching process. Based on the etched pellets the telinite and collotelinite have superficial similarities in the unetched portion of Figure 8A. Etching emphasizes the fundamental differences in maceral structure, with the telinite band shown to largely consist of cell walls and the collotelinite band showing a more uniform structure. Telinite is also seen in Figure 8B and in the lower half of Figure 8C. The upper half of the Figure 8C vitrinite particle consists of corpogelinite cell fillings surrounded by telinite cell walls. Similarly, that in Figure 8D has bands of corpogelinite with thin cell walls. The latter bands are separated by *ca.* 10-μm bands, probably originating from detrovitrinite.

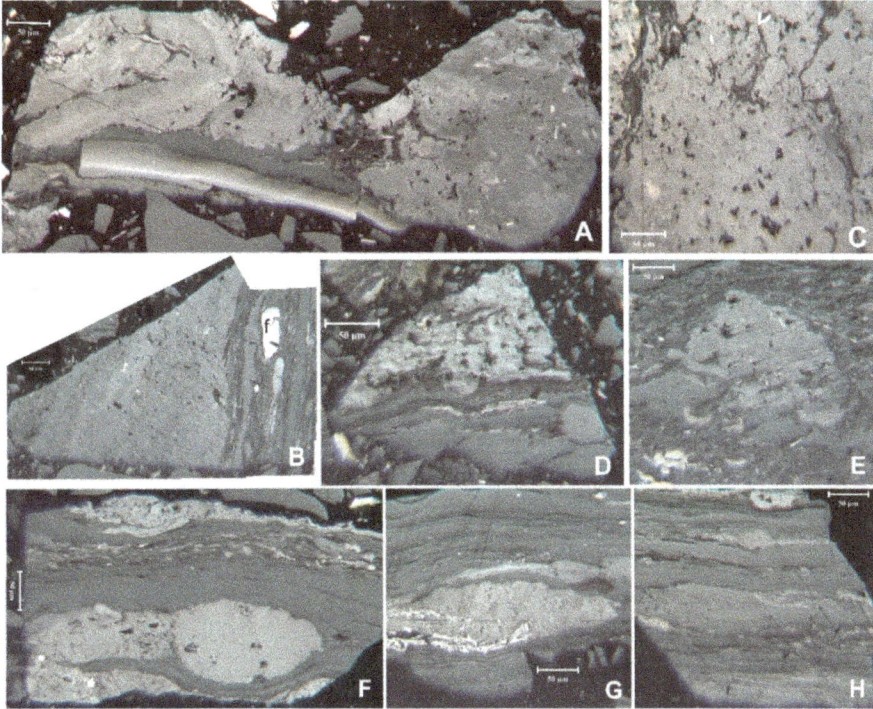

**Figure 7.** Macrinite in the coal. (**A**) macrinite with inertodetrinite inclusions on right; macrinite on left may have replaced sporinite, as noted in the Hower *et al.* [4] publication of the image, or it may just have a shape coincident with sporinite; note that Hower and Ruppert [55] showed an example of funginite replacing a megaspore in a Pennsylvanian coal; (**B**) assemblage of macrinite (ma), semifusinite (sf), and fusinite (f); (**C**) agglomerated rounded macrinite reminiscent of fecal pellet macrinite seen in younger coals [49]; from Hower *et al.* [4]; (**D**) agglomerated rounded macrinite reminiscent of fecal pellet macrinite; (**E**) macrinite with inertodetrinite inclusions but also with agglomerated rounded macrinite reminiscent of fecal pellet macrinite; (**F**) macrinite (ma) with semifusinite (sf); (**G**) macrinite (ma) with fusinite; (**H**) fecal pellet macrinite (fp), Reflected light, oil immersion; (**A–E**) sample 5510; (**F**) sample 5028; (**G,H**) sample 5023. Scale bar: 50 μm.

Petrographically, the relatively high vitrinite contents, especially telovitrinite which peaks in the middle bench, are likely a function of the same wet conditions that promoted arborescent lycopods, pteridospermalean and cordaitalean gymnosperms, as well as the relatively slower rates of decomposition present in the Pennsylvanian coal bed. A direct correspondence between high vitrinite contents and high percentages of arborescent lycopod spores, especially *Lycospora*, is a common feature in many Early and Middle Pennsylvanian coal beds in the Appalachian and Eastern Interior, USA basins [69–74]. The decrease in fusinite + semifusinite from the bottom bench to the middle bench, followed by a large increase to the top bench, is a further indication of the cycle of relative wet and dry conditions in the Gray Hawk coal.

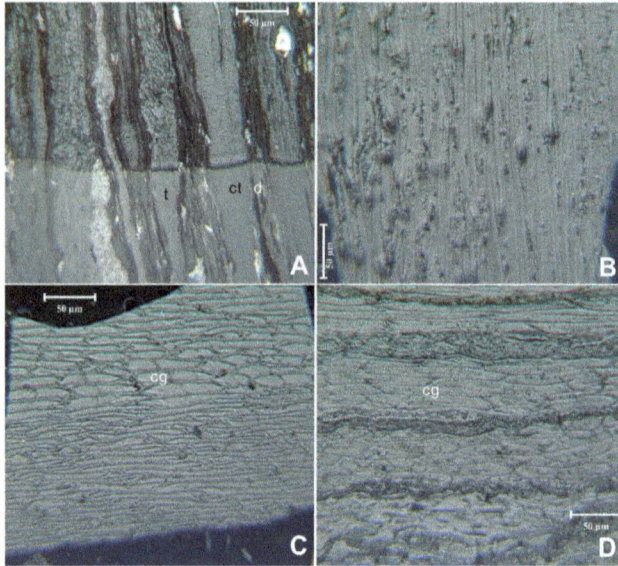

**Figure 8.** Vitrinite in the coal. (**A**) etched vitrinite, inertinite, and liptinite showing contrast between telinite (t), collotelinite (ct), and detrovitrinite (d) bands; (**B**) etched telinite; (**C**) etched corpogelinite (cg) cell filling with telinite cell walls; (**D**) etched corpogelinite (cg) cell filling with telinite cell walls, Reflected light, oil immersion; (A) sample 5509; (B) sample 5510; (C) and (D) 5509.

*4.3. Palynology*

Palynomorph data is reported as percentages of the total count in Table 6, according to parent plant affinity. Palynomorph/parent plant affinities are based on excellent summaries by [75–77]. Forms identified in addition to those observed in the point counts are marked with an X in Table 6.

The occurrence of a few, stratigraphically constrained, spore taxa indicate a late Early Pennsylvanian (late Langsettian) age for the Gray Hawk coal bed. These taxa include *Schulzospora rara, Radiizonates aligerans, R. striatus, Secarisporites remotus, Endosporites globifomis, Laevigatosporites* spp. and *Granasporites medius* [74]. The Gray Hawk coal palynoflora is strongly dominated by *Lycospora* spp. (>85% in all three samples), which is typical of coal beds in this stratigraphic interval. Indeed, many, if not most, Early Pennsylvanian coal beds in the Appalachian Basin are dominated by *Lycospora* spp., which is the dispersed spore genus of several of the large lycopsid trees (e.g., *Lepidodendron, Lepidophloios*) that were important elements of Early Pennsylvanian mire floras [71–74,78]. *Lycospora pellucida* and *L. pusilla* are abundant in all three benches (Table 6). *Lycospora granulate* is relatively abundant in the bottom two benches, exceeding the concentration of *L. pusilla* in the bottom bench, but undergoes a precipitous drop in concentration to the top bench, sample 5508. Small ferns, particularly represented by *Granulatisporites parvus*, and calamites (*Calamospora breviradiata*) in the bottom bench, are present in lesser abundance than the lycopsids. Cordaites, primarily *Florinites* spp., occur in all three benches, but reach significant abundances in the bottom and middle benches. Common palynomorph taxa recovered from the Gray Hawk coal bed are shown in Figure 9.

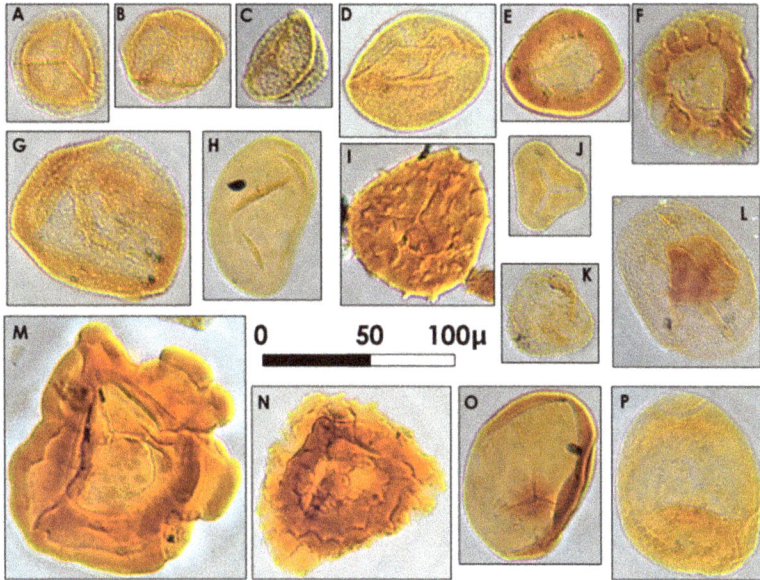

**Figure 9.**  Common palynomorphs in the Gray Hawk coal bed.   (**A**) *Lycospora pellucida;* (**B**) *Lycospora pusilla;* (**C**) *Lycospora granulata;* (**D**) *Granasporites medius;* (**E**) *Densosporites annulatus;* (**F**) *Radiizonates aligerans;* (**G**) *Crassispora kosankei;* (**H**) *Laevigatosporites minor;* (**I**) *Camptotriletes bucculentus;* (**J**) *Granulatisporites adnatoides;* (**K**) *Pilosisporites triquetrus;* (**L**) *Florinites similis;* (**M**) *Reticulatisporites reticulatus;* (**N**) *Cristatisporites connexus;* (**O**) *Calamospora breviradiata;* (**P**) *Schulzospora rara.* All images were collected using Nomarski interference contrast illumination (DIC).

Abundant *Lycospora* spp. spores indicate that arborescent lycopsids, notably *Lepidophloios* and *Lepidodendron*, were contributors to the peat that formed the Gray Hawk coal bed. The proportion of *Granasporites medius* in the samples indicates that *Diaphorodendron* and/or *Synchysidendron*, was also present, though likely in proportionally fewer numbers than *Lepidophloios* or *Lepidodendron*. Mire-centered lycopsids had developed reproductive and vegetative strategies for growth and development in very wet areas. The megasporangium of *Lepidophloios* (*Lepidocarpon*) and *Lepidodendron* (*Achlamydocarpon*) were both boat-shaped, facilitating dispersal in areas of standing water. Lycopsid root systems (*Stigmaria*) radiated laterally from the base of the plant, rather than downward, to provide stability. This is a common feature of many modern tree-size plants growing in wet, soft substrates. Furthermore, the rootlets emanating from *Stigmarian* axes were very slender in design, and would have been incapacitated by firm substrates [78].

Other plant groups, ferns, seed ferns, calamites and cordaites, are represented by minor quantities of palynomorphs in the Gray Hawk coal bed. The same supersaturated substrates that would have favored the proliferation of arborescent lycopods may have also favored pteridospermatophytean and cordaitalean gymnospermous trees [79]. Both taxa produced fewer palynomorphs than the lycopsid taxa, and may have been, at least to a degree, arthropod pollenated [68]. The presence of coprolitic macrinite in the samples supports the supposition that saprophytic and/or grazing arthropods occurred in the mire ecosystem; these and others may have served as pollinators for the early gymnosperms.

**Table 6.** Palynology of Gray Hawk benches. Note: X indicates that the palynomorph was observed to occur in trace amounts.

| Sample | 5508 | 5509 | 5510 |
|---|---|---|---|
| Lycospora pellucida | 46.8 | 37.2 | 41.2 |
| L. pusilla | 32.8 | 30.4 | 22.4 |
| L. granulata | 3.2 | 22.0 | 25.6 |
| L. orbicula | X | X | X |
| L. micropapillata | 3.2 | 0.4 | 3.2 |
| L. rotunda | X | X | X |
| L. torquifer | X | X | |
| Granasporites medius | 3.2 | 3.2 | X |
| Crassispora kosankei | X | X | X |
| **Total Lycopsid Trees** | **89.2** | **93.2** | **92.4** |
| Densosporites annulatus | X | X | 0.4 |
| D. triangularis | X | | |
| D. sphaerotriangularis | 0.8 | 0.4 | X |
| Cristatisporites indignabundulus | X | X | X |
| C. connexus | X | X | |
| Cingulizonates loricatus | X | X | |
| Radiizonates aligerans | X | X | X |
| R. striatus | | X | |
| Cirratriradites saturni | X | X | |
| Endosporites globiformis | X | X | X |
| E. ornatus | | X | |
| Spencerisporites radiatus | X | X | X |
| **Total Small Lycopsids** | **0.8** | **0.4** | **0.4** |
| Punctatisporites minutus | 0.4 | X | 1.2 |
| Laevigatosporites minimus | | X | |
| **Total Tree Ferns** | **0.4** | **0.0** | **1.2** |

| Sample | 5508 | 5509 | 5510 |
|---|---|---|---|
| Granulatisporites parvus | 5.6 | 2.4 | 2.4 |
| G. adnatoides | X | X | X |
| G. piroformis | X | X | X |
| G. granulatus | X | | |
| Cyclogranisporites microgranus | X | | |
| C. minutus | | X | |
| Lophotriletes microsaetosus | 0.8 | 0.4 | |
| L. commissuralis | X | X | X |
| L. granoornatus | X | X | |
| Leiotriletes subadnatoides | 1.2 | X | X |
| L. adnatus | X | X | |
| L. pridyi | X | | |
| Pilosisporites triquetrus | X | X | |
| P. aculeolatus | X | | X |
| Savitrisporites majus | X | | |
| Camptotriletes bucculentus | X | X | X |
| C. corrugatus | X | X | |
| Verrucosisporites verrucosus | X | X | X |
| V. donarii | X | X | |
| Convolutispora florida | X | X | |
| Reticulitriletes reticulocingulum | | X | |
| R. falsus | X | | |
| Microreticulatisporites concavus | | 0.8 | 0.4 |
| Raistrickia abdita | X | X | X |
| R. saetosa | | | X |
| Apiculatisporis variocomeus | X | X | |
| Apiculatasporites spinulistratus | X | X | |
| Punctatisporites punctatus | X | | |
| P. aerarius | X | X | |
| **Total Small Ferns** | **7.6** | **3.6** | **2.8** |

| Sample | 5508 | 5509 | 5510 |
|---|---|---|---|
| Schulzospora rara | | 1.6 | |
| **Total Seed Ferns** | **0.0** | **1.6** | **0.0** |
| Calamospora breviradiata | 0.8 | X | 0.4 |
| C. microrugosa | X | X | |
| C. pallida | X | | X |
| C. parva | | X | |
| Laevigatosporites minor | 0.4 | 0.8 | 2.0 |
| L. vulgaris | | | 0.4 |
| Reticulatisporites reticulatus | | X | 0.4 |
| R. muricatus | X | X | |
| **Total Calamites** | **1.2** | **0.8** | **3.2** |
| Florinites florini | 0.8 | 0.4 | X |
| F. mediapudens | X | X | |
| F. similis | X | | |
| F. miloti | | | X |
| **Total Cordaites** | **0.8** | **0.4** | **0.0** |
| Potonieisporites elegans | | | X |
| **Total Conifers** | **0.0** | **0.0** | **0.0** |
| Ahrensisporites guerickei | X | | |
| Scarisporites remotus | X | X | |
| Reinschospora magnifica | X | | |
| Dictyotriletes bireticulatus | X | | X |
| Grumosisporites varrireticulatus | X | | |
| **Total Unknown Affinity** | **0.0** | **0.0** | **0.0** |

### 4.4. Geochemistry

Table 7 lists the percentages of major-element oxides and concentrations of trace elements in the newly-collected samples (5508, 5509 and 5510). Due to the low ash yields, the major-element oxide contents are very low relative to those in coals of other areas (e.g., Dai *et al.* [80]). Iron content is highest in the upper bench (48.24 wt % $Fe_2O_3$) falling to 5.70 wt % in the lowermost bench with concomitant increases in silica, alumina and titanium. Calcium concentration is highest in the middle bench.

Compared to average values for world hard coals reported by Ketris and Yudovich [81], Ge and As are slightly enriched in the coal with concentration coefficients between 2-5 (CC, ratio of trace element concentration in studied samples *vs.* averages for world hard coals); however, the concentrations of other trace elements are either close to or much lower than world averages (Figure 10).

With the exception of As (28.04 µg/g), most elements of environmental concern in the coal are low in concentrations, such as Be (2.53 µg/g on average for the three benches), Cr (5.65 µg/g), As (28.04 µg/g), Se (2.02 µg/g), Cd (0.05 µg/g), Mo (0.54 µg/g), Hg (174 ng/g), Tl (0.23 µg/g), Pb (12.34 µg/g) and U (0.31 µg/g). Arsenic and Hg are also enriched in the top bench (sample 5508). This is also the bench with the highest pyritic sulfur content (Table 1), and, as discussed more fully below, the highest proportion of pyrite both within the LTA and the whole-coal material. Such an occurrence suggests an association with the sulfide minerals, in accordance with the observations of many other authors [82–90].

A number of elements (Li, Be, Sc, V, Cr, Co, Ni, Cu, Zn, Ga, Ge, Se, Zr, Nb, Th, U and REY) are relatively enriched in the lower bench. Enrichment of $Zr + TiO_2$ in the basal lithotype of eastern Kentucky coals has been noted in other studies, for example Hower and Bland [8]. Considering the known association of U and Th with zircons, more abundant zircon in the basal bench may explain the greater abundance of U and Th in the coal. The lowermost bench (sample 5510) has the highest ash yield (Table 4) and also (as described below) the highest proportion of illite and illite/smectite among the clay minerals. Vanadium and Cr are known to be associated with clay minerals [91,92], and in some cases with organic matter [93]. These two elements in the present study are thus probably associated with the more abundant clay minerals, especially illite and illite/smectite, in the lower bench sample.

Germanium, enriched in both the top and, in particular, the bottom bench, has been found to be enriched in the same position in many other coals [94,95]. The element is generally associated with organic matter [96], and its enrichment is usually attributed to the leaching of granite by hydrothermal solutions and the Ge-rich solutions discharged into the peat swamp [80,96]. Copper and Zn are associated with traces of chalcopyrite and sphalerite, respectively, as described more fully below.

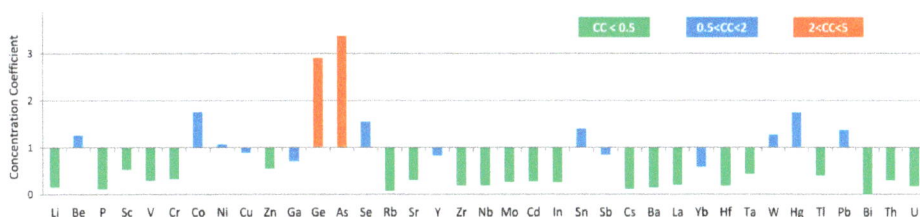

**Figure 10.** Concentration coefficients (CC) of trace elements in the Gray Hawk coals, normalized by average trace element concentrations in world hard coals [81].

The $Al_2O_3/TiO_2$ ratio is a useful provenance indicator for sedimentary rocks [97] and sediments associated with coal deposits [63]. Typical $Al_2O_3/TiO_2$ ratios are 3-8, 8-21, and 21-70 for sediments derived from mafic, intermediate, and felsic igneous rocks, respectively [97]. The three benches of the Gray Hawk coals have $Al_2O_3/TiO_2$ ratios of 16.8 (sample 5508), 22.1 (sample 5509), and 17.7 (sample 5510), indicating a sediment-source region with intermediate and felsic compositions; this is also supported by the relationship of Cr/Th-Sc/Th (Figure 11). However, the relationship of Co/Th *vs.* La/Sc plot (Figure 11) indicates a minor input of intermediate-mafic materials.

Although the rare metals (Ga; rare earth elements and yttrium, REY or REE if Y is not included) in the coal samples are relatively low in concentration compared to world averages, they are, as along with Ge, highly enriched in the coal ashes (Table 8). Based on the cut-off grade (higher than 1000 µg/g) for beneficial recovery of REY from coal combustion products as proposed by Seredin and Dai [98], the Gray Hawk coals have great potential for industrial use. The REY$_{def, rel}$-Coutl graph, which was proposed by Seredin and Dai [98] to evaluate of REY-rich ashes as raw materials (Figure 12), where the X-axis is the outlook coefficient (Coutl, ratio of the relative amount of critical REY metals in the REY sum to the relative amount of excessive REY) and the Y-axis is the percentage of critical elements in total REY (REY$_{def, rel}$, %), shows that the REY in the three coal bench samples (ash basis) fall in Area III or between Areas II (promising) and III (highly promising).

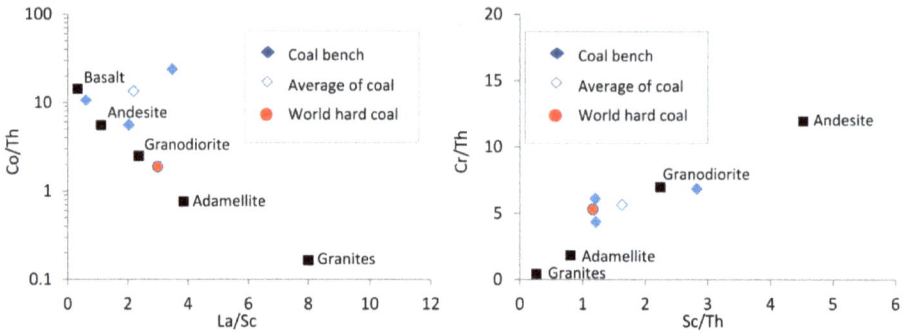

**Figure 11.** Relationship of Co/Th against La/Sc [99] and Cr/Th against Sc/Th [100] for the Gray Hawk coal samples. Data of different igneous rocks are from Condie [99].

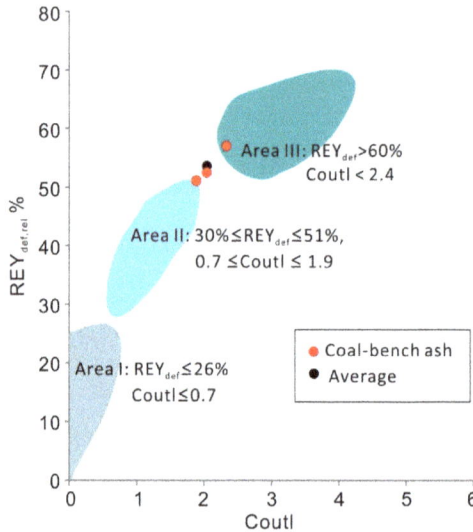

**Figure 12.** The REY$_{def, rel}$-Coutl plot for Gray Hawk coal ashes. Area I, unpromising, Area II, promising, and Area III, highly promising.

**Table 7.** Concentrations of major-element oxides and trace elements in the samples 5508, 5509 and 5510, as well as their comparison with averages for world hard coals (Unit for loss-on-ignition and major-element oxides is %; Unit for trace elements is µg/g and for Hg is ng/g).

| Sample | $SiO_2$ | $TiO_2$ | $Al_2O_3$ | $Fe_2O_3$ | MgO | CaO | $Na_2O$ | $K_2O$ | $P_2O_5$ | $SO_3$ | LOI | Cl | Sc | V | Cr | Co | Ni | Cu |
|---|---|---|---|---|---|---|---|---|---|---|---|---|---|---|---|---|---|---|
| 5508 | 27.97 | 0.92 | 15.4 | 47.64 | 0.54 | 2.91 | 0.37 | 0.98 | 0.23 | 2.08 | 96.96 | 2608 | 0.92 | 4.61 | 3.31 | 4.22 | 9.56 | 6.44 |
| 5509 | 36.16 | 1.35 | 29.85 | 18.18 | 0.88 | 6.43 | 1.08 | 0.78 | 0.33 | 3.91 | 98.58 | 3571 | 0.49 | 3.63 | 2.51 | 9.73 | 9.32 | 11.18 |
| 5510 | 52.98 | 1.74 | 30.9 | 5.7 | 0.78 | 2.36 | 0.88 | 2.41 | 0.18 | 1.31 | 95.93 | 2743 | 5.49 | 20.54 | 13.36 | 20.71 | 43.1 | 29.74 |
| Average | 37.48 | 1.29 | 24.70 | 25.97 | 0.73 | 4.04 | 0.76 | 1.28 | 0.25 | 2.54 | 97.16 | 2993 | 1.96 | 8.42 | 5.65 | 10.53 | 18.24 | 14.25 |
| World | nd | nd | nd | nd | nd | nd | nd | nd | nd | nd | nd | 340 | 3.7 | 28 | 17 | 6 | 17 | 16 |

| Sample | Zn | Ga | Ge | As | Se | Rb | Sr | Y | Zr | Nb | Mo | Cd | In | Sn | Sb | Cs | Ba | La | Ce | Pr |
|---|---|---|---|---|---|---|---|---|---|---|---|---|---|---|---|---|---|---|---|---|
| 5508 | 7.66 | 3.12 | 1.72 | 71.11 | 2.12 | 0.83 | 30.47 | 4.9 | 4.81 | 0.47 | 0.8 | 0.03 | 0.01 | 1.21 | 0.38 | 0.09 | 24.69 | 1.87 | 3.66 | 0.54 |
| 5509 | 8.08 | 1.98 | 0.24 | 1.93 | 1.07 | bdl | 28.45 | 4.62 | 3.83 | 0.3 | 0.5 | 0.01 | 0.01 | 3.42 | 0.06 | 0.02 | 14.46 | 1.7 | 4.17 | 0.62 |
| 5510 | 36.73 | 9.03 | 23.83 | 2.48 | 3.18 | 3.44 | 33.44 | 13.03 | 13.02 | 1.72 | 0.24 | 0.15 | 0.01 | 0.99 | 2.6 | 0.33 | 25.28 | 3.41 | 7.21 | 1.04 |
| Average | 15.42 | 4.25 | 6.97 | 28.04 | 2.02 | 1.21 | 30.51 | 6.92 | 6.6 | 0.74 | 0.54 | 0.05 | 0.01 | 1.95 | 0.84 | 0.13 | 21.13 | 2.21 | 4.77 | 0.7 |
| World | 28 | 6 | 2.4 | 8.3 | 1.3 | 18 | 100 | 8.4 | 36 | 4 | 2.1 | 0.2 | 0.04 | 1.4 | 1 | 1.1 | 150 | 11 | 23 | 3.4 |

| Sample | Nd | Sm | Eu | Gd | Tb | Dy | Ho | Er | Tm | Yb | Lu | Hf | Ta | W | Tl | Pb | Bi | Th | U |
|---|---|---|---|---|---|---|---|---|---|---|---|---|---|---|---|---|---|---|---|
| 5508 | 2.39 | 0.56 | 0.14 | 0.64 | 0.12 | 0.68 | 0.15 | 0.47 | 0.06 | 0.35 | 0.06 | 0.15 | 0.06 | 1.68 | 0.61 | 15.67 | bdl | 0.76 | 0.16 |
| 5509 | 2.86 | 0.7 | 0.15 | 0.75 | 0.11 | 0.65 | 0.12 | 0.37 | 0.04 | 0.27 | 0.03 | 0.1 | bdl | 1.3 | bdl | 3.91 | bdl | 0.41 | 0.11 |
| 5510 | 4.74 | 1.25 | 0.32 | 1.54 | 0.31 | 2.23 | 0.46 | 1.47 | 0.19 | 1.39 | 0.18 | 0.46 | 0.41 | 0.6 | bdl | 19.28 | bdl | 1.95 | 0.82 |
| Average | 3.18 | 0.79 | 0.19 | 0.92 | 0.17 | 1.07 | 0.22 | 0.7 | 0.09 | 0.59 | 0.08 | 0.21 | 0.13 | 1.26 | 0.23 | 12.34 | 0 | 0.94 | 0.31 |
| World | 12 | 2.2 | 0.43 | 2.7 | 0.31 | 2.1 | 0.57 | 1 | 0.3 | 1 | 0.2 | 1.2 | 0.3 | 0.99 | 0.58 | 9 | 1.1 | 3.2 | 1.9 |

nd, no data; bdl; below detection limit.

**Table 8.** Rare metals (Ga, Ge and REY) in the ash of Gray Hawk coal (µg/g).

| Sample | Ga | Ge | La | Ce | Pr | Nd | Sm | Eu | Gd | Tb | Dy | Ho | Er | Tm | Yb | Lu | Y | REY | REO |
|---|---|---|---|---|---|---|---|---|---|---|---|---|---|---|---|---|---|---|---|
| 5508 | 99.8 | 54.9 | 59.6 | 117 | 17.2 | 76.3 | 17.8 | 4.31 | 20.4 | 3.77 | 21.9 | 4.89 | 15.0 | 1.88 | 11.1 | 1.76 | 157 | 529 | 635 |
| 5509 | 135 | 16.1 | 116 | 284 | 42.1 | 195 | 47.4 | 10.1 | 51.2 | 7.45 | 44.0 | 8.33 | 24.9 | 2.64 | 18.6 | 2.37 | 315 | 1168 | 1402 |
| 5510 | 215 | 569 | 81.4 | 172.11 | 24.9 | 113 | 29.7 | 7.70 | 36.8 | 7.34 | 53.2 | 11.0 | 35.1 | 4.50 | 33.2 | 4.26 | 311 | 925 | 1110 |
| Average | 143 | 175 | 85.77 | 192 | 28.26 | 129 | 31.7 | 7.30 | 35.9 | 6.04 | 38.1 | 7.75 | 23.8 | 2.84 | 19.6 | 2.64 | 254 | 865 | 1038 |

The classification of REY used in the present study is based on Seredin and Dai [98] and includes light (LREY: La, Ce, Pr, Nd and Sm), medium (MREY: Eu, Gd, Tb, Dy and Y), and heavy (HREY: Ho, Er, Tm, Yb and Lu) REY. Accordingly, in comparison with the upper continental crust (UCC) [101], three enrichment types are identified: L-type (light-REY; $La_N/Lu_N > 1$), M-type (medium-REY; $La_N/Sm_N < 1$, $Gd_N/Lu_N > 1$), and H-type (heavy-REY; $La_N/Lu_N < 1$) [98]. The REY in samples 5508 and 5509, however, have similar distribution patterns; both are weakly fractionated (Figure 13) but are slightly enriched in medium- and heavy-REY. Sample 5510 is distinctively characterized by a heavy-REY enrichment type (Figure 13). The three samples do not show anomalies of Eu, Ce and Y (Figure 13). However, the REE distribution patterns of the coal samples present in this study would have been expected to have a light-REY enrichment type because the sediment-source region is mainly of felsic to intermediate composition. The REY in the present coal samples may have been subjected to acid natural waters or epithermal solutions, which can lead to the medium-REY [98,102] and heavy-REY [98,103] enrichment types, respectively.

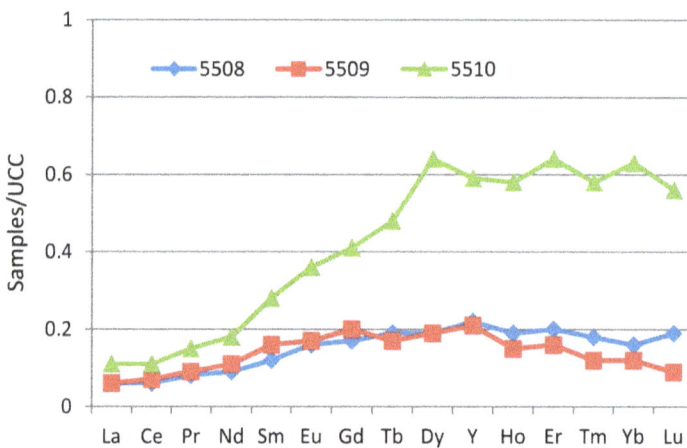

**Figure 13.** Distribution patterns of REY in the Gray Hawk coals. REY are normalized by Upper Continental Crust [101].

The concentration of Ge in the Gray Hawk coal ash is 18-times higher than that in the world hard coal ashes reported by Ketris and Yudovich [81]. For comparison, the Ge concentration in the ashes of coal-hosted Ge ore deposits including Wulantuga, Lincang, and Spetzugli, are 2820, 3902 and 4906 µg/g, respectively [104]. Germanium is currently being industrially extracted as a raw material from these three Ge-bearing coal deposits. Gallium concentrations in the coal-ash samples are 103, 139 and 222 µg/g, respectively, much higher than the average value for world hard coal ash (36 µg/g) [81]. The average concentration of Ga in fly ash derived from the coal-hosted Al-Ga ore deposit in Jungar, Inner Mongolia, China, is 92 µg/g [105].

### 4.5. Mineralogy

The proportion of LTA for each of the coal bench samples studied is slightly higher than the respective high-temperature ash yield (Figure 14). The difference is in part due to dehydration of the clay minerals, oxidation of the pyrite, and conversion of the bassanite-forming components to either anhydrite or lime during the (high-temperature) ashing process. Different mineral assemblages occur in the LTA residues of the three coal bench samples (Table 9). The LTA of the uppermost bench mainly contains kaolinite, pyrite, illite and iron III hydroxyl-sulfate, with minor proportions of quartz, and szomolnokite, and traces of jarosite, anatase and bassanite. Minerals in the LTA of the middle bench are mainly kaolinite, with minor proportions of illite, pyrite, quartz, chlorite and bassanite, along with traces of iron III hydroxyl-sulfate, jarosite, anatase and calcite. The LTA of the lower bench is composed of kaolinite, illite, mixed-layer illite/smectite, quartz, and traces of iron III hydroxyl-sulfate, pyrite, basanite and anatase.

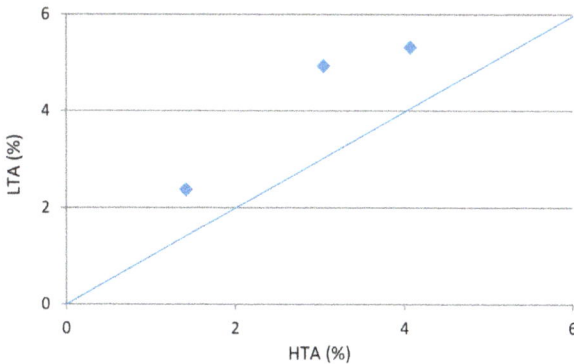

**Figure 14.** Relationship between low-temperature and high-temperature ash yields.

**Table 9.** Minerals in the coal low-temperature ashes (LTA) (**A**) and in whole-coal samples (**B**) wt %.

| Sample | LTA | Quartz | Kaolinite | Illite | I/S | Chlorite | Pyrite | Iron-HS | Jarosite | Szomolnokite | Anatase | Calcite | Bassanite |
|---|---|---|---|---|---|---|---|---|---|---|---|---|---|
| | | | | | **A-Low-Temperature Ashes** | | | | | | | | |
| 5508 | 5.3 | 6.7 | 29.2 | 15.6 | | | 27.2 | 11.8 | 1.6 | 4.5 | 0.6 | | 2.7 |
| 5509 | 2.4 | 5.3 | 63.6 | 9.3 | 5.1 | | 6.0 | 1.7 | 0.8 | | 1.2 | 0.3 | 6.7 |
| 5510 | 4.9 | 17.1 | 45.1 | 19.2 | 12.8 | | 0.3 | 1.6 | | | 0.8 | | 3.0 |
| | | | | | **B-Whole Coal** | | | | | | | | |
| 5508 | - | 0.36 | 1.55 | 0.83 | 0.00 | 0.00 | 1.44 | 0.63 | 0.08 | 0.24 | 0.03 | 0.00 | 0.14 |
| 5509 | - | 0.13 | 1.53 | 0.22 | 0.00 | 0.12 | 0.14 | 0.04 | 0.02 | 0.00 | 0.03 | 0.01 | 0.16 |
| 5510 | - | 0.84 | 2.21 | 0.94 | 0.63 | 0.00 | 0.01 | 0.08 | 0.00 | 0.00 | 0.04 | 0.00 | 0.15 |

Iron-HS, Iron III hydroxyl-sulfate.

The bassanite in coal LTAs is commonly thought to be formed as an artifact of the plasma-ashing process, derived from interaction of non-mineral Ca and sulfur released from the maceral components during oxidation [106–108]. In some cases, however, the bassanite may be produced when sulfuric acid from oxidation of pyrite during storage of the coal reacts with calcite (if present) to form gypsum, which is then partly dehydrated during the low-temperature ashing process [109].

SEM-EDS studies show that the kaolinite mainly occurs in cell-fillings, indicating an authigenic origin (Figure 15A,B), although in some cases, it also occurs as massive material in the collodetrinite, most likely of detrital origin. (Figure 15E). Illite is distributed along bedding planes (Figure 15C,D), or occurs in lath form (Figure 15D,E), respectively indicating an authigenic origin and derivation from detrital materials of terrigenous origin [110,111]. Chlorite occurs in association with illite along bedding planes, suggesting a detrital origin. Quartz occurs as detrital grains in collodetrinite (Figure 15E,F). Pyrite, mainly concentrated in the top bench, occurs as framboidal and dispersed fine-grained particles in the collodetrinite (Figure 16A,B), both occurrences indicative of an authigenic origin. Anatase occurs in cell-fillings of the coal-forming plants (Figure 16C), indicating an authigenic origin. Traces of sphalerite and chalcopyrite, which are the carriers of relatively high Zn and Cu concentrations in the samples, occur in the collodetrinite of sample 5510 (Figure 15B).

A trace of an Fe-bearing phase without S, possibly an iron oxide or hydroxy-oxide (such as hematite or goethite), was detected by SEM-EDS in sample 5510 but is below the detection limit of the XRD technique. The modes of occurrence of this Fe-bearing phase, including fracture-fillings (Figure 17A), cell-fillings (Figure 17C,D), and fine particles distributed in the collodetrinite (Figure 17D,E), indicate an epigenetic origin.

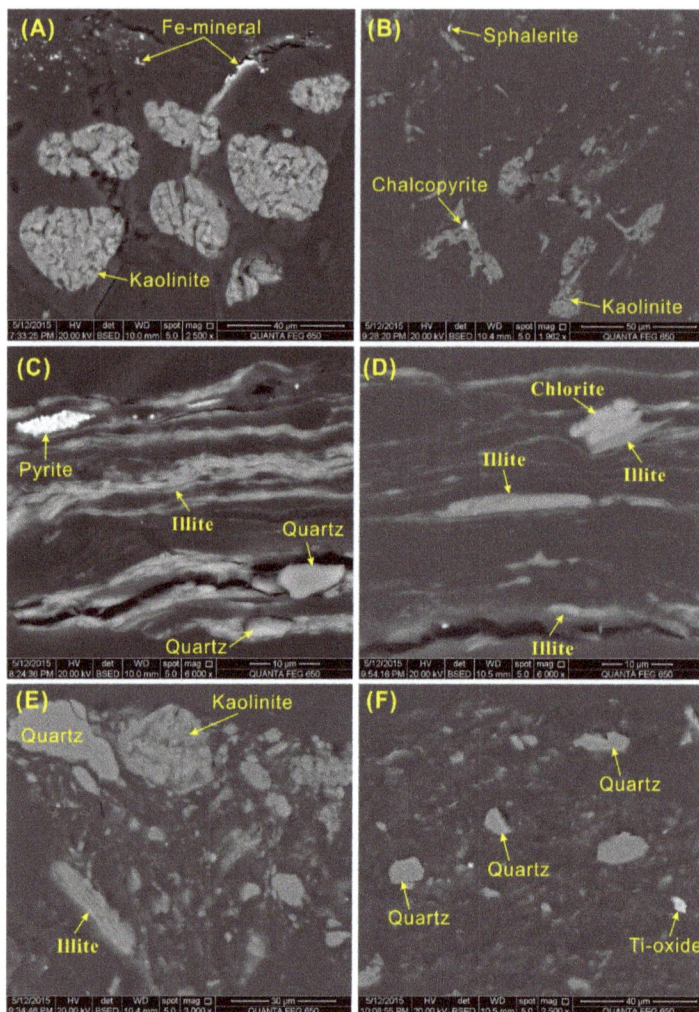

**Figure 15.** Back-scattered electron images of kaolinite, illite and quartz in the coal. (**A**) Kaolinite in cell-fillings, sample 5508; (**B**) Kaolinite in cell-fillings, and chalcopyrite and sphalerite in collodetrinite of sample 5510; (**C**) Illite distributed along bedding planes, as well as pyrite and quartz in collodetrinite, sample 5508; (**D**) Illite and chlorite occurring in lath form in collodetrinite, sample 5510; (**E**) Discrete quartz, massive kaolinite, and lath-like illite in collodetrinite, sample 5510; (**F**) Discrete quartz in collodetrinite, sample 5510.

(A)

(B)

(C)

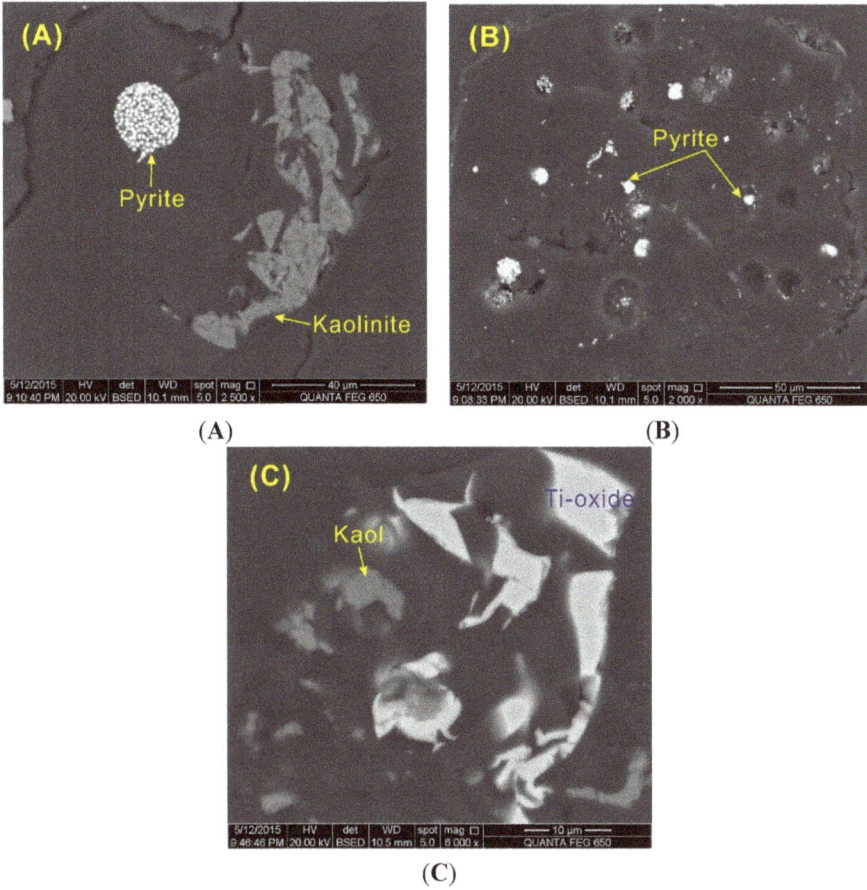

**Figure 16.** Back-scattered electron images of pyrite, rutile, and kaolinite in the coal. (**A**) Framboidal pyrite and cell-filling pyrite in sample 5509; (**B**) Cell-filling and dispersed pyrite in sample 5509; (**C**) Cell-filling anatase and kaolinite in sample 5510.

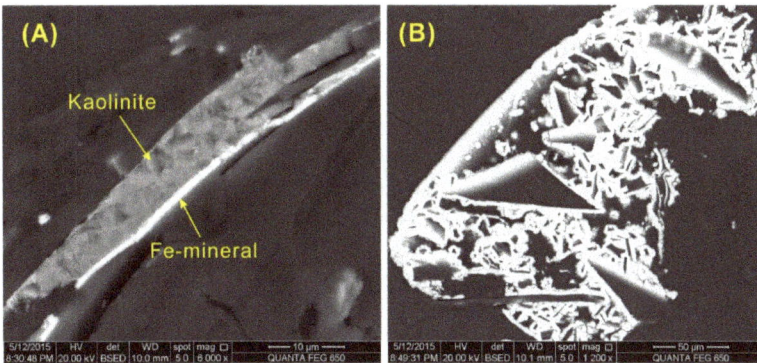

**Figure 17.** *Cont.*

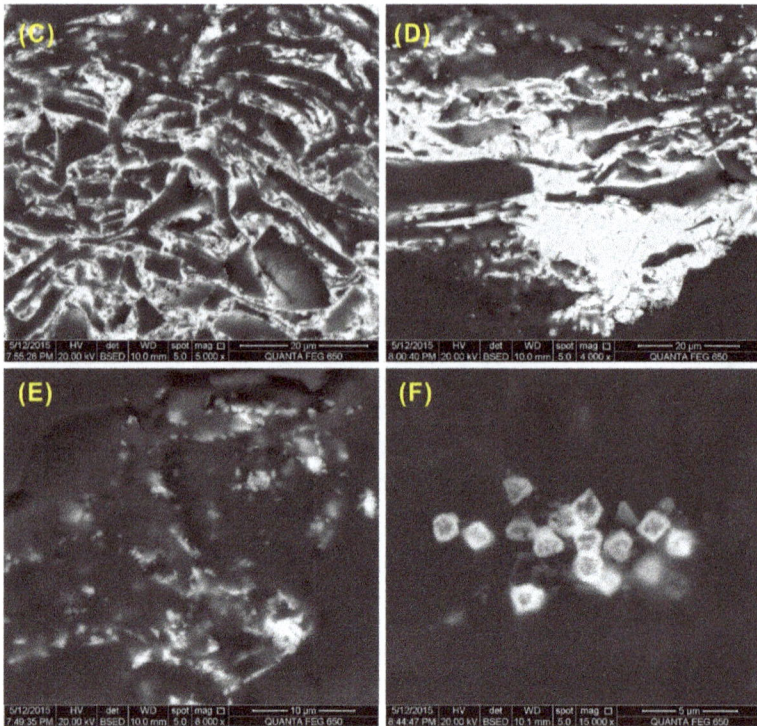

**Figure 17.** Back-scattered electron images of iron oxide or hydroxy-oxide (such as hematite or goethite; light-colored) in sample 5508. (**A**) Fe-mineral and kaolinite in fractures. (**B–D**), Fe-mineral in fractures. (**E–F**), Fe-mineral organic matter matrix.

## 5. Summary

The Langsettian-age high volatile A bituminous Gray Hawk coal is a relatively thin, low-S, low-ash coal that has been mined near the western margin of the Eastern Kentucky coalfield. The maceral assemblages are dominated by the vitrinite macerals collotelinite, telinite, and vitrodetrinite; and in lesser amounts, inertinite macerals fusinite + semifusinite, fecal pellet-derived macrinite, and micrinite; as well as the liptinites sporinite and cutinite. The decrease in fusinite + semifusinite from the bottom to middle benches followed by an increase in fusinite + semifusinite in the top bench is an indicator of the cycle of relatively dry *vs.* wet conditions during deposition of the Gray Hawk peat. The palynomorph assemblages are dominated by varieties of *Lycospora*.

The element associations in the coals (e.g., $Al_2O_3/TiO_2$, Cr/Th-Sc/Th) suggest a sediment source region dominantly comprising intermediate and felsic compositions. High As and Hg concentrations in the upper bench coincide with an increase in pyrite concentration, confirming a sulfide association.

On an ash basis, the Gray Hawk coal is significantly enriched in Ge relative to world hard coals [81], although the concentration of 569 ppm in the ash of sample 5510 is lower than that of the commercial coal-hosted Ge deposits at Lincang and Wulantuga in China and Spetuzgli in Russia. The Ga concentration on an ash-basis in the Gray Hawk coal exceeds the concentration in the fly ash from the Al-Ga ore at Jungar, Inner Mongolia, China. The total rare earth elements plus yttrium (REY) concentrations are in excess of 1000 ppm (ash basis), making them of interest for potential industrial use.

**Acknowledgments:** The analyses of trace elements and mineral compositions were supported by the National Key Basic Research Program of China (No. 2014CB238902) and the National Natural Science Foundation of China (No. 41420104001). Thanks are given to Hongjian Song for his help with the low temperature ashing of coal samples.

**Author Contributions:** James C. Hower collected samples 5508-5510. James C. Hower and Jennifer M.K. O'Keefe were responsible for the petrology and the interpretation of the petrology. Cortland F. Eble was responsible for the palynology. Shifeng Dai, Peipei Wang, Panpan Xie and Jingjing Liu were responsible for the geochemical analyses at CUMT—Beijing. Colin R. Ward and David French were responsible for mineralogy.

**Conflicts of Interest:** The authors declare no conflict of interest.

## References

1. Eble, C.F.; Hower, J.C. Palynologic, petrographic, and geochemical characteristics of the Manchester coal bed in eastern Kentucky. *Int. J. Coal Geol.* **1995**, *27*, 249–278. [CrossRef]
2. Hower, J.C.; Misz-Keenan, M.; O'Keefe, J.M.K.; Mastalerz, M.; Eble, C.F.; Garrison, T.M.; Johnston, M.N.; Stucker, J.D. Macrinite forms in Pennsylvanian coals. *Int. J. Coal Geol.* **2013**, *116–117*, 172–181. [CrossRef]
3. Hower, J.C.; O'Keefe, J.M.K.; Eble, C.F.; Raymond, A.; Valentim, B.; Volk, T.J.; Richardson, A.R.; Satterwhite, A.B.; Hatch, R.S.; Stucker, J.D.; *et al.* Notes on the origin of inertinite macerals in coal: Evidence for fungal and arthropod transformations of degraded macerals. *Int. J. Coal Geol.* **2011**, *86*, 231–240. [CrossRef]
4. Crowder, D.F. *Geology of the Parrot Quadrangle, Kentucky*; U.S. Geologic Quadrangle Map GQ–236; Geological Survey: Washington, D.C., USA, 1963.
5. Weir, G.W.; Mumma, M.D. *Geologic Map of the McKee Quadrangle, Jackson and Owsley Counties, Kentucky*; Geologic Quadrangle Map GQ–236; U.S. Geological Survey: Washington, D.C., USA, 1973.
6. Greb, S.F.; Eble, C.F.; Chesnut, D.R., Jr. Comparison of the Eastern and Western Kentucky coal fields (Pennsylvanian), USA—Why are coal distribution patterns and sulfur contents so different in these coal fields? *Int. J. Coal Geol.* **2002**, *50*, 89–118. [CrossRef]
7. Rimmer, S.M.; Hower, J.C.; Moore, T.A.; Esterle, J.S.; Walton, R.L.; Helfrich, C.T. Petrography and palynology of the Blue Gem coal bed, southeastern Kentucky, USA. *Int. J. Coal Geol.* **2000**, *42*, 159–184. [CrossRef]
8. Hower, J.C.; Bland, A.E. Geochemistry of the Pond Creek coal bed, Eastern Kentucky coalfield. *Int. J. Coal Geol.* **1989**, *11*, 205–226. [CrossRef]
9. Helfrich, C.T.; Hower, J.C. Palynologic and petrographic variation in the Pond Creek coal bed, Pike County, Kentucky. *Org. Geochem.* **1991**, *17*, 153–159. [CrossRef]
10. Hower, J.C.; Pollock, J.D.; Griswold, T.B. Structural controls on petrology and geochemistry of the Pond Creek coal bed, Pike and Martin Counties, eastern Kentucky. In *Geology in Coal Resource Utilization*; Peters, D.C., Ed.; American Association of Petroleum Geologists, Energy Minerals Division: Tulsa, OK, USA, 1991; pp. 413–427.
11. American Society for Testing and Materials (ASTM) International. *Test Method for Moisture in the Analysis Sample of Coal and Coke*; ASTM D3173-11; ASTM International: West Conshohocken, PA, USA, 2011.
12. American Society for Testing and Materials (ASTM) International. *Test Method for Ash in the Analysis Sample of Coal and Coke*; ASTM D3174-11; ASTM International: West Conshohocken, PA, USA, 2011.
13. American Society for Testing and Materials (ASTM) International. *Test Method for Volatile Matter in the Analysis Sample of Coal and Coke*; ASTM D3175-11; ASTM International: West Conshohocken, PA, USA, 2011.
14. American Society for Testing and Materials (ASTM) International. *Test Methods for Total Sulfur in the Analysis Sample of Coal and Coke*; ASTM D3177-02; ASTM International: West Conshohocken, PA, USA, 2002.
15. American Society for Testing and Materials (ASTM) International. *Standard Test Method for Forms of Sulfur in Coal*; ASTM D2492-02; ASTM International: West Conshohocken, PA, USA, 2002.
16. American Society for Testing and Materials (ASTM) International. *Standard Practice for Ultimate Analysis of Coal and Coke*; ASTM D3176-15; ASTM International: West Conshohocken, PA, USA, 2015.
17. American Society for Testing and Materials (ASTM) International. *Standard Test Method for Gross Calorific Value of Coal and Coke*; ASTM D5865-13; ASTM International: West Conshohocken, PA, USA, 2013.
18. Dai, S.; Wang, X.; Zhou, Y.; Hower, J.C.; Li, D.; Chen, W.; Zhu, X. Chemical and mineralogical compositions of silicic, mafic, and alkali tonsteins in the late Permian coals from the Songzao Coalfield, Chongqing, Southwest China. *Chem. Geol.* **2011**, *282*, 29–44. [CrossRef]

19. Li, X.; Dai, S.; Zhang, W.; Li, T.; Zheng, X.; Chen, W. Determination of As and Se in coal and coal combustion products using closed vessel microwave digestion and collision/reaction cell technology (CCT) of inductively coupled plasma mass spectrometry (ICP-MS). *Int. J. Coal Geol.* **2014**, *124*, 1–4. [CrossRef]
20. Taylor, J.C. Computer programs for standard less quantitative analysis of minerals using the full powder diffraction profile. *Powder Diffr.* **1991**, *6*, 2–9. [CrossRef]
21. Rietveld, H.M. A profile refinement method for nuclear and magnetic structures. *J. Appl. Crystallogr.* **1969**, *2*, 65–71. [CrossRef]
22. Ward, C.R.; Spears, D.A.; Booth, C.A.; Staton, I.; Gurba, L.W. Mineral matter and trace elements in coals of the Gunnedah Basin, New South Wales, Australia. *Int. J. Coal Geol.* **1999**, *40*, 281–308. [CrossRef]
23. Ward, C.R. Mineralogical analysis in hazard assessment. In *Geological Hazards—The Impact to Mining*; Doyle, R., Moloney, J., Eds.; Coalfield Geology Council of New South Wales: Newcastle, Australia, 2001; pp. 81–88.
24. Ruan, C.-D.; Ward, C.R. Quantitative X-ray Powder Diffraction analysis of clay minerals in Australian coals using Rietveld methods. *Appl. Clay Sci.* **2002**, *21*, 227–240. [CrossRef]
25. International Committee for Coal Petrology (ICCP). The new vitrinite classification (ICCP System 1994). *Fuel* **1998**, *77*, 349–358.
26. International Committee for Coal Petrology (ICCP). The new inertinite classification (ICCP System 1994). *Fuel* **2001**, *80*, 459–471.
27. Eble, C.F.; Gastaldo, R.A.; Demko, T.M.; Liu, Y. Coal compositional changes along a mire interior to mire margin transect in the Mary Lee coal bed, Warrior Basin, Alabama, USA. *Int. J. Coal Geol.* **1994**, *26*, 43–62. [CrossRef]
28. Hower, J.C.; Pollock, J.D. Petrology of the River Gem Coal Bed, Whitley County, Kentucky. *Int. J. Coal Geol.* **1989**, *11*, 227–245. [CrossRef]
29. Sakulpitakphon, T.; Hower, J.C.; Schram, W.H.; Ward, C.R. Tracking Mercury from the Mine to the Power Plant: Geochemistry of the Manchester Coal Bed, Clay County, Kentucky. *Int. J. Coal Geol.* **2004**, *57*, 127–141. [CrossRef]
30. Mardon, S.M.; Hower, J.C. Impact of coal properties on coal combustion by-product quality: Examples from a Kentucky power plant. *Int. J. Coal Geol.* **2004**, *59*, 153–169. [CrossRef]
31. Chou, C.-L. Sulfur in coals: A review of geochemistry and origins. *Int. J. Coal Geol.* **2012**, *100*, 1–13. [CrossRef]
32. Hower, J.C.; Rimmer, S.M. Coal rank trends in the Central Appalachian coalfield: Virginia, West Virginia, and Kentucky. *Org. Geochem.* **1991**, *17*, 161–173. [CrossRef]
33. Taylor, E.L.; Taylor, T.N.; Krings, M. *Paleobotany: The Biology and Evolution of Fossil Plants*; Elsevier: Burlington, MA, USA, 2009.
34. Krings, M. Pilzreste auf und den Fiedern zweier Pteridospermen auf und in den Stefan von Blanzy-Montceau (Zentralmassiv, Frankreich). *Geol. Saxon.* **2001**, *46–47*, 189–196.
35. Stach, E. The origin of fusain. *Gluckauf* **1927**, *63*, 759.
36. Evans, W.P. The formation of fusain from a comparatively recent angiosperm. *N. Z. J. Sci. Technol.* **1929**, *11*, 262–268.
37. Scott, A.C. Observations on the nature and origin of fusain. *Int. J. Coal Geol.* **1989**, *12*, 443–475. [CrossRef]
38. Scott, A.C. The pre-quaternary history of fire. *Palaeogeogr. Palaeoclim. Palaeoecol.* **2000**, *164*, 281–329. [CrossRef]
39. Scott, A.C. Coal petrology and the origin of coal macerals: A way ahead? *Int. J. Coal Geol.* **2002**, *50*, 119–134. [CrossRef]
40. Scott, A.C.; Glasspool, I.J. Charcoal reflectance as a proxy for the emplacement temperature of pyroclastic flow deposits. *Geology* **2005**, *33*, 589–592. [CrossRef]
41. Scott, A.C.; Glasspool, I.J. The diversification of Paleozoic fire systems and fluctuations in atmospheric oxygen concentration. *Proc. Natl. Acad. Sci. USA* **2006**, *103*, 10861–10865. [CrossRef] [PubMed]
42. Scott, A.C.; Glasspool, I.J. Observations and experiments on the origin and formation of inertinite group macerals. *Int. J. Coal Geol.* **2007**, *70*, 53–66. [CrossRef]
43. Scott, A.C.; Jones, T.P. The nature and influence of fire in Carboniferous ecosystems. *Palaeogeogr. Palaeoclim. Palaeoecol.* **1994**, *106*, 91–112. [CrossRef]

44. Scott, A.C.; Cripps, J.A.; Collinson, M.E.; Nichols, G.J. The taphonomy of charcoal following a recent heathland fire and some implications for the interpretation of fossil charcoal deposits. *Palaeogeogr. Palaeoclim. Palaeoecol.* **2000**, *164*, 1–31. [CrossRef]

45. McParland, L.C.; Collinson, M.E.; Scott, A.C.; Steart, D.C.; Grassineau, N.V.; Gibbons, S.J. Ferns and fires: Experimental charring of ferns compared to wood and implications for paleobiology, paleoecology, coal petrology, and isotope geochemistry. *Palaios* **2007**, *22*, 528–538. [CrossRef]

46. Hudspith, V.; Scott, A.C.; Collinson, M.E.; Pronina, N.; Beeley, T. Evaluating the extent to which wildfire history can be interpreted from inertinite distribution in coal pillars: An example from the Late Permian, Kuznetsk Basin, Russia. *Int. J. Coal Geol.* **2012**, *89*, 13–25. [CrossRef]

47. Hudspith, V.A.; Belcher, C.M.; Yearsley, J.M. Charring temperatures are driven by the fuel types burned in a peatland wildfire. *Front Plant Sci.* **2014**, *5*. [CrossRef] [PubMed]

48. Hudspith, V.A.; Rimmer, S.M.; Belcher, C.M. Latest Permian chars may derive from wildfires, not coal combustion. *Geology* **2014**, *42*, 879–882. [CrossRef]

49. Hower, J.C.; O'Keefe, J.M.K.; Wagner, N.J.; Dai, S.; Wang, X.; Xue, W. An investigation of Wulantuga coal (Cretaceous, Inner Mongolia) macerals: Paleopathology of faunal and fungal invasions into wood and the recognizable clues for their activity. *Int. J. Coal Geol.* **2013**, *114*, 44–53. [CrossRef]

50. O'Keefe, J.M.K.; Bechtel, A.; Christanis, K.; Dai, S.; DiMichele, W.A.; Eble, C.F.; Esterle, J.S.; Mastalerz, M.; Raymond, A.L.; Valentim, B.V.; *et al.* On the fundamental difference between coal rank and coal type. *Int. J. Coal Geol.* **2013**, *118*, 58–87. [CrossRef]

51. Chipman, M.L.; Hudspith, V.; Higuera, P.E.; Duffy, P.A.; Kelly, R.; Oswald, W.W. Spatiotemporal patterns of tundra fires: Late-Quaternary charcoal records from Alaska. *Biogeosci. Discuss.* **2015**, *12*, 3177–3209. [CrossRef]

52. Pujana, R.R.; Massini, J.L.G.; Brizuela, R.R.; Burrieza, H.P. Evidence of fungal activity in silicified gymnosperm wood from the Eocene of southern Patagonia (Argentina). *Geobios* **2009**, *42*, 639–647. [CrossRef]

53. Tanner, L.H.; Lucas, S.G. Degraded wood in the upper Triassic Petrified Forest Formation (Chinle Group), Northern Arizona: Differentiating Fungal Rot from Arthropod Boring. In *The Triassic System*; Tanner, L.H., Spielmann, J.A., Lucas, S.G., Eds.; New Mexico Museum of Natural History and Science Bulletin: Albuquerque, NM, USA, 2013; Volume 61, pp. 582–588.

54. Hower, J.C.; O'Keefe, J.M.K.; Watt, M.A.; Pratt, T.J.; Eble, C.F.; Stucker, J.D.; Richardson, A.R.; Kostova, I.J. Notes on the origin of inertinite macerals in coals: Observations on the importance of fungi in the origin of macrinite. *Int. J. Coal Geol.* **2009**, *80*, 135–143. [CrossRef]

55. Hower, J.C.; Ruppert, L.F. Splint coals of the Central Appalachians: Petrographic and geochemical facies of the Peach Orchard No. 3 split coal bed, southern Magoffin County, Kentucky. *Int. J. Coal Geol.* **2011**, *85*, 268–273. [CrossRef]

56. O'Keefe, J.M.K.; Hower, J.C.; Finkelman, R.B.; Drew, J.W.; Stucker, J.D. Petrographic, geochemical, and mycological aspects of Miocene coals from the Nováky and Handlová mining districts, Slovakia. *Int. J. Coal Geol.* **2011**, *87*, 268–281. [CrossRef]

57. Richardson, A.R.; Eble, C.F.; Hower, J.C.; O'Keefe, J.M.K. A critical re-examination of the petrology of the No. 5 Block coal in eastern Kentucky with special attention to the origin of inertinite macerals in the splint lithotypes. *Int. J. Coal Geol.* **2012**, *98*, 41–49. [CrossRef]

58. Edwards, D.; Selden, P.A.; Richardson, J.B.; Axe, L. Coprolites as evidence of plant-animal interaction in Siluro-Devonian terrestrial ecosystems. *Nature* **1995**, *377*, 329–331. [CrossRef]

59. Edwards, D.; Selden, P.A.; Axe, L. Selective Feeding in an Early Devonian Terrestrial Ecosystem. *Palaios* **2012**, *27*, 509–522. [CrossRef]

60. Honegger, R.; Axe, L.; Edwards, D. Bacterial Epibionts and Endolichenic Actinobacteria and Fungi in the Lower Devonian Lichen *Chlorolichenomycites salopensis*. *Fungal Biol.* **2013**, *117*, 512–518. [CrossRef] [PubMed]

61. Cichan, M.A.; Taylor, T.N. Wood-borings in *Premnoxlon*: Plant-animal interactions in the Carboniferous. *Palaeogeogr. Palaeoclim. Palaeoecol.* **1982**, *39*, 123–127. [CrossRef]

62. Taylor, T.N.; Scott, A.C. Interactions of plants and animals during the Carboniferous. *BioScience* **1983**, *33*, 488–493. [CrossRef]

63. Dai, S.; Liu, J.; Ward, C. R.; Hower, J.C.; Xie, P.; Jiang, Y.; Hood, M.M.; O'Keefe, J.M.K.; Song, H. Petrological, geochemical, and mineralogical compositions of the low-Ge coals from the Shengli Coalfield, China: A

comparative study with Ge-rich coals and a formation model for the Wulantuga Ge ore deposit. *Ore Geol. Rev.* **2015**, *71*, 318–349. [CrossRef]

64. Raymond, A.; Cutlip, P.; Sweet, M. Rates and processes of terrestrial nutrient cycling in the Paleozoic: The world before beetles, termites, and flies. In *Evolutionary Paleoecology: The Ecological Context of Macroevolutionary Change*; Allmon, W.D., Bottjer, D.J., Eds.; Columbia University Press: New York, NY, USA, 2001; pp. 235–284.

65. Baxendale, R.W. Plant-bearing coprolites from North American Pennsylvanian coal balls. *Paleontology* **1979**, *22*, 537–548.

66. Lesnikowska, A.D. Evidence of Herbivory in Tree-Fern Petioles from the Calhoun Coal (Upper Pennsylvanian) of Illinois. *Palaios* **1990**, *5*, 76–80. [CrossRef]

67. Labandiera, C.C. Early History of Arthropod and Vascular plant associations. *Annu. Rev. Earth Planet. Sci.* **1998**, *26*, 329–377. [CrossRef]

68. Labandiera, C.C.; Kvaček, J.; Mostovski, M.B. Pollination drops, pollen, and insect pollination of Mesozoic gymnosperms. *Taxon* **2007**, *56*, 663–695. [CrossRef]

69. Peppers, R.A. Comparison of Miospore Assemblages in the Pennsylvanian System of the Illinois Basin with those in the Upper Carboniferous of Western Europe. In *Proceeding of the Ninth International Congress for Carboniferous Geology and Stratigraphy*; Sutherland, P.K., Manger, W.L., Eds.; Southern Illinois University Press: Carbondale, IL, USA, 1985; Volume 2, pp. 483–502.

70. Peppers, R.A. Palynological correlation of major Pennsylvanian (Middle and Upper Carboniferous) chronostratigraphic boundaries in the Illinois and other coal basins. *Geol. Soc. Am. Mem.* **1996**, *188*, 111.

71. Eble, C.F. Applications of coal palynology to biostratigraphic and paleoecologic analyses of Pennsylvanian coal beds. In *Predictive Stratigraphic Analysis—Concepts and Application*; Cecil, C.B., Edgar, T.N., Eds.; United States Geological Survey Bulletin: Reston, VA, USA, 1994; Volume 2110, pp. 28–32.

72. Eble, C.F. Palynostratigraphy of selected Middle Pennsylvanian coal beds in the Appalachian Basin. In *Elements of Pennsylvanian Stratigraphy, Central Appalachian Basin*; Rice, C.L., Ed.; Geological Society of America Special Paper: Boulder, CO, USA, 1994; Volume 294, pp. 55–68.

73. Eble, C.F. Paleoecology of Pennsylvanian coal beds in the Appalachian Basin. In *Palynology: Principles and Applications*; Jansonius, J., McGregor, D.C., Eds.; American Association of Stratigraphic Palynologists Foundation: Salt Lake City, UT, USA, 1996; Volume 3, pp. 1143–1156.

74. Eble, C.F. Lower and lower Middle Pennsylvanian coal palynofloras, southwestern Virginia. *Int. J. Coal Geol.* **1996**, *31*, 67–114. [CrossRef]

75. Ravn, R.L. Palynostratigraphy of the Lower and Middle Pennsylvanian coals of Iowa. *Iowa Geol. Survey Tech. Paper* **1986**, *7*, 245.

76. Traverse, A. *Paleopalynology*; Unwin Hyman, Ltd.: London, UK, 1988; p. 600.

77. Balme, B.E. Fossil *in situ* spores and pollen grains: An annotated catalogue. *Rev. Palaeobota. Palynol.* **1995**, *87*, 81–323. [CrossRef]

78. Dimichele, W.A.; Phillips, T.L. Paleobotanical and paleoecological constraints on models of peat formation in the Late Carboniferous of Euramerica. *Palaeogeogr. Palaeoclim. Palaeoecol.* **1994**, *106*, 39–90. [CrossRef]

79. Rothwell, G.W. Cordaixylon dumusum (Cordaitales). II. Reproductive biology, phenology, and growth ecology. *Int. J. Plant Sci.* **1993**, *154*, 572–586. [CrossRef]

80. Dai, S.; Ren, D.; Chou, C.-L.; Finkelman, R.B.; Seredin, V.V.; Zhou, Y. Geochemistry of trace elements in Chinese coals: A review of abundances, genetic types, impacts on human health, and industrial utilization. *Int. J. Coal Geol.* **2012**, *94*, 3–21. [CrossRef]

81. Ketris, M.P.; Yudovich, Y.E. Estimations of Clarkes for carbonaceous biolithes: World average for trace element contents in black shales and coals. *Int. J. Coal Geol.* **2009**, *78*, 135–148. [CrossRef]

82. Dai, S.; Li, T.; Seredin, V.V.; Ward, C.R.; Hower, J.C.; Zhou, Y.; Zhang, M.; Song, X.; Song, W.; Zhao, C. Origin of minerals and elements in the Late Permian coals, tonsteins, and host rocks of the Xinde Mine, Xuanwei, eastern Yunnan, China. *Int. J. Coal Geol.* **2014**, *121*, 53–78. [CrossRef]

83. Dai, S.; Luo, Y.; Seredin, V.V.; Ward, C.R.; Hower, J.C.; Zhao, L.; Liu, S.; Tian, H.; Zou, J. Revisiting the late Permian coal from the Huayingshan, Sichuan, southwestern China: Enrichment and occurrence modes of minerals and trace elements. *Int. J. Coal Geol.* **2014**, *122*, 110–128. [CrossRef]

84. Erarslan, C.; Örgün, Y.; Bozkurtoğlu, E. Geochemistry of trace elements in the Keşan coal and its effect on the physicochemical features of ground- and surface waters in the coal fields, Edirne, Thrace Region, Turkey. *Int. J. Coal Geol.* **2014**, *133*, 1–12. [CrossRef]

85. Tian, C.; Zhang, J.; Zhao, Y.; Gupta, R. Understanding of mineralogy and residence of trace elements in coals via a novel method combining low temperature ashing and float-sink technique. *Int. J. Coal Geol.* **2014**, *131*, 162–171. [CrossRef]

86. Sutcu, E.C.; Karayigit, A.I. Mineral matter, major and trace element content of the Afşin–Elbistan coals, Kahramanmaraş, Turkey. *Int. J. Coal Geol.* **2015**, *144–145*, 111–129. [CrossRef]

87. Brownfield, M.E.; Affolter, R.H.; Cathcart, J.D.; Johnson, S.Y.; Brownfield, I.K.; Rice, C.A.; Zielinski, R.A. Geologic setting and characterization of coal and the modes of occurrence of selected elements from the Franklin coal zone, Puget Group, John Henry No. 1 mine, King County, Washington. *Int. J. Coal Geol.* **2005**, *63*, 247–275. [CrossRef]

88. Yudovich, Y.E.; Ketris, M.P. Arsenic in coal: A review. *Int. J. Coal Geol.* **2005**, *61*, 141–196. [CrossRef]

89. Hower, J.C.; Campbell, J.L.; Teesdale, W.J.; Nejedly, Z.; Robertson, J.D. Scanning proton microprobe analysis of mercury and other trace elements in Fe-sufides from a Kentucky coal. *Int. J. Coal Geol.* **2008**, *75*, 88–92. [CrossRef]

90. Kolker, A. Minor element distribution in iron disulfides in coal: A geochemical review. *Int. J. Coal Geol.* **2012**, *94*, 32–43. [CrossRef]

91. Finkelman, R.B. Mode of occurrence of potentially hazardous elements in coal: Levels of confidence. *Fuel Proc. Technol.* **1994**, *39*, 21–34. [CrossRef]

92. Swaine, D.J. Why trace elements are important. *Fuel Proc. Technol.* **2000**, *65*, 21–66. [CrossRef]

93. Liu, J.; Yang, Z.; Yan, X.; Ji, D.; Yang, Y.; Hu, L. Modes of occurrence of highly-elevated trace elements in superhigh-organic-sulfur coals. *Fuel* **2015**, *156*, 190–197. [CrossRef]

94. Hower, J.C.; Ruppert, L.F.; Williams, D.A. Controls on Boron and Germanium Distribution in the Low-Sulfur Amos Coal Bed: Western Kentucky Coalfield, USA. *Int. J. Coal Geol.* **2002**, *53*, 27–42. [CrossRef]

95. Yudovich, Y.E. Notes on the marginal enrichment of Germanium in coal beds. *Int. J. Coal Geol.* **2003**, *56*, 223–232. [CrossRef]

96. Seredin, V.V.; Finkelman, R.B. Metalliferous coals: A review of the main genetic and geochemical types. *Int. J. Coal Geol.* **2008**, *76*, 253–289. [CrossRef]

97. Hayashi, K.I.; Fujisawa, H.; Holland, H.D.; Ohmoto, H. Geochemistry of ~1.9 Ga sedimentary rocks from northeastern Labrador, Canada. *Geochim. Cosmochim. Acta* **1997**, *61*, 4115–4137. [CrossRef]

98. Seredin, V.V.; Dai, S. Coal deposits as potential alternative sources for lanthanides and yttrium. *Int. J. Coal Geol.* **2012**, *94*, 67–93. [CrossRef]

99. Condie, K.C. Chemical composition and evolution of the upper continental crust: contrasting results from surface samples and shales. *Chem. Geol.* **1993**, *104*, 1–37. [CrossRef]

100. Condie, K.C.; Wronkiewicz, D.J. The Cr/Th ratio in Precambrian pelites from the Kaapvaal Craton as an index of craton evolution. *Earth. Planet. Sci. Lett.* **1990**, *97*, 256–267. [CrossRef]

101. Taylor, S.R.; McLennan, S.M. *The Continental Crust: Its Composition and Evolution*; Blackwell: London, UK, 1985; p. 312.

102. Johanneson, K.H.; Zhou, X. Geochemistry of the rare earth element in natural terrestrial waters: A review of what is currently known. *Chin. J. Geochem.* **1997**, *16*, 20–42. [CrossRef]

103. Michard, A.; Albarède, F. The REE content of some hydrothermal fluids. *Chem. Geol.* **1986**, *55*, 51–60.

104. Dai, S.; Seredin, V.V.; Ward, C.R.; Jiang, J.; Hower, J.C.; Song, X.; Jiang, Y.; Wang, X.; Gornostaeva, T.; Li, X.; *et al.* Composition and modes of occurrence of minerals and elements in coal combustion products derived from high-Ge coals. *Int. J. Coal Geol.* **2014**, *121*, 79–97. [CrossRef]

105. Seredin, V.V. From coal science to metal production and environmental protection: A new story of success. *Int. J. Coal Geol.* **2012**, *90–91*, 1–3. [CrossRef]

106. Frazer, F.W.; Belcher, C.B. Quantitative determination of the mineral matter content of coal by a radio-frequency oxidation technique. *Fuel* **1973**, *52*, 41–46. [CrossRef]

107. Ward, C.R. Analysis and significance of mineral matter in coal seams. *Int. J. Coal Geol.* **2002**, *50*, 135–168. [CrossRef]

108. Ward, C.R.; Matulis, C.E.; Taylor, J.C.; Dale, L.S. Quantification of mineral matter in the Argonne Premium coals using interactive Rietveld-based X-ray diffraction. *Int. J. Coal Geol.* **2001**, *46*, 67–82. [CrossRef]

109. Rao, C.P.; Gluskoter, H.J. Occurrence and distribution of minerals in Illinois coals. *Illinois State Geol. Surv. Circ.* **1973**, *476*, 1–56.

110. Dai, S.; Hower, J.C.; Ward, C.R.; Guo, W.; Song, H.; O'Keefe, J.M.K.; Xie, P.; Hood, M.M.; Yan, X. Elements and phosphorus minerals in the middle Jurassic inertinite-rich coals of the Muli Coalfield on the Tibetan Plateau. *Int. J. Coal Geol.* **2015**, *144–145*, 23–47. [CrossRef]

111. Dai, S.; Wang, P.; Ward, C.R.; Tang, Y.; Song, X.; Jiang, J.; Hower, J.C.; Li, T.; Seredin, V.V.; Wagner, N.J.; *et al.* Elemental and mineralogical anomalies in the coal-hosted Ge ore deposit of Lincang, Yunnan, southwestern China: Key role of $N_2$–$CO_2$-mixed hydrothermal solutions. *Int. J. Coal Geol.* **2015**. [CrossRef]

*Article*

# Petrology and Geochemistry of the Harlan, Kellioka, and Darby Coals from the Louellen 7.5-Minute Quadrangle, Harlan County, Kentucky

**Michelle N. Johnston [1], James C. Hower [1,\*], Shifeng Dai [2], Peipei Wang [2], Panpan Xie [2] and Jingjing Liu [2]**

[1]   Center for Applied Energy Research, University of Kentucky, 2540 Research Park Drive, Lexington, KY 40511, USA; mnjohn5@g.uky.edu

[2]   State Key Laboratory of Coal Resources and Safe Mining, China University of Mining and Technology (Beijing), Beijing 100083, China; daishifeng@gmail.com (S.D.); wangpeipei110@163.com (P.W.); xiepanpan90@163.com (P.X.); liujj.cumtb@gmail.com (J.L.)

\*   Correspondence: james.hower@uky.edu; Tel.: +1-859-257-0261

Academic Editor: Antonio Simonetti

Received: 9 November 2015; Accepted: 4 December 2015; Published: 11 December 2015

**Abstract:** The Harlan, Kellioka, and Darby coals in Harlan County, Kentucky, have been among the highest quality coals mined in the Central Appalachians. The Middle Pennsylvanian coals are correlative with the Upper Elkhorn No. 1 to Upper Elkhorn No. 3½ coals to the northwest of the Pine Mountain thrust fault. Much of the mining traditionally was controlled by captive, steel-company-owned mines and the coal was part of the high volatile A bituminous portion of the coking coal blend. Overall, the coals are generally low-ash and low-sulfur, contributing to their desirability as metallurgical coals. We did observe variation both in geochemistry, such as individual lithologies with significant $P_2O_5$/Ba + Sr/Rare earth concentrations, and in maceral content between the lithotypes in the mine sections.

**Keywords:** Pennsylvanian; mining history; coking coal; coal quality

## 1. Introduction

Harlan County, Kentucky, has had a long, colorful, and, at times, violent mining history [1–3]. None of that would have been the case without a base of extensive reserves of high quality coal, much of it directed towards the metallurgical coal market and mined at steel-company-owned mines [4,5].

In this investigation, we examined the petrology and chemistry of coals in Harlan County southeast of Pine Mountain, on the Pine Mountain thrust sheet (Figure 1). In particular, we appraised, in ascending order, the Harlan, Kellioka, and Darby coals of the Pikeville Formation of the Middle Pennsylvanian Breathitt Group (Figure 2). These coals were traditionally some of the better coking coal reserves. The study coals are the approximate correlatives of the Upper Elkhorn No. 1 and Upper Elkhorn No. 2, Upper Elkhorn No. 3 or Van Lear, and the Upper Elkhorn No. 3½ coals, respectively, to the northwest of Pine Mountain [6]. The underlying Path Fork coal, the correlative of the Blue Gem and Pond Creek coals on the northwest side of the Pine Mountain thrust fault, was investigated by Hatton *et al.* [7]. Below the Path Fork, the Grundy Formation Hance coal, correlative of the Manchester and many other coals on the northwest side of the Pine Mountain thrust fault, was studied by Esterle and Ferm [8] as well as Hubbard *et al.* [9]. The study areas for the Path Fork and Hance coals, however, although on the thrust sheet, were to the southwest of the present study area.

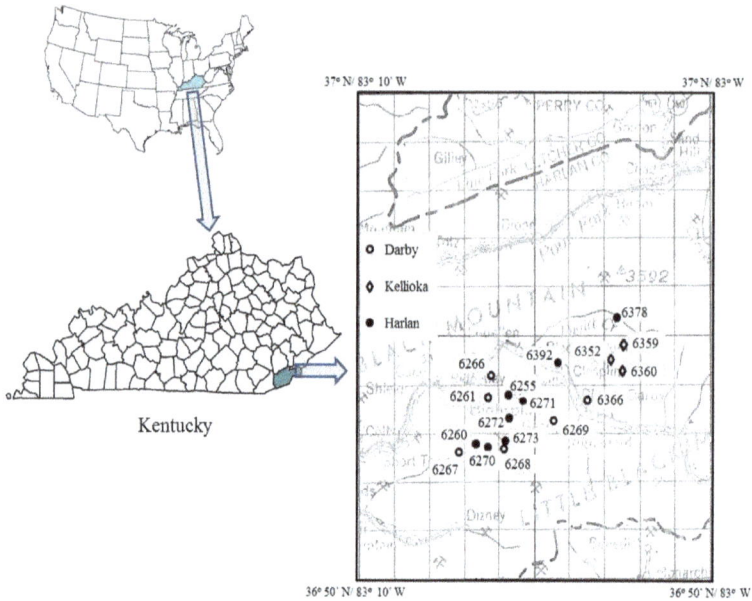

**Figure 1.** Location of the sample sites in Harlan County, Kentucky. For multiple-bench/multiple-lithotype samples, the site is designated by the sample number of the accompanying whole-coal sample.

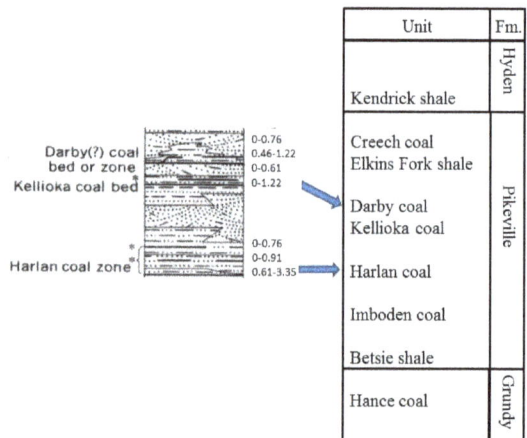

**Figure 2.** Geologic section of the study interval in the Louellen 7.5-minute quadrangle [10].

The environments of deposition for Kentucky coals have been investigated by the Kentucky Geological Survey and the University of Kentucky Center for Applied Energy Research [7,9,11–14]; John Ferm and his students at various universities, most recently (1980–1999) at the University of Kentucky (for example: Esterle and Ferm [8]); and outside researchers [15,16]. With the exception of Esterle and Ferm [8], Hatton *et al.* [7], and Hubbard *et al.* [9], the studies emphasized settings to the northwest of Pine Mountain. Much of the mining of the coals preceded the 1980s and 1990s studies noted above. As such, the current examination of coals collected in the 1980s represents one of the few detailed studies of Harlan County coals.

*Minerals* **2015**, *5*, 894–918

## 2. Methods

Samples were collected both at mines and from company-supplied cores in the 1980s (The samples from the 1980s represent the last widespread availability of mine samples and the serendipitous availability of core samples. Aside from shifts in mining, the CAER sampling interests shifted to other coals) (locations on Figure 1). The coals were collected as whole coal and bench samples, excluding rock partings greater than about 1-cm thick. The proximate and sulfur analyses were conducted at the University of Kentucky Center for Applied Energy Research (CAER) following ASTM procedures. Major oxides were analyzed at the CAER by X-ray fluorescence following procedures outlined by Hower and Bland [17]. All of the latter analyses were done shortly after sampling. Inductively coupled plasma mass spectrometry (X series II ICP-MS, ThermoFisher, Waltham, MA, USA), in pulse counting mode (three points per peak), was used to determine trace elements in the coal ash samples obtained from raw coals at 815 °C. The ICP-MS analyses were conducted at the China University of Mining and Technology (Beijing) on ash samples provided by the CAER. For ICP-MS analysis, samples were digested using an UltraClave Microwave High Pressure Reactor (Milestone, Milano, Italy) [18]. The digestion reagents for each 50-mg coal ash sample are 2-mL 65% $HNO_3$ and 5-mL 40% HF [18]. The Guaranteed-Reagent $HNO_3$ and HF for sample digestion were further purified by sub-boiling distillation. Arsenic and Selenium were determined by ICP-MS using collision cell technology (CCT) in order to avoid disturbance of polyatomic ions [19]. Multi-element standards (Inorganic Ventures: CCS-1, CCS-4, CCS-5, and CCS-6; NIST 2685b and Chinese standard reference GBW 07114) were used for calibration of trace element concentrations. The method detection limit (MDL) for each of the trace elements, calculated as three times the standard deviation of the average from the blank samples ($n = 10$), is listed in Table 1.

**Table 1.** Method detection limit (MDL, µg/L) of inductively coupled plasma mass spectrometry (ICP-MS) for the Harlan, Kellioka, and Darby coal ashes.

| Elements | MDL | Elements | MDL | Elements | MDL |
|----------|--------|----------|--------|----------|--------|
| Li | 0.0090 | Zr | 0.0173 | Gd | 0.0113 |
| Be | 0.0103 | Nb | 0.0567 | Tb | 0.0071 |
| Sc | 0.0067 | Mo | 0.1052 | Dy | 0.0198 |
| V | 0.0062 | Ag | 0.0028 | Ho | 0.0044 |
| Cr | 0.0293 | Cd | 0.0019 | Er | 0.0089 |
| Co | 0.0032 | Sn | 0.0176 | Tm | 0.0061 |
| Ni | 0.0219 | Sb | 0.0037 | Yb | 0.0050 |
| Cu | 0.0481 | In | 0.0028 | Lu | 0.0042 |
| Zn | 0.1547 | Cs | 0.0088 | Hf | 0.0129 |
| Ga | 0.0016 | Ba | 0.2019 | Ta | 0.1194 |
| Ge | 0.0015 | La | 0.0165 | W | 0.1465 |
| As | 0.1799 | Ce | 0.0071 | Tl | 0.2137 |
| Se | 0.2291 | Pr | 0.0122 | Pb | 0.0124 |
| Rb | 0.7208 | Nd | 0.0247 | Bi | 0.0331 |
| Sr | 0.0440 | Sm | 0.0229 | Th | 0.0677 |
| Y | 0.0144 | Eu | 0.0049 | U | 0.88 |

Maceral analysis, originally done shortly after the sampling, was re-examined for this study following the ICCP nomenclature [20,21]. The petrology was done using Leitz Orthoplan microscopes with oil-immersion, reflected-light, 50-x objectives on particulate pellets prepared to a final 0.05-µm-alumina polish. The reflectance was measured using a 547-nm bandpass filter and a 9-µm-diameter measuring spot with a photomultiplier calibrated against a series of glass reflectance standards in the range of the coal reflectances.

## 3. Discussion and Results

### *3.1. Proximate and Sulfur Analysis*

The Harlan coal is the thickest of the three coals investigated; with two of the sections exceeding 2.72 m. The Harlan ash yield is higher and more variable than in the Kellioka and Darby coals (Table 2), discussed below. Several samples exceed 20% ash yield, with sample 6400 having 53% ash yield, sufficient to classify it as carbonaceous shale. On the whole-coal basis, the Harlan coal is a low- to medium-S coal, also higher than in the other two coals; exceeding 2% S in the 23.8%-ash-yield sample 6384. With the relative increase in sulfur compared to the other two coals, we can infer that the Harlan peat was subjected to a more significant marine influence. The nature and extent of such an influence is difficult to discern with just three detailed sections. With the exception of this study, the Harlan coal has not been studied in the same detail as some other eastern Kentucky coals (see Introduction). The mined Harlan coal, destined for the metallurgical market, was beneficiated prior to shipment from the facility; therefore, much of the high-ash and high-S coal was not included in those shipments.

**Table 2.** Thickness (cm), proximate analysis (%), total sulfur and forms of sulfur (%) in the Harlan, Kellioka, and Darby coals. T—total; py—pyritic; org—organic; sulf—sulfate; wc—whole coal; nd—not determined.

| Coal | Sample | Bench | Thickness (cm) | Mois | Ash | S (t) | S (py) | S (sulf) | S (org) |
|------|--------|-------|----------------|------|-----|-------|--------|----------|---------|
| | 6255 | wc | 111.60 | 2.11 | 7.64 | 1.10 | 0.25 | 0.00 | 0.85 |
| | 6256 | 1/4 (top) | 16.70 | 2.41 | 18.66 | 1.30 | 0.37 | 0.02 | 0.91 |
| | 6257 | 2/4 | 9.60 | 2.29 | 5.94 | 1.73 | 0.67 | 0.02 | 1.04 |
| | 6258 | 3/4 | 28.40 | 2.20 | 11.34 | 1.85 | 0.84 | 0.03 | 0.98 |
| | 6259 | 4/4 | 57.00 | 2.20 | 8.40 | 1.08 | 0.26 | 0.01 | 0.81 |
| | 6260 | wc | 140.21 | 1.98 | 8.88 | 1.19 | 0.30 | 0.01 | 0.88 |
| | 6270 | wc | 121.92 | 1.73 | 13.54 | 1.07 | nd | nd | nd |
| | 6271 | wc | 126.49 | 1.59 | 10.75 | 1.22 | nd | nd | nd |
| | 6272 | wc | 126.49 | 1.79 | 14.54 | 1.35 | nd | nd | nd |
| | 6273 | wc | 134.11 | 1.76 | 35.91 | 0.91 | nd | nd | nd |
| | 6378 | wc | 272.80 | 2.08 | 9.19 | 0.91 | 0.20 | 0.00 | 0.71 |
| | 6379 | 1/9 (top) | 27.43 | 2.03 | 9.36 | 0.89 | 0.11 | 0.01 | 0.77 |
| | 6380 | 2/9 | 11.58 | 1.43 | 29.52 | 0.53 | 0.09 | 0.00 | 0.44 |
| | 6381 | 3/9 | 32.92 | 2.07 | 8.84 | 1.34 | 0.30 | 0.01 | 1.03 |
| | 6382 | 4/9 | 17.68 | 1.80 | 4.96 | 0.79 | 0.08 | 0.01 | 0.70 |
| Harlan | 6383 | 5/9 | 51.51 | 2.13 | 5.46 | 1.51 | 0.44 | 0.02 | 1.05 |
| | 6384 | 6/9 | 13.41 | 2.57 | 23.80 | 2.06 | 1.23 | 0.06 | 0.77 |
| | 6385 | 7/9 | 3.35 | 1.49 | 16.96 | 0.49 | 0.04 | 0.00 | 0.45 |
| | 6386 | 8/9 | 41.15 | 2.07 | 3.41 | 0.64 | 0.08 | 0.00 | 0.56 |
| | 6387 | 9/9 | 50.29 | 2.21 | 7.67 | 1.19 | 0.34 | 0.04 | 0.81 |
| | 6392 | wc | 292.61 | 1.88 | 16.72 | 1.09 | 0.29 | 0.00 | 0.80 |
| | 6393 | 1/11 (top) | 35.05 | 1.72 | 7.65 | 0.71 | 0.05 | 0.00 | 0.66 |
| | 6394 | 2/11 | 6.71 | 1.44 | 8.02 | 0.73 | 0.06 | 0.00 | 0.67 |
| | 6395 | 3/11 | 23.77 | 1.70 | 5.93 | 1.27 | 0.24 | 0.01 | 1.02 |
| | 6396 | 4/11 | 6.40 | 2.68 | 25.88 | 1.09 | 0.36 | 0.00 | 0.73 |
| | 6397 | 5/11 | 17.98 | 1.69 | 4.74 | 0.98 | 0.12 | 0.00 | 0.86 |
| | 6398 | 6/11 | 3.05 | 1.17 | 5.48 | 0.80 | 0.01 | 0.00 | 0.79 |
| | 6399 | 7/11 | 49.38 | 1.85 | 4.24 | 0.95 | 0.12 | 0.00 | 0.83 |
| | 6400 | 8/11 | 20.42 | 1.91 | 53.04 | 1.37 | 0.73 | 0.04 | 0.60 |
| | 6401 | 9/11 | 6.71 | 1.77 | 23.13 | 0.58 | 0.09 | 0.00 | 0.49 |
| | 6402 | 10/11 | 43.89 | 1.95 | 4.51 | 0.77 | 0.09 | 0.00 | 0.68 |
| | 6403 | 11/11 | 52.12 | 2.25 | 5.87 | 1.25 | 0.41 | 0.03 | 0.81 |

Table 2. *Cont.*

| Coal | Sample | Bench | Thickness (cm) | Mois | Ash | S (t) | S (py) | S (sulf) | S (org) |
|------|--------|-------|----------------|------|-----|-------|--------|----------|---------|
| Kellioka | 6352 | wc | 86.87 | 2.41 | 5.44 | 0.72 | 0.11 | 0.00 | 0.61 |
| | 6353 | 1/6 (top) | 18.29 | 2.60 | 3.14 | 0.93 | 0.21 | 0.02 | 0.70 |
| | 6354 | 2/6 | 16.76 | 2.60 | 3.22 | 1.13 | 0.39 | 0.03 | 0.71 |
| | 6355 | 3/6 | 8.23 | 2.32 | 2.27 | 0.52 | 0.05 | 0.00 | 0.47 |
| | 6356 | 4/6 | 11.58 | 1.99 | 2.00 | 0.56 | 0.05 | 0.00 | 0.51 |
| | 6357 | 5/6 | 19.81 | 1.79 | 3.80 | 0.52 | 0.03 | 0.00 | 0.49 |
| | 6358 | 6/6 | 12.19 | 2.06 | 15.62 | 0.60 | 0.09 | 0.01 | 0.50 |
| | 6359 | wc | 107.19 | 2.07 | 4.19 | 0.67 | 0.10 | 0.00 | 0.57 |
| | 6360 | wc | 113.03 | 2.09 | 4.86 | 0.97 | 0.29 | 0.01 | 0.67 |
| | 6361 | 1/5 (top) | 19.81 | 2.25 | 3.75 | 1.06 | 0.26 | 0.03 | 0.77 |
| | 6362 | 2/5 | 15.85 | 2.00 | 4.50 | 0.92 | 0.14 | 0.02 | 0.76 |
| | 6363 | 3/5 | 12.80 | 2.14 | 2.93 | 0.65 | 0.05 | 0.00 | 0.60 |
| | 6364 | 4/5 | 43.59 | 1.87 | 3.98 | 0.64 | 0.03 | 0.01 | 0.60 |
| | 6365 | 5/5 | 21.03 | 2.13 | 10.26 | 1.14 | 0.36 | 0.04 | 0.74 |
| Darby | 6261 | wc | 92.05 | 2.59 | 2.18 | 0.56 | 0.06 | 0.00 | 0.50 |
| | 6262 | 1/4 (top) | 13.41 | 2.56 | 1.77 | 0.53 | 0.04 | 0.01 | 0.48 |
| | 6263 | 2/4 | 12.19 | 5.23 | 1.93 | 0.48 | 0.04 | 0.01 | 0.43 |
| | 6264 | 3/4 | 49.38 | 3.46 | 1.29 | 0.55 | 0.04 | 0.00 | 0.51 |
| | 6265 | 4/4 | 17.07 | 3.55 | 1.80 | 0.60 | 0.05 | 0.01 | 0.54 |
| | 6266 | wc | 99.06 | 1.94 | 2.21 | 0.54 | nd | nd | nd |
| | 6267 | wc | 164.59 | 1.96 | 3.04 | 0.70 | nd | nd | nd |
| | 6268 | wc | 108.20 | 2.20 | 2.67 | 0.63 | nd | nd | nd |
| | 6269 | wc | 102.11 | 1.85 | 4.06 | 0.53 | nd | nd | nd |
| | 6366 | wc | 76.71 | 3.03 | 2.21 | 0.63 | 0.04 | 0.01 | 0.58 |
| | 6367 | 1/5 (top) | 3.35 | 2.54 | 7.44 | 0.56 | 0.07 | 0.02 | 0.47 |
| | 6368 | 2/5 | 13.41 | 2.11 | 2.02 | 0.53 | 0.05 | 0.01 | 0.47 |
| | 6369 | 3/5 | 4.57 | 2.23 | 3.28 | 0.53 | 0.04 | 0.01 | 0.48 |
| | 6370 | 4/5 | 41.45 | 3.12 | 1.76 | 0.54 | 0.05 | 0.02 | 0.47 |
| | 6371 | 5/5 | 14.33 | 3.29 | 2.78 | 0.59 | 0.04 | 0.02 | 0.53 |

The ash yield of Harlan coals varies somewhat in our samples depending upon the decisions made about sampling benches. For example, in retrospect, sample 6400 perhaps should not have been included in the whole-coal sample although it would have been part of the mined section along with other partings and portions of the roof and floor. As noted above, coal beneficiation would eliminate many of the higher mineral matter particles, producing a low-ash product.

With the exception of the lower bench at both of the benched sites, the ash yield of the Kellioka coals is less than 5%; sulfur content is generally low, exceeding 0.9% only in the top two benches at both sites and the basal bench at site 6360. The Darby has a low-ash, low-S content, with the exception of the thin top lithotype (sample 6367) at site 6366 with 7.44% ash yield and 36.4% total vitrinite (ash-free basis).

## 3.2. Petrology

The petrology of the coals is presented on Table 3. The lithologic profile of the Harlan coal is shown in Figure 3. Despite some similarities between nearby sites, particularly between seam sections 6378 and 6392, the continuity is not as great as we have seen in studies of the Pond Creek and Blue Gem coals [22,23], the Fire Clay coal [12,24,25], although significant short-distance, few-hundred-meter variation is known to occur in other economically important eastern Kentucky coals [23,26].

The total vitrinite in the whole Harlan coals ranges from 58% to 75% (mineral-included basis), the lowest being in sample 6273 owing to the high mineral content. There is a wide variety of maceral distributions among the bench/lithotype samples. This is well illustrated in the low-mineral matter samples 6399 and 6398 of the 6392 sequence. The sample 6399 bright clarain, bench 7 of 11, has 81.6% total vitrinite. In contrast, sample 6398, the thin, 3.05-cm durain directly overlying 6399, has 31.2% vitrinite, 31.6% inertinite, and 36.6% liptinite.

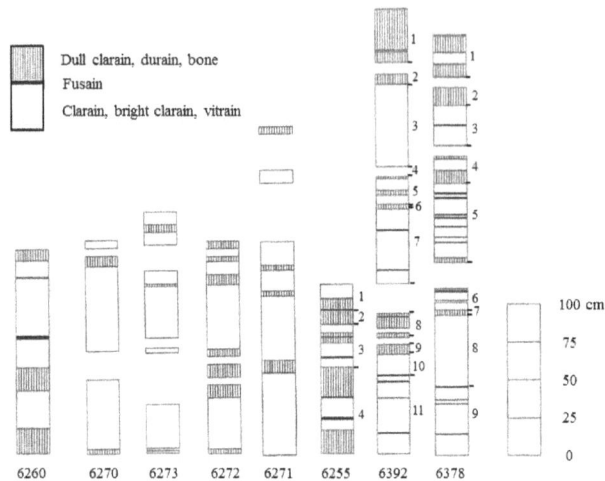

**Figure 3.** Lithologic sections of the Harlan coal. For the 6255, 6378, and 6392 sequences, the tick marks along the right edge indicate the boundaries of the sampled intervals and the associated numbers represent the bench number. The blank spaces between coal benches indicate non-coal rock intervals. See the tables for the correlation between the bench and sample numbers.

With a few exceptions, such as sample 6357 with less than 41% total vitrinite and sample 6364 with less than 48% total vitrinite, the Kellioka coal samples have over 60% vitrinite. The highest vitrinite is found in the relatively high-S upper benches. The inertinite assemblages in the Kellioka coals are dominated by varying amounts of fusinite, semifusinite, micrinite, and inertodetrinite. Macrinite is most abundant in the low-vitrinite lithologies 6357 and 6364, at 6.4% and 4.6% (mineral-free basis), with 2.2% and 2.4% macrinite found in samples 6355 and 6356, respectively, the lithologies with 63%–65% vitrinite. The liptinite assemblages in the Kellioka coals are dominated by sporinite with lesser amounts of resinite and cutinite.

The whole-coal Darby samples all have at least 68.9% total vitrinite (ash-free basis) and none of the other lithotypes have less than 55% vitrinite. The inertinite is generally a function of varying amounts of fusinite, semifusinite, micrinite, and inertodetrinite. Macrinite exceeds 1.5% only in the whole coal sample 6267 and the lithotype sample 6367. The liptinite assemblages in the Darby samples are dominated by sporinite with lesser amounts of resinite and minor amounts of cutinite.

*3.3. Elemental Geochemistry*

Table 4 lists the concentrations of major-element oxides and trace elements in the samples from the Harlan, Kellioka, and Darby coals. Compared to average values for world hard coals reported by Ketris and Yudovich [26] and based on the enrichment classification of elements in coal outlined by Dai *et al.* [27], only the averages of Co in the Harlan coals and As in the Kellioka coals are slightly enriched, with CC (CC = ratio of element concentration in investigated coals *vs.* world hard coals) 2.10 and 2.02, respectively. Lithium and Cu in the Harlan coals have CC of 1.60 and 1.84, respectively. The average concentrations of other trace elements in the three coals are either close to or depleted relative to the averages of the same elements for the world hard coals (Figure 4). Particularly, the concentrations of quite a number of trace elements in Kellioka and Darby coals are depleted (Figure 4). According to Dai *et al.* [27], elemental concentrations in coal can be classified as six levels relative to the averages for world coals, unusually enriched (CC > 100), significantly enriched (10 < CC < 100), enriched(5 < CC < 10); slightly enriched (2 < CC < 5); close to the average values for world hard coals (0.5 < CC < 2), and depleted (CC < 0.5).

Table 3. Petrological compositions (volume%) and maximum vitrinite reflectance (%) of coals from Harlan, Kellioka, and Darby.

| Coal | Sample | Bench | T | CT | VD | CD | CG | G | T-V | F | SF | Mic | Mac | Sec | Fun | ID | T-I | Sp | Cut | Res | Alg | LD | Sub | Ex | T-L | Sil | Sul | Car | Oth | T-M | $R_{o,max}$ |
|---|---|---|---|---|---|---|---|---|---|---|---|---|---|---|---|---|---|---|---|---|---|---|---|---|---|---|---|---|---|---|---|
| Harlan | 6255 | wc | 3 | 40.2 | 3.4 | 15.2 | 3 | 1.2 | 66 | 3.2 | 3.8 | 3.8 | 0.2 | 0.4 | 0 | 4.6 | 16 | 12.6 | 1 | 2.8 | 0 | 0.2 | 0 | 0 | 16.6 | 0.8 | 0.6 | 0 | 0 | 1.4 | 0.95 |
| | 6256 | 1/4 | 3.6 | 39 | 7 | 18.4 | 1.6 | 2.6 | 72.2 | 2.2 | 3 | 1.8 | 0.2 | 0.4 | 0 | 1.6 | 9.2 | 9.2 | 0.2 | 2.4 | 0 | 0.2 | 0 | 0 | 12 | 5.8 | 0.8 | 0 | 0 | 6.6 | 0.91 |
| | 6257 | 2/4 | 0.6 | 26.2 | 1.2 | 29.8 | 0.2 | 0.4 | 58.4 | 4.2 | 4.4 | 4.2 | 2.8 | 0 | 0 | 7.2 | 22.8 | 12.4 | 0 | 4.8 | 0 | 0 | 0 | 0 | 17.2 | 1.6 | 1.6 | 0 | 0 | 1.6 | 0.96 |
| | 6258 | 3/4 | 2.2 | 59.8 | 1.6 | 8.4 | 0.6 | 1.8 | 74.4 | 2.2 | 1.2 | 1.2 | 0.4 | 0.4 | 0 | 4.8 | 9.8 | 7.6 | 1.2 | 1.6 | 0 | 0 | 0 | 0.2 | 10.6 | 1.8 | 3.4 | 0 | 0 | 5.2 | 0.93 |
| | 6259 | 4/4 | 2 | 61 | 1.2 | 6.4 | 0.2 | 0.4 | 71.2 | 7 | 4.8 | 1.2 | 0.6 | 0.4 | 0 | 2.6 | 16.6 | 8.8 | 0.6 | 1.6 | 0 | 0 | 0 | 0.1 | 11 | 0.2 | 1 | 0.0 | 0 | 1.2 | 0.93 |
| | WA-H1 | | 2.2 | 54.4 | 2.2 | 10.7 | 0.5 | 1.1 | 71.1 | 4.8 | 3.6 | 1.5 | 0.7 | 0.3 | 0 | 3.4 | 14.3 | 8.9 | 0.6 | 2.0 | 0 | 0 | 0 | 0 | 11.6 | 1.4 | 1.6 | 0.0 | 0.0 | 3.1 | 0.93 |
| | 6260 | wc | 0.4 | 63.8 | 1.6 | 7.4 | 0.2 | 1.2 | 74.6 | 4.4 | 3.8 | 0.6 | 0.6 | 0.2 | 0 | 4.2 | 13.8 | 6.6 | 1 | 1.4 | 0 | 0 | 0 | 0 | 9 | 1.4 | 1.2 | 0 | 0 | 2.6 | 0.88 |
| | 6270 | wc | 0.6 | 60 | 2.8 | 7 | 0.4 | 1.2 | 72 | 6 | 3.4 | 0.8 | 0.8 | 0 | 0 | 2.6 | 13.6 | 5.4 | 0.6 | 1.8 | 0 | 0 | 0 | 0 | 7.8 | 5 | 1.6 | 0 | 0 | 6.6 | 0.9 |
| | 6271 | wc | 1 | 57.6 | 1.8 | 10.6 | 1 | 1.4 | 73.4 | 4 | 2.4 | 1.2 | 1.8 | 0 | 0 | 1.8 | 11.2 | 7 | 0.2 | 0.8 | 0 | 0.4 | 0 | 0 | 8.4 | 5.8 | 0.8 | 0.4 | 0 | 7 | 0.87 |
| | 6272 | wc | 1 | 52.2 | 0.8 | 7.4 | 1 | 1 | 64.8 | 6 | 2.4 | 2.4 | 2.2 | 0.8 | 0 | 3.6 | 17.4 | 10 | 0.2 | 1.8 | 0 | 0.4 | 0 | 0 | 12 | 4.4 | 1.2 | 0 | 0.2 | 5.8 | 0.9 |
| | 6273 | wc | 3.6 | 35.2 | 7 | 9.4 | 1.6 | 1.6 | 58 | 4.6 | 0.2 | 1.6 | 0.8 | 0.4 | 0 | 2 | 9.6 | 4 | 1.2 | 1.8 | 0 | 0.4 | 0 | 0.2 | 7.4 | 23.4 | 1.4 | 0 | 0.2 | 25 | 0.89 |
| | 6378 | wc | 1.2 | 58.2 | 1 | 8 | 2.4 | 4 | 74.8 | 3.6 | 2.8 | 2.8 | 0.6 | 0.6 | 0 | 1.4 | 11.4 | 8.8 | 1 | 0.6 | 0 | 0.2 | 0 | 0 | 10.8 | 2.4 | 0.4 | 0 | 0.2 | 3 | 0.95 |
| | 6379 | 1/9 | 2.8 | 25 | 0.6 | 13 | 1.4 | 2.4 | 45.2 | 5.8 | 13.4 | 4.4 | 9 | 0.6 | 0 | 7.4 | 40.6 | 7.2 | 1 | 2.2 | 0 | 0 | 0 | 0 | 10.4 | 3.4 | 0.2 | 0 | 0.2 | 3.8 | 0.93 |
| | 6380 | 2/9 | 2.8 | 22.2 | 3.2 | 9.2 | 1.8 | 1.8 | 39.8 | 6.8 | 7.8 | 5.4 | 3.8 | 1.8 | 0 | 7.8 | 33.4 | 9.8 | 0 | 8.4 | 0 | 0 | 0 | 0 | 18.2 | 7.8 | 0.2 | 0 | 0.6 | 8.6 | 0.92 |
| | 6381 | 3/9 | 1 | 55.6 | 1.2 | 7.4 | 3.8 | 3.8 | 70.6 | 2.4 | 3.4 | 3.2 | 1.2 | 0.6 | 0 | 3 | 13.6 | 8.2 | 1.6 | 1.4 | 0 | 0 | 0 | 0 | 11.2 | 2.2 | 0.2 | 0 | 0.4 | 4.6 | 0.91 |
| | 6382 | 4/9 | 2 | 46.4 | 2 | 11.2 | 1 | 1.2 | 64.6 | 2 | 4.6 | 2.6 | 2 | 0.4 | 0 | 4.4 | 16 | 14 | 0.6 | 3.4 | 0 | 0 | 0 | 0 | 18 | 0.6 | 0.4 | 0 | 0.4 | 1.4 | 0.92 |
| | 6383 | 5/9 | 2.8 | 57.8 | 0 | 10 | 1.2 | 1.2 | 73.2 | 5.2 | 3.6 | 5 | 0 | 0 | 0 | 2.8 | 17 | 7.2 | 0.4 | 0.6 | 0.4 | 0 | 0 | 0 | 8.6 | 0.6 | 0.6 | 0 | 0 | 1.2 | 0.93 |
| | 6384 | 6/9 | 3.6 | 38.4 | 6.4 | 12.2 | 1.2 | 2.2 | 64 | 0.4 | 2.6 | 5 | 0.4 | 0.4 | 0.2 | 3.8 | 9.4 | 5.4 | 0.8 | 2 | 0 | 0.4 | 0 | 0 | 8.2 | 13.4 | 5 | 0 | 0 | 18.4 | 0.91 |
| | 6385 | 7/9 | 4.2 | 13 | 3.4 | 13.2 | 0.6 | 1.8 | 36.2 | 11.6 | 14.8 | 1.8 | 3.2 | 0.6 | 0 | 13 | 45 | 14 | 0.2 | 2.8 | 0 | 0 | 0 | 0 | 17 | 1.2 | 0.6 | 0 | 0 | 1.8 | 0.92 |
| | 6386 | 8/9 | 1 | 60 | 0.2 | 16.4 | 1 | 1 | 79.6 | 2 | 2 | 2.6 | 0 | 0 | 0 | 1.4 | 6.2 | 8 | 0.2 | 2.8 | 0 | 0 | 0 | 0 | 12 | 1.8 | 0.2 | 0.2 | 0 | 2.2 | 0.95 |
| | 6387 | 9/9 | 1.2 | 59.8 | 1.2 | 9.2 | 0.8 | 0.8 | 73 | 3.8 | 4 | 3.4 | 0.2 | 0.2 | 0 | 2.2 | 13.6 | 8.2 | 0.4 | 1.6 | 0 | 0.4 | 0 | 0 | 10.6 | 1.8 | 1 | 0.4 | 0 | 2.8 | 0.95 |
| | WA-H2 | | 1.9 | 50.6 | 1.2 | 11.1 | 1.2 | 1.7 | 67.6 | 3.8 | 4.5 | 3.6 | 1.6 | 0.3 | 0 | 3.5 | 17.3 | 8.3 | 0.8 | 2.1 | 0.1 | 0.1 | 0 | 0 | 11.3 | 2.6 | 1.0 | 0 | 0.1 | 3.7 | 0.93 |
| | 6392 | wc | 4.4 | 51.8 | 1.2 | 11 | 1.2 | 0.8 | 70.2 | 3.2 | 4.6 | 1.4 | 1 | 1 | 0 | 4 | 14.4 | 8 | 1 | 1 | 0 | 0.4 | 0 | 0 | 9.4 | 4.6 | 1 | 0 | 0 | 6 | 0.85 |
| | 6393 | 1/11 | 0.8 | 22.8 | 3.4 | 23.2 | 0.8 | 1.4 | 52.4 | 3.4 | 11.2 | 2.6 | 5 | 1.8 | 0 | 7.2 | 31.2 | 10.4 | 0 | 0.8 | 0 | 0.6 | 0 | 0 | 11.8 | 1.8 | 0 | 0 | 0 | 4.6 | 0.82 |
| | 6394 | 2/11 | 1.2 | 25 | 1 | 18.8 | 0.2 | 0.2 | 46.4 | 4.4 | 13.8 | 3.4 | 1.8 | 1 | 0 | 9.6 | 34 | 14 | 0.4 | 3.4 | 0 | 0 | 0 | 0 | 17.8 | 1.8 | 0 | 0.4 | 0 | 1.8 | 0.83 |
| | 6395 | 3/11 | 2.2 | 57.6 | 0.8 | 10.8 | 0.2 | 1.4 | 74 | 2.6 | 5 | 2 | 1 | 1 | 0 | 3.4 | 13.4 | 6.4 | 3 | 1.2 | 0 | 0 | 0 | 0.2 | 10.8 | 9.4 | 0.6 | 0.4 | 0 | 1.8 | 0.82 |
| | 6396 | 4/11 | 6.4 | 46.8 | 7 | 8.8 | 2.4 | 0.4 | 71.8 | 4 | 3.4 | 1.8 | 0.2 | 0.4 | 0 | 1.2 | 10.6 | 2.4 | 1 | 1.6 | 0 | 0.8 | 0 | 0.2 | 5 | 9.4 | 2.8 | 0.4 | 0 | 12.6 | 0.78 |
| | 6397 | 5/11 | 1 | 51.6 | 0.2 | 10.2 | 0.2 | 2.2 | 65.4 | 3 | 4.8 | 1.8 | 2.2 | 0.4 | 0 | 5.6 | 17.8 | 10.4 | 1 | 3.2 | 0.4 | 0 | 0 | 0 | 15.8 | 0.4 | 0.2 | 0.2 | 0.2 | 1.0 | 0.81 |
| | 6398 | 6/11 | 0.2 | 16.2 | 0.4 | 12.8 | 0.2 | 0.6 | 31.2 | 2.2 | 3.4 | 1.8 | 4.2 | 0.4 | 0 | 14 | 31.6 | 29.2 | 1 | 7.4 | 0 | 0 | 0 | 0 | 36.6 | 0.8 | 0.2 | 0 | 0 | 0.6 | 0.86 |
| | 6399 | 7/11 | 4 | 58.6 | 0 | 17.6 | 0.2 | 1.2 | 81.6 | 3.2 | 3.2 | 2 | 0.2 | 0 | 0 | 1.6 | 10.2 | 5.2 | 0.6 | 0.6 | 0 | 0 | 0 | 0 | 6.4 | 0.8 | 1 | 0 | 0 | 1.8 | 0.8 |
| | 6400 | 8/11 | 5.4 | 26.2 | 5.8 | 10.2 | 1.6 | 0.4 | 49.6 | 2.6 | 3.2 | 0.2 | 0.4 | 0 | 0 | 1.6 | 7.6 | 2.4 | 0 | 0.2 | 0 | 0 | 0 | 0 | 2.6 | 35.4 | 4.8 | 0 | 0.2 | 40.2 | 0.78 |
| | 6401 | 9/11 | 1.8 | 47.6 | 0.4 | 7.8 | 0 | 0.8 | 58.8 | 4.4 | 2.8 | 0.4 | 0 | 0.2 | 0 | 9.8 | 17.8 | 13 | 1.2 | 2 | 0 | 0.8 | 0 | 0 | 17 | 6.4 | 0 | 0 | 0 | 6.4 | 0.82 |
| | 6402 | 10/11 | 2.4 | 66.2 | 0.4 | 5.2 | 0 | 2.4 | 76.8 | 0.6 | 0.2 | 1.6 | 0 | 0.2 | 0 | 1.8 | 4.4 | 11 | 2 | 1.8 | 0 | 1.6 | 0 | 0 | 16.4 | 2.2 | 0 | 0.2 | 0 | 2.4 | 0.87 |
| | 6403 | 11/11 | 2.4 | 64.2 | 0.4 | 9.8 | 0.2 | 2.2 | 79 | 3.8 | 5 | 2.2 | 0.2 | 0.4 | 0 | 1.4 | 10 | 7.8 | 2 | 1 | 0 | 0.6 | 0 | 0 | 10.4 | 0.2 | 0.2 | 0.2 | 0 | 0.6 | 0.86 |
| | WA-H3 | | 2.7 | 51.4 | 1.4 | 12.7 | 0.5 | 1.6 | 70.2 | 2.8 | 4.2 | 1.9 | 1.0 | 0.4 | 0 | 3.1 | 13.5 | 8.1 | 0.8 | 1.3 | 0 | 0.5 | 0 | 0 | 11.0 | 4.4 | 0.2 | 0.1 | 0.0 | 5.3 | 0.83 |
| | WA-H | | 2.0 | 52.3 | 2.2 | 10.0 | 1.3 | 1.5 | 69.3 | 4.2 | 3.2 | 2.0 | 1.1 | 0.3 | 0 | 3.1 | 13.9 | 8.0 | 0.7 | 1.6 | 0 | 0.2 | 0 | 0 | 10.5 | 5.1 | 1.0 | 0.1 | 0.1 | 6.3 | 0.90 |

Table 3. Cont.

| Coal | Sample | Bench T | CT | VD | CD | CG | G | T-V | F | SF | Mic | Mac | Sec | Fun | ID | T-I | Sp | Cut | Res | Alg | LD | Sub | Ex | T-L | Sil | Sul | Car | Oth | T-M | $R_{o,max}$ |
|---|---|---|---|---|---|---|---|---|---|---|---|---|---|---|---|---|---|---|---|---|---|---|---|---|---|---|---|---|---|---|
| | 6352 | wc 1.0 | 54.4 | 1.4 | 15.4 | 0.6 | 0.6 | 73.4 | 4.0 | 4.6 | 2.0 | 1.2 | 0.4 | 0 | 4.0 | 16.2 | 7.4 | 0.8 | 0.8 | 0 | 0 | 0 | 0 | 9.0 | 1.4 | 0 | 0 | 0 | 1.4 | 0.94 |
| | 6353 | 1/6 0.0 | 69.4 | 0.8 | 10.2 | 0.6 | 1.6 | 82.6 | 1.4 | 0.8 | 3.2 | 0 | 0.2 | 0 | 6.4 | 7.0 | 0.2 | 1.2 | 0.2 | 0 | 0 | 0 | 0 | 8.4 | 0.8 | 1.8 | 0 | 0 | 2.6 | 0.94 |
| | 6354 | 2/6 1.4 | 57.9 | 0.8 | 13.4 | 0.4 | 1.4 | 75.4 | 3.2 | 3.8 | 2.8 | 0.2 | 0.0 | 0 | 2.6 | 12.6 | 7.8 | 1.0 | 1.0 | 0 | 0 | 0 | 0 | 9.8 | 0.6 | 1.6 | 0 | 0 | 2.2 | 0.94 |
| | 6355 | 3/6 0.6 | 44.6 | 0.0 | 17.2 | 0.4 | 0.6 | 63.4 | 5.2 | 11.6 | 2.0 | 2.4 | 0.0 | 0 | 4.6 | 25.8 | 9.8 | 0.2 | 0.6 | 0 | 0 | 0 | 0 | 10.6 | 0.2 | 0 | 0 | 0 | 0.2 | 0.93 |
| | 6356 | 4/6 0.8 | 47.8 | 0.0 | 14.2 | 0.4 | 1.0 | 64.2 | 4.8 | 6.4 | 3.6 | 2.2 | 0.2 | 0 | 6.6 | 23.8 | 7.8 | 1.0 | 2.2 | 0 | 0.6 | 0 | 0 | 11.6 | 0.0 | 0.0 | 0.2 | 0 | 0.4 | 0.92 |
| | 6357 | 5/6 4.0 | 22.2 | 0.8 | 9.4 | 0.8 | 3.4 | 40.6 | 7.0 | 12.8 | 5.2 | 6.4 | 1.6 | 0 | 11.6 | 44.6 | 9.8 | 0.0 | 4.4 | 0 | 0.4 | 0 | 0 | 14.6 | 0.0 | 0.0 | 0.0 | 0 | 0.2 | 0.96 |
| | 6358 | 6/6 3.2 | 58.4 | 2.2 | 3.8 | 0.0 | 1.7 | 68.6 | 3.0 | 2.4 | 4.6 | 0.4 | 0.0 | 0 | 2.2 | 12.6 | 6.6 | 1.0 | 2.4 | 0 | 0.2 | 0 | 0 | 10.2 | 8.2 | 0.4 | 0.0 | 0.2 | 8.6 | 0.91 |
| | WA-K1 | 1.8 | 49.6 | 0.8 | 10.9 | 0.5 | 1.7 | 65.4 | 4.1 | 6.1 | 3.7 | 2.1 | 0.4 | 0 | 4.9 | 21.3 | 8.1 | 0.7 | 1.9 | 0 | 0.2 | 0 | 0 | 11.0 | 1.5 | 0.7 | 0.0 | 0.0 | 2.3 | 0.94 |
| Kellioka | 6359 | wc 2.2 | 50.6 | 0.6 | 7.8 | 0.0 | 1.6 | 62.8 | 3.2 | 6.6 | 3.8 | 2.6 | 0.2 | 0 | 6.0 | 22.2 | 10.2 | 0.0 | 2.8 | 0 | 0.0 | 0 | 0 | 13.0 | 1.8 | 0.2 | 0 | 0 | 2.0 | 0.90 |
| | 6360 | wc 1.2 | 55.8 | 1.6 | 8.6 | 0.4 | 1.2 | 68.8 | 6.0 | 4.2 | 4.6 | 2.2 | 0.2 | 0 | 5.2 | 22.4 | 6.2 | 0.0 | 1.0 | 0 | 0.2 | 0 | 0 | 7.8 | 0.8 | 0.2 | 0 | 0 | 1.0 | 0.96 |
| | 6361 | 1/5 3.4 | 65.2 | 0.8 | 8.4 | 0.2 | 2.0 | 80.0 | 2.2 | 4.2 | 7.0 | 0.0 | 0.2 | 0 | 1.0 | 11.6 | 3.6 | 0.6 | 0.4 | 0.2 | 0.4 | 0 | 0 | 5.2 | 1.6 | 0.2 | 0.4 | 0 | 3.2 | 0.95 |
| | 6362 | 2/5 1.2 | 48.6 | 1.6 | 16.2 | 0.2 | 2.0 | 69.8 | 3.0 | 4.2 | 3.4 | 0.0 | 0.4 | 0 | 6.0 | 15.6 | 10.8 | 0.4 | 1.0 | 0 | 0.6 | 0 | 0 | 12.8 | 1.6 | 0.2 | 0.0 | 0 | 1.8 | 0.94 |
| | 6363 | 3/5 0.6 | 43.4 | 0.0 | 18.2 | 0.6 | 2.0 | 63.4 | 3.0 | 8.4 | 5.6 | 0.0 | 0.0 | 0 | 6.0 | 24.4 | 8.8 | 0.2 | 1.4 | 0 | 0.8 | 0 | 0 | 10.8 | 0.4 | 0.4 | 0.6 | 0 | 1.8 | 0.96 |
| | 6364 | 4/5 2.6 | 26.8 | 0.2 | 15.2 | 0.8 | 2.0 | 47.6 | 5.4 | 16.0 | 5.0 | 4.6 | 0.6 | 0 | 10.6 | 42.2 | 6.6 | 0.2 | 2.0 | 0 | 0.8 | 0 | 0 | 9.6 | 0.6 | 0.4 | 0.6 | 0 | 0.6 | 0.95 |
| | 6365 | 5/5 4.4 | 49.8 | 3.4 | 12.2 | 1.4 | 4.4 | 75.6 | 1.6 | 1.8 | 4.0 | 0.2 | 0.6 | 0 | 2.2 | 9.4 | 4.8 | 1.6 | 1.0 | 1.2 | 0.8 | 0 | 0 | 9.4 | 3.0 | 2.6 | 0.1 | 0 | 5.6 | 0.94 |
| | WA-K2 | WA 2.7 | 42.7 | 1.1 | 13.9 | 0.7 | 2.3 | 63.4 | 3.6 | 8.3 | 5.0 | 1.9 | 0.4 | 0 | 5.8 | 25.0 | 6.6 | 0.4 | 1.3 | 0.3 | 0.7 | 0 | 0 | 9.4 | 1.3 | 0.8 | 0.1 | 0 | 2.2 | 0.95 |
| | WA-K | 1.8 | 50.6 | 1.1 | 11.3 | 0.4 | 1.5 | 66.8 | 4.2 | 6.0 | 3.8 | 2.0 | 0.3 | 0 | 5.2 | 21.4 | 7.7 | 0.5 | 1.6 | 0.1 | 0.2 | 0 | 0 | 10.0 | 1.4 | 0.4 | 0.1 | 0 | 1.8 | 0.94 |
| | 6261 | wc 6.8 | 54.2 | 0 | 11.2 | 1.8 | 3.8 | 77.8 | 4.2 | 2 | 2.4 | 0.2 | 0 | 0 | 3.4 | 12.2 | 6.8 | 1.6 | 0.8 | 0 | 0.2 | 0 | 0 | 9.4 | 0.2 | 0.2 | 0.2 | 0 | 0.6 | 0.93 |
| | 6262 | 1/4 1.6 | 40.2 | 0.2 | 16.6 | 0.4 | 1.6 | 60.6 | 4.2 | 8 | 4 | 0.2 | 0.2 | 0 | 7 | 23.6 | 12 | 0.6 | 2.4 | 0 | 0.4 | 0 | 0 | 15.4 | 0 | 0.4 | 0 | 0 | 0.4 | 0.86 |
| | 6263 | 2/4 4 | 55.4 | 0 | 11.2 | 0.4 | 2.4 | 73.4 | 9 | 2.8 | 4.8 | 0 | 0 | 0 | 3.6 | 20.2 | 5.6 | 1 | 0.6 | 0 | 0.2 | 0 | 0 | 6.4 | 0 | 0 | 0 | 0 | 0 | 0.9 |
| | 6264 | 3/4 2.8 | 57.2 | 0.4 | 13.2 | 0.2 | 4 | 77.8 | 4.2 | 3 | 0.4 | 0 | 0 | 0 | 2.2 | 9.8 | 9.2 | 1 | 1.8 | 0 | 0.2 | 0 | 0.2 | 12.4 | 0 | 0 | 0 | 0 | 0 | 0.93 |
| | 6265 | 4/4 3.6 | 64.2 | 0 | 5.4 | 0.8 | 8.8 | 82.8 | 2.4 | 0.4 | 2.2 | 0 | 0 | 0 | 0.8 | 5.8 | 7.8 | 1.2 | 1.2 | 0 | 0.8 | 0 | 0.2 | 11 | 0.0 | 0.4 | 0 | 0 | 0.4 | 0.92 |
| | WA-D1 | 2.8 | 55.8 | 0.3 | 12.1 | 0 | 4.6 | 76.0 | 3.8 | 3.3 | 1.4 | 0 | 0 | 0 | 2.7 | 11.3 | 9.4 | 1.0 | 1.8 | 0 | 0.4 | 0 | 0.1 | 12.6 | 0.0 | 0.2 | 0 | 0 | 0.4 | 0.92 |
| | 6266 | wc 1 | 64.6 | 0.2 | 9.2 | 0 | 3.2 | 78.2 | 1.4 | 4.4 | 0.2 | 1 | 0 | 0 | 3.4 | 10.4 | 6 | 1.8 | 2.4 | 0 | 0.4 | 0 | 0 | 10.6 | 0.8 | 0 | 0 | 0 | 0.8 | 0.93 |
| | 6267 | wc 2 | 57.6 | 0.4 | 9.2 | 0.2 | 2.2 | 71.6 | 2.4 | 5.2 | 2.2 | 3.2 | 0 | 0 | 4 | 17 | 7 | 0.8 | 1 | 0 | 0 | 0 | 0 | 8.8 | 1.8 | 0.8 | 0 | 0 | 2.6 | 0.92 |
| Darby | 6268 | wc 7.6 | 49.2 | 0.4 | 11.8 | 0.2 | 2.2 | 71.8 | 2.4 | 9.4 | 0.8 | 0.6 | 0.4 | 0 | 4.6 | 18.2 | 7 | 0.4 | 1.2 | 0 | 0 | 0 | 0 | 8.6 | 1.4 | 0 | 0 | 0 | 1.4 | 0.92 |
| | 6269 | wc 1 | 55.8 | 0.6 | 13.6 | 0 | 2.2 | 73.2 | 2.6 | 5.8 | 0.8 | 1.4 | 0 | 0 | 2 | 12.6 | 9.6 | 0.8 | 1.8 | 0 | 0.2 | 0 | 0 | 12.4 | 1.6 | 0.2 | 0 | 0 | 1.8 | 0.9 |
| | 6366 | wc 5.4 | 48.8 | 0 | 11.8 | 0.2 | 2.6 | 68.8 | 2.6 | 10 | 3 | 1.2 | 0.2 | 0 | 4 | 20.4 | 6.6 | 0.6 | 3 | 0 | 0.4 | 0 | 0 | 10.6 | 0.2 | 0 | 0 | 0 | 0.2 | 0.99 |
| | 6367 | 1/5 0.4 | 17.4 | 1.2 | 15 | 0 | 1.6 | 35.6 | 1.6 | 25.8 | 2.4 | 5.6 | 0 | 0 | 4.6 | 40 | 14.8 | 0 | 7.2 | 0 | 0.2 | 0 | 0 | 22.2 | 2 | 0 | 0.2 | 0 | 2.2 | 0.95 |
| | 6368 | 2/5 0.8 | 41.8 | 0 | 18 | 0 | 0.4 | 61 | 0.2 | 12.8 | 1.6 | 0.6 | 0.2 | 0 | 9 | 24.4 | 9.4 | 0 | 4.8 | 0 | 0.2 | 0 | 0 | 14.4 | 0.2 | 0.2 | 0 | 0 | 0.2 | 0.95 |
| | 6369 | 3/5 3.8 | 38 | 0.6 | 10.6 | 0 | 2.6 | 55.6 | 5.4 | 10.4 | 1.4 | 0.8 | 0.6 | 0 | 5.4 | 24 | 12.8 | 0.4 | 5.4 | 0 | 1.6 | 0 | 0 | 20.2 | 0.2 | 0 | 0 | 0 | 0.2 | 0.93 |
| | 6370 | 4/5 4.4 | 65 | 0 | 10 | 0.4 | 0 | 79.8 | 2.4 | 2.6 | 0.4 | 0 | 0.4 | 0 | 2.8 | 8.6 | 7.6 | 1 | 1 | 0 | 2 | 0 | 0 | 11.6 | 0 | 0 | 0 | 0 | 0 | 1 |
| | 6371 | 5/5 5.6 | 69 | 0.6 | 6.6 | 0.4 | 1.4 | 83.6 | 2.6 | 2 | 3.2 | 0 | 0 | 0 | 2.6 | 10.4 | 2.4 | 1.2 | 0.6 | 0 | 1 | 0 | 0 | 5.2 | 0.8 | 0 | 0 | 0 | 0.8 | 0.97 |
| | WA-D2 | 3.8 | 58.0 | 0.2 | 11.0 | 0.3 | 0.6 | 73.9 | 2.6 | 5.7 | 1.3 | 0 | 0.3 | 0 | 4.1 | 14.0 | 7.6 | 0.8 | 2.1 | 0 | 1.4 | 0 | 0 | 11.9 | 0.3 | 0 | 0 | 0 | 0.3 | 0.98 |
| | WA-D | 3.8 | 55.5 | 0.3 | 11.2 | 0.4 | 2.7 | 73.9 | 2.6 | 5.7 | 1.5 | 1.0 | 0.1 | 0 | 3.5 | 14.5 | 7.5 | 1.0 | 1.8 | 0 | 0.4 | 0 | 0 | 10.6 | 0.8 | 0.2 | 0 | 0 | 1.0 | 0.94 |

T, telinite; CT, collotelinite; VD, vitrodetrinite; CD, collodetrinite; CG, corpogelinite; G, gelinite; T-V, total vitrinite; F, fusinite; SF, semifusinite; Mic, micrinite; Mac, macrinite; Sec, secretinite; Fun, funginite; ID, inertodetrinite; T-I, total inertinite; Sp, sporinite; Cut, cutinite; Res, resinite; Alg, alginate; LD, liptodetrinite; Sub, suberinite; Ex, exsudatinite; T-L, total liptinite; Sil, silicate; Sul, sulfide; Car, carbonate; Oth, others; T-M, total mineral; WA-H1, weighted average based the thickness of bench interval (samples 6256 to 6259); WA-H2, weighted average of samples 6379 to 6387; WA-H3, weighted average of samples 6393 to 6403; WA-H, weighted average of all the Harlan coal samples collected; WA-K1, weighted average of samples 6353 to 6358; WA-K2, weighted average of samples 6361 to 6365; WA-K, weighted average of all Kellioka samples collected; WA-D1, weighted average of samples 6262 to 6265; WA-D2, weighted average of samples 6367 to 6371; WA-D, weighted average of all Darby samples collected.

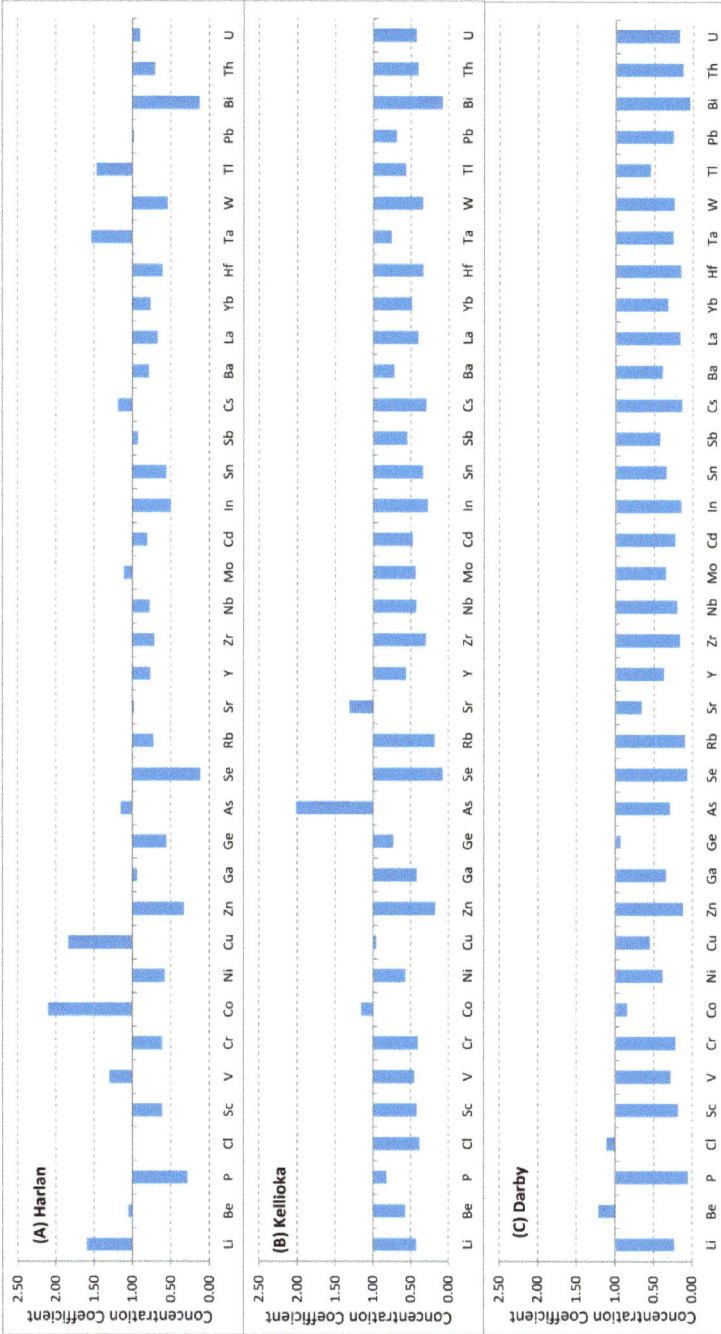

**Figure 4.** Concentration coefficients of trace elements in the coals studied. (**A**) Harlan; (**B**) Kellioka; (**C**) Darby. Concentration coefficients (CC) are the ratio of the trace-element concentrations in the coal samples *vs.* world hard coals reported by Ketris and Yudovich [26].

**Table 4.** Percentages of major-element oxides and chlorine (%) and concentrations of trace elements (µg/g) in coals from Harlan, Kellioka, and Darby (on whole coal basis). Bdl—below detection limit; nd—ot determined.

| Coal | Sample | Bench | SiO₂ | TiO₂ | Al₂O₃ | Fe₂O₃ | MgO | CaO | Na₂O | K₂O | P₂O₅ | Li | Be | Cl | Sc | V | Cr | Co | Ni | Cu | Zn | Ga | Ge |
|---|---|---|---|---|---|---|---|---|---|---|---|---|---|---|---|---|---|---|---|---|---|---|---|
| Harlan | 6255 | wc | 3.82 | 0.123 | 2.67 | 0.58 | 0.073 | 0.12 | 0.001 | 0.254 | 0.009 | 17.2 | 1.58 | 0.02 | 1.69 | 26.2 | 10.8 | 6.4 | 10.8 | 22.7 | 7.59 | 3.79 | 0.59 |
| | 6256 | 1/4 | 9.99 | 0.185 | 6.09 | 1.1 | 0.285 | 0.08 | 0.044 | 0.883 | 0.019 | 36.6 | 2.16 | 0.02 | 3 | 75.7 | 26.7 | 8.69 | 22 | 43.3 | 35.5 | 8.95 | 1.44 |
| | 6257 | 2/4 | 2.47 | 0.096 | 1.74 | 1.35 | 0.036 | 0.17 | bdl | 0.082 | 0.006 | 8.69 | 0.65 | 0.02 | 0.94 | 15.5 | 7.54 | 2.15 | 8.25 | 13.4 | 4.2 | 2.14 | 0.25 |
| | 6258 | 3/4 | 4.99 | 0.144 | 3.79 | 1.75 | 0.113 | 0.1 | 0.014 | 0.442 | 0.013 | 29.4 | 0.62 | 0.03 | 1.55 | 37.5 | 14.7 | 3.28 | 12.4 | 27.5 | 8.3 | 5.33 | 0.42 |
| | 6259 | 4/4 | 4.15 | 0.136 | 3.04 | 0.56 | 0.081 | 0.15 | 0.012 | 0.286 | 0.01 | 20.6 | 2.68 | bdl | 1.75 | 34.0 | 12.7 | 10.7 | 15.4 | 30.0 | 9.7 | 4.85 | 1.05 |
| | WA-H1 | | 5.09 | 0.142 | 3.575 | 1.011 | 0.116 | 0.13 | 0.016 | 0.397 | 0.012 | 24.2 | 1.9 | 0.02 | 1.82 | 39.5 | 14.8 | 7.8 | 15 | 29.9 | 12.7 | 5.35 | 0.88 |
| | 6260 | wc | 4.34 | 0.129 | 3.25 | 0.59 | 0.106 | 0.18 | 0.028 | 0.277 | 0.011 | 19.8 | 3.59 | bdl | 1.88 | 35.7 | 13.2 | 9.06 | 16.1 | 23.9 | 10.7 | 5.0 | 2.14 |
| | 6270 | wc | 6.62 | 0.164 | 4.61 | 0.73 | 0.269 | 0.57 | 0.12 | 0.536 | 0.017 | 26.8 | 4.18 | 0.02 | 1.19 | 43.3 | 17.8 | 23.7 | 25 | 35.4 | 19.5 | 6.85 | 1.39 |
| | 6271 | wc | 5.02 | 0.122 | 3.32 | 1.36 | 0.169 | 0.33 | 0.07 | 0.377 | 0.016 | 19.1 | 1.76 | bdl | 3.11 | 32.4 | 16 | 44.4 | 45.3 | 44.9 | 21.1 | 5.05 | 0.89 |
| | 6272 | wc | 7.18 | 0.172 | 4.75 | 1.1 | 0.276 | 0.43 | 0.131 | 0.552 | 0.017 | 27.3 | 1.59 | 0.02 | 3.31 | 34.1 | 16.9 | 19.9 | 26.3 | 29.8 | 18 | 6.83 | 1.28 |
| | 6273 | wc | 17.6 | 0.388 | 11.4 | 3.4 | 0.854 | 0.51 | 0.297 | 1.66 | 0.037 | 64.5 | 4.13 | 0.02 | 8.05 | 112 | 45.4 | 16.6 | 32.7 | 74.9 | 61.3 | 17.5 | 2.35 |
| | 6378 | wc | 4.63 | 0.115 | 2.9 | 0.77 | 0.127 | 0.3 | 0.052 | 0.31 | 0.017 | 16.7 | 0.86 | bdl | 0.38 | 25.6 | 10.8 | 2.91 | 7.62 | 19.0 | 10.1 | 3.78 | 1.46 |
| | 6379 | 1/9 | 5.53 | 0.183 | 2.76 | 0.44 | 0.068 | 0.18 | bdl | 0.16 | 0.056 | 16.6 | 2.26 | bdl | 3.46 | 30.7 | 14.3 | 5.39 | 8.4 | 28.2 | 9.16 | 5.58 | 5.05 |
| | 6380 | 2/9 | 19.7 | 0.466 | 7.46 | 0.57 | 0.22 | 0.01 | 0.033 | 1.032 | 0.051 | 27.0 | 1.29 | bdl | 1.5 | 51.1 | 32.3 | 3.16 | 7.35 | 25.7 | 10.9 | 10.7 | 0.85 |
| | 6381 | 3/9 | 4.03 | 0.093 | 3.12 | 0.91 | 0.088 | 0.22 | 0.063 | 0.328 | 0.019 | 17.0 | 0.47 | bdl | 1.11 | 26.4 | 11.4 | 3.18 | 7.42 | 22.8 | 8.46 | 4.01 | 0.39 |
| | 6382 | 4/9 | 2.31 | 0.094 | 1.93 | 0.31 | 0.039 | 0.14 | 0.041 | 0.08 | 0.034 | 10.9 | 0.47 | bdl | 1.33 | 16.6 | 7.62 | 1.77 | 6.04 | 14.8 | 3.75 | 2.41 | 0.29 |
| | 6383 | 5/9 | 1.96 | 0.045 | 1.67 | 1.27 | 0.051 | 0.35 | 0.033 | 0.077 | 0.042 | 5.56 | 1.82 | bdl | 1.42 | 14 | 6.51 | 1.86 | 4.41 | 6.78 | 7.64 | 3.55 | 0.33 |
| | 6384 | 6/9 | 12.1 | 0.243 | 7.58 | 2.46 | 0.355 | 0.08 | 0.071 | 0.936 | 0.015 | 36.4 | 2.01 | bdl | 7.22 | 63.8 | 37.0 | 24.6 | 83.4 | 95.8 | 31.9 | 9.19 | 1.43 |
| | 6385 | 7/9 | 10.3 | 0.412 | 5.39 | 0.38 | 0.124 | 0.15 | 0.017 | 0.257 | 0.01 | 2.25 | 0.67 | bdl | 4.42 | nd | nd | nd | nd | nd | nd | 3.99 | 0.44 |
| | 6386 | 8/9 | 1.32 | 0.03 | 1.19 | 0.44 | 0.069 | 0.22 | 0.06 | 0.077 | 0.002 | 6.42 | 0.32 | bdl | 0.82 | 7.21 | nd | 0.14 | nd | 0.59 | 4.03 | 1.28 | 0.16 |
| | 6387 | 9/9 | 3.41 | 0.084 | 2.44 | 1.02 | 0.12 | 0.3 | 0.045 | 0.273 | 0.006 | 19.3 | 1.38 | 0.02 | 1.83 | 22.8 | 9.56 | 3.45 | 8.49 | 18.9 | 11.4 | 3.49 | 1.96 |
| | WA-H2 | | 4.32 | 0.111 | 2.713 | 0.895 | 0.099 | 0.24 | 0.043 | 0.252 | 0.025 | 14.2 | 1.23 | bdl | 1.94 | 22.5 | 10.4 | 3.71 | 9.78 | 18.8 | 9.16 | 4.01 | 1.24 |
| | 6392 | wc | 8.77 | 0.177 | 5.3 | 1.11 | 0.296 | 0.23 | 0.194 | 0.68 | 0.012 | bdl | bdl | bdl | nd | nd | nd | nd | nd | nd | nd | nd | nd |
| | 6393 | 1/11 | 4.03 | 0.128 | 2.9 | 0.24 | 0.056 | 0.09 | 0.068 | 0.138 | 0.006 | 23.2 | 10.37 | bdl | 1.29 | 16.9 | 9.69 | 4.83 | 8.65 | 20.3 | 5.2 | 5.19 | 5.68 |
| | 6394 | 2/11 | 4.22 | 0.068 | 3.2 | 0.22 | 0.042 | 0.11 | 0.05 | 0.115 | 0.01 | 24.5 | 1.05 | bdl | 1.31 | 27.2 | 12.1 | 3.53 | 9.59 | 14.6 | 2.97 | 4.91 | 0.86 |
| | 6395 | 3/11 | 2.60 | 0.053 | 2.14 | 0.67 | 0.057 | 0.2 | 0.068 | 0.137 | 0.007 | 15.3 | 0.48 | bdl | 0.97 | 23.6 | 7.19 | 4.03 | 9.41 | 21.4 | 6.46 | 3.18 | 0.49 |
| | 6396 | 4/11 | 12.1 | 0.318 | 9.32 | 1.06 | 0.483 | 0.81 | 0.405 | 1.521 | 0.022 | 48.6 | 0.96 | bdl | 1.81 | 83.6 | 35.2 | 6.02 | 17.6 | 62.4 | 38.7 | 11.3 | 0.75 |
| | 6397 | 5/11 | 2.15 | 0.038 | 0.67 | 1.44 | 0.048 | 0.29 | 0.011 | 0.101 | 0.003 | 2.36 | 0.25 | bdl | 0.73 | 23.7 | 4.66 | 1.37 | 4.32 | 3.1 | 9.11 | 0.94 | 2.12 |
| | 6398 | 6/11 | 1.96 | 0.04 | 0.65 | 2.41 | 0.037 | 0.32 | 0.001 | 0.083 | 0.002 | 12.1 | 0.33 | bdl | 0.81 | 18.3 | 8.64 | 8.47 | 7.86 | 14.1 | 3.93 | 2.8 | 2.06 |
| | 6399 | 7/11 | 1.27 | 0.027 | 0.45 | 2.04 | 0.025 | 0.38 | bdl | 0.056 | 0.032 | 9.79 | 0.77 | bdl | 0.71 | 11.4 | 6.71 | 0.4 | 1.22 | 7.38 | 2.85 | 1.89 | 1.7 |
| | 6400 | 8/11 | 21.1 | 0.37 | 6.96 | 21.14 | 0.408 | 5.56 | 0.102 | 0.892 | 0.032 | 29.1 | 3.03 | bdl | 6.95 | 161 | 56.2 | 21.5 | 57.1 | 112 | 73.2 | 10 | 13.22 |
| | 6401 | 9/11 | 10.3 | 0.238 | 7.7 | 2.09 | 0.448 | 1.42 | 0.605 | 0.62 | 0.018 | bdl | bdl | bdl | nd | 55.1 | 23.1 | 7.4 | 25 | 32.4 | 14.3 | nd | nd |
| | 6402 | 10/11 | 1.99 | 0.047 | 1.35 | 0.7 | 0.087 | 0.12 | 0.048 | 0.177 | 0.004 | 11.7 | 0.95 | bdl | 0.84 | 14.8 | 5.7 | 2.75 | 6.88 | 16.2 | 16.1 | 2.03 | 0.27 |
| | 6403 | 11/11 | 2.59 | 0.062 | 1.75 | 0.92 | 0.111 | 0.15 | 0.064 | 0.23 | 0.005 | 15.3 | 1.19 | bdl | 1.02 | 18.5 | 7.28 | 3.38 | 8.66 | 20.1 | 18.8 | 2.56 | 0.35 |
| | WA-H3 | | 4.13 | 0.09 | 2.174 | 2.616 | 0.104 | 0.64 | 0.057 | 0.237 | 0.008 | 15.9 | 2.31 | bdl | 1.43 | 30.2 | 11.735 | 4.38 | 10.8 | 24.1 | 15.8 | 3.51 | 2.51 |
| | WA-H | | 6.50 | 0.157 | 4.24 | 1.287 | 0.226 | 0.33 | 0.092 | 0.503 | 0.016 | 22.3 | 2.1 | bdl | 2.25 | 36.5 | 15.245 | 12.6 | 18.1 | 29.4 | 16.9 | 5.61 | 1.34 |

**Table 4.** *Cont.*

| Coal | Sample | Bench | $SiO_2$ | $TiO_2$ | $Al_2O_3$ | $Fe_2O_3$ | MgO | CaO | $Na_2O$ | $K_2O$ | $P_2O_5$ | Li | Be | Cl | Sc | V | Cr | Co | Ni | Cu | Zn | Ga | Ge |
|---|---|---|---|---|---|---|---|---|---|---|---|---|---|---|---|---|---|---|---|---|---|---|---|
| | 6352 | wc | 2.80 | 0.091 | 1.54 | 0.5 | 0.079 | 0.23 | 0.04 | 0.138 | 0.043 | 6.27 | 1.31 | nd | 1.7 | 13.1 | 7 | 5.26 | 9.26 | 16.3 | 7.64 | 2.72 | 1.58 |
| | 6353 | 1/6 | 1.42 | 0.036 | 0.92 | 0.54 | 0.034 | 0.07 | 0.02 | 0.097 | 0.002 | 3.32 | 1.34 | 0.03 | 1.29 | 7.34 | 3.56 | 2.12 | 6.26 | 7.37 | 2.7 | 1.83 | 0.22 |
| | 6354 | 2/6 | 1.25 | 0.036 | 0.84 | 0.88 | 0.031 | 0.09 | 0.026 | 0.07 | 0.002 | 2.81 | 0.22 | 0.04 | 0.76 | 6.51 | 3.45 | 1.85 | 6.85 | 10.9 | 1.78 | 1.11 | 0.14 |
| | 6355 | 3/6 | 0.71 | 0.028 | 0.58 | 0.23 | 0.033 | 0.67 | 0.033 | 0.004 | 0.001 | 1.01 | 0.12 | 0.04 | 0.44 | 2.75 | 1.93 | 1.77 | 6.79 | 9.11 | 1.67 | 0.55 | 0.08 |
| | 6356 | 4/6 | 0.83 | 0.039 | 0.63 | 0.26 | 0.031 | 0.17 | 0.033 | 0.008 | 0.001 | 1.61 | 0.2 | 0.03 | 0.75 | 4.42 | 3.31 | 4.37 | 10.6 | 18.4 | 1.88 | 1.07 | 0.15 |
| | 6357 | 5/6 | 1.97 | 0.092 | 1.03 | 0.25 | 0.061 | 0.37 | 0.037 | 0.005 | 0.005 | 4.1 | 0.44 | 0.03 | 1.43 | 7.67 | 5.16 | 5.44 | 8.47 | 21.5 | 3.7 | 1.09 | 0.11 |
| | 6358 | 6/6 | 8.91 | 0.221 | 4.76 | 0.59 | 0.232 | 0.08 | 0.024 | 0.81 | 0.01 | 17.6 | 7.66 | 0.04 | 5.55 | 50.9 | 20.2 | 6.94 | 15.3 | 32.7 | 21.3 | 13.2 | 14.7 |
| | WA-K1 | | 2.42 | 0.074 | 1.398 | 0.48 | 0.067 | 0.21 | 0.029 | 0.15 | 0.004 | 4.9 | 1.54 | 0.03 | 1.66 | 12.5 | 6.04 | 3.77 | 8.77 | 16.5 | 5.15 | 2.9 | 2.19 |
| Kellioka | 6359 | wc | 1.97 | 0.069 | 1.14 | 0.6 | 0.047 | 0.2 | 0.028 | 0.085 | 0.057 | 6.72 | 1.07 | nd | 0.95 | 10.4 | 7.16 | 2.35 | 5.77 | 10.7 | 3.49 | 2.07 | 1.58 |
| | 6360 | wc | 2.27 | 0.078 | 1.31 | 0.73 | 0.055 | 0.22 | 0.033 | 0.103 | 0.065 | 5.51 | 0.99 | nd | 1.91 | 10.6 | 6.25 | 15.4 | 13.9 | 15.6 | 4.49 | 2.28 | 1.55 |
| | 6361 | 1/5 | 1.71 | 0.038 | 1.01 | 0.69 | 0.057 | 0.1 | 0.028 | 0.121 | 0.004 | 4.46 | 0.71 | 0.03 | 2.32 | 10 | 4.85 | 19.07 | 8.47 | 14.9 | 3.34 | 2.16 | 0.85 |
| | 6362 | 2/5 | 2.29 | 0.058 | 1.29 | 0.5 | 0.067 | 0.11 | 0.042 | 0.141 | 0.006 | 5.01 | 0.26 | 0.03 | 0.51 | 9.7 | 5.06 | 1.37 | 4.57 | 10.7 | 2.76 | 1.68 | 0.18 |
| | 6363 | 3/5 | 0.75 | 0.027 | 0.72 | 0.57 | 0.081 | 0.75 | 0.038 | 0.007 | 0.017 | 1.07 | 0.11 | 0.03 | 0.23 | 3.16 | 1.56 | 1.15 | 4.11 | 8.14 | 1.64 | 0.49 | 0.08 |
| | 6364 | 4/5 | 1.96 | 0.093 | 1.1 | 0.29 | 0.038 | 0.31 | 0.051 | 0.004 | 0.146 | 3.77 | 0.35 | 0.03 | 1.39 | 8.54 | 6.90 | 3.44 | 7.03 | 17.4 | 2.58 | 1.47 | 0.13 |
| | 6365 | 5/5 | 5.21 | 0.147 | 3.3 | 0.89 | 0.108 | 0.12 | 0.019 | 0.446 | 0.043 | 17.2 | 2.72 | 0.04 | 2.65 | 52.4 | 19.4 | 16.3 | 29.4 | 31.5 | 7.52 | 7.42 | 8.94 |
| | WA-K2 | | 2.43 | 0.081 | 1.477 | 0.533 | 0.063 | 0.26 | 0.038 | 0.126 | 0.068 | 6.2 | 0.81 | 0.03 | 1.53 | 16.5 | 8.0 | 8.02 | 10.8 | 17.6 | 3.55 | 2.62 | 1.90 |
| | WA-K | | 2.38 | 0.079 | 1.373 | 0.569 | 0.062 | 0.23 | 0.034 | 0.12 | 0.047 | 5.94 | 1.14 | 0.01 | 1.55 | 12.6 | 6.89 | 6.95 | 9.7 | 15.3 | 4.86 | 2.52 | 1.76 |
| Darby | 6261 | wc | 0.90 | 0.033 | 0.71 | 0.24 | 0.044 | 0.18 | 0.026 | 0.056 | 0.004 | 2.71 | 4.76 | 0.05 | 1.06 | 6.55 | 3.42 | 10.5 | 9.81 | 10.7 | 1.79 | 2.6 | 4.72 |
| | 6262 | 1/4 | 0.77 | 0.052 | 0.62 | 0.15 | 0.023 | 0.14 | 0.01 | 0.008 | 0.004 | 2.12 | 19.67 | 0.04 | 1.66 | 6.32 | 3.78 | 16.5 | 9.87 | 8.02 | 1.76 | 10.7 | 11.0 |
| | 6263 | 2/4 | 0.64 | 0.028 | 0.68 | 0.22 | 0.048 | 0.26 | 0.041 | 0.011 | 0.003 | nd | 0.48 | 0.03 | nd | nd | nd | nd | nd | nd | nd | nd | nd |
| | 6264 | 3/4 | 0.39 | 0.015 | 0.49 | 0.2 | 0.031 | 0.13 | 0.033 | 0.006 | 0.002 | 2.44 | 7.41 | 0.04 | 0.33 | 2.55 | 1.5 | 5.69 | 7.58 | 7.51 | 1.36 | 0.43 | 0.11 |
| | WA-D1 | 4/4 | 0.54 | 0.023 | 0.542 | 0.19 | 0.029 | 0.117 | 0.014 | 0.056 | 0.003 | 2.57 | 5.18 | 0.05 | 0.09 | 8.28 | 3.47 | 19 | 12 | 16.7 | 1.78 | 3.21 | 10.3 |
| | 6266 | wc | 0.89 | 0.031 | 0.73 | 0.26 | 0.05 | 0.14 | 0.051 | 0.017 | 0.004 | 2.41 | 0.78 | 0.04 | 1.07 | 5.89 | 2.57 | 10.3 | 8.91 | 9.56 | 1.52 | 2.74 | 4.11 |
| | 6267 | wc | 1.35 | 0.04 | 0.83 | 0.43 | 0.06 | 0.18 | 0.063 | 0.054 | 0.004 | 2.47 | 1.23 | 0.05 | 0.26 | 9.21 | 4.56 | 1.59 | 5.41 | 3.86 | 5.48 | 1.15 | 4.70 |
| | 6268 | wc | 1.20 | 0.035 | 0.78 | 0.3 | 0.052 | 0.17 | 0.059 | 0.082 | 0.003 | 3.08 | 4.68 | 0.04 | 0.66 | 7.71 | 3.51 | 3.63 | 4.79 | 10.5 | 3.53 | 2.42 | 2.03 |
| | 6269 | wc | 2.00 | 0.053 | 1.18 | 0.38 | 0.102 | 0.15 | 0.058 | 0.071 | 0.005 | 4.83 | 2.12 | 0.04 | 0.47 | 13.1 | 6.71 | 4.29 | 5.59 | 9.47 | 2.78 | 2.64 | 0.49 |
| | 6366 | wc | 1.05 | 0.034 | 0.75 | 0.23 | 0.024 | 0.06 | 0.026 | 0.129 | 0.002 | 4.23 | 0.5 | nd | 0.49 | 6.6 | 3.14 | 7.02 | 8.82 | 14.1 | 4.84 | 2.69 | 1.44 |
| | 6367 | 1/5 | 4.38 | 0.098 | 2.07 | 0.5 | 0.124 | 0.02 | bdl | 0.038 | 0.005 | 3.06 | 2.9 | 0.04 | 4.59 | 72.2 | 15.4 | 5.69 | 18.8 | 8.13 | 18.5 | 4.85 | 1.59 |
| | 6368 | 2/5 | 1.03 | 0.046 | 0.61 | 0.2 | 0.018 | 0.08 | 0.021 | 0.247 | 0.003 | 7.75 | 0.12 | nd | 0.86 | 6.06 | 2.45 | 1.03 | 2.16 | 4.09 | 2.19 | 0.86 | 0.3 |
| | 6369 | 3/5 | 1.89 | 0.098 | 0.98 | 0.18 | 0.02 | 0.04 | 0.043 | 0.008 | 0.004 | 2.77 | 0.18 | nd | 1.13 | 8.27 | 3.71 | 1.34 | 4.1 | 6.43 | 3.61 | 0.71 | 0.16 |
| | 6370 | 4/5 | 0.70 | 0.024 | 0.64 | 0.25 | 0.021 | 0.06 | 0.038 | 0.019 | 0.002 | 1.42 | 0.14 | nd | 0.45 | 5.15 | 2.04 | 0.75 | 1.83 | 4.26 | 2.84 | 0.68 | 0.11 |
| | 6371 | 5/5 | 1.35 | 0.029 | 0.95 | 0.29 | 0.031 | 0.04 | 0.013 | 0.067 | 0.003 | 2.50 | 0.25 | 0.17 | 0.65 | 3.92 | 2.33 | 0.83 | 2.62 | 2.14 | 9.08 | 1.23 | 0.22 |
| | WA-D2 | 5/5 | 1.11 | 0.036 | 0.77 | 0.26 | 0.027 | 0.06 | 0.029 | 0.036 | 0.003 | 4.35 | 0.28 | 0.055 | 0.78 | 8.18 | 2.85 | 1.06 | 2.91 | 4.39 | 4.61 | 1.0 | 0.23 |
| | WA-D | | 1.13 | 0.036 | 0.79 | 0.29 | 0.048 | 0.13 | 0.042 | 0.06 | 0.003 | 3.05 | 2.44 | 0.038 | 0.66 | 7.71 | 3.63 | 5.08 | 6.53 | 8.84 | 3.32 | 2.04 | 2.24 |

Table 4. Cont.

| Coal | Sample | Bench | As | Se | Rb | Sr | Zr | Nb | Mo | Cd | In | Sn | Sb | Cs | Ba | Hf | Ta | W | Tl | Pb | Bi | Th | U |
|---|---|---|---|---|---|---|---|---|---|---|---|---|---|---|---|---|---|---|---|---|---|---|---|
| | 6255 | wc | 10 | 0.1 | 10.2 | 54.1 | 22.7 | 2.79 | 2.09 | 0.11 | 0.017 | 0.66 | 0.78 | 0.86 | 67.4 | 0.65 | 1.31 | 0.5 | 0.94 | 6.43 | 0.13 | 1.75 | 1.11 |
| | 6256 | 1/4 | 5.23 | 0.15 | 31.8 | 44.9 | 31.6 | 4.16 | 3.08 | 0.17 | 0.031 | 1.17 | 2.36 | 3.31 | 109 | 0.98 | 0.32 | 0.92 | 0.81 | 13.32 | 0.14 | 2.77 | 2.62 |
| | 6257 | 2/4 | 37.8 | 0.17 | 1.37 | 45.3 | 17.2 | 1.95 | 3.24 | 0.09 | 0.013 | 0.56 | 0.37 | 0.21 | 33.4 | 0.51 | 0.19 | 0.33 | 2.43 | 4.43 | 0.11 | 1.17 | 0.79 |
| | 6258 | 3/4 | 22.6 | 0.13 | 14.1 | 45.9 | 27.3 | 2.89 | 3.96 | 0.12 | 0.021 | 0.88 | 1.11 | 1.46 | 64.8 | 0.8 | 0.27 | 0.42 | 1.51 | 9.34 | 0.17 | 1.89 | 2.77 |
| | 6259 | 4/4 | 6.33 | 0.1 | 7.09 | 47.9 | 23.4 | 2.75 | 2.05 | 0.13 | 0.016 | 0.62 | 1.31 | 0.75 | 60.5 | 0.65 | 0.5 | 0.64 | 0.42 | 7.19 | 0.1 | 2.05 | 1.57 |
| | WA-H1 | | 13 | 0.12 | 12.1 | 46.7 | 25.1 | 2.93 | 2.79 | 0.13 | 0.019 | 0.76 | 1.34 | 1.27 | 66.6 | 0.73 | 0.39 | 0.6 | 0.93 | 8.42 | 0.12 | 2.04 | 1.97 |
| | 6260 | wc | 9.54 | 0.05 | 8.41 | 64.9 | 23.8 | 2.79 | 2.34 | 0.13 | 0.017 | 0.67 | 0.97 | 1.02 | 66.7 | 0.66 | 0.38 | 0.58 | 0.74 | 6.39 | 0.13 | 2.16 | 1.7 |
| | 6270 | wc | 9.44 | 0.15 | 10 | 350 | 30.2 | 3.69 | 2.74 | 0.18 | 0.024 | 0.92 | 1 | 1.79 | 307 | 0.88 | 0.57 | 0.83 | 0.72 | 6.86 | 0.16 | 0.69 | 2.23 |
| | 6271 | wc | 10.6 | 0.27 | 6.59 | 46.4 | 23.1 | 2.82 | 3.08 | 0.18 | 0.017 | 0.71 | 1.1 | 1.02 | 66.6 | 0.67 | 0.36 | 0.6 | 0.68 | 10.54 | 0.1 | 3.08 | 1.79 |
| | 6272 | wc | 20 | 0.24 | 23.1 | 108 | 31.3 | 3.98 | 2.39 | 0.15 | 0.025 | 0.91 | 1.19 | 1.62 | 144 | 0.9 | 0.46 | 0.7 | 1.24 | 9 | 0.17 | 3.46 | 2.36 |
| | 6273 | wc | 7.96 | 0.32 | 52.9 | 170 | 67.9 | 8.54 | 2.38 | 0.53 | 0.062 | 2.04 | 1.43 | 4.51 | 307 | 1.86 | 0.95 | 1.04 | 1.49 | 25.38 | 0.38 | 8.16 | 4.32 |
| | 6378 | wc | 4.2 | 0.17 | 3.74 | 75.8 | 20.5 | 2.15 | 1.7 | 0.1 | 0.014 | 0.55 | 0.7 | 0.74 | 64.7 | 0.59 | 0.17 | 0.27 | 0.3 | 5.35 | 0.11 | 0.42 | 0.99 |
| | 6379 | 1/9 | 6.2 | 0.18 | 4.81 | 131 | 38.3 | 4.47 | 1.21 | 0.09 | 0.025 | 0.92 | 1.05 | 0.52 | 85.3 | 1.04 | 0.39 | 0.54 | 0.19 | 7.84 | 0.2 | 2.88 | 1.55 |
| | 6380 | 2/9 | 2.08 | 0.21 | 21.2 | 122 | 99.7 | 9.14 | 0.76 | 0.2 | 0.039 | 1.9 | 0.41 | 2.58 | 140 | 2.74 | 0.72 | 0.62 | 0.16 | 11.47 | 0.25 | 1.43 | 1.74 |
| | 6381 | 3/9 | 11.1 | 0.08 | 5.82 | 84.3 | 15.6 | 1.71 | 2.82 | 0.1 | 0.019 | 0.57 | 0.8 | 0.79 | 119 | 0.46 | 0.15 | 0.29 | 0.6 | 6.23 | 0.12 | 1.09 | 0.68 |
| | 6382 | 4/9 | 1.76 | 0.04 | 1.25 | 128 | 17 | 1.86 | 1.59 | 0.11 | 0.013 | 0.42 | 0.54 | 0.18 | 96.7 | 0.49 | 0.17 | 0.21 | 0.07 | 4.59 | 0.11 | 1.06 | 0.52 |
| | 6383 | 5/9 | 22.3 | 0.15 | 12.5 | 23.2 | 9.36 | 1.17 | 3.36 | 0.04 | 0.011 | 0.35 | 0.5 | 0.9 | 51.4 | 0.25 | 0.13 | 0.3 | 0.12 | 3.74 | 0.06 | 1.26 | 0.74 |
| Harlan | 6384 | 6/9 | 14.5 | 0.14 | 14.8 | 618 | 40.9 | 5.52 | 3.38 | 0.41 | 0.035 | 1.22 | 1.92 | 1.01 | 499 | 1.06 | 0.53 | 0.93 | 1.33 | 30.58 | 0.28 | 6.17 | 4.2 |
| | 6385 | 7/9 | 0.91 | 0.27 | 39.3 | 6.6 | 70.2 | 8.78 | 0.82 | 0 | 0.012 | 1.34 | 0.36 | 0.55 | 84.8 | 1.89 | 1.65 | 0.65 | -0.06 | 0 | 0.01 | 4.07 | 1.3 |
| | 6386 | 8/9 | 1.35 | 0.07 | 1.6 | 25 | 5.46 | 0.76 | 2.01 | 0.09 | 0.007 | 0.2 | 0.15 | 0.14 | 18.1 | 0.17 | 0.32 | 0.2 | 0.02 | 9.12 | 0.05 | 0.55 | 0.27 |
| | 6387 | 9/9 | 14.1 | 0.14 | 8.03 | 91.3 | 15.1 | 1.55 | 1.62 | 0.12 | 0.011 | 0.49 | 0.86 | 1.21 | 95 | 0.47 | 0.14 | 0.36 | 1.05 | 5.47 | 0.11 | 1.49 | 0.81 |
| | WA-H2 | | 10.8 | 0.13 | 8.16 | 101 | 21.1 | 2.37 | 2.2 | 0.11 | 0.016 | 0.58 | 0.69 | 0.81 | 99.1 | 0.59 | 0.27 | 0.37 | 0.42 | 7.57 | 0.12 | 1.64 | 0.98 |
| | 6392 | wc | nd | nd | nd | nd | nd | nd | nd | nd | nd | nd | nd | nd | nd | nd | nd | nd | nd | nd | nd | nd | nd |
| | 6393 | 1/11 | 1.5 | 0.08 | 3.05 | 27.3 | 25.1 | 2.8 | 0.8 | 0.07 | 0.014 | 0.55 | 0.67 | 0.3 | 75.5 | 0.67 | 0.26 | 0.31 | 0.11 | 6.84 | 0.15 | 1.37 | 0.75 |
| | 6394 | 2/11 | 2 | 0.07 | 2.04 | 23.1 | 13.1 | 2.01 | 0.63 | 0.03 | 0.012 | 0.36 | 0.55 | 0.25 | 34.1 | 0.4 | 0.15 | 0.17 | 0.06 | 7.83 | 0.08 | 0.9 | 0.71 |
| | 6395 | 3/11 | 11.8 | 0.08 | 1.98 | 79.3 | 9.36 | 1.1 | 3.46 | 0.09 | 0.011 | 0.36 | 1.48 | 0.23 | 78.9 | 0.29 | 0.15 | 0.44 | 0.82 | 7.18 | 0.09 | 1.13 | 0.9 |
| | 6396 | 4/11 | 5.12 | 0.61 | 23.3 | 96.7 | 56.8 | 6.04 | 3.68 | 0.32 | 0.049 | 1.74 | 1.75 | 3.11 | 191 | 1.73 | 0.5 | 0.4 | 0.41 | 20.85 | 0.6 | 0.83 | 1.98 |
| | 6397 | 5/11 | 2.47 | 0.06 | 4.88 | 10.5 | 6.98 | 0.73 | 1.33 | 0.06 | 0.003 | 0.19 | 0.24 | 0.33 | 110 | 0.18 | 0.06 | 0.12 | 0.33 | 9.07 | 0.01 | 0.38 | 0.35 |
| | 6398 | 6/11 | 3.6 | 0.03 | 11.2 | 60 | 7.92 | 0.91 | 3.04 | 0.04 | 0.009 | 0.17 | 0.8 | 0.73 | 54.4 | 0.18 | 0.09 | 0.2 | 0.14 | 6.66 | 0.05 | 0.6 | 0.89 |
| | 6399 | 7/11 | 1.85 | 0.04 | 2.87 | 36.8 | 5.68 | 0.66 | 4.01 | 0.04 | 0.007 | 0.21 | 0.08 | 0.58 | 42.5 | 0.16 | 0.21 | 0.12 | 0.01 | 3.24 | 0.06 | 0.36 | 0.25 |
| | 6400 | 8/11 | 35.6 | 0.32 | 42.3 | 115 | 61.3 | 7.89 | 25.97 | 0.89 | 0.041 | 1.53 | 4.6 | 3.11 | 519 | 1.5 | 0.82 | 1.98 | 19.36 | 53.6 | 0.17 | 5.17 | 7.97 |
| | 6401 | 9/11 | nd | nd | 36.1 | 794 | 87.2 | 6.25 | nd | nd | nd | nd | nd | nd | 665 | nd | nd | nd | nd | nd | nd | nd | nd |
| | 6402 | 10/11 | 10.8 | 0.09 | 5.34 | 57 | 5.84 | 0.91 | 1.12 | 0.06 | 0.011 | 0.83 | 0.67 | 0.54 | 58.2 | 0.26 | 0.08 | 0.3 | 0.62 | 6.66 | 0.04 | 1.12 | 0.82 |
| | 6403 | 11/11 | 14.4 | 0.08 | 5.34 | 67.5 | 9.64 | 1.17 | 1.44 | 0.12 | 0.014 | 1.64 | 0.81 | 0.65 | 70.2 | 0.34 | 0.18 | 0.37 | 0.87 | 7.71 | 0.08 | 1.51 | 1.09 |
| | WA-H3 | | 9.58 | 0.1 | 7.56 | 54.8 | 15.4 | 1.88 | 3.96 | 0.15 | 0.014 | 0.81 | 0.97 | 0.76 | 104 | 0.45 | 0.23 | 0.41 | 1.94 | 10.54 | 0.09 | 1.34 | 1.32 |
| | WA-H | | 9.56 | 0.15 | 13 | 97.4 | 25.5 | 3.09 | 2.33 | 0.16 | 0.02 | 0.78 | 0.92 | 1.31 | 118 | 0.73 | 0.46 | 0.54 | 0.85 | 8.77 | 0.14 | 2.25 | 1.71 |

Table 4. Cont.

| Coal | Sample | Bench | As | Se | Rb | Sr | Zr | Nb | Mo | Cd | In | Sn | Sb | Cs | Ba | Hf | Ta | W | Tl | Pb | Bi | Th | U |
|------|--------|-------|----|----|----|----|----|----|----|----|----|----|----|----|----|----|----|----|----|----|----|----|----|
| | 6352 | wc | 3.12 | 0.09 | 6.12 | 106 | 10.5 | 1.49 | 0.91 | 0.08 | 0.012 | 0.46 | 0.53 | 0.69 | 85.5 | 0.38 | 0.14 | 0.36 | 0.26 | 4.46 | 0.11 | 1.4 | 0.77 |
| | 6353 | 1/6 | 16.2 | 0.06 | 4.85 | 67.9 | 5.64 | 0.72 | 1.21 | 0.04 | 0.005 | 0.21 | 0.45 | 0.44 | 63.5 | 0.17 | 0.07 | 0.54 | 0.51 | 3.37 | 0.05 | 0.76 | 0.5 |
| | 6354 | 2/6 | 39.7 | 0.05 | 3.2 | 85.7 | 3.7 | 0.72 | 0.94 | 0.05 | 0.006 | 0.17 | 0.15 | 0.23 | 73.8 | 0.17 | 0.13 | 0.11 | 0.89 | 2.9 | 0.07 | 0.72 | 0.49 |
| | 6355 | 3/6 | 0.74 | 0.03 | 0.05 | 117 | 3.29 | 0.56 | 0.6 | 0.02 | 0.004 | 0.15 | 0.05 | 0.01 | 91.9 | 0.15 | 0.06 | 0.08 | 0.08 | 1.62 | 0.02 | 0.33 | 0.15 |
| | 6356 | 4/6 | 0.47 | 0.04 | 0.3 | 116 | 6.96 | 0.84 | 0.56 | 0.03 | 0.006 | 0.25 | 0.08 | 0.03 | 102 | 0.19 | 0.08 | 0.13 | 0.12 | 1.72 | 0.05 | 0.51 | 0.26 |
| | 6357 | 5/6 | 0.68 | 0.06 | 0.06 | 198 | −0.01 | 2.06 | 0.32 | 0.05 | 0.015 | 0.91 | 0.1 | 0.02 | 104 | 0.49 | 0.29 | 0.14 | 0.12 | 3.1 | 0.1 | 1.56 | 0.62 |
| | 6358 | 6/6 | 2.12 | 0.13 | 3.55 | 587 | 38.2 | 4.96 | 0.79 | 0.16 | 0.046 | 1.18 | 2.71 | 0.25 | 543 | 0.94 | 0.43 | 0.64 | 1.39 | 13.78 | 0.31 | 3.5 | 1.91 |
| | WA-K1 | | 11.6 | 0.06 | 2.2 | 185 | 8.5 | 1.62 | 0.75 | 0.06 | 0.013 | 0.5 | 0.54 | 0.18 | 150 | 0.35 | 0.18 | 0.28 | 0.53 | 4.29 | 0.1 | 1.25 | 0.66 |
| Kellioka | 6359 | wc | 7.31 | 0.12 | 2.51 | 72 | 13 | 1.6 | 0.79 | 0.08 | 0.008 | 0.5 | 0.54 | 0.24 | 74 | 0.41 | 0.21 | 0.35 | 0.16 | 3.3 | 0.06 | 0.72 | 0.81 |
| | 6360 | wc | 49.8 | 0.14 | 2.33 | 129 | 14.8 | 1.94 | 1.13 | 0.17 | 0.011 | 0.53 | 0.56 | 0.25 | 107 | 0.46 | 0.43 | 0.38 | 0.45 | 14.55 | 0.09 | 1.9 | 1.05 |
| | 6361 | 1/5 | 37.3 | 0.08 | 3.98 | 102 | 6.34 | 1.19 | 1.63 | 0.14 | 0.009 | 0.22 | 0.77 | 0.23 | 93.2 | 0.21 | 0.15 | 0.76 | 0.25 | 2.71 | 0.06 | 0.91 | 1.09 |
| | 6362 | 2/5 | 23.9 | 0.07 | 1.77 | 69.5 | 9.2 | 1.11 | 1.02 | 0.04 | 0.008 | 0.36 | 0.16 | 0.28 | 64.1 | 0.27 | 0.09 | 0.11 | 0.44 | 3.23 | 0.07 | 0.29 | 0.46 |
| | 6363 | 3/5 | 0.79 | 0.07 | 0.11 | 100 | 4.73 | 0.56 | 0.94 | 0.03 | 0.004 | 0.18 | 0.1 | 0.01 | 67.6 | 0.14 | 0.05 | 0.1 | 0.08 | 2.08 | 0.06 | 0.14 | 0.21 |
| | 6364 | 4/5 | 0.84 | 0.06 | 0 | 276 | 0 | 2.01 | 0.62 | 0.05 | 0.012 | 0.39 | 0.15 | 0.01 | 186 | 0.53 | 0.21 | 0.18 | 0.09 | 3.46 | 0.1 | 0.63 | 0.63 |
| | 6365 | 5/5 | 8.37 | 0.14 | 11.9 | 118 | 24.9 | 3.3 | 1.23 | 0.17 | 0.021 | 0.71 | 1.83 | 1 | 106 | 0.71 | 0.26 | 0.47 | 0.58 | 10.68 | 0.14 | 1.59 | 1.29 |
| | WA-K2 | | 11.8 | 0.08 | 3.17 | 167 | 7.56 | 1.82 | 1 | 0.09 | 0.012 | 0.39 | 0.57 | 0.27 | 124 | 0.43 | 0.17 | 0.32 | 0.26 | 4.48 | 0.09 | 1.23 | 0.76 |
| | WA-K | | 16.8 | 0.1 | 3.27 | 132 | 10.9 | 1.69 | 0.92 | 0.1 | 0.011 | 0.48 | 0.55 | 0.33 | 108 | 0.41 | 0.23 | 0.34 | 0.33 | 6.22 | 0.09 | 1.3 | 0.81 |
| | 6261 | wc | 1.2 | 0.09 | 1.65 | 82.8 | 4.93 | 0.79 | 0.46 | 0.03 | 0.006 | 0.17 | 0.66 | 0.18 | 58 | 0.17 | 0.06 | 0.27 | 0.15 | 2 | 0.04 | 0.47 | 0.28 |
| | 6262 | 1/4 | 0.74 | 0.09 | 0.18 | 75.9 | 1.74 | 1.47 | 0.17 | 0.02 | 0.005 | 0.29 | 0.97 | 0.02 | 44 | 0.28 | 0.1 | 0.28 | 0.2 | 1.94 | 0.04 | 0.69 | 0.24 |
| | 6264 | 3/4 | 0.63 | 0.05 | 0.16 | 101 | 1.21 | 0.29 | 0.49 | 0.02 | 0.004 | 0.16 | 0.05 | 0.02 | 86.5 | 0.07 | 0.03 | 0.1 | 0.02 | 1.02 | 0.03 | 0.17 | 0.05 |
| | 6265 | 4/4 | 1.24 | 0.08 | 0.55 | 29.8 | 4.24 | 0.76 | 0.46 | 0.04 | 0.006 | 0.21 | 1.43 | 0.04 | 17.3 | 0.13 | 0.04 | 0.45 | 0.17 | 1.93 | 0.04 | 0.04 | 0.26 |
| | WA-D1 | | 0.78 | 0.06 | 0.25 | 81.8 | 1.95 | 0.59 | 0.43 | 0.02 | 0.005 | 0.19 | 0.5 | 0.04 | 64.6 | 0.12 | 0.04 | 0.21 | 0.08 | 1.37 | 0.03 | 0.23 | 0.13 |
| | 6266 | wc | 1.42 | 0.1 | 0.68 | 31.4 | 5.53 | 0.81 | 0.48 | 0.04 | 0.004 | 0.16 | 0.6 | 0.11 | 21.4 | 0.16 | 0.09 | 0.23 | 0.61 | 1.88 | 0.03 | 0.66 | 0.32 |
| | 6267 | wc | 8.64 | 0.09 | 4.62 | 44.6 | 7.01 | 1.05 | 0.78 | 0.06 | 0.005 | 0.21 | 0.45 | 0.29 | 56.5 | 0.21 | 0.08 | 0.36 | 0.2 | 2.84 | 0.03 | 0.18 | 0.54 |
| | 6268 | wc | 2.82 | 0.07 | 3.37 | 74.9 | 5.79 | 0.8 | 0.57 | 0.04 | 0.006 | 0.16 | 0.22 | 0.25 | 71.9 | 0.17 | 0.09 | 0.18 | 0.24 | 2.26 | 0.05 | 0.29 | 0.25 |
| Darby | 6269 | wc | 1.33 | 0.11 | 1.54 | 99.2 | 8.74 | 0.7 | 0.95 | 0.07 | 0.009 | 0.24 | 0.34 | 0.16 | 83.8 | 0.26 | 0.05 | 0.17 | 0.9 | 4.38 | 0.08 | 0.45 | 0.6 |
| | 6366 | wc | 1.51 | 0.05 | 0.33 | 63.4 | 5.87 | 0.74 | 0.98 | 0.07 | 0.005 | 0.2 | 0.14 | 0.07 | 52.9 | 0.17 | 0.08 | 0.21 | 0.13 | 1.41 | 0.05 | 0.41 | 0.23 |
| | 6367 | 1/5 | 3.47 | 0.32 | 6.42 | 29.3 | 16.9 | 3.37 | 0.65 | 0.09 | 0.014 | 0.73 | 1.69 | 0.5 | 39.7 | 0.48 | 0.16 | 0.34 | 0.29 | 6.75 | 0.1 | 1.1 | 1.52 |
| | 6368 | 2/5 | 0.85 | 0.09 | 3.83 | 7.72 | 8.74 | 1.03 | 1.33 | 0.03 | 0.003 | 0.26 | 0.25 | 0.22 | 15.1 | 0.22 | 0.1 | 0.55 | 0.2 | 1.71 | 0.02 | 0.65 | 0.22 |
| | 6369 | 3/5 | 1.07 | 0.05 | 2.77 | 3.01 | 20.8 | 2.11 | 1.18 | 0.06 | 0.003 | 0.38 | 0.09 | 0.22 | 12.5 | 0.57 | 0.3 | 0.31 | 0.15 | 2.38 | 0.01 | 1.26 | 0.36 |
| | 6370 | 4/5 | 1.34 | 0.05 | 0.46 | 41.7 | 4.07 | 0.54 | 1.27 | 0.03 | 0.004 | 4.32 | 0.63 | 0.09 | 54.5 | 0.13 | 0.09 | 0.14 | 0.24 | 1.31 | 0.04 | 0.42 | 0.14 |
| | 6371 | 5/5 | 3.94 | 0.09 | 1.07 | 140 | 4.66 | 0.69 | 0.87 | 0.06 | 0.026 | 0.21 | 0.13 | 0.09 | 141.6 | 0.16 | 0.1 | 0.38 | 0.34 | 1.64 | 0.06 | 0.6 | 0.31 |
| | WA-D2 | | 1.81 | 0.07 | 1.56 | 51.3 | 6.54 | 0.87 | 1.17 | 0.04 | 0.00829 | 2.46 | 0.49 | 0.11 | 60.7 | 0.19 | 0.11 | 0.27 | 0.25 | 1.74 | 0.04 | 0.57 | 0.26 |
| | WA-D | | 2.44 | 0.08 | 1.75 | 66.2 | 5.79 | 0.79 | 0.73 | 0.04 | 0.006 | 0.47 | 0.42 | 0.15 | 58.7 | 0.18 | 0.08 | 0.24 | 0.32 | 2.23 | 0.04 | 0.41 | 0.33 |

Because the three coals have relatively low ash yields and most of the trace elements in the coals have inorganic affinity, the concentration of trace elements in ashes of the three coals were compared to the averages of the same elements for the world coal ash reported by Ketris and Yudovich [26]. Elements including Li, Co, Cu, As, and Ta in the Harlan coal ashes, elements P, Co, Cu, As, Sr, Ba, Ta, and Pb in the Kellika coal ashes, and elements Be, Co, Ni, Cu, Ga, Ge, Sr, Y, Mo, Sn, Sb, Ba, and Tl in the Darby coal ashes are relatively enriched (Figure 5).

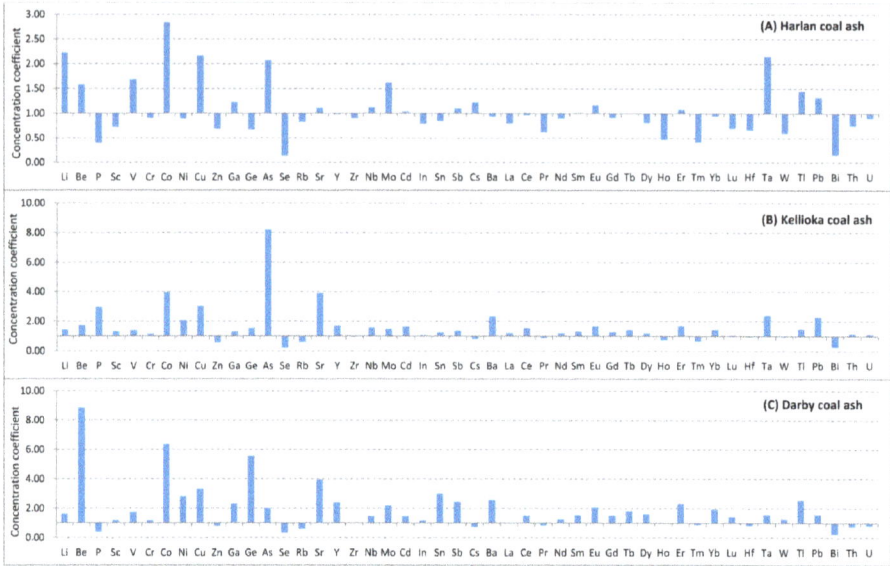

**Figure 5.** Concentration coefficients of trace elements in the coal ashes studied. (**A**) Harlan; (**B**) Kellioka; (**C**) Darby. Concentration coefficients (CC) are the ratio of the trace-element concentrations in the coal ash samples *vs.* world coal ash reported by Ketris and Yudovich [26].

The correlation coefficient ($r = 0.59$) of Li and ash yield in the Harlan coals indicates an inorganic affinity. Further, lithium positively correlated to Mg ($r = 0.74$), SiO$_2$ ($r = 0.61$), Al$_2$O$_3$ ($r = 0.740$), and K$_2$O ($r = 0.84$), indicatingit is mainly associated with clay minerals (e.g., kaolinite, mixed-layer illite/smectite, or illite).

The Cu in the Harlan coals is positively correlated to Ash ($r = 0.87$), Al$_2$O$_3$ ($r = 0.77$), SiO$_2$ ($r = 0.80$), and K$_2$O ($r = 0.76$), but has a weak correlation coefficient with total sulfur ($r = 0.37$), indicating Cu mainly occurs in clay minerals. The correlation coefficient of Co and ash is 0.41, indicating that Co has a dominant inorganic association and a small proportion may be associated with organic matter.

The concentrations of As in Kellioka coals and coal ashes are 16.8 and 378 μg/g respectively, much higher than their averages for world hard coals and coal ashes (9 and 46 μg/g, respectively) [26]. The adverse effects on environment of arsenic in Kellioka coals should be of concern. The correlation coefficient of As-St ($r = 0.77$) and As-Fe$_2$O$_3$ ($r = 0.63$), and Fe-St ($r = 0.85$) (Figure 6) of the Kellioka coals indicate that As is mainly associated with pyrite.

With exceptions of Li, Cu, and Co in the Harlan coals, and As in the Kellioka coals, the remaining elements in the three coals are either close to or lower than the averages for world hard coals [26], and most of them have an inorganic affinity (Figure 7). However, some elements including Be, Ga, Ge, Sr, Mo, Sn, and W have different modes of occurrence in the three coals. For example:

The correlation coefficient of Be and ash yield in the Harlan and Darby coals ($r = 0.17$, and $r = -0.14$, respectively) show an organic-inorganic mixed affinity (also see Figure 8). Be in the Kellioka coals, however, showed inorganic affinity ($r = 0.95$; Figure 8).

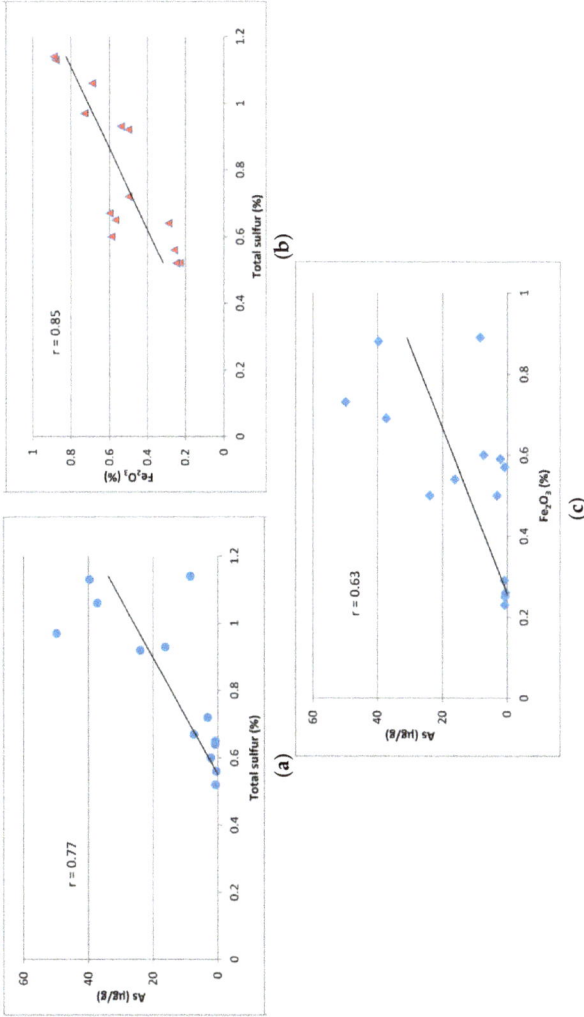

**Figure 6.** Relation of (**a**) Arsenic-total sulfur, (**b**) $Fe_2O_3$-total sulfur, and (**c**) arsenic-$Fe_2O_3$ in Kellioka coals.

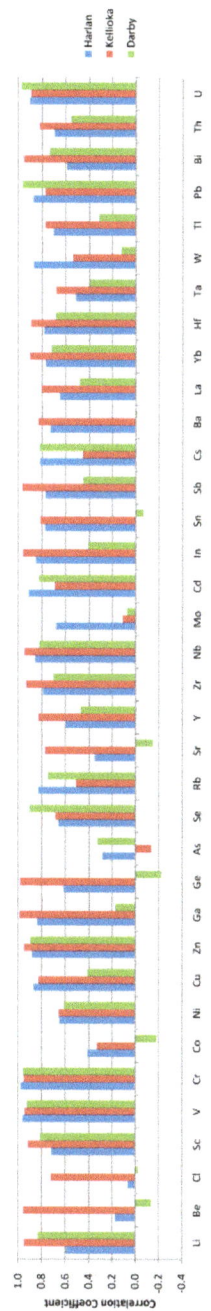

**Figure 7.** Correlation coefficient of trace elements and ash yield of the coals in Harlan, Kellioka, and Darby.

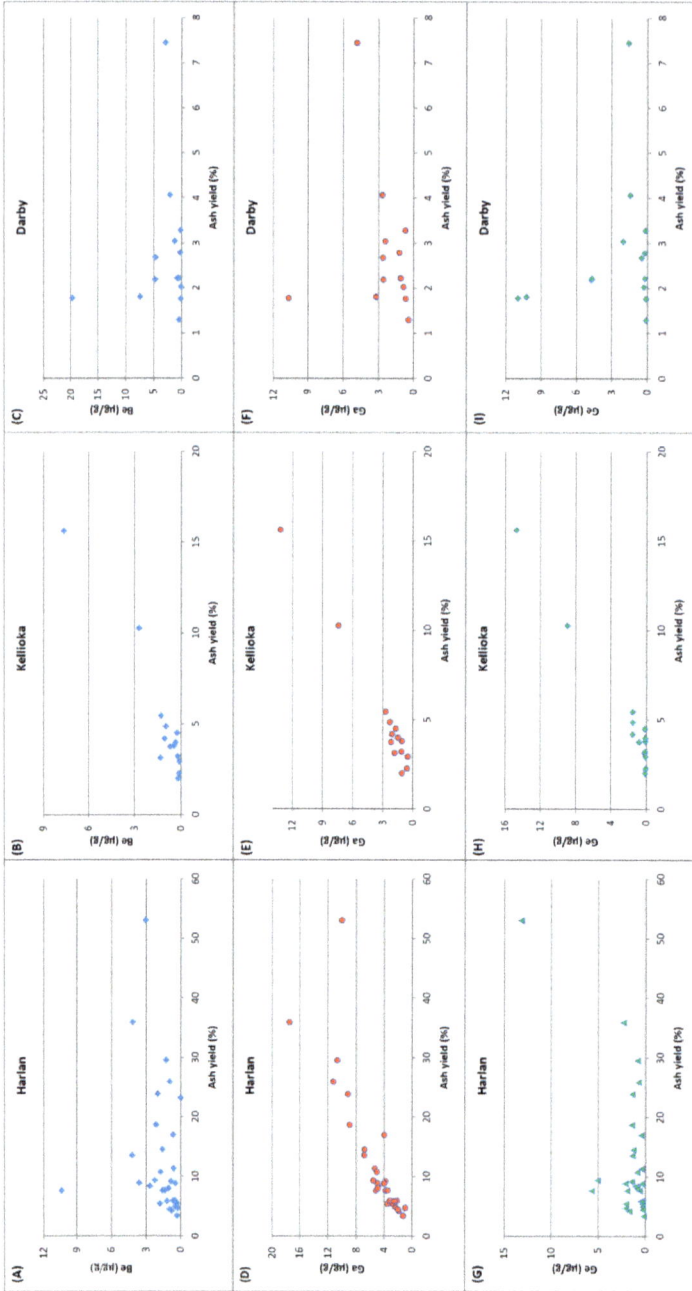

Figure 8. *Cont.*

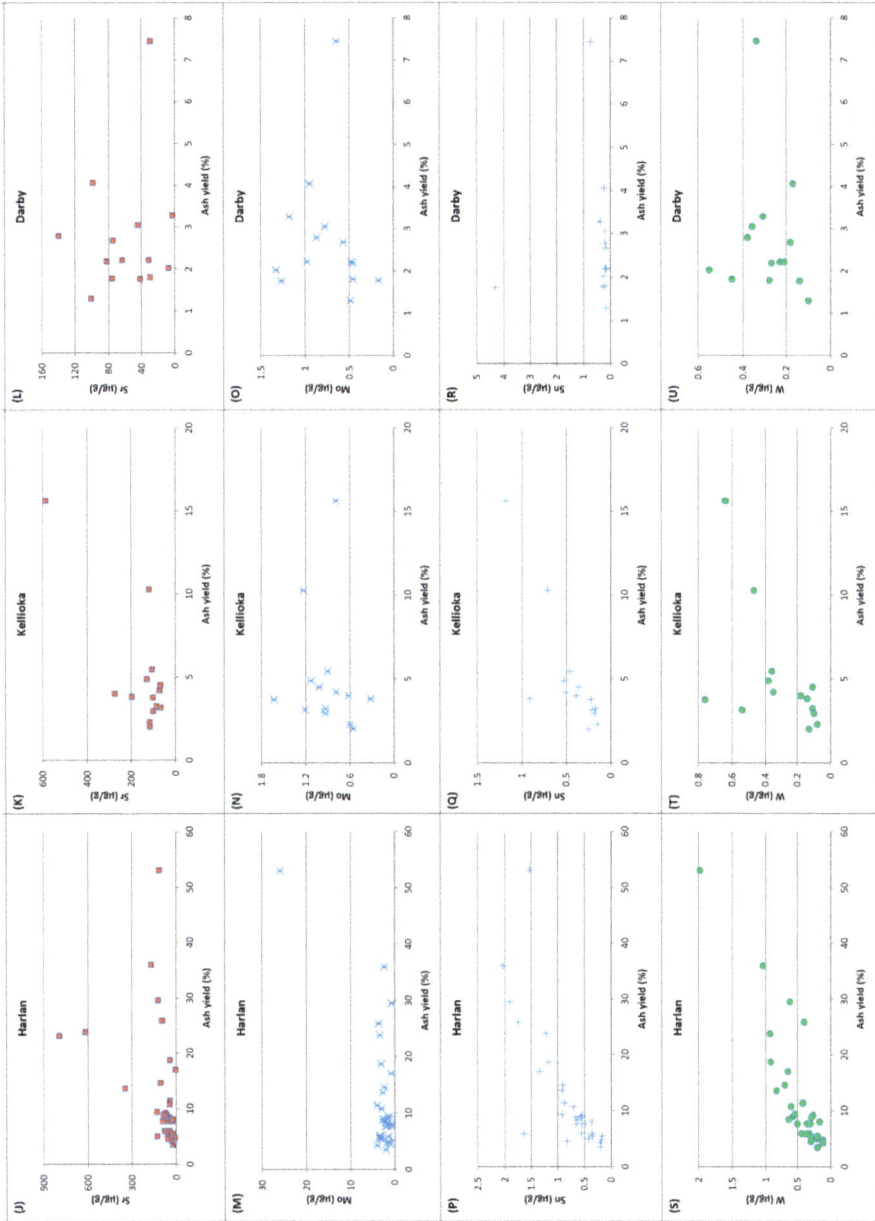

**Figure 8.** Relations of ash yield and some selected trace elements in the coals in Harlan, Kellioka, and Darby.

Gallium in the Harlan and Kellioka coals shows an inorganic affinity ($r$ = 0.84 and $r$ = 0.99 respectively; Figure 8D,E), but in the Darby coals it has an organic-inorganic mixed affinity ($r$ = 0.17; Figure 8F).

Germanium, Sr, and Mo in the three coals show an organic-inorganic mixed affinity (Figure 8). Although the correlation coefficient of Ge and ash yield in Kellioka coals is high ($r$ = 0.98; Figure 8H), there are only two points fall in the area with Ge concentration higher than 8 μg/g. However, twelve scattered points fall in the area of Ge concentration of less than 1.6μg/g, showing an organic-inorganic mixed affinity. The correlation coefficient of Sr and ash yield in the Kellioka is also high ($r$ = 0.77), the scattered points in the Figure 8K (only one point with high Sr concentration, 587μg/g) also indicate a mixed affinity.

Tin and W show an inorganic affinity in the Harlan and Kellioka coals but have an inorganic-organic mixed affinity in the Darby coals (Figure 8).

Although the average concentrations of most of trace elements in the three coals are not enriched relative to the averages of the world coals, some trace elements are relatively enriched in some benches of each coal seam. For example, see Sections 3.3.1–3.3.3 below.

### 3.3.1. The Harlan coals

The Harlan geochemistry has hints of the high values of certain minor element associations noted in other coals, such as $TiO_2$ + Zr, V + Cr, and Ba + Sr (such as the Darby for the latter association, see below). The $TiO_2$ + Zr has been found to be associated with detrital minerals in the basal benches of some coals [17,28]; V + Cr, possibly in association with clay minerals, can be enriched in the top bench; and Ba and Sr can be associated with phosphates and carbonates [28,29]. The fourth benches in both sections 6378 and 6392, bench 6 of 9 of section 6378, and bench 8 of 11 of section 6392 have some of the higher Cr and V values. In all cases, these benches underlie a parting, an event nearly as significant as the final demise of the coal [13], therefore, also an event likely to be marked by the same geochemical indicators as the top of the coal.

The Ba + Sr content exceeds 6000 ppm in sample 6401, but this is considerably lower than the high values encountered in the Darby coal (see below). Certain benches in the 6378 section also have >1000 ppm (ash basis) Ba and/or Sr, in some cases corresponding with $P_2O_5$ > 0.5% (ash basis). The Rare earth elements + Y (REY) content is not high by what might be considered to be potential commercial standards (perhaps 900 ppm on the ash basis) [30]. We note, however, that the samples with REY >600 ppm correspond to the samples with $P_2O_5$ >0.5%, not surprising since the REY are often found in phosphate minerals.

### 3.3.2. The Kellioka coals

The high-$Fe_2O_3$ content generally occurs in the top two benches, the higher pyritic S lithologies. These are also the benches with the highest As concentrations, up to 1231 ppm As and 0.39% $S_{py}$ (both on ash basis) in sample 6354. The third benches from the top at both sites, samples 6355 and 6363, are the highest CaO samples. Very little mineral matter is evident in microscopic examination and carbonates are not among the microscopic minerals. The concentrations of Sr, Ba, and REY are generally highest in the same samples, for example >11000 ppm Sr + Ba and 929 ppm REY in sample 6364, corresponding to a phosphate concentration of 3.50% (all on the ash basis).

### 3.3.3. The Darby coals

The samples generally have a relatively high amount of Ba + Sr, with sample 6264 (sample 3 of 4 from the 6261 series) exceeding 14560 ppm and sample 6370 (sample 4 of 5 from the 6366 series) having >11000 ppm Ba + Sr (both on the ash basis). Such levels of Ba + Sr might be attributable to associations with carbonates or phosphates. Neither sample has the highest REY content of the Darby samples, nearly 1700 ppm in sample 6262 (bench 1 or 4 from the 6261 series). Without further microbeam-based mineralogy studies, we cannot be certain about the association.

Vanadium and Cr, known in other coals to be associated with clays and frequently observed in the uppermost lithotype of many coal beds [31,32], are highest in the top lithology of both bench suites. Germanium and Ga are relatively high in the upper and lower benches of the 6261 series. Germanium is known to be enriched in coal lithotypes bordering the roof, floor, or partings [33].

### 3.4. Rare Earth Elements and Yttrium

The classification of rare earth elements and yttrium (REY, or REE if yttrium is not included) used in the present study is based on Seredin and Dai [30] and includes light (LREY: La, Ce, Pr, Nd, and Sm), medium (MREY: Eu, Gd, Tb, Dy, and Y), and heavy (HREY: Ho, Er, Tm, Yb, and Lu) REY. Accordingly, normalized to the upper continental crust (UCC; Taylor and McLennan [34]), three enrichment types are identified [30]: L-type (light-REY; $La_N/Lu_N > 1$), M-type (medium-REY; $La_N/Sm_N < 1$, $Gd_N/Lu_N > 1$), and H-type (heavy REY; $La_N/Lu_N < 1$).

The concentrations of rare earth elements (Table 5) in the three coals are lower than the averages for the world coals [26]; Figure 4), but their concentrations in coal ashes are close to the average for the world coal ash (Ketris and Yudovich [26]; Figure 5). The three coal seams have different REY distribution patterns:

(1) With the exceptions of some samples (samples 6387, 6386, and 6383 in Figure 9C; samples in Figure 9D; samples 6397 and 6398 in Figure 9E; samples 6399, 6402, and 6403 in Figure 9F), the REY in the Harlan coals are characterized by M-type enrichment.

(2) The Kellioka coal samples do not show much fractionation among the L-, M-, and H-REY, with the exception of sample 6358, which has a distinct H-REY enrichment type (Figure 10).

(3) With a few exceptions of samples 6264, 6370, 6371, and 6369, which a slight M-REY enrichment, the Darby coal samples are enriched in heavy REY relative to the upper continental crust [35] (Figure 11).

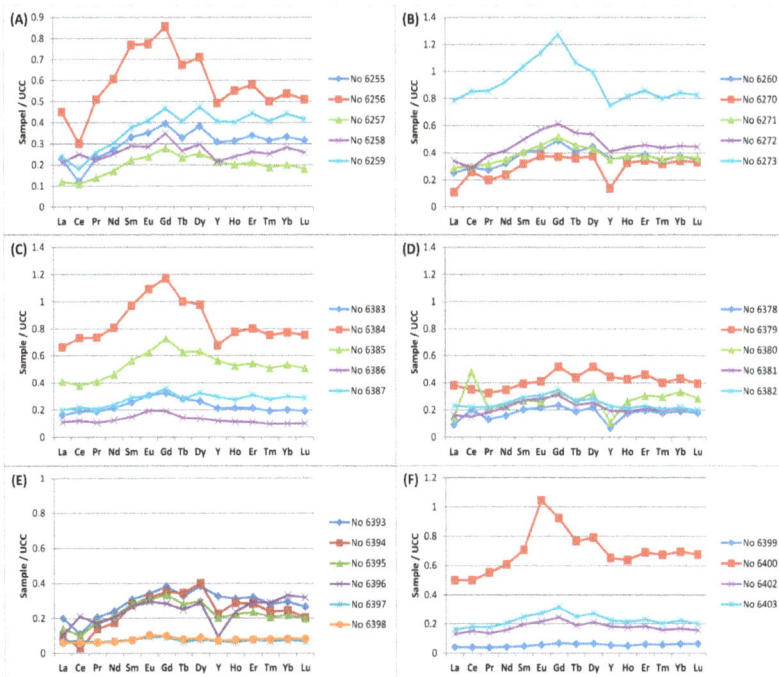

**Figure 9.** Distribution patterns of REY in Harlan coals. REY concentrations are normalized by those in the Upper Continental Crust [35].

**Table 5.** Concentrations of rare earth elements and yttrium (µg/g) in coals from Harlan, Kellioka, and Darby (on whole coal basis).

| Coal | Sample | Bench | La | Ce | Pr | Nd | Sm | Eu | Gd | Tb | Dy | Y | Ho | Er | Tm | Yb | Lu |
|---|---|---|---|---|---|---|---|---|---|---|---|---|---|---|---|---|---|
| Harlan | 6255 | wc | 6.82 | 7.82 | 1.65 | 7.01 | 1.49 | 0.31 | 1.5 | 0.21 | 1.34 | 6.77 | 0.25 | 0.78 | 0.1 | 0.73 | 0.1 |
| | 6256 | 1/4 | 13.6 | 19 | 3.61 | 15.76 | 3.46 | 0.68 | 3.25 | 0.43 | 2.49 | 10.8 | 0.44 | 1.33 | 0.17 | 1.18 | 0.16 |
| | 6257 | 2/4 | 3.63 | 7.25 | 0.99 | 4.44 | 1.01 | 0.21 | 1.06 | 0.15 | 0.89 | 4.72 | 0.16 | 0.49 | 0.06 | 0.44 | 0.06 |
| | 6258 | 3/4 | 6.38 | 16 | 1.56 | 6.53 | 1.3 | 0.25 | 1.32 | 0.17 | 1.04 | 4.72 | 0.19 | 0.6 | 0.08 | 0.62 | 0.08 |
| | 6259 | 4/4 | 7.06 | 11.4 | 1.84 | 7.88 | 1.69 | 0.36 | 1.77 | 0.26 | 1.66 | 8.94 | 0.32 | 1.02 | 0.13 | 0.97 | 0.13 |
| | WA-H1 | | 7.58 | 13.4 | 1.96 | 8.42 | 1.8 | 0.37 | 1.82 | 0.25 | 1.56 | 7.78 | 0.29 | 0.91 | 0.12 | 0.87 | 0.12 |
| | 6260 | wc | 7.55 | 18.3 | 1.94 | 8.33 | 1.8 | 0.37 | 1.86 | 0.26 | 1.56 | 7.86 | 0.29 | 0.89 | 0.11 | 0.83 | 0.11 |
| | 6270 | wc | 3.24 | 16.6 | 1.42 | 6.21 | 1.43 | 0.33 | 1.41 | 0.23 | 1.31 | 3.04 | 0.26 | 0.79 | 0.11 | 0.75 | 0.11 |
| | 6271 | wc | 8.55 | 19.3 | 2.25 | 9.09 | 1.83 | 0.4 | 1.96 | 0.29 | 1.51 | 7.67 | 0.3 | 0.86 | 0.12 | 0.82 | 0.12 |
| | 6272 | wc | 10.1 | 18.5 | 2.69 | 10.8 | 2.23 | 0.5 | 2.33 | 0.35 | 1.87 | 8.99 | 0.35 | 1.05 | 0.14 | 0.99 | 0.14 |
| | 6273 | wc | 23.3 | 54.1 | 6.06 | 24 | 4.65 | 1 | 4.84 | 0.68 | 3.48 | 16.39 | 0.65 | 1.97 | 0.26 | 1.85 | 0.26 |
| | 6378 | wc | 2.59 | 12.7 | 0.92 | 4.12 | 0.91 | 0.19 | 0.89 | 0.12 | 0.76 | 1.43 | 0.14 | 0.45 | 0.06 | 0.42 | 0.06 |
| | 6379 | 1/9 | 11.4 | 22.4 | 2.3 | 9.02 | 1.77 | 0.36 | 1.97 | 0.28 | 1.81 | 9.74 | 0.34 | 1.06 | 0.13 | 0.95 | 0.13 |
| | 6380 | 2/9 | 4.41 | 30.5 | 1.48 | 6.11 | 1.21 | 0.23 | 1.26 | 0.17 | 1.13 | 2.42 | 0.21 | 0.7 | 0.1 | 0.73 | 0.09 |
| | 6381 | 3/9 | 4.7 | 9.47 | 1.29 | 5.68 | 1.21 | 0.25 | 1.17 | 0.15 | 0.89 | 4.2 | 0.15 | 0.48 | 0.06 | 0.47 | 0.06 |
| | 6382 | 4/9 | 6.85 | 14.3 | 1.55 | 6.45 | 1.32 | 0.27 | 1.32 | 0.17 | 0.99 | 4.99 | 0.17 | 0.53 | 0.07 | 0.48 | 0.06 |
| | 6383 | 5/9 | 4.9 | 12.2 | 1.34 | 5.6 | 1.15 | 0.27 | 1.23 | 0.18 | 0.92 | 4.66 | 0.17 | 0.49 | 0.06 | 0.44 | 0.06 |
| | 6384 | 6/9 | 19.7 | 46.8 | 5.21 | 21.01 | 4.37 | 0.96 | 4.45 | 0.64 | 3.42 | 14.8 | 0.62 | 1.84 | 0.25 | 1.7 | 0.24 |
| | 6385 | 7/9 | 12.2 | 24 | 2.91 | 11.95 | 2.54 | 0.55 | 2.76 | 0.4 | 2.2 | 12.4 | 0.42 | 1.25 | 0.17 | 1.17 | 0.16 |
| | 6386 | 8/9 | 3.34 | 7.42 | 0.76 | 3.23 | 0.67 | 0.17 | 0.72 | 0.09 | 0.48 | 2.63 | 0.09 | 0.25 | 0.03 | 0.22 | 0.03 |
| | 6387 | 9/9 | 5.97 | 14.2 | 1.46 | 6.22 | 1.31 | 0.27 | 1.35 | 0.18 | 1.14 | 6.49 | 0.22 | 0.71 | 0.09 | 0.66 | 0.09 |
| | WA-H2 | | 6.56 | 15.6 | 1.62 | 6.72 | 1.39 | 0.3 | 1.45 | 0.2 | 1.15 | 5.76 | 0.21 | 0.65 | 0.08 | 0.6 | 0.08 |
| | 6393 | 1/11 | 5.87 | 7.06 | 1.47 | 6.28 | 1.39 | 0.3 | 1.45 | 0.21 | 1.34 | 7.17 | 0.25 | 0.74 | 0.09 | 0.65 | 0.09 |
| | 6394 | 2/11 | 2.8 | 2.23 | 0.97 | 4.49 | 1.25 | 0.28 | 1.33 | 0.22 | 1.4 | 4.97 | 0.23 | 0.65 | 0.08 | 0.54 | 0.07 |
| | 6395 | 3/11 | 4.22 | 6.68 | 1.24 | 5.48 | 1.27 | 0.27 | 1.27 | 0.18 | 1.05 | 4.43 | 0.18 | 0.54 | 0.07 | 0.48 | 0.06 |
| | 6396 | 4/11 | 3.02 | 13.5 | 1.19 | 5.33 | 1.2 | 0.26 | 1.08 | 0.16 | 1.02 | 2.05 | 0.19 | 0.67 | 0.1 | 0.73 | 0.1 |
| | 6397 | 5/11 | 2 | 4.32 | 0.44 | 1.81 | 0.34 | 0.08 | 0.35 | 0.04 | 0.27 | 1.48 | 0.05 | 0.17 | 0.02 | 0.16 | 0.02 |
| | 6398 | 6/11 | 1.66 | 3.53 | 0.43 | 1.67 | 0.33 | 0.09 | 0.37 | 0.05 | 0.3 | 1.57 | 0.06 | 0.18 | 0.03 | 0.18 | 0.03 |
| | 6399 | 7/11 | 1.25 | 2.39 | 0.27 | 1.1 | 0.21 | 0.05 | 0.26 | 0.04 | 0.23 | 1.19 | 0.04 | 0.14 | 0.02 | 0.14 | 0.02 |
| | 6400 | 8/11 | 15.1 | 31.9 | 3.91 | 15.78 | 3.18 | 0.92 | 3.51 | 0.49 | 2.76 | 14.34 | 0.51 | 1.59 | 0.22 | 1.53 | 0.22 |
| | 6402 | 10/11 | 3.86 | 9.64 | 0.96 | 4.1 | 0.88 | 0.19 | 0.93 | 0.12 | 0.74 | 3.99 | 0.14 | 0.42 | 0.05 | 0.37 | 0.05 |
| | 6403 | 11/11 | 4.93 | 11.6 | 1.25 | 5.34 | 1.12 | 0.24 | 1.2 | 0.16 | 0.95 | 4.95 | 0.17 | 0.53 | 0.07 | 0.49 | 0.06 |
| | WA-H3 | | 4.56 | 9.26 | 1.18 | 4.97 | 1.06 | 0.25 | 1.13 | 0.16 | 0.94 | 4.71 | 0.17 | 0.53 | 0.07 | 0.49 | 0.07 |
| | WA-H | | 7.35 | 16.9 | 1.97 | 8.15 | 1.69 | 0.37 | 1.74 | 0.25 | 1.41 | 6.4 | 0.26 | 0.81 | 0.11 | 0.76 | 0.11 |

**Table 5.** *Cont.*

| Coal | Sample | Bench | La | Ce | Pr | Nd | Sm | Eu | Gd | Tb | Dy | Y | Ho | Er | Tm | Yb | Lu |
|---|---|---|---|---|---|---|---|---|---|---|---|---|---|---|---|---|---|
| | 6352 | wc | 4.22 | 9.87 | 1.01 | 4.22 | 0.9 | 0.2 | 1 | 0.14 | 0.96 | 5.68 | 0.18 | 0.59 | 0.08 | 0.55 | 0.07 |
| | 6353 | 1/6 | 2.29 | 5.52 | 0.63 | 2.85 | 0.7 | 0.17 | 0.88 | 0.14 | 0.95 | 6.49 | 0.19 | 0.62 | 0.08 | 0.55 | 0.08 |
| | 6354 | 2/6 | 2.54 | 6.01 | 0.63 | 2.68 | 0.55 | 0.12 | 0.59 | 0.07 | 0.43 | 2.1 | 0.08 | 0.23 | 0.03 | 0.2 | 0.03 |
| | 6355 | 3/6 | 1.95 | 4.11 | 0.41 | 1.7 | 0.33 | 0.08 | 0.35 | 0.04 | 0.27 | 1.34 | 0.05 | 0.14 | 0.02 | 0.13 | 0.02 |
| | 6356 | 4/6 | 2.81 | 6.09 | 0.55 | 2.23 | 0.42 | 0.1 | 0.47 | 0.06 | 0.36 | 1.89 | 0.07 | 0.2 | 0.03 | 0.17 | 0.02 |
| | 6357 | 5/6 | 4.99 | 10.8 | 0.97 | 3.86 | 0.74 | 0.16 | 0.82 | 0.11 | 0.65 | 3.53 | 0.12 | 0.37 | 0.05 | 0.34 | 0.05 |
| | 6358 | 6/6 | 9.41 | 19.4 | 2.32 | 8.89 | 1.61 | 0.39 | 1.92 | 0.31 | 2.03 | 12.37 | 0.43 | 1.4 | 0.19 | 1.36 | 0.2 |
| Kellioka | WA-K1 | | 3.99 | 8.7 | 0.91 | 3.7 | 0.74 | 0.17 | 0.85 | 0.12 | 0.79 | 4.69 | 0.16 | 0.5 | 0.06 | 0.46 | 0.06 |
| | 6359 | wc | 2.93 | 8.56 | 0.87 | 3.5 | 0.71 | 0.18 | 0.78 | 0.12 | 0.68 | 2.69 | 0.13 | 0.41 | 0.06 | 0.37 | 0.05 |
| | 6360 | wc | 5.67 | 12.5 | 1.37 | 5.42 | 1.08 | 0.26 | 1.2 | 0.18 | 1.01 | 5.47 | 0.2 | 0.6 | 0.08 | 0.56 | 0.08 |
| | 6361 | 1/5 | 2.52 | 6.07 | 0.77 | 3.34 | 0.81 | 0.21 | 0.93 | 0.18 | 1.17 | 7.46 | 0.24 | 0.76 | 0.1 | 0.69 | 0.1 |
| | 6362 | 2/5 | 1.56 | 6.14 | 0.52 | 2.28 | 0.49 | 0.11 | 0.48 | 0.06 | 0.38 | 1.2 | 0.07 | 0.21 | 0.03 | 0.21 | 0.03 |
| | 6363 | 3/5 | 0.39 | 2.32 | 0.1 | 0.44 | 0.09 | 0.03 | 0.12 | 0.01 | 0.08 | 0.4 | 0.01 | 0.04 | 0.01 | 0.04 | 0.01 |
| | 6364 | 4/5 | 7.08 | 13.8 | 1.38 | 5.56 | 1.05 | 0.23 | 1.12 | 0.15 | 0.87 | 4.57 | 0.16 | 0.48 | 0.06 | 0.42 | 0.06 |
| | 6365 | 5/5 | 9.8 | 22.7 | 2.2 | 8.9 | 1.78 | 0.36 | 1.89 | 0.26 | 1.67 | 9.46 | 0.32 | 1.03 | 0.13 | 0.96 | 0.14 |
| | WA-K2 | | 5.26 | 11.7 | 1.16 | 4.75 | 0.96 | 0.21 | 1.03 | 0.15 | 0.91 | 5.04 | 0.17 | 0.54 | 0.07 | 0.5 | 0.07 |
| | WA-K | | 4.41 | 10.3 | 1.06 | 4.32 | 0.88 | 0.2 | 0.97 | 0.14 | 0.87 | 4.71 | 0.17 | 0.53 | 0.07 | 0.49 | 0.07 |
| | 6261 | wc | 2.08 | 5.4 | 0.52 | 2.32 | 0.55 | 0.13 | 0.67 | 0.1 | 0.7 | 5.13 | 0.14 | 0.46 | 0.06 | 0.42 | 0.06 |
| | 6262 | 1/4 | 3.4 | 6.99 | 0.71 | 3.15 | 0.75 | 0.19 | 0.98 | 0.16 | 1.24 | 10.23 | 0.26 | 0.84 | 0.11 | 0.76 | 0.1 |
| | 6264 | 3/4 | 1.43 | 3.99 | 0.36 | 1.64 | 0.36 | 0.09 | 0.4 | 0.05 | 0.31 | 1.9 | 0.06 | 0.17 | 0.02 | 0.14 | 0.02 |
| | 6265 | 4/4 | 0.56 | 3.09 | 0.24 | 1.19 | 0.34 | 0.08 | 0.43 | 0.08 | 0.64 | 1.41 | 0.14 | 0.46 | 0.06 | 0.41 | 0.06 |
| | WA-D1 | | 1.57 | 4.3 | 0.39 | 1.8 | 0.42 | 0.1 | 0.5 | 0.07 | 0.54 | 3.19 | 0.11 | 0.34 | 0.04 | 0.3 | 0.04 |
| | 6266 | wc | 2.02 | 5.13 | 0.57 | 2.45 | 0.57 | 0.15 | 0.62 | 0.11 | 0.68 | 4.41 | 0.14 | 0.42 | 0.06 | 0.38 | 0.06 |
| | 6267 | wc | 1.61 | 5.51 | 0.53 | 2.25 | 0.49 | 0.13 | 0.54 | 0.09 | 0.57 | 1.37 | 0.12 | 0.36 | 0.05 | 0.31 | 0.04 |
| | 6268 | wc | 1.38 | 4.76 | 0.45 | 1.93 | 0.43 | 0.11 | 0.47 | 0.08 | 0.45 | 2.06 | 0.09 | 0.27 | 0.04 | 0.25 | 0.04 |
| Darby | 6269 | wc | 1.77 | 6 | 0.6 | 2.59 | 0.59 | 0.14 | 0.61 | 0.1 | 0.63 | 2.27 | 0.12 | 0.39 | 0.05 | 0.36 | 0.05 |
| | 6366 | wc | 1.57 | 3.68 | 0.45 | 2.04 | 0.47 | 0.11 | 0.52 | 0.08 | 0.48 | 2.83 | 0.09 | 0.29 | 0.04 | 0.26 | 0.04 |
| | 6367 | 1/5 | 3.89 | 12.47 | 1.18 | 5.51 | 1.5 | 0.35 | 1.77 | 0.29 | 1.99 | 11.8 | 0.39 | 1.26 | 0.17 | 1.23 | 0.16 |
| | 6368 | 2/5 | 3.15 | 7.52 | 0.77 | 3.26 | 0.71 | 0.18 | 0.86 | 0.14 | 0.84 | 7.77 | 0.18 | 0.54 | 0.07 | 0.46 | 0.07 |
| | 6369 | 3/5 | 4.18 | 8.95 | 0.85 | 3.33 | 0.64 | 0.15 | 0.72 | 0.09 | 0.5 | 2.82 | 0.09 | 0.29 | 0.04 | 0.27 | 0.04 |
| | 6370 | 4/5 | 1.67 | 4.02 | 0.47 | 2 | 0.42 | 0.12 | 0.43 | 0.06 | 0.32 | 1.65 | 0.06 | 0.18 | 0.02 | 0.16 | 0.02 |
| | 6371 | 5/5 | 1.85 | 3.82 | 0.59 | 2.61 | 0.62 | 0.16 | 0.63 | 0.1 | 0.57 | 3.02 | 0.11 | 0.31 | 0.04 | 0.27 | 0.04 |
| | WA-D2 | | 2.21 | 5.25 | 0.6 | 2.56 | 0.57 | 0.15 | 0.62 | 0.09 | 0.54 | 3.48 | 0.11 | 0.32 | 0.04 | 0.29 | 0.04 |
| | WA-D | | 1.78 | 5 | 0.51 | 2.24 | 0.51 | 0.13 | 0.57 | 0.09 | 0.57 | 3.09 | 0.11 | 0.36 | 0.05 | 0.32 | 0.04 |

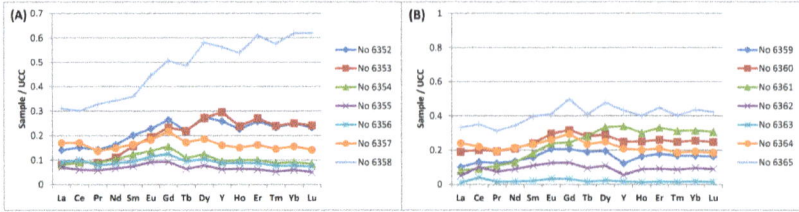

**Figure 10.** Distribution patterns of REY in Kellioka coals. REY concentrations are normalized by those in the Upper Continental Crust [35].

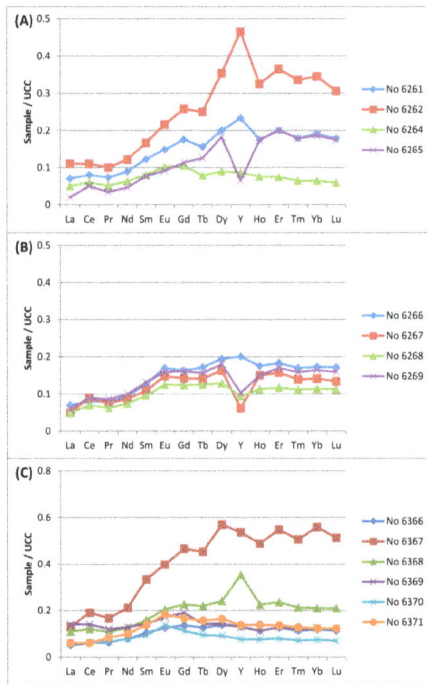

**Figure 11.** Distribution patterns of REY in Darby coals. REY concentrationsare normalized by those in the Upper Continental Crust [35].

Geochemical influences were likely to have been complex. Aside from the expected terrigenous influx at the time of deposition, the region was subject to the influence of hydrothermal fluids during diagenesis. This is most notable on the footwall side of the Pine Mountain thrust fault where an enhanced coal rank compared to correlative coals on the thrust sheet (as we are studying here), albeit all within the high volatile A bituminous rank range, are accompanied by enhanced levels of Cl and Hg and other trace metals [34]. While not previously demonstrated, it is possible that the coals on the Pine Mountain thrust sheet could have been similarly influenced, if not from fluids squeezed out in advance of the Pine Mountain thrust sheet, then by fluid flow influenced by thrust faults to the southeast in Virginia. Examination of the $Al_2O_3$ *vs.* $TiO_2$ plot (Figure 12) provides a view of another aspect of mineral influx. The Harlan benches have a much wider distribution than the Kellioka or Darby benches, having both higher and lower $Al_2O_3$ and lower $TiO_2$ than the other coals. Among the Darby samples with the highest $TiO_2$, bench samples 6262 and 6368 have strikingly different REY distributions than any of the other benches among the three coals. In particular, the Y concentration *versus* the UCC baseline value is high. Relative to other Darby benches, the $P_2O_5$ is also high, suggesting that an influx of Y- (and REY) bearing phosphates could have accompanied the $TiO_2$ influx. $TiO_2$-mineral/Phosphate/Zircon sediments are common in detrital (often the basal coal) lithotypes [17,25,26,36–43].

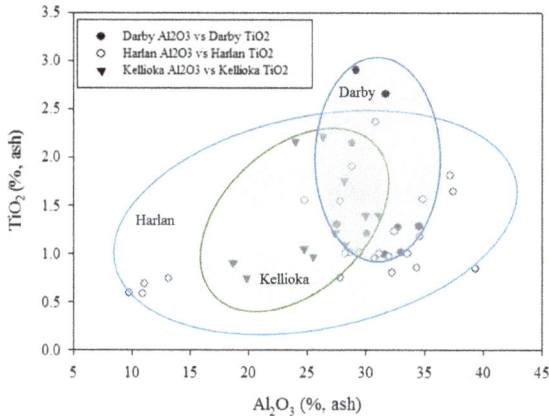

**Figure 12.** Distribution of $Al_2O_3$-$TiO_2$ in the coals from in Harlan, Kellioka, and Darby.

## 4. Summary

The Harlan, Kellioka, and Darby coals have traditionally been among the more important coal resources in Harlan County, Kentucky.

The Harlan coal is the thickest and has the highest ash and sulfur content of the three coals in the study. In practice, the ash and sulfur content could be reduced by beneficiation. An enrichment of $TiO_2$ + Zr in the basal lithotype and V + Cr in the top lithotype and in lithologies immediately below partings is similar to occurrences seen in other eastern Kentucky coals. The lithotypes with REY >600 ppm correspond to concentrations of $P_2O_5$ > 0.5%. Much of the Harlan coal has an M-type REY distribution pattern (after Seredin and Dai [30]).

The Kellioka coal generally has less than 5% ash yield and a sulfur content <0.9% in most lithotypes. Pyritic S is highest in the uppermost two lithotypes and, in the 6360 section, in the basal lithotype. Concentrations of Sr + Ba > 11000 ppm, accompanied by 929 ppm REY, occur in a lithotype with 3.50% $P_2O_5$. The Kellioka samples generally do not have REY patterns corresponding to the Seredin and Dai [30] distributions.

The Darby coal is generally low-ash and low-sulfur. The Ba + Sr content exceeds 14,560 ppm (ash basis) in one sample and has relatively high values in other lithotypes. While a carbonate or phosphate association might be the source of the elements, there is no direct mineral evidence for such an association in this coal. The highest REY content, nearly 1700 ppm, does not correspond to the highest Sr + Ba. A few of the Darby samples show an M-type distribution, with most samples enriched in heavy REY elements. As with the Harlan coal, the V + Cr is highest in the uppermost lithotype in both benched sections.

**Acknowledgments:** The trace elements analysis was supported by the National Key Basic Research Program of China (No. 2014CB238902) and the National Natural Science Foundation of China (No. 41420104001).

**Author Contributions:** James C. Hower was a participant in the original sampling. Michelle N. Johnston and James C. Hower re-did the petrology using the ICCP 1994 nomenclature. Shifeng Dai, Peipei Wang, Panpan Xie, and Jingjing Liu were responsible for the ICP-MS chemistry. James C. Hower and Shifeng Dai were responsible for the writing of the manuscript.

**Conflicts of Interest:** The authors declare no conflict of interest.

## References

1.  Bragg, B.; Reece, F. *Which Side Are You On?*; Sony/ATV Music Publishing LLC (Current Copyright Owner): Nashville, TN, USA, 1931.
2.  Woolley, B. *We Be Here When the Morning Comes*; University Press of Kentucky: Lexington, KY, USA, 1975.
3.  Portelli, A. *They Say in Harlan County: An Oral History*; Oxford University Press: Oxford, UK, 2010.
4.  Caudill, H.M. *Theirs Be the Power—The Moguls of Eastern Kentucky*; University of Illinois Press: Urbana, IL, USA, 1983.
5.  Shifflett, C.A. *Coal Towns: Life, Work, and Culture in Company Towns of Southern Appalachia, 1880–1960*; University of Tennessee Press: Knoxville, TN, USA, 1991.
6.  Rice, C.L.; Smith, J.H. *Correlation of Coal Beds, Coal Zones, and Key Stratigraphic Units, Pennsylvanian Rocks of Eastern Kentucky*; U.S. Geological Survey Map MF-1188; U.S. Geological Survey: Washington, DC, USA, 1980.
7.  Hatton, A.R.; Hower, J.C.; Helfrich, C.T.; Pollock, J.D.; Wild, G.D. Lithologic succession in the Path Fork coal bed (Breathitt Formation, Middle Pennsylvanian), southeastern Kentucky. *Org. Geochem.* **1992**, *18*, 301–311. [CrossRef]
8.  Esterle, J.S.; Ferm, J.C. Relationship between petrographic and chemical properties and coal seam geometry, Hance seam, Breathitt Formation, southeastern Kentucky. *Int. J. Coal Geol.* **1986**, *6*, 199–214. [CrossRef]
9.  Hubbard, T.E.; Miller, T.R.; Hower, J.C.; Ferm, J.C.; Helfrich, C.F. The Upper Hance coal bed in southeastern Kentucky: Palynologic, geochemical, and petrographic evidence for environmental succession. *Int. J. Coal Geol.* **2002**, *49*, 177–194. [CrossRef]
10. Froelich, A.J. *Geologic Map of the Louellen Quadrangle, Southeastern Kentucky*; U.S. Geological Survey Map GQ-1060; U.S. Geological Survey: Washington, DC, USA, 1975.
11. Chesnut, D.R. Geologic framework for the coal-bearing rocks of the Central Appalachian Basin. *Int. J. Coal Geol.* **1996**, *31*, 55–66. [CrossRef]
12. Greb, S.F.; Eble, C.F.; Hower, J.C. Depositional history of the Fire Clay coal bed (Late Duckmantian), eastern Kentucky, USA. *Int. J. Coal Geol.* **1999**, *40*, 255–280. [CrossRef]
13. Greb, S.F.; Eble, C.F.; Hower, J.C.; Andrews, W.M. Multiple-bench architecture and interpretations of original mire phases in Middle Pennsylvanian coal seams: Examples from the Eastern Kentucky coal field. *Int. J. Coal Geol.* **2002**, *49*, 147–175. [CrossRef]
14. Greb, S.F.; Eble, C.F.; Chesnut, D.R., Jr. Comparison of the Eastern and Western Kentucky coal fields (Pennsylvanian), USA—Why are coal distribution patterns and sulfur contents so different in these coal fields? *Int. J. Coal Geol.* **2002**, *50*, 89–118. [CrossRef]
15. Aitken, J.F.; Flint, S.S. The application of high-resolution sequence stratigraphy to fluvial systems: A case study from the Upper Carboniferous Breathitt Group, eastern Kentucky, USA. *Sedimentology* **1995**, *42*, 3–30. [CrossRef]
16. Aitken, J.F.; Flint, S.S. *Variable Expressions of Interfluvial Sequence Boundaries in the Breathitt Group (Pennsylvanian), Eastern Kentucky, USA*; Geological Society Special Publication: London, UK, 1996; pp. 193–206.
17. Hower, J.C.; Bland, A.E. Geochemistry of the Pond Creek coal bed, Eastern Kentucky coalfield. *Int. J. Coal Geol.* **1989**, *11*, 205–226. [CrossRef]
18. Dai, S.; Wang, X.; Zhou, Y.; Hower, J.C.; Li, D.; Chen, W.; Zhu, X. Chemical and mineralogical compositions of silicic, mafic, and alkali tonsteins in the late Permian coals from the Songzao Coalfield, Chongqing, Southwest China. *Chem. Geol.* **2011**, *282*, 29–44. [CrossRef]
19. Li, X.; Dai, S.; Zhang, W.; Li, T.; Zheng, X.; Chen, W. Determination of As and Se in coal and coal combustion products using closed vessel microwave digestion and collision/reaction cell technology (CCT) of inductively coupled plasma mass spectrometry (ICP-MS). *Int. J. Coal Geol.* **2014**, *124*, 1–4. [CrossRef]
20. International Committee for Coal and Organic Petrology. The new vitrinite classification (ICCP system 1994). *Fuel* **1998**, *77*, 349–358.
21. International Committee for Coal and Organic Petrology. The new inertinite classification (ICCP system 1994). *Fuel* **2001**, *80*, 459–471.
22. Hower, J.C.; Pollock, J.D.; Griswold, T.B. Structural Controls on Petrology and Geochemistry of the Pond Creek Coal Bed, Pike and Martin Counties, Eastern Kentucky. In *Geology in Coal Resource Utilization, American Association of Petroleum Geologists*; Peters, D.C., Ed.; Energy Minerals Division: Tulsa, OK, USA, 1991; pp. 413–427.

23. Rimmer, S.M.; Hower, J.C.; Moore, T.A.; Esterle, J.S.; Walton, R.L.; Helfrich, C.T. Petrography and palynology of the Blue Gem coal bed, southeastern Kentucky, USA. *Int. J. Coal Geol.* **2000**, *42*, 159–184. [CrossRef]

24. Eble, C.F.; Hower, J.C.; Andrews, W.M., Jr. Paleoecology of the Fire Clay coal bed in a portion of the Eastern Kentucky coal field. *Palaeogeogr. Palaeoclimatol. Palaeoecol.* **1994**, *106*, 287–305. [CrossRef]

25. Hower, J.C.; Andrews, W.M., Jr.; Wild, G.D.; Eble, C.F.; Dulong, F.T.; Salter, T.L. Coal quality trends for the Fire Clay coal bed, southeastern Kentucky. *J. Coal Qual.* **1994**, *13*, 13–26.

26. Ketris, M.P.; Yudovich, Y.E. Estimates of Clarkes for carbonaceous biolithes: World average for trace elements contents in black shales and coals. *Int. J. Coal Geol.* **2009**, *78*, 135–148. [CrossRef]

27. Dai, S.; Seredin, V.V.; Ward, C.R.; Hower, J.C.; Xing, Y.; Zhang, W.; Song, W.; Wang, P. Enrichment of U-Se-Mo-Re-V in coals preserved within marine carbonate successions: Geochemical and mineralogical data from the Late Permian Guiding Coalfield, Guizhou, China. *Miner. Deposita* **2015**, *50*, 159–186. [CrossRef]

28. Hower, J.C.; Taulbee, D.N.; Rimmer, S.M.; Morrell, L.G. Petrographic and geochemical anatomy of lithotypes from the Blue Gem coal bed, southeastern Kentucky. *Energy Fuels* **1994**, *8*, 719–728. [CrossRef]

29. Hower, J.C.; Rimmer, S.M.; Bland, A.E. Geochemistry of the Blue Gem coal bed, Knox County, Kentucky. *Int. J. Coal Geol.* **1991**, *18*, 211–231. [CrossRef]

30. Seredin, V.V.; Dai, S. Coal deposits as a potential alternative source for lanthanides and yttrium. *Int. J. Coal Geol.* **2012**, *94*, 67–93. [CrossRef]

31. Zubović, P. Physico-Chemical Properties of Certain Minor Elements as Controlling Factors in Their Distribution in Coal. In *Coal Science*; Given, P.H., Ed.; American Chemical Society Advances in Chemistry Series: Washington, DC, USA, 1966; pp. 211–231.

32. Hower, J.C.; Greb, S.F.; Cobb, J.C.; Williams, D.A. Discussion on origin of vanadium in coals: Parts of the Western Kentucky (USA) No. 9 coal rich in vanadium: Special Publication No. 125, 1997, 273–286. *J. Geol. Soc. Lond.* **2000**, *157*, 1257–1259. [CrossRef]

33. Yudovich, Y.E. Notes on the marginal enrichment of Germanium in coal beds. *Int. J. Coal Geol.* **2003**, *56*, 223–232. [CrossRef]

34. Sakulpitakphon, T.; Hower, J.C.; Schram, W.H.; Ward, C.R. Tracking Mercury from the Mine to the Power Plant: Geochemistry of the Manchester Coal Bed, Clay County, Kentucky. *Int. J. Coal Geol.* **2004**, *57*, 127–141. [CrossRef]

35. Taylor, S.R.; McLennan, S.M. The Continental Crust: Its Composition and Evolution. Blackwell: London, UK, 1985; p. 312.

36. Hower, J.C.; Pollock, J.D. Petrology of the River Gem Coal Bed, Whitley County, Kentucky. *Int. J. Coal Geol.* **1989**, *11*, 227–245. [CrossRef]

37. Hower, J.C.; Riley, J.T.; Thomas, G.A.; Griswold, T.B. Chlorine in Kentucky coals. *J. Coal Qual.* **1991**, *10*, 152–158.

38. Hower, J.C.; Hiett, J.K.; Wild, G.D.; Eble, C.F. Coal resources, production, and quality in the Eastern Kentucky coalfield: Perspectives on the future of steam coal production. *Nonrenew. Resour.* **1994**, *3*, 216–236. [CrossRef]

39. Hower, J.C.; Ruppert, L.F.; Eble, C.F.; Graham, U.M. Geochemical and palynological indicators of the paleoecology of the River Gem coal bed, Whitley County, Kentucky. *Int. J. Coal Geol.* **1996**, *31*, 135–149. [CrossRef]

40. Andrews, W.M., Jr.; Hower, J.C.; Hiett, J.K. Investigations of the Fire Clay coal bed, southeastern Kentucky, in the vicinity of sandstone washouts. *Int. J. Coal Geol.* **1994**, *26*, 95–115. [CrossRef]

41. Mardon, S.M.; Hower, J.C. Impact of coal properties on coal combustion by-product quality: Examples from a Kentucky power plant. *Int. J. Coal Geol.* **2004**, *59*, 153–169. [CrossRef]

42. Dai, S.; Wang, P.; Ward, C.R.; Tang, Y.; Song, X.; Jiang, J.; Hower, J.C.; Li, T.; Seredin, V.V.; Wagner, N.J.; *et al.* Elemental and mineralogical anomalies in the coal-hosted Ge ore deposit of Lincang, Yunnan, Southwestern China: Key role of $N_2$-$CO_2$-mixed hydrothermal solutions. *Int. J. Coal Geol.* **2014**. [CrossRef]

43. Dai, S.; Hower, J.C.; Ward, C.R.; Guo, W.; Song, H.; O'Keefe, J.M.K.; Xie, P.; Hood, M.M.; Yan, X. Elements and phosphorus minerals in the middle Jurassic inertinite-rich coals of the Muli Coalfield on the Tibetan Plateau. *Int. J. Coal Geol.* **2015**, *144–145*, 23–47. [CrossRef]

**minerals**

MDPI

*Article*

# Major and Trace Element Geochemistry of Coals and Intra-Seam Claystones from the Songzao Coalfield, SW China

Lei Zhao [1,2,]*, Colin R. Ward [3], David French [3] and Ian T. Graham [3]

[1] State Key Laboratory of Coal Resources and Safe Mining, China University of Mining and Technology (Beijing), Beijing 100083, China
[2] College of Geoscience and Survey Engineering, China University of Mining and Technology (Beijing), Beijing 100083, China
[3] School of Biological, Earth and Environmental Sciences, University of New South Wales, Sydney NSW 2052, Australia; c.ward@unsw.edu.au (C.R.W.), d.french@unsw.edu.au (D.F.); i.graham@unsw.edu.au (I.T.G.)
* Correspondence: lei.zhao@y7mail.com; Tel.: +86-10-6234-1868

Academic Editor: Antonio Simonetti
Received: 19 October 2015; Accepted: 25 November 2015; Published: 3 December 2015

**Abstract:** Silicic, mafic and alkali intra-seam tonsteins have been known from SW China for a number of years. This paper reports on the geochemical compositions of coals and tonsteins from three seam sections of the Songzao Coalfield, SW China, and evaluates the geological factors responsible for the chemical characteristics of the coal seams, with emphasis on the influence from different types of volcanic ashes. The roof and floor samples of the Songzao coal seams mostly have high $TiO_2$ contents, consistent with a high $TiO_2$ content in the detrital sediment input from the source region, namely mafic basalts from the Kangdian Upland on the western margin of the coal basin. The coals from the Songzao Coalfield generally have high ash yields and are highly enriched in trace elements including Nb, Ta, Zr, Hf, rare earth elements (REE), Y, Hg and Se; some variation occurs among different seam sections due to input of geochemically different volcanic ash materials. The geochemistry of the Songzao coals has also been affected by the adjacent tonstein/K-bentonite bands. The relatively immobile elements that are enriched in the altered volcanic ashes also tend to be enriched in the adjacent coal plies, possibly due to leaching by groundwaters. The coals near the alkali tonstein bands in the Tonghua and Yuyang sections of the Songzao Coalfield are mostly high in Nb, Ta, Zr, Hf, Th, U, REE and Y. Coal samples overlying the mafic K-bentonite in the Tonghua section are high in V, Cr, Zn and Cu. The Datong coal, which has neither visible tonstein layers nor obvious volcanogenic minerals, has high $TiO_2$, V, Cr, Ni, Cu and Zn concentrations in the intervals between the coal plies affected by mafic and alkaline volcanic ashes. This is consistent with the suggestion that a common source material was supplied to the coal basin, derived from the erosion of mafic basaltic rocks of the Kangdian Upland. Although the Songzao coal is generally a high-sulfur coal, most of the chalcophile trace elements show either poor or negative correlations with total iron sulfide contents. The absence of traditional pyrite-metal associations may reflect wide variations in the concentrations of these elements in individual pyrite/marcasite components, or simply poor retention of these elements in the pyrite/marcasite of the relevant coals.

**Keywords:** geochemistry; coal; rare earth elements; volcanic ash; Late Permian

## 1. Introduction

The trace element geochemistry of a particular coal is the result of the interaction of the original peaty material with water- and/or air-borne detrital input, and solutions that circulated within the coal basin [1–3], influenced in different ways by the botanical, biochemical and geological factors that acted throughout the long-term process of coal formation [1,4,5]. Among all the factors, incorporation of volcanic ash or influence of volcanic ash layers may have a significant impact not only on the mineralogy, but also on the geochemical characteristics of the individual layers within the coal seams [6,7].

Altered volcanic ash layers are widespread in the Permian strata of southwestern (SW) China [8–10]. Although the geochemistry of tonsteins in the Late Permian coals of SW China indicates an origin from silicic volcanic ash fallout [11], alkali tonsteins that developed in the early part of the Late Permian in SW China have also been reported [10,12,13]. The enrichment of rare metals in coal and its host rocks (e.g., roof and floor strata) in southwestern China, caused by alkali volcanic ashes, has attracted much attention in recent years [3,14–16].

Dai *et al.* [17] indicated that coal from one Songzao coal seam (the No. 11 seam), which contains no visible tonsteins, is significantly enriched in some alkaline elements such as Nb, Ta, Zr, Hf and rare earth elements (REE), and suggested that these geochemical anomalies can be mainly attributed to synsedimentary alkaline volcanic ashes. Another study by Dai *et al.* [18] distinguished three types of tonstein bands (silicic, mafic and alkali) in the Songzao Coalfield based on their distinctive chemical compositions. In a recent study of three individual seam sections in the Songzao Coalfield [19], the modes of occurrence and origin of the mineral assemblages in the volcanic-influenced coal seams were more fully investigated. The present study discusses the modes of occurrence of the trace elements in the coal and associated non-coal strata from the same coal seams. It also provides an opportunity to evaluate the geological factors responsible for the chemical characteristics of coal seams that have been influenced by different types of volcanic ashes. Importantly, the concentrations of rare earth elements and Y (REY, or REE if Y is not included) in the Songzao coals are comparable to those of conventional rare-metal ore deposits and thus the coals are potential raw sources of these metals.

## 2. Geologic Setting

The Songzao Coalfield is located in SW Chongqing, and encompasses eight different mines (Figure 1). The coal reserves of the Songzao Coalfield are estimated to be 811 Mt as of 2003 [20], accounting for 42.6% of the total coal reserves in Chongqing [18]. The coals of the Songzao Coalfield are mostly high-sulfur anthracites and, in a few cases, medium-sulfur coals (e.g., No. 8 Coal), and are rich in methane. The coalfield is located on the northwestern flanks of the Jiudianya, Jiulongshan and Sangmuchang anticlines.

The Longtan Formation (Late Permian) is the coal-bearing sequence in the coalfield, deposited in a tidal flat system along the western margin of an epicontinental sea [21]. The Longtan Formation consists of limestone, sandstone, silty mudstone, mudstone, coal seams and tuffaceous sediments (Figure 2). The Kangdian Upland to the west was the major sediment source for the coalfield [10,15,22]. The coal-bearing sequence contains 6–11 coal seams, among which the No. 8 coal is workable throughout the entire coalfield, and the Nos. 6, 7, 8, 11 and 12 are locally workable. The Longtan Formation is disconformably underlain by the Maokou Formation, an Early Permian shallow marine limestone unit.

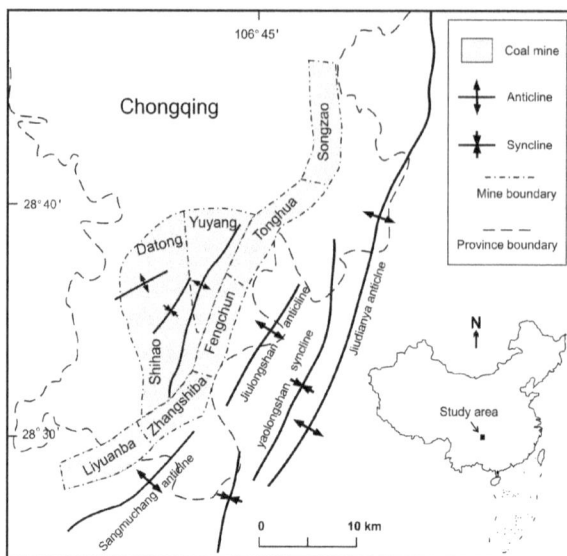

**Figure 1.** Locality map of the Songzao Coalfield, indicating the mining areas (grey) (after [19]).

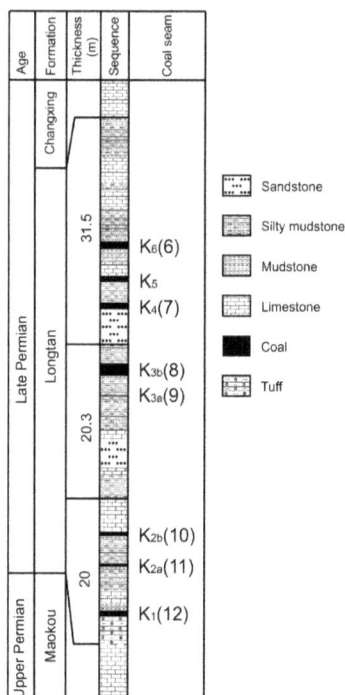

**Figure 2.** Sedimentary sequence of the Songzao Coalfield, showing the location of the coal seams (after [19]).

## 3. Sampling and Methods

A total of 24 coal, associated mudrock and intra-seam claystone samples (channel samples) from three seam sections (Datong, Tonghua, and Yuyang) were used for this investigation, as well as for a previous study of the Songzao coal seams [19]. Three series of samples were taken at the underground working faces of the Datong (No. 7 coal), Tonghua (No. k2b coal), and Yuyang (No. 11 coal) mines, respectively (Figure 1).

Each sample was ground to fine powder (about 200 mesh) using a zirconia mill, and split into representative portions for further analyses. All the coal samples were ashed at 815 °C, following procedures described by Standards Australia [23]; the resultant ashes and the ground non-coal samples were analyzed by X-ray fluorescence (XRF) spectrometry to determine the concentrations of major elements. The coal samples were also subjected to low-temperature oxygen-plasma ashing and both the coal mineral residues (low-temperature ash or LTA) and the non-coal rock samples analyzed by X-ray diffraction (XRD) techniques, using Siroquant software (Sietronics Pty Ltd., Belconnen, ACT, Australia) for quantitative mineralogical analysis. Samples were also examined in polished section using scanning electron microscopy combined with energy-dispersive spectrometry (SEM-EDS) techniques, to identify the modes of mineral occurrence. The XRD, XRF and SEM procedures are discussed further by Zhao *et al.* [19].

Concentrations of most trace elements in the coal and rock samples were determined by inductively coupled plasma-mass spectrometry/ optical emission spectrometry (ICP-MS/OES). Prior to the ICP-MS/OES analysis, two separate digestion procedures were carried-out, to accomplish total decomposition of the samples and to ensure that the total element content in each sample was reflected in the resultant digests. One procedure involved ashing the coal and rock samples at 450 °C. The ashes were then subjected to microwave dissolution in a mixed acid (HCl, HF and $HNO_3$). The other procedure involved fusion with a mixture of lithium tetraborate ($Li_2B_4O_7$) and lithium metaborate ($LiBO_2$) flux, done on the relevant samples without an ashing process. Following microwave-assisted acid digestion or borate fusion, the resultant digests were analyzed by ICP-MS/OES, and the determined values were calculated as concentrations in the original coal or rock samples. ICP-MS techniques for determination of trace elements in coal and associated rock samples have been discussed by Dai *et al.* [18]. Arsenic and selenium in the samples were analyzed using the ICP-MS technique, following the method described by Li *et al.* [24].

Fluorine in the samples was determined using a pyrohydrolysis/fluoride ion-selective electrode technique, following procedures described by Chinese National Standard GB/T 4633-1997 [25]. Mercury in the samples was analyzed using a Milestone DMA-80 Hg analyzer (Milestone, Milan, Italy); the detection limit of Hg is 0.005 ng and the linearity of the calibration is in the range 0–1000 ng.

## 4. Results and Discussion

### 4.1. Coal Characteristics

Table 1 lists the proximate analysis, total and pyritic sulfur contents of selected samples, and the mean maximum vitrinite reflectance value of the coal samples, as well as clay mineralogy obtained from the <2 µm fractions of all coal LTAs and non-coal strata, as discussed by Zhao *et al.* [19]. The Songzao coals have medium to high ash yield and varying sulfur percentages. The coal is mainly semi-anthracite under the classification of the American Society for Testing and Materials (ASTM), based on the volatile matter value and fixed carbon percentages [26].

**Table 1.** Proximate analysis, total and pyritic sulfur (selected samples), mean maximum vitrinite reflectance value of the Songzao coal samples (%, air-dried basis, unless indicated), as well as oriented-aggregate X-ray diffraction (XRD) data for clay minerals (wt % of <2 μm fraction) in all coal low-temperature ash (LTAs) and non-coal strata (all data from Zhao *et al.*, 2013 [19]).

| Sample | Thickness (cm) | Ash Yield | $VM_{daf}$ | $FC_{daf}$ | Total Sulfur | Pyritic Sulfur | $Rv_{max}$ | Clay Minerals | | |
|---|---|---|---|---|---|---|---|---|---|---|
| | | | | | | | | Kao (+ Chl) | I | E |
| dt-7-0 | - | - | - | - | - | - | - | 71 | 7 | 23 |
| dt-7-1 | 15 | 20.6 | 10.0 | 90.0 | 6.73 | 5.00 | 2.22 | 86 | 0 | 14 |
| dt-7-2 | 22 | 18.2 | 10.3 | 89.7 | 4.89 | - | 2.17 | 100 | 0 | 0 |
| dt-7-3 | 12 | 19.4 | 9.4 | 90.6 | 4.62 | 4.10 | 2.36 | 86 | 0 | 14 |
| dt-7-4 | 24 | 23.2 | 8.0 | 92.0 | 3.73 | - | 2.33 | 78 | 4 | 18 |
| dt-7-5 | 20 | 35.4 | 9.5 | 90.5 | 3.21 | - | 2.29 | 80 | 7 | 12 |
| dt-7-6 | - | - | - | - | - | - | - | 69 | 26 | 5 |
| th-k2b-0 | - | - | - | - | - | - | - | 11 | 25 | 64 |
| th-k2b-1 | 30 | 32.3 | 8.1 | 91.9 | 2.97 | - | 2.40 | 76 | 1 | 22 |
| th-k2b-2 | 7 | 43.7 | 6.9 | 93.1 | 0.78 | - | 2.42 | 89 | 0 | 10 |
| th-k2b-3 | - | - | - | - | - | - | - | 22 | 28 | 50 |
| th-k2b-4 | 11 | 36.6 | 11.1 | 88.1 | 10.10 | - | 2.30 | 78 | 0 | 22 |
| th-k2b-5 | - | - | - | - | - | - | - | 66 | 10 | 25 |
| th-k2b-6 | 6 | 36.6 | 10.7 | 89.3 | 1.22 | - | 2.42 | 90 | 0 | 10 |
| th-k2b-7 | - | - | - | - | - | - | - | 40 | 35 | 26 |
| yy-11-0 | - | - | - | - | - | - | - | 27 | 45 | 28 |
| yy-11-1 | 10 | 35.9 | 7.0 | 93.0 | 8.11 | - | 2.09 | 100 | 0 | 0 |
| yy-11-2 | 7 | 24.1 | 8.9 | 91.2 | 8.13 | - | 2.25 | 100 | 0 | 0 |
| yy-11-3 | 7 | 24.2 | 10.0 | 90.0 | 13.39 | 11.0 | 2.31 | 100 | 0 | 0 |
| yy-11-4 | 14 | 23.6 | 9.5 | 90.5 | 10.08 | 8.50 | 2.40 | 100 | 0 | 0 |
| yy-11-5 | 8 | 41.3 | 7.7 | 92.3 | 3.03 | 2.40 | 2.28 | 93 | 0 | 7 |
| yy-11-6 | - | - | - | - | - | - | - | 88 | 0 | 12 |
| yy-11-7 | 11 | 31.5 | 8.4 | 91.6 | 1.64 | - | 2.25 | 93 | 0 | 7 |
| yy-11-8 | - | - | - | - | - | - | - | 41 | 26 | 33 |

VM, volatile matter; FC, fixed carbon; daf, dry ash-free basis; $R_{v,max}$, mean maximum vitrinite reflectance; Kao, kaolinite; Chl, chlorite; I, illite; E, expandable clays.

According to Lyons *et al.* [27] and Spears [28], volcanogenic claystones are referred to as tonsteins or K-bentonites, respectively, when kaolinite or mixed-layer illite/smectite (I/S) exceeds 50% of the respective clay mineral assemblages. Claystone sample th-k2b-3 is thus a K-bentonite, and claystones th-k2b-5 and yy-11-6, which have >50% kaolinite, are tonsteins.

*4.2. Geochemical Associations in Coal Samples*

Major and trace element data for the Songzao coal and non-coal samples are given in Table 2. In general, the Songzao coals have relatively high concentrations of most trace elements compared to the respective averages for worldwide coals [29]. This is especially so for the lithophile elements, which usually show a positive correlation with the ash yield of coal [30,31]. The relationship between the mineralogical data and major-element ash chemistry for the Songzao coals has been previously evaluated by Zhao *et al.* [19].

The geochemical results from the Songzao coal samples were evaluated using cluster analysis, to identify groups of associated trace elements and major element oxides. The major element oxide percentages determined by XRF analysis of the coal ashes were recalculated to give the percentages of those oxides in the whole coal, before the cluster analysis was undertaken.

Hierarchical clustering was performed using Pearson correlation coefficients. The likely organic/mineral affinity of the elements in the Songzao coals is indicated by the statistical correlation of the different trace element concentrations with the ash yield. Elements with a strong inorganic affinity would be expected to show a positive correlation to the ash percentage, and those with a strong organic affinity would show a negative correlation. Elements that are the most strongly correlated are linked first, and then elements or element groups with decreasing correlation, until a complete dendrogram is achieved.

**Table 2.** Major oxide and trace element analyses of the Songzao coal and associated non-coal samples from three seam sections (Major oxides in wt %, trace elements in ppm, unless otherwise indicated. All data are on a whole-coal basis. Major element oxides recalculated from the data by X-ray fluorescence (XRF) analysis. Trace elements determined by ICP-MS/OES analysis).

| Sample | Major Element Oxide (wt %) and Trace Element (ppm) | | | | | | | | | | | | | | | | | | | | | | |
|---|---|---|---|---|---|---|---|---|---|---|---|---|---|---|---|---|---|---|---|---|---|---|---|
| | SiO2 | Al2O3 | TiO2 | Fe2O3 | MgO | CaO | Na2O | K2O | P2O5 | MnO | Li | Be | F | Sc | V | Cr | Co | Ni | Cu | Zn | Ga | Ge | As |
| dt-7-0 | 36.9 | 24.9 | 3.82 | 9.04 | 0.66 | 0.50 | 0.870 | 1.042 | 0.226 | 0.148 | 197 | 4.9 | 354 | 62.0 | 360 | 168 | 65.4 | 101 | 192 | 213 | 44.0 | 2.2 | 9.3 |
| dt-7-1 | 6.9 | 4.8 | 0.30 | 7.35 | 0.06 | 0.44 | 0.079 | 0.073 | 0.048 | 0.006 | 57 | 2.1 | 71 | 10.0 | 47 | 22 | 11.6 | 19 | 35 | 12 | 8.0 | 1.7 | 3.9 |
| dt-7-2 | 5.3 | 4.0 | 0.20 | 5.63 | 0.08 | 1.77 | 0.040 | 0.056 | 0.029 | 0.011 | 57 | 1.6 | 64 | 11.4 | 108 | 23 | 15.8 | 15 | 60 | 16 | 6.1 | 2.0 | 3.0 |
| dt-7-3 | 7.1 | 5.0 | 0.26 | 5.11 | 0.11 | 1.11 | 0.053 | 0.118 | 0.011 | 0.008 | 69 | 1.4 | 41 | 11.5 | 97 | 24 | 20.4 | 23 | 56 | 49 | 8.6 | 1.7 | 2.9 |
| dt-7-4 | 10.2 | 6.8 | 0.68 | 3.99 | 0.13 | 0.52 | 0.114 | 0.255 | 0.015 | 0.004 | 70 | 6.0 | 76 | 18.9 | 157 | 56 | 23.1 | 41 | 98 | 251 | 10.1 | 1.1 | 4.4 |
| dt-7-5 | 17.0 | 7.6 | 0.59 | 4.36 | 0.32 | 3.37 | 0.108 | 0.351 | 0.027 | 0.015 | 107 | 1.4 | 96 | 25.2 | 154 | 50 | 35.5 | 213 | 71 | 34 | 13.0 | 1.1 | 2.5 |
| dt-7-6 | 42.1 | 31.5 | 5.99 | 1.41 | 0.36 | 0.21 | 1.697 | 1.203 | 0.067 | 0.010 | 235 | 4.6 | 309 | 65.5 | 433 | 297 | 22.9 | 55 | 152 | 85 | 51.4 | 2.7 | <1 |
| th-k2b-0 | 41.3 | 16.1 | 1.79 | 11.19 | 0.72 | 0.64 | 0.907 | 1.965 | 0.114 | 0.050 | 53 | 5.6 | 1697 | 47.6 | 279 | 137 | 51.5 | 125 | 88 | 81 | 31.2 | 1.8 | 36.7 |
| th-k2b-1 | 15.6 | 7.7 | 0.51 | 4.20 | 0.72 | 1.81 | 0.181 | 0.343 | 0.037 | 0.020 | 80 | 3.5 | 533 | 33.7 | 335 | 65 | 43.3 | 91 | 334 | 24 | 14.6 | 1.7 | 2.9 |
| th-k2b-2 | 23.1 | 15.2 | 1.79 | 1.00 | 0.25 | 0.75 | 0.290 | 0.395 | 0.056 | 0.006 | 185 | 4.2 | 536 | 48.4 | 457 | 103 | 9.6 | 39 | 305 | 43 | 24.5 | 2.5 | 0.8 |
| th-k2b-3 | 39.6 | 28.6 | 4.09 | 4.17 | 0.53 | 0.25 | 2.016 | 1.300 | 0.051 | 0.025 | 186 | 12.8 | 1014 | 65.8 | 385 | 201 | 32.8 | 140 | 260 | 110 | 53.3 | 4.5 | 1.0 |
| th-k2b-4 | 8.3 | 4.9 | 0.32 | 13.80 | 1.46 | 3.36 | 0.145 | 0.177 | 0.040 | 0.033 | 50 | 2.2 | 326 | 19.5 | 170 | 32 | 18.7 | 76 | 81 | 42 | 9.0 | 1.1 | 5.4 |
| th-k2b-5 | 42.5 | 31.0 | 1.44 | 0.88 | 0.55 | 0.22 | 1.205 | 1.375 | 0.031 | 0.007 | 211 | 9.4 | 1288 | 33.6 | 27 | 8 | 3.5 | 34 | <2 | 45 | 75.3 | 3.1 | <1 |
| th-k2b-6 | 13.7 | 7.2 | 0.44 | 7.24 | 1.71 | 3.70 | 0.080 | 0.174 | 0.067 | 0.040 | 97 | 1.6 | 365 | 26.0 | 160 | 41 | 25.7 | 70 | 69 | 48 | 12.7 | 1.5 | 5.2 |
| th-k2b-7 | 42.2 | 31.4 | 5.28 | 2.39 | 0.65 | 0.24 | 2.188 | 1.359 | 0.050 | 0.014 | 233 | 6.8 | 1184 | 60.9 | 502 | 234 | 21.7 | 111 | 124 | 86 | 59.1 | 4.4 | <1 |
| yy-11-0 | 50.4 | 20.3 | 1.84 | 5.26 | 0.89 | 0.28 | 1.260 | 2.459 | 0.112 | 0.029 | 39 | 5.8 | 1635 | 45.1 | 182 | 39 | 25.6 | 42 | 144 | 119 | 39.8 | 2.0 | 2.3 |
| yy-11-1 | 21.6 | 2.3 | 0.37 | 10.56 | 0.02 | 0.12 | 0.014 | 0.040 | 0.012 | 0.011 | 18 | 17.4 | 58 | 16.6 | 33 | 15 | 15.4 | 28 | 18 | 13 | 5.6 | 3.8 | 4.8 |
| yy-11-2 | 11.0 | 2.1 | 0.28 | 9.17 | 0.06 | 0.61 | 0.022 | 0.041 | 0.007 | 0.005 | 15 | 6.8 | 70 | 10.2 | 24 | 15 | 17.2 | 22 | 17 | 10 | 6.2 | 3.6 | 5.3 |
| yy-11-3 | 5.1 | 2.1 | 0.17 | 13.49 | 0.09 | 0.92 | 0.019 | 0.024 | 0.009 | 0.006 | 14 | 7.8 | 69 | 6.1 | 18 | 11 | 13.1 | 6 | 23 | 11 | 6.7 | 3.4 | 6.9 |
| yy-11-4 | 7.7 | 3.9 | 0.27 | 13.91 | 0.15 | 0.96 | 0.015 | 0.059 | 0.019 | 0.009 | 30 | 5.9 | 98 | 6.7 | 22 | 15 | 107 | 68 | 92 | 11 | 8.5 | 3.7 | 6.9 |
| yy-11-5 | 21.0 | 13.6 | 0.59 | 3.53 | 0.30 | 0.95 | 0.186 | 0.334 | 0.027 | 0.010 | 110 | 4.9 | 530 | 16.5 | 26 | 14 | 4.2 | 10 | 7 | 42 | 26.5 | 3.8 | 2.3 |
| yy-11-6 | 41.6 | 31.8 | 1.42 | 0.37 | 0.36 | 0.20 | 0.575 | 0.809 | 0.018 | 0.008 | 251 | 14.0 | 1225 | 27.7 | 11 | 5 | 9.8 | 6 | <2 | 20 | 62.4 | 7.9 | <1 |
| yy-11-7 | 16.8 | 8.0 | 0.24 | 2.26 | 0.42 | 2.17 | 0.093 | 0.159 | 0.050 | 0.021 | 59 | 7.3 | 343 | 15.1 | 23 | 13 | 3.7 | 16 | 3 | 9 | 12.2 | 2.0 | 1.6 |
| yy-11-8 | 43.0 | 25.6 | 2.94 | 4.81 | 0.96 | 0.31 | 0.739 | 2.949 | 0.073 | 0.024 | 75 | 9.3 | 2587 | 52.8 | 300 | 89 | 30.6 | 65 | 123 | 182 | 55.0 | 3.4 | 16.6 |

Table 2. *Cont.*

| Sample | Se | Rb | Cs | Sr | Y | Zr | Nb | Mo | Ag | Cd | Sn | Sb | Te | Ba | Hf | Ta | W | Hg (ppb) | Tl | Pb | Bi | Th | U |
|---|---|---|---|---|---|---|---|---|---|---|---|---|---|---|---|---|---|---|---|---|---|---|---|
| | | | | | | | | | | | | | | Trace Element (ppm) | | | | | | | | | |
| dt-7-0 | 3.43 | 24.05 | 1.94 | 613 | 56 | 536 | 69.0 | 1.3 | 1.34 | 0.22 | 4.65 | 0.70 | 1.14 | 216.9 | 26.7 | 6.48 | 1.76 | 108 | <0.02 | 16.6 | 0.18 | 20.6 | 6.1 |
| dt-7-1 | 8.63 | 1.67 | 0.18 | 206 | 13 | 56 | 5.6 | 1.7 | 0.12 | 0.09 | 1.06 | 0.47 | 0.16 | 18.2 | 3.1 | 0.50 | 0.37 | 459 | 0.06 | 16.9 | 0.35 | 6.3 | 2.5 |
| dt-7-2 | 9.17 | 1.51 | 0.18 | 213 | 18 | 49 | 3.8 | 1.0 | 0.10 | 0.10 | 1.00 | 0.53 | 0.23 | 14.6 | 3.2 | 0.48 | 0.21 | 450 | 0.03 | 13.8 | 0.33 | 6.7 | 1.9 |
| dt-7-3 | 12.32 | 3.16 | 0.44 | 133 | 17 | 67 | 10.0 | 2.4 | 0.16 | 0.46 | 1.79 | 1.05 | 0.29 | 17.6 | 4.3 | 0.64 | 1.09 | 852 | 0.02 | 16.5 | 0.48 | 10.3 | 2.3 |
| dt-7-4 | 15.30 | 6.06 | 0.73 | 120 | 23 | 99 | 10.4 | 1.7 | 0.22 | 0.58 | 1.65 | 0.86 | 0.11 | 36.6 | 6.2 | 0.96 | 0.54 | 1102 | 0.04 | 46.6 | 0.32 | 10.8 | 2.6 |
| dt-7-5 | 13.05 | 8.48 | 1.07 | 292 | 32 | 221 | 12.4 | 1.5 | 0.35 | 0.22 | 3.31 | 1.14 | 0.20 | 45.1 | 9.1 | 1.34 | 0.49 | 679 | 0.06 | 10.1 | 0.48 | 11.8 | 3.7 |
| dt-7-6 | 6.46 | 21.32 | 1.17 | 841 | 49 | 611 | 107.5 | 1.5 | 1.69 | 0.22 | 5.85 | 0.78 | 1.06 | 266.2 | 30.9 | 8.25 | 2.24 | 855 | <0.02 | 6.8 | 0.13 | 25.1 | 5.6 |
| th-k2b-0 | 8.21 | 45.79 | 2.94 | 719 | 38 | 521 | 73.0 | 20.4 | 1.47 | 0.34 | 5.96 | 1.27 | 0.80 | 173.3 | 28.5 | 7.64 | 0.97 | 414 | 0.03 | 21.7 | 0.30 | 25.2 | 10.2 |
| th-k2b-1 | 10.81 | 7.22 | 1.00 | 285 | 77 | 162 | 20.2 | 2.0 | 0.81 | 0.25 | 4.77 | 0.50 | 0.36 | 35.0 | 9.9 | 1.83 | 0.58 | 906 | 0.07 | 22.8 | 0.38 | 17.6 | 4.2 |
| th-k2b-2 | 5.28 | 8.54 | 1.09 | 301 | 42 | 283 | 51.0 | 1.2 | 0.92 | 0.15 | 4.12 | 0.39 | 0.18 | 57.3 | 17.2 | 4.01 | 1.56 | 196 | 0.03 | 10.0 | 0.42 | 23.0 | 4.6 |
| th-k2b-3 | 10.64 | 22.92 | 1.44 | 1134 | 37 | 581 | 82.8 | 2.2 | 1.60 | 0.31 | 6.18 | 1.15 | 0.58 | 214.5 | 30.4 | 7.35 | 3.17 | 369 | 0.05 | 30.2 | 0.17 | 21.5 | 4.7 |
| th-k2b-4 | 15.62 | 3.68 | 0.53 | 311 | 70 | 334 | 41.6 | 3.5 | 0.76 | 0.39 | 4.94 | 0.86 | 0.29 | 20.9 | 15.1 | 1.42 | 0.80 | 781 | 0.03 | 23.3 | 0.41 | 8.3 | 4.3 |
| th-k2b-5 | 3.29 | 23.10 | 1.65 | 825 | 37 | 857 | 355.6 | 1.1 | 5.16 | 0.34 | 13.16 | 0.75 | 0.70 | 183.8 | 66.8 | 31.71 | 1.87 | 457 | <0.02 | 9.2 | 0.63 | 76.8 | 9.3 |
| th-k2b-6 | 11.93 | 3.97 | 0.59 | 432 | 101 | 665 | 38.1 | 6.0 | 0.76 | 0.34 | 5.69 | 0.78 | 0.41 | 23.7 | 26.6 | 2.04 | 1.16 | 361 | 0.03 | 11.7 | 0.39 | 13.7 | 5.0 |
| th-k2b-7 | 3.07 | 23.97 | 1.60 | 971 | 53 | 739 | 106.6 | 2.7 | 1.67 | 0.29 | 6.93 | 0.72 | 0.27 | 204.1 | 36.1 | 7.79 | 3.07 | 148 | 0.03 | 10.6 | 0.19 | 24.7 | 5.0 |
| yy-11-0 | 9.70 | 43.50 | 2.55 | 729 | 41 | 531 | 88.7 | 1.5 | 1.47 | 0.22 | 5.71 | 0.92 | 0.28 | 167.5 | 30.5 | 7.00 | 2.89 | 196 | 0.03 | 22.8 | 0.07 | 25.8 | 4.4 |
| yy-11-1 | 7.88 | 0.73 | 0.14 | 72 | 13 | 106 | 7.3 | 3.2 | 0.17 | 0.09 | 1.44 | 0.43 | 0.15 | 9.1 | 5.6 | 0.91 | 0.70 | 917 | 0.50 | 8.9 | 0.24 | 4.9 | 2.9 |
| yy-11-2 | 12.05 | 0.76 | 0.14 | 99 | 14 | 85 | 6.4 | 3.5 | 0.14 | 0.06 | 1.25 | 0.39 | <0.1 | 9.0 | 4.6 | 0.77 | 0.20 | 764 | 0.46 | 7.8 | 0.18 | 5.8 | 1.8 |
| yy-11-3 | 14.01 | 0.51 | 0.10 | 112 | 16 | 68 | 4.9 | 5.8 | 0.12 | 0.12 | 0.87 | 0.43 | <0.1 | 5.2 | 3.7 | 0.64 | 0.38 | 599 | 0.37 | 20.0 | 0.09 | 4.0 | 1.3 |
| yy-11-4 | 12.57 | 1.11 | 0.20 | 95 | 22 | 123 | 11.9 | 4.1 | 0.21 | 0.11 | 1.26 | 0.41 | <0.1 | 7.4 | 6.6 | 1.29 | 0.13 | 861 | 0.57 | 16.6 | 0.13 | 7.1 | 2.6 |
| yy-11-5 | 7.06 | 6.04 | 0.99 | 239 | 56 | 537 | 134.2 | 2.5 | 2.58 | 0.28 | 8.55 | 0.55 | 0.18 | 37.5 | 32.2 | 12.69 | 1.55 | 443 | 0.11 | 21.0 | 0.30 | 27.2 | 6.5 |
| yy-11-6 | 0.24 | 12.20 | 1.98 | 418 | 40 | 529 | 403.8 | 2.5 | 6.28 | 0.13 | 14.48 | 0.78 | 0.43 | 84.4 | 31.7 | 33.27 | 4.22 | 93 | <0.02 | 1.2 | 0.17 | 29.2 | 4.8 |
| yy-11-7 | 6.28 | 3.72 | 0.65 | 204 | 75 | 773 | 57.4 | 1.8 | 1.12 | 0.16 | 6.12 | 0.36 | 0.23 | 19.6 | 37.8 | 3.82 | 0.24 | 203 | 0.02 | 8.2 | 0.11 | 23.4 | 7.8 |
| yy-11-8 | 10.85 | 57.98 | 4.01 | 794 | 69 | 1068 | 155.7 | 3.4 | 2.46 | 0.40 | 8.60 | 1.01 | 0.47 | 174.5 | 50.3 | 12.22 | 7.57 | 175 | 0.09 | 30.0 | 0.24 | 37.7 | 10.3 |

Major element data from Zhao *et al.* [19].

Associations of elements in the Songzao coals are broadly indicated by the resulting dendrogram (Figure 3). The six main groups, and also the statistical correlation coefficients between selected elements and ash yield, are shown in Table 3. Apart from the inter-correlation among elements in the same group, each group may also include elements of different sub-groups that have different correlations with $Al_2O_3$, $CaO$, or the abundance of particular minerals in the coals. The possible modes of occurrence of the different elements can be inferred based on the correlation of their concentrations with particular mineralogical abundances, with both of the element and mineral contents recalculated to a whole-coal basis prior to the correlation analysis.

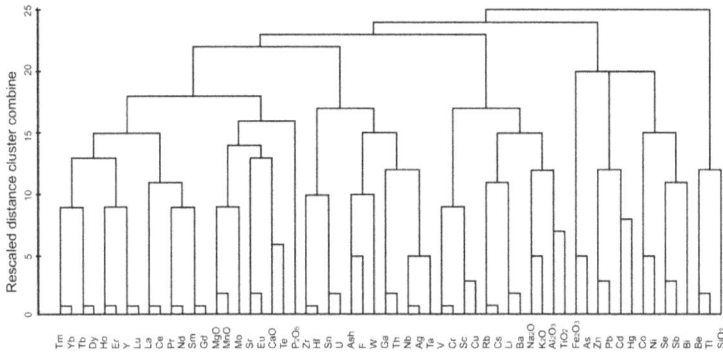

**Figure 3.** Dendrogram developed from cluster analysis on the geochemical data of the coals from three seam sections in the Songzao Coalfield (cluster method, centroid clustering; interval, Pearson correlation; transform values, maximum magnitude of 1).

**Table 3.** Broad classification of elements according to the results from cluster analysis, and also correlation coefficients ($R$, in parentheses) between concentrations of individual elements (E) and $Al_2O_3$, $CaO$ or the sum of pyrite and marcasite in the coals.

| Group | Element (Correlation Coefficient) | Element or Mineral to which the Elements in the Group Correlate |
|---|---|---|
| Group A | $Al_2O_3$ (1), $TiO_2$ (0.79), Rb (0.82), Cs (0.86), Li (0.92), Ba (0.87), $Na_2O$ (0.9), $K_2O$ (0.86), Sc (0.73), Cr (0.63), V(0.59), Cu (0.47) | $R(E-Al_2O_3)$ |
| Group B | Hf (0.63), Sn (0.72), U (0.69), F (0.8), W (0.74), Ga (0.97), Th (0.91), Nb (0.75), Ag (0.76), Ta (0.73), Zr (0.5) | $R(E-Al_2O_3)$ |
| Group C | Tm (0.44), Yb (0.47), Tb (0.52), Dy (0.5), Ho (0.47), Er (0.49), Y (0.46), Lu (0.43), La (0.59), Ce (0.57), Pr (0.55), Nd (0.53), Sm (0.51), Gd (0.59) | $R(E-Al_2O_3)$ |
| Group D | CaO (1), MgO (0.8), MnO (0.86), Sr (0.77), Eu (0.42), Te (0.66), $P_2O_5$ (0.52), Mo (0.19) | $R(E-CaO)$ |
| Group E | $Fe_2O_3$ (0.95), As (0.89), Hg (0.32), Zn (−0.32), Pb (−0.06), Cd (−0.37), Co (0.35), Ni (−0.17), Se (0.43), Sb (−0.3), Bi (−0.58) | $R(E-(Py + Mar))$ |
| Group F | Be (−0.28), Tl (−0.16), $SiO_2$ (0.84) | $R(E-Al_2O_3)$ |

Group A includes $TiO_2$, V, Cr, Sc, Cu, Rb, Cs, Li, Ba, $Na_2O$, $K_2O$ and $Al_2O_3$. With the exception of Sc, Cr, V and Cu, elements in Group A are all strongly correlated with $Al_2O_3$, with high correlation coefficients ($R > 0.79$). On the other hand, slightly lower correlation coefficients generally exist between these elements and the ash yield. Most of the elements in this group probably have a common source.

Group B includes Zr, Hf, Sn, U, F, W, Ga, Th, Nb, Ag and Ta. With the exception of Ga and Th, these elements have relatively strong correlations with $Al_2O_3$, with correlation coefficients in the range of 0.5–0.79. Ga and Th stand out in this group, as they have greater affinity with $Al_2O_3$ than other elements ($R = 0.97$ and 0.91, respectively). The comparison of Ga against $Al_2O_3$ is shown in Figure 4A. Ga and Th are clustered in this group because of their close association with other elements in the group, for example Th with U ($R = 0.87$), and Ga with Nb ($R = 0.83$). The elements in this group also probably have a common source.

Group C includes all the REE except Eu. The correlation coefficients between the REE in this group and the $Al_2O_3$ concentration in the coals are in the range of 0.44 to 0.59. Higher correlation coefficients generally exist between these elements and the ash yield ($R = 0.6$–0.68). This may indicate that the REE are mainly moderately associated with ash yield in the Songzao coals.

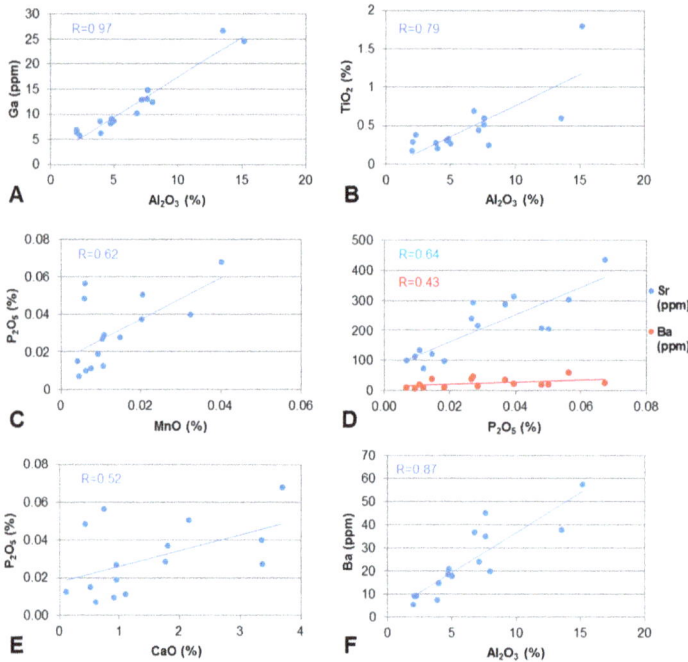

**Figure 4.** Correlations between selected elements in the Songzao coal samples: (**A**) Ga against $Al_2O_3$; (**B**) $TiO_2$ against $Al_2O_3$; (**C**) $P_2O_5$ against MnO; (**D**) Sr and Ba against $P_2O_5$; (**E**) $P_2O_5$ against CaO; and (**F**) Ba against $Al_2O_3$. Relevant correlation coefficients ($R$), obtained from linear regression analysis, are also shown in each case.

Group D includes MgO, MnO, Mo, Sr, Eu, CaO, Te and $P_2O_5$. With the exception of Mo, all the elements in this group are strongly or relatively strongly correlated with CaO, and thus have a carbonate affinity. $P_2O_5$ and Sr are probably associated with aluminophosphate minerals, which were detected by EDS analysis in the coals [19]. Molybdenum, although clustered in this group, also has an affinity with the sum of the abundances of pyrite and marcasite (correlation coefficient of 0.66), expressed on a whole-coal basis.

Group E includes $Fe_2O_3$, As, Zn, Pb, Cd, Hg, Co, Ni, Se, Sb and Bi. These elements have either no or a negative correlation with $Al_2O_3$ ($R$ in the range of $-0.73$ to 0.39) or with ash yield ($R$ in the range of $-0.36$ to 0.3). The elements in this group are mainly chalcophile elements. However, only As is significantly correlated with pyrite, having a correlation coefficient of 0.89. Hg and Se are weakly correlated with the sum of the abundances of pyrite and marcasite in the coals, with correlation coefficients of 0.32 and 0.43, respectively.

Group F includes Be, Tl and $SiO_2$, which do not have obvious correlation with each other. No overall correlation exists between Be and quartz (or $SiO_2$), although an elevated Be concentration is present in one quartz-rich coal sample (yy-11-1).

*4.3. Associations of Major Elements in the Coals*

The major elements in the Songzao coals are dominated by $SiO_2$, $Al_2O_3$ and $Fe_2O_3$ (Table 2). The main carriers of these elements are quartz, clay minerals and pyrite [19]. The concentration of $TiO_2$ is relatively high in the Songzao coals. EDS analysis indicates that $TiO_2$ occurs in the clay minerals, as anatase crystals, and as finely disseminated submicron particles of anatase or Ti-bearing phases with possibly poor crystallinity [19]. A positive correlation exists between $TiO_2$ and $Al_2O_3$ ($R = 0.79$) in the ash chemistry (Figure 4B). The $TiO_2/Al_2O_3$ ratio of the Songzao coal ashes ranges from 0.03 to 0.16. As discussed by Ward *et al.* [32], part of the $TiO_2/Al_2O_3$ ratio might be attributed to the incorporation of Ti in the aluminosilicate (e.g., kaolinite) structure. The intimate association of anatase (and fine Ti-bearing phases) and clay minerals in the Songzao coals, as noted in some cases by Zhao *et al.* [19], indicates that kaolinite and $TiO_2$ may have been co-precipitated. $TiO_2$ in some of the coals also occurs as separate masses of anatase replacing probable volcanic components [19]. The high proportions of $TiO_2$ in the Songzao coals may partly reflect high Ti contents in the sediment input to the original peat swamp, derived from the mafic basaltic rocks of the Kangdian Upland on the western margin of the coal basin [18]. The occurrence of the volcanic components replaced by $TiO_2$ also suggests that Ti can be introduced, or at least moved around, in solutions permeating through the peat/coal after deposition, perhaps as a more soluble $Ti(OH)_4$ component or as organo-metallic complexes.

$Na_2O$ shows great variability in the Songzao coals. Several coal and non-coal samples in the Datong and Tonghua sections are high in $Na_2O$, which is mainly attributed to the presence of albite. High proportions of $Na_2O$ in a few of the coal samples are also attributed to the presence of Na-rich illite and Na-I/S, rather than the more common K-illite and K-I/S in the Songzao coals [19].

High correlations of MnO-CaO ($R = 0.86$) and MnO-MgO (0.96) indicate that the MnO in the Songzao coals may be closely associated with carbonates (calcite, dolomite and ankerite). Dai *et al.* [17] noted the presence of fine-grained alabandite (MnS) (about 1 μm) of hydrothermal origin in the Songzao No. 11 coal, which was probably the most important carrier of manganese in those samples. However, alabandite was not observed in the Songzao coals during the present study. Siderite is also present in some of the coals [19], but the concentration of Mn in the siderite was below the detection limit of the SEM-EDS system used for that study.

$P_2O_5$ shows a positive correlation with MnO (Figure 4C). As discussed further below, $P_2O_5$ also shows significant correlations with Sr (Figure 4D) and Ca (Figure 4E), but no correlation with Ba (Figure 4D). This indicates that $P_2O_5$ mainly occurs in aluminophosphates of the goyazite and probably crandallite groups. Although no correlation exists between $P_2O_5$ and Ba, the presence of gorceixite was indicated by the SEM study [19]. Ba in the Songzao coals may have additional sources other than gorceixite (e.g., barite), which were not identified by either XRD or microscope studies.

*4.4. Selected Elements in the Roof, Floor and Claystone Samples*

The concentration of $TiO_2$ is as high as 4.09% in the K-bentonite band of the Tonghua section (sample th-k2b-3), much higher than that in the tonsteins (samples th-k2b-5 and yy-11-6) of the Songzao coal seam. The $TiO_2/Al_2O_3$ ratio has been compared with that found in volcanic rocks to identify sediments with a possible volcanic component in the coal-bearing sequences, or to indicate the possible composition of the parent magma in many studies [18,33–37]. In the study of Spears and Kanaris-Sotiriou [38], tonsteins with $TiO_2/Al_2O_3$ values of <0.02 and >0.07 are grouped to indicate parent magmas of acid and mafic compositions, respectively; those with values in between are thought to represent intermediate ash materials.

A comparison of the $TiO_2$ and $Al_2O_3$ percentages in the Songzao non-coal samples is potted in Figure 5A. On this diagram, the $TiO_2$-rich bentonite, sample th-k2b-3, plots in the mafic field while the other two tonsteins plot between the two lines indicating $TiO_2/Al_2O_3$ values of 0.02 and 0.07. The key parameters for the claystones were also investigated using the magma source discrimination diagram of Winchester and Floyd [39] (Figure 5B). On this diagram, the bentonite (th-k2b-3) falls in the alkali basalt field, while the two tonsteins fall in the trachyte and basanite/nephelinite fields.

As discussed above, Zr, Nb, and Y are relatively mobile in the volcanic ash layers investigated, potentially making the use of discrimination diagrams such as that of Winchester and Floyd [39] less reliable as provenance indicators. The actual $Zr/TiO_2$ ratios for the claystones may also be less than those shown in the plot, due to possible contamination of the samples from the zirconia grinding mill used during sample preparation. This, coupled with decreased Zr, Nb and Y concentrations in the studied claystones due to leaching, may therefore have affected to some extent the fields in which the samples plot. Nevertheless, mafic characteristics for claystone th-k2b-3 and intermediate characteristics for claystones th-k2b-5 and yy-11-6 are confirmed by their $TiO_2/Al_2O_3$ ratios (Figure 5A). It is thus tentatively concluded that the bentonite and the two tonsteins were derived from mafic and alkali ashes, respectively.

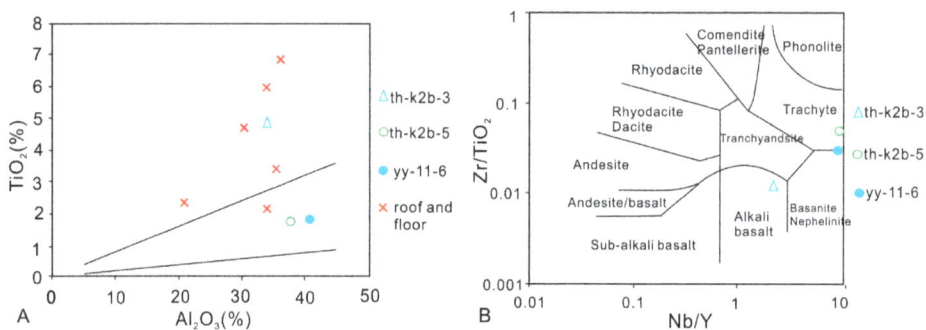

**Figure 5.** Plots of elements for the Songzao non-coal samples. (**A**) Comparison of $TiO_2$ and $Al_2O_3$ concentrations. The upper and lower diagonal lines represent $TiO_2/Al_2O_3$ values of 0.07 and 0.02, respectively. (**B**) Plot of $Zr/TiO_2$ against $Nb/Y$ ratios using the magma source discrimination diagram of Winchester and Floyd [39].

As indicated in Figure 5A, the roof and floor samples generally have high $TiO_2$ contents. This reflects a high Ti content in the detrital sediment input from the source region, probably mafic basalts from the Kangdian Upland on the western margin of the coal basin.

Two groups of elements are enriched in the mafic K-bentonite and the alkali tonsteins, respectively, not only relative to the coal plies, but to the roof and floor samples in the respective seam sections (Figure 6). The high field strength elements, including Nb, Ta, Zr, Hf, REE and Y, are enriched in the alkali tonsteins. Another group of elements (V, Cr, Co, Cu and Ni), most of which are transition elements, are concentrated in the mafic K-bentonite band (Figure 6).

The relatively immobile elements in tonsteins are considered to be present in resistate phases, such as ilmenite (Ti) and zircon (Zr and Hf), and diagenetic minerals such as kaolinite (Al) and anatase (Ti) [40]. Spears and Rice [34] suggested that Ga and Th can be accommodated in kaolinite, and U and probably Y in zircon. Zircon in coal may also contain Nb, Ta and Th [6]. These minerals, if incorporated as volcanogenic components within the coal, may be responsible for enrichment of the immobile elements.

High concentrations of Nb, Ta, Zr, Hf and REE have been reported in alkali tonsteins, relative to silicic and mafic tonsteins [10,18]. Dai *et al.* [17] suggested that alkaline volcanic ash is responsible for the enrichment of elements such as Nb, Zr, Ga, Hf and REE in the No. 11 coal of the Songzao Coalfield, where only clayey micro-sized bands were observed.

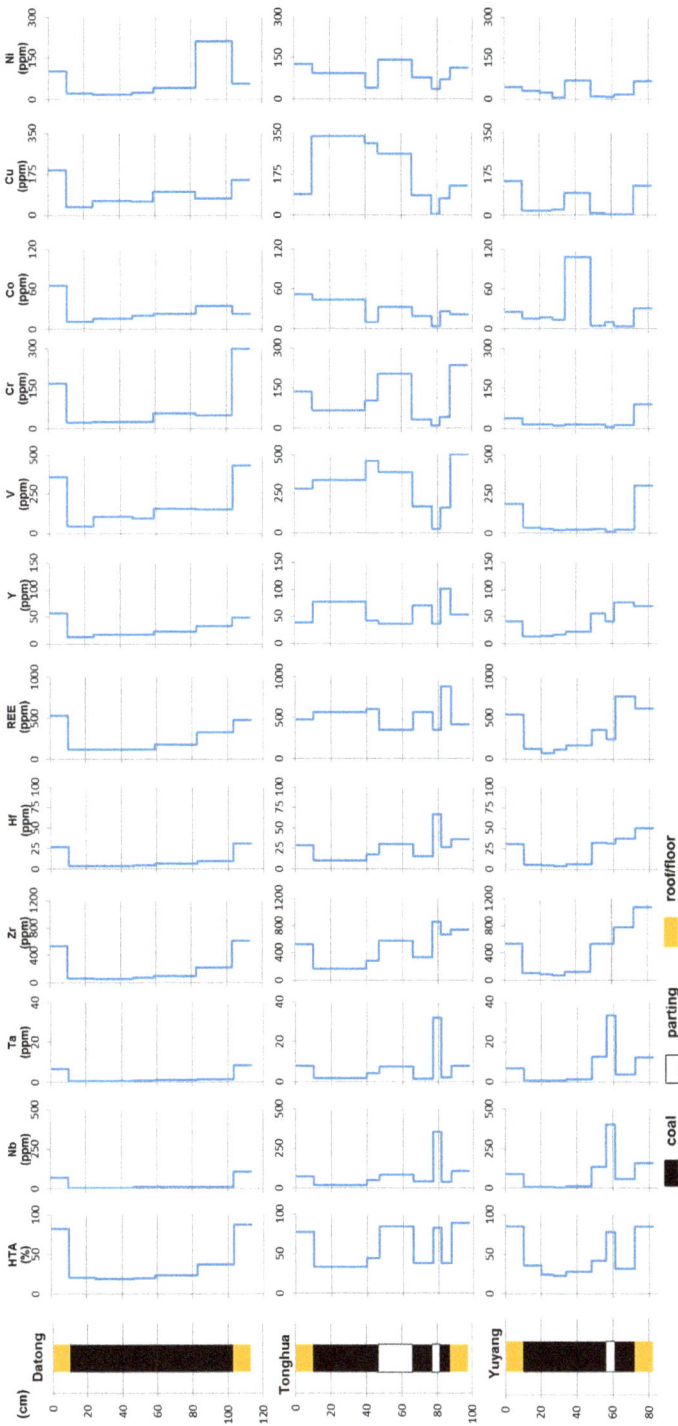

**Figure 6.** Plots showing vertical variation of percentage of high temperature ash and selected trace element concentrations in the Songzao seam sections.

*4.5. Selected Trace Elements in Coal Samples*

As indicated by the cluster analysis, many elements are associated with $Al_2O_3$ or the ash yield of the coals with different degrees of affinity. Apart from the ash yield, other factors may also control the concentrations of different trace elements in the different coal seams.

4.5.1. Sr and Ba

The concentration of $P_2O_5$ shows positive correlations with the concentrations of Sr and Ba in the coal samples ($R$ = 0.64 and 0.43 respectively). The Sr and Ba in the coals probably occur in aluminophosphates (goyazite and gorceixite), also confirmed by SEM-EDS analysis for some coal samples [19]. The moderate correlation between Ba and $P_2O_5$ may indicate that some of the Ba had sources other than gorceixite. Some Ba also occurs in authigenic rhabdophane, as indicated by EDS analysis [19]. A significant positive correlation, however, is shown in the plot of Ba against $Al_2O_3$ (Figure 4F). This indicates that a large proportion of the Ba in the Songzao coals is associated with aluminosilicates, probably clay minerals, or substituting for calcium in the plagioclase structure, with only minor Ba occurring in the aluminophosphates (mainly gorceixite and rhabdophane). Alternatively, the correlation between Ba and $Al_2O_3$ may indicate a common source.

Additionally, the occurrence of goyazite-, gorceixite- or crandallite-group minerals and elevated concentrations of F, P and Sr in other coals and associated rocks is often indicative of volcanic input [41–43].

4.5.2. V, Cr, Cu, Co and Ni

The concentrations of V, Cr, Cu, Co and Ni are generally high in the Datong and Tonghua coals. Cu, for example, has the highest concentration in the coals, being 457, 103 and 334 ppm, respectively. These elements are also correlated with each other. The correlations among V, Cr and Cu are especially strong, with $R$ being 0.96 for V and Cr, and 0.92 for V and Cu (Figure 7A). A significant correlation between V and Cr was also observed in coals from the Gunnedah Basin, Australia, in a study by Ward *et al.* [32], who suggested that a common magmatic source may be reflected. Glick and Davis [44] suggested that Cr may have an association with illite in a large number of US coals.

Copper in the Songzao coals has a moderate correlation with $Al_2O_3$ ($R$ = 0.47). However, along with V and Cr, Cu shows a significant positive correlation with the sum of illite and I/S (Figure 7B) ($R$ = 0.88, 0.84 and 0.76 respectively). The correlations are apparent in those samples where the proportion of illite +I/S is greater than 5% (whole coal basis). Siroquant may have difficulties in quantification of illite and I/S when in small proportions (<5%, whole coal basis), and this may account for the poor correlations between illite + I/S and these elements.

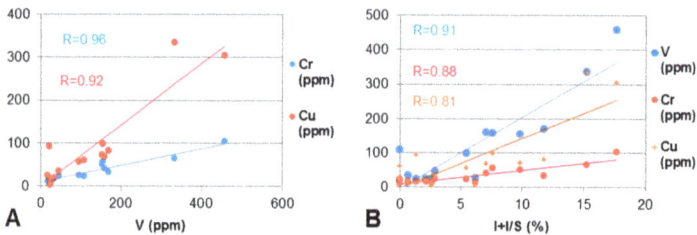

**Figure 7.** Correlations between selected elements in the Songzao coal samples: (**A**) Cr and Cu against V; and (**B**) V, Cr and Cu against the sum of illite and I/S, on a whole-coal basis. Relevant correlation coefficients ($R$), obtained from linear regression analysis, are also shown in each case.

Cu in coal has been reported to be associated with sulfides [45–47] and carbonates [48], and occassionally to occur as Cu sulfides and oxides [49]. Organically bound Cu has also been suggested in some coals [50,51]. However, neither of these associations is indicated in the Songzao coals. Cu is not only associated with illite and I/S, but also appears to have a similar pattern of variation to V and Cr (Figure 7).

The correlations between V, Cr and Cu and the sum of illite and I/S may indicate a common source of clastic material supplied to the coal basin, which was in turn probably derived from the mafic basaltic rocks of the Kangdian Upland in southwestern China. Vanadium, Cr and Cu are especially concentrated in the Tonghua coals, with the coals of the upper section being more enriched in these elements than those in the lower section. This is mostly likely related to the underlying mafic bentonite. Leaching of the original mafic ash may have led to higher concentrations of these elements in the underlying coal than in coals without such an influence or affected by alkaline ashes. As physio-chemical conditions change, re-precipitation of these elements within the overlying coal layers may also occur, due to the leaching of these elements from the mafic bentonite by upwelling fluids.

### 4.5.3. Chalcophile Elements

The plot of As against the proportion of iron sulfides (the sum of pyrite and marcasite) shows a relatively strong and consistent correlation (Figure 8A). The correlation trend indicates the presence of approximately 0.3 ppm of As per 1% of iron sulfides, or 30 ppm of As present in the iron sulfides themselves. This ratio is comparable with that in the high sulfur Greta coals of the Sydney Basin, Australia (10 ppm in pyrite) (unpublished data), but much lower than that in the coals from the Gunnedah Basin, Australia (1000 ppm in pyrite) as discussed by Ward *et al.*, [32]. Further evaluation of Figure 8A indicates that the correlation line intersects the $y$ (As) axis, indicating a value of 1.4 ppm As when the pyrite concentration is zero. This may be due to the presence of small proportions of organically-associated As in the coal samples. Alternatively, this may also be due to the presence of arsenate ($AsO_4^{3-}$), formed during pyrite oxidation during storage under ambient conditions. The presence of arsenate has been indicated by XAFS studies in a range of US bituminous coals [52,53].

As indicated in Figure 8B, a positive correlation also exists between Mo and the total iron sulfides in the coals, with a correlation coefficient of 0.66. Figure 8B shows the presence of approximately 0.2 ppm of Mo with 1% of iron sulfides, or 20 ppm of Mo present in the iron sulfides themselves. An association of Mo with iron sulfides is observed in some coal deposits [54,55] but not in others [32]. LA-ICP-MS studies of coals from the Black Warrior Basin, USA, by Diehl *et al.* [56], indicated that varying but significant concentrations of Mo (<10–582 ppm) and As (<100–27400 ppm) are present in pyrite within the coals of that basin. The results of the present study show an overall consistent relationship between As and Mo to the proportion of iron sulfides in the Songzao coals. However, there appears to be some variation in Mo concentration in these sulfides, as expressed by the scatter of individual points on the graph.

Mercury and Se are only broadly correlated with the iron sulfide content (Figure 8C,D). Both of these elements have been reported to be commonly associated with pyrite in other coals, but the degree of scatter found in the present study is relatively high. The concentration of Hg in the individual coal plies of the Songzao Coalfield is mostly in the range of 0.3 to 0.9 ppm, with the highest value being 1.1 ppm. This is much higher than the average Hg concentration of Chinese coals, which is 0.163 ppm [57]. A relatively high correlation coefficient exists between Tl and iron sulfides (Figure 8E). However, this relationship is dominated by the presence of several samples with very high iron sulfide proportions (>12.4%, on a whole coal basis). The same iron sulfide-rich coals contain the highest concentrations of Ge (Figure 8F), despite poor correlation between Ge and the iron sulfides in other coals of the samples studied.

Other chalcophile elements, such as Sb, Pb, Co, Ni, Cu and Zn, show either poor or negative correlations with total iron sulfides, although different degrees of positive correlation have also been reported in other coal deposits. This may reflect the variation of the concentrations of these elements in individual pyrite/marcasite (including Tl and Ge), or simply poor retention of those elements in the pyrite/marcasite of the Songzao coals. Co, Ni, Cu and Zn could also be associated with the clay minerals. For example, Cu in the Songzao coals is correlated with the sum of illite and I/S, as discussed above.

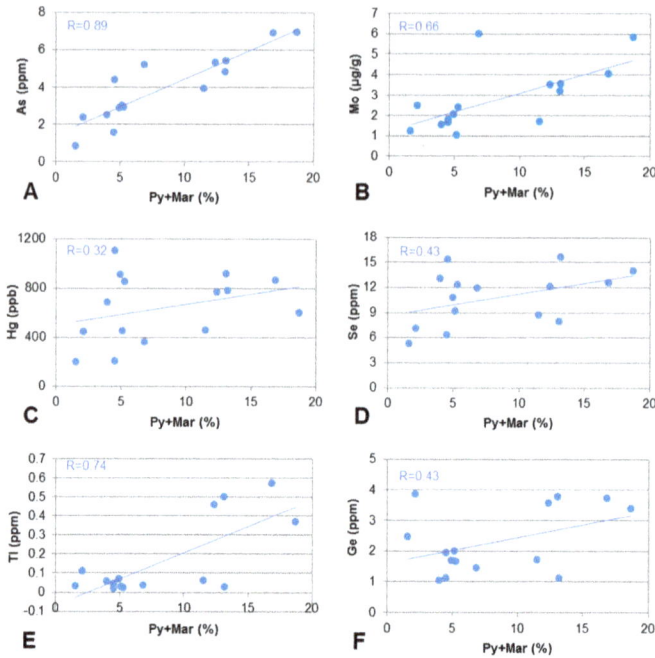

**Figure 8.** Correlation of selected elements (As, Mo, Hg, Se, Tl and Ge) with the sum of pyrite and marcasite in the Songzao coal samples, on a whole-coal basis: (**A**) As against iron sulfides (the sum of pyrite and marcasite); (**B**) Mo against iron sulfides; (**C**) Hg against iron sulfides; (**D**) Se against iron sulfides; (**E**) Tl against iron sulfides; and (**F**) Ge against iron sulfides. Relevant correlation coefficients (*R*), obtained from linear regression analysis, are also shown in each case.

### 4.5.4. Nb, Ta, Zr, Hf, Th, U and REE

Niobium, Ta, Zr, Hf, Th and U, also referred to as high field strength elements, are enriched in all the coals in the Tonghua section and two coals near the alkali tonstein band in the Yuyang section (Table 2). The variation in Zr, however, may also be derived from contamination of the coals from the zirconia grinding mill used during sample preparation.

The concentrations of Nb, Ta, Zr, Hf, REE and Y are notably higher in the coal ply (yy-11-5) overlying the alkali tonstein in the Yuyang section, and in the coal plies (th-k2b-6 and yy-11-7) under the alkali tonsteins in the Tonghua and Yuyang sections (Figure 6). It is also worth noting that the REE and Y concentrations in both the overlying and underlying coal plies are higher than those in the tonsteins of the Tonghua and Yuyang sections.

Elevated concentrations of trace elements in coals near tonsteins are relatively common and volcanic minerals or volcanic glasses are probably the sources for the elevated element concentrations in such coals [6]. For example, the enrichment of Zr, Nb, Th and Ce in coals directly above and below tonsteins in the C coal bed of the Emery Coal Field, Utah was reported by Crowley *et al.* [6]. Crowley *et al.* [6] suggested that the mechanism of enrichment for some elements in the coal was leaching of volcanic ash by groundwater and subsequent incorporation in organic matter or authigenic minerals, or, alternatively, the incorporation of volcanic ash in the original peat material. Leaching of the volcanic ash by ground water was used by Hower *et al.* [7] to explain the high concentrations of Zr, Y and REE in the coal directly underlying a tonstein in the Fire Clay coal bed, Kentucky. Similar observations suggesting enrichment of elements due to leaching of volcanic ash beds in coals were also made by Wang [58].

Although Nb, Ta and the REE are relatively immobile in most low-temperature environments, Zielinski [40] suggested that the high leaching efficiency in acid coal-forming swamps may explain the significant mobility of these elements during the alteration of volcanic ash to tonsteins. Zielinski [40] indicated that Zr and Hf were the most resistant to mobilization. In the present study, although Nb and Ta are significantly enriched in the alkali tonsteins, no significant elevation of their concentrations was observed in the adjacent coal samples (Figure 6).

The positive correlations between the high field strength elements and $Al_2O_3$ in coal most likely indicate a common source, namely the original volcanic ash. In the present study, Nb was also detected in anatase by EDS in the Tonghua coals, and in fine Zr-phases (<0.5 μm), probably zircon, in the tonsteins [19]. The fine Zr-phases in the tonsteins are probably authigenic, and similar phases may also occur in the coal samples [19]. If such material is present in the coals, it may possibly have been overlooked during SEM examination due to the fine particle size. Primary minerals of volcanogenic origin (e.g., zircon) mainly made-up of these elements, however, were not observed in the coals of the present study, and thus are probably not the main carrier of the elements in question.

Although the presence of REE mineral veins in a Tonghua coal (th-k2b-4) [19] may lead to high REE concentrations, the occurrence of such minerals may not account for the elevated REE in all the coal samples adjacent to the tonsteins. The main carrier of REE in the coals is fine-grained authigenic REE-phosphates, probably rhabdophane, which is also indicated by SEM data [19].

### 4.6. Distribution and Affinity of REE and Y

The REY content ($\sum REE + Y$) of the Songzao coals from the three sections ranges between 70 and 874 ppm (Table 4). The maximum REY concentration in each section appears to occur in the coal immediately above the floor strata (Table 4). The highest REY concentration in the Datong section is in the lowermost coal ply, which may be related to its high ash yield; in both the Tonghua and Yuyang sections, the coal plies near alkali tonsteins have elevated REY contents. However, all the altered volcanic ash bands have REE and Y concentrations lower than those in the respective overlying and underlying coal samples.

The correlation coefficients between the concentrations of individual REY and the ash yield are mainly in the range of 0.6 to 0.7, except for Eu, which has a correlation coefficient with ash of 0.58 (Figure 9). The REE exhibit less significant correlations with $Al_2O_3$ ($R = 0.4$–0.6). This indicates that the REE generally have similar mineral affinities. Ce may have a slightly greater organic affinity than the other REE. A stronger organic affinity for heavy REE has been observed in some coal deposits [59,60], while the light REE exhibit greater organic affinity in other coals [61].

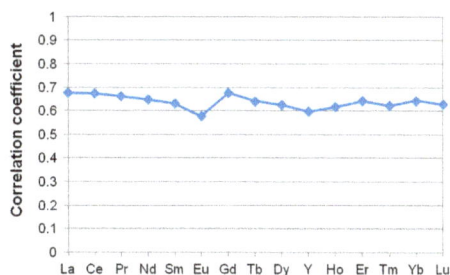

**Figure 9.** Correlation coefficients between mean individual REE and Y with ash yield in the Songzao coal samples.

Table 4. Rare earth elements and Y (REY) in the Songzao coal samples and associated strata (REY concentrations in ppm, on whole-coal basis).

| Element | Sample | | | | | | | | | | | | | | |
|---|---|---|---|---|---|---|---|---|---|---|---|---|---|---|---|
| | dt-7-0 | dt-7-1 | dt-7-2 | dt-7-3 | dt-7-4 | dt-7-5 | dt-7-6 | th-k2b-0 | th-k2b-1 | th-k2b-2 | th-k2b-3 | th-k2b-4 | th-k2b-5 | th-k2b-6 | th-k2b-7 |
| La | 85.56 | 20.33 | 19.49 | 18.38 | 32.03 | 61.68 | 85.92 | 88.63 | 92.56 | 121.1 | 59.46 | 100 | 58.17 | 151.5 | 68.12 |
| Ce | 187.1 | 43.26 | 36.89 | 38.05 | 61.1 | 125.4 | 172.2 | 194.2 | 203.3 | 239.2 | 115.1 | 206.6 | 117.2 | 316.3 | 131.9 |
| Pr | 24.32 | 5.35 | 4.53 | 4.68 | 6.82 | 14.65 | 21.57 | 22.78 | 24.27 | 27.36 | 16.45 | 24.54 | 13.88 | 39 | 17.82 |
| Nd | 91.54 | 19.59 | 17.81 | 18.4 | 25.51 | 57.94 | 77.68 | 81.41 | 91.4 | 99.71 | 63.58 | 90.02 | 49.82 | 153.4 | 69.01 |
| Sm | 19.43 | 3.59 | 3.73 | 3.43 | 4.4 | 10.24 | 12.66 | 10.67 | 18.37 | 19.69 | 10.7 | 16.85 | 10.38 | 32.14 | 16.1 |
| Eu | 4.96 | 0.88 | 1.05 | 0.68 | 0.82 | 2.08 | 2.59 | 1.44 | 3.73 | 3.26 | 2.38 | 3.75 | 2.65 | 5.81 | 2.68 |
| Gd | 23.72 | 3.05 | 4 | 3.37 | 5.08 | 9.95 | 20.11 | 16.8 | 19.17 | 20.19 | 16.37 | 15.99 | 19.48 | 25.83 | 22.52 |
| Tb | 2.72 | 0.46 | 0.63 | 0.52 | 0.74 | 1.29 | 2.29 | 1.73 | 2.78 | 2.2 | 2.4 | 2.64 | 2.9 | 3.5 | 2.98 |
| Dy | 14.52 | 2.58 | 3.37 | 3.02 | 4.25 | 6.6 | 13.08 | 10.29 | 14.99 | 10.45 | 13.74 | 14.32 | 17.82 | 18.98 | 15.28 |
| Y | 56.21 | 12.85 | 17.64 | 16.87 | 23.29 | 32.44 | 48.98 | 38.27 | 77.11 | 42.29 | 36.78 | 70.26 | 36.54 | 101.4 | 52.67 |
| Ho | 2.79 | 0.51 | 0.7 | 0.61 | 0.92 | 1.33 | 2.48 | 1.98 | 3.17 | 1.92 | 2.51 | 3.19 | 3.44 | 3.92 | 2.66 |
| Er | 7.36 | 1.41 | 1.8 | 1.7 | 2.59 | 3.87 | 6.96 | 5.72 | 8.36 | 5.49 | 6.48 | 8.71 | 9.26 | 10.4 | 6.78 |
| Tm | 0.99 | 0.2 | 0.24 | 0.25 | 0.35 | 0.51 | 0.96 | 0.77 | 1.23 | 0.73 | 0.76 | 1.36 | 1.21 | 1.45 | 0.88 |
| Yb | 6.45 | 1.27 | 1.48 | 1.53 | 2.26 | 3.42 | 5.75 | 4.69 | 7.27 | 4.76 | 4.71 | 8.19 | 7.34 | 8.79 | 5.43 |
| Lu | 0.96 | 0.19 | 0.21 | 0.24 | 0.32 | 0.52 | 0.88 | 0.67 | 1.07 | 0.67 | 0.64 | 1.29 | 1.05 | 1.36 | 0.77 |
| REY | 529 | 116 | 114 | 112 | 170 | 332 | 474 | 480 | 569 | 599 | 352 | 568 | 351 | 874 | 416 |
| REO | 634 | 139 | 136 | 134 | 205 | 398 | 569 | 576 | 683 | 719 | 422 | 681 | 421 | 1049 | 499 |
| $REO_a$ | 777 | 675 | 747 | 691 | 884 | 1124 | 652 | 742 | 2115 | 1645 | 501 | 1861 | 512 | 2866 | 562 |
| Eu/Eu* | 1.07 | 1.24 | 1.27 | 0.93 | 0.8 | 0.96 | 0.73 | 0.48 | 0.93 | 0.76 | 0.81 | 1.07 | 0.81 | 0.95 | 0.64 |
| Ce/Ce* | 0.93 | 0.94 | 0.9 | 0.93 | 0.94 | 0.95 | 0.91 | 0.98 | 0.98 | 0.95 | 0.84 | 0.95 | 0.94 | 0.94 | 0.86 |
| Y/Y* | 0.67 | 0.85 | 0.87 | 0.94 | 0.9 | 0.83 | 0.65 | 0.64 | 0.85 | 0.71 | 0.47 | 0.79 | 0.35 | 0.89 | 0.62 |
| $(La/Lu)_N$ | 0.95 | 1.13 | 0.98 | 0.81 | 1.07 | 1.27 | 1.05 | 1.41 | 0.93 | 1.92 | 0.997 | 0.83 | 0.59 | 1.19 | 0.95 |
| $(La/Sm)_N$ | 0.66 | 0.85 | 0.78 | 0.8 | 1.09 | 0.9 | 1.02 | 1.25 | 0.76 | 0.92 | 0.83 | 0.89 | 0.84 | 0.71 | 0.63 |
| $(Gd/Lu)_N$ | 2.08 | 1.34 | 1.58 | 1.17 | 1.34 | 1.62 | 1.93 | 2.11 | 1.51 | 2.52 | 2.17 | 1.04 | 1.56 | 1.6 | 2.47 |
| Type | H-M | L-M | H-M | H-M | L | L-M | L | L | H-M | L-M | M | M | M | L-M | H-M |

**Table 4.** *Cont.*

| Element | Sample | | | | | | | | |
|---|---|---|---|---|---|---|---|---|---|
| | yy-11-0 | yy-11-1 | yy-11-2 | yy-11-3 | yy-11-4 | yy-11-5 | yy-11-6 | yy-11-7 | yy-11-8 |
| La | 98.11 | 24.25 | 11.69 | 20.84 | 30.15 | 69.42 | 44.3 | 160.9 | 122.1 |
| Ce | 215.3 | 47.44 | 23.26 | 41 | 60.41 | 128.6 | 83.98 | 304.5 | 244.6 |
| Pr | 25.64 | 4.75 | 2.56 | 4.5 | 6.91 | 13.99 | 8.33 | 32.81 | 24.22 |
| Nd | 94.73 | 15.57 | 9.08 | 15 | 24.36 | 47.5 | 26.09 | 114 | 70.8 |
| Sm | 17.4 | 2.69 | 1.81 | 2.75 | 4.28 | 8.35 | 5.03 | 19.46 | 12.26 |
| Eu | 3.15 | 0.4 | 0.28 | 0.37 | 0.56 | 1.09 | 0.81 | 2.18 | 2.69 |
| Gd | 20.98 | 2.78 | 1.54 | 2.38 | 4.53 | 8.92 | 9.93 | 17.58 | 26.77 |
| Tb | 2.02 | 0.36 | 0.32 | 0.46 | 0.67 | 1.59 | 1.22 | 2.84 | 3.11 |
| Dy | 11.04 | 2.16 | 2.13 | 2.72 | 4.05 | 9.54 | 7.45 | 15.31 | 18.37 |
| Y | 41.25 | 12.77 | 14.24 | 15.65 | 21.85 | 55.61 | 40.32 | 75.1 | 68.58 |
| Ho | 2.09 | 0.47 | 0.48 | 0.58 | 0.85 | 2.09 | 1.48 | 3.03 | 3.39 |
| Er | 6.11 | 1.43 | 1.37 | 1.64 | 2.48 | 5.88 | 4.19 | 8.32 | 9.75 |
| Tm | 0.85 | 0.21 | 0.21 | 0.23 | 0.35 | 0.82 | 0.54 | 1.2 | 1.42 |
| Yb | 5.33 | 1.47 | 1.3 | 1.4 | 2.2 | 5 | 3.25 | 7.42 | 9.52 |
| Lu | 0.78 | 0.22 | 0.21 | 0.22 | 0.33 | 0.74 | 0.44 | 1.13 | 1.3 |
| REY | 545 | 117 | 70 | 110 | 164 | 359 | 237 | 766 | 619 |
| REO | 654 | 140 | 85 | 132 | 197 | 431 | 285 | 919 | 743 |
| $REO_a$ | 768 | 390 | 353 | 545 | 835 | 1044 | 365 | 2917 | 871 |
| Eu/Eu* | 0.76 | 0.68 | 0.78 | 0.67 | 0.6 | 0.59 | 0.49 | 0.55 | 0.63 |
| Ce/Ce* | 0.98 | 1 | 0.97 | 0.96 | 0.95 | 0.94 | 0.99 | 0.95 | 1.02 |
| Y/Y* | 0.65 | 0.96 | 1.07 | 0.95 | 0.9 | 0.95 | 0.92 | 0.84 | 0.66 |
| $(La/Lu)_N$ | 1.35 | 1.19 | 0.61 | 1.01 | 0.98 | 0.998 | 1.07 | 1.52 | 0.999 |
| $(La/Sm)_N$ | 0.85 | 1.35 | 0.97 | 1.13 | 1.06 | 1.25 | 1.32 | 1.24 | 1.49 |
| $(Gd/Lu)_N$ | 2.28 | 1.08 | 0.63 | 0.91 | 1.16 | 1.01 | 1.89 | 1.31 | 1.73 |
| Type | L-M | L | H | L | H | H | L | L | H |

REY, sum of rare earth elements and yttrium; REO, sum of oxides of rare earth elements and yttrium; Subscript a indicates value on ash basis; Subscript N indicates values normalized by the average REE content of Upper Continental Crust (Taylor and McLennan, 1985). $Eu/Eu* = 2Eu_N/(Sm_N + Gd_N)$; $Ce/Ce* = 2Ce_N/(La_N + Pr_N)$; $Y/Y* = 2Y/(Dy_N + Ho_N)$.

High correlation coefficients (0.71 to 0.85) exist between the concentrations of individual REY and $P_2O_5$. This indicates that some REY probably occur as phosphate phases, such as rhabdophane. Thus, the high REE concentration of the coal underlain by the bentonite in the Tonghua section may be attributed to authigenic rhabdophane and REE-hydroxides/oxyhydroxides and REE-carbonates occurring as fracture infillings in the coal. The low REE concentrations in the partings are probably due to leaching by groundwater during parting formation [61,62].

The REY concentrations in each coal and non-coal sample were also normalized against the Upper Continental Crust (UCC) [63], in order to obtain a more clear indication of the distribution patterns (Figure 10). Seredin and Dai [64] classified the distribution of REY into three enrichment types, namely L-type (light REY type enrichment), M-type (medium REY enrichment type) and H-type (heavy REY enrichment type). All the three types, as well as a mixed type, of REE enrichment based on this classification occur in the Songzao coal and non-coal samples (Table 4).

The REY enrichment in most of the Datong and Tonghua coals is dominated by a mixed type, either L-M or H-M type. The Yuyang coals have either L-type or H-type enrichment. Nevertheless, the dominance of M-type for the Datong and Tonghua coals indicates a different REY source from that of the Yuyang coals. As discussed by Seredin and Dai [64], an M-type of REY plot, normalized to the UCC, may be due to the circulation of acid natural waters, including acid hydrothermal solutions with high REY concentrations, in the coal basin.

Despite the diversity in the enrichment types, relatively flat REY distribution patterns tend to exist for the coals in the upper part of the Datong (Figure 10A) and Yuyang (Figure 10E) sections. The non-coal samples, including the claystone partings, show no obvious Ce anomalies, pronounced negative Y anomalies, and either no obvious or negative Eu anomalies (Figure 10B,D,F).

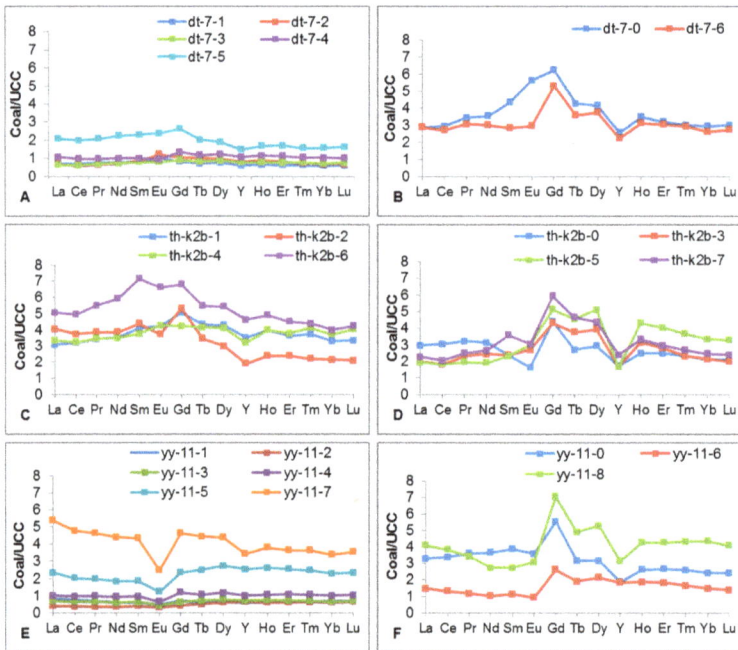

**Figure 10.** Distribution patterns of REE in the three seam sections. REE are normalized to Upper Continental Crust (UCC) (data from [63]). (**A**) Coal samples in the Datong section; (**B**) rock samples in the Datong section; (**C**) coal samples in the Tonghua section; (**D**) rock samples in the Tonghua section; (**E**) coal samples in the Yuyang section; and (**F**) rock samples in the Yuyang section.

*4.7. Potential Industrial Value of REY in Coal Ashes*

The average REY concentrations in the Datong, Tonghua and Yuyang coals ashes are 704 µg/g (or 0.84% $REY_2O_3$), 1737 µg/g (or 2.09% $REY_2O_3$), and 911 µg/g (or 1.09% $REY_2O_3$), respectively. The REY concentrations of the latter two coal ashes are higher than the typical REY cut-off-grade (0.1% $REY_2O_3$) in coal combustion wastes for by-product recovery [64].

In order to evaluate the potential industrial value of REY in the Songzao coal ashes, the $REY_{def, rel}$-Coutl graph proposed by Seredin and Dai [64] is also adopted in the present study (Figure 11). The *y*-axis is the percentage of critical elements (Nd, Eu, Tb, Dy, Y and Er) in the total REY ($REY_{def, rel}$), and the *x*-axis is the ratio of the amount of critical REY metals to the relative amount of excessive REY (Ce, Ho, Tm, Yb and Lu) in total REY. The Songzao coal ashes fall mostly in area II of the graph, which indicates that the coal ashes can be regarded as promising REY raw materials.

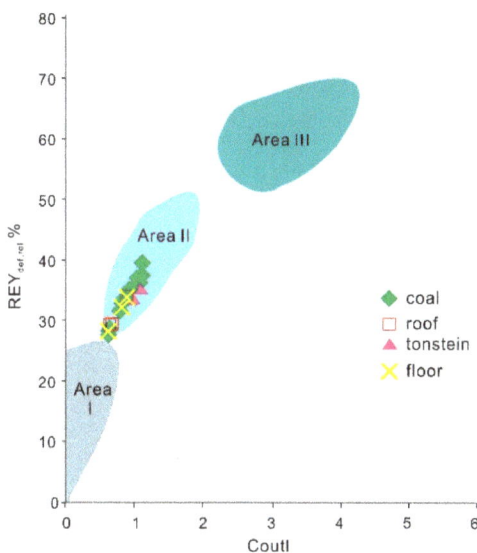

**Figure 11.** $REY_{def, rel}$-Coutl plot for the ashes of Songzao coals and associated rocks: Area I, unpromising; Area II, promising; and Area III, highly promising (adapted from [64]).

## 5. Conclusions

The geochemistry of coals from the Songzao Coalfield has been affected by the adjacent tonstein or K-bentonite bands. Coals near the alkali tonsteins in the Tonghua and Yuyang sections are high in Nb, Ta, Hf, Ga, Th, U, REE and Y. Relative to mafic and felsic volcanic ashes, alkaline volcanic ash is enriched with REE and Y, and tends to lead to elevated REE and Y concentrations in the adjacent coals. Both of the coals occurring near the alkaline tonsteins in the Tonghua and Yuyang sections are enriched with REE and Y, which may be attributed to: (1) abundant fine-grained authigenic rhabdophane in the coal, which was probably precipitated from leachates derived from the overlying bentonite; and (2) REE-hydroxides/oxyhydroxides and REE-carbonates occurring as fracture infillings in the coal, which probably crystallized from ascending hydrothermal fluids carrying high REE concentrations.

Coal samples overlying the mafic bentonite in the Tonghua section are high in $TiO_2$, V, Cr, Zn and Cu. Leaching of the original mafic ash may have led to higher concentrations of these elements in the adjacent coals than in coals without such influence, or in coals affected by alkaline ashes. The $TiO_2$ mainly occurs as anatase, poorly-crystallized Ti-phases, and kaolinite in some of the coal and claystone samples.

The coals in the Datong section, which have neither visible tonstein layers nor obvious volcanogenic minerals, have high concentrations of $TiO_2$, V, Cr, Ni, Cu and Zn in intervals between coals affected by mafic and alkaline volcanic ashes. This is consistent with the suggestion that a common source material was supplied to the coal basin, derived from the mafic basaltic rocks of the Kangdian Upland to the west.

In the Songzao coals, only As and Mo show positive correlations with iron sulfides. No definitive correlations have been found between other chalcophile trace elements (e.g., Sb, Pb, Co, Ni, Cu and Zn) and iron sulfides. This may reflect wide variations of the concentrations of those elements in individual pyrite/marcasite components (including Tl and Ge), or simply poor retention of these elements in the pyrite/marcasite of the Songzao coals. Lead, Cu, Sn and Sb are positively correlated with $Al_2O_3$, rather than pyrite, probably indicating a common source for those elements.

**Acknowledgments: Acknowledgments:** This research was partly supported by the National Key Basic Research Program of China (No. 2014CB238904), the National Natural Science Foundation of China (Nos. 41302128 and 41420104001), and the Fundamental Research Funds for the Central Universities of China (2014QM01). Portions of the present work were published in thesis form in fulfillment of the requirements for the PhD for Lei Zhao from the University of New South Wales, Australia [65].

**Author Contributions: Author Contributions:** Lei Zhao carried out this project in partial fulfillment of her PhD program at the University of New South Wales, Australia, under the supervision of Colin Ward, co-supervised by David French, and Ian Graham. Colin Ward, David French, and Ian Graham have provided major contributions to the design of the work and gave extensive suggestions on data analyses and interpretation, as well as on the English language editing of the paper.

**Conflicts of Interest:** The authors declare no conflict of interest.

## References

1. Swaine, D.J. *Trace Elements in Coal*; Butterworths: London, UK, 1990; p. 278.
2. Kolker, A.; Finkelman, R.B. Potentially hazardous elements in coal: Modes of occurrence and summary of concentration data for coal components. *Coal Prep.* **1998**, *19*, 133–157. [CrossRef]
3. Seredin, V.V.; Finkelman, R.B. Metalliferous coals: A review of the main genetic and geochemical types. *Int. J. Coal Geol.* **2008**, *76*, 253–289. [CrossRef]
4. Bouška, V. *Geochemistry of Coal*; Elsevier: Amsterdam, The Nederland, 1981; p. 284.
5. Hower, J.C.; Rimmer, S.M.; Bland, A.E. Geochemistry of the Blue Gem coal bed, Knox County, Kentucky. *Int. J. Coal Geol.* **1991**, *18*, 211–231. [CrossRef]
6. Crowley, S.S.; Stanton, R.W.; Ryer, T.A. The effects of volcanic ash on the maceral and chemical composition of the C coal bed, Emery Coal Field, Utah. *Org. Geochem.* **1989**, *14*, 315–331. [CrossRef]
7. Hower, J.C.; Ruppert, L.F.; Eble, C.F. Lanthanide, yttrium, and zirconium anomalies in the Fire Clay coal bed, eastern Kentucky. *Int. J. Coal Geol.* **1999**, *39*, 141–153. [CrossRef]
8. Zhou, Y.; Tang, D.; Ren, Y. Characteristics of zircons from volcanic ash-derivd tonsteins in Late Permian coalfields of eastern Yunnan, China. *Acta Sedimentol. Sin.* **1992**, *10*, 28–38. (In Chinese)
9. Zhou, Y.; Ren, Y.; Tang, D.; Bohor, B. Characteristics of zircons from volcanic ash-derived tonsteins in Late Permian coal fields of eastern Yunnan, China. *Int. J. Coal Geol.* **1994**, *25*, 243–264. [CrossRef]
10. Zhou, Y.; Bohor, B.F.; Ren, Y. Trace element geochemistry of altered volcanic ash layers (tonsteins) in Late Permian coal-bearing formations of eastern Yunnan and western Guizhou provinces, China. *Int. J. Coal Geol.* **2000**, *44*, 305–324. [CrossRef]
11. Zhou, Y.; Ren, Y.; Bohor, B.F. Origin and distribution of tonsteins in Late Permian coal seams of southwestern China. *Int. J. Coal Geol.* **1982**, *2*, 49–77. [CrossRef]
12. Zhou, Y.; Ren, Y. Element geochemistry of volcanic ash derived tonsteins in Late Permian coal-bearing formation of eastern Yunnan and western Guizhou, China. *Acta Sedimentol. Sin.* **1994**, *12*, 123–132. (In Chinese)
13. Zhou, Y.P. The alkali pyroclastic tonsteins of early stage of Late Permian age in southwestern China. *Coal Geol. Explor.* **1999**, *27*, 5–9. (In Chinese)

14. Dai, S.; Zhou, Y.; Zhang, M.; Wang, X.; Wang, J.; Song, X.; Jiang, Y.; Luo, Y.; Song, Z.; Yang, Z.; *et al.* A new type of Nb(Ta)-Zr(Hf)-REE-Ga polymetallic deposit in the Late Permian coal-bearing strata, eastern Yunnan, southwestern China: Possible economic significance and genetic implications. *Int. J. Coal Geol.* **2010**, *83*, 55–63. [CrossRef]

15. Dai, S.; Chekryzhov, I.Y.; Seredin, V.V.; Nechaev, V.P.; Graham, I.T.; Hower, J.C.; Ward, C.R.; Ren, D.; Wang, X. Metalliferous coal deposits in east Asia (Primorye of Russia and south China): A review of geodynamic controls and styles of mineralization. *Gondwana Res.* **2015**, in press. [CrossRef]

16. Seredin, V.V.; Dai, S.; Sun, Y.; Chekryzhov, I.Y. Coal deposits as promising sources of rare metals for alternative power and energy-efficient technologies. *Appl. Geochem.* **2013**, *31*, 1–11. [CrossRef]

17. Dai, S.; Zhou, Y.; Ren, D.; Wang, X.; Li, D.; Zhao, L. Geochemistry and mineralogy of the Late Permian coals from the Songzao Coalfield, Chongqing, southwestern China. *Sci. China Ser. D Earth Sci.* **2007**, *50*, 678–688. [CrossRef]

18. Dai, S.; Wang, X.; Zhou, Y.; Hower, J.C.; Li, D.; Chen, W.; Zhu, X.; Zou, J. Chemical and mineralogical compositions of silicic, mafic, and alkali tonsteins in the Late Permian coals from the Songzao Coalfield, Chongqing, southwest China. *Chem. Geol.* **2011**, *282*, 29–44. [CrossRef]

19. Zhao, L.; Ward, C.R.; French, D.; Graham, I.T. Mineralogical composition of Late Permian coal seams in the Songzao Coalfield, southwestern China. *Int. J. Coal Geol.* **2013**, *116–117*, 208–226. [CrossRef]

20. Li, W. A preliminary evaluation of coal resources in the Songzao Coalfield. *Inf. Sci. Technol. Econ.* **2007**, *17*, 148–150, (In Chinese).

21. Dai, S.; Wang, X.; Chen, W.; Li, D.; Chou, C.-L.; Zhou, Y.; Zhu, C.; Li, H.; Zhu, X.; Xing, Y.; *et al.* A high-pyrite semianthracite of Late Permian age in the Songzao Coalfield, southwestern China: Mineralogical and geochemical relations with underlying mafic tuffs. *Int. J. Coal Geol.* **2010**, *83*, 430–445. [CrossRef]

22. China Coal Geology Bureau. *Sedimentary Environments and Coal Accumulation of Late Permian Coal Formation in Western Guizhou, Southern Sichuan, and Eastern Yunnan, China*; Chongqing University Press: Chongqing, China, 1996; p. 216. (In Chinese)

23. Standards Australia. *Coal and Coke—Analysis and Testing Part 3: Proximate Analysis of Higher Rank Coal. Australian Standard 1038.22*; Standards Australia: Strathfield, Australia, 2000.

24. Li, X.; Dai, S.; Zhang, W.; Li, T.; Zheng, X.; Chen, W. Determination of As and Se in coal and coal combustion products using closed vessel microwave digestion and collision/reaction cell technology (CCT) of inductively coupled plasma mass spectrometry (ICP-MS). *Int. J. Coal Geol.* **2014**, *124*, 1–4. [CrossRef]

25. Beijing Research Institute of Coal Chemistry. *Chinese National Standard GB/T 4633-1997, Determination of Fluorine in Coal*; Standand Press of China: Beijing, China, 1997.

26. American Society for Testing and Materials (ASTM). *ASTM Standard D388. Classification of Coals By Rank*; ASTM International: West Conshohocken, PA, USA, 2012.

27. Lyons, P.C.; Spears, D.A.; Outerbridge, W.F.; Congdon, R.D.; Evans, H.T. Euramerican tonsteins: Overview, magmatic origin, and depositional-tectonic implications. *Palaeogeogr. Palaeoclimatol. Palaeoecol.* **1994**, *106*, 113–134. [CrossRef]

28. Spears, D.A. The origin of tonsteins, an overview, and links with seatearths, fireclays and fragmental clay rocks. *Int. J. Coal Geol.* **2012**, *94*, 22–31. [CrossRef]

29. Ketris, M.P.; Yudovich, Y.E. Estimations of clarkes for carbonaceous biolithes: World averages for trace element contents in black shales and coals. *Int. J. Coal Geol.* **2009**, *78*, 135–148. [CrossRef]

30. Finkelman, R.B. Modes of occurrence of environmentally-sensitive trace elements in coal. In *Environmental Aspect of Trace Elements in Coal*; Swaine, D.J., Goodarzi, F., Eds.; Kluwer Academic Publishers: Amsterdam, The Nederland, 1995.

31. Ren, D.; Zhao, F.; Dai, S.; Zhang, J.; Luo, K. *Geochemistry of Trace Elements in Coal*; Science Press: Beijing, China, 2006; p. 556. (In Chinese)

32. Ward, C.R.; Spears, D.A.; Booth, C.A.; Staton, I.; Gurba, L.W. Mineral matter and trace elements in coals of the Gunnedah Basin, New South Wales, Australia. *Int. J. Coal Geol.* **1999**, *40*, 281–308. [CrossRef]

33. Price, N.B.; Duff, P.M.D. Mineralogy and chemistry of tonsteins from Carboniferous sequences in Great Britain. *Sedimentology* **1969**, *13*, 45–69. [CrossRef]

34. Spears, D.A.; Rice, C.M. An Upper Carboniferous tonstein of volcanic origin. *Sedimentology* **1973**, *20*, 281–294. [CrossRef]

35. Spears, D.A.; Kanaris-Sotiriou, R. Titanium in some Carboniferous sediments from Great Britain. *Geochim. Cosmochim. Acta.* **1976**, *40*, 345–351. [CrossRef]

36. Addison, R.; Harrison, R.K.; Land, D.H.; Young, B.R.; Davis, A.E.; Smith, T.K. Volcanogenic tonsteins from Tertiary coal measures, east Kalimantan, Indonesia. *Int. J. Coal Geol.* **1983**, *3*, 1–30. [CrossRef]

37. Burger, K.; Zhou, Y.; Ren, Y. Petrography and geochemistry of tonsteins from the 4th Member of the Upper Triassic Xujiahe Formation in southern Sichuan province, China. *Int. J. Coal Geol.* **2002**, *49*, 1–17. [CrossRef]

38. Spears, D.A.; Kanaris-Sotiriou, R. A geochemical and mineralogical investigation of some British and other European tonsteins. *Sedimentology* **1979**, *26*, 407–425. [CrossRef]

39. Winchester, J.A.; Floyd, P.A. Geochemical discrimination of different magma series and their differentiation products using immobile elements. *Chem. Geol.* **1977**, *20*, 325–343. [CrossRef]

40. Zielinski, R.A. Element mobility during alteration of silicic ash to kaolinite-a study of tonstein. *Sedimentology* **1985**, *32*, 567–579. [CrossRef]

41. Brownfield, M.E.; Affolter, R.H.; Cathcart, J.D.; Johnson, S.Y.; Brownfield, I.K.; Rice, C.A. Geologic setting and characterization of coals and the modes of occurrence of selected elements from the Franklin coal zone, Puget Group, John Henry No. 1 mine, King County, Washington, USA. *Int. J. Coal Geol.* **2005**, *63*, 247–275. [CrossRef]

42. Rao, P.D.; Walsh, D.E. Nature and distribution of phosphorus minerals in Cook Inlet coals, Alaska. *Int. J. Coal Geol.* **1997**, *33*, 19–42. [CrossRef]

43. Zhao, L.; Ward, C.R.; French, D.; Graham, I.T. Mineralogy of the volcanic-influenced Great Northern coal seam in the Sydney Basin, Australia. *Int. J. Coal Geol.* **2012**, *94*, 94–110. [CrossRef]

44. Glick, D.C.; Davis, A. Variability in the inorganic element content of U.S. coals including results of cluster analysis. *Org. Geochem.* **1987**, *11*, 331–342. [CrossRef]

45. Spears, D.A.; Martinez-Tarazona, M.R. Geochemical and mineralogical characteristics of a power station feed-coal, Eggborough, England. *Int. J. Coal Geol.* **1993**, *22*, 1–20. [CrossRef]

46. Querol, X.; Cabrera, L.; Pickel, W.; López-Soler, A.; Hagemann, H.W.; Fernández-Turiel, J.L. Geological controls on the coal quality of the Mequinenza subbituminous coal deposit, northeast Spain. *Int. J. Coal Geol.* **1996**, *29*, 67–91. [CrossRef]

47. Kolker, A. Minor element distribution in iron disulfides in coal: A geochemical review. *Int. J. Coal Geol.* **2012**, *94*, 32–43. [CrossRef]

48. Dai, S.; Chou, C.-L.; Yue, M.; Luo, K.; Ren, D. Mineralogy and geochemistry of a Late Permian coal in the Dafang Coalfield, Guizhou, China: Influence from siliceous and iron-rich calcic hydrothermal fluids. *Int. J. Coal Geol.* **2005**, *61*, 241–258. [CrossRef]

49. Finkelman, R.B. *Mode of Occurrence of Trace Elements in Coal*; University of Maryland: College Park, MD, USA, 1980.

50. Miller, R.N.; Given, P.H. The association of major, minor and trace inorganic elements with lignites. I. Experimental approach and study of a North Dakota lignite. *Geochim. Cosmochim. Acta* **1986**, *50*, 2033–2043. [CrossRef]

51. Querol, X.; Klika, Z.; Weiss, Z.; Finkelman, R.B.; Alastuey, A.; Juan, R.; López-Soler, A.; Plana, F.; Kolker, A.; Chenery, S.R.N. Determination of element affinities by density fractionation of bulk coal samples. *Fuel* **2001**, *80*, 83–96. [CrossRef]

52. Huggins, F.E.; Shah, N.; Zhao, J.; Lu, F.; Huffman, G.P. Nondestructive determination of trace element speciation in coal and coal ash by XAFS spectroscopy. *Energy Fuels* **1993**, *7*, 482–489. [CrossRef]

53. Kolker, A.; Huggins, F.E.; Palmer, C.A.; Shah, N.; Crowley, S.S.; Huffman, G.P.; Finkelman, R.B. Mode of occurrence of arsenic in four US coals. *Fuel Proc. Technol.* **2000**, *63*, 167–178. [CrossRef]

54. Lindahl, P.C.; Finkelman, R.B. Factors influencing major, minor, and trace element variations in U.S. Coals. In *Mineral Matter and Ash in Coal*; Vorres, K.S., Ed.; American Chemical Society: Washington, DC, USA, 1986; Volume 301, pp. 61–69.

55. Spears, D.A.; Zheng, Y. Geochemistry and origin of elements in some UK coals. *Int. J. Coal Geol.* **1999**, *38*, 161–179. [CrossRef]

56. Diehl, S.F.; Goldhaber, M.B.; Hatch, J.R. Modes of occurrence of mercury and other trace elements in coals from the Warrior field, Black Warrior Basin, northwestern Alabama. *Int. J. Coal Geol.* **2004**, *59*, 193–208. [CrossRef]

57. Dai, S.; Ren, D.; Chou, C.-L.; Finkelman, R.B.; Seredin, V.V.; Zhou, Y. Geochemistry of trace elements in Chinese coals: A review of abundances, genetic types, impacts on human health, and industrial utilization. *Int. J. Coal Geol.* **2012**, *94*, 3–21. [CrossRef]

58. Wang, X. Geochemistry of Late Triassic coals in the Changhe mine, Sichuan Basin, southwestern China: Evidence for authigenic lanthanide enrichment. *Int. J. Coal Geol.* **2009**, *80*, 167–174. [CrossRef]

59. Eskenazy, G.M. Aspects of the geochemistry of rare earth elements in coal: An experimental approach. *Int. J. Coal Geol.* **1999**, *38*, 285–295. [CrossRef]

60. Dai, S.; Zou, J.; Jiang, Y.; Ward, C.R.; Wang, X.; Li, T.; Xue, W.; Liu, S.; Tian, H.; Sun, X.; *et al.* Mineralogical and geochemical compositions of the Pennsylvanian coal in the Adaohai mine, Daqingshan coalfield, Inner Mongolia, China: Modes of occurrence and origin of diaspore, gorceixite, and ammonian illite. *Int. J. Coal Geol.* **2012**, *94*, 250–270. [CrossRef]

61. Dai, S.; Li, D.; Chou, C.-L.; Zhao, L.; Zhang, Y.; Ren, D.; Ma, Y.; Sun, Y. Mineralogy and geochemistry of boehmite-rich coals: New insights from the haerwusu surface mine, Jungar Coalfield, Inner Mongolia, China. *Int. J. Coal Geol.* **2008**, *74*, 185–202. [CrossRef]

62. Dai, S.; Ren, D.; Chou, C.-L.; Li, S.; Jiang, Y. Mineralogy and geochemistry of the No. 6 coal (Pennsylvanian) in the Junger coalfield, Ordos Basin, China. *Int. J. Coal Geol.* **2006**, *66*, 253–270. [CrossRef]

63. Taylor, S.R.; McLennan, S.M. *The Continental Crust: Its Composition and Evolution*; Blackwell: Oxford, UK, 1985; p. 312.

64. Seredin, V.V.; Dai, S. Coal deposits as potential alternative sources for lanthanides and yttrium. *Int. J. Coal Geol.* **2012**, *94*, 67–93. [CrossRef]

65. Zhao, L. Mineralogy and geochemistry of Permian coal seams of the Sydney Basin, Australia, and the Songzao Coalfield, SW China. Ph.D. Thesis, University of New South Wales, Australia, 2012, unpublished.

*Article*

# Geochemical Characteristics of Trace Elements in the No. 6 Coal Seam from the Chuancaogedan Mine, Jungar Coalfield, Inner Mongolia, China

**Lin Xiao [1,2,]***, **Bin Zhao [1]**, **Piaopiao Duan [3]**, **Zhixiang Shi [2]**, **Jialiang Ma [1]** and **Mingyue Lin [1]**

[1]   Key Laboratory of Resource Exploration Research of Hebei Province, Hebei University of Engineering, Handan 056038, China; zhaobin@hebeu.edu.cn (B.Z.); majialiang@hebeu.edu.cn (J.M.); linmingyue@hebeu.edu.cn (M.L.)
[2]   Hebei Collaborative Innovation Center of Coal Exploitation, Hebei University of Engineering, Handan 056038, China; shizhixiang@hebeu.edu.cn
[3]   Department of Resources and Earth Science, China University of Mining and Technology, Xuzhou 221008, China; duanpiaopiao@cumt.edu.cn
*   Correspondence: xiaolin@hebeu.edu.cn; Tel.: +86-310-8579-315

Academic Editors: Shifeng Dai and Dimitrina Dimitrova
Received: 19 November 2015; Accepted: 23 February 2016; Published: 30 March 2016

**Abstract:** Fourteen samples of No. 6 coal seam were obtained from the Chuancaogedan Mine, Jungar Coalfield, Inner Mongolia, China. The samples were analyzed by optical microscopic observation, X-ray diffraction (XRD), scanning electron microscope equipped with an energy-dispersive X-ray spectrometer (SEM-EDS), inductively coupled plasma mass spectrometry (ICP-MS) and X-ray fluorescence spectrometry (XRF) methods. The minerals mainly consist of kaolinite, pyrite, quartz, and calcite. The results of XRF and ICP-MS analyses indicate that the No. 6 coals from Chuancaogedan Mine are higher in $Al_2O_3$, $P_2O_5$, Zn, Sr, Li, Ga, Zr, Gd, Hf, Pb, Th, and U contents, but have a lower $SiO_2/Al_2O_3$ ratio, compared to common Chinese coals. The contents of Zn, Sr, Li, Ga, Zr, Gd, Hf, Pb, Th, and U are higher than those of world hard coals. The results of cluster analyses show that the most probable carrier of strontium in the coal is gorceixite; Lithium mainly occurs in clay minerals; gallium mainly occurs in inorganic association, including the clay minerals and diaspore; cadmium mainly occurs in sphalerite; and lead in the No. 6 coal may be associated with pyrite. Potentially valuable elements (e.g., Al, Li, and Ga) might be recovered as byproducts from coal ash. Other harmful elements (e.g., P, Pb, and U) may cause environmental impact during coal processing.

**Keywords:** mode of occurrence; cluster analysis; minerals in coal; Jungar Coalfield

## 1. Introduction

Coal is the main fossil fuel resource and energy source in China. China's energy consumption has grown and will continue to grow along with its economic growth [1]. In the process of coal utilization, the recovery of valuable elements from fly ash, as well as the impact on environment from harmful trace elements, have become important research topics [2,3]. In some cases, Ge, Ga, Li and U can be enriched to higher levels than usual economic grades [4], while As, Pb, Hg, and F are potentially toxic. Furthermore, the modes of occurrence of trace elements in coal control the stratus of their emission in coal combustion processes [5]. Moreover, Sr, Ba, B, and V have great significance to the environment regarding coal formation [5].

The valuable trace elements associated with Jungar coal have been reported by many authors [1,6–12], and the toxic trace elements have been investigated by several researchers [13–16]. However, these previous studies were essentially focused on the northern and central parts of the Jungar Coalfield, such as the Heidaigou Opencut Mine, the Haerwusu Opencut Mine, and the Guanbanwusu Mine (Figure 1) [4,5,7,17,18].

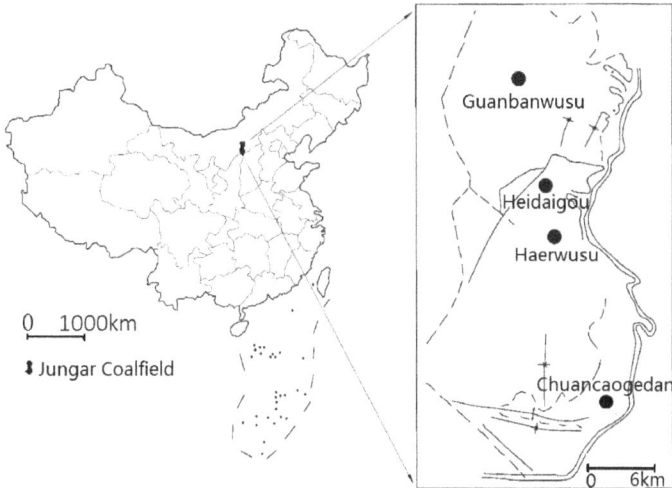

**Figure 1.** Locations of the Chuancaogedan Mine, Guanbanwusu Mine, Heidaigou Opencut Mine, and Haerwusu Opencut Mine in the Jungar Coalfield.

In this study, the No. 6 coal seam from the Chuancaogedan Mine was chosen because this mine is located on the southern edge of the Jungar Coalfield (Figure 1). Compared with the northern and central parts, Chuancangedan is further away from the provenance—the Yinshan Upland [4,17,18]. In this paper, the concentrations and modes of occurrence of the trace elements of No. 6 coal from the Chuancaogedan Mine are reported. The results provide new data on trace element enrichment in coal.

## 2. Geological Setting

The Jungar coalfield is located in the southern Yinshan Oldland (Figure 1), and is one of several Late Palaeozoic coal-bearing basins in this region [19]. The Jungar coalfield is ~64 km long (N–S) by ~26 km wide (W–E), with a total area of 1700 km². This coalfield was sustained by dynamic tectonic activities, and the formation, sedimentation, and evolution of the Jungar coalfield was controlled by the tectonic processes of the Central Asian Orogenic Belt (CAOB) [20,21]. The Permo-Carboniferous denudation processes of the basin development started from the Cambrian-Ordovician periods, then gradually progressed to the Middle and Late Palaeozoic [22]. During the period from the end of the Early Permian to the Late Permian, intermediate-felsic lavas erupted in the CAOB and this lava sequence might have provided an important source of minerals to the coal formations in the region.

The Taiyuan Formation, with a total thickness of 21–95 m, is mainly composed of grey and greyish-white quartzose sandstone, mudstone, siltstone, and coal which is interbedded with dark-grey mudstone, siltstone, limestone, and thin-bedded quartzose sandstone. The Taiyuan Formation was formed in paralic delta and tidal flat-barrier complex environments (Figure 2).

The No. 6 coal seam, the main minable seam, is located in the second section of the Taiyuan Formation. The thickness of the No. 6 coal varies from 0.30 to 16.80 m (11.61 m on average), with 0 to 7 partings. The partings consist mainly of mudstone. The floor and the roof are mainly composed of mudstone, sandy mudstone, and siltstone.

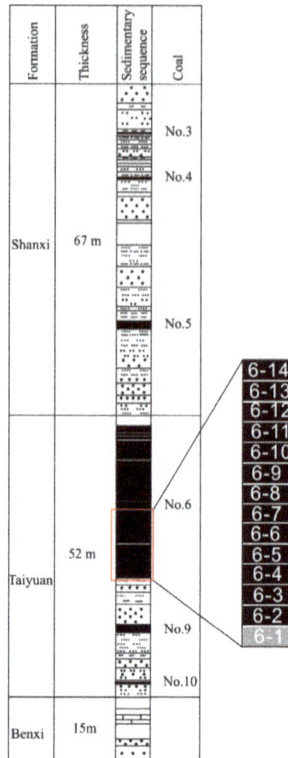

**Figure 2.** Lithostratigraphical column of the Jungar Coalfield and lithological column of the sampling profile.

## 3. Samples and Methods

Fourteen bench samples were taken from the workface at the Chuancaogedan Mine, following the Chinese Standard Method GB 482-2008 [23]. Every coal bench sample was cut over a column that was 10-cm wide, 10-cm deep, and 50-cm thick. All the collected samples were immediately stored in plastic bags to minimize contamination and oxidation. From bottom to top, the 14 bench samples were identified as 6-1 (roof) to 6-14 (Figure 2)

The mineralogical composition was determined on the raw coal samples by coal-petrography microscopy (Leica DM 4500P microscope (Leica Microsystems, Solms, Germany) (at a magnification of 500×) equipped with a Craic QDI 302™ spectrophotometer, CRAIC, San Dimas, CA, USA). Low-temperature ashing of coal was performed on an EMITECH K1050× plasma asher (Quorum, Ashford, UK). The temperature for low-temperature ashing was kept lower than 200 °C (75 W power). X-ray diffraction (XRD) analyses on the resultant low-temperature ashes and the parting samples were performed on a D/max-2500/PC powder diffractometer (Rigaku, Tokyo, Japan) with Ni-filtered Cu-Kα radiation and a scintillation detector. The XRD patterns were recorded over a 2θ interval of 2.6°–70°, with a step size of 0.01°.

A scanning electron microscope (HITACHI UHR FE-SEM, SU8220, HITACHI, Tokyo, Japan) equipped with an energy-dispersive X-ray spectrometer (SEM-EDS) was used to study the distribution characteristics of the minerals, and the distribution patterns of some elements of interest in the coal.

All of the samples were crushed and ground to pass 200 mesh (75 μm) for elemental analysis. X-ray fluorescence spectrometry (XRF) was used to determine the oxides of the major elements in the coal ash (815 °C), including $Na_2O$, $MgO$, $Al_2O_3$, $SiO_2$, $P_2O_5$, $K_2O$, $CaO$, $TiO_2$, $MnO$, and $Fe_2O_3$ [24].

Inductively coupled plasma mass spectrometry (ICP-MS) was applied to determine the trace element contents in the coal samples. For the ICP-MS analysis, microwave digestion of an approximate 200-mg sample ($\varnothing$ <40 μm) was weighed into PTFE (Poly Tetra Fluoro Ethylene) vessels; 2 mL of HF (50%) + 5 mL of $HNO_3$ (65%) + 2 mL of $H_2O_2$ (30%) were added, and microwave digestion was performed for 1 h at a temperature of 210 °C. This solution was then transferred into 125-mL FEP (Fluorinated Ethylene Propylene) bottles that were filled with 100 g of deionized water [24].

## 4. Results and Discussion

### 4.1. Minerals in the Coal

The XRD results from the low temperature ashes, optical microscopic observations, and SEM-EDS data show that the minerals in the No. 6 coal from Chuancaogedan are mainly composed of clay (kaolinite) pyrite, quartz and calcite (Figures 3–6).

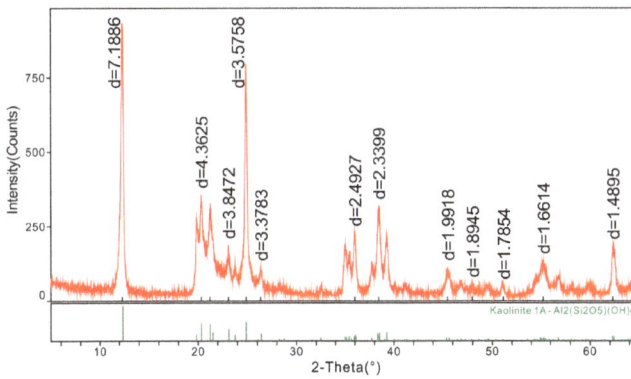

**Figure 3.** Identification of minerals in the X-ray diffraction (XRD) pattern of the low temperature ash (LTA) of Sample 6-10.

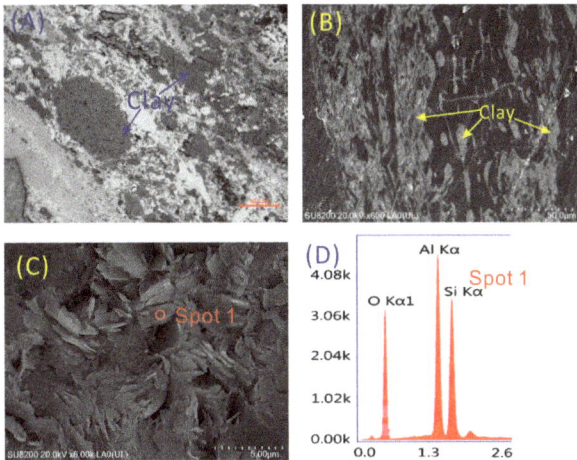

**Figure 4.** Clay minerals in Sample 6-1. (**A**) Lumpy clay with microgranular texture (reflected light); (**B**) Cell-filling clay minerals (scanning electron microscopy, SEM); (**C**) Crystalline kaolinite (SEM) and energy-dispersive X-ray spectrometry (EDS) spectrum from Spot 1 (**D**).

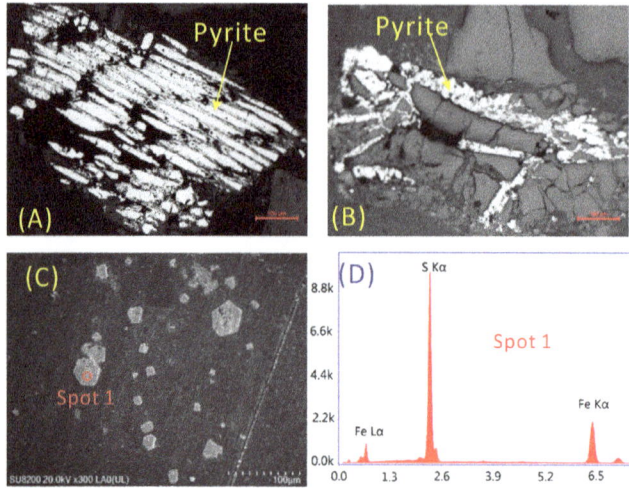

**Figure 5.** Pyrite in the Sample 6-13. (**A**) Pyritized cell filling (reflected light); (**B**) Fracture-filling pyrite (reflected light); (**C**) Crystals of pyrite (SEM) and EDS spectra of it (**D**).

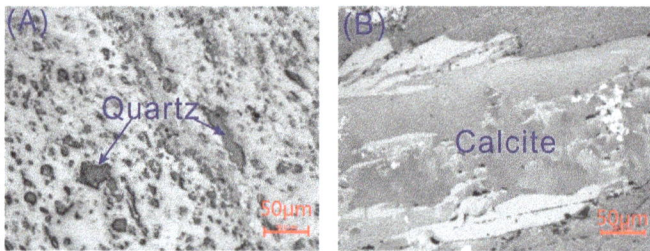

**Figure 6.** Quartz (**A**) and calcite (**B**) in the Sample 6-3 (reflected light).

The clay minerals mainly occur as lumps and cell-fillings with microgranular surfaces (Figure 4A) in telinite and fusinite (Figure 4B). This is common in many other coals and closely associated strata, and may indicate formation by authigenic processes. The results from XRD (Figure 3) and SEM-EDS studies (Figure 4C,D) show that the clay minerals consist mainly of kaolinite. Moreover, the data also indicates that kaolinite is well crystallized.

Pyrite is one of the most common sulfides occurring in coal, especially in coals formed in marine influenced depositional environments. Pyrite in the No. 6 coal predominantly occurs in pyritized cells (Figure 5A) and as fracture-fillings (Figure 5B); euhedral pyrite crystals are also found in the coal (Figure 5C,D), showing that pyrite can be of both syngenetic and epigenetic origin.

Quartz is commonly distributed in the macerals as irregular particles (Figure 6A) in this coal, suggesting a terrigenous detrital origin. Calcite mainly occurs as fracture-infillings (Figure 6B) in the No. 6 coal, indicating an epigenetic origin.

### 4.2. Major Element Contents in the No. 6 Coal

The contents of major oxides ($Na_2O$, $MgO$, $Al_2O_3$, $SiO_2$, $P_2O_5$, $K_2O$, $CaO$, $TiO_2$, $MnO$, and $Fe_2O_3$, on dry coal basis) in the No. 6 coal from Chuancaogedan Mine, in comparison to the average values of coals from Guanbanwusu, Haerwusu, Heidaigou and averages for Chinese coals, are listed in Table 1. Although the oxides of major elements in the No. 6 coal from the Chuancaogedan Mine are dominated by $SiO_2$ (8.09% on average) and $Al_2O_3$ (6.76% on average) (Table 1), the $SiO_2/Al_2O_3$ ratio of the No. 6 coal (1.20 on average) is lower than the common values of Chinese coals. Because quartz is virtually absent, the clay minerals are the major carrier of Si in the coal [25].

**Table 1.** Contents of major elements (re-calculated as oxides; in %, on dry coal basis) and total sulfur in No. 6 coal from Chuancaogedan.

| Sample | LOI | Na$_2$O | MgO | Al$_2$O$_3$ | SiO$_2$ | SiO$_2$/Al$_2$O$_3$ | P$_2$O$_5$ | K$_2$O | CaO | TiO$_2$ | MnO | Fe$_2$O$_3$ | St |
|---|---|---|---|---|---|---|---|---|---|---|---|---|---|
| 6-1 (roof) | 20.24 | 0.008 | 0.000 | 34.98 | 43.30 | 1.24 | 0.025 | 0.072 | 0.06 | 0.85 | 0.0000 | 0.18 | 0.07 |
| 6-2 | 95.21 | 0.008 | 0.013 | 1.96 | 2.29 | 1.17 | 0.004 | 0.019 | 0.06 | 0.07 | 0.0001 | 0.29 | 0.60 |
| 6-3 | 89.53 | 0.004 | 0.029 | 2.45 | 2.24 | 0.91 | 0.011 | 0.004 | 0.09 | 0.04 | 0.0005 | 5.48 | 4.37 |
| 6-4 | 96.00 | 0.002 | 0.013 | 1.64 | 1.87 | 1.14 | 0.005 | 0.004 | 0.08 | 0.07 | 0.0003 | 0.25 | 0.64 |
| 6-5 | 93.09 | 0.003 | 0.024 | 2.84 | 2.89 | 1.02 | 0.253 | 0.012 | 0.08 | 0.08 | 0.0003 | 0.23 | 0.75 |
| 6-6 | 91.29 | 0.003 | 0.011 | 3.69 | 4.23 | 1.15 | 0.125 | 0.003 | 0.10 | 0.09 | 0.0003 | 0.31 | 0.34 |
| 6-7 | 94.56 | 0.002 | 0.013 | 2.22 | 2.54 | 1.14 | 0.043 | 0.003 | 0.10 | 0.04 | 0.0005 | 0.41 | 0.74 |
| 6-8 | 76.45 | 0.010 | 0.069 | 9.10 | 13.91 | 1.53 | 0.006 | 0.077 | 0.06 | 0.13 | 0.0005 | 0.16 | 1.16 |
| 6-9 | 84.03 | 0.006 | 0.033 | 6.78 | 7.93 | 1.17 | 0.037 | 0.019 | 0.21 | 0.16 | 0.0004 | 0.62 | 0.73 |
| 6-10 | 85.25 | 0.003 | 0.019 | 6.24 | 6.96 | 1.12 | 0.389 | 0.007 | 0.23 | 0.14 | 0.0004 | 0.54 | 0.81 |
| 6-11 | 84.12 | 0.003 | 0.016 | 6.68 | 7.59 | 1.14 | 0.265 | 0.021 | 0.12 | 0.27 | 0.0010 | 0.66 | 0.87 |
| 6-12 | 87.32 | 0.002 | 0.019 | 4.69 | 4.79 | 1.02 | 0.796 | 0.005 | 0.76 | 0.10 | 0.0006 | 1.06 | 1.43 |
| 6-13 | 88.58 | 0.002 | 0.013 | 4.77 | 5.16 | 1.08 | 0.368 | 0.010 | 0.19 | 0.16 | 0.0003 | 0.50 | 1.00 |
| 6-14 | 82.97 | 0.006 | 0.024 | 6.57 | 7.57 | 1.15 | 0.038 | 0.038 | 0.15 | 0.29 | 0.0010 | 2.26 | 2.30 |
| Av. | 81.47 | 0.004 | 0.021 | 6.76 | 8.09 | 1.20 | 0.169 | 0.021 | 0.17 | 0.18 | 0.0004 | 0.93 | 1.13 |
| Guanbanwusu [17] | nd | 0.020 | 0.110 | 9.34 | 6.97 | 0.74 | 0.126 | 0.120 | 0.83 | 0.43 | 0.0140 | 0.73 | nd |
| Haerwusu [17] | nd | 0.070 | <0.110 | 8.89 | 6.19 | 0.70 | 0.100 | 0.100 | 1.33 | 0.47 | 0.0100 | 0.56 | nd |
| Heidaigou [17] | nd | 0.010 | 3.660 | 10.56 | 8.04 | 0.76 | 0.016 | 0.210 | 0.44 | 0.74 | 0.0060 | 0.93 | nd |
| China [26] | nd | 0.160 | 0.220 | 5.98 | 8.47 | 1.42 | 0.090 | 0.190 | 1.23 | 0.33 | 0.0200 | 4.85 | nd |

Av., average; St, total sulfur; LOI, loss on ignition; nd, no data.

The average content of $Al_2O_3$ is higher than that in other Chinese coals, because abundant kaolinite and boehmite are present in these coals. In addition, Chuancaogedan coal displays higher $SiO_2$, $P_2O_5$, and $Fe_2O_3$ contents than those of Guanbanwusu, Haerwusu, and Heidaigou coals; this is maybe due to higher quartz and pyrite contents in the former than in the latter coal.

### 4.3. Trace Elements in the No. 6 Coal

The contents of trace elements in the coal samples, in comparison to the average values for Guanbanwusu, Haerwusu, and Heidaigou coals, as well as other Chinese coals [26] and world hard coals [27], are listed in Table 2. The abundance of trace elements in the No. 6 coal from the Chuancaogedan Mine, in comparison to the average values for Chinese and world hard coals, is shown in Figure 7.

Compared to world hard coals (Figure 7A), the only trace element with a CC > 5 (the CC (concentration coefficient) which is the ratio of the element concentration in Chuancaogedan and world hard coals) in the coals is Sr. Elements with a weak enrichment (2 < CC < 5) include Li, Ga, Zn, Zr, Gd, Hf, Pb, Th and U. Co, Ni, Rb, and Cs, which have lower concentrations than those of world hard coals (CC < 0.5). Beryllium, Sc, V, Cr, Nb, Mo, Ba, Ta, W, and Bi (0.5 < CC < 2) are close to the levels of abundance found in average world hard coals.

Compared to Chinese coals (Figure 7B), only Sr has a CC > 5, and Zn shows a weak enrichment (2 < CC < 5). Caesium is lower compared to world hard coals (CC < 0.5). The concentrations of the remaining elements are close to those found in average world hard coals.

Compared to the Chinese coals [28] (Table 2), Lithium, Sr, and Bi are enriched in the No. 6 Coal, Sc, Co, Ni, Cu, Rb, Nb, Cs, Ba, and Ta are depleted.

The coals from Chuancaogedan Mine contain more Zn, Sr, Cd, and Ba than Guanbanwusu, Haerwusu and Heidaigou coals. The remaining elements in the Chuancaogedan Mine samples display lower contents compared to the coals from Guanbanwusu, Haerwusu and Heidaigou Mines.

**Table 2.** Concentrations of trace elements in the No. 6 coal from Chuancaogedan (µg/g, on dry coal basis).

| Sample | Li | Be | Sc | V | Cr | Co | Ni | Cu | Zn | Ga | Rb | Sr | Zr | Nb | Mo | Cd | Cs | Ba | Hf | Ta | W | Pb | Bi | Th | U |
|---|---|---|---|---|---|---|---|---|---|---|---|---|---|---|---|---|---|---|---|---|---|---|---|---|---|
| 6-1 | 253 | 0.7 | 3.2 | 24.5 | 6.5 | 0.4 | 0.8 | 16.3 | 13.5 | 30.4 | 4.7 | 80.3 | 209 | 34.4 | 2.2 | 0.1 | 0.60 | 19.3 | 7.4 | 2.45 | 5.8 | 19.2 | 1.0 | 12.5 | 6.1 |
| 6-2 | 28.9 | 2.2 | 3.0 | 19.4 | 6.8 | 0.9 | 1.2 | 5.1 | 24.3 | 8.8 | 0.4 | 31.3 | 192 | 3.0 | 1.3 | 0.2 | 0.02 | 27.2 | 4.4 | 0.19 | 0.4 | 5.40 | 0.2 | 15.3 | 3.6 |
| 6-3 | 16.8 | 1.5 | <0.5 | 9.12 | 5.2 | 1.6 | 1.5 | 16.9 | 83.1 | 4.5 | 0.2 | 48.4 | 48.7 | 1.6 | 7.7 | 0.1 | 0.02 | 32.0 | 1.2 | 0.08 | 1.4 | 122 | 0.2 | 2.8 | 0.9 |
| 6-4 | 23.1 | 1.5 | <0.5 | 9.00 | 6.0 | 0.6 | 1.0 | 4.1 | 26.2 | 3.3 | 0.2 | 47.7 | 39.7 | 1.5 | 1.0 | 0.1 | 0.01 | 5.08 | 0.9 | 0.10 | 0.6 | 3.51 | 0.2 | 2.0 | 0.4 |
| 6-5 | 34.2 | 1.5 | 2.2 | 30.0 | 10.5 | 1.0 | 1.9 | 10.8 | 125 | 15.2 | 0.7 | 1623 | 407 | 11.3 | 3.0 | 0.7 | 0.04 | 112 | 9.9 | 0.27 | 0.9 | 15.2 | 0.3 | 23.8 | 5.2 |
| 6-6 | 33.2 | 1.5 | 0.2 | 11.7 | 6.4 | 1.0 | 2.1 | 7.5 | 160.5 | 8.7 | 0.4 | 759.7 | 7.6 | 56.7 | 3.1 | 1.2 | 0.5 | 0.0 | 60.1 | 30.3 | 54.1 | 5.5 | 17.8 | 3.0 | 0.5 |
| 6-7 | 23.1 | 1.3 | <0.5 | 12.6 | 6.2 | 1.6 | 3.7 | 4.6 | 159 | 8.9 | 0.5 | 309 | 21.5 | 1.1 | 1.4 | 0.3 | 0.01 | 25.0 | 0.6 | 0.07 | 1.2 | 7.16 | 0.2 | 1.1 | 0.4 |
| 6-8 | 88.6 | 2.0 | 5.2 | 33.8 | 8.9 | 2.0 | 5.3 | 19.2 | 83.6 | 13.1 | 1.6 | 5737 | 186 | 17.8 | 2.7 | 0.2 | 0.13 | 377 | 4.9 | 1.13 | 2.1 | 36.1 | 0.8 | 17.2 | 7.9 |
| 6-9 | 57.9 | 1.1 | 3.1 | 20.1 | 6.0 | 1.5 | 2.3 | 8.3 | 180 | 18.0 | 1.0 | 375 | 161 | 8.2 | 3.2 | 0.2 | 0.10 | 46.2 | 3.4 | 0.44 | 2.1 | 14.2 | 0.3 | 8.6 | 7.6 |
| 6-10 | 72.9 | 2.0 | 2.4 | 18.2 | 7.5 | 1.2 | 3.3 | 15.2 | 217 | 13.1 | 0.3 | 920 | 123 | 4.1 | 2.1 | 0.1 | 0.04 | 34.8 | 2.9 | 0.27 | 0.6 | 18.8 | 0.4 | 7.0 | 4.5 |
| 6-11 | 49.1 | 2.5 | 2.0 | 17.0 | 8.8 | 1.2 | 3.5 | 11.3 | 102 | 12.9 | 1.2 | 1075 | 104 | 7.7 | 2.0 | 0.1 | 0.19 | 212 | 3.0 | 0.68 | 1.4 | 14.4 | 0.6 | 15.2 | 3.6 |
| 6-12 | 33.9 | 1.5 | 1.4 | 26.0 | 8.0 | 2.0 | 6.4 | 19.7 | 561 | 19.4 | 0.4 | 1759 | 122 | 5.9 | 4.4 | 2.7 | 0.03 | 90.9 | 3.0 | 0.21 | 0.8 | 27.4 | 0.6 | 8.4 | 5.6 |
| 6-13 | 40.4 | 2.3 | 1.5 | 21.1 | 12.8 | 1.3 | 4.8 | 9.7 | 81.1 | 13.8 | 0.5 | 1915 | 88.7 | 3.8 | 2.1 | 0.2 | 0.04 | 73.9 | 2.3 | 0.25 | 0.6 | 14.1 | 0.3 | 8.8 | 2.8 |
| 6-14 | 60.6 | 1.5 | 0.9 | 39.0 | 22.6 | 2.7 | 11.2 | 28.6 | 37.2 | 13.9 | 1.8 | 108 | 160 | 7.7 | 4.2 | 0.2 | 0.17 | 39.8 | 3.8 | 0.48 | 1.0 | 25.3 | 0.6 | 8.9 | 2.5 |
| Av. | 56.6 | 1.6 | 2.1 | 20.2 | 8.6 | 1.3 | 3.4 | 12.3 | 134.2 | 12.9 | 1.0 | 1037 | 132 | 7.6 | 2.7 | 0.4 | 0.10 | 81.0 | 3.4 | 0.47 | 1.4 | 23.1 | 0.4 | 9.3 | 3.6 |
| Guanbanwusu [17] | 175 | 1.64 | 6.87 | 38.3 | 16.2 | 1.28 | 2.76 | 13.3 | 29.1 | 12.9 | 2.99 | 703 | 143 | 11.1 | 1.83 | 0.11 | 0.15 | 62 | 3.96 | 0.85 | 1.1 | 26.5 | 0.49 | 12.9 | 3.74 |
| Haerwusu [17] | 116 | 2.8 | 7 | 27 | 10 | 1.3 | 2.3 | 13 | 40 | 18 | 1.3 | 350 | 268 | 13 | 1.6 | 0.06 | 0.07 | 41 | 7.2 | 0.9 | 1.7 | 30 | 0.5 | 17 | 3.7 |
| Heidaigou [17] | 38 | 2.3 | 8.4 | 32 | 15 | 2.1 | 5.6 | 16 | 17 | 45 | 2 | 423 | 234 | 15 | 3.1 | 0.13 | 0.35 | 56 | 8 | 1 | 1.8 | 36 | 0.8 | 18 | 3.9 |
| Clarke value [28] | 20 | 2.8 | 22 | 135 | 100 | 25 | 75 | 55 | 70 | 15 | 90 | 375 | 165 | 20 | 1.5 | 0.2 | 3 | 425 | 8 | 2 | 1.5 | 12.5 | 0.17 | 9.6 | 2.7 |
| China [26] | 31.8 | 2.1 | 4.2 | 35.1 | 15.4 | 7.1 | 13.7 | 17.5 | 41.4 | 6.6 | 9.3 | 140 | 89.5 | 9.4 | 3.1 | 0.3 | 1.1 | 159 | 3.7 | 0.6 | 1.1 | 15.1 | 0.8 | 5.8 | 2.4 |
| World [27] | 14 | 2 | 3.7 | 28 | 17 | 6 | 17 | 16 | 28 | 6 | 18 | 100 | 36 | 4 | 2.1 | 0.2 | 1.1 | 150 | 1.2 | 0.3 | 0.99 | 9 | 1.10.8 | 3.2 | 1.9 |

Av., average.

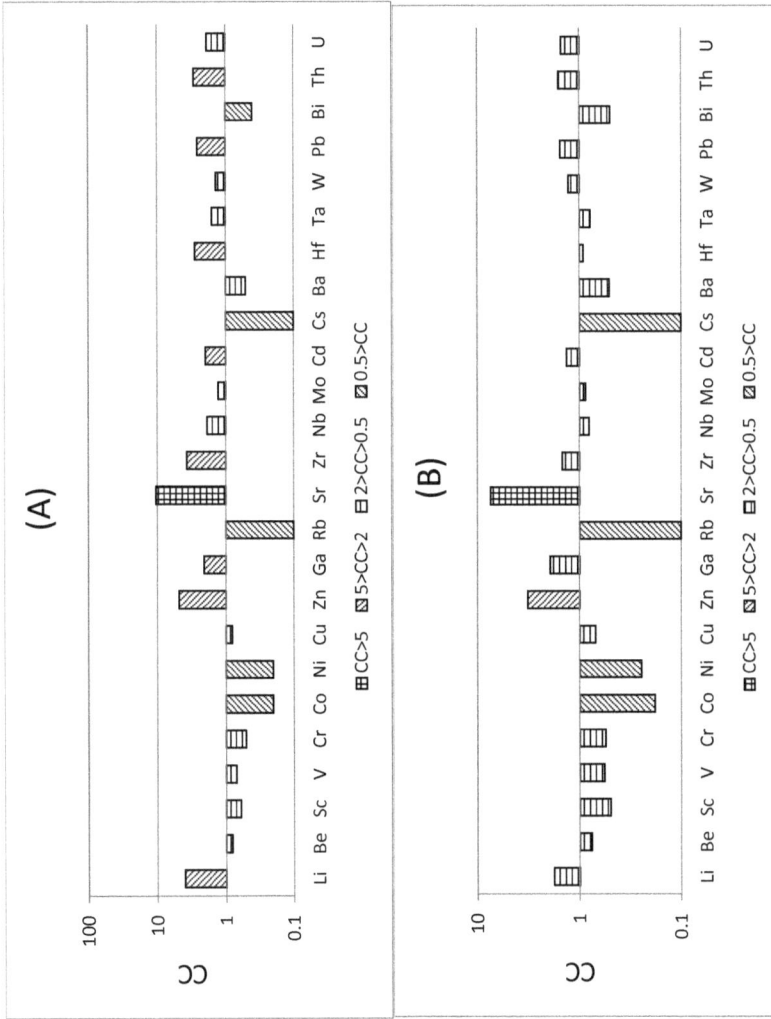

**Figure 7.** (**A**) Concentrations coefficients (CC) of elements in the Chuancaogedan coal *vs.* world coals; (**B**) CC of elements in the Chuancaogedan coals *vs.* Chinese coals.

*4.4. Paragenetic Association of Trace Elements in the No. 6 Coal*

4.4.1. Affinity of the Elements

Trace elements bound in the organic matter of coal volatilize more easily during combustion than these bound in the inorganic matter, which tend to remain in the ash. Therefore, ash yield and trace element contents in coal often display a close relationship [29–31].

Four groups (Groups 1 to 4) of elements have been identified in the No. 6 coal from the present based on their correlation coefficients with ash yield (Table 3).

**Table 3.** Element affinities between the concentration of each element in the coal and ash yield or selected elements.

| Correlation With Ash Yield |
| --- |
| Group 1: $r_{ash}$ = 0.8–1.0 Li (0.99), Ta (0.85), Bi (0.85), $Al_2O_3$ (0.99), $SiO_2$ (0.99), $TiO_2$ (0.96) |
| Group 2: $r_{ash}$ = 0.5–0.8 Sc (0.59), V (0.59), Co (0.67), Ni (0.56), Cu (0.69), W (0.65), U (0.66) $Na_2O$ (0.5), Cs (0.76), Ga (0.53), Rb (0.75), Sr (0.63), Nb (0.73), Ba (0.69), $K_2O$ (0.74) |
| Group 3: $r_{ash}$ = 0.3–0.5 Cr (0.35) |
| Group 4: $r_{ash}$ = −0.3–0.3 Zn (0.11), $Fe_2O_3$ (−0.12), MnO (−0.29), CaO (−0.14), Pb (0.19), Th (0.27), $SO_3$ (−0.17), $P_2O_5$ (−0.15), MgO (−0.17), Hf (0.14), Zr (0.11), Mo (0.28), Cd (−0.02), Be (0.22) |
| **Aluminosilicate Affinity** |
| $r_{Al–Si}$ > 0.8 $TiO_2$, Li, Ga, Rb, Nb, Cs, Ta, W |
| $r_{Al–Si}$ = 0.5–0.8 $K_2O$, Bi, $Na_2O$ |
| $r_{Al–Si}$ = 0.3–0.5 Sc, Cu, Hf, U |
| **Correlation Coefficients Between Selected Elements** |
| V-Cr 0.7, V-Co 0.5, V-Ni 0.7, V-MnO 0.27, V-Cu 0.7 Cr-Co 0.6, Cr-Ni 0.8, Cr-MnO 0.54, Cr-Cu 0.6 Co-Ni 0.88, Co-MnO 0.5, Co-Cu 0.70 Ni-MnO 0.71, Ni-Cu 0.73, MnO-Cu 0.5, Sr-Ba 0.90 Li-$K_2O$ 0.77, Li-Bi 0.77, Li-W 0.93, Li-Ta 0.97, Li-Nb 0.94, Li-Rb 0.95, Li-Ga 0.81 Th-Sc 0.68, Th-V 0.64, Th-Zr 0.89, Th-Hf 0.87, Cd-$SO_3$ 0.5 Ga-S −0.25, Cd-Zn 0.90 |

$r_{ash}$: correlation of elements with ash yield; figures in the brackets: correlation coefficients.

Group 1 includes $Al_2O_3$, $SiO_2$, $TiO_2$, Li, Ta, and Bi, which are strongly correlated with the ash yield ($r_{ash}$ = 0.8–1.0). Silicon and Al are major constituents of the aluminosilicate minerals (kaolinite) [11]. The correlation coefficient between ash yield and $Al_2O_3$, $SiO_2$ is 0.99. Li, $TiO_2$, Ta, and Bi have high correlation coefficients with $SiO_2$ and $Al_2O_3$.

Group 2 includes elements with a relatively high inorganic affinity. The elements in this group (Sc, V, Co, Ni, Cu, W, U, $Na_2O$, Cs, Ga, Rb, Sr, Nb, Ba, and $K_2O$) are strongly correlated with the ash yield, with correlation coefficients between 0.50 and 0.80.

Group 3 includes only Cr, which has a correlation coefficient with the ash yield of 0.35.

Group 4 includes $Fe_2O_3$, MnO, CaO, Pb, Th, $SO_3$, $P_2O_5$, MgO, Hf, Zr, Mo, Cd, and Be. These elements have correlation coefficients with the ash yield that range from −0.30 to 0.30, indicating an intermediate affinity.

4.4.2. Cluster Analysis

The elemental associations in the Chuancaogedan coals were studied by cluster analysis. Four groups of elemental association were identified (Figure 8), referred to as Groups 1, 2, 3 and 4.

Group 1. This group includes $Al_2O_3$, $SiO_2$, Li, Rb, $TiO_2$, Cs, Nb, and Ta (Figure 8). All of the elements in this group have high positive correlation coefficients with the ash yield, ranging from 0.53 to

0.99 (Table 3). $SiO_2$ and $Al_2O_3$ are major constituents of the ash-forming minerals (clay minerals) [18]. All of the elements in Group 1 are lithophile elements that probably occur in aluminosilicate minerals.

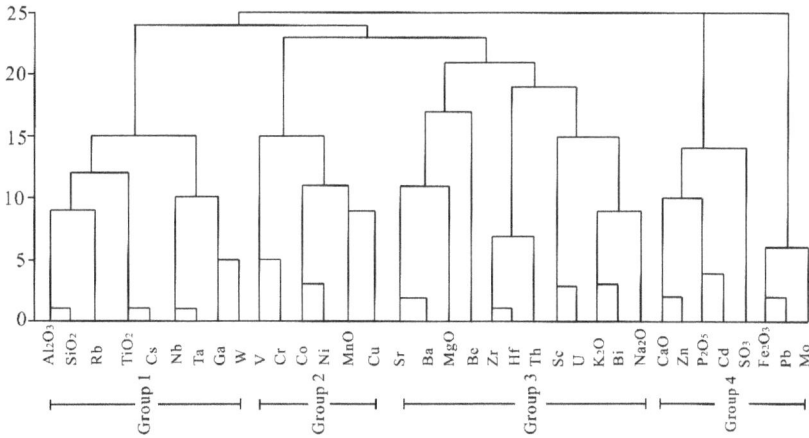

**Figure 8.** Cluster analyses of analytical results on 14 samples.

Group 2. This group includes the elements V, Cr, Co, Ni, MnO, and Cu. All of the elements in this group, except MnO, have relatively high correlation coefficients with the ash yield, ranging from 0.34 to 0.67. With the exception of V-MnO (0.27), the correlation coefficients between the pairs of elements in this association are higher than 0.50. The elements in this group, except Cu, are lithophile elements that are probably associated with the clay minerals.

Group 3. This group consists of Sr, Ba, MgO, Be, Zr, Hf, Th, Sc, U, $K_2O$, Bi, and $Na_2O$. With the exception of $K_2O$, Bi, and $Na_2O$, they have low correlation coefficients with the ash yield (Table 3), and these elements are probably associated with unidentified traces of sulfide minerals. $K_2O$, Bi, and $Na_2O$ have high correlation coefficients with $SiO_2$ and $Al_2O_3$, and these elements probably occur in clay minerals and diaspore.

Group 4. This group includes CaO, Zn, $P_2O_5$, Cd, $SO_3$, $Fe_2O_3$, Pb, and Mo. All of these elements have negative correlation coefficients with the ash yield, possibly because they occur in phosphate minerals (gorceixite and fluorapatite).

*4.5. Elevated Trace Elements in the Coal*

4.5.1. Strontium

The concentration of Sr in the samples varies considerably, from 31.3 to 1915 µg/g, with a weighted average of 1037 µg/g. This level is much higher than that in common Chinese coals (140 µg/g on average) [26] and world hard coals (114 µg/g on average) [27]. The correlation coefficient between Sr and Ba is high at 0.90 (Table 3). The most probable carrier of Ba in the coal is gorceixite, so Sr may be also contained in this mineral [18].

4.5.2. Lithium

The arithmetic average lithium content reaches 56.6 µg/g in the No. 6 coal, and is much higher than that in common Chinese coals and world hard coals. The highest content of Li is 253 µg/g (Table 2). Sun *et al.* suggest that the cut-off grade of Li for economic recovery should be 120 µg/g [32]. However, the Li content of the coal from the Chuancaogedan Mine does not reach this level.

The modes of occurrence of lithium in coal have not been fully studied. Li mainly occurs in granite pegmatite deposits, alkali feldspar granite deposits and salt lake deposits [23,25,32]. Researchers thought that Li in coal generally occurred in clay minerals and partially in mica and tourmaline [33,34]. The high correlation coefficient between Li and ash indicates that a large proportion of the Li occurs in the inorganic matter [18]. Lithium is positively correlated with $Al_2O_3$, $SiO_2$, and $K_2O$, with correlation coefficients of 0.99, 0.99 and 0.77, respectively (Table 3), indicating that Li may also be associated with kaolinite, chlorite, and possibly with illite.

Lithium is also positively correlated with some lithophile elements, including Rb, Nb, Bi, Ga, Ta, and W (Table 3), with correlation coefficients of 0.95, 0.94, 0.77, 0.81, 0.97, and 0.93, respectively, and it is further confirmed that Li occurs in aluminosilicate minerals. The distribution of the Li content in the coal benches, except sample 6-1, is uniform. The high content of Li in sample 6-1 may be caused by a parting within the seam.

### 4.5.3. Gallium

Gallium in coal is generally related to clay minerals [35–40]. The Ga content of the Chuancaogedan coals (12.8 µg/g) is much higher than that of other Chinese coals and world hard coals.

It can be deduced that the Ga mainly occurs in inorganic association, including the clay minerals and diaspore, based on the positive correlations of Ga-ash ($r = 0.81$), Ga-$Al_2O_3$ ($r = 0.82$), and Ga-$SiO_2$ ($r = 0.80$). Gallium may replace Zn in sphalerite, but the low-sulfur content and the negative correlation coefficient of Ga-S ($r = -0.25$) indicate that Ga is not related to sulfide in the Chuancaogedan coals.

### 4.5.4. Zirconium

The average concentration of Zr is 132 µg/g, which is higher than that of common Chinese (89.5 µg/g) [26] and world coals (36 µg/g) [27]. The correlation coefficients of Zr-$Al_2O_3$ and Zr-$SiO_2$ are relatively low, 0.24 and 0.25, respectively. Dai *et al.* found that the content of Zr in the coals from the northern and central Jungar Coalfield is higher than that of common Chinese coals, and pointed out that the major carrier of Zr is zircon [18].

### 4.5.5. Cadmium

Cadmium is one of the toxic trace elements in coal. The average Cd content of the No. 6 coal (0.4 µg/g on average) is higher than that of common Chinese coals and world hard coals. Cd is a chalcophile element. Cd is positively correlated with S and Zn (Table 3), with correlation coefficients of 0.5 and 0.9, respectively, indicating that the Cd in the No. 6 coal mainly occurs in sphalerite.

### 4.5.6. Lead

The average concentration of Pb is 23.1 µg/g, which is higher than that of common Chinese (15.1 µg/g) [26] and world coals (7.8 µg/g) [27].

Pb in coal mainly occurs in galena or is associated with other sulfide minerals. The relationships between Pb and $Al_2O_3$ and $SiO_2$ in the No. 6 coal are poor, with correlation coefficients of 0.05 and 0.10, respectively. The correlation coefficient with $Fe_2O_3$ is high, suggesting that the occurrence of Pb in the No. 6 coal may be associated with pyrite.

### 4.5.7. Thorium

The average Th content of the No. 6 coal (9.3 µg/g) is higher than that of common Chinese coals and world hard coals. Thorium in the Chuancaogedan coals has high correlation coefficients with Sc (0.68), V (0.64), Zr (0.89), and Hf (0.87), probably indicating the same source for these elements. The relatively low correlation coefficients of Th-$Al_2O_3$ (0.21), Th-$SiO_2$ (0.23), and Th-ash (0.20) indicate that Th may occur in accessory minerals in the coal and probably in the organic matter as well.

## 5. Conclusions

1.  The No. 6 coal from Chuancaogedan Mine is significantly enriched in Zn and Sr and is slightly enriched in Li, Ga, Zr, Gd, Hf, Pb, Th, and U compared with world hard coals. The major elements exhibit enrichment in $Al_2O_3$ (6.76%) and $P_2O_5$ (0.169%), but with a lower $SiO_2/Al_2O_3$ ratio (1.20), compared to Chinese hard coals. The contents of Zn, Sr, Li, Ga, Zr, Gd, Hf, Pb, Th, and U are higher than those of world hard coals. Aluminum, Li, and Ga could be recovered as the byproducts from coal ash, and P, Pb, and U may be harmful to the environment during coal processing.

2.  The elements in the No. 6 coal may be classified into four groups of association according to their modes of occurrence. Group 1 includes $Al_2O_3$, $SiO_2$, Li, Rb, $TiO_2$, Cs, Nb, and Ta. Group 2 includes the elements V, Cr, Co, Ni, MnO, and Cu. Group 3 consists of Sr, Ba, MgO, Be, Zr, Hf, Th, Sc, U, $K_2O$, Bi, and $Na_2O$. Group 4 includes CaO, Zn, $P_2O_5$, Cd, $SO_3$, $Fe_2O_3$, Pb, and Mo. Most of the elements in Group 1 and Group 2 are strongly correlated with the ash yield, but the elements of the remaining two associations have negative or weak correlation coefficients with the ash yield.

3.  The most probable carriers of Sr in the coal are barite and gorceixite. Lithium is mainly associated with kaolinite and possibly with illite. Gallium mainly occurs in inorganic association, including the clay minerals and diaspore, but is not related to sulfide. Zirconium occurs in association with sulfide minerals. Cadmium mainly occurs in sphalerite. Lead in the No. 6 coal may be associated with pyrite. Thorium may occur in accessory minerals in the coal and probably in the organic matter as well.

**Acknowledgments:** This research was supported by the National Natural Science Foundation of China (Nos. 41330317, 41402138, and 41511011206).

**Author Contributions:** Lin Xiao and Bin Zhao conceived and designed the experiments; Zhixiang Shi and Jialiang Ma performed the experiments; Lin Xiao, Piaopiao Duan, and Mingyue Lin analyzed the data; Lin Xiao wrote the paper.

**Conflicts of Interest:** The authors declare no conflict of interest.

## References

1.  Sun, Y.Z.; Duan, P.P.; Li, X.W.; Wang, J.X.; Deng, X.L. Advance of mining technology for coals under buildings in China. *World J. Eng.* **2012**, *9*, 213–220. [CrossRef]
2.  Neupane, G.; Donahoe, R. Leachability of elements in alkaline and acidic coal fly ash samples during batch and column leaching tests. *Fuel* **2013**, *104*, 758–770. [CrossRef]
3.  Jankowski, J.; Ward, C.R.; French, D.; Groves, S. Mobility of trace elements from selected Australian fly ashes and its potential impact on aquatic ecosystems. *Fuel* **2006**, *85*, 243–256. [CrossRef]
4.  Sun, Y.Z.; Zhao, C.L.; Zhang, J.Y.; Yang, J.J.; Zhang, Y.Z.; Yuan, Y.; Xu, J.; Duan, D.J. Concentrations of valuable elements of the coals from the Pingshuo Minging District, Ningwu Coalfield, Northern China. *Energy Explor. Exploit.* **2013**, *31*, 727–744. [CrossRef]
5.  Zhao, C.L.; Sun, Y.Z.; Xiao, L.; Qin, S.J.; Wang, J.X.; Duan, D.J. The occurrence of barium in Jurassic coal in the Huangling 2 mine, Ordos Basin, northern China. *Fuel* **2014**, *128*, 428–432. [CrossRef]
6.  Sun, Y.Z.; Zhao, C.L.; Li, Y.H.; Wang, J.X.; Zhang, J.Y.; Jin, Z.; Lin, M.Y.; Kalkreuth, W. Further information of the associated Li deposits in the No. 6 coal seam at Junger Coalfield, Inner Mongolia, Northern China. *Acta Geol. Sin. Engl.* **2013**, *87*, 1097–1108.
7.  Wang, W.F.; Qin, Y.; Liu, X.H.; Zhao, J.L.; Wang, Y.Y.; Wu, G.D.; Liu, J.T. Distribution, occurrence and enrichment causes of gallium in coals from the Ningdong Coalfield, Inner Mongolia. *Sci. China Earth Sci.* **2011**, *41*, 181–196. (In Chinese)
8.  Dai, S.; Ren, D.; Li, S. Discovery of the superlarge gallium ore deposit in Jungar, Inner Mongolia, North China. *Chin. Sci. Bull.* **2006**, *5*, 2243–2252. [CrossRef]

9. Chu, G.; Xiao, L.; Jin, Z.; Lin, M.; Blokhin, M. The relationship between trace element concentrations and coal-forming environments in the No. 6 Coal Seam, Haerwusu Mine, China. *Energy Explor. Exploit.* **2015**, *33*, 99–104. [CrossRef]
10. Dai, S.; Li, D.; Chou, C.L.; Zhao, L.; Zhang, Y.; Ren, D.; Ma, Y.; Sun, Y. Mineralogy and geochemistry of boehmite-rich coals: New insights from the Haerwusu Surface Mine, Jungar Coalfield, Inner Mongolia, China. *Int. J. Coal Geol.* **2008**, *74*, 185–202. [CrossRef]
11. Dai, S.; Zhao, L.; Peng, S.; Chou, C.L.; Wang, X.; Zhang, Y.; Li, D.; Sun, Y. Abundances and distribution of minerals and elements in high-alumina coal fly ash from the Jungar Power Plant, Inner Mongolia, China. *Int. J. Coal Geol.* **2010**, *81*, 320–332. [CrossRef]
12. Dai, S.; Li, T.; Jiang, Y.; Ward, C.R.; Hower, J.C.; Sun, J.; Liu, J.; Song, H.; Wei, J.; Li, Q.; *et al.* Mineralogical and geochemical compositions of the Pennsylvanian coal in the Hailiushu Mine, Daqingshan Coalfield, Inner Mongolia, China: Implications of sediment-source region and acid hydrothermal solutions. *Int. J. Coal Geol.* **2015**, *137*, 92–110. [CrossRef]
13. Wang, X.; Dai, S.; Sun, Y.; Li, D.; Zhang, W.; Zhang, Y.; Luo, Y. Modes of occurrence of fluorine in the late paleozoic No. 6 coal from the Haerwusu surface mine, Inner Mongolia, China. *Fuel* **2011**, *90*, 248–254. [CrossRef]
14. Li, S.S.; Ren, D.Y. Analysis of annmalous high concentration of lead and selenium and their origin in the main minable coal seam in the Ningdong Coalfield. *J. China Univ. Min. Technol.* **2006**, *35*, 612–615. (In Chinese)
15. Liu, D.M.; Yang, Q.; Tang, D.Z. A study of abundances and distribution of ash yield, sulfur, phosphorus and chlorine content of the coals from Ordos Basin. *Earth Sci. Front.* **1996**, *6*, 53–59. (In Chinese)
16. Xu, J.; Sun, Y.Z.; Kalkreuth, W. Characteristics of trace elements of the No. 6 Coal in the Guanbanwusu Mine, Junger Coalfield, Inner Mongolia. *Energy Explor. Exploit.* **2011**, *29*, 827–842. [CrossRef]
17. Dai, S.F.; Zou, J.H.; Jiang, Y.F.; Ward, C.L.; Wang, X.B.; Li, T.; Xue, W.F.; Liu, S.D.; Tian, H.M.; Sun, X.H.; *et al.* Mineralogical and geochemical compositions of the Pennsylvanian coal in the Adaohai Mine, Daqingshan Coalfield, Inner Mongolia, China: Modes of occurrence and origin of diaspore, gorceixite, and ammonian illite. *Int. J. Coal Geol.* **2012**, *94*, 250–270. [CrossRef]
18. Dai, S.F.; Jiang, Y.F.; Ward, C.R.; Gu, L.D.; Seredin, V.V.; Liu, H.D.; Zhou, D.; Wang, X.B.; Sun, Y.Z.; Zou, J.H.; *et al.* Mineralogical and geochemical compositions of the coal in the Guanbanwusu Mine, Inner Mongolia, China: Further evidence for the existence of an Al (Ga and REE) ore deposit in the Jungar Coalfield. *Int. J. Coal Geol.* **2012**, *98*, 10–40. [CrossRef]
19. Dai, S.F.; Ren, D.Y.; Chou, C.L.; Li, S.S.; Jiang, Y.F. Mineralogy and geochemistry of the No. 6 Coal (Pennsylvanian) in the Jungar Coalfield, Ordos Basin, China. *Int. J. Coal Geol.* **2006**, *66*, 253–270. [CrossRef]
20. Xiao, W.J.; Kröner, A.; Windley, B. Geodynamic evolution of Central Asia in the Paleozoic and Mesozoic. *Int. J. Earth Sci.* **2009**, *98*, 1185–1188. [CrossRef]
21. Yang, T.N.; Li, J.Y.; Zhang, J.; Hou, K.J. The Altai-Mongolia terrane in the Central Asian Orogenic Belt (CAOB): A peri-Gondwana one? Evidence from zircon U–Pb, Hf isotopes and REE abundance. *Precambrian Res.* **2011**, *187*, 79–98. [CrossRef]
22. Jian, P.; Kröner, A.; Windley, B.F.; Zhang, Q.; Zhang, W.; Zhang, L. Episodic mantle melting-crustal reworking in the late Neoarchean of the northwestern North China Craton: Zircon ages of magmatic and metamorphic rocks from the Yinshan Block. *Precambrian Res.* **2012**, *222*, 230–254. [CrossRef]
23. China Coal Research Institute (CCRI) Coal Analysis Laboratory. *GB/T 482-2008 Sampling of Coal in Seam*; Standardization Administration of the People's Republic of China: Beijing, China, 2008. (In Chinese)
24. Zhao, C.L.; Duan, D.J.; Li, Y.H.; Zhang, J.Y. Rare earth elements in No. 2 coal of Huangling mine, Huanglong Coalfield, China. *Energy Explor. Exploit.* **2012**, *30*, 803–818. [CrossRef]
25. Sun, Y.Z.; Zhao, C.L.; Qin, S.J.; Xiao, L.; Li, Z.S.; Lin, M.Y. Occurrence of some vluable elements in the unique "high-aluminium coals" from the Jungar Coalfield, China. *Ore Geol. Rev.* **2016**, *72*, 659–668. [CrossRef]
26. Dai, S.F.; Ren, D.Y.; Chou, C.-L.; Finkelman, R.B.; Seredin, V.V.; Zhou, Y.P. Geochemistry of trace elements in Chinese coals: A review of abundances, genetic types, impacts on human health, and industrial utilization. *Int. J. Coal Geol.* **2012**, *94*, 3–21. [CrossRef]
27. Ketris, M.P.; Yudovich, Y.E. Estimations of clarkes for carbonaceous biolithes: World average for trace element contents in black shales and coals. *Int. J. Coal Geol.* **2009**, *78*, 135–148. [CrossRef]
28. Taylor, S.R.; McLennan, S.M. *The Continental Crust: Its Composition and Evolution*; Blackwell Oxford: Oxford, UK, 1985.

29. Liu, G.J.; Yang, P.Y.; Wang, G.L. Geochemistry of elements from the No. 3 coal seam of Shanxi Formation in the Yanzhou mining district. *Geochimica* **2003**, *32*, 255–262. (In Chinese).

30. Sun, Y.Z.; Zhao, C.L.; Li, Y.H.; Wang, J.X.; Liu, S.M. Li distribution and mode of occurrences in Li-bearing coal seam #6 from the Guanbanwusu Mine, Inner Mongolia, Northern China. *Energy Explor. Exploit.* **2012**, *30*, 109–130.

31. Wang, J.; Yamada, O.; Nakazato, T.; Zhang, Z.G.; Suzuki, Y.; Sakanishi, K. Statistical analysis of the concentrations of trace elements in a wide diversity of coals and its implications for understanding elemental modes of occurrence. *Fuel* **2008**, *87*, 2211–2222. [CrossRef]

32. Sun, Y.Z.; Zhao, C.L.; Li, Y.H.; Wang, J.X. Minimum mining grade of the selected trace elements in Chinese coal. *J. China Coal Soc.* **2014**, *39*, 744–748. (In Chinese).

33. Qin, S.J.; Zhao, C.L.; Li, Y.H.; Zhang, Y. Review of coal as a promising source of lithium. *Int. J. Oil Gas Coal Technol.* **2015**, *9*, 215–229. [CrossRef]

34. Finkelman, R.B. Modes of occurrence of environmentally sensitive trace elements of coal. In *Environmental Aspects of Trace Elements of Coal*; Swaine, D.J., Goodarzi, F., Eds.; Kluwer Academic Publishers: Dordrecht, The Netherlands, 1995.

35. Qin, S.J.; Sun, Y.Z.; Li, Y.H.; Wang, J.X.; Zhao, C.L.; Gao, K. Coal deposits as promising alternative sources for gallium. *Earth Sci. Rev.* **2015**, *150*, 95–101. [CrossRef]

36. Seredin, V.V.; Dai, S. Coal deposits as a potential alternative source for lanthanides and yttrium. *Int. J. Coal Geol.* **2012**, *94*, 67–93. [CrossRef]

37. Seredin, V.V.; Finkelman, R.B. Metalliferous coals: A review of the main genetic and geochemical types. *Int. J. Coal Geol.* **2008**, *76*, 253–289. [CrossRef]

38. Hower, J.C.; Ruppert, L.F.; Eble, C.F. Lanthanide, yttrium, and zirconiumanomalies in the fire clay coal bed, Eastern Kentucky. *Int. J. Coal Geol.* **1999**, *39*, 141–153. [CrossRef]

39. Wang, W.; Qin, Y.; Sang, S.; Jiang, B.; Zhu, Y.; Guo, Y. Sulfur variability and element eochemistry of the No. 11 coal seam from the Antaibao mining district, China. *Fuel* **2007**, *86*, 777–784. [CrossRef]

40. Dai, S.; Chekryzhov, I.Y.; Seredin, V.V.; Nechaev, V.P.; Graham, I.T.; Hower, J.C.; Ward, C.R.; Ren, D.; Wang, X. Metalliferous coal deposits in East Asia (Primorye of Russia and South China): A review of geodynamic controls and styles of mineralization. *Gondwana Res.* **2016**, *29*, 60–82. [CrossRef]

![minerals logo]

*minerals*

MDPI

*Article*

# Modes of Occurrence and Abundance of Trace Elements in Pennsylvanian Coals from the Pingshuo Mine, Ningwu Coalfield, Shanxi Province, China

**Ning Yang, Shuheng Tang \*, Songhang Zhang and Yunyun Chen**

School of Energy Resources, China University of Geosciences, Beijing 100083, China;
yangning@cugb.edu.cn (N.Y.); zhangsh@cugb.edu.cn (S.Z.); chenyy@cugb.edu.cn (Y.C.)
* Correspondence: tangsh@cugb.edu.cn; Tel.: +86-10-8232-2005

Academic Editor: Shifeng Dai
Received: 29 January 2016; Accepted: 15 April 2016; Published: 27 April 2016

**Abstract:** The Pingshuo Mine is an important coal mine of the Ningwu coalfield in northern Shanxi Province, China. To investigate the mineralogy and geochemistry of Pingshuo coals, core samples from the mineable No. 4 coals were collected. The minerals, major element oxides, and trace elements were analyzed by scanning electron microscopy (SEM), LTA-XRD in combination with Siroquant software, X-ray fluorescence (XRF), inductively coupled plasma mass spectrometry (ICP-MS) and ICP-CCT-MS (As and Se). The minerals in the Pennsylvanian coals from the Pingshuo Mine dominantly consist of kaolinite and boehmite, with minor amounts of siderite, anatase, goyazite, calcite, apatite and florencite. Major-element oxides including $SiO_2$ (9.54 wt %), $Al_2O_3$ (9.68 wt %), and $TiO_2$ (0.63 wt %), as well as trace elements including Hg (449.63 ng/g), Zr (285.95 µg/g), Cu (36.72 µg/g), Ga (18.47 µg/g), Se (5.99 µg/g), Cd (0.43 µg/g), Hf (7.14 µg/g), and Pb (40.63 µg/g) are enriched in the coal. Lithium and Hg present strong positive correlations with ash yield and $SiO_2$, indicating an inorganic affinity. Elements Sr, Ba, Be, As and Ga have strong positive correlations with CaO and $P_2O_5$, indicating that most of these elements may be either associated with phosphates and carbonates or have an inorganic–organic affinity. Some of the Zr and Hf may occur in anatase due to their strong positive correlations with $TiO_2$.

**Keywords:** Pennsylvanian coals; minerals; trace elements; Ningwu Coalfield

## 1. Introduction

Coal is the most abundant fossil fuel in China, and it is a reliable long-term fuel source for China and other countries, including Turkey and South Africa. With the increasing use of coal in China, a large amount of pollutants is produced, not only in the form of gas emissions but also as ash residues. However, many valuable elements in the coal and coal ash are not yet extracted and used, with the exception of Ge [1–3]. Studies on the geochemistry of elements in coals serve as the basis for the environmental impacts of coals and the efficient use of valuable elements. Many previous investigations have studied the geochemistry and mineralogy of coal deposits around the world, such as the Guanbanwusu and Haerwusu Surface Mines in the Junger Coalfield, northern China [4,5], the Donglin Coal Mine and Xinde Mine, southwestern of China [6,7], Yili Basin, northwestern China [8], the Mariza-east lignite deposit, Bulgaria [9], the Çan coals, Çanakkale, Turkey [10] and Gray Hawk Coal, Eastern Kentucky, USA [11].

As the province with the highest coal production in China, Shanxi has produced up to 505 Mt of raw coal in 2014, which accounts for more than a quarter of China's total coal production. To further understand hazardous elements in coals and in coal combustion products, many coal production areas have been investigated. In this paper, we reported the data on the mineralogy and elemental geochemistry of the No. 4 Coal in the Pingshuo mine, Ningwu Coalfield, Shanxi Province, China.

## 2. Geological Setting

The Pingshuo mine district covers an area of 396 km$^2$. The total coal reserves are estimated at 13 billion tons. The mine district is part of the Ningwu Coalfield, which has an area of 2761 km$^2$ and is located in the northern region of Shanxi Province, north China, in the Ningwu Syncline Basin (Figure 1). To the west of the basin are the Lvliang Mountains and to the east is Wutai Mountain. The base rocks of the basin are metamorphic rocks of the Archean Group [12].

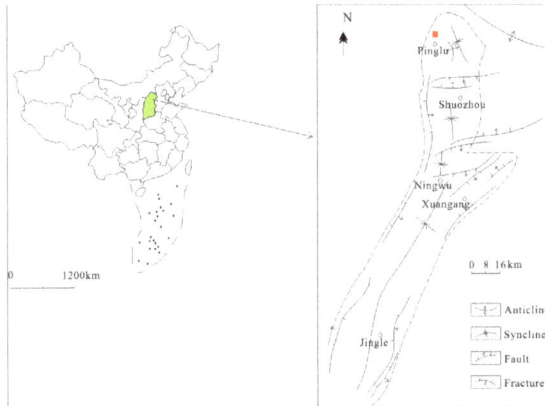

**Figure 1.** Location of the Pingshuo Mine in the Ningwu Coalfield, Shanxi Province, China.

Coal-bearing strata in the study area occur in the Taiyuan Formation of Upper Carboniferous age and the Shanxi Formation of Lower Permian age [13]. In ascending stratigraphic order, the coal-bearing strata in the area are No. 11, 10, 9, 5 and 4 of the Taiyuan Formation and No. 3 of the Shanxi Formation (Figure 2a).

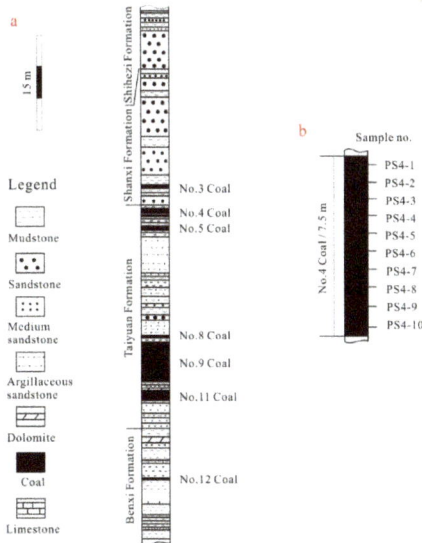

**Figure 2.** Generalized stratigraphic column: (**a**) of the Late Pennsylvanian-Permian coal measures in the Ningwu coalfield and sampling profiles; and (**b**) of the No. 4 Coal in the Pingshuo Coal Mine.

### 3. Samples and Analytical Procedures

To examine the mineralogical and geochemical composition of the Ningwu coals, a total of ten samples were collected from the No. 4 Coal in the Pingshuo Coal Mine, Ningwu Coalfield. The cumulative thickness of the No. 4 Coal is approximately 7.5 m. From top to bottom, the ten samples are labeled PS4-1 to PS4-10 (Figure 2b). All samples were air-dried, sealed in polyethylene bags to prevent oxidation, and splits were ground to pass 200-mesh and were stored in brown glass bottles for chemical analyses.

Proximate analyses ($A_{ad}$, $M_{ad}$, and $V_{daf}$,) of the coal samples were conducted in accordance with ASTM standards (ASTM D3173-11 [14], ASTM D3175-11 [15], and ASTM D3174-11 [16], respectively). Total sulfur was determined according to the ASTM D3177-02 standard [17].

A scanning electron microscope (SEM) was used to study the surface characteristics and the distribution of minerals in the coal. The accelerating voltage was 20 KV, and the beam current was $10^{-10}$ A.

LTAs (low-temperature ashes) of the powdered coal samples were produced by an EMITECH K1050X plasma asher (Quorum, Lewes, UK) prior to XRD analysis. XRD analysis of the LTAs was performed on a D/max-2500/PC powder diffractometer (Rigaku, Tokyo, Japan) with Ni-filtered Cu-K radiation and a scintillation detector. Each XRD pattern was recorded over a 2θ interval of 2.6°–70°, with a step size of 0.01°. X-ray diffractograms of the LTAs and non-coal samples were subjected to quantitative mineralogical analysis using the Siroquant™ interpretation software system (Sietronics, Mitchell, Australia). More analytical details are given by Dai *et al.* [18,19] and Wang *et al.* [20].

X-ray fluorescence (XRF) spectrometry (ARL ADVANT'XP+) was performed to determine the major-element oxides (*i.e.*, $SiO_2$, $TiO_2$, $Al_2O_3$, $Fe_2O_3$, $MgO$, $CaO$, $MnO$, $Na_2O$, $K_2O$, and $P_2O_5$) of the high-temperature coal ash samples.

Trace elements, except for As, Se, Hg and F, were determined by inductively coupled plasma mass spectrometry (ICP-MS). For ICP-MS analysis, coal samples were digested using an UltraClave Microwave High Pressure Reactor (Milestone, Sorisole, Italy). The basic load for the digestion tank was composed of 330-mL distilled $H_2O$, 30-mL 30% $H_2O_2$, and 2-mL 98% $H_2SO_4$. Initial nitrogen pressure was set at 50 bars and the highest temperature was set at 240 °C for 75 min. The reagents for 50-mg sample digestion were 2 mL 40% HF, 5 mL 65% $HNO_3$ and 1 mL 30% $H_2O_2$. Multi-element standards were used for calibration of trace element concentrations. More details are given in Dai *et al.* [21]. Arsenic and Se were analyzed by a collision/reaction cell technology of inductively coupled plasma mass spectrometry (ICP-CCT-MS), as described by Li *et al.* [22]. Fluorine was determined by the ion-selective electrode (ISE) method. Mercury was determined using a Milestone DMA-80 Hg Analyzer (Milestone).

The quantitative analysis of minerals and determinations of elements were completed at the State Key Laboratory of Coal Resources and Safe Mining of China University of Mining and Technology (Beijing, China).

### 4. Results and Discussion

*4.1. Coal Chemistry*

The results of the proximate analysis and the total sulfur of the Late Pennsylvanian coal samples are presented in Table 1. Ash yields of the Pingshuo No. 4 coal range from 12.96% to 32.14%, with an average of 21.42%, indicating a medium ash coal according to Chinese National Standards (GB/T 15224.1-2004, which shows 10.01%–16.00% ash yield is for low-ash coal, 16.01%–29.00% ash yield for medium-ash coal, and >29.00% ash yield for high-ash coal) [23]. The ash yields range irregularly from bottom to top through the coal-seam section (Figure 3).

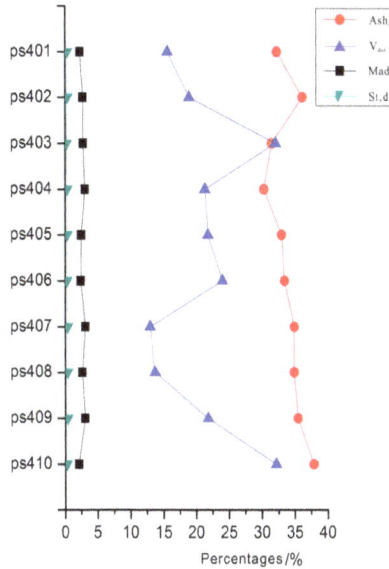

**Figure 3.** Variation of total sulfur and proximate analysis through the Pingshuo coal section.

**Table 1.** Proximate analysis and total sulfur in the Pingshuo coals (%).

| Sample | Proximate Analysis | | | $S_{t,d}$ |
|---|---|---|---|---|
| | $M_{ad}$ | $V_{daf}$ | $A_d$ | |
| PS4-1 | 2.28 | 32.19 | 15.56 | 0.33 |
| PS4-2 | 2.74 | 36.11 | 18.91 | 0.36 |
| PS4-3 | 2.78 | 31.43 | 32.02 | 0.25 |
| PS4-4 | 3.05 | 30.27 | 21.33 | 0.33 |
| PS4-5 | 2.49 | 32.95 | 21.78 | 0.34 |
| PS4-6 | 2.41 | 33.40 | 24.00 | 0.32 |
| PS4-7 | 3.12 | 34.89 | 12.96 | 0.43 |
| PS4-8 | 2.66 | 34.88 | 13.69 | 0.41 |
| PS4-9 | 3.07 | 35.47 | 21.80 | 0.43 |
| PS4-10 | 2.12 | 37.88 | 32.14 | 0.31 |
| Average | 2.67 | 33.95 | 21.42 | 0.30 |

M, moisture; V, volatile matter; A, ash yield; St, total sulfur; ad, air-dry basis; d, dry basis; daf, dry and ash-free basis.

The volatile matter yields of the No. 4 coal vary from 30.27% to 37.88% through the coal-seam section, with a mean of 33.95%, suggesting that the Late Pennsylvanian coals in the Pingshuo coal mine are medium-high volatile bituminous coals based on Chinese Standard MT/T 849-2000 (28.01% to 37.00% for medium-high volatile coal, 37.01% to 50.00% for high volatile coal and >29.00% for super high volatile coal) [24].

The coals from the Pingshuo coal mine have a moisture content of 2.12% to 3.12%, with an average of 2.67%, indicating a low-medium rank coal, in accordance with MT/T 850-2000 (≤5% for low moisture coal, 5% to 15% for medium moisture coal, and >15% for high moisture coal) [25].

The total sulfur of the No. 4 coals ranges from 0.25% to 0.43%, averaging 0.30%, which corresponds to a super-low sulfur coal according to Chinese National Standards (GB/T 15224.2-2004) (<0.5% for super-low sulfur coal, 0.51%–0.9% for low-sulfur coal and 0.9%–1.50% for medium-sulfur coal) [26].

*4.2. Minerals in Coal*

The mineral phase percentages were calculated on a coal ash basis from the XRD results obtained for the low temperature ashes and are reported in Table 2. The results show that minerals in the Pingshuo coals mainly consist of kaolinite, followed by boehmite (averaging 5.09%), siderite (1.35%), anatase (0.29%), goyazite (0.8%), calcite (0.07%), apatite (0.01%) and florencite (0.33%) (Figure 4).

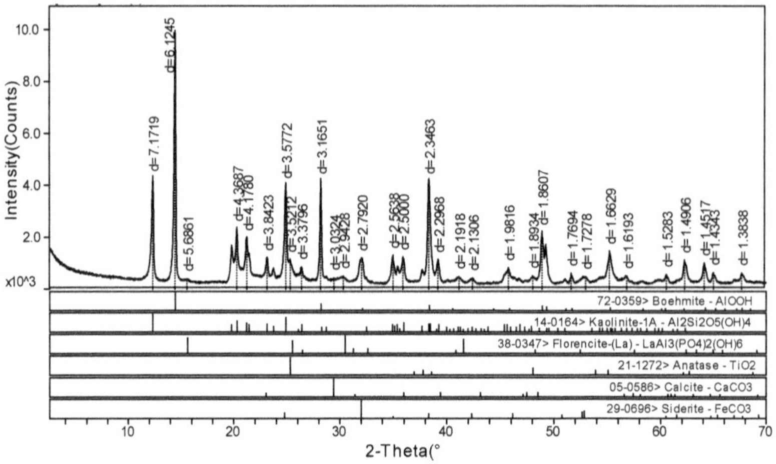

**Figure 4.** XRD patterns of coal samples (PS4-9).

**Table 2.** Mineral contents in coal samples from the Pingshuo Mine measured by LTA-XRD (%).

| Samples | PS4-1 | PS4-2 | PS4-3 | PS4-4 | PS4-5 | PS4-6 | PS4-7 | PS4-8 | PS4-9 | PS4-10 |
|---|---|---|---|---|---|---|---|---|---|---|
| LTA yield | 15.04 | 15.25 | 30.58 | 20.62 | 20.81 | 23.11 | 13.15 | 21.03 | 28.89 | 13.91 |
| Kaolinite | 13.72 | 11.16 | 27.82 | 12.47 | 12.53 | 11.49 | 7.72 | 13.36 | 14.30 | 12.79 |
| Boehmite | 0.92 | 0.05 | 1.87 | 5.77 | 7.89 | 11.44 | 4.47 | 6.04 | 11.61 | 0.83 |
| Arsenopyrite | 0.33 | | | | | | | | | |
| Rutile | 0.08 | | | | 0.06 | | | | | |
| Siderite | | 3.54 | 0.31 | | | | 0.16 | 0.88 | 1.54 | 1.70 |
| Apatite | | 0.27 | | | 0.27 | | | | | |
| Hexahydrite | | 0.12 | | | | | | | | |
| Goyazite | | 0.08 | | 2.25 | 0.08 | | | | | |
| Anatase | | 0.03 | 0.58 | | | | 0.02 | | 0.11 | 0.69 |
| Florencite | | | | | 0.02 | | | | 0.49 | |
| Calcite | | | | | | | 0.08 | | 0.06 | |
| Quartz | | | | | | | | | | 0.29 |

Kaolinite is a very common clay mineral in coal [27,28]. As presented in Table 2, kaolinite is the most abundant mineral in the Pingshuo coals, with abundance varying from 7.72% to 27.82% (13.74% on average). Kaolinite occurs as infillings of cells or fractures (Figure 5a) and as thin-layered or flocculent forms (Figure 5b) in the No. 4 Coal.

The percentage of boehmite in Pingshuo coals varies from 0.05% to 11.61%, with an average of 5.09%. As for the variation of content in the coal profile, boehmite is lower in the top and bottom portions than in the middle portion. Some trace elements (Ga, F) occur in boehmite according to a previous study [5].

Other minerals are detected in only a few samples, and there are no obvious vertical trends of their distribution.

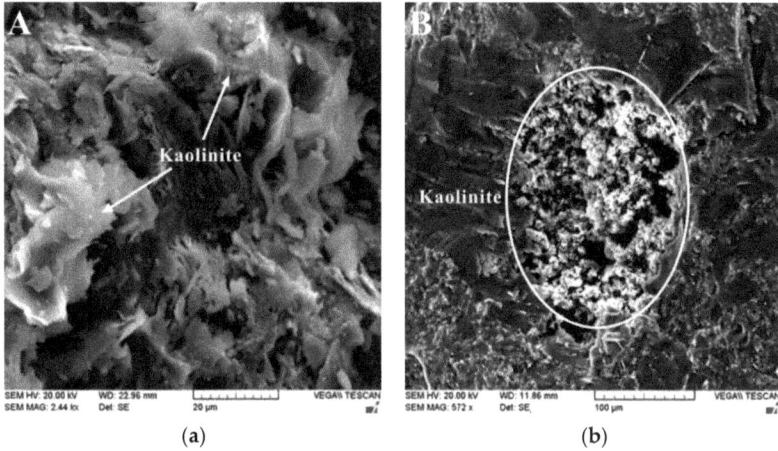

**Figure 5.** Minerals in the Pingshuo coals (SEM, secondary electron images): (**a**) kaolinite in thin-layered forms; and (**b**) flocculent kaolinite.

### 4.3. Abundance of Elements in Pennsylvanian Coals

The percentages of major-element oxides and concentrations of trace elements in the Pennsylvanian coal samples from the Pingshuo Mine, in comparison with the average values of Chinese coals [29] or world hard coals [30], are listed in Table 3.

**Table 3.** Contents of major-element oxides and trace elements in Pennsylvanian coals from the Pingshuo Mine (LOI, oxides in %, elements in µg/g) (whole-coal basis).

| Elemental Contents | Samples | | | | | | | | | | | |
|---|---|---|---|---|---|---|---|---|---|---|---|---|
| | PS4-1 | PS4-2 | PS4-3 | PS4-4 | PS4-5 | PS4-6 | PS4-7 | PS4-8 | PS4-9 | PS4-10 | Average | Coal [b] |
| LOI | 84.9 | 84.8 | 69.4 | 79.4 | 79.2 | 76.9 | 86.9 | 79.0 | 86.1 | 71.1 | 79.8 | - |
| $SiO_2$ | 8.17 | 7.15 | 16.3 | 9.84 | 9.47 | 9.08 | 5.18 | 5.86 | 7.61 | 16.7 | 9.54 | 8.47 [a] |
| $TiO_2$ | 0.42 | 0.57 | 0.81 | 0.90 | 0.69 | 0.79 | 0.43 | 0.36 | 0.93 | 0.41 | 0.63 | 0.33 [a] |
| $Al_2O_3$ | 6.63 | 6.61 | 14.0 | 8.10 | 10.4 | 13.2 | 6.15 | 6.44 | 11.2 | 14.1 | 9.68 | 5.98 [a] |
| $Fe_2O_3$ | 0.06 | 3.33 | 0.36 | 0.53 | 0.58 | 0.42 | 0.67 | 0.59 | 1.13 | 0.33 | 0.80 | 4.85 [a] |
| $Na_2O$ | 0.01 | 0.01 | 0.03 | 0.02 | 0.02 | 0.02 | 0.01 | 0.01 | 0.02 | 0.03 | 0.02 | 0.16 [a] |
| $K_2O$ | 0.01 | 0.02 | 0.18 | 0.09 | 0.17 | 0.10 | 0.05 | 0.06 | 0.11 | 0.06 | 0.08 | 0.19 [a] |
| CaO | 0.09 | 0.25 | 0.10 | 0.28 | 0.11 | 0.10 | 0.12 | 0.10 | 0.17 | 0.11 | 0.14 | 1.23 [a] |
| MgO | 0.02 | 0.15 | 0.06 | 0.06 | 0.08 | 0.06 | 0.07 | 0.08 | 0.10 | 0.05 | 0.07 | 0.22 [a] |
| $P_2O_5$ | 0.02 | 0.21 | 0.02 | 0.86 | 0.02 | 0.03 | 0.04 | 0.02 | 0.19 | 0.14 | 0.15 | 0.09 [a] |
| MnO | - | 0.06 | - | - | - | - | 0.01 | - | 0.01 | - | 0.03 | 0.02 [a] |
| Li | 4.82 | 12.2 | 27.4 | 10.6 | 3.95 | 11.5 | 12.6 | 10.2 | 3.12 | 42.4 | 13.9 | 14.0 |
| Be | 2.66 | 1.88 | 2.49 | 4.16 | 2.48 | 1.94 | 1.70 | 1.55 | 1.60 | 0.87 | 2.13 | 2.00 |
| F | 81.5 | 105 | 123 | 384 | 194 | 203 | 116 | 11 | 241 | 117 | 167 | 140 |
| Sc | 2.08 | 13.3 | 2.55 | 4.84 | 1.75 | 1.59 | 2.74 | 1.62 | 2.03 | 1.56 | 3.40 | 3.00 |
| V | 18.7 | 24.7 | 46.6 | 42.3 | 42.4 | 35.8 | 39.2 | 26.4 | 40.3 | 21.9 | 33.8 | 21.0 |
| Cr | 5.09 | 6.99 | 12.4 | 12.9 | 9.21 | 21.9 | 8.75 | 16.0 | 16.3 | 12.1 | 12.0 | 12.0 |
| Co | 2.83 | 1.52 | 0.68 | 0.66 | 1.11 | 0.88 | 1.29 | 1.29 | 1.04 | 1.08 | 1.24 | 7.00 |
| Ni | 4.31 | 4.54 | 3.99 | 1.57 | 2.28 | 4.02 | 1.94 | 2.31 | 3.13 | 5.46 | 3.36 | 14.0 |
| Cu | 85.6 | 54.2 | 39.4 | 39.3 | 31.1 | 33.8 | 24.2 | 23.0 | 24.3 | 12.4 | 36.7 | 13.0 |
| Zn | 65.3 | 26.8 | 19.6 | 16.6 | 14.7 | 13.5 | 18.8 | 12.3 | 23.5 | 12.5 | 22.4 | 35.0 |
| Ga | 11.4 | 21.5 | 19.8 | 32.4 | 25.0 | 16.1 | 19.4 | 18.8 | 10.9 | 9.62 | 18.5 | 9.00 |
| Ge | 1.16 | 1.23 | 0.41 | 1.93 | 0.80 | 0.48 | 0.73 | 0.76 | 0.52 | 0.38 | 0.84 | 2.78 |
| As | 1.71 | 1.87 | 0.94 | 2.54 | 0.83 | 0.77 | 0.92 | 0.71 | 0.75 | 0.79 | 1.18 | 5.00 |
| Se | 3.56 | 4.85 | 5.68 | 6.60 | 5.75 | 9.73 | 5.77 | 4.08 | 6.46 | 7.45 | 5.99 | 2.00 |
| Hg | 0.18 | 0.13 | 0.08 | 0.09 | 0.34 | 0.09 | 0.07 | 0.09 | 0.37 | 0.31 | 0.18 | 0.15 |
| Rb | 0.28 | 0.19 | 3.77 | 1.70 | 3.88 | 1.82 | 0.81 | 1.27 | 1.96 | 1.55 | 1.72 | 8.00 |

**Table 3.** *Cont.*

| Elemental Contents | Samples | | | | | | | | | | | |
|---|---|---|---|---|---|---|---|---|---|---|---|---|
| | PS4-1 | PS4-2 | PS4-3 | PS4-4 | PS4-5 | PS4-6 | PS4-7 | PS4-8 | PS4-9 | PS4-10 | Average | Coal [b] |
| Sr | 74.5 | 710 | 35.6 | 2908 | 57.7 | 52.1 | 166 | 70.8 | 557 | 264 | 229 | 423 |
| Zr | 107 | 250 | 691 | 276 | 365 | 184 | 282 | 144 | 405 | 157 | 286 | 52.0 |
| Mo | 2.10 | 1.51 | 0.77 | 1.34 | 2.01 | 1.82 | 1.80 | 2.07 | 1.63 | 1.04 | 1.61 | 4.00 |
| Cd | 0.33 | 0.43 | 0.91 | 0.39 | 0.50 | 0.25 | 0.39 | 0.25 | 0.57 | 0.24 | 0.43 | 0.20 |
| Sn | 3.60 | 3.12 | 4.79 | 3.23 | 5.52 | 2.73 | 1.52 | 1.90 | 3.52 | 2.13 | 3.21 | 2.00 |
| Sb | 0.29 | 0.38 | 0.27 | 0.27 | 0.53 | 0.19 | 0.41 | 0.32 | 0.46 | 0.14 | 0.33 | 2.00 |
| Cs | 0.02 | 0.02 | 0.13 | 0.19 | 0.07 | 0.08 | 0.03 | 0.06 | 0.09 | 0.13 | 0.08 | 1.00 |
| Ba | 16.2 | 64.9 | 25.0 | 213 | 26.7 | 20.4 | 30.0 | 20.8 | 54.6 | 28.5 | 50.0 | 56.0 |
| Hf | 2.74 | 6.02 | 15.6 | 6.75 | 9.59 | 4.84 | 7.51 | 3.87 | 10.4 | 4.06 | 7.14 | 2.40 |
| Ta | 0.47 | 1.12 | 1.41 | 1.36 | 3.33 | 1.02 | 0.77 | 0.62 | 1.27 | 0.88 | 1.22 | 0.70 |
| W | 5.02 | 1.20 | 1.40 | 1.86 | 1.66 | 1.53 | 0.55 | 0.84 | 1.14 | 0.84 | 1.60 | 2.00 |
| Tl | 0.05 | 0.06 | 0.04 | 0.05 | 0.10 | 0.05 | 0.09 | 0.09 | 0.13 | 0.09 | 0.07 | 0.40 |
| Pb | 31.4 | 38.2 | 39.5 | 40.6 | 97.3 | 40.0 | 32.1 | 24.6 | 49.2 | 13.3 | 40.6 | 13.0 |
| Bi | 0.21 | 0.41 | 0.54 | 0.56 | 0.61 | 0.55 | 0.33 | 0.35 | 0.54 | 0.32 | 0.44 | 0.80 |
| Th | 5.52 | 13.1 | 4.85 | 20.6 | 12.0 | 2.76 | 14.3 | 10.0 | 13.8 | 1.02 | 9.78 | 6.00 |
| U | 1.87 | 3.33 | 4.56 | 5.40 | 5.86 | 3.85 | 4.12 | 2.84 | 4.40 | 2.08 | 3.83 | 3.00 |

LOI, loss on ignition; [a] Chinese average coals value by Dai *et al.* [29]; [b] world hard coals by Ketris and Yudovich [30].

The major elements in the Pingshuo coals are dominated by $SiO_2$ and $Al_2O_3$. The main carriers of these elements are quartz and clay minerals [31,32]. Average values of major-element oxides in high-temperature ashes of Pingshuo coal samples are as follows: $SiO_2$ (9.54 wt %), $Al_2O_3$ (9.68 wt %), $Fe_2O_3$ (0.80 wt %), $TiO_2$ (0.63 wt %), CaO (0.14 wt %), $K_2O$ (0.08 wt %), MgO (0.07 wt %), $Na_2O$ (0.02 wt %), $P_2O_5$ (0.15 wt %) and MnO (0.27 wt %). Compared with the average values of Chinese coals [29], the Pennsylvanian coals from the Pingshuo Mine contain higher proportions of $SiO_2$, $Al_2O_3$, $TiO_2$, and $P_2O_5$ and lower proportions of $Fe_2O_3$, CaO, $K_2O$, MgO and $Na_2O$.

The $SiO_2/Al_2O_3$ ratios range from 0.68 to 1.23, with an average of 0.99 for the Pennsylvanian coals. This range is lower than those of other Chinese coals (1.42) [25] and lower than the theoretical $SiO_2/Al_2O_3$ ratio of kaolinite (1.18), in accordance with the occurrence of boehmite and the lack of quartz in the coal.

In comparison with Chinese coals [29] or world hard coals [30], Zr (CC = 5.50) is substantially higher in the Pennsylvanian coals from the Pingshuo Mine. Copper (CC = 2.82), Ga (CC = 2.05), Se (CC = 2.99), Cd (CC = 2.14), Hf (CC = 2.97) and Pb (CC = 3.13) are slightly higher than the average values of Chinese coals [29], whereas Li (CC = 0.99), Be (CC = 1.07), F (CC = 1.20), Sc (CC = 1.13), V (CC = 1.61), Cr (CC = 1.00), Zn (CC = 0.64), Hg (CC = 1.20), Sr (CC = 1.16), Sn (CC = 1.60), Ba (CC = 0.89), Ta (CC = 1.75), W (CC = 0.80), Bi (CC = 0.55), Th (CC = 1.63) and U (CC = 1.27) are very close to those of Chinese coals, and the remaining trace elements are lower than the averages of Chinese and world hard coals (CC < 0.5) (CC, concentration coefficient, the ratio of trace element concentration in coal samples investigated *vs.* averages for Chinese or world hard coals [33]) (Figure 6).

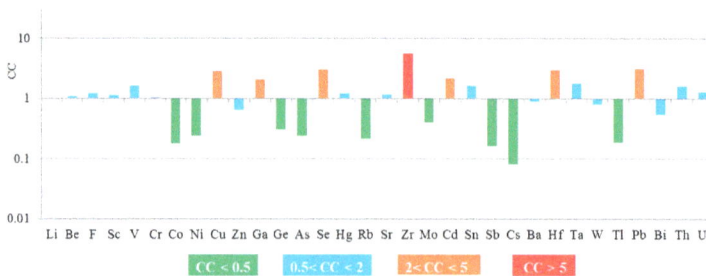

**Figure 6.** Concentration coefficients (CC) of trace elements in the Pingshuo coals.

*4.4. Geochemical Associations in the Coal Samples*

The modes of occurrence of trace elements in the Pingshuo coals were preliminarily investigated using cluster analysis and correlation coefficients, which are effective indirect methods for coal geochemistry [34]. Pearson correlation coefficients were used in hierarchical clustering. Elements that are the most strongly correlated are linked first, followed by elements or element groups with decreasing correlations, until a complete dendrogram is achieved, as reported by Zhao *et al.* [35].

Associations of trace elements in the Pingshuo coals are indicated by the hierarchical clustering dendrogram (Figure 7). Elements in the Pingshuo coals are divided into six groups. Apart from the inter-correlation among elements in the same group, each group contains elements of different sub-groups that have different correlations with ash yield and some oxides in the coals. The modes of occurrence of trace elements could be deduced based on the cluster analysis results and correlation coefficients.

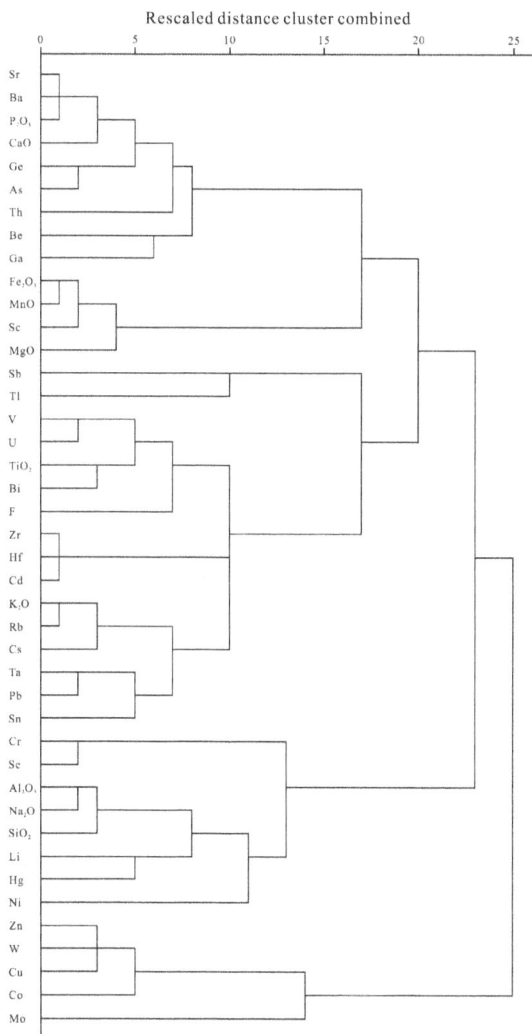

**Figure 7.** Cluster analysis of the geochemical data of the Pennsylvanian Pingshuo coal samples.

Group 1 includes $Al_2O_3$, $SiO_2$, $Na_2O$, Li, Hg, Ni, Se, and Cr (Figure 7). Elements in Group 1 are all strongly correlated with $Al_2O_3$ or $SiO_2$, with high correlation coefficients ($r > 0.60$, ranging from 0.62 to 0.81). In addition, these elements are slightly correlated with ash yields ($r$Li-ash = 0.72, $r$Hg-ash = 0.56, $r$Ni-ash = 0.54, $r$Se-ash = 0.53, and $r$Cr-ash = 0.55). Most of the elements in this group probably have the same modes of occurrence.

Group 2 includes $P_2O_5$, CaO, Ba, Sr, As, Th, Be, and Ga (Figure 7). Their correlation coefficients with $P_2O_5$ are $r$Be = 0.68, $r$Ga = 0.61, $r$As = 0.78, $r$Sr = 0.99, $r$Ba = 0.99, and $r$Th = 0.66, indicating that $P_2O_5$ and these elements are probably associated with phosphate minerals (goyazite, apatite, and florencite in coal samples), which were detected by previous investigations [35–37]. Ba, Sr, As, and Th also have strong correlations with CaO, and they have a carbonate affinity.

Group 3 includes $Fe_2O_3$, MgO, MnO and Sc (Figure 7). Elements in this group have a strong correlation with each other ($r > 0.92$). $Fe_2O_3$ has strong correlations with total sulfur in the coal samples (Figure 8), indicating most of elements in Group 3 probably occur in sulfides in the Pingshuo coals.

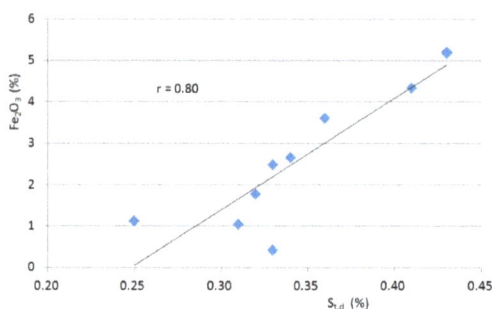

**Figure 8.** Correlations between $Fe_2O_3$ and total sulfur in the Pingshuo coals.

Group 4 includes three subgroups. Subgroup 1 includes $TiO_2$, Bi, V, U, F, Zr, Hf and Cd. The correlation coefficients of trace elements in this subgroup with $TiO_2$ are larger than 0.72 ($r$Bi = 0.87, $r$V = 0.74, and $r$U = 0.72), indicating that $TiO_2$ and these elements probably occur in anatase. Subgroup 2 includes $K_2O$, Rb and Cs. Elements in this group have strong correlations with $K_2O$ ($r$Rb = 0.99 and $r$Cs = 0.86). Subgroup 3 includes Ta, Pb and Sn. These elements are weakly correlated with ash yield ($r$ varies from −0.19 to 0.36), $SiO_2$ ($r$ varies from −0.30 to 0.32) or $Al_2O_3$ ($r$ varies from −0.01 to 0.32).

Group 5 includes Zn, W, Cu, Co and Mo (Figure 7). These metal elements present negative correlations with ash ($r$ ranging from −0.81 to −0.29), $SiO_2$ ($r$ is from −0.81 to −0.18) and $Al_2O_3$ ($r$ ranging from −0.60 to −0.39).

Group 6 includes Sb and Tl (Figure 7). Their correlation coefficient is 0.55, and there are no obvious correlations with ash, $SiO_2$, $Al_2O_3$ or other elements.

### 4.5. Modes of Occurrence of Some Trace Elements in Coals

Based on the cluster analysis, many elements are associated with $SiO_2$, $Al_2O_3$ and other elements in the coals with different degrees of affinity. Apart from these correlations, ash yield and other factors may also affect the abundance and modes of occurrence of trace elements in coals.

### 4.5.1. Li and Hg

Lithium (Li) is a very important element used for batteries and in other industries [38]. Many authors have investigated Li enrichment [38,39]. The average Li content in coals is 14 µg/g globally and 31.8 µg/g in China [29]. The Li contents of the Pingshuo coals range from 3.12 to 42.40 µg/g, averaging 13.87 µg/g (Table 3), close to the world hard coal content (CC = 0.99). Modes of occurrence of Li in coal have not been fully addressed in other studies because of its low atomic number and its reduced level of toxicity to the environment compared to many other elements [4]. Limited previous studies show that Li in coals is generally associated with aluminosilicate minerals [4,38,39] and organic matter. Lithium in the Pingshuo coals presents strong positive correlations with ash yield, $SiO_2$ and $Al_2O_3$ (Figure 9), indicating that Li may be associated with aluminosilicate minerals in the Pingshuo coals.

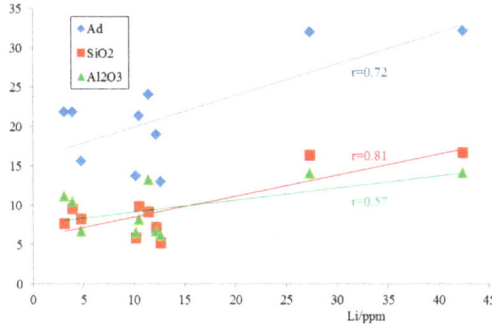

**Figure 9.** Correlations between Li and ash yield (Ad), $SiO_2$ and $Al_2O_3$.

The coal samples from Pingshuo are enriched in mercury (Hg) (CC = 1.20). The Hg concentrations vary between 0.07 and 0.37 µg/g, with an average value of 0.18 µg/g. There have been many reports on the modes of occurrence of Hg in coals [40,41], and most of the results agree that mercury in coals has a sulfide affinity, usually in pyrite [42–44], getchellite [45], or organic matter [46]. However, none of these associations are indicated in the Pingshuo coals. There is no obvious correlation between Hg and total sulfur contents in the coal samples (Figure 10), and the correlation coefficients between Hg and ash yield, $SiO_2$ and Li are 0.56, 0.62 and 0.78, respectively, indicating that Hg is associated with aluminosilicate minerals other than sulfides.

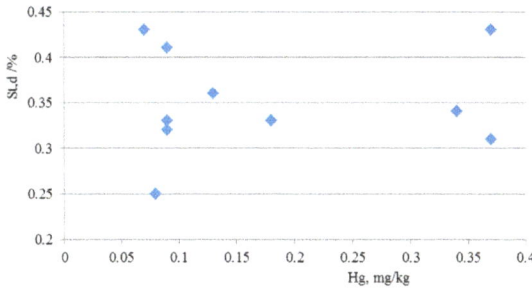

**Figure 10.** Correlations between Hg and total sulfur.

### 4.5.2. Sr, Ba, Be, As and Ga

The concentrations of Sr and Ba in coal samples are very close to the averages for Chinese or world hard coals (CC = 1.16 and 0.89), with average contents of 489.50 and 50.02 µg/g, respectively. The concentration of $P_2O_5$ shows positive correlations with Sr and Ba in the coal samples (Figure 11), with correlation coefficients of 0.99, indicating that most of the Sr and Ba in Pingshuo coals may occur in aluminophosphates (goyazite and gorceixite), as previously reported [35,47]. Sr and Ba also have positive correlations with CaO in the samples (Figure 11), indicating that the Sr and Ba in the coals exhibit some carbonate affinity.

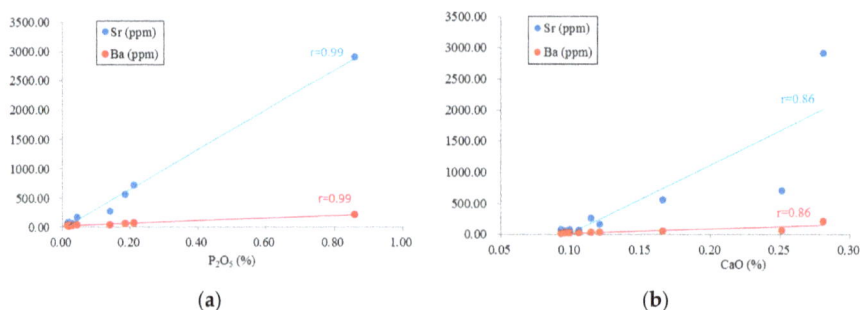

**Figure 11.** Correlations of Sr and Ba with: $P_2O_5$ (**a**); and CaO (**b**).

The concentration of beryllium (Be) in the Pingshuo coals is close to the average content of Chinese coals (CC = 1.07). Modes of occurrence of Be in the coal have been addressed in previous studies [9,41], and Be has a mixed inorganic–organic affinity in the Yimin Ge-rich coals [48]. The correlation coefficients for Be-$P_2O_5$ ($r$ = 0.68), Be-Sr ($r$ = 0.71) and Be-Ba ($r$ = 0.73) in the present study indicate that the Be has a mixed inorganic–organic affinity.

A wide variety of As-bearing phases has been observed in the Guizhou high-As coals, including pyrite, Fe-As oxides, and As-bearing clays [49,50]. Arsenic in the Chongqing coals is correlated with $Fe_2O_3$, suggesting a pyrite affinity [6,44]. The correlation coefficient between As and ash yield is 0.20, which indicates As in the Pingshuo coals has an inorganic–organic affinity. Arsenic has a positive correlation with CaO and $P_2O_5$ ($r$CaO = 0.79, $r$$P_2O_5$ = 0.78), which suggests that some As may be affiliated with phosphates and carbonates in the Pingshuo coals.

According to previous research, Gallium (Ga) is generally related to clay minerals in coal [39,51]. The modes of Ga occurrence in the Guanbanwusu, Haerwusu and Heidaigou coals are slightly different. Ga in the Haerwusu coals occurs mainly in boehmite and organic matter [27]. The correlation coefficient between Ga and ash yield is −0.15, suggesting that Ga has mixed affinities in coal samples. The strong positive correlation between Ga and $P_2O_5$ ($r$ = 0.61) suggests that Ga mainly occurs in inorganic associations.

### 4.5.3. Zr and Hf

The concentration of Zirconium (Zr) in Pingshuo coals is considerably higher than the average content of Chinese coals (CC = 5.50), whereas that of hafnium (Hf) is slightly higher (CC = 2.97). Zr and Hf show similar distributions in the coal samples (Figure 12), and their correlation coefficient is 0.99, indicating that they have nearly the same modes of occurrence in the Pingshuo coals.

Zircon is the most common zirconium mineral; therefore, Zr is believed to be at least partly due to the probable presence of this heavy mineral in these samples [5]. The fine Zr-phases (<0.5 μm), probably zircon, were detected in anatase [39,51]. Zr and Hf have strong positive correlations with $TiO_2$ and $K_2O$ (Figure 13), indicating that some of Zr and Hf may occur in zircons included in the anatase.

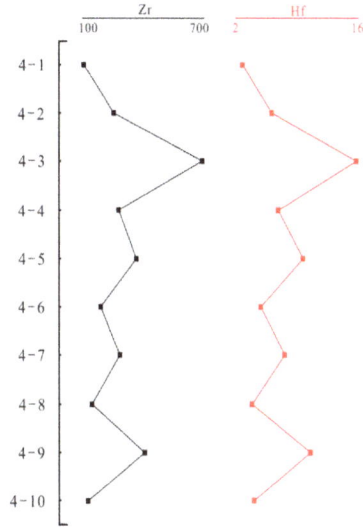

**Figure 12.** Vertical variation of Zr and Hf in the profile of the Pingshuo coals.

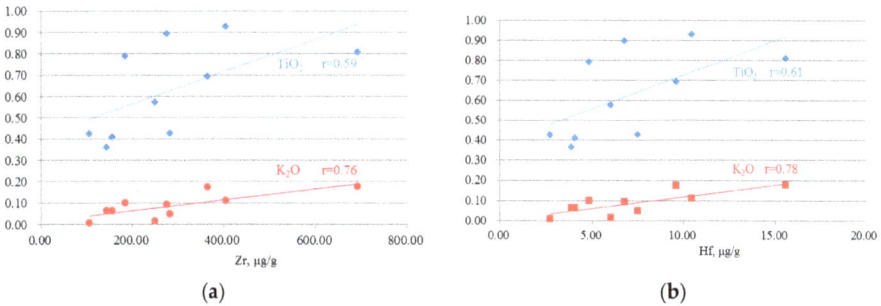

**Figure 13.** Correlations of Zr and Hf with: TiO$_2$ (**a**); and K$_2$O (**b**).

## 5. Conclusions

The Pingshuo coals of the Ningwu Coalfield have a medium-ash yield (averaging 21.42%) and an ultra-low-sulfur content (0.30%), with mean volatile matter and moisture contents of 33.95% and 2.67%, respectively.

The minerals in the Pingshuo coals mainly consist of kaolinite, followed by boehmite (averaging 5.09%), siderite (1.35%), anatase (0.29%), goyazite (0.8%), calcite (0.07%), apatite (0.01%) and florencite (0.33%).

The Pennsylvanian coals from the Pingshuo Mine contain higher proportions of SiO$_2$, Al$_2$O$_3$, TiO$_2$, and P$_2$O$_5$ and lower proportions of Fe$_2$O$_3$, CaO, K$_2$O, MgO and Na$_2$O compared with the average values for common Chinese coals [26].

Based on the CC values, Hg and Zr concentrations are considerably higher in the Pennsylvanian coals from the Pingshuo Mine. Concentrations of elements Cu, Ga, Se, Cd, Hf and Pb are slightly higher, and the those of Li, Be, F, Sc, V, Cr, Zn, Sr, Sn, Ba, Ta, W, Bi, Th and U are very close to common Chinese and world hard coals. The remaining trace elements are much lower than the averages for common Chinese and world hard coals.

The modes of occurrence of trace elements in the Pingshuo coals were preliminarily investigated by using cluster analysis and correlation analysis. Elements Li and Hg show strong positive correlations with ash yield and $SiO_2$, indicating that Li and Hg occur in mineral phases. Elements Sr, Ba, Be, As and Ga have strong positive correlations with CaO and $P_2O_5$, indicating that most of these elements may be affiliated with phosphates and carbonates or have an inorganic–organic affinity in the Pingshuo coals. Elements Zr and Hf have strong positive correlations with $TiO_2$ and $K_2O$, indicating part of them occur in anatase.

**Acknowledgments:** This work was supported by the National Key Basic Research Program of China (No. 2014CB238901) and the Key Program of the National Natural Science Foundation of China (No. 41330317). The authors are grateful to Shifeng Dai for his experimental and technical assistance. Special thanks are given to anonymous reviewers for their useful suggestions and comments.

**Author Contributions:** Ning Yang conceived the overall experimental strategy and analyzed testing data of coal samples. Shuheng Tang guided all experiments. Songhang Zhang and Yunyun Chen processed pictures. All authors participated in writing the manuscript.

**Conflicts of Interest:** The authors declare no conflict of interest.

## References

1. Seredin, V.V.; Finkelman, R.B. Metalliferous coals: A review of the main genetic and geochemical types. *Int. J. Coal Geol.* **2008**, *76*, 253–289. [CrossRef]
2. Seredin, V.V.; Dai, S.F.; Sun, Y.Z.; Chekryzhov, I.Y. Coal deposits as promising sources of rare metals for alternative power and energy-efficient technologies. *Appl. Geochem.* **2013**, *31*, 1–11. [CrossRef]
3. Dai, S.F.; Seredin, V.V.; Ward, C.R.; Jiang, J.H.; Hower, J.C.; Song, X.L.; Jiang, Y.F.; Wang, X.B.; Gornostaeva, T.; Li, X.; *et al.* Composition and modes of occurrence of minerals and elements in coal combustion products derived from high-Ge coals. *Int. J. Coal Geol.* **2014**, *121*, 79–97. [CrossRef]
4. Dai, S.F.; Jiang, Y.F.; Ward, C.R.; Gu, L.; Seredin, V.V.; Liu, H.D.; Zhou, D.; Wang, X.B.; Sun, Y.Z.; Zou, J.H.; *et al.* Mineralogical and geochemical compositions of the coal in the Guanbanwusu Mine, Inner Mongolia, China: Further evidence for the existence of an Al (Ga and REE) ore deposit in the Jungar Coalfield. *Int. J. Coal Geol.* **2012**, *98*, 10–40. [CrossRef]
5. Dai, S.F.; Li, D.; Chou, C.-L.; Zhao, L.; Zhang, Y.; Ren, D.Y.; Ma, Y.W.; Sun, Y.Y. Mineralogy and geochemistry of boehmite-rich coals: New insights from the Haerwusu Surface Mine, Jungar Coalfield, Inner Mongolia, China. *Int. J. Coal Geol.* **2008**, *74*, 185–202. [CrossRef]
6. Chen, J.; Chen, P.; Yao, D.X.; Liu, Z.; Wu, Y.S.; Liu, W.Z.; Hu, Y.B. Mineralogy and geochemistry of late Permian coals from the Donglin Coal Mine in the Nantong coalfield in Chongqing, southwestern China. *Int. J. Coal Geol.* **2015**, *149*, 24–40. [CrossRef]
7. Dai, S.F.; Li, T.; Seredin, V.V.; Ward, C.R.; Hower, J.C.; Zhou, Y.P.; Zhang, M.Q.; Song, X.L.; Song, W.J.; Zhao, C.L. Origin of minerals and elements in the late Permian coals, tonsteins, and host rocks of the Xinde Mine, Xuanwei, eastern Yunnan, China. *Int. J. Coal Geol.* **2014**, *121*, 53–78. [CrossRef]
8. Dai, S.F.; Yang, J.Y.; Ward, C.R.; Hower, J.C.; Liu, H.D.; Garrison, T.M.; French, D.; O'Keefe, J.M.K. Geochemical and mineralogical evidence for a coal-hosted uranium deposit in the Yili Basin, Xinjiang, northwestern China. *Ore Geol. Rev.* **2015**, *70*, 1–30. [CrossRef]
9. Eskenazy, G.M.; Valceva, S.P. Geochemistry of beryllium in the Mariza-east lignite deposit (Bulgaria). *Int. J. Coal Geol.* **2003**, *55*, 47–58. [CrossRef]
10. Gürdal, G. Abundances and modes of occurrence of trace elements in the Çan coals (Miocene), Çanakkale-Turkey. *Int. J. Coal Geol.* **2011**, *87*, 157–173. [CrossRef]
11. Hower, J.C.; Eble, C.F.; O'Keefe, J.M.K.; Dai, S.F.; Wang, P.P.; Xie, P.P.; Liu, J.J.; Ward, C.R.; French, D. Petrology, palynology, and geochemistry of Gray Hawk Coal (Early Pennsylvanian, Langsettian) in Eastern Kentucky, USA. *Minerals* **2015**, *5*, 592–622. [CrossRef]
12. Tang, M.Z.; Liu, X.L. Geology and sedimentary environmental analysis of bauxite deposits in Ningwu, Shanxi. *J. Geol. Miner. Resour. North China* **1996**, *11*, 580–585. (In Chinese)
13. Liu, D.M.; Yang, Q.; Tang, D.Z.; Kang, X.D.; Huang, W.H. Geochemistry of sulfur and elements in coals from the Antaibao surface mine, Pingshuo, Shanxi Province, China. *Int. J. Coal Geol.* **2001**, *46*, 51–64. [CrossRef]

14. American Society for Testing and Materials (ASTM) International. *Test Method for Moisture in the Analysis Sample of Coal and Coke*; ASTM D3173-11; ASTM International: West Conshohocken, PA, USA, 2011.

15. American Society for Testing and Materials (ASTM) International. *Test Method for Volatile Matter in the Analysis Sample of Coal and Coke*; ASTM D3175-11; ASTM International: West Conshohocken, PA, USA, 2011.

16. American Society for Testing and Materials (ASTM) International. *Test Method for Ash in the Analysis Sample of Coal and Coke from Coal*; ASTM D3174-11; ASTM International: West Conshohocken, PA, USA, 2011.

17. American Society for Testing and Materials (ASTM) International. *Test Methods for Total Sulfur in the Analysis Sample of Coal and Coke*; ASTM D3177-02; ASTM International: West Conshohocken, PA, USA, 2011.

18. Dai, S.F.; Li, T.J.; Jiang, Y.F.; Ward, C.R.; Hower, J.C.; Sun, J.H.; Liu, J.J.; Song, H.J.; Wei, J.P.; Li, Q.Q.; *et al.* Mineralogical and geochemical compositions of the Pennsylvanian coal in the Hailiushu Mine, Daqingshan Coalfield, Inner Mongolia, China: Implications of sediment-source region and acid hydrothermal solutions. *Int. J. Coal Geol.* **2015**, *137*, 92–110. [CrossRef]

19. Dai, S.F.; Zou, J.H.; Jiang, Y.F.; Ward, C.R.; Wang, X.B.; Li, T.; Xue, W.F.; Liu, S.D.; Tian, H.M.; Sun, X.H.; *et al.* Mineralogical and geochemical compositions of the Pennsylvanian coal in the Adaohai Mine, Daqingshan Coalfield, Inner Mongolia, China: Modes of occurrence and origin of diaspore, gorceixite, and ammonian illite. *Int. J. Coal Geol.* **2012**, *94*, 250–270. [CrossRef]

20. Wang, X.B.; Wang, R.X.; Wei, Q.; Wang, P.P.; Wei, J.P. Mineralogical and geochemical characteristics of late Permian coals from the Mahe Mine, Zhaotong Coalfield, Northeastern Yunnan, China. *Minerals* **2015**, *5*, 380–396. [CrossRef]

21. Dai, S.F.; Wang, X.B.; Zhou, Y.P.; Hower, J.C.; Li, D.H.; Chen, W.M.; Zhu, X.W.; Zou, J.H. Chemical and mineralogical compositions of silicic, mafic, and alkali tonsteins in the late Permian coals from the Songzao Coalfield, Chongqing, Southwest China. *Chem. Geol.* **2011**, *282*, 29–44. [CrossRef]

22. Li, X.; Dai, S.F.; Zhang, W.G.; Li, T.; Zheng, X.; Chen, W. Determination of As and Se in coal and coal combustion products using closed vessel microwave digestion and collision/reaction cell technology (CCT) of inductively coupled plasma mass spectrometry (ICP-MS). *Int. J. Coal Geol.* **2014**, *124*, 1–4.

23. China Coal Research Institute. *GB/T 15224.1-2004, Classification for Quality of Coal—Part 1: Ash*; Chinese National Standard. General Administration of Quality Supervision, Inspection and Quarantine of the People's Republic of China: Beijing, China, 2004. (In Chinese)

24. China Coal Research Institute. *MT/T849-2000, Classification for Volatile Matter of Coal*; Chinese National Standard. General Administration of Quality Supervision, Inspection and Quarantine of the People's Republic of China: Beijing, China, 2000. (In Chinese)

25. China Coal Science Research Institute Beijing Coal Chemical Research Branch. *MT/T850-2000, Classification for Total Moisture in Coal*; Chinese National Standard. General Administration of Quality Supervision, Inspection and Quarantine of the People's Republic of China: Beijing, China, 2000. (In Chinese)

26. China Coal Research Institute Beijing Coal Chemical Research Branch. *GB/T 15224.2-2004, Classification for Coal Quality—Part 2: Sulfur Content*; Chinese National Standard. General Administration of Quality Supervision, Inspection and Quarantine of the People's Republic of China: Beijing, China, 2004. (In Chinese)

27. Dai, S.F.; Ren, D.Y.; Chou, C.-L.; Li, S.S.; Jiang, Y.F. Mineralogy and geochemistry of the No. 6 Coal (Pennsylvanian) in the Junger Coalfield, Ordos Basin, China. *Int. J. Coal Geol.* **2006**, *66*, 253–270. [CrossRef]

28. Dai, S.F.; Luo, Y.B.; Seredin, V.V.; Ward, C.R.; Hower, J.C.; Zhao, L.; Liu, S.D.; Zhao, C.L.; Tian, H.M.; Zou, J.H. Revisiting the late Permian coal from the Huayingshan, Sichuan, southwestern China: Enrichment and occurrence modes of minerals and trace elements. *Int. J. Coal Geol.* **2014**, *122*, 110–128. [CrossRef]

29. Dai, S.F.; Ren, D.Y.; Chou, C.-L.; Finkelman, R.B.; Seredin, V.V.; Zhou, Y.P. Geochemistry of trace elements in Chinese coals: A review of abundances, genetic types, impacts on human health, and industrial utilization. *Int. J. Coal Geol.* **2012**, *94*, 3–21. [CrossRef]

30. Ketris, M.P.; Yudovich, Y.E. Estimations of clarkes for carbonaceous biolithes: World average for trace element contents in black shales and coals. *Int. J. Coal Geol.* **2009**, *78*, 135–148. [CrossRef]

31. Yang, J.Y. Concentrations and modes of occurrence of trace elements in the Late Permian coals from the Puan Coalfield, southwestern Guizhou, China. *Environ. Geochem. Health* **2006**, *28*, 567–576. [CrossRef] [PubMed]

32. Tang, S.S.; Sun, S.L.; Qin, Y.; Jiang, Y.F.; Wang, W.F. Distribution characteristics of sulfur and the main harmful trace elements in China's coal. *Acta Geol. Sin. Engl. Ed.* **2008**, *82*, 722–730.

33. Dai, S.F.; Seredin, V.V.; Ward, C.R.; Hower, J.C.; Xing, Y.W.; Zhang, W.G.; Song, W.J.; Wang, P.P. Enrichment of U-Se-Mo-Re-V in coals preserved within marine carbonate successions: Geochemical and mineralogical data from the Late Permian Guiding Coalfield, Guizhou, China. *Miner. Depos.* **2015**, *50*, 159–186. [CrossRef]

34. Dai, S.F.; Wang, X.B.; Seredin, V.V.; Hower, J.C.; Ward, C.R.; O'Keefe, J.M.K.; Huang, W.H.; Li, T.; Li, X.; Liu, H.D.; *et al.* Petrology, mineralogy, and geochemistry of the Ge-rich coal from the Wulantuga Ge ore deposit, Inner Mongolia, China: New data and genetic implications. *Int. J. Coal Geol.* **2012**, *90–91*, 72–99.

35. Zhao, L.; Ward, C.R.; French, D.; Graham, I.T. Major and trace element geochemistry of coals and intra-seam claystones from the Songzao Coalfield, SW China. *Minerals* **2015**, *5*, 870–893.

36. Hower, J.C.; Eble, C.F.; Pierce, B.S. Petrography, geochemistry and palynology of the Stockton coal bed (Middle Pennsylvanian), Martin County, Kentucky. *Int. J. Coal Geol.* **1996**, *31*, 195–215. [CrossRef]

37. Dai, S.F.; Hower, J.C.; Ward, C.R.; Guo, W.M.; Song, H.J.; O'Keefe, J.M.K.; Xie, P.P.; Hood, M.M.; Yan, X.Y. Elements and phosphorus minerals in the middle Jurassic inertinite-rich coals of the Muli Coalfield on the Tibetan Plateau. *Int. J. Coal Geol.* **2015**, *144–145*, 23–47. [CrossRef]

38. Yang, N.; Tang, S.H.; Zhang, S.H.; Chen, Y.Y. Mineralogical and geochemical compositions of the No. 5 Coal in Chuancaogedan Mine, Junger Coalfield, China. *Minerals* **2015**, *5*, 788–800. [CrossRef]

39. Dai, S.F.; Liu, J.J.; Ward, C.R.; Hower, J.C.; Xie, P.P.; Jiang, Y.F.; Hood, M.M.; O'Keefe, J.M.K.; Song, H.D. Petrological, geochemical, and mineralogical compositions of the low-Ge coals from the Shengli Coalfield, China: A comparative study with Ge-rich coals and a formation model for coal-hosted Ge ore deposit. *Ore Geol. Rev.* **2015**, *71*, 318–349. [CrossRef]

40. Dai, S.F.; Wang, P.P.; Ward, C.R.; Tang, Y.G.; Song, X.L.; Jiang, J.H.; Hower, J.C.; Li, T.; Seredin, V.V.; Wagner, N.J.; *et al.* Elemental and mineralogical anomalies in the coal-hosted Ge ore deposit of Lincang, Yunnan, southwestern China: Key role of $N_2$-$CO_2$-mixed hydrothermal solutions. *Int. J. Coal Geol.* **2015**, *152*, 19–46. [CrossRef]

41. Kolker, A. Minor element distribution in iron disulfides in coal: A geochemical review. *Int. J. Coal Geol.* **2012**, *94*, 32–43. [CrossRef]

42. Diehl, S.F.; Goldhaber, M.B.; Hatch, J.R. Modes of occurrence of mercury and other trace elements in coals from the warrior field, Black Warrior Basin, Northwestern Alabama. *Int. J. Coal Geol.* **2004**, *59*, 193–208.

43. Kolker, A.; Senior, C.; Alphen, C.V.; Koenig, A.; Geboy, N. Mercury and trace element distribution in density separates of a South African Highveld (#4) coal: Implications for mercury reduction and preparation of export coal. *Int. J. Coal Geol.* **2016**. [CrossRef]

44. Dai, S.F.; Zeng, R.S.; Sun, Y.Z. Enrichment of arsenic, antimony, mercury, and thallium in a Late Permian anthracite from Xingren, Guizhou, Southwest China. *Int. J. Coal Geol.* **2006**, *66*, 217–226. [CrossRef]

45. Kolker, A.; Senior, C.L.; Quick, J.C. Mercury in coal and the impact of coal quality on mercury emissions from combustion systems. *Appl. Geochem.* **2006**, *21*, 1821–1836. [CrossRef]

46. Li, J.; Zhuang, X.G.; Querol, X.; Font, O.; Izquierdo, M.; Wang, Z.M. New data on mineralogy and geochemistry of high-Ge coals in the Yimin coalfield, Inner Mongolia, China. *Int. J. Coal Geol.* **2014**, *125*, 10–21. [CrossRef]

47. Zhao, L.; Ward, C.R.; French, D.; Graham, I.T. Mineralogical composition of Late Permian coal seams in the Songzao Coalfield, southwestern China. *Int. J. Coal Geol.* **2013**, *116–117*, 208–226. [CrossRef]

48. Belkin, H.E.; Zheng, B.S.; Finkelman, R.B. Geochemistry of coals causing arsenismin southwest China. In *4th International Symposium on Environmental Geochemistry*; US Geological Survey Open-File Report; U.S. Geological Survey: Reston, VA, USA, 1997.

49. Ding, Z.H.; Zheng, B.S.; Long, J.P.; Belkin, H.E.; Finkelman, R.B.; Chen, C.G.; Zhou, D.; Zhou, Y. Geological and geochemical characteristics of high arsenic coals from endemic arsenosis areas in southwestern Guizhou Province. *Appl. Geochem.* **2001**, *16*, 1353–1360. [CrossRef]

50. Chou, C.-L. Abundances of sulfur, chlorine, and trace elements in Illinois Basin coals, USA. In Proceedings of the 14th Annual International Pittsburgh Coal Conference & Workshop, Taiyuan, China, 23–27 September 1997; Section 1, pp. 76–87.

51. Dai, S.F.; Liu, J.J.; Ward, C.R.; Hower, J.C.; French, D.; Jia, S.H.; Hood, M.M.; Garrison, T.M. Mineralogical and geochemical compositions of Late Permian coals and host rocks from the Guxu Coalfield, Sichuan Province, China, with emphasis on enrichment of rare metals. *Int. J. Coal Geol.* **2015**. [CrossRef]

*Article*

# Radioactivity of Natural Nuclides ($^{40}$K, $^{238}$U, $^{232}$Th, $^{226}$Ra) in Coals from Eastern Yunnan, China

Xin Wang [1,3], Qiyan Feng [1,*], Ruoyu Sun [2] and Guijian Liu [1,2,*]

1 School of Environment Science and Spatial Informatics, China University of Mining and Technology, Xuzhou 221000, China; wangxinhello@126.com
2 Chinese Academy of Sciences Key Laboratory of Crust-Mantle Materials and the Environments, School of Earth and Space Sciences, University of Science and Technology of China, Hefei 230026, China; rousun1986@gmail.com
3 College of Life Sciences, Huai Bei Normal University, Huaibei 235000, China
* Correspondence: fqycumt@126.com (Q.F.); lgj@ustc.edu.cn (G.L.); Tel.: +86-516-8358-1303 (Q.F.); Fax: +86-551-6362-1485 (G.L.)

Academic Editors: Shifeng Dai and Mostafa Fayek
Received: 29 June 2015 ; Accepted: 18 September 2015 ; Published: 30 September 2015

**Abstract:** The naturally occurring primordial radionuclides in coals might exhibit high radioactivity, and can be exported to the surrounding environment during coal combustion. In this study, nine coal samples were collected from eastern Yunnan coal deposits, China, aiming at characterizing the overall radioactivity of some typical nuclides (*i.e.*, $^{40}$K, $^{238}$U, $^{232}$Th, $^{226}$Ra) and assessing their ecological impact. The mean activity concentrations of $^{238}$U, $^{232}$Th, $^{40}$K and $^{226}$Ra are 63.86 (17.70–92.30 Bq·kg$^{-1}$), 23.76 (11.10–37.10 Bq·kg$^{-1}$), 96.84 (30.60–229.30 Bq·kg$^{-1}$) and 28.09 Bq·kg$^{-1}$ (3.10–61.80 Bq·kg$^{-1}$), respectively. Both $^{238}$U and $^{232}$Th have high correlations with ash yield of coals, suggesting their inorganic origins. The overall environmental effect of natural radionuclides in studied coals is considered to be negligible, as assessed by related indexes (*i.e.*, radium equivalent activity, air-adsorbed dose rate, annual effective dose, and external hazard index). However, the absorbed dose rates values are higher than the average value of global primordial radiation and the Chinese natural gamma radiation dose rate.

**Keywords:** coal; radioactivity; nuclide; Yunnan; China

## 1. Introduction

Combustion of coals that are enriched in radioactive nuclides, particularly those of U series, possibly magnifies the background levels of radioactive nuclides in the surrounding environment and increases human exposure risks [1–4]. High concentrations of radioactive nuclides in coals have been reported worldwide. For example, Barber and Giorgio (1977) [5] found that the radioactive level of $^{226}$Ra in a coal sample from Illinois, USA, could reach up to 1500 Bq·kg$^{-1}$. Measurement of $^{226}$Ra on the lignite originated from Kotili of Xanthi Prefecture, Northern Greece, showed a radioactivity as high as 2600 Bq·kg$^{-1}$, which is more than one order of magnitude higher than that (110–260 Bq·kg$^{-1}$) of lignite commonly used in coal-fired power plants [6,7]. High levels of radioactive nuclides in Chinese coals are rarely reported. Jiang *et al.* (1989) [8] reported that the weighted means of $^{226}$Ra, $^{232}$Th and $^{40}$K in Chinese coals are 36 (2–2300 Bq·kg$^{-1}$), 30 (4–110 Bq·kg$^{-1}$) and 104 Bq·kg$^{-1}$ (5–750 Bq·kg$^{-1}$), respectively. Liu *et al.* (2007) [9] calculated a weighted mean of 79.5 ± 45, 73.9 ± 53, 40.3 ± 34 and 152.4 ± 21 Bq·kg$^{-1}$ for $^{238}$U, $^{226}$Ra, $^{232}$Th and $^{40}$K, respectively, from ~1000 samples collected from several Chinese coal mines. However, Dai *et al.* found that high concentration of U occurs in some coals from southwestern China, e.g., Linchang and Yanshan in Yunnan province [10,11] and Guiding in Guizhou province [12].

The coal deposits in eastern Yunnan province, China, have been identified to enrich high levels of natural radionuclides [13–17]. However, their spatial variation, modes of occurrences, and controlling factors are rarely illustrated. In this study, the radioactive nuclides in coal deposits of this area are reassessed, in an effort to generalize a basic knowledge on their potential environmental impacts.

## 2. Geological Setting

Eastern Yunnan province is located at the southern part of the Yangzi Craton. This region is composed of intermediate massif and suture zones, and has a complicated geological setting due to extensive development of faulting and folding systems [18–21]. In addition to coal deposits, this area is also abundant in metal resources, such as gold, antimony, germanium, gallium, arsenic, mercury, and thallium. The coal-forming periods in eastern Yunnan include the Early Carboniferous, Late Permian, Late Triassic and Neogene [18,19]. The main coal-bearing strata are the Upper Carboniferous Wanshoushan Formation, the Lower Permian Longtan Formation and the Miocene Xiaolongtan Formation. Different ranks of coal, from gas coal (subbituminous coal) to anthracite, have been developed. The frequent eruptions of volcanoes during coal-forming periods and the alternated marine and terrestrial facies have been suggested to be the main reasons causing the geochemical and mineralogical anomalies of regional coals [22,23].

## 3. Sampling and Analysis

Nine coal samples from eight active coalmines (Maoergou, Kelang, Changsheng, Xujiayuan, Shizhuang, Xingying, Tuobai and Gongqing) in eastern Yunnan were sampled (Figure 1 and Table 1) according to the Chinese Standard Method GB/T 482-2008 [24]. The samples include anthracite from Zhaotong (coalmines of Maoergou, Changsheng, Xujiayuan and Shizhuang), lignite from Kunming (Kelang coalmine) and coking (bituminous) coals from Qujing (coal mines of Xingying and Gongqing) and Honghezhou (Tuobai coal mine) (Table 1), on basis of GB5751-86 [25]. Each sample was taken by cutting 10-cm wide and 10-cm deep into the coalface to represent the full height or a section of the coal. Upon collection, all samples were stored in polyethylene bags to preclude contamination and weathering.

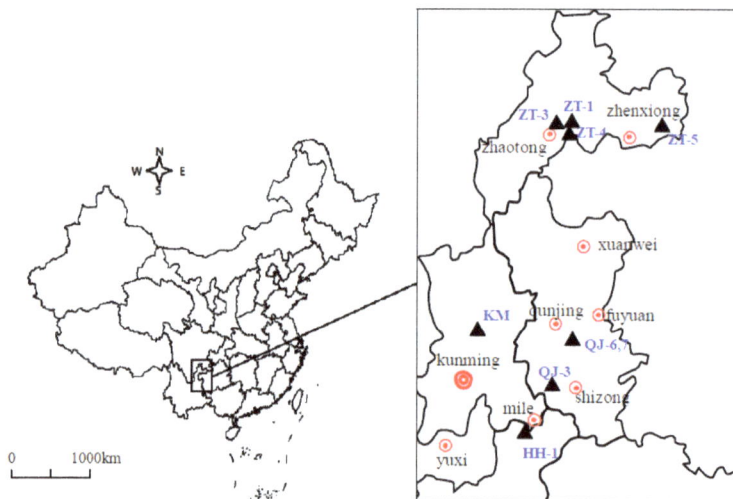

**Figure 1.** Sampling locations (filled triangle) of the coalmines in eastern Yunnan, China (HH: Honghezhou; KM: Kunming; QJ: Qujing; ZT: Zhaotong).

*Minerals* **2015**, *5*, 637–646

Table 1. Proximate analysis (%), sulfur (%), radioactive elements (μg/g) of the eastern Yunnan coal.

| Sample ID | Coal Mine | Coal Rank | Proximate Analysis | | | $S_{t,d}$ | U | Th | Ra |
|---|---|---|---|---|---|---|---|---|---|
| | | | $M_{ad}$ | $A_d$ | $V_{daf}$ | | | | |
| ZT-1 | Maoergou | Anthracite | 1.20 | 29.24 | 12.66 | 4.00 | 5.50 | 9.65 | 1.87 |
| KM | Kelang | Lignite | 9.97 | 15.46 | 56.34 | 1.95 | 7.32 | 2.71 | 2.49 |
| ZT-3 | Changsheng | Anthracite | 0.61 | 4.71 | 5.67 | 0.83 | 0.36 | 1.06 | 0.12 |
| ZT-4 | Xujiayuan | Anthracite | 0.47 | 9.47 | 7.00 | 1.68 | 0.54 | 1.93 | 0.18 |
| ZT-5 | Shizhuang | Anthracite | 0.82 | 32.12 | 12.42 | 3.22 | 3.72 | 7.59 | 1.26 |
| QJ-3 | Xingying | Coking coal | 0.62 | 14.04 | 19.96 | 0.46 | 1.54 | 3.46 | 0.52 |
| HH-1 | Tuobai | Coking coal | 0.55 | 22.81 | 17.99 | 4.15 | 4.76 | 8.41 | 1.62 |
| QJ-6 | Gongqing | Coking coal | 0.69 | 10.85 | 22.64 | 1.74 | 1.64 | 7.47 | 1.52 |
| QJ-7 | Gongqing | Coking coal | 0.75 | 19.81 | 20.48 | 0.93 | 1.74 | 3.78 | 0.59 |

M, moisture; A, ash yield; V, volatile matter; $S_t$, total sulfur; ad, air-dry basis; d, dry basis; daf, dry and ash-free basis.

Coal samples were air-dried, pulverized using a jaw crusher and passed through a 100-mesh sieve before analysis. Proximate analysis (ash yield, volatile matter, moisture) and total sulfur were determined according to ASTM standards [26–29]. Powdered samples were digested using mixed acids ($HNO_3$:$HClO_4$:HF = 3:1:1) before U and Th measurement by inductively coupled-plasma mass spectrometry (ICP-MS). The method detection limit is 0.002 ng/mL for U and 0.003 ng/mL for Th. The uncertainty of elemental concentration is within ± 10% as evaluated by soil standard GBW 07114 [30] (GBW: national standard substance). The standard deviation for replicate sample digests is less than 10%.

Before radioactive measurement, powdered samples were dried in a temperature controlled furnace at 70 ± 1 °C for 24 h to remove moisture. After cooling down to room temperature, ~500 g of samples were sealed in gas-light, Rn impermeable cylindrical polyethylene containers (Inner Diameter = 40 mm). The samples were sealed and stored for 40 days in order to reach a radioactive equilibrium between $^{226}$Ra and its daughters. A GEM60P4-83 high-purity Ge gamma ray spectrometer (ORTEC, Oak Ridge, TN, USA) shielded from background radiation using Pb bricks was employed to measure the most prominent γ-ray energies of the radionuclides in equilibrium with $^{238}$U, $^{226}$Ra, $^{40}$K and $^{232}$Th. The gamma spectrometry measurements were made with an energy resolution of 1.95 keV at the 1.332 Mev of $^{60}$Co and the relative efficiency of 60%. The activity concentrations of $^{238}$U was determined through its daughter product $^{234}$Th, $^{226}$Ra by its daughter products $^{214}$Bi and $^{214}$Pb and $^{232}$Th by its daughter products $^{228}$Ac and $^{208}$Tl [31]. The activity concentration of $^{40}$K was obtained from its single gamma ray lines of 1460.83 keV. The detector output was connected to a spectroscopy amplifier. The counting time for each sample and background was 10,000 s. The energy and absolute efficiency calibrations of the spectrometers were carried out using calibration sources with energy range covering the studied nuclides. The activity concentrations (*As*) for the natural radionuclides is calculated as:

$$A_S = \frac{N}{\varepsilon P_r M t} \tag{1}$$

where $N$ is the net counting rate for a specific gamma line corrected for background, $\varepsilon$ is the detector efficiency of a specific gamma ray, $Pr$ is the absolute transition probability of gamma decay, $M$ is the mass of the sample (kg), and $t$ is counting time (s). The national standard reference materials (GBW04127 [32], GBW04325 [33] and GBW04326 [34]) were used for quality assurance, and the typically obtained values are within 5% of certified values. The statistic uncertainties of activity concentrations for radionuclides are estimated less then 2%.

## 4. Results and Discussion

### 4.1. Coal Basic Parameters

The studied coals have ash yields ranging from 4.71% to 32.12% (mean = 17.61%), and total sulfur ranging from 0.46% to 4.15% (mean = 2.11%), classifying as low ash, medium-high sulfur coals. The volatile matter values are generally low for most coal samples, varying from 5.67% to 23.28%, with the exception of one sample (KM) of up to 56.34%. According to the proximate analysis and on the basis of the Chinese Standard GB5751-86 [25], the studied coal samples are classified into different ranks: lignite, coking coal (bituminous coal) and anthracite (Table 1).

### 4.2. Concentrations of U and Th in Coals

The highest U concentration is observed in Kelang mine (KM) of 7.32 µg/g, whereas the lowest U concentration in Changsheng mine (ZT-3) of 0.36 µg/g (Figure 1). The average U concentration in studied coals is 3.01 µg/g, which is slightly higher than that in Chinese coals (2.71 µg/g) [35] and world coals (2.40 µg/g) [36], but is significantly lower than that in coals from Yunnan, China (7.08 µg/g) reported by Li *et al.* [37]. For Th in coals, the highest concentration is 9.65 µg/g in Maoergou mine (ZT-1), and the lowest concentration is 1.06 µg/g in Changsheng mine (ZT-3). The average Th concentration in studied coals is 5.12 µg/g, which is slightly higher than that in Chinese coals (4.93 µg/g) [35] and world coals (3.3 µg/g) [36], but is still lower than that in coals from Yunnan, China (5.83 µg/g) reported by Li *et al.* [37].

### 4.3. Levels of Radioactivity in Coals

The radioactivity of these radionuclides in studied coals is summarized in Table 2. The mean radioactivity of $^{238}$U, $^{226}$Ra, $^{232}$Th and $^{40}$K are 63.86, 28.09, 23.76 and 96.84 Bq·kg$^{-1}$, respectively. The largest contributor to the total radioactivity is $^{40}$K for these coal samples.

**Table 2.** Comparison of radioactivity (Bq·kg$^{-1}$) of nuclides in coals of eastern Yunnan, China with previous publications.

| Sample ID | $^{238}$U | $^{232}$Th | $^{226}$Ra | $^{40}$K |
|---|---|---|---|---|
| ZT-1 | 84.90 | 36.00 | 46.40 | 165.00 |
| KM | 68.80 | 17.00 | 61.80 | 229.30 |
| ZT-3 | 17.70 | 10.30 | 3.10 | 30.60 |
| ZT-4 | 29.80 | 15.80 | 4.50 | 78.70 |
| ZT-5 | 92.30 | 37.10 | 31.30 | 134.20 |
| QJ-3 | 65.10 | 32.30 | 13.00 | 46.10 |
| HH-1 | 78.20 | 33.70 | 40.20 | 65.00 |
| QJ-6 | 59.30 | 11.10 | 37.80 | 54.90 |
| QJ-7 | 78.60 | 20.50 | 14.60 | 67.80 |
| mean | 63.86 ± 25 | 23.76 ± 11 | 28.09 ± 20 | 96.84 ± 66 |
| Yunnan [a] | 36.60 ± 31 | 16.50 ± 15 | 39.20 ± 4 | 36.10 ± 35 |
| China [a] | 64.90 ± 32 | 37.50 ± 18 | 49.40 ± 31 | 106.00 ± 27 |
| Poland [b] | 23.50 | 14.30 | 18.10 | 129.90 |
| Australia [c] | 25.00 | 24.00 | 21.00 | 75.00 |

[a] Liu *et al.* (2007) [12]; [b] Bemet *et al.* (2002) [38]; [c] Fardyet *et al.* (1989) [39].

For comparison, the radioactivity of these radionuclides in coals from Yunnan province, China, and other countries are shown in Table 2. The average radioactivity of $^{238}$U and $^{40}$K in studied coals are 63.86 ± 25 and 96.84 ± 66 Bq·kg$^{-1}$, respectively. Both values are comparable to those in Chinese coals, but are elevated by a factor of 1.7–2.7 as compared to those previously measured in Yunnan coals [11], Radioactivity of $^{232}$Th and $^{226}$Ra in both previously measured Yunnan coals and this study are slightly lower than those in Chinese coals [12]. It is noted that radionuclides in specific coals can reach to rather high levels such as $^{238}$U (84.90 Bq·kg$^{-1}$), $^{232}$Th (36.00 Bq·kg$^{-1}$) and $^{40}$K (165.00 Bq·kg$^{-1}$) in Maoergou Mine (ZT-1), and $^{238}$U (68.80 Bq·kg$^{-1}$), $^{226}$Ra (61.80 Bq·kg$^{-1}$) and $^{40}$K (229.30 Bq·kg$^{-1}$) in Kelang Mine (KM) (Table 2). Australia coals generally have low levels of radioactivity of nuclides, while Poland coals have relatively high radioactivity of $^{40}$K [38,39].

### 4.4. Factors Influencing Radioactivity of Nuclides

Coles *et al.* (1978) [40] identified radionuclides associated with organic matter and sulfides are more easily volatilized than those with affinity to alumino-silicate minerals. The correlation analysis of these radionuclides with coal ash yield is used to determine their preliminary modes of occurrence (Figure 2). As shown in Figure 2, the $^{238}$U has a positive correlation with ash yield ($R^2 = 0.81$), indicating a dominant inorganic affinity. However, the correlation between U and ash yield is insignificant albeit positive ($R^2 = 0.24$). Uranium can reside in coal matrix in a highly disperse, organic state, or as discrete minerals such as brannerite, uraninite and coffinite, depending on the coal-forming environment [10,40,41]. Seredin and Finkelman (2008) [41] and Dai *et al.* (2014) [10] showed that U in U-bearing coal deposits worldwide is mainly associated with the organic matter, and that only a small proportion of the U occurs in U-bearing minerals. The inferred modes of occurrence for $^{238}$U and U by statistic correction method are not consistent in present study, which need further confirmation.

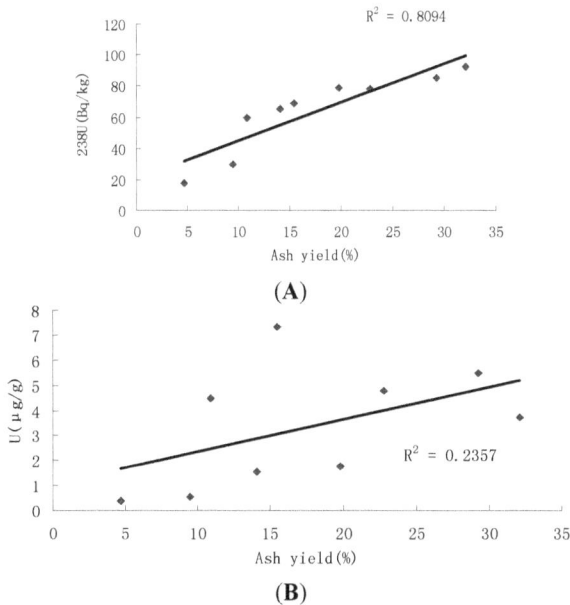

**Figure 2.** Correlations of $^{238}$U (A) and U (B) with ash yield.

Thorium, Ra and K are commonly considered as refractory elements and associated with alumino-silicate ash of coal [40]. Thorium is generally associated with zircon ($ZrSiO_4$) [42], a highly chemical-resistant mineral. Zircons in coals can have both volcanic and authigenic origins [43,44]. Besides, Th can be adsorbed by clay minerals in coals [45]. The positive correlations (Figure 3) of $^{232}$Th radioactivity ($R^2 = 0.73$) and Th ($R^2 = 0.58$) with ash yield demonstrate a dominant inorganic affinity of Th. It has been shown that the radioactivity of $^{226}$Ra in Indian coal slag is two times higher than that in feed coal [35,36], implying a dominant inorganic Ra state. A highly positive correlation of Ra with ash yield has been observed in our study ($R^2 = 0.50$). However, the correlative of $^{226}$Ra with ash yield is weak (Figure 4).

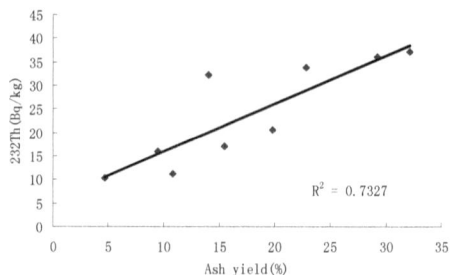

**(A)**

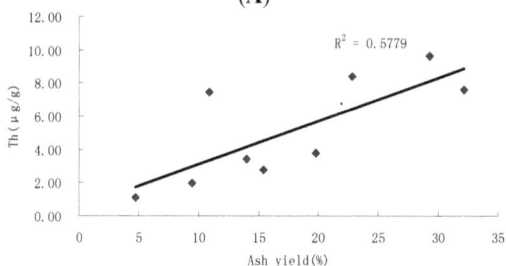

**(B)**

**Figure 3.** Correlations of $^{232}$Th (**A**) and Th (**B**) with ash yield.

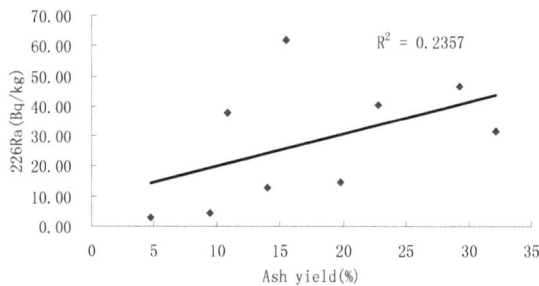

**(A)**

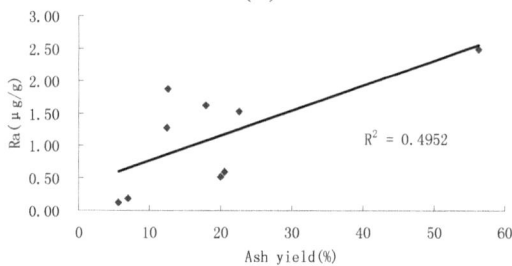

**(B)**

**Figure 4.** Correlations of $^{226}$Ra (**A**) and Ra (**B**) with ash yield.

*4.5. Radioactive Impacts of Nuclides*

The assessment of the radioactive impacts of nuclides in coals has an important environmental significance, since coal is abundantly used in power generation. Emissions from coal-fired power plants in gaseous and particulate forms containing nuclides probably discharge and accumulate in the surrounding environment, thus causing radiation exposures to the local population [46,47]. Inhalation of fly ash emitted from the stacks and the ingestion of foodstuffs receiving atmospheric deposition are the main pathways to increase human radiation exposure [48].

It has been hypothesized that 370 Bq·kg$^{-1}$ of $^{226}$Ra or 259 Bq·kg$^{-1}$ of $^{232}$Th or 4810 Bq·kg$^{-1}$ of $^{40}$K could produce the same gamma dose rate [31,49]. The Ra equivalent radioactivity, the internal exposure index ($I_{Ra}$) and the external hazard index ($I_r$) can be calculated according to Beretka and Mathew (1985) [50] and GB6566-2010 [51] as:

$$Ra_{eq} = C_{Ra} + 1.43C_{Th} + 0.077C_K \tag{2}$$

$$I_{Ra} = \frac{C_{Ra}}{200} \leqslant 1 \tag{3}$$

$$I_r = \frac{C_{Ra}}{370} + \frac{C_{Th}}{259} + \frac{C_K}{4810} \leqslant 1 \tag{4}$$

where $C_{Ra}$, $C_{Th}$ and $C_K$, are the radioactivity of $^{226}$Ra, $^{232}$Th and $^{40}$K, respectively. The maximum value of $Ra_{eq}$ in construction materials is set to 370 Bq·kg$^{-1}$ for safe use [52], the maximum values of $I_{Ra}$ and $I_r$ must be lower than 1.0 according to GB6566-2010 [48]. The calculated values of $Ra_{eq}$ for studied coals vary from 19.97 to 109.43 Bq·kg$^{-1}$ (Table 3), with an average of 68.84 Bq·kg$^{-1}$, which is lower than the recommended limit. The calculated values of $I_r$ and $I_{Ra}$ are generally below than 0.3 significantly lower than unity.

**Table 3.** Radium equivalent activity (*Raeq*), external hazard index (*$I_r$*), internal exposure index (*$I_{Ra}$*), air absorbed dose rates (D (nGy/h)) and annual effective dose (AED) of radionuclides in coals from eastern Yunnan, China.

| Sample ID | *Raeq* | *$I_r$* | *$I_{Ra}$* | D (nGy/h) | AED |
|-----------|--------|---------|------------|-----------|-------|
| ZT-1 | 110.59 | 0.30 | 0.23 | 50.06 | 61.40 |
| KM | 103.77 | 0.28 | 0.31 | 48.38 | 59.33 |
| ZT-3 | 20.19 | 0.05 | 0.02 | 8.93 | 10.95 |
| ZT-4 | 33.15 | 0.09 | 0.02 | 14.90 | 18.28 |
| ZT-5 | 94.69 | 0.26 | 0.16 | 42.47 | 52.08 |
| QJ-3 | 62.74 | 0.17 | 0.07 | 27.44 | 33.65 |
| HH-1 | 93.40 | 0.25 | 0.20 | 41.64 | 51.06 |
| QJ-6 | 57.90 | 0.16 | 0.19 | 26.46 | 32.45 |
| QJ-7 | 49.14 | 0.13 | 0.07 | 21.95 | 26.92 |
| mean | 69.51 | 0.19 | 0.14 | 31.36 | 38.46 |

The total outdoor air absorbed dose rate (nGy/h) due to terrestrial gamma rays at 1 m above the ground is calculated from $^{226}$Ra, $^{232}$Th and $^{40}$K. The conversion factors used to calculate the absorbed dose rates are given as follows [53]:

$$D(nGy/h) = 0.462C_{Ra} + 0.604C_{Th} + 0.0417C_K \tag{5}$$

The calculated absorbed dose rates range from 8.93 to 50.06 nGy/h with a mean value of 31.36 nGy/h (Table 3), which is lower than the average value of global primordial radiation of 59 nGy/h [53] and the Chinese natural gamma radiation dose rate of 63.0 nGy/h [53].

To estimate the annual effective dose rates, the conversion coefficient from the absorbed dose in air to the effective dose (0.7 nGy/h) and the outdoor occupancy factor (20%) recommended by UNSCEAR (2000) [53] are used. The annual effective dose rate is calculated by the following formula [52]:

$$\text{Effective dose rate } (\mu Sv/year) = D \ (nGy/h) \ \times \ 8760 \ (h/year) \ \times \ 0.7 \times \ 10^6 \mu Sv/10^9 nGy \times 0.2 \quad (6)$$

The calculated effective dose rates vary from 10.95 to 61.40 μSv/year with an average value of 38.46 μSv/year.

## 5. Conclusions

This study provides an initial assessment on the natural radionuclides in coals from eastern Yunnan, China. The mean specific activity concentrations of radionuclides $^{238}$U, $^{226}$Ra, $^{232}$Th and $^{40}$K are 63.86, 28.09, 23.76 and 96.84 Bq· kg$^{-1}$ in coals, respectively. Radioactivity of $^{238}$U and $^{40}$K in studied coal are elevated by a factor of 1.7–2.7 as compared to previously measured Yunnan coals. Radioactivity of $^{238}$U and $^{40}$K in all Yunnan coals is slightly lower than Chinese coals. Radioactivity of $^{238}$U and $^{232}$Th is closely correlated with coal ash, suggesting their inorganic origins. The environmental effect of natural radionuclides is considered to be negligible as evaluated by indexes of $Ra_{eq}$, $I_R$ and $I_r$. The absorbed dose rates D (nGy/h) values are lower than the average value of global primordial radiation and the Chinese natural gamma radiation dose rate.

**Acknowledgments: Acknowledgments:** This work was supported by the National Basic Research Program of China (973 Program, 2014CB238903), the National Natural Science Foundation of China (No. 41173032 and 41373110). We acknowledge editors and reviewers for polishing the language of the paper and for in-depth discussion.

**Author Contributions: Author Contributions:** Guijian Liu and Qiyan Feng contributed to the conception of the study. Guijian Liu, Qiyan Feng, Ruoyu Sun and Xin Wang contributed significantly to analysis and manuscript preparation; Xin Wang performed the data analyses and wrote the manuscript; Ruoyu Sun helped perform the analysis with constructive discussions. All authors read and approved the manuscript.

**Conflicts of Interest: Conflicts of Interest:** The authors declare no conflict of interest.

## References

1. Papastefanou, C. Escaping radioactivity from coal-fired power plants (CPPs) due to coal burning and the associated hazards: A review. *J. Environ. Radioact.* **2010**, *101*, 191–200. [CrossRef] [PubMed]
2. Eisenbud, M.; Petrow, H.G. Radioactivity in the atmospheric effluents of power plants that use fossil fuels. *Science* **1964**, *144*, 288–289. [CrossRef] [PubMed]
3. Jaworowski, Z.; Bilkiewicz, J.; Zylicz, E. $^{226}$Ra in contemporary and fossil snow. *Health Phys.* **1971**, *20*, 449–450. [PubMed]
4. Kirchner, H.; Merz, E.; Schiffers, A. Radioaktive emissionen aus mit rheinischer braunkohle befeuerten kraftwerksanlagen. *Braunkohle* **1974**, *11*, 340–345.
5. Barber, D.E.; Giorgio, H.R. Gamma-ray activity in bituminous, sub-bituminous and lignite coals. *Health Phys.* **1977**, *32*, 83–88. [CrossRef] [PubMed]
6. Papastefanou, C.; Charalambous, S. On the radioactivity of fly ashes from coal power plants. *Z. Naturforschung A* **1979**, *34*, 533–537. [CrossRef]
7. Papastefanou, C.; Charalambous, S. Hazards from radioactivity of fly ash from Greek coal power plants (CPP). In Proceedings of the Fifth International Congress of the International Radiation Protection Association (IRPA), Jerusalem, Israel, 9–14 March 1980; Volume 3, pp. 161–165.
8. Jiang, X.W.; Liu, Q.S.; Li, R.X.; Bai, G.; Lin, Z.C.; Liu, X.H.; Pan, S.; Gan, L.; Zhu, L. Level of natural radionuclides in coal in China. *Radiat. Prot.* **1989**, *9*, 181–188, (In Chinese).
9. Liu, F.D.; Pan, Z.Q.; Liu, S.L.; Chen, L.; Wang, C.H.; Liao, H.T.; Wu, Y.H.; Wang, N.P. Investigation and analysis of the content of natural radionuclides at coal mines in China. *Radiat. Prot.* **2007**, *27*, 171–180, (In Chinese).

10. Dai, S.; Wang, P.; Ward, C.R.; Tang, Y.; Song, X.; Jiang, J.; Hower, J.C.; Li, T.; Seredin, V.V.; Wagner, N.J.; *et al.* Elemental and mineralogical anomalies in the coal-hosted Ge ore deposit of Lincang, Yunnan, southwestern China: Key role of $N_2$-$CO_2$-mixed hydrothermal solutions. *Int. J. Coal Geol.* **2014**. [CrossRef]

11. Dai, S.; Ren, D.; Zhou, Y.; Chou, C.L.; Wang, X.; Zhao, L.; Zhu, X. Mineralogy and geochemistry of a superhigh-organic-sulfur coal, Yanshan Coalfield, Yunnan, China: Evidence for a volcanic ash component and influence by submarine exhalation. *Chem. Geol.* **2008**, *255*, 182–194. [CrossRef]

12. Dai, S.; Seredin, V.V.; Ward, C.R.; Hower, J.C.; Xing, Y.; Zhang, W.; Song, W.; Wang, P. Enrichment of U-Se-Mo-Re-V in coals preserved within marine carbonate successions: Geochemical and mineralogical data from the Late Permian Guiding Coalfield, Guizhou, China. *Miner. Depos.* **2015**, *50*, 159–186. [CrossRef]

13. Huang, W.H.; Wan, H.; Finkelman, R.B.; Tang, X.; Zhao, Z. Distribution of uranium in the main coalfields of China. *Energy Explor. Exploit.* **2012**, *30*, 819–835. [CrossRef]

14. Xi, W.S. The survey of uranium in coal in Yunnan province. *Coal Geol. China* **1992**, *4*, 22, (In Chinese).

15. Zhou, Y.P. The survey of some trace elements and toxic elements in Yunnan province. *Sci. Technol. Yunnan Coal* **1985**, *3*, 2–8, (In Chinese).

16. Yu, Y.L. The survey of natural radioactivity level of coal mine in Yunnan. *Chin. J. Radio Health* **2007**, *16*, 196–198, (In Chinese).

17. Xiong, Z.W.; Yu, Y.L.; You, M.; Guo, C.L.; Zhou, S.K.; Yu, Z.X. Analysis of environment contamination from concomitant radioactivity of coal mine source. *J. China Coal Soc.* **2007**, *32*, 762–766, (In Chinese ).

18. Li, J.; Cao, D.Y.; Lin, Y.C.; Tao, Z.G.; Zhang, S.; Wang, J.; Yao, Z. Study on tectonic analysis of coal controlled structure mode in Yunnan province. *Coal Sci. Technol.* **2011**, *39*, 100–103, (In Chinese).

19. Tao, Z.G.; Cao, D.Y.; Li, J.; Shi, X.Y.; Lin, Y.C.; Wang, J. Study on coalfield structural framwork and coal measures hosting pattern in eastern Yunnan area. *Coal Geol. China* **2011**, *23*, 56–59, (In Chinese).

20. Zhou, Y.P. Distribution of arsenic in coal of Yunnan province, China and its controlling factors. *Int. J. Coal Geol.* **1992**, *20*, 85–98. [CrossRef]

21. Zhou, Y.P.; Bohor, B.F.; Ren, Y.L. Trace element geochemistry of altered volcanic ash layers (tonsteins) in Late Permian coal-bearing formations of eastern Yunnan and western Guizhou Provinces, China. *Int. J. Coal Geol.* **2000**, *44*, 305–324. [CrossRef]

22. Wang, X.C.; Zhang, Y.C.; Pan, R.Q.; Liu, C.R. *Sedimentary Environments and Coal Accumulation of Late Permian Coal Formation in Western Guizhou, Southern Sichuan, and Eastern Yunnan, China*; Chongqing University Press: Chongqing, China, 1996; pp. 124–155. (In Chinese)

23. Miao, Q. *Study on Gas Geology Condition of Coal Mines in Yunnan*; Coal Industry Publishing House: Beijing, China, 2013; pp. 61–75. (In Chinese)

24. National Coal Standardization Technical Committee. *Sampling of Coal Seams*; GB/T482-2008 C.S.; Standard Press of China: Beijing, China, 2008. (In Chinese)

25. China's Ministry of Coal Industry. *Chinese Standard for Coal Classification*; GB5751-86 C.S.; Standard Press of China: Beijing, China, 1986. (In Chinese)

26. American Society for Testing and Materials (ASTM). *Test Method for Moisture in the Analysis Sample of Coal and Coke*; D3173-11, A.S.; American Society for Testing and Materials International: West Conshohocken, PA, USA, 2011.

27. American Society for Testing and Materials (ASTM). *Test Method for ash in the Analysis Sample of Coal and Coke*; D3174-11, A.S.; American Society for Testing and Materials International: West Conshohocken, PA, USA, 2011.

28. American Society for Testing and Materials (ASTM). *Test Method for Volatile Matter in the Analysis Sample of Coal and Coke*; D3175-11, A.S.; American Society for Testing and Materials International: West Conshohocken, PA, USA, 2011.

29. American Society for Testing and Materials (ASTM). *Test Methods for Total Sulfur in the Analysis Sample of Coal and Coke*; D3177-02, A.S. (Reapproved 2007); American Society for Testing and Materials International: West Conshohocken, PA, USA, 2011.

30. National Standard Substance: Standard Substance for Rock Composition Analysis, No. GBW07114.

31. Cevik, U.; Damla, N.; Koz, B.; Kaya, S. Radiological characterization around the Afsin-Elbistan coal-fired power plant in Turkey. *Energy Fuels* **2008**, *22*, 428–432. [CrossRef]

32. National Standard Substance: Natural Uranium and Thorium Ore Standard Material, No. GBW04127.

33. National Standard Substance: Thorium Powder Standard Radioactive Source, No. GBW04325.

34. National Standard Substance: Potassium-40 Powder Standard Radioactive Source, No. GBW04326.
35. Ren, D.Y.; Zhao, F.H.; Wang, Y.Q.; Yang, S.J. Distributions of minor and trace elements in Chinese coals. *Int. J. Coal Geol.* **1999**, *40*, 109–118. [CrossRef]
36. Ketris, M.P.; Yudovich, Y.E. Estimations of clarkes for Carbonaceous biolithes: World averages for trace element contents in black shales and coals. *Int. J. Coal Geol.* **2009**, *78*, 135–148. [CrossRef]
37. Li, D.H.; Tang, Y.G.; Chen, K.; Deng, T.; Cheng, F.P.; Liu, D. Distribution of twelve toxic trace elements in coals from Southwest China. *J. China Univ. Min. Technol.* **2006**, *35*, 15–20.
38. Bem, H.; Wieczorkowski, P.; Budzanowski, M. Evaluation of technologically enhanced natural radiation near the coal-fired power plants in the Lodz region of Poland. *J. Environ. Radioact.* **2002**, *61*, 191–201. [CrossRef]
39. Fardy, J.J.; McOrist, G.D.; Farrar, Y.J. Neutron activation analysis and radioactivity measurements of Australian coals and fly ashes. *J. Radioanal. Nucl. Chem.* **1989**, *133*, 217–226. [CrossRef]
40. Coles, D.G.; Ragainl, R.C.; Ondov, J.M. Behavior of natural radionuclides in western coal-fired power plants. *Environ. Sci. Technol.* **1978**, *12*, 442–446. [CrossRef]
41. Seredin, V.V.; Finkelman, R.B. Metalliferous coals: A review of the main genetic and geochemical types. *Int. J. Coal Geol.* **2008**, *76*, 253–289. [CrossRef]
42. Kirby, H.W. *Geochemistry of the Naturally Occuring Radioactive Series*; National Technical Information Service: Alexandria, VA, USA, 1974; p. 82.
43. Finkelman, R.B. *Modes of Occurrence of Trace Elements in Coal*; United States Geological Survey Open-File Report; United States Geological Survey: Reston, VA, USA, 1981; Volume 81–99, p. 322.
44. Seredin, V.V. Metalliferous coals: Formation conditions and outlooks for development. In *Coal Resources of Russia*; Geoinformmark: Moscow, Russia, 2004; Volume 6, pp. 452–519. (In Russian)
45. Arbuzov, S.I.; Volostnov, A.V.; Rikhvanov, L.P.; Mezhibor, A.M.; Ilenok, S.S. Geochemistry of radioactive elements (U, Th) in coal and peat of northern Asia (Siberia, Russian Far East, Kazakhstan, and Mongolia). *Int. J. Coal Geol.* **2011**, *86*, 318–328. [CrossRef]
46. McBride, J.P.; Moore, R.E.; Witherspoon, J.P.; Blanco, R.E. Radiological impact of airborne effluents of coal and nuclear plants. *Science* **1978**, *202*, 1045–1050. [CrossRef] [PubMed]
47. Zhou, C.; Liu, G.; Wu, S.; Lam, P.K. The environmental characteristics of usage of coal gangue in bricking-making: A case study at Huainan, China. *Chemosphere* **2014**, *95*, 274–280. [CrossRef] [PubMed]
48. Papastefanou, C. Radiation impact from lignite burning due to $^{226}$Ra in Greek coal-fired power plants. *Health Phys.* **1996**, *70*, 187–191. [CrossRef] [PubMed]
49. Mishra, U.C. Environmental impact of coal industry and thermal power plants in India. *J. Environ. Radioact.* **2004**, *72*, 35–40. [CrossRef]
50. Beretka, J.; Mathew, P.J. Natural radioactivity of Australian building materials, industrial wastes and by-products. *Health Phys.* **1985**, *48*, 87–95. [CrossRef] [PubMed]
51. China Building Materials Academy. *Building Materials Radionuclide Limited*; GB/6566-2010, C.S.; Standard Press of China: Beijing, China, 2010. (In Chinese)
52. Organisation for Economic Co-operation and Development (OECD). *Exposure to Radiation from the Natural Radioactivity in Building Materials*; OECD Nuclear Energy Agency: Paris, France, 1979.
53. United Nations Scientific Committee on the Effects of Atomic Radiation (UNSCEAR). *Sources and Effects of Ionizing Radiation: Sources*; United Nations Publications: New York, NY, USA, 2000.

*minerals*

MDPI

*Article*

# Morphology and Composition of Microspheres in Fly Ash from the Luohuang Power Plant, Chongqing, Southwestern China

**Huidong Liu [1,2,\*], Qi Sun [1], Baodong Wang [1], Peipei Wang [3] and Jianhua Zou [3,4]**

[1]  National Institute of Clean-and-Low-Carbon Energy, Beijing 102209, China; sunqi@nicenergy.com (Q.S.); wangbaodong@nicenergy.com (B.W.)
[2]  College of Chemistry, Dalian University of Technology, Dalian 116024, China
[3]  College of Geoscience and Surveying Engineering, China University of Mining and Technology, Beijing 100083, China; wangpeipei1100@gmail.com (P.W.); zoujianhua1200@gmail.com (J.Z.)
[4]  Chongqing Institute of Geology and Mineral Resources, Chongqing 400042, China
\*  Correspondence: liuhuidong@nicenergy.com; Tel.: +86-10-5733-6190

Academic Editor: Thomas Kerestedjian
Received: 9 December 2015; Accepted: 29 January 2016; Published: 1 April 2016

**Abstract:** In order to effectively raise both utilization rate and additional value of fly ash, X-Ray diffraction (XRD), scanning electron microscope (SEM) and energy-dispersive X-Ray spectrometer (EDS) were used to investigate the morphology, and chemical and mineral composition of the microspheres in fly ash from the Luohuang coal-fired power plant, Chongqing, southwestern China. The majority of fly ash particles are various types of microspheres, including porous microsphere, plerospheres (hollow microspheres surrounding sub-microspheres or mineral fragments) and magnetic ferrospheres. Maghemite ($\gamma$-Fe$_2$O$_3$) crystals with spinel octahedron structure regularly distribute on the surfaces of ferrospheres, which explained the source of their strong magnetism that would facilitate the separation and classification of these magnetic ferrospheres from the fly ash. Microspheres in Luohuang fly ash generally are characterized by an elemental transition through their cross-section: the inner layer consists of Si and O; the chemical component of the middle layer is Si, Al, Fe, Ti, Ca and O; and the Fe-O mass (maghemite or hematite) composes the outer layer (ferrosphere). Studies on composition and morphological characteristics of microspheres in fly ash would provide important information on the utilization of fly ash, especially in the field of materials.

**Keywords:** porous microspheres; plerospheres; maghemite ($\gamma$-Fe$_2$O$_3$); magnetic ferrospheres; elemental differentiation

## 1. Introduction

Coal accounts for over 74% of the present energy consumption of China and will be the primary energy for the foreseeable future [1]. With the rapid economic growth and enormous energy demand in China over the past thirty years, both the coal consumption and the emission of coal-fired fly ash have increased continuously. Environmental pollution and induced human health problems relevant to coal combustion, such as endemic arsenosis, fluorosis, selenosis, and lung cancer, are serious hazards in some districts of China [1–8]. There is no doubt that the fly ash stock in China will keep growing for decades to come. Thus, effective utilization of fly ash is urgently needed.

A number of studies showed that some coals are enriched in rare metals (such as Ge, Ga, Nb, Zr, Au and rare earth elements) that can be potentially utilized from coal combustion products [9–14]. Research on coal combustion byproducts is of great significance not only environmentally but also economically. The Luohuang power plant, one of the largest thermal power plants in southwestern China, consumes about six million tons of coal each year, with an annual fly ash output of about two million tons. Nearly

90% of this fly ash was sold at a fairly low price as raw material for concrete and cement production after a rough particle size classification. Along with recent deceleration of hydropower and real estate construction and reduced consumption of fly ash in China, numerous coal-fired power plants including the Luohuang plant have been facing increasing environmental and economic pressure.

In addition to valuable rare metals that could be potentially extracted from coal combustion by-production, the microspheres separated from fly ash of power plants are valuable industrial products, owing to their particular chemical and mechanical properties including density, hydrophobicity, thermo conductivity and stability [15–18]. Fly ash microspheres have been widely used to create functional materials, such as thermoset plastics, special concrete, nylon, coating material, high-density polyethylene (HDPE), and others [19–21]. After a high-temperature (1200–1700 °C) thermochemical transformation of the organic matter and mineral constituents in coal during combustion, several morphological types of microspheres may form [22], such as cenospheres, porous microspheres and plerospheres. Plerospheres, as identified and described by Fisher *et al.* [23] and Goodarzi *et al.* [24], are hollow microspheres filled with finer microspheres or mineral particles. Meanwhile, the probable element differentiations in fly ash would result in the formation of microspheres with various chemical compositions, such as iron-rich or alumino-silicate microspheres [25,26]. Based on the magnetic difference or density variation, several types of fly ash microspheres can be extracted out for the varying application scenarios mentioned above.

Fly ash samples from 15 Chinese coal-fired power plants contained between 10% and 80% microspheres [27]. A primary particle count under optical microscope indicated that the percentage of microspheres in Luohuang fly ash exceeds 80% (Table S1). Enrichment of microspheres may allow Luohuang fly ash to be used efficiently and have higher economic value. Chemical and mineral composition and the interior microstructure characteristics of fly ash microspheres were investigated in the present study. Detailed information revealed by this study may not only benefit the eventual extraction of microspheres, but also expand or deepen the utilization of fly ash.

## 2. Samples Collection and Analysis

The Luohuang power plant, affiliated with the Huaneng Group, one of the largest coal-fired thermal power enterprises in China, is located in the town of Luohuang, Jiangjin district, approximately 40 km SW of the center of Chongqing city. This plant is mainly fueled with the high ash, medium-high to high sulfur anthracite from the Songzao Coalfield in Chongqing, and other provinces (e.g., Ningxia) occasionally. All six subcritical W-type flame pulverized coal furnaces are applied in this plant. $NO_x$ selective catalytic reduction (SCR) technology is adopted, with liquid ammonia as the reducing agent, using imported catalysts. The total flue gas denitration efficiency is above 85%, namely a $NO_x$ emission concentration below 200 mg/N· m$^3$. Limestone-gypsum wet flue gas desulfurization (WFGD) system is also applied to control the emission of $SO_2$, with a desulfurization efficiency of 96%. The annual output of the desulfurization gypsum reaches one million tons. Six sets of double-room four-field horizontal electrostatic precipitators (ESP) are used to capture the fly ash from the flue gas, with a collection efficiency of 99.7%. Fly ash captured by all four electric fields of each ESP system is transported by a pneumatic ash pipeline to an ash silo, without separation of fine and coarse ash. Fly ash from silos is subsequently classified into three levels according to particle size for sale.

Sampling continued for 13 days by collecting one approximately one-kilogram fly ash sample each day. All 13 fly ash samples were collected through a tap on the ash pipeline connecting to the ESP system. Considering the possible particle size differentiation of fly ash along with the pneumatic transportation distance, sampling point was set close to ash buckets of the ESP.

X-Ray fluorescence spectrometry (XRF, ARL ADVANT' XP+, Thermo Fisher, Washington, D.C., USA) was used to determine the concentrations of major elements in these fly ash samples after high-temperature ashing (HTA, at 815 °C, following the Chines Standard GB/T 212-2008 [28]) as outlined by Dai *et al.* [29]. One gram fly ash (HTA treated) was homogenously mixed with ten grams lithium borate flux (50% $Li_2B_4O_7$ + 50% $LiBO_2$), and then was fully fused in an automated fusion furnace (CLAISSE TheBee-10, Claisse, Quebec, QC, Canada). Finally, a glass-like disk (diameters 35 mm) was obtained for the XRF analysis.

XRD (X-Ray diffraction) analysis of all these fly ash samples was then performed by a D/max-2500/PC powder diffractometer (Rigaku, Tokyo, Japan), equipped with a Ni-filtered Cu-Kα radiation source and a scintillation detector. All samples were scanned within a 2θ interval of 2.6°–70°, with a step size of 0.01°. Based on the X-Ray diffractograms acquired, JADE 6.5 (MDI, Burbank, CA, USA) and Siroquant™ (Canberra, Australia) were applied to identify and quantify the mineral phases in the sample, respectively. Siroquant™ was developed by Taylor [30] on the basis of diffractogram profiling principles presented by Rietveld [31]. Detailed practices of this technique on coal-related materials were further described by Ward *et al.* [32] and Dai *et al.* [33,34]. Metakaolin and tridymite were consistent in representing the amorphous or glassy material in fly ash in the Siroquant quantitative analysis [35]. In this study, however, tridymite was preferred in Siroquant quantitative analysis, considering its better-fitted value compared to that of metakaolin.

XRF (Table S2) and XRD (Table S3) analyses indicate that the chemical and mineral compositions of these samples are quite stable. Thus, one sample whose major elements concentration belongs to the medium level of the 13 fly ash samples was chosen for further study.

The particle size distribution of the selected fly ash sample was analyzed by laser particle analyzer (Malvern Mastersizer 2000, Malvern Instruments, Malvern, UK) in conjunction with a dispersal device (Hydro G, Malvern Instruments).

After sample splitting, approximately one gram of fly ash was made into pellet, polished, and then coated with carbon in a sputtering coater (Q150T ES, Quorum Technologies, Lewes, UK). A Field Emission-Scanning Electron Microscope (FE-SEM, FEI Quanta™ 650 FEG, FEI, Hillsboro, OR, USA), equipped with an energy-dispersive X-Ray spectrometer (EDS, Genesis Apex 4, EDAX Inc., Mahwah, NJ, USA), was applied to observe the microstructure of the microspheres in the polished section, as well as to evaluate the distribution of some elements. The working distance of the SEM-EDS was 10 mm, with beam voltage 20.0 kV, aperture 6, and spot sizes 3–5.5 nm. Images were acquired through a backscatter electron detector or a secondary electron detector. For more details on the FE SEM-EDS working conditions, see Dai *et al.* [33,34,36,37].

## 3. Results and Discussion

### 3.1. The Particle Size Distribution of the Fly Ash

As shown in Figure 1, the particle size of the Luohuang fly ash is between 0.5 and 400 μm, with $D_{50}$ (medium diameter) of 50.7 μm, $D_{10}$ of 6.10 μm and $D_{90}$ of 178.38 μm, similar to those previously reported by Mardon *et al.* [38], Vassilev *et al.* [39] and Dai *et al.* [40].

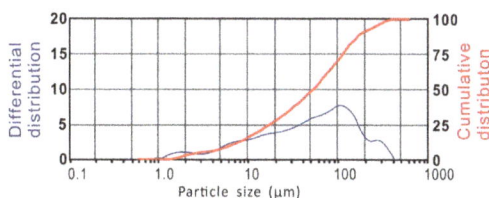

**Figure 1.** Particle size distribution of the Luohuang fly ash.

### 3.2. Major Elements of the Fly Ash

An obvious characteristic of the chemical composition of the Luohuang fly ash is the enrichment (14.09%) of $Fe_2O_3$, which can be attributed to the relatively high percentage of pyrite in the feed coal. As reported by Zhao *et al.* [41] and Dai *et al.* [42], the Songzao coals contain 2.15–10.65 wt % $Fe_2O_3$, with an average of 7.63%; however, the average $Fe_2O_3$ value in common Chinese coals is 5.78% [43].

The value of loss on ignition (LOI) was used to represent the content of unburned carbon (4.01%). Concentration of major elements in fly ash was given in the form of oxides by XRF. Oxides of major elements, including $SiO_2$ (48.27%), $Al_2O_3$ (21.59%), $Fe_2O_3$ (14.09%), and $CaO$ (5.72%), account for 93.4% of the inorganic matter of the fly ash (Table 1).

**Table 1.** Loss on ignition (LOI, %) and the concentrations of major elements (%) in the Luohuang fly ash.

| LOI | Na$_2$O | MgO | Al$_2$O$_3$ | SiO$_2$ | P$_2$O$_5$ | SO$_3$ | K$_2$O | CaO | TiO$_2$ | MnO | Fe$_2$O$_3$ |
|------|---------|------|-------------|---------|------------|--------|--------|------|---------|------|-------------|
| 4.01 | 0.77 | 1.21 | 21.59 | 48.27 | 0.13 | 0.78 | 1.21 | 5.72 | 1.78 | 0.09 | 14.09 |

*3.3. Mineral Composition of the Fly Ash*

During the high-temperature (over than 1400 °C) combustion in the pulverized coal furnaces of the Luohuang plant, minerals in coal including kaolinite, illite-montmorillonite mixed layer, pyrite, calcite, siderite, and even anatase and quartz (to an extent at least) have melted. Some newly-formed minerals such as mullite, hematite, maghemite, anhydrite, and maybe a proportion of quartz, were formed by recrystallization as the molten mass cooled down. However, there are still large percentages of major elements existing in fly ash as an amorphous substance, or so-called glass (Table 2). In the XRD pattern (Figure 2), the section where the baseline of the curve is raised up (2θ from 13° to 38°) indicates the existence of glass. Percentage of active silicon and aluminum in glass is one of the most important factors affecting the pozzolanic activity of fly ash [44]. High percentage of glass (79.5%) made the use of large quantities of the Luohuang fly ash in cement and concrete production possible. Ferromagnetic matter separated from coal fly ash could be utilized in coal cleaning circuits as a dense medium [45], or be used for special concrete [46]. Maghemite ($\gamma$-$Fe_2O_3$, strong magnetic) accounts for 70.8% of the iron-bearing independent minerals, maghemite plus hematite ($\alpha$-$Fe_2O_3$, weak magnetic), which will facilitate the separation of ferromagnetic matter from the Luohuang fly ash.

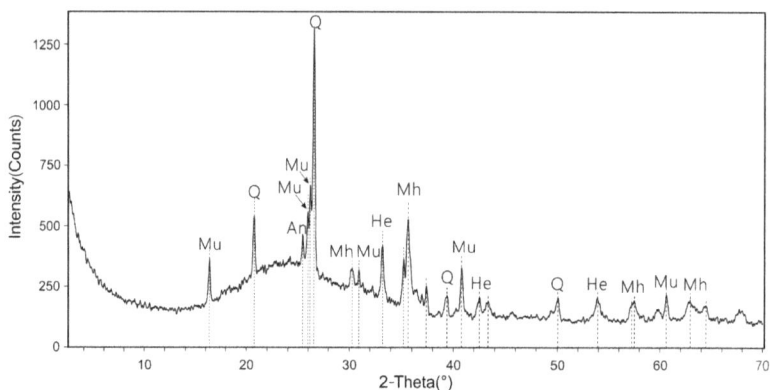

**Figure 2.** The X-Ray diffraction (XRD) pattern of the fly ash sample from the Luohuang power plant. Q, quartz; Mu, mullite; He, hematite; Mh, maghemite; An, anhydrite.

The content of quartz (5.8%) in the fly ash is much lower than that in the feed coal (18.8%, on ash basis), which indicates that quartz, at least partially, melted during the coal combustion at around 1400 °C. $FeS_2$ (pyrite), $CaO$ (calcite) and some alkali metals (e.g., K and Na) in coal are likely to react with the clay, quartz and other minerals in coal to form a low-temperature eutectic mixture [47,48], which can melt at a temperature much lower than the melting points of single minerals (e.g., 1750 °C of quartz). Anatase (melting point of 1850 °C) existing in feed coal is not detected in the fly ash, which can be attributed to the same reason.

**Table 2.** Mineral compositions of the Luohuang fly ash determined by X-Ray diffraction and Siroquant technologies.

| Mineral Phase | Percentage (%) |
|:---:|:---:|
| Glass | 79.5 |
| Mullite | 8.3 |
| Maghemite | 4.6 |
| Quartz | 5.1 |
| Hematite | 1.9 |
| Anhydrite | 0.7 |

### 3.4. Morphology and Composition of Microspheres in the Fly Ash

Under the scanning electron microscope (SEM), a few irregular mineral fragments, cohesive bodies and debris of microspheres can be observed in the Luohuang fly ash (Figure 3A); the majority are spherical particles called "microspheres" (Figure 3A,B,E). Microspheres smaller than 10 μm are uniformly spherical (Figure 3E,F). Porous microspheres (Figure 3D) are found in the Luohuang fly ash. The presence of these pores further increases the specific surface area of the fly ash and enhances its adsorption ability. During the cooling of the molten drop, trapped gas was emitted and gave rise to pores on the surfaces of the porous microspheres. Plerospheres (Figure 3E,F) with a larger diameter (e.g., 100 μm) enclosing sub-microspheres or mineral fragments (mostly <10 μm) are rather common. Fine spheres have better mechanical properties and a higher chemical reactivity [49–52]. The value of Luohuang fly ash might be increased by crushing plerospheres to release the sub-microspheres.

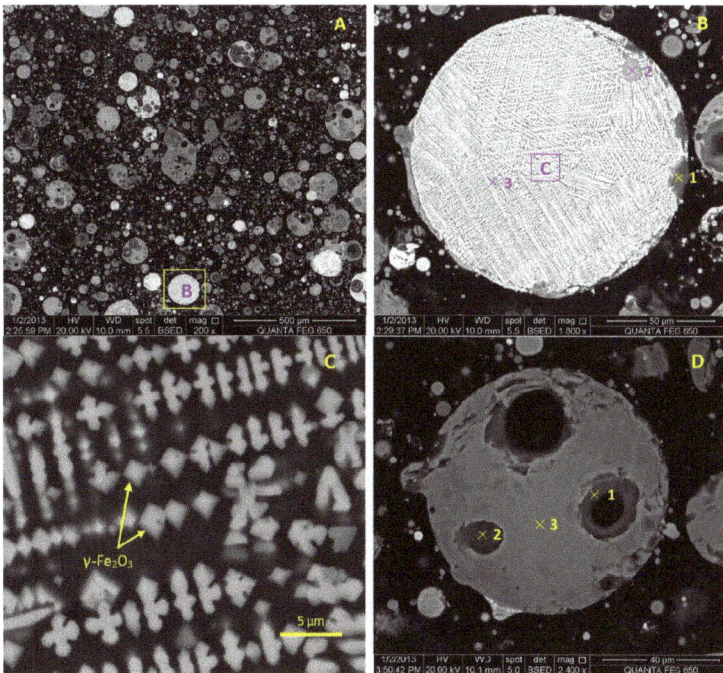

**Figure 3.** *Cont.*

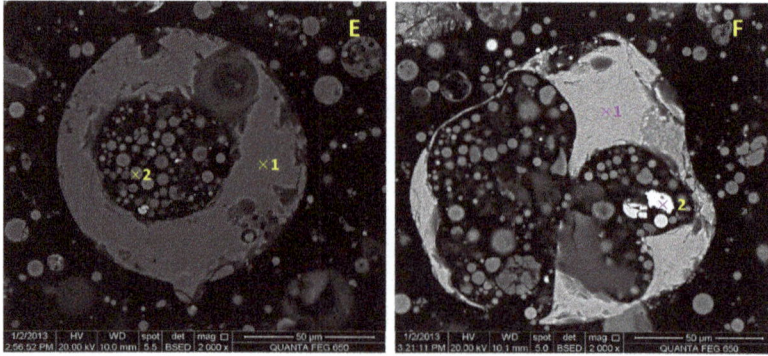

**Figure 3.** Scanning electron microscope (SEM) backscattered electron images and energy-dispersive X-Ray spectrometer (EDS) analyses of microspheres in the Luohuang fly ash: (**A**) Overall view of the Luohuang fly ash; (**B**) Enlargement of the area marked in (A), magnetic microsphere; (**C**) Enlargement of the area marked in (B), maghemite ($\gamma$-$Fe_2O_3$) crystals with spinel structure; (**D**) Porous microsphere; (**E**) Plerosphere; and (**F**) Cracked plerosphere containing a particle of metallic iron.

A total of 11 individual microspheres with various particle sizes (30–250 μm) of different types (ferrospheres, porous and plerospheres) and 26 detection points were analyzed using SEM and EDS. Backscattered electron images of several typical microspheres and their EDS analysis data are listed in Figure 3 and Table 3, respectively. In the SEM backscattered electron images, the higher the atomic number of an element, the brighter it appears. As shown in Figure 3A, the majority of the widespread bright microspheres or spots are ferrospheres (Figure 3B,C,F), microspheres coated with iron oxides. According to Figure 3B and the EDS analysis results (Table 3), iron oxides are common on the surfaces of ferrospheres and have a dendritic form [53]. In Figure 3C, an enlargement of the area marked in Figure 3B, further shows a spinel octahedron structure of the iron oxide crystals. Additionally, as discussed in Section 3.3., maghemite is the primary iron-bearing mineral in the Luohuang fly ash. It is therefore concluded that these iron oxide crystals are maghemite. Likewise, these ferrospheres are magnetic microspheres. In view of the promising applications of magnetic microspheres in the composites of magnetic materials, magnetic media, adsorbents, catalysts and ion exchangers [22], separating and classifying magnetic microspheres from the fly ash will bring additional value.

**Table 3.** Data from energy-dispersive X-Ray spectrometer (EDS) analyses of the Luohuang microspheres shown in Figure 3.

| Detection Spot * | Elements (at. %) Detected by EDS | | | | | | | | |
|---|---|---|---|---|---|---|---|---|---|
| | Si | Al | Fe | Ca | Ti | Na | Mg | K | O ** |
| B1 | 44.82 | 1.64 | 5.14 | 0.83 | - | - | - | - | 47.57 |
| B2 | 26.51 | 9.28 | 10.09 | 4.24 | 0.49 | - | 0.44 | - | 48.93 |
| B3 | 4.40 | 6.50 | 40.07 | 1.15 | 0.31 | - | 0.94 | - | 46.63 |
| D1 | 46.54 | 1.63 | - | - | - | - | - | - | 51.82 |
| D2 | 47.19 | 0.98 | - | - | - | - | - | - | 51.83 |
| D3 | 24.02 | 15.97 | 3.92 | 1.48 | 2.50 | 1.07 | 0.67 | 1.07 | 49.29 |
| E1 | 26.84 | 16.24 | 1.88 | 1.92 | 2..22 | 0.58 | 1.89 | 1.35 | 47.09 |
| E2 | 29.87 | 22.19 | 4.30 | - | 0.99 | 0.47 | 1.16 | 1.07 | 47.59 |
| F1 | 22.68 | 6.60 | 23.71 | 0.40 | - | - | - | - | 46.61 |
| F2 | 1.75 | - | 98.25 | - | - | - | - | - | - |

\* Detection spot B1 means the 1st detection spot in Figure 3B, D3 means the 3rd detection spot in Figure 3D; and so on for the other spots. \*\* The value of oxygen is obtained through calculation rather than physical detection, and is therefore not fully credible.

Additionally, metallic iron is detected in a cracked plerosphere (Figure 3F). It may be generated from the disoxidation of $Fe^{2+}$ or $Fe^{3+}$ in a strong reducing environment before the burst of plerospheres.

As discussed above, variance of brightness in the backscattered electron images can reflect the chemical change on the whole. Brightness of zones at different depths (surface or inside) of microspheres varies obviously. In EDS data, detection spots distributed at different depths of microspheres (Figure 3D) reveal that these microspheres show the characteristic of elemental transition through their cross-section. The inner layer with the lowest brightness consists of Si and O (Detection spot B1, D1 and D2, as shown in Figure 3 and Table 3). The chemical component of the middle layer is Si, Al, Fe, Ti, Ca and O; this layer appears brighter than the inner one but darker than the iron-bearing minerals. For ferrospheres, the Fe-O mass (maghemite or hematite) would compose the outer layer. This indicates that an elemental differentiation may have occurred during the formation process of these microspheres.

## 4. Conclusions

Minerals including mullite (8.3%), quartz (5.1%), maghemite (4.6%), hematite (1.9%), and anhydrite (0.7%) are detected in the fly ash from the Luohuang plant in Chongqing, southwest China. Amorphous aluminosilicate glass accounts for 79.5% of the fly ash. This contributes to the prominent pozzolanic activity of Luohuang fly ash in cement and concrete production.

The majority of particles in the fly ash from the Luohuang plant are various types of microspheres: porous microspheres, plerospheres (hollow microspheres surrounding sub-microspheres or mineral fragments) and magnetic ferrospheres. Maghemite ($\gamma$-$Fe_2O_3$) crystals with spinel octahedron structure occur on the surfaces of microspheres, displaying a dendritic or fabric framework. Separating and classifying these magnetic ferrospheres from the Luohuang fly ash would generate considerable additional value. Microspheres in Luohuang fly ash generally show a characteristic of elemental transition through their cross-section: the inner layer consists of Si and O; the chemical component of the middle layer is Si, Al, Fe, Ti, Ca and O; and the Fe-O mass (maghemite or hematite) composes the outer layer (plerospheres). This indicates that an elemental differentiation occurred during the formation process of the microspheres. The above investigations on the composition and morphological characteristics of microspheres in fly ash provide important information on the utilization of the Luohuang fly ash, especially in the field of materials.

**Supplementary Materials:** The following are available online at www.mdpi.com/2075-163X/6/2/30/s1.

**Acknowledgments:** This research was totally supported by the National Key Basic Research Program of China (No. 2014CB238900), the National Natural Science Foundation of China (Nos. 41420104001, 41272182 and 41502162), and Innovative Research Team in University (No. IRT13099). Special thanks are given to Dai Shifeng, Zhao Lei, and Weijiao Song for their valuable advice and corrections on the manuscript.

**Author Contributions:** Huidong Liu designed and operated the largest share of the experiment. Qi Sun and Baodong Wang made many important modifications and provided some good suggestion on the structure of this paper. Peipei Wang performed the majority of tests and analyses. Jianhua Zou contributed greatly to the sample collection and preparation. Huidong Liu wrote the paper.

**Conflicts of Interest:** The authors declare no conflict of interest.

## References

1. Dai, S.; Ren, D.; Chou, C.L.; Finkelman, R.B.; Seredin, V.V.; Zhou, Y. Geochemistry of trace elements in Chinese coals: A review of abundances, genetic types, impacts on human health, and industrial utilization. *Int. J. Coal Geol.* **2012**, *94*, 3–21. [CrossRef]
2. Belkin, H.E.; Finkelman, R.B.; Zheng, B.S.; Zhou, D.X. Human health effects of domestic combustion of coal: A causal factor for arsenosis and fluorosis in rural China. In Proceedings of the Air Quality Conference, McLean, VA, USA, 1–4 December 1998.

3.  Belkin, H.E.; Finkelman, R.B.; Zheng, B.S. Geochemistry of fluoride-rich coal related to endemic fluorosis in Guizhou Province, China. In Proceedings of the Pan-Asia Pacific Conference on Fluoride and Arsenic Research, Shenyang, China, 16–20 August 1999.

4.  Belkin, H.E.; Finkelman, R.B.; Zheng, B.S. Human health effects of domestic combustion of coal: A causal factor for arsenic poisoning and fluorosis in rural China. *Am. Geophys. Union* **1999**, *80*, 377–378.

5.  Belkin, H.E.; Zheng, B.S.; Zhou, D.; Finkelman, R.B. Chronic arsenic poisoning from domestic combustion of coal in rural China: A case study of the relationship between earth materials and human health. *Environ. Geochem.* **2008**. [CrossRef]

6.  Finkelman, R.B.; Orem, W.; Castranova, V.; Tatu, C.A.; Belkin, H.E.; Zheng, B.S.; Lercha, H.E.; Maharaje, S.V.; Bates, A.L. Health impacts of coal and coal use: Possible solutions. *Int. J. Coal Geol.* **2002**, *50*, 425–443. [CrossRef]

7.  Dai, S.; Ren, D.; Ma, S. The cause of endemic fluorosis in western Guizhou Province, Southwest China. *Fuel* **2004**, *83*, 2095–2098. [CrossRef]

8.  Dai, S.; Li, W.; Tang, Y.; Zhang, Y.; Feng, P. The sources, pathway, and preventive measures for fluorosis in Zhijin County, Guizhou, China. *Appl. Geochem.* **2007**, *22*, 1017–1024. [CrossRef]

9.  Dai, S.; Chou, C.L.; Yue, M.; Luo, K.; Ren, D. Mineralogy and geochemistry of a Late Permian coal in the Dafang Coalfield, Guizhou, China: Influence from siliceous and iron-rich calcic hydrothermal fluids. *Int. J. Coal Geol.* **2005**, *61*, 241–258. [CrossRef]

10. Dai, S.; Ren, D.; Tang, Y.; Yue, M.; Hao, L. Concentration and distribution of elements in Late Permian coals from western Guizhou Province, China. *Int. J. Coal Geol.* **2005**, *61*, 119–137. [CrossRef]

11. Dai, S.; Zhou, Y.; Ren, D.; Wang, X.; Li, D.; Zhao, L. Geochemistry and mineralogy of the Late Permian coals from the Songzo Coalfield, Chongqing, southwestern China. *Sci. China Ser. D Earth Sci.* **2007**, *50*, 678–688. [CrossRef]

12. Seredin, V.V.; Dai, S. The occurrence of gold in fly ash derived from high-Ge coal. *Miner. Deposita* **2014**, *49*, 1–6. [CrossRef]

13. Seredin, V.V.; Dai, S. Coal deposits as potential alternative sources for lanthanides and yttrium. *Int. J. Coal Geol.* **2012**, *94*, 67–93. [CrossRef]

14. Seredin, V.V.; Dai, S.; Sun, Y.; Chekryzhov, I.Y. Coal deposits as promising sources of rare metals for alternative power and energy-efficient technologies. *Appl. Geochem.* **2013**, *31*, 1–11. [CrossRef]

15. Page, A.L.; Elseewi, A.A.; Straughan, I.R. Physical and chemical properties of fly ash from coal-fired power plants with reference to environmental impacts. In *Residue Reviews*; Springer: New York, NY, USA, 1979; pp. 83–120.

16. Sear, L.K.A. *Properties and Use of Coal Fly Ash: A Valuable Industrial by-Product*; Thomas Telford: London, UK, 2001; p. 220.

17. Chávez, V.A.; Arizmendi, M.A.; Vargas, G.; Almanza, J.M.; Alvarez, Q.J. Ultra-low thermal conductivity thermal barrier coatings from recycled fly-ash cenospheres. *Acta Mater.* **2011**, *59*, 2556–2562. [CrossRef]

18. Hwang, J.Y.; Sun, X.; Li, Z. Unburned carbon from fly ash for mercury adsorption: I. Separation and characterization of unburned carbon. *J. Miner. Mater. Charact. Eng.* **2002**, *1*, 39–60. [CrossRef]

19. Drozhzhin, V.S.; Danilin, L.D.; Pikulin, I.V.; Khovrin, A.N.; Maximova, N.V.; Regiushev, S.A.; Pimenov, V.G. Functional materials on the basis of cenospheres. In Proceedings of the 2005 World of Coal Ash Conference, Lexington, KY, USA, 11–15 April 2005.

20. Wasekar, P.A.; Pravin, G.K.; Shashank, T.M. Effect of cenosphere concentration on the mechanical, thermal, rheological and morphological properties of nylon 6. *J. Miner. Mater. Charact. Eng.* **2012**, *11*, 807–812. [CrossRef]

21. Deepthi, M.V.; Sharma, M.; Sailaja, R.R.N.; Anantha, P.; Sampathkumaran, P.; Seetharamu, S. Mechanical and thermal characteristics of high density polyethylene-fly ash cenospheres composites. *Mater. Des.* **2010**, *31*, 2051–2060. [CrossRef]

22. Sharonova, O.M.; Anshits, N.N.; Oruzheinikov, A.I.; Akimochkina, G.V.; Salanov, A.N.; Nizovshii, A.I.; Semenova, O.N.; Anshits, A.G. Composition and morphology of magnetic microspheres in power plant fly ash of coal from the Ekibastuz and Kuznetsk basins. *Chem. Sustain.* **2003**, *11*, 639–648.

23. Fisher, G.L.; Chang, D.P.Y.; Brummer, M. Fly ash collected from electrostatic precipitators: Microcrystalline structures and the mystery of the spheres. *Science* **1976**, *192*, 553–555. [CrossRef] [PubMed]

24. Goodarzi, F.; Sanei, H. Plerosphere and its role in reduction of emitted fine fly ash particles from pulverized coal-fired power plants. *Fuel* **2009**, *88*, 382–386. [CrossRef]

25. Zhao, Y.; Zhang, J.; Gao, Q.; Guo, X.; Zheng, C. Chemical composition and evolution mechanism of ferrospheres in fly ash from coal combustion. *Proc. CSEE* **2006**, *26*, 82. (In Chinese).

26. Goodarzi, F. Characteristics and composition of fly ash from Canadian coal-fired power plants. *Fuel* **2006**, *85*, 1418–1427. [CrossRef]

27. Zhou, Z.; Zhu, Z.; Huang, Z.; Tang, Y.; Shi, Q. Determination of content and properties of microspheres in fly ash. *Fly Ash* **1997**, *6*, 13–15. (In Chinese).

28. Standardization Administration of P.R. China; General Administration of Quality Supervision, Inspection and Quarantine of the P.R. China. *Chinese National Standard GB/T 212-2008, Proximate Analysis of Coal*; Standand Press of China: Beijing, China, 2008; pp. 3–5. (In Chinese)

29. Dai, S.; Wang, X.; Zhou, Y.; Hower, J.C.; Li, D.; Chen, W.; Zhu, X. Chemical and mineralogical compositions of silicic, mafic, and alkali tonsteins in the Late Permian coals from the Songzao Coalfield, Chongqing, Southwest China. *Chem. Geol.* **2011**, *282*, 29–44. [CrossRef]

30. Taylor, S.R.; McLennan, S.M. *The Continental Crust: Its Composition and Evolution*; Blackwell: London, UK, 1985; p. 312.

31. Rietveld, H.M. A profile refinement method for nuclear and magnetic structures. *J. Appl. Crystallogr.* **1969**, *2*, 65–71. [CrossRef]

32. Ward, C.R.; Spears, D.A.; Booth, C.A.; Staton, I.; Gurba, L.W. Mineral matter and trace elements in coals of the Gunnedah Basin, New South Wales, Australia. *Int. J. Coal Geol.* **1999**, *40*, 281–308. [CrossRef]

33. Dai, S.; Liu, J.; Ward, C.R.; Hower, J.C.; French, D.; Jia, S.; Hood, M.M.; Garrison, T.M. Mineralogical and geochemical compositions of Late Permian coals and host rocks from the Guxu Coalfield, Sichuan Province, China, with emphasis on enrichment of rare metals. *Int. J. Coal Geol.* **2015**. [CrossRef]

34. Dai, S.; Li, T.; Seredin, V.V.; Ward, C.R.; Hower, J.C.; Zhou, Y.; Zhang, M.; Song, X.; Song, W.; Zhao, C. Origin of minerals and elements in the Late Permian coals, tonsteins, and host rocks of the Xinde Mine, Xuanwei, eastern Yunnan, China. *Int. J. Coal Geol.* **2014**, *121*, 53–78. [CrossRef]

35. Ward, C.R.; French, D. Determination of glass content and estimation of glass composition in fly ash using quantitative X-Ray diffractometry. *Fuel* **2006**, *85*, 2268–2277. [CrossRef]

36. Dai, S.; Liu, J.; Ward, C.R.; Hower, J.C.; Xie, P.; Jiang, Y.; Hood, M.M.; O'Keefee, J.M.K.; Song, H. Petrological, geochemical, and mineralogical compositions of the low-Ge coals from the Shengli Coalfield, China: A comparative study with Ge-rich coals and a formation model for coal-hosted Ge ore deposit. *Ore Geol. Rev.* **2015**, *71*, 318–349. [CrossRef]

37. Dai, S.; Hower, J.C.; Ward, C.R.; Guo, W.; Song, H.; O'Keefe, J.M.K.; Xie, P.; Hood, M.M.; Yan, X. Elements and phosphorus minerals in the middle Jurassic inertinite-rich coals of the Muli Coalfield on the Tibetan Plateau. *Int. J. Coal Geol.* **2015**, *144–145*, 23–47. [CrossRef]

38. Mardon, S.M.; Hower, J.C.; O'Keefe, J.M.K.; Marks, M.N.; Hedges, D.H. Coal combustion by-product quality at two stoker boilers: Coal source *vs.* fly ash collection system design. *Int. J. Coal Geol.* **2008**, *75*, 248–254. [CrossRef]

39. Vassilev, S.V.; Vassileva, C.G.; Karayigit, A.I.; Bulut, Y.; Alastuey, A.; Querol, X. Phase-mineral and chemical composition of composite samples from feed coals, bottom ashes and fly ashes at the Soma power station, Turkey. *Int. J. Coal Geol.* **2005**, *61*, 35–63. [CrossRef]

40. Dai, S.; Zhao, L.; Peng, S.; Chou, C.L.; Wang, X.; Zhang, Y.; Li, D.; Sun, Y. Abundances and distribution of minerals and elements in high-alumina coal fly ash from the Jungar Power Plant, Inner Mongolia, China. *Int. J. Coal Geol.* **2010**, *81*, 320–332. [CrossRef]

41. Zhao, L.; Ward, C.R.; French, D.; Graham, I.T. Mineralogical composition of Late Permian coal seams in the Songzao Coalfield, southwestern China. *Int. J. Coal Geol.* **2013**, *116*, 208–226. [CrossRef]

42. Dai, S.; Wang, X.; Chen, W.; Li, D.; Chou, C.L.; Zhou, Y.; Zhu, C.; Li, H.; Zhua, X.; Xing, Y.; *et al.* A high-pyrite semianthracite of Late Permian age in the Songzao Coalfield, southwestern China: Mineralogical and geochemical relations with underlying mafic tuffs. *Int. J. Coal Geol.* **2010**, *83*, 430–445. [CrossRef]

43. Dai, S.; Li, D.; Chou, C.L.; Zhao, L.; Zhang, Y.; Ren, D.; Ma, Y.; Sun, Y. Mineralogy and geochemistry of boehmite-rich coals: New insights from the Haerwusu Surface Mine, Jungar Coalfield, Inner Mongolia, China. *Int. J. Coal Geol.* **2008**, *74*, 185–202. [CrossRef]

44. Diaz, E.I.; Allouche, E.N.; Eklund, S. Factors affecting the suitability of fly ash as source material for geopolymers. *Fuel* **2010**, *89*, 992–996. [CrossRef]

45. Groppo, J.; Honaker, R. Economical recovery of fly ash-derived magnetics and evaluation for coal cleaning. In Proceedings of the WOCA 09 The World of Coal Ash, Lexington, KY, USA, 4–7 May 2009.

46. Huang, Y.; Qian, J.; Zhang, J. Research on the building electromagnetic wave absorber mixing high-iron fly ash. *J. China Coal Soc.* **2010**, *35*, 135–139. (In Chinese).

47. Huffman, G.P.; Huggins, F.E. Reactions and transformations of coal mineral matter at elevated temperatures. *Am. Chem. Soc. Symp. Ser.* **1986**, *301*, 100–113.

48. Huffman, G.P.; Huggins, F.E.; Shah, N.; Shah, A. Behavior of basic elements during coal combustion. *Prog. Energy Combust. Sci.* **1990**, *16*, 243–251. [CrossRef]

49. Nascimento, R.S.V.; D'Almeida, J.R.M.; Abreu, E.S.V. *Proceedings of the Third Ibero-American Polymer Symposium*; Wiley-VCH: Vigo, Spain, 1992; pp. 531–532.

50. D'Almeida, J.R.M. An analysis of the effect of the diameters of glass microspheres on the mechanical behavior of glass-microsphere/epoxy-matrix composites. *Compos. Sci. Technol.* **1999**, *59*, 2087–2091. [CrossRef]

51. Blissett, R.S.; Rowson, N.A. A review of the multi-component utilisation of coal fly ash. *Fuel* **2012**, *97*, 1–23. [CrossRef]

52. Seames, W.S. An initial study of the fine fragmentation fly ash particle mode generated during pulverized coal combustion. *Fuel Process. Technol.* **2003**, *81*, 109–125. [CrossRef]

53. Sokol, E.V.; Kalugin, V.M.; Nigmatulina, E.N.; Volkova, N.I.; Frenkel, A.E.; Maksimova, N.V. Ferrospheres from fly ashes of Chelyabinsk coals: Chemical composition, morphology and formation conditions. *Fuel* **2002**, *81*, 867–876. [CrossRef]

*Article*

# Modes of Occurrence of Fluorine by Extraction and SEM Method in a Coal-Fired Power Plant from Inner Mongolia, China

**Guangmeng Wang †, Zixue Luo \*,†, Junying Zhang † and Yongchun Zhao**

State Key Laboratory of Coal Combustion, Huazhong University of Science and Technology,
Wuhan 430074, China; guangmeng66@163.com (G.W.); jyzhang@hust.edu.cn (J.Z.); yczhao@hust.edu.cn (Y.Z.)
* Correspondence: luozixue@hust.edu.cn; Tel.: +86-27-8754-2417 (ext. 8318); Fax: +86-27-8754-5526
† These authors contributed equally to this work.

Academic Editor: Shifeng Dai
Received: 19 October 2015; Accepted: 18 November 2015; Published: 2 December 2015

**Abstract:** In this study, an extraction method and environmental scanning electron microscopy (SEM) are employed to reveal the changes in the occurrence mode of fluorine in a coal-fired power plant in Inner Mongolia, China. The different occurrence states of fluorine during coal combustion and emission show that fluorine in coal mainly assumes insoluble inorganic mineral forms. The results illustrate that the three typical occurrence modes in coal are $CaF_2$, $MgF_2$ and $AlF_3$. The fluorine in fly ash can be captured by an electrostatic precipitator (EPS) or a bag filter. In contrast, the gaseous fluorine content in flue gas is only in the range of several parts per million; thus, it cannot be used in this study. The occurrence mode of fluorine in bottom ash and slag is inorganic villiaumite (e.g., soluble NaF, KF and insoluble $CaF_2$) which is difficult to break down even at high temperatures. The occurrence mode of fluorine with the highest content in fly ash is physically adsorbed fluorine along the direction of the flue gas flow. The insoluble inorganic mineral fluoride content in fly ash is also high, but the gradually increasing fluorine content in fly ash is mainly caused by physical adsorption. Fluorine in the coal-fired power plant discharges mostly as solid products; however, very little fluorine emitted into the environment as gas products (HF, $SiF_4$) cannot be captured. The parameters used in this study may provide useful references in developing a monitoring and control system for fluorine in coal-fired power plants.

**Keywords:** extraction; SEM; occurrence mode of fluorine; coal-fired power plant; combustion products

## 1. Introduction

In recent years, the toxic elements in coal have been research highlights in many countries, especially in China, the largest coal-consuming country in the world. The fluorine in coal is a toxic trace element, with a general content of 100–300 ppm. The average fluorine concentration level in coal in China is 130 ppm [1], 48% higher than that in the rest of the world (the average level in the world is 88 ppm [2]). The reason China is higher than the rest of the world is that the amount in the sample determined by researchers in China is limited, or different determination method was used. Although the concentration of fluorine in coal is on a ppm level, the emission needs to be paid special attention as the coal consumption increases. A gas product of coal combustion: HF, poses the most harm to humans and animals.

There have been a number of studies on modes of fluorine in coal. Eskenazy [3] illustrated the occurrence modes of fluorine in coal by examining the relationship of fluorine with ash yield, as well as the oxides in fly ash. Martinez [4] analyzed the modes of fluorine in Austrian coals. Dai [5] studied the correlation between fluorine and minerals to evaluate the occurrence modes of fluorine in different coal types. He discovered mineralogical and geochemical compositions of the coal in Inner Mongolia, China [6], and made a case study of factors on controlling compositions of coals [7]. He also discussed elements and phosphorus minerals in the middle Jurassic inertinite-rich coals including the existence of trace elements [8]. Wang [9] employed the continuous extraction/density separation method to reveal the occurrence modes of fluorine in Chinese coal. Liu [10] investigated the correlation between fluorine and alkalis to identify the occurrence states of fluorine in Guizhou coal. The conventional analytical method to assess the fluorine occurrence states in coal mainly uses correlation analysis, but a relatively simple and effective extraction method has been adopted in recent years.

The extraction method [11] has been proved suitable for occurrence studies on sulfur, selenium and other elements. Such a method was also used to analyze fluorine in soil and for chemical extraction for trace elements in coal [12–14]. However, this method has been seldom used for analyzing fluorine in coal [15], especially for tracking the changes in fluorine occurrence modes in power plant boilers. In this study, an extraction method and scanning electron microscopy (SEM) are employed to determine the fluorine changes during coal combustion in a coal-fired power plant. During coal combustion, the fluorine in coal remains in slag and bottom ash or adheres to fly ash. This research also provides a reference for the study on the change regularity of the occurrence modes of fluorine in coal-fired power plants.

## 2. Method and Sampling

Several types of samples, including pulverized coal, slag, flue gas desulfurization gypsum and fly ash, in different parts of the boiler were collected from the Shangdu coal-fired power plant in Inner Mongolia, China. All samples were dried and filtered through a 200-mesh sieve. The extraction method means that fluorine from solid samples was leached by solvent. The fluorine content in the extraction solution was determined using the fluoride ion-selective electrode method (GB/T 4633-1997 [16]). SEM was used to analyze the fluorine mineral forms in the samples. Figure 1 shows the experimental procedure. We adopted 4 g coal sample (or 2 g other samples) and 50 mL extraction solution (deionized water, 0.1 mol/L NaOH solution, 1 mol/L $HNO_3$ solution, 0.1 mol/L KCl solution, 0.05 mol/L $CaCl_2$ solution, respectively). Additionally, these samples are adopted with 30 min magnetic stirring and 1 h extraction, then filtered and combined with 25 mL liquid supernatant to adjust their pH concentration. Determination of fluoride by the ion-selective electrode should be obtained in the end.

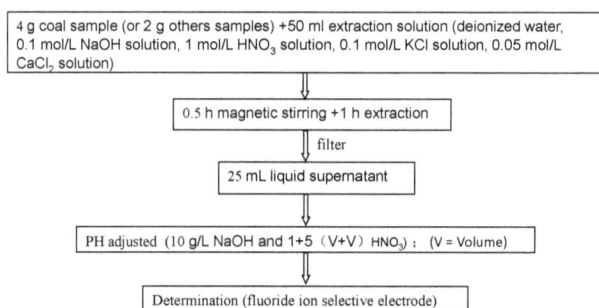

**Figure 1.** Scheme for extraction of fluorine in the samples (referred to Qi [17]).

## 3. Results and Discussion

### 3.1. Occurrence Modes of Fluorine in Coal

The extraction data are presented in Table 1. The results show that water-soluble fluorine content is 4.51 µg/g with an extraction rate of 1.33%. Extraction rate means that the leaching amount accounted for the proportion of the total content in the sample. It can be expressed: extraction rate = leaching amount/total amount. The alkali-soluble fluorine content is 9.74 µg/g with an extraction rate of 2.88%. The acid-soluble fluorine content is 39.92 µg/g with an extraction rate of 11.81%. The acid solution has the highest extraction rate among solutions. In salt-soluble fluorine, the calcium chloride is 4.89 µg/g with an extraction rate of 1.45%, and potassium chloride is 4.16 µg/g with an extraction rate of 1.23%. The water-soluble fluorine forms, namely NaF, KF and NH$_4$F, mainly come from soluble fluoride and adsorbed fluoride. The extraction rate of water-soluble fluorine indicates that free-state F$^-$ is adsorbed on mineral particle surfaces, and fluorine is among the mineral particles in pore water; however, water-soluble fluorine is not the main occurrence mode of fluorine in coal. Alkali-soluble fluorine is fluoride adsorbed by the clay minerals in coal, and OH$^-$ can replace the fluorine adsorbed by the clay minerals, thereby increasing the OH$^-$ concentration in the solution. OH$^-$ can also displace a certain amount of fluorine in the mineral lattice [18–20], which is ionic with an isomorphism. This process can be expressed by the following formula:

$$\text{Minerals–OH} + \text{F}^- \overset{\text{Replacement}}{\Leftrightarrow} \text{Minerals–F} + \text{OH}^-$$

**Table 1.** Extraction results from the coal-fired power plant.

| Items | Water-Soluble | | Alkali-Soluble | | Acid-Soluble | | CaCl$_2$-Soluble | | KCl-Soluble | |
|---|---|---|---|---|---|---|---|---|---|---|
| | Con (µg/g) | E-Rate (%) | Con (µg/g) | E-Rate (%) | Con (µg/g) | E-Rate (%) | Con (µg/g) | E-Rate (%) | Con (µg/g) | E-Rate (%) |
| YM | 4.51 | 1.33 | 9.74 | 2.88 | 39.92 | 11.81 | 4.16 | 1.23 | 4.89 | 1.45 |
| LZ | 2.45 | 3.03 | 2.77 | 3.42 | 4.33 | 5.34 | 2.36 | 2.91 | 2.17 | 2.68 |
| QH | 6.36 | 11.78 | 6.49 | 12.02 | 34.22 | 63.38 | 6.06 | 11.22 | 6.76 | 12.52 |
| HHC | 57.97 | 21.08 | 71.0 | 25.82 | 85.21 | 30.99 | 48.69 | 17.71 | 63.64 | 23.14 |
| HHX | 349.37 | 42.87 | 383.52 | 47.06 | 426.15 | 52.29 | 324.70 | 39.85 | 365.30 | 44.82 |

Con = Content; E-rate = Extraction rate; YM = coal; LZ = slag; QH = ash of sample before ESP; HHC = sample of ESP ash; HHX = sample of bag filter ash.

The fluorine in coal is mainly insoluble fluoride. Unlike water and alkali, acid can dissolve insoluble fluoride, including CaF$_2$, MgF$_2$, AlF$_3$, FeF$_3$ and Ca$_5$(PO$_4$)$_3$F. Iron and sulfur contents are high in fluorine-containing minerals (Figure 2a). In addition, the fluorine content of calcium chloride solution is slightly below that of potassium, and it is on the ppm level as CaF$_2$ is in the coal. The proximate and ultimate analysis of the coal are shown in Table 2.

**Table 2.** Proximate and ultimate analysis of coal.

| Proximate (wt %) | | | Ultimate (wt %) | | | | |
|---|---|---|---|---|---|---|---|
| Moisture | Ash | Volatile | C | H | O | N | S |
| 16.83 | 13.57 | 48.65 | 68.84 | 5.12 | 22.73 | 1.51 | 1.32 |

(a)

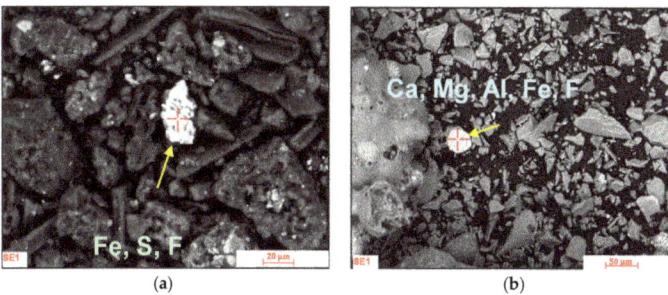

(b)

**Figure 2.** *Cont.*

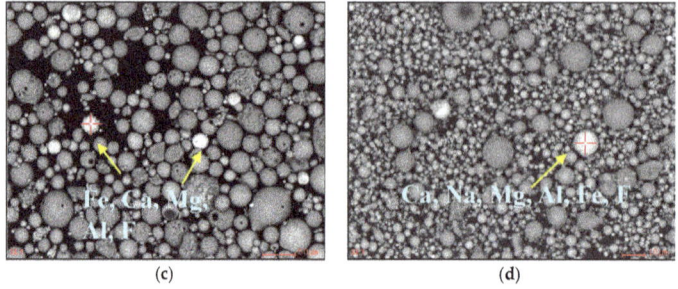

**Figure 2.** SEM analysis of coal and products of combustion: (**a**) is the backscattered image of YM; (**b**) is the backscattered image of LZ; (**c**) is the backscattered image of QH; (**d**) is the backscattered image of HHX. YM = coal; LZ = slag; QH = ash of sample before ESP; HHX = sample of bag filter ash.

*3.2. Occurrence Modesof Fluorine in LZ*

As shown in Figure 3, the extraction rate of the fluorine in LZ minimally changed, and only the acid-soluble fluorine content slightly decreased. Adsorbed fluoride, organic fluorides, and some mineral fluorine are released into the flue gas at elevated temperatures during coal combustion. Some fluorides were retained in the LZ, such as soluble fluorine (NaF, KF) and insoluble fluorine ($CaF_2$, $MgF_2$, $FeF_3$, $AlF_3$). Water-soluble fluorine and salt-soluble ($CaCl_2$ and KCl) fluorine mainly came from NaF and KF; alkali-soluble fluorine was derived from NaF, KF and the mineral lattice. Acid-soluble fluorine primarily resulted from soluble (NaF, KF) and insoluble fluorine ($CaF_2$, $MgF_2$, $FeF_3$ and $AlF_3$). Those fluorides are difficult to break down even at high temperatures. It is obvious that soluble and insoluble fluorine are the dominative occurrence modes in LZ. Figure 2b shows the SEM results of LZ. The outcomes also showed that $CaF_2$, $MgF_2$, $FeF_3$ and $AlF_3$ are the main occurrence states.

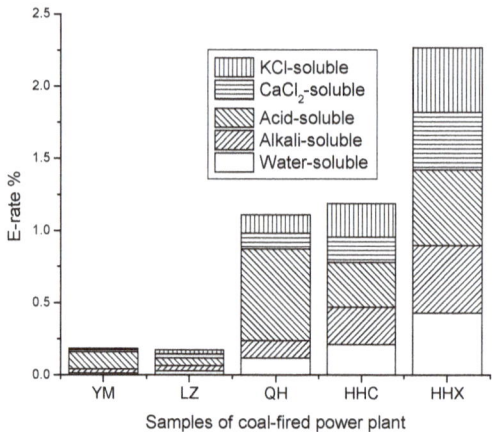

**Figure 3.** Different extraction rate of each sample.

*3.3. Occurrence Modesof Fluorine in QH*

As shown in Figure 3, the extraction rate of fluorine in QH, *i.e.*, 63.38%, is significantly higher than those in coal and slag, especially the extraction rate of acid-soluble fluorine. Some minerals are decomposed during coal combustion, and metal oxides (*i.e.*, CaO, MgO and $Al_2O_3$) are subsequently generated. There are some metal oxides (*i.e.*, CaO, MgO, $Al_2O_3$ and $Fe_2O_3$) in the fly ash [21]. These metal oxides react with gaseous fluorine (mostly HF) in flue gas to produce insoluble fluorides ($CaF_2$, $MgF_2$, $FeF_3$ and $AlF_3$) and other soluble fluorides (NaF, KF). Figure 2c shows that the Ca, Mg, Al and Fe contents in fluorine-containing minerals are high. Therefore, adsorbed fluorine and insoluble fluoride (*i.e.*, $CaF_2$, $MgF_2$ and $AlF_3$) are the main occurrence modes in QH.

### 3.4. Occurrence Modes of Fluorine in HHC and HHX

Among QH, HHC and HHX, HHX has the smallest particle diameter, whereas QH has the largest particle diameter. A smaller particle size indicates stronger fluoride absorption. The extraction fluorines in HHC and HHX, which are electrostatic precipitator ash and bag filter ash, respectively, are significantly more than those in YM, LZ and QH. The proportion of acid-soluble fluorine in HHC and HHX is less than that in QH. The fluorine in QH is mainly $CaF_2$, $MgF_2$, $FeF_3$ and $AlF_3$, while the fluorine in HHC and HHX is physically adsorbed fluorine, and a certain amount of fluorines are soluble fluorine (NaF, KF) in HHC and HHX. As shown in Figure 3, fluorines ordered from the fluorine with the lowest content on the basis of their extraction rates are as follows: acid-soluble, alkali-soluble, KCl-soluble, water-soluble and $CaCl_2$-soluble fluorines. Acid-soluble fluorine accounts for the largest proportion of extraction-soluble fluorine. Thus, physically adsorbed fluorine and insoluble inorganic mineral fluoride ($CaF_2$, $MgF_2$ and $AlF_3$) are the main occurrence modes of fluorine in HHC and HHX, according to the experimental results. The results are also consistent with Figure 2c,d.

### 3.5. The Changes of Fluorine in the Power Plant

Figure 4 shows that the water-soluble fluorine content rapidly increases along the flue gas flow, and the insoluble fluorine content can be estimated from acid-soluble fluorine and water-soluble fluorine. The change in insoluble fluorine content is represented by curve B in Figure 4, and the insoluble fluorine contents in HHC and HHX are 27.24 and 76.78 µg/g, respectively. The gradually increasing fluorine content in fly ash is caused by physical adsorption.

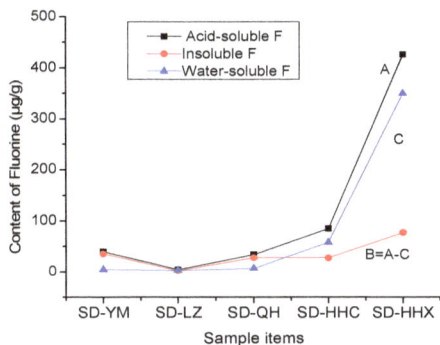

**Figure 4.** Transformation of fluorine during coal combustion in a coal-fired power plant.

Fluorine content in YM is 169 ppm (Table 3), and migration takes place during coal combustion. The fluorine is immigrated into LZ, QH, HHC, HHX, SG and flue gas flow. As shown in Figure 5, only 3.02% of fluorine remains in LZ, and a total of 23% immigrates into QH (8.05%), HHC (14.96%) and HHX (1.84%) during coal combustion. The rest of the fluorine is released into the flue gas flow, and only 1.24% is emitted into the environment, and 72.74%of the fluorine is absorbed by flue gas desulfurization gypsum. Fluorine in the coal-fired power plant discharges mostly as solid products; however, only tiny amounts of fluorine (1.24%) emitted into environment as gas products (HF, $SiF_4$) cannot be captured.

**Table 3.** Fluorine content of different samples in the coal-fired power plant.

| Sample | YM | LZ | QH | HHC | HHX | SG | Gas |
|---|---|---|---|---|---|---|---|
| P (%) | 100 | 3.02 | 8.05 | 14.96 | 1.84 | 72.74 | 1.24 |
| S (ppm) | 169 | 81 | 54 | 275 | 815 | 2730 | - |

$P = X_i/X$ ($X_i$ = total fluorine in a by-product of 1 kg coal, $X$ = total fluorine in 1 kg coal); S = fluorine content of a sample; Gas = gaseous fluorine emitted into environment; SG = flue gas desulfurization gypsum.

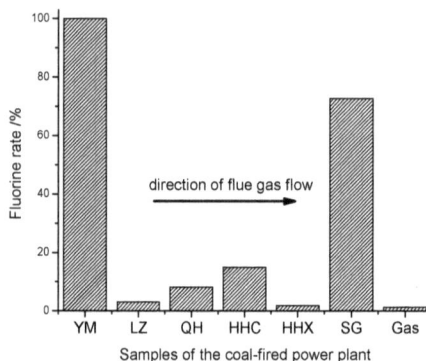

**Figure 5.** Migration of fluorine during coal combustion in the coal-fired power plant.

## 4. Conclusions

The different occurrence modes of fluorine during combustion and emission are as follows: (1) the fluorine in YM mainly exists in insoluble inorganic mineral forms, such as $CaF_2$, $MgF_2$ and $AlF_3$; (2) the fluorine occurrence modes in LZ are inorganic villiaumite, such as soluble NaF, KF, and insoluble $CaF_2$, which are difficult to break down even at high temperatures; and (3) physically adsorbed fluorine and insoluble inorganic mineral fluoride($CaF_2$, $MgF_2$ and $AlF_3$) are the main occurrence modes of fluorine in HHC and HHX. The high fluorine content in fly ash is mainly attributed to physically adsorbed fluorine along the direction of the flue gas flow. The insoluble inorganic mineral fluoride content in fly ash is also high, but the gradually increasing fluorine content in fly ash is mainly caused by physical adsorption. Fluorine in the coal-fired power plant discharges mostly as solid products; however, very little fluorine as HF and $SiF_4$ released into air cannot be captured.

**Acknowledgments:** The research is supported by the National Key Basic Research Program of China (No. 2014CB238904) and National Science Foundation of China (No. 51576082). The authors also thank the Foundation of State Key Laboratory of Coal Combustion (No. FSKLCCB1411).

**Author Contributions:** Zixue Luo and Junying Zhang conceived and designed the experiments; Guangmeng Wang performed the experiments; Zixue Luo and Guangmeng Wang analyzed the data; Yongchun Zhao contributed reagents and materials; Guangmeng Wang and Zixue Luo wrote the paper.

**Conflicts of Interest:** The authors declare no conflict of interest.

## References

1. Dai, S.; Ren, D.; Chou, C.L.; Finkelman, R.B.; Seredin, V.V.; Zhou, Y. Geochemistry of trace elements in chinese coals: A review of abundances, genetic types, impacts on human health, and industrial utilization. *Int. J. Coal Geol.* **2012**, *94*, 3–21. [CrossRef]
2. Ketris, M.P.; Yudovich, Y.E. Estimations of Clarkes for Carbonaceous biolithes: World averages for trace element contents in black shales and coals. *Int. J. Coal Geol.* **2009**, *78*, 135–148. [CrossRef]
3. Greta, E.; Dai, S.; Li, X. Fluorine in Bulgarian coals. *Int. J. Coal Geol.* **2013**, *105*, 16–23. [CrossRef]

4. Martinez-Tarazona, M.R.; Suarez-Fernandez, G.P.; Cardin, J.M. Fluorine in Asturian coals. *Fuel* **1994**, *73*, 1209–1213. [CrossRef]

5. Dai, S.; Li, D.; Chou, C.L.; Zhao, L.; Zhang, Y.; Ren, D.; Ma, Y.; Sun, Y. Mineralogy and geochemistry of boehmite-rich coals: New insights from the Haerwusu Surface Mine, Jungar Coalfield, Inner Mongolia, China. *Int. J. Coal Geol.* **2008**, *74*, 185–202. [CrossRef]

6. Dai, S.; Jiang, Y.; Ward, C.R.; Gu, L.; Seredin, V.V.; Liu, H.; Zhou, D.; Wang, X.; Sun, Y.; Zou, J.; *et al.* Mineralogical and geochemical compositions of the coal in the Guanbanwusu Mine, Inner Mongolia, China: Further evidence for the existence of an Al (Ga and REE) ore deposit in the Jungar Coalfield. *Int. J. Coal Geol.* **2012**, *98*, 10–40. [CrossRef]

7. Dai, S.; Zhang, W.; Seredin, V.V.; Ward, C.R.; Hower, J.C.; Wang, X.; Li, X.; Song, W.; Zhao, L.; Kang, H.; *et al.* Factors controlling geochemical and mineralogical compositions of coals preserved within marine carbonate successions: A case study from the Heshan Coalfield, southern China. *Int. J. Coal Geol.* **2013**, *109–110*, 77–100. [CrossRef]

8. Dai, S.; Hower, J.C.; Ward, C.R.; Guo, W.; Song, H.; O'Keefe, J.M.K.; Xie, P.; Hood, M.M.; Yan, X. Elements and phosphorus minerals in the middle Jurassic inertinite-rich coals of the Muli Coalfield on the Tibetan Plateau. *Int. J. Coal Geol.* **2015**, *144–145*, 23–47. [CrossRef]

9. Wang, X.; Dai, S.; Sun, Y.; Li, D.; Zhang, W.; Zhang, Y. Modes of occurrence of fluorine in the late paleozoic No. 6 coal from the Haerwusu surface mine, Inner Mongolia, China. *Fuel* **2011**, *90*, 248–254. [CrossRef]

10. Liu, X.; Zheng, C.; Liu, J.; Zhang, J.Y.; Song, D.Y. Analysis on fluorine speciation in coals from Guizhou province. *Chin. J. Electr. Eng.* **2008**, *28*, 46–51. (In Chinese).

11. Derher, G.B.; Finkelman, R.B. Selenium mobilization in a surface coal mine. *Environ. Geol. Water Sci.* **1992**, *19*, 155–167. [CrossRef]

12. Cavender, P.F.; Spears, D.A. Analysis of forms of sulfur within coal, and minor and trace element associations with pyrite by ICP analysis of extraction solutions. *Coal Sci. Technol.* **1995**, *24*, 1653–1656.

13. Spears, D.A. The determination of trace element distributions in coals using sequential chemical leaching—A new approach to an old method. *Fuel* **2013**, *114*, 31–37. [CrossRef]

14. Liu, J.; Yang, Z.; Yan, X.; Ji, D.; Yang, Y.; Hu, L. Modes of occurrence of highly-elevated trace elements in superhigh-organic-sulfur coals. *Fuel* **2015**, *156*, 190–197. [CrossRef]

15. Lessing, R. Fluorine in coal. *Nature* **1934**, *134*, 699–700. [CrossRef]

16. Beijing Research Institute of Coal Chemistry. *GB/T 4633-1997, Determination of Fluorine in Coal*; Standards Press of China: Beijing, China, 1997.

17. Qi, Q.J.; Liu, J.Z.; Wang, J.R.; Cao, X.Y.; Zhou, J.H.; Cen, K.F. Extracting experimental research on occurrence modes of fluorine in coal. *J. Liaoning Tech. Univ.* **2003**, *22*, 577–579. (In Chinese).

18. Elrashidi, M.A.; Lindsay, W.L. Chemical equilibria of fluorine in soils. *Soil Sci.* **1986**, *141*, 274–280. [CrossRef]

19. Larsen, S.; Widdowson, A.E. Soil fluorine. *J. Soil Sci.* **1971**, *22*, 210–221. [CrossRef]

20. Elrashidi, M.A.; Lindsay, W.L. Solubility of aluminum fluoride, fluorite, and fluorophlogopite minerals in soils. *Soil Sci. Soc. Am. J.* **1986**, *50*, 594–598. [CrossRef]

21. Koukouzas, N.; Ketikidis, C.; Itskos, G. Heavy metal characterization of CFB-derived coal fly ash. *Fuel Process. Technol.* **2011**, *92*, 441–446. [CrossRef]

*Article*

# Minerals in the Ash and Slag from Oxygen-Enriched Underground Coal Gasification

**Shuqin Liu \*, Chuan Qi, Shangjun Zhang and Yunpeng Deng**

School of Chemistry and Environmental Engineering, China University of Mining and Technology (Beijing), Beijing 100083, China; 18813089528@163.com (C.Q.); 13520219887@126.com (S.Z.); 15101142344@126.com (Y.D.)
* Correspondence: 13910526026@163.com; Tel.: +86-10-6233-9156

Academic Editor: Thomas N. Kerestedjian
Received: 30 October 2015; Accepted: 21 January 2016; Published: 30 March 2016

**Abstract:** Underground coal gasification (UCG) is a promising option for the recovery of low-rank and inaccessible coal resources. Detailed mineralogical information is essential to understand underground reaction conditions far from the surface and optimize the operation parameters during the UCG process. It is also significant in identifying the environmental effects of UCG residue. In this paper, with regard to the underground gasification of lignite, UCG slag was prepared through simulation tests of oxygen-enriched gasification under different atmospheric conditions, and the minerals were identified by X-Ray diffraction (XRD) and a scanning electron microscope coupled to an energy-dispersive spectrometer (SEM-EDS). Thermodynamic calculations performed using FactSage 6.4 were used to help to understand the transformation of minerals. The results indicate that an increased oxygen concentration is beneficial to the reformation of mineral crystal after ash fusion and the resulting crystal structures of minerals also tend to be more orderly. The dominant minerals in 60%-$O_2$ and 80%-$O_2$ UCG slag include anorthite, pyroxene, and gehlenite, while amorphous substances almost disappear. In addition, with increasing oxygen content, mullite might react with the calcium oxide existed in the slag to generate anorthite, which could then serve as a calcium source for the formation of gehlenite. In 80%-$O_2$ UCG slag, the iron-bearing mineral is transformed from sekaninaite to pyroxene.

**Keywords:** underground coal gasification; coal ash; mineralogy; oxygen-enriched gasification

## 1. Introduction

Coal, the main energy resource in China, accounts for approximately 70% of the primary energy resource structure. To avoid environmental pollution, the development of methods for clean and efficient utilization of coal has become necessary in recent years. During coal combustion and gasification at high temperatures, the reactivity differences of organic matters in coal can be almost ignored while the transformation behavior of minerals becomes important to the stability of the process. The reactions of inorganic minerals during combustion and gasification include a series of complicated physical and chemical changes that eventually form ash and slag with complex compositions [1]. Hence, detailed information about the mineralogical properties of coal ash is essential to optimize the operation parameters during coal utilization. It is also significant for improving the coal utilization efficiency and determining the influence of the solid wastes on the environment.

Underground coal gasification (UCG) is the process of *in situ* conversion of coal directly into combustible gaseous products. A sketch of the UCG process is shown in Figure 1. The first step of UCG is to choose a proper location and then design and construct an underground reactor. Boreholes are drilled from the surface to the coal bed, followed by a horizontal channel connecting the boreholes along the bottom of the coal bed. After a gasifier is prepared, the coal at one end of the channel is ignited, and gasification agents such as air, oxygen, steam, or their mixtures with different oxygen ratios are injected into the reactor. Accompanied by a series of coal reactions including pyrolysis, reduction, and oxidation, the fire moves along the channel towards the production borehole where the coal gas is collected by a pipeline. Unlike traditional coal mining and ground coal gasification technologies, UCG is carried out in the underground coal bed without physical coal mining, transportation, or coal preparation, which is regarded as supplementary to the coal mining method. The composition and heat value of the product gas depend on the initial gas injected, the position for gas injection, and the temperature profile of the coal bed.

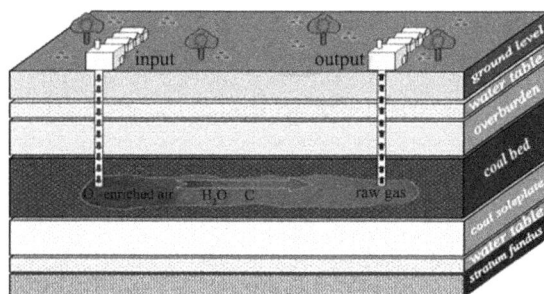

**Figure 1.** Sketch of the underground coal gasification (UCG) process.

Because UCG is always performed within the coal bed, several hundred meters beneath the surface, only the injection and production parameters can be determined. It is particularly difficult to determine the actual reaction conditions, especially the temperature field distribution and thermal equilibrium of the underground gasifier. Therefore, it is necessary to investigate the relationship between gasification technology and mineralogical characteristics of ash and slag through UCG simulation experiments. On the other hand, the potential for groundwater pollution from UCG-generated residues has also been a concern in recent years. The leaching behavior of toxic elements from solid residues is closely related to the characteristics of UCG ash and slag.

Plenty of research has been published on the conversion of minerals during coal combustion, but a low number of papers have reported on the transformation of minerals during coal gasification. Most of the study of ash chemistry during the high-temperature gasification process focused on the investigation of ash deposition and slag formation, as well as on the difficulties found in industrial gasifiers regarding fluidized bed gasification and entrained flow gasification [2–5]. However, few papers focus on the formation mechanism of ash and slag during the UCG process.

Mineral matter in coal can be classified as external minerals and inherent minerals according to their origin, with distinct differences in composition and form. At elevated temperatures, mineral transformation occurs, including chemical reactions between the clay minerals, carbonate minerals, pyrite, and quartz in the coal. It was discovered that the transformation temperature of external minerals was relatively lower than that of inherent minerals [6] and that the reaction rate and degree were greatly affected by temperature. Furthermore, the transformation processes varied in different gasification atmospheres. For instance, the softening temperature of minerals in the gasification condition was found to be lower than that in the combustion condition [7].

In addition to temperature and atmosphere, the furnace type is another important influence factor for ash formation behaviors [8]. The thermal conversion of minerals occurs at high temperatures (>800 °C) [9–12], including the transformation of clay minerals, carbonate minerals, pyrite, and quartz. During coal gasification, external minerals were fragmented into fine particles in the thermal conversion process, thus determining the particle size distribution of the fly ash. The formation of ash or slag is also enhanced by the cracking and thermal decomposition of external minerals as well as the reaction between external minerals and other minerals/gaseous substances [13]. A large number of studies show that pyrite and calcium carbonate break up when heated, while quartz and clay minerals do not. In addition, the breaking mechanism of some other types of external minerals is controversial and is affected by the residence time and heating rate [14–18]. The fusible minerals, such as the carbonate, sulfate minerals, and feldspar contained in inherent or external minerals, tend to become "solvent minerals" at high temperatures, which in turn promote the melting and slagging of gasification residues [19].

There are few reports on the formation and properties of UCG slag. In the USA, a series of residues and rocks were sampled from the UCG test site near Centralia, Washington [20]. X-Ray diffraction (XRD) and a scanning electron microscope coupled to an electron microprobe (SEM-EPMA) were used to analyze the mineralogical characteristics of the samples, and a moderate temperature reaction was confirmed. Reduced iron reacted with clay minerals to form a solid solution of aluminum-rich hercynite ($FeAl_2O_4$), which then serve as the precursor and react with $SiO_2$ to form sekaninaite ($Fe_2Al_4Si_5O_{18}$) at higher temperatures.

In this study, based on the first UCG field test of lignite in Inner Mongolia, laboratory UCG simulation tests were performed to prepare UCG residues in different atmospheres. XRD and a scanning electron microscope coupled to an energy-dispersive spectrometer (SEM-EDS) were used to identify the composition and microstructure of the typical minerals formed and existing in UCG ash and slag. Thermodynamic calculations using FactSage 6.4 (Thermfact/CRCT, Montreal, QC, Canada; GTT-Technologies, Aachen, Germany) were also carried out to investigate the transformation of minerals at elevated temperatures during the UCG process.

## 2. Experimental and Modeling

### 2.1. Geological Setting of the Coalfield

The strata in the study area are from the Lower Jining Group in the Mesoarchean ($Ar_2J^1$), the Oligocene Huerjing Formation ($E_3h$), the Miocene Hannuoba Formation ($N_1h$), the Pliocene Baogedawula Formation ($N_2b$), and the Holocene. The coal-bearing strata are a set of sedimentary sequences of terrestrial clastic rocks formed in lake and swamp facies and overlain by the Quaternary strata. In addition, the Jining Group is the basement of the coal-bearing strata, and the major lithology is granite at depths of 202.95–565.25 m.

The main coal bed in this area occurs in the Lower Huerjing Formation ($E_3h^1$). It is 7.05 m thick on average and minable in most coalfields, with 0–12 partings of accumulative thickness in 1.79 m. The dip angle of the coal bed is less than 5°. The roof of the coal bed is siltstone and dark grey carbonaceous mudstone with clastic organic debris, while the bottom of the coal bed is a thin layer of mudstone close to the lowest basement of granite and gneiss. The coal is identified as lignite, with an ignition point of 268 °C and the net heating value ranging from 13.37 to 16.72 MJ/kg [21].

### 2.2. Coal Samples

The test lignite is from a neighboring coal mine of the UCG field test area, also in the Gonggou coal field located in Ulanqab, Inner Mongolia, China. The coal bed has an average depth of 280 m.

Proximate analysis, including moisture, ash, volatile matter, and fixed carbon, was determined in accordance with Chinese Standards GB/T 212-2008 [22]. Ultimate analysis, including carbon, hydrogen, nitrogen, and total sulfur, was measured following Chinese Standards GB/T 476-2008 [23], GB/T 19227-2008 [24] and GB/T 214-2007 [25], respectively. The results of the ultimate and proximate analyses of the coal sample are shown in Table 1, indicating that the lignite is higher in volatile materials, ash and moisture. With a sulfur content lower than 1%, the test lignite is considered a low-sulfur coal according to Chinese standards GB/T 15224.2-2010 [26].

The ash composition of the test lignite was conducted using GB/T 1574-2007 [27], which expressed in oxide percentages (Table 2). The ash fusion test is performed according to Chinese standards GB/T 219-2008 [28].

**Table 1.** Proximate and ultimate analysis of the test coal (%).

| Proximate Analysis/% | | | | Ultimate Analysis/% | | | | |
|---|---|---|---|---|---|---|---|---|
| $M_{ad}$ | $A_{ad}$ | $V_{ad}$ | $FC_{ad}$ | $C_{ad}$ | $H_{ad}$ | $O_{ad}$ | $N_{ad}$ | $S_{t,ad}$ |
| 11.50 | 29.10 | 28.47 | 30.93 | 43.70 | 3.11 | 11.59 | 0.57 | 0.65 |

M, moisture; A, ash; V, volatile matter; FC, fixed carbon; C, carbon; H, hydrogen; O, oxygen; N, nitrogen; St, total sulfur; ad, air dried basis.

**Table 2.** Ash composition of lignite (wt %).

| Oxide | $SiO_2$ | $Al_2O_3$ | $Fe_2O_3$ | $TiO_2$ | CaO | MgO | $K_2O$ | $Na_2O$ | $MnO_2$ | $SO_3$ | $P_2O_5$ |
|---|---|---|---|---|---|---|---|---|---|---|---|
| Content | 59.04 | 18.02 | 5.83 | 0.93 | 5.80 | 2.71 | 2.74 | 1.26 | 1.26 | 2.50 | 0.20 |

It is obvious that the lignite is enriched in $SiO_2$, $Al_2O_3$ and CaO. Among them, $SiO_2$ accounts for more than half of the composition, followed by $Al_2O_3$ with a ratio of 18.02%. The content of $Fe_2O_3$ and $SO_3$ are relatively high in this sample and account for 5.83% and 2.50%, respectively. The analysis results of the coal ash fusibility in a weak reducing atmosphere are exhibited in Table 3. It can be seen that the coal ash is medium melting, beginning to soften at approximately 1200 °C and starting to flow at approximately 1270 °C.

**Table 3.** Coal ash fusibility (°C).

| Deformation Temperature (DT) | Softening Temperature (ST) | Hemispherical Temperature (HT) | Flow Temperature (FT) |
|---|---|---|---|
| 1160 | 1200 | 1230 | 1270 |

### 2.3. Underground Coal Gasification (UCG) Simulation Facility

The UCG simulation facility, as shown in Figure 2, consists of four parts: the gasifier gas supply system, the gas cleaning system, the sampling, and monitoring systems. The whole gasifier is designed in the shape of a cylinder with external dimensions of 7.4 m in length and 3.5 m in diameter. The shell of the gasifier is made of special steel used for pressure vessels, and the design pressure is 1.6 MPa. The hearth of the reactor is cast with refractory material in the shape of a rectangular prism (5.0 m × 1.6 m × 1.6 m), the working temperature of which can be up to 1800 °C. Five inlet or outlet pipes, 33 measurement points for temperature and pressure, and four observation holes for a closed circuit industrial television (CCTV) are installed in the reactor. Gas composition analysis is performed by gas chromatography (GC-2014, Shimadzu, Kyoto, Japan). Data are collected online during the test using a distributed control system (DCS) (Honeywell, Morristown, NJ, USA) and saved to a hard disk. Various curves for parameters can be transferred to the screen from the DCS.

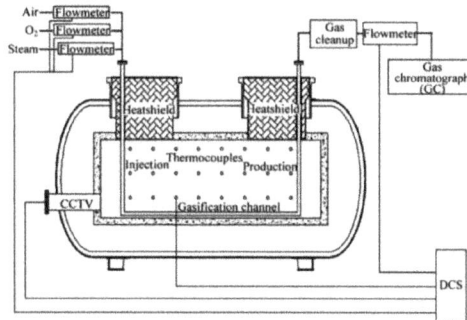

**Figure 2.** A schematic diagram of an Underground Coal Gasification (UCG) simulation facility.

## 2.4. Coal Bed Simulation

The coal bed layout is shown in Figure 3. A soil layer with a thickness of 50 mm was arranged at the bottom of the hearth of the test gasifier to act as the bottom layers. Then, coal lumps with dimensions of 0.4 m × 0.8 m × 0.8 m were piled into the internal hearth to create a coal bed of 0.8 m in length, 0.8 m in width and 3.5 m in height. The gaps between the coal blocks were filled with coal slurry, which is a mixture of coal powder and clay. Then, a square channel with dimensions of 0.1 m × 0.1 m was drilled into the bottom of the coal bed and connected to the vertical inlet and outlet holes. In the following step, the coal bed was covered by a 2-cm sand layer and then cast with a thin layer of cement. Finally, the gap between the cement and the hearth was tightly filled with soil to prevent gas leakage.

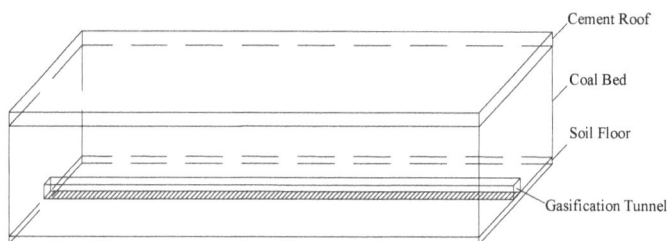

**Figure 3.** Schematic diagram of coal bed layout.

To monitor and control the temperature profile in the gasified coal bed during the test, 63 K-type armored thermocouples were installed in the coal bed and arranged in three levels, 0.15-m, 0.3-m or 0.4-m below the top of coal bed. In each level, 21 thermocouples were laid with an average distribution.

## 2.5. Experimental Procedure

The UCG method of controlled moving injection point was used to perform gasification in order to change the oxygen concentration in the injection gas. After a leak test of the UCG simulation facility, the gasifier was prepared for ignition. First, air was introduced into the channel through the injection hole. Then, an electric igniter was placed inside the horizontal channel and used to ignite the coal. The temperature was monitored and collected in real time, and the gas from the production hole was analyzed every 20 min. If the temperature in the coal bed increased to 600 °C and the concentration of $CO_2$ in the effluent gas exceeded 20%, this suggested that the ignition was successful. Then, the gas mixture containing 40%-$O_2$ and 60%-$N_2$ was injected into the gasifier to produce gas. With the expansion of the cavity, the high-temperature zone, or fire face, moves towards the production hole. When the gas quality grew poor, as determined by a gas heat value lower than 4.18 MJ/m³, we switched to the second injection hole nearest the fire face, and a mixture of 60%-$O_2$ and 40%-$N_2$ was injected. The entire duration lasted for 80 h, including 10 h for ignition and preheating, 30 h for 40%-$O_2$ gasification, 20 h for 60%-$O_2$ gasification, and 20 h for 80%-$O_2$ gasification. After the three-stage simulation test, $N_2$ was continuously injected to cool the gasifier.

## 2.6. Sampling and Analysis of UCG Slag

### 2.6.1. Sampling of UCG Slag

After the temperature inside the furnace reached room temperature, the gasifier cover was opened. The sandy soil above the coal bed was removed carefully, and slag formed under different UCG conditions was collected near the injection point for analysis. Pictures of the UCG cavity and UCG slag in different atmospheres are shown in Figure 4.

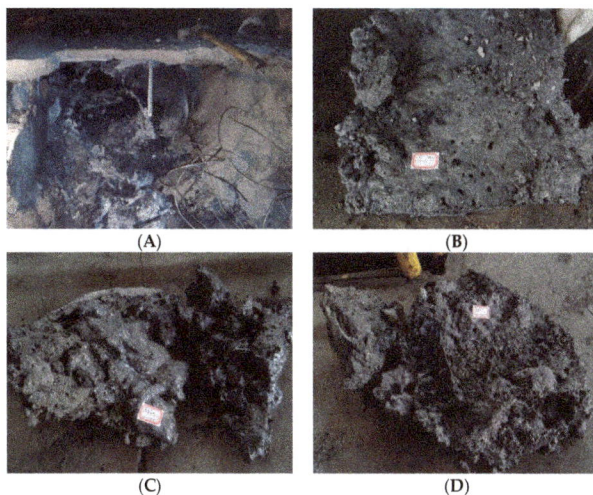

**Figure 4.** UCG cavity and UCG slag in different atmospheres. (**A**) UCG cavity; (**B**) 40%-$O_2$ slag; (**C**) 60%-$O_2$ slag; (**D**) 80%-$O_2$ slag.

### 2.6.2. X-Ray Diffraction (XRD) Analysis

The raw coal, ash, and slag were crushed, ground, and sieved before analysis. Samples below 75 μm were selected for XRD analysis, which was performed on a powder diffractometer (D/max-2500/pc XRD, Rigaku, Tokyo, Japan) with Ni-filtered Cu-Kα radiation and a scintillation detector. The XRD pattern was recorded over a 2θ range of 2.6°–70° at a scan rate of 3°/min and a step size of 0.01°. Jade 6.5 software (MDI, Livermore, CA, USA) was used to analyze the XRD curve for qualitative analysis. X-Ray diffractograms of the LTAs (low temperature ash) and partings were subjected to quantitative mineralogical analysis using Siroquant™ (Sietronics, Mitchell, Australia), a commercial interpretation software developed by Taylor (1991), based on the principles for diffractogram profiling set out by Rietveld (1969). Further details indicating the use of this technique for coal-related materials are given by Ward *et al.* [29] and Dai *et al.* [30].

### 2.6.3. Scanning Electron Microscope and Energy-Dispersive Spectrometer (SEM-EDS) Analysis

The raw coal and ash and slag samples were firstly crushed and ground, and particles in the range of 1–3 mm were collected. The sample particles were then mixed with ethyl α-cyanoacrylate and the mixture was polished and mounted on standard aluminum SEM stubs using sticky electronic-conductive carbon tabs. A field emission-scanning electron microscope (FE-SEM, Quanta™ 650 FEG, FEI, Hillsboro, OR, USA), in conjunction with an energy-dispersive X-Ray (EDAX) spectrometer (EDS, Apex 4, Genesis, NJ, USA), was used to study the morphology of the minerals and to determine the distribution of some elements. The samples were prepared under low-vacuum SEM conditions. The analytical conditions were as follows: working distance (WD) 10 mm, beam voltage 20 kV, aperture 6, and micron spot size 5. The images were captured via a retractable solid state backscatter electron detector (SSBSED). More details of FE-SEM-EDS are described by Dai *et al.* [31].

### 2.7. FactSage Thermochemical Modeling

FactSage® (Thermfact/CRCT, Montreal, QC, Canada; GTT-Technologies, Aachen, Germany) was introduced in 2001 as the fusion of two well-known software packages in the field of computational thermochemistry: F*A*C*T/FACT-Win and ChemSage. The thermochemistry models can be used to analyze equilibrium conditions for reactions occurring between inorganic and/or organic materials, as well as providing insight into the mineral formation and slag formation speciation. The database can assist in understanding, as well as predicting, what can and will happen with specific coal and mineral sources inside the gasification process [32]. In this study, thermodynamic equilibrium modeling was accomplished by the "Equilib module" in FactSage 6.4, which is the Gibbs energy minimization workhorse of FactSage. It calculates the concentrations of chemical species when specific elements or compounds react or partially react to reach a state of chemical equilibrium [33].

For the calculations the equilibrium module has been employed together with the databases FToxid and FactPS. Additionally, the solution phases of FToxid-SLAG and FToxid-oPyr have been selected. In order to simulate the gasification process as close as possible to the actual gasification process, the temperature is from 0 to 1500 °C in 100 °C intervals and the pressure is atmospheric pressure. The calculations were conducted based on the mineral composition of coal and the amount of coal, oxygen, and nitrogen consumed to produce 1 N·m$^3$ of gas. Since the amount of ash in coal is the sum of the mineral composition of coal such as $SiO_2$, $Al_2O_3$, $Fe_2O_3$, $TiO_2$, $CaO$, $MgO$, $K_2O$, $Na_2O$, $MnO_2$, $SO_3$ and $P_2O_5$, the input into FactSage as shown in Table 4 is done in elemental form *i.e.*, carbon (C), hydrogen (H), nitrogen (N), sulfur (S), oxygen (O), and other mineral elements.

**Table 4.** Input into calculations.

| Element | 40%-$O_2$ Gasification (g) | 60%-$O_2$ Gasification (g) | 80%-$O_2$ Gasification (g) |
|---------|---------------------------|---------------------------|---------------------------|
| C | 184.65 | 247.98 | 308.67 |
| H | 13.14 | 17.65 | 21.97 |
| O | 397.15 | 536.51 | 628.44 |
| N | 459.39 | 277.83 | 123.59 |
| S | 2.75 | 3.69 | 4.59 |
| Si | 33.89 | 45.51 | 56.65 |
| Al | 11.75 | 15.78 | 19.64 |
| Fe | 5.03 | 6.75 | 8.41 |
| Ti | 0.68 | 0.91 | 1.13 |
| Ca | 5.11 | 6.87 | 8.55 |
| Mg | 1.99 | 2.67 | 3.32 |
| K | 2.79 | 3.75 | 4.66 |
| Na | 1.14 | 1.53 | 1.91 |
| Mn | 0.04 | 0.06 | 0.07 |
| P | 0.13 | 0.17 | 0.21 |

## 3. Results and Discussion

### 3.1. Gas Composition under Oxygen-Enriched Gasification Conditions

Increasing the oxygen concentration of the gasification agent is an effective way to improve the quality of the product gas. Table 5 provides detailed information about coal gas composition and gas heat value under oxygen-enriched gasification conditions during the UCG simulation test. The gas heat values and the contents of CO and $H_2$ in the gas gradually increased with the increase in oxygen concentrations. The yield of CO and $H_2$ mainly depends on the rate and extent of the reduction reaction between C and $CO_2/H_2O$ (g), which is dominated by the temperature. Coal combustion was enhanced when a larger amount of oxygen was injected and the reduction process was strengthened, accompanied by an increase in the temperature field in the gasifier. This suggests that during UCG, an optimum oxygen concentration in the injection agent could be found to yield the highest combustible composition for certain coal types and typical reaction conditions. In addition, it was found that the methane content remains at a lower level and is less affected by the oxygen concentration.

**Table 5.** Gas composition under oxygen-enriched gasification conditions (V%).

| Oxygen (%) | $H_2$ | CO | $CH_4$ | $CO_2$ | $N_2$ | $O_2$ | Heat Value MJ/N·m$^3$ |
|---|---|---|---|---|---|---|---|
| 40 | 21.06 | 15.04 | 1.50 | 25.85 | 35.67 | 0.72 | 4.70 |
| 60 | 30.11 | 24.86 | 1.25 | 26.75 | 16.81 | 0.13 | 6.82 |
| 80 | 31.33 | 27.29 | 1.03 | 31.69 | 8.36 | 0.21 | 7.18 |

## 3.2. Distribution of Temperature Field in the Coal Bed

During the UCG process, the transformation of the organic and inorganic components of the coal can be divided into three steps. The first step is the drying and pyrolysis of coal below 600 °C, which involves the release of water and volatile matter and the crystal transformation of some minerals. The second step is the reduction reaction of char with $CO_2$ and $H_2O$ (g) at temperatures ranging from 600 to 900 °C. The final step is the oxidation of residual carbon above 900 °C. The real reaction temperature is much higher than the theoretical temperature because of the thermal storage in the simulated coal bed.

Temperature profiles of the coal bed under different oxygen-enriched conditions are displayed in Figures 5–7 and were constructed from thermocouple data. It is clear that the high-temperature area is narrow in 40%-$O_2$ and 60%-$O_2$ conditions because their reduction and oxidation reactions occur in smaller areas. However, when the oxygen concentration was increased to 80%, the temperature field in the reaction area significantly increased. In the oxidation zone near the injection hole, the temperature was increased remarkably. In addition, most of the monitoring points exceeded 600 °C, and the maximum temperature of the central area of oxidation increased from 1200 to 1400 °C. It is inferred that the reduction area is enlarged and the reduction process is enhanced.

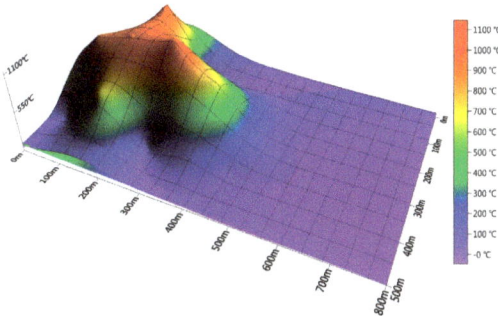

**Figure 5.** Temperature profile distribution during UCG simulation test (40%-$O_2$).

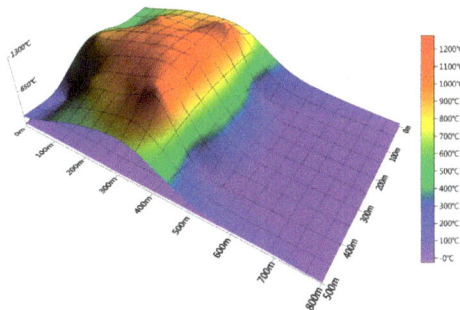

**Figure 6.** Temperature profile distribution during UCG simulation test (60%-$O_2$).

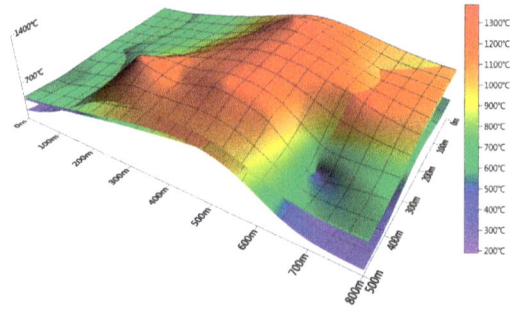

**Figure 7.** Temperature profile distribution during UCG simulation test (80%-$O_2$).

### 3.3. Minerals in the UCG Ash and Slag

To further understand the mineral transformation behavior that occurs during the UCG process, X-Ray diffraction analysis was carried out to identify the typical minerals present in the UCG residual samples. The XRD patterns of raw coal and UCG slag at different atmospheres are summarized in Figure 8, and the quantitative analysis results using Siroquant are listed in Table 6. The major minerals found in raw coal include quartz (melting point (*Tm*): 1723 °C), illite, and clay minerals (mostly kaolinite). Minor amounts of pyrite and chlorite are also observed. In the 40%-$O_2$ slag, high-temperature quartz, anorthite (*Tm*: 1550 °C), mullite (*Tm*: 1900 °C), sekaninaite (*Tm*: 1200 °C), and massive amorphous substance materials (49%) become the major minerals. The mineral compositions of 60%-$O_2$ slag and 80%-$O_2$ slag are similar, and the dominant minerals involve high-temperature quartz, anorthite, gehlenite (*Tm*: 1500 °C), and pyroxene. Furthermore, the amount of quartz and anorthite in 80%-$O_2$ slag is less than that in 60%-$O_2$ slag, while gehlenite is formed in great quantities. Clay and iron minerals in coal have not been found in UCG slag, which suggests that they have been transformed to anorthite and pyroxene during oxygen-enriched gasification.

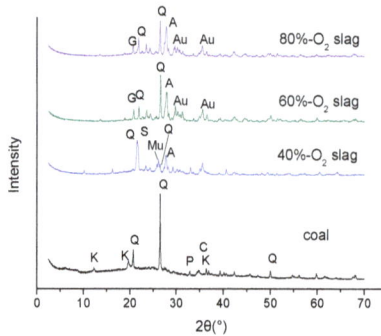

**Figure 8.** XRD patterns of coal and UCG slag. Q-Quartz; Py-Pyroxene; A-Anorthite; S-Sekaninaite; G-Gehlenite; P-Pyrite; C-Chlorite; K-Kaolinite; Mu-Mullite.

In addition to X-Ray diffraction analysis, SEM-EDS examination was also performed to investigate the typical minerals present in the UCG slag. A great deal of amorphous materials were identified in the 40%-$O_2$ slag based on XRD analysis, which was also proven by SEM images, shown in Figure 9A, in which a large amount of porous and melted materials could be observed. It has been reported that during gasification, the decrease in crystallization intensity of the minerals with increasing temperature is not only due to the decomposition of some mineral phases, but also because of the formation of molten liquid(SLAG) [32]. Based on the previous information for temperature distribution in the

40%-$O_2$ gasification condition, the temperatures in the oxidation zone are in the range of 1100 to 1200 °C, which is close to the ash fusion point of the test coal. Therefore, it is concluded that under this gasification condition, the minerals in coal might melt and then form slag after cooling down, leading to an obvious disappearance of crystal minerals. Simultaneously, a small portion of the crystal minerals could be encapsulated by the melting material.

**Table 6.** Mineral composition of UCG slag by XRD analysis and Siroquant.

| Mineral Composition | Raw Coal | 40%-$O_2$ Slag | 60%-$O_2$ Slag | 80%-$O_2$ Slag |
|:---:|:---:|:---:|:---:|:---:|
| Quartz | 29.6 | 1.5 | 9.3 | 1.8 |
| Illite | 43.3 | — | — | — |
| Kaolinite | 21.5 | — | — | — |
| Chlorite | 1.4 | — | — | — |
| Pyrite | 4 | — | — | — |
| Anorthite | — | 13.7 | 59.7 | 45.4 |
| Pyroxene | — | — | 25.7 | 26.3 |
| Gehlenite | — | — | 1.3 | 26.4 |
| Sekaninaite | — | 11.5 | — | — |
| Mullite | — | 22.3 | — | — |
| Amorphous | — | 49.0 | 3.6 | — |

Note: "—", less than 1%.

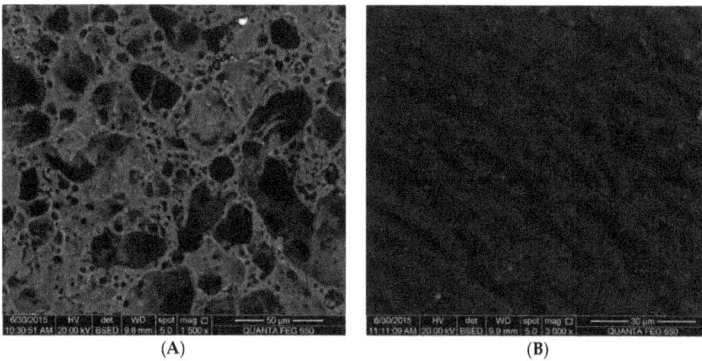

**Figure 9.** Melting material and unburned carbon in 40%-$O_2$ slag, SEM back scattering images. (**A**) Porous and melting materials; (**B**) Unburned carbon.

The mineral transformations with increasing temperature are displayed in Figure 10. It is indicated that in the 40%-$O_2$ gasification condition, mineral melting occurs at temperatures lower than 900 °C. With further increasing temperature, the slag content continuously increases. Massive high-temperature quartz mineral slag ($SiO_2$(SLAG)) and pyroxene mineral solid solution (oPyr(solution)) are generated at temperatures ranging from 900 to 1200 °C. The difference between the thermodynamic calculation and the sample analysis should be attributed to the ideal state adopted in the equilibrium calculations. However, both sample analysis and thermodynamic simulation show that the melting temperature of coal minerals is significantly lower than the ash fusion temperature of coal in UCG reduction conditions. Because there is a big difference between the modeling results of mineral transformation in these oxygen-enriched conditions with the experimental results, the results in 60%-$O_2$ and 80%-$O_2$ are not given here. The real gasification reaction is always limited by the reaction kinetic, mass transport, unknown reactions, and interfaces, especially in underground coal simulation conditions.

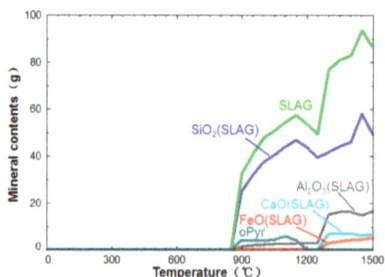

**Figure 10.** Thermodynamic simulation of mineral transformation in 40%-$O_2$ gasification condition.

SEM images of 60%-$O_2$ slag and 80%-$O_2$ slag are shown in Figures 11 and 12 respectively. In 60%-$O_2$ slag, a small amount of amorphous glass beads could still be found, which is in agreement with the XRD quantitative analysis. However, amorphous material is hardly observed in 80%-$O_2$ slag, and a large amount of crystals appear in the shape of rod-like stacks (Figure 12A). A possible reason for this change could be explained as follows: with an increase in the oxygen concentration in the injection gas, the reaction temperature continuously increases, causing melting minerals to react further and produce new crystal minerals at temperatures over 1200 °C, contributing to the remarkable increase in the anorthite and pyroxene contents in the 60%-$O_2$ slag and 80%-$O_2$ slag. In other words, oxygen-enriched gasification is beneficial to the regeneration of typical minerals.

In the SEM image of 40%-$O_2$ slag, unburned carbon in the shape of plant cells was observed, and the whole micro-morphology is comparatively complicated (Figure 9B). In comparison, the SEM image of the 60%-$O_2$ slag seems to be more homogeneous and is shown to have a wheat head formation (Figure 11B), which has been previously noted in the study of surface gasification ash by Matjie [34]. The homogeneous phenomenon is even more obvious for the 80%-$O_2$ slag, where crystal minerals are regularly arranged in lamellar stacks formation. The transformation of micro-morphology from disordered, porous, and melting minerals to homogeneous and orderly crystals indicates that the crystal structure of minerals tends to be more orderly with increases in the oxygen content from 40% to 80%.

Mullite is found in the 40%-$O_2$ slag, while it disappears in the 60%-$O_2$ and 80%-$O_2$ slag. Instead, massive anorthite is formed in the 60-$O_2$ slag. It is suggested that mullite reacts with calcium oxide contained in the slag to generate anorthite at temperatures over 1130 °C [35]. In addition, the alkali metals in coal may inhibit the formation of mullite at high temperatures [36]. These factors lead to the reduction and disappearance of mullite with the increase in the oxygen concentration during UCG process.

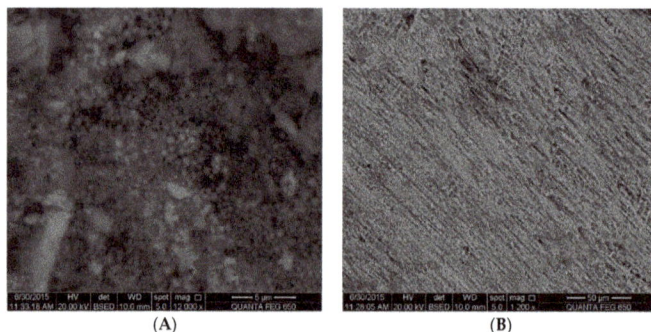

**Figure 11.** Amorphous glass beads and wheat head formation crystals in 60%-$O_2$ slag, SEM back-scattering images. (**A**) Amorphous glass beads; (**B**) Wheat head formation crystals.

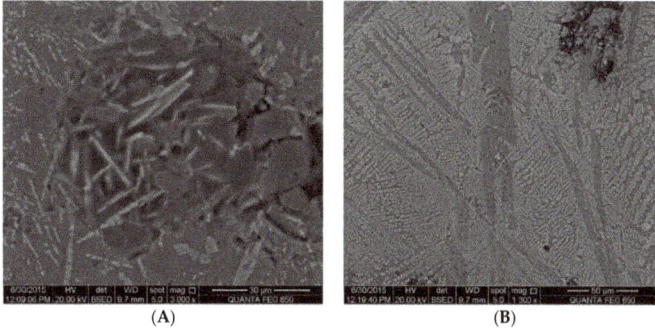

**Figure 12.** Rod-like stacks crystals and lamellar stacks crystals in 80%-$O_2$ slag, SEM back scattering images. (**A**) Rod-like stacks crystals; (**B**) Lamellar stacks crystals.

As shown in the SEM image in Figure 13, as a whole, 80%-$O_2$ slag is mainly composed of two type of materials, phase "A" and phase "B." From the Energy-Dispersive Spectrometer (EDS) quantitative analysis, as listed in Table 7, it is inferred that phase "A" contains anorthite crystals and phase "B" is the solid solution of gehlenite and pyroxene. The previous XRD analysis showed that in the 80%-$O_2$ slag, the anorthite content is reduced, while gehlenite is formed in great quantities. Therefore, it can be concluded that anorthite forms in great quantities at 1200 °C and tends to melt as the temperature increases, so its crystal content gradually decreases until it finally disappears at 1400 °C. It is also reported that gehlenite is formed between 1200 and 1400 °C and begins to decrease above 1400 °C [37]. Thus, it is assumed that anorthite may provide a calcium source for the formation of gehlenite, which also accounts for the reduction of anorthite in the 80%-$O_2$ slag.

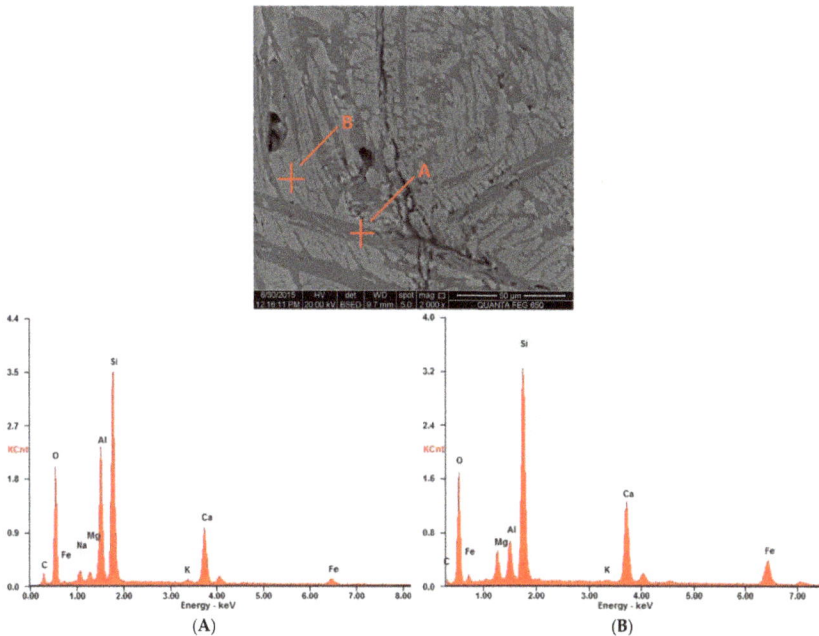

**Figure 13.** Scanning electron microscope and energy-dispersive spectrometer (SEM-EDS) analysis of minerals in 80%-$O_2$ slag. (**A**) Anorthite crystals; (**B**) Solid solution of gehlenite and pyroxene.

**Table 7.** Energy-dispersive spectrometer (EDS) quantitative results of anorthite crystals and gehlenite and pyroxene solid solution in 80%-$O_2$ slag (wt %).

| Element | C | O | Na | Mg | Al | Si | K | Ca | Fe |
|---------|-----|------|-----|-----|------|------|-----|------|------|
| A | 6.9 | 35.7 | 1.7 | 1.0 | 13.7 | 26.6 | 0.6 | 10.8 | 3.0 |
| B | 7.2 | 32.7 | — | 3.4 | 4.4 | 24.5 | 0.4 | 14.8 | 12.6 |

A: Anorthite crystals; B: Solid solution of gehlenite and pyroxene. Note: "—", undetectable.

Based on the result of SEM-EDS analysis in Figure 14, sekaninaite is proven to exist in 40%-$O_2$ slag. Moreover, iron oxide is also found in the form of $Fe_3O_4$, as concluded from the EDS quantitative results (Table 8). However, the iron-bearing mineral in 60%-$O_2$ slag and 80%-$O_2$ slag is mainly pyroxene on the basis of the SEM-EDS results, which is in agreement with the previous XRD quantitative analysis result. For the UCG residue, the existence of sekaninaite ($Fe_2Al_4Si_5O_{18}$) has been observed and proven to be the product of the reaction between $SiO_2$ and hercynite ($FeAl_2O_4$) at high temperatures [20]. It has been reported that under oxygen-enriched gasification conditions, the iron-bearing mineral tends to react with aluminosilicate to form pyroxene [38]. Therefore, the reaction mechanism of iron-bearing minerals at high temperatures could be concluded to be iron mineral oxidizing to form magnetite ($Fe_3O_4$) and then converting to $Fe^{2+}$ in hercynite during the gasification process (in a reductive atmosphere). Hercynite reacts with $SiO_2$ to form sekaninaite, and then sekaninaite is further oxidized to produce pyroxene with the increase in oxygen concentration.

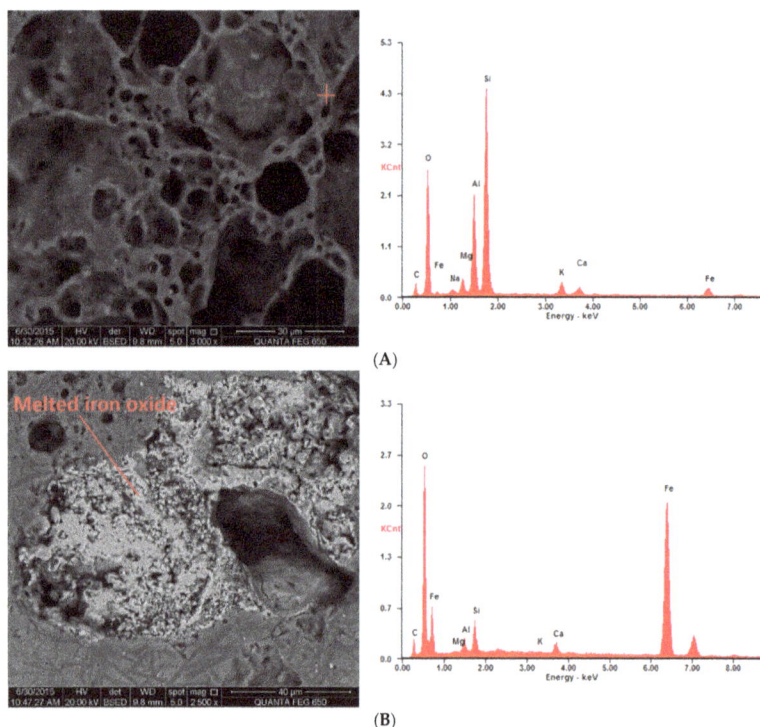

(A)

(B)

**Figure 14.** *Cont.*

(C)

**Figure 14.** SEM-EDS analysis of Fe-bearing minerals in UCG slag. (**A**) Sekaninaite in 40%-$O_2$ slag, EDS analysis; (**B**) Melting iron oxide in 40%-$O_2$ slag, EDS analysis; (**C**) Pyroxene crystals in 60%-$O_2$ slag, EDS analysis.

**Table 8.** EDS quantitative results of iron-containing minerals in UCG slag.

| Element | C | O | Na | Mg | Al | Si | K | Ca | Fe | Ti |
|---------|------|------|-----|-----|------|------|-----|------|------|------|
| A | 10.1 | 37.9 | 0.6 | 1.9 | 11.8 | 28.3 | 2.7 | 1.8 | 5.0 | — |
| B | 6.0 | 24.3 | — | 0.2 | 1.6 | 3.3 | 0.2 | 1.4 | 63.3 | — |
| C | 23.9 | 18.4 | 0.5 | 1.0 | 2.9 | 11.5 | 0.7 | 26.8 | 2.1 | 12.2 |

A, Sekaninaite in 40%-$O_2$ slag; B, Melting iron oxide in 40%-$O_2$ slag; C, Pyroxene crystals in 60%-$O_2$ slag. Note: "—", undetectable.

## 4. Conclusions

(1) The typical minerals in the 40%-$O_2$ UCG slag include anorthite, mullite, sekaninaite, and approximately 49% amorphous substances. The mineral compositions of the 60%-$O_2$ slag and 80%-$O_2$ UCG slag are similar, and the dominant minerals involve anorthite, pyroxene, and gehlenite, while the amorphous substance almost disappears.

(2) In micro-appearance, the whole micro-morphology of the 40%-$O_2$ slag is comparatively complicated, with unburned carbon in the form of plant cells and a large amount of porous and melting material observed. In contrast, the 60%-$O_2$ slag seems to be homogeneous and is shown to have a wheat head formation. The homogeneous phenomenon is even more obvious in the 80%-$O_2$ slag, with mineral crystals regularly arranged in lamellar stacks. It is inferred that the increased oxygen concentration during UCG is beneficial to the reformation of the mineral crystals and that the crystal structure of the minerals tends to be more orderly when the oxygen content increases from 40% to 80%.

(3) Mullite may react with the calcium oxide contained in slag to generate anorthite when the oxygen concentration is higher than 40%, which contributes to the disappearance of mullite and the remarkable increase of anorthite in the 60%-$O_2$ slag. Anorthite may serve as a calcium source for the formation of gehlenite, which also accounts for the reduction of anorthite in the 80%-$O_2$ slag.

(4) Sekaninaite is proven to exist in the low-oxygen-concentration slag; however, the iron-bearing mineral in higher-oxygen-concentration slag is mainly pyroxene. The reaction mechanism of iron-bearing minerals at high temperatures could be assumed to be iron mineral oxidizing to magnetite ($Fe_3O_4$) and then converting to $Fe^{2+}$ of hercynite in a reducing atmosphere. The hercynite then reacts with $SiO_2$ to form sekaninaite. Finally, with the increase in the oxygen concentration, sekaninaite is further oxidized to produce pyroxene.

**Acknowledgments:** The authors are grateful for the financial support provided by the National Basic Research Program of China (973 Program) (NO. 2014CB238905), the Natural Science Foundation of China (NO. 51476185), and the Fundamental Research Funds for the Central Universities (No. 2009QH13). The authors wish to thank Shifeng Dai for his great supports to X-Ray diffraction (XRD) and scanning electron microscope (SEM) analysis.

**Author Contributions:** Shuqin Liu conceived, designed, and performed the experiments and contributed reagents; Chuan Qi and Yunpeng Deng performed the experiments, performed the X-Ray diffraction (XRD) determination and thermodynamic simulation, observed the slag using SEM-EDS, and analyzed the data; Shangjun Zhang analyzed the data. All authors participated in writing the manuscript.

**Conflicts of Interest:** The authors declare no conflict of interest.

## References

1. Li, W.; Bai, J. Mineral composition and characterization of coals and coal ashes. In *Chemistry of Ash from Coal*; Science Press: Beijing, China, 2013; pp. 1–7. (In Chinese)

2. Srinivasachar, S.; Helble, J.J.; Boni, A.A. An experimental study of the inertial deposition of ash under coal combustion conditions. In *Twenty-Third Symposium (International) on Combustion*; The Combustion Institute: Pittsburgh, PA, USA, 1991; Volume 23, pp. 1305–1312.

3. Brooker, D. Chemistry of deposit formation in a coal gasification syngas cooler. *Fuel* **1993**, *72*, 665–670. [CrossRef]

4. Marinov, V.; Marinov, S.P.; Lazarov, L.; Stefanova, M. Ash agglomeration during fluidized bed gasification of high sulphur content lignites. *Fuel Process. Technol.* **1992**, *31*, 181–191. [CrossRef]

5. McLennan, A.R.; Bryant, G.W.; Bailey, C.W.; Stanmore, B.R.; Wall, T.F. An experimental comparison of the ash formed from coals containing pyrite and siderite mineral in oxidizing and reducing conditions. *Energy Fuels* **2000**, *14*, 308–315. [CrossRef]

6. Raask, E. *Mineral Impurities in Coal Combustion: Behavior, Problems, and Remedial Measures*; Hemisphere Publishing Corporation: Washington, DC, USA, 1985; pp. 231–239.

7. Wagner, N.J.; Coertzen, M.; Matjie, R.H.; van Dyk, J.C. Coal gasification. In *Applied Coal Petrology*; Suárez-Ruiz, I., Crelling, J.C., Eds.; Elsevier Publications: Amsterdam, The Netherlands, 2008; pp. 119–144.

8. Koyama, S.; Morimoto, T.; Ueda, A.; Matsuoka, H. A microscopic study of ash depositsin a two-stage entrained-bed coal gasifier. *Fuel* **1996**, *75*, 459–465. [CrossRef]

9. Slade, R.C.T.; Davies, T.W. Evolution of structural changes during flash calcination of kaolinite. A $^{29}$Si and $^{27}$Al nuclear magnetic resonance spectroscopy study. *J. Mater. Chem.* **1991**, *1*, 361–364. [CrossRef]

10. Srinivasachar, S.; Helble, J.J.; Boni, A.A. Mineral behavior during coal combustion 1. Pyrite transformations. *Prog. Energy Combust. Sci.* **1990**, *16*, 281–292. [CrossRef]

11. Ranjan, S.; Sridhar, S.; Fruehan, R.J. Reaction of FeS with simulated slag and atmosphere. *Energy Fuels* **2010**, *24*, 5002–5007. [CrossRef]

12. Hu, G.; Dam-Johansen, K.; Wedel, S.; Hansen, J.P. Decomposition and oxidation of pyrite. *Prog. Energy Combust. Sci.* **2006**, *32*, 295–314. [CrossRef]

13. Miller, S.F.; Schobert, H.H. Effect of the occurrence and composition of silicate and aluminosilicate compounds on ash formation in pilot-scale combustion of pulverized coal and coal-water slurry fuels. *Energy Fuels* **1994**, *8*, 1197–1207. [CrossRef]

14. Raask, E. Creation, capture and coalescence of mineral species in coal flames. *J. Inst. Energy* **1984**, *57*, 231–239.

15. Helble, J.J.; Srinivasachar, S.; Boni, A.A. Factors influencing the transformation of minerals during pulverized coal combustion. *Prog. Energy Combust. Sci.* **1990**, *16*, 267–279. [CrossRef]

16. Brink, H.M.; Eenkhoorn, S.; Weeda, M. The behaviour of coal mineral carbonates in a simulated coal flame. *Fuel Process. Technol.* **1996**, *47*, 233–243. [CrossRef]

17. Yan, L.; Gupta, R.P.; Wall, T.F. The implication of mineral coalescence behaviour on ash formation and ash deposition during pulverised coal combustion. *Fuel* **2001**, *80*, 1333–1340. [CrossRef]

18. Yan, L.; Gupta, R.P.; Wall, T.F. Ash formation from excluded minerals including consideration of mineral-mineral associations. *Energy Fuels* **2007**, *21*, 461–467.

19. Jing, N.J.; Wang, Q.H.; Cheng, L.M.; Luo, Z.Y.; Cen, K.F.; Zhang, D.K. Effect of temperature and pressure on the mineralogical and fusion characteristics of Jincheng coal ash in simulated combustion and gasification environments. *Fuel* **2013**, *104*, 647–655. [CrossRef]

20. McCarthy, G.J.; Stevenso, R.J.; Oliver, R.L. Mineralogical characterization of the residues from the Tono I UCG experiment. In Proceedings of the Fourteenth Annual Underground Coal Gasification Symposium, Chicago, IL, USA, 15–18 August 1988; pp. 41–50.

21. Chen, Z.W.; Liu, J.S.; Pan, Z.J.; Connell, L.D.; Elsworth, D. Influence of the effective stress coefficient and sorption-induced strain on the evolution of coal permeability: Model development and analysis. *Int. J. Greenh. Gas Control* **2012**, *8*, 101–110. [CrossRef]

22. Standardization Administration of the People's Republic of China. *Proximate Analysis of Coal 2008*; Chinese Standard GB/T 212-2008. Standardization Administration of the People's Republic of China: Beijing, China (In Chinese).

23. Standardization Administration of the People's Republic of China. *Determination of Carbon and Hydrogen in Coal 2008*; Chinese Standard GB/T 476-2008. Standardization Administration of the People's Republic of China: Beijing, China. (In Chinese)

24. Standardization Administration of the People's Republic of China. *Determination of Nitrogen in Coal 2008*; Chinese Standard GB/T 214-2007. Standardization Administration of the People's Republic of China: Beijing, China. (In Chinese)

25. Standardization Administration of the People's Republic of China. *Determination of Total Sulfur in Coal 2008*; Chinese Standard GB/T 212-2008. Standardization Administration of the People's Republic of China: Beijing, China. (In Chinese)

26. Standardization Administration of the People's Republic of China. *Classification for Quality of Coal. Part 2: Sulfur, 2010*; Chinese Standard GB/T 15224, 2–2010. Standardization Administration of the People's Republic of China: Beijing, China. (In Chinese)

27. Standardization Administration of the People's Republic of China. *Test Method for Analysis of Coal Ash 2008*; Chinese Standard GB/T 1574-2007. Standardization Administration of the People's Republic of China: Beijing, China. (In Chinese)

28. Standardization Administration of the People's Republic of China. *Determination of Fusibility of Coal Ash 2008*; Chinese Standard GB/T 219-2008. Standardization Administration of the People's Republic of China: Beijing, China. (In Chinese)

29. Ward, C.R.; Spears, D.A.; Booth, C.A.; Staton, I.; Gurba, L.W. Mineral matter and trace elements in coals of the Gunnedah Basin, New South Wales, Australia. *Int. J. Coal Geol.* **1999**, *40*, 281–308. [CrossRef]

30. Dai, S.; Wang, P.; Ward, C.R.; Tang, Y.; Song, X.; Jiang, J.; Hower, J.C.; Li, T.; Seredin, V.V.; Wagner, N.J.; *et al.* Elemental and mineralogical anomalies in the coal-hosted Ge ore deposit of Lincang, Yunnan, southwestern China: Key role of $N_2$–$CO_2$-mixed hydrothermal solutions. *Int. J. Coal Geol.* **2015**, *152*, 19–46. [CrossRef]

31. Dai, S.; Liu, J.; Ward, C.R.; Hower, J.C.; French, D.; Jia, S.; Hood, M.M.; Garrison, T.M. Mineralogical and geochemical compositions of Late Permian coals and host rocks from the Guxu Coalfield, Sichuan Province, China, with emphasis on enrichment of rare metals. *Int. J. Coal Geol.* **2015**. [CrossRef]

32. Van Dyk, J.C.; Melzer, S.; Sobiecki, A. Mineral matter transformation during Sasol-Lurgi fixed bed dry bottom gasification-utilization of HT-XRD and FactSage modelling. *Miner. Eng.* **2006**, *19*, 1126–1135. [CrossRef]

33. Bale, C.W.; Chartrand, P.; Degterov, S.A.; Eriksson, G.; Hack, K.; Manfoud, R.B.; Melancon, J.; Pelton, A.D.; Peterson, S. FactSage thermochemical software and databases. *Calphad* **2002**, *26*, 189–228. [CrossRef]

34. Matjie, R.H.; Li, Z.S.; Ward, C.R.; French, D. Chemical composition of glass and crystalline phases in coarse coal gasification ash. *Fuel* **2008**, *87*, 857–869. [CrossRef]

35. Wu, X.J.; Zhang, Z.X.; Zhou, T.; Chen, Y.S.; Chen, G.Y.; Lu, C.; Huang, F.B. Ash fusion characteristics and mineral evolvement of blended ash under gasification condition. *J. Combust. Sci. Technol.* **2010**, *16*, 511–512.

36. Grim, R.E. *Clay Mineralogy*, 2nd ed.; McGraw-Hill Book Company: New York, NY, USA, 1968; p. 173.

37. Li, W.; Bai, J. Mineral transformation during thermal conversion of coals. In *Chemistry of Ash from Coal*; Science Press: Beijing, China, 2013; p. 52. ( In Chinese)

38. Huang, Z.Y.; Li, Y.; Zhao, J.; Zhou, Z.J.; Zhou, J.H.; Cen, K.F. Ash fusion regulation mechanism of coal with melting point and different ash compositions. *J. Fuel Chem. Technol.* **2012**, *40*, 1038–1043.

*minerals*

MDPI

Article

# Leaching Behavior and Potential Environmental Effects of Trace Elements in Coal Gangue of an Open-Cast Coal Mine Area, Inner Mongolia, China

Liu Yang [1,*], Jianfei Song [1], Xue Bai [2], Bo Song [1], Ruduo Wang [1], Tianhao Zhou [1], Jianli Jia [1] and Haixia Pu [3]

[1]   College of Geoscience and Surveying Engineering, China University of Mining and Technology, Beijing 100083, China; songjianfeiii@gmail.com (J.S.); songbo529@gmail.com (B.S.); ruduowang@gmail.com (R.W.); tian675677299@gmail.com (T.Z.); jjl@cumtb.edu.cn (J.J.)
[2]   Branch of Resources and Environment, China National Institute of Standardization, Beijing 100088, China; baixue@cnis.gov.cn
[3]   Institute of Geographic Sciences and Natural Resources Research, Chinese Academy of Sciences, Beijing 100101, China; puhx.14b@igsnrr.ac.cn
*   Correspondence: 108889@cumtb.edu.cn; Tel.: +86-10-6233-1180

Academic Editors: Shifeng Dai, William Skinner and Anna H. Kaksonen
Received: 28 February 2016; Accepted: 6 May 2016; Published: 27 May 2016

**Abstract:** In order to better understand the role of coal gangue in potential environmental and ecological risks, the leaching behavior of trace elements from coal gangue has been investigated in an open-cast coal mine, Inner Mongolia, China. Four comparative column leaching experiments were conducted to investigate the impacts of leaching time, pH values and sample amount on the leaching behavior of trace elements. Enrichment factors (EF), maximum leached amount ($L_{am}$), maximum leachability ($L_{rm}$), effects range low (ERL) and effects range median (ERM) were employed to evaluate potential environmental and ecological hazards resulting from the leaching behavior of environment-sensitive trace elements from coal gangue. Leaching time and sample amount display important effects on trace element concentrations, leached amounts and leachability. The pH values exhibit a weak influence on the leaching behavior of the selected trace elements (e.g., As, V, Cr, Co, Ni, Cu, Zn, Se, Cd, Sn, Pb and Hg). The coal gangue are enriched in As, Co, Se and Pb and, in particular, show higher environmental pollution levels of As and Se (EF > 2). $L_{am}$ values suggest that all of the elements investigated do not show potential risk to soils and vegetation, but have a high hazard risk for ground water. Elements including Ni, As, Cr and Zn are inclined to show high or moderate biological toxicity.

**Keywords:** coal gangue; environment-sensitive trace elements; column leaching; leaching behavior; environmental hazard; ecological risk

## 1. Introduction

In recent years, coal has accounted for 74% of China's total primary energy consumption and will continue to be the major energy source in the next decades [1]. However, environmental hazards were caused as a result of the release and dispersal of harmful trace elements contained in coal, coal gangue and coal combustion residues [2–5]. The rapid development of coal mining in China over the last twenty years has led to a huge coal gangue accumulation in coalfields. According to the incomplete statistics of 2010, there were more than 4.5 billion tons of coal gangue, which covered approximately 1.5 thousands square kilometers of land in China [6]. This could result in substantial environmental hazard and ecological risk, such as soil and water pollution and ecological deterioration, if reasonable precautions are not taken [7,8].

Coal gangue, a mixture of rocks derived from coal bed's roof, floor, partings and coal itself, is mainly produced from coal mining and, in some cases, is enriched in some toxic trace elements [9–11]. Trace elements are defined as elements with concentrations lower than 0.1% in coal gangue [12]; they have become a hot topic due to the complex changes of their particles, inability to decompose in natural processes and high toxicity to ecosystems following their release and dispersal into the atmosphere, soils, water and vegetation through the pathways of leaching, weathering and spontaneous combustion [6,9,13,14]. Leaching has proven to be one of the primary pathways for trace elements entering into the ecosystem. Extensive studies on the leaching behavior of trace elements from coal, coal fly ash and bottom ash have been conducted [13,15–17]. However, studies focused on the leaching behavior of trace elements from coal gangue are relatively rare [6,18], especially from coal gangue piles. Nevertheless, "leaching behavior" and its impact factors are not clearly defined in these current investigations.

Time is one of the important impact factors on the leaching behavior of trace elements from coal gangue [4,19,20]. The integrative efforts to determine the leaching behavior of the selected trace elements from coal gangue piles indicated that it is a very complex process in terms of the leaching pathways, which might be influenced by various factors. Moreover, the potential environmental impacts of the resulting leachates from coal gangue should be given more attention and studied extensively using quantitative assessment methods due to their emergent potential pollution and toxicity.

Therefore, this study was conducted to focus on investigations on the leaching behavior of trace elements from coal gangue piles and their environmental effects. Based on the comparative column leaching experiments, the environmental and ecological risks generated by the trace elements of the resulting leachates from coal gangue piles were evaluated in detail by different semi-quantitative methods. The leaching behavior of the trace elements from coal gangue piles, in this study, mainly refers to the leached concentrations of trace elements, leached amount, leachability, maximum leached amount ($L_{am}$) and maximum leachability ($L_{rm}$). The roles of the impact factors, including leaching time, pH of the leaching solution and sample amount, were simultaneously analyzed.

Twenty-six trace elements in coal, proposed by Swaine [12], could lead to potential environmental impacts, including As, Cr, Cd, Hg, Pb, Se, B, Mn, Ni, Cu, V, Zn, Co, Sn, Cl, F, Mo, Bo, P, Th, U, Ba, I, Ra, Sb and Tl. Particularly, elements As, Cd, Cr, Hg, Pb and Se are of most environmental importance [12], while elements Ni, Cu, V, Zn, Co and Sn have generally been analyzed in other leaching experiments [6,8]. Twelve trace elements, including As, V, Cr, Co, Ni, Cu, Zn, Se, Cd, Sn, Pb and Hg, were therefore selected for investigation in this study, to assess their leaching behavior in coal gangue piles and the corresponding environmental and ecological hazards levels.

## 2. Materials and Methods

### 2.1. Study Area

The Wulantuga open-cast germanium coal mine area (43°56′57.86″ N, 115°54′37.36″ E), covering an area of 2.2 km² [21,22], is located in the southwest of the Shengli Coalfield (with a total area of 342 km² [21]) in northeastern Inner Mongolia, northern China (Figure 1). The germanium-rich coal in the open-cast mine has been mined since 1997. The production of raw coal was 7.3 million tons in 2014, resulting in approximately 0.7 million tons of coal gangue accumulation [23]. The coal properties and the geological setting have previously been described in a great detail [21,22,24–26].

The study area has a semi-arid continental climate of the middle temperate zone. The annual average temperature is 0–3 °C, and the mean annual rainfall is approximately 276.3 mm. The soil type is chestnut soil, and the vegetation type is typical grassland, with 50% coverage. There is a seasonal river named the Xilin River flowing northward, 12 km from the east of the open-cast coal mine area, which has no direct hydraulic connection to the coal mine. The terrain of the Shengli Coalfield is gentle, with a slope of no more than 7° and an elevation of 1061–1196 m. The germanium coal mine area belongs to an approximate level-slight inclination monocline structure, with a formation dip of less than 5°.

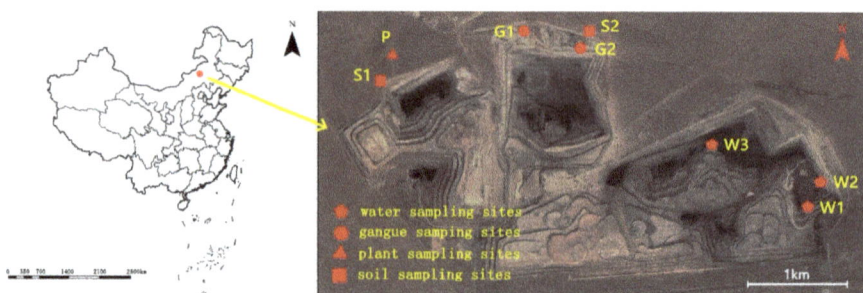

**Figure 1.** Study area and sample location in the open-cast germanium coal-mine area, Inner Mongolia, China.

*2.2. Sample Collection*

To investigate the mineral and chemical compositions in coal gangue and their potential environmental effects, coal gangue samples were collected in the open-cast germanium coal mine in July 2014, as well as the soils, water and plant samples around the mine area (Figure 1).

Coal gangue samples were selected from two sites of coal gangue piles in study area (Figure 1). Three samples were taken by hand and were immediately stored in individual sealed plastic bags in order to avoid any contamination and oxidation. Soil samples were collected from two sites in the north of the open-cast coal mine area using a geotome at a depth of 0–15 cm of each layer (Figure 2), and three sampling points were set in each layer, according to the sample collection methods described in detail by Jia *et al.* [23]. Background soil samples were taken from the grassland located approximately 15 km to the northeast of Xilinhaote city. All of the soil samples were also stored in sealed plastic bags in a portable freezer to minimize possible changes and contaminations. Three water samples were taken using a glass water sampler with 2 L capacity from the water pools in the open-cast germanium coal mine area. Two background samples were collected from the Jiuquwan Reservoir, which is situated approximately 20 km from the southeast of the coal mine area. The samples were immediately put into polyethylene terephthalate (PET) bottles and stored in the portable freezer. Four plant samples (*Filifolium Kitam, Artemisia lavandulaefolia, Allium tuberosum Rottler* and *Leymus chinensis (Trin.) Tzvel*) were selected in the grassland located in the north of the coal mine area. The leaves and trunks of plant samples were collected and immediately stored in sealed plastic bags.

*2.3. Analytical Methods*

The contents of major-element oxides, including $SiO_2$, $TiO_2$, $Al_2O_3$, $Fe_2O_3$, $MgO$, $CaO$, $MnO$, $Na_2O$, $K_2O$ and $P_2O_5$, in coal gangue samples were analyzed by X-ray fluorescence spectrometry (XRF, ARL ADVANT'XP+, ThermoFisher, Waltham, MA, USA) as outlined by Dai *et al.* [27]. The mineralogical compositions were determined on a D/max-2500/PC powder diffractometer with Ni-filtered Cu-Kα radiation and a scintillation detector. Each XRD pattern was recorded over a 2θ interval of 2.6°–70°, with a step size of 0.01° [27]. The selected environmentally-sensitive trace elements were As, V, Cr, Co, Ni, Cu, Zn, Se, Cd, Sn, Pb and Hg. The trace element contents of V, Cr, Co, Ni, Cu, Zn, Cd, Sn and Pb in the coal gangue, soil, water and plant samples and those in the resulting leachates of the following experiments of coal gangue were all determined by inductively-coupled plasma mass spectrometry (X series II ICP-MS, ThermoFisher), according to the procedures described in detail by Dai *et al.* [28]. Arsenic and Se were determined by ICP-MS using collision cell technology (CCT), as described by Li *et al.* [29]. The concentration of Hg was determined by a Milestone DMA-80 Hg analyzer (Milestone, Sorisole, Italy). The detection limit of Hg is 0.005 ng; the relative standard deviation from eleven runs on Hg standard reference is 1.5%; and the linearity of the calibration is in the range 0–1000 ng [30]. The handling methods of soil samples were described in detail by Jia *et al.* [24]. Four plant samples were mixed and were cleaned by deionized water. They were dried in a drying oven (60 °C) and crushed to 100 mesh size for testing.

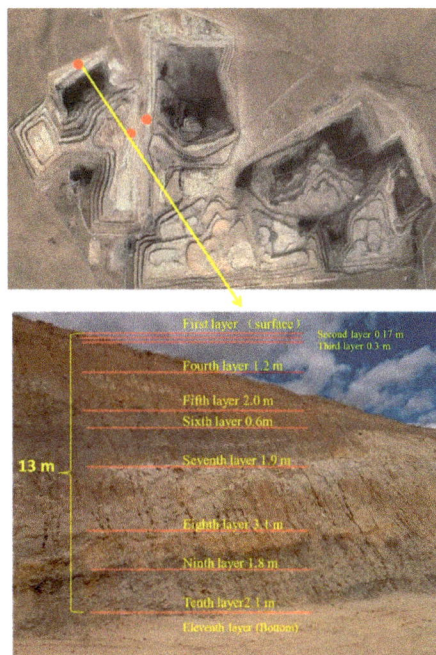

**Figure 2.** The distribution of soil sampling points and sections in the mining area [23].

*2.4. Leaching Experiments*

Coal gangue samples were air-dried and crushed to 200 mesh. They were blended by the method of repeated division into four equal portions to mix thoroughly (four times) and prepared for the leaching experiments (5 kg). Then, three 30-g sub-samples and one 45-g sub-sample were obtained by an analytical balance of 0.01-mg precision, accurate to four decimal places.

To investigate the impacts of different pH values and sample quantities on the leaching behavior of the selected trace elements in the coal gangue, four column leaching experimental groups were determined in this study (Table 1). The pH of the rainfall in the study area is approximately 6.60–8.19, slightly alkaline, with no acid rain. The concentration ratio between $SO_4^{2-}$ and $NO_3^-$ in the rainfall is approximately 2.3–29.0, suggesting sulfate precipitation [19,31]. According to the rainfall characteristics, distilled water (pH = 7.0 ± 0.3), acidic solution (pH = 6.0 ± 0.3, using distilled water with $H_2SO_4$) and alkaline solution (pH = 8.0 ± 0.3, using distilled water with NaOH) were prepared for the leaching experimental groups. In this study, Experiments I, II and III were set for detecting the effects of solution pH on leaching behavior, with the same sample weight of 30-g, and different solutions of acid, alkaline and neutral pH, respectively. All of the test utensils were soaked in a 14% $HNO_3$ solution for 24 h and rinsed by distilled water before the leaching experiments [19].

Each coal gangue sample was transferred into a fixed glass column, which was 30 mm in internal diameter and 50 cm in length (Figure 3). Quartz sand (10 g, particle size < 0.83 mm) was packed into the bottom of the column in Experiments I, III and IV, respectively, to prevent fine particle loss during leaching. It was also packed at the top of the sample to make the solution disperse uniformly. However, a small amount of absorbent cotton, instead of quartz sand, was used at the top and bottom of the column in Experiment II. This was to prevent the quartz sand from reacting with the alkaline solution, in which absorbent cotton could not be dissolved. The four solutions were controlled in terms of influx into the columns at room temperature. The experiments lasted for 90 h. The resulting leachates were sampled once every 3 h and then put into 50-mL volumetric tubes for element analysis. Thirty samples were obtained from each experimental group. As a matter of convenience for displaying and analyzing the results, the concentrations of the 30 samples were averaged over 10 time units, *i.e.*, 0–9 h, 9–18 h, 18–27 h, 27–36 h, 36–45 h, 45–54 h, 54–63 h, 63–72 h, 72–81 h and 81–90 h.

**Table 1.** Test setting data for the four column leaching experimental groups.

| Group | Weight of Samples (g) | pH of Solution | Test Purpose |
|-------|----------------------|----------------|--------------|
| I | 30 | $6.0 \pm 0.3$ | acid solution |
| II | 30 | $8.0 \pm 0.3$ | alkaline solution |
| III | 30 | $7.0 \pm 0.3$ | neutral solution |
| IV | 45 | $7.0 \pm 0.3$ | different weight of samples |

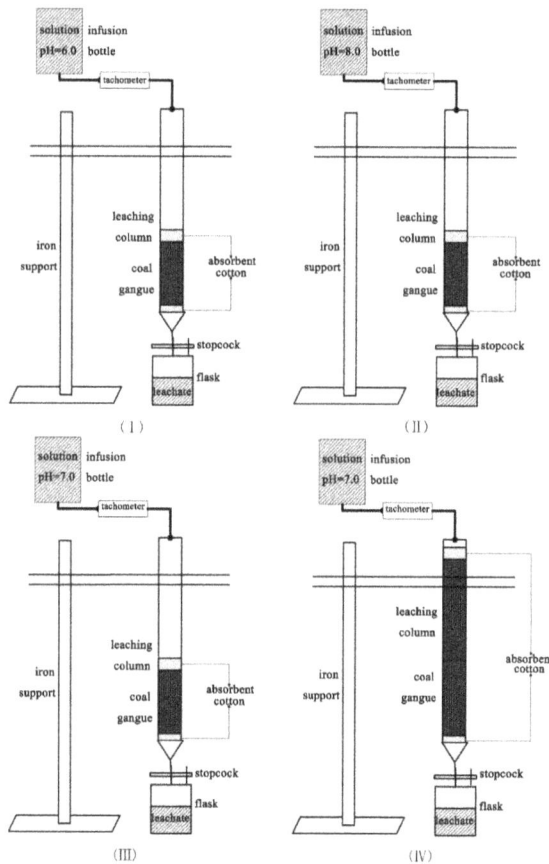

**Figure 3.** Installations of the leaching experiments: I, II, III and IV. Experiment I, 30-g samples, pH= 6.0 ± 0.3; Experiment II, 30-g samples, pH = 8.0 ± 0.3; Experiment III, 30-g samples, pH = 7.0 ± 0.3; Experiment III, 45-g samples, pH = 7.0 ± 0.3.

The "leachability $(L_r)$", "leached concentration $(C_l)$" and "maximum leached concentration $(C_{lm})$" were used to analyze the leaching behavior of the coal gangue during the four experimental groups. $L_r$ (%) is calculated by Equation (1), and $C_{lm}$ (µg/L) was calculated by Equation (2):

$$L_r = \frac{100C_{lx} \times V_x}{M_x} \tag{1}$$

$$C_{lm} = Max(C_{lx}) \tag{2}$$

where $C_{lx}$ (µg/L) and $V_x$ (L) represent the concentration of an element and the volume of the resulting leachates, respectively, during a period of leaching time. $M_x$ (µg) is the total mass of an element in the coal gangue samples.

### 2.5. Environmental Effect Indicators

To compare the concentrations of an element in coal gangue with black shales, Clarke values were used as a practical method of assessing trace element enrichment [6]. The enrichment factor (EF) was then an important parameter to evaluate the pollution level of an element [6,32–34]. EF values of elements in coal gangue, coal, soil, water and vegetation of the Wulantuga open-cast coal mine were applied to assess the pollution degrees of the 12 trace elements. The EF value is expressed as follows:

$$EF = \frac{A_i/C}{B_i/D} \tag{3}$$

where $A_i$ (µg/g) is an element's concentration; $B_i$ (µg/g) is the background value of an element (Clarke value); $C$ (µg/g) is the reference element concentration; $D$ (µg/g) is the Clarke value of the reference element. Scandium is usually used as the reference element due to its stable concentration, and it was also applied in this study.

"Maximum leached amount" $(L_{am})$ and "maximum leachability" $(L_{rm})$ are important indicators for trace element environmental risk assessment, which could provide valuable information of the maximum extent of element migrating ability [4,33]. The value of $L_{am}$ is defined here as the ratio of all amounts of an element in the leachates of 90 h to the mass of the coal gangue sample. The value of $L_{rm}$ is based on the ratio of the amount of the trace element in the leachates after 90 h to the mass of the trace element in the coal gangue sample.

Aimed at further evaluating the environmental threat of the 12 environmentally-sensitive elements, the concentration limits of the trace elements in groundwater, soil and food were also compared in this research. Furthermore, the effects range low (ERL) and effects range median (ERM), defined in international sediment quality guidelines (SQGs), were adopted in this study to assess the ecological risk, namely the biological toxicity of trace elements in coal gangue. ERL values are the concentrations below which adverse effects on sediment-dwelling fauna would be unlikely, and ERM values were, in contrast, the concentrations above which adverse effects are likely [5,34]. The existence of biological toxicity could be divided into three levels, *i.e.*, <ERL, very low toxicity, >ERL and <ERM, middle toxicity, and >ERM, probable toxicity [35].

The environmental quality standard for soils, the quality standard for groundwater and limits of contaminants in foods were also applied to be compared with the $L_{am}$ values of the trace elements in soil samples, water samples and vegetation samples, respectively, in the study. Unfortunately, there are no specific limits, standards or references of the trace element concentrations for plants. Therefore, the concentrations in vegetation samples had to be compared to the limits of contaminants in foods, the alternative comparing reference, for evaluating the environmental impacts of coal gangue leaching on vegetation.

## 3. Results

### 3.1. Chemistry and Mineralogy in Coal Gangue

Proximate and sulfur analyses were performed on the coal gangue sample. Proximate analysis on coal or/and coal gangue samples is long-term, well-established term in coal industry and covers the determination of moisture, volatile matter and ash yield in samples [36]. These data are expressed as percentages of the air-dried coal gangue (on an air dry basis; Table 2), and these include the air-dried moisture, but do not involve the surface moisture of the samples [36–39].

**Table 2.** Contents (wt %) of coal gangue samples in the Wulantuga open-cast coal mine area.

| Proximate and Sulfur Analysis | | | Chemical Compositions | | | | | | | | | |
|---|---|---|---|---|---|---|---|---|---|---|---|---|
| $A_{ad}$ | $M_{ad}$ | $S_{td}$ | $SiO_2$ | $Al_2O_3$ | $Fe_2O_3$ | $K_2O$ | MgO | $TiO_2$ | CaO | $Na_2O$ | $P_2O_5$ | MnO |
| 90.30 | 9.70 | 0.07 | 70.26 | 19.23 | 3.76 | 3.15 | 1.31 | 1.10 | 0.50 | 0.40 | 0.04 | 0.04 |

$A_{ad}$, ash yield, air dry basis; $M_{ad}$, moisture content, air dry basis; $S_{td}$, total sulfur, dry basis.

The coal gangue samples are dominated by $SiO_2$ (70.26%) and $Al_2O_3$ (19.23%), followed by $Fe_2O_3$ (3.76%) and $K_2O$ (3.15%) (Table 2), along with trace percentages of MgO (1.31%), $TiO_2$ (1.10%), CaO (0.50%) and $Na_2O$ (0.40%).

The XRD patterns showed that the main mineral phases found in the coal gangue samples were montmorillonite, illite mixed layer, kaolinite and quartz. The dominant minerals of montmorillonite and kaolinite indicated that the coal gangue in the Wulantuga open-cast coal mine area had a high expansibility.

In comparison to the world Clarke values, the concentration of the coal gangue was enriched in Co (the average concentration equaled 53.92 µg/g in study area) and depleted in Se, Cd and Hg. The remains of the selected 12 trace elements were close to the concentrations of the world Clarke values (Table 3).

### 3.2. Leaching Characteristics of Coal Gangue

#### 3.2.1. Leaching Time

Eight different trend curves for the leached concentration of 12 elements, which were changed by leaching time, could be identified in the four leaching tests (Tables 4–7). Among them, a sharp drop followed by a steady curve was the major trend, which accounted for more than one third of the concentration time changing curves. This trend curve was observed for elements Co, Ni, Se, Cu, Zn and Pb. Descending curves and curves with a short rise followed by a large decline accounted for 20.83% and 14.58% of the trends, respectively. Elements Cr, As and Cu exhibited descending curves. Short rise, large decline curves were observed for V and As. These three trends, in general, showed higher extractable concentrations of the elements in the initial leaching phase, but decreased sharply or gradually as the leaching time goes on. Steady descending curves (14.58% of all of the trends), *i.e.*, concentrations of the elements decreased slightly during the whole leaching time, were observed for the elements Cd, Sn and Hg. A wave-like curve in decreasing order of significance accounted for 4.17%, including Zn. Double wave-like curves contributed to 2.08%, showing a slight and steady decrease order for elements of Cd and Hg. A curve with a slight and steady rise followed by a decline (2.08% of all of the trends) was observed for Sn. Based on these trends, the elements could be divided into two categories. Elements Co, Ni, Se, Cu, Zn, Pb, Cr, As and V were in the category with a sharp initial decrease in the leaching phase, followed by a steady decline. Elements Cd, Hg and Sn were in the category with characteristics of slight and steady waves, decreasing throughout the leaching phases.

**Table 3.** Concentrations of the 12 elements in coal gangue, soil, water and vegetation in the Wulantuga open-cast coal mine area. SQGs, sediment quality guidelines.

| Elements | Coal Gangue (µg/g) | | | | Soil (µg/g) | | Water (µg/L) | | Vegetation (µg/g) | | Clarke Value[2] | ERL-ERM (µg/g) | % of Samples amongst Ranges of SQGs | | |
|---|---|---|---|---|---|---|---|---|---|---|---|---|---|---|---|
| | Min–Max | AC | World Coal Gangue[1] | EF | AC | EF | AC | EF | AC | EF | | | <ERL | >ERL and <ERM | >ERM |
| As | 7.35–28.53 | 17.94 | 10–80 | 4.86 | 30.87 | 9.08 | 8.30 | 41.07 | 3.91 | 3.78 | 1.8 | 8.2–70 | 50 | 50 | 0 |
| V | 107.33–107.35 | 107.34 | 100–400 | 0.39 | 99.74 | 0.39 | 1.93 | 0.13 | 18.99 | 0.24 | 135 | | | | |
| Cr | 72.48–94.46 | 83.47 | 50–160 | 0.41 | 92.82 | 0.49 | 1.10 | 0.10 | 161.53 | 2.81 | 100 | 81–370 | 50 | 50 | 0 |
| Co | 29.64–78.21 | 53.92 | 10–30 | 1.05 | 12.55 | 0.27 | 0.59 | 0.21 | 2.68 | 0.19 | 25 | | | | |
| Ni | 28.38–52.88 | 40.63 | 40–140 | 0.26 | 23.56 | 0.17 | 1.63 | 0.19 | 15.54 | 0.36 | 75 | 20.9–51.6 | 0 | 50 | 50 |
| Cu | 34.58–44.97 | 39.77 | 35–150 | 0.35 | 32.44 | 0.31 | 5.01 | 0.81 | 25.65 | 0.81 | 55 | 34–270 | 0 | 100 | 0 |
| Zn | 127.41–155.31 | 141.36 | 60–300 | 0.99 | 104.48 | 0.79 | 17.55 | 2.23 | 30.54 | 0.76 | 70 | 150–410 | 50 | 50 | 0 |
| Se | 0.36–1.16 | 0.76 | 3–30 | 7.41 | 0.53 | 5.61 | 3.51 | 625.26 | 0.45 | 15.67 | 0.05 | | | | |
| Cd | 0.36–0.41 | 0.39 | 2–12 | 0.95 | 0.44 | 1.17 | 0.01 | 0.45 | 0.14 | 1.22 | 0.2 | | | | |
| Sn | 3.98–4.01 | 4.00 | 2–10 | 0.98 | 4.46 | 1.18 | −0.02 | 0.09 | 0.35 | 0.30 | 2 | | | | |
| Pb | 26.40–28.39 | 27.39 | 10–40 | 1.07 | 24.00 | 1.02 | −0.08 | 0.06 | 5.99 | 0.83 | 12.5 | | | | |
| Hg | 0.09–0.17 | 0.13 | 0.2–0.6 | 0.79 | 0.08 | 0.53 | Nd | 0.00 | 0.03 | 0.65 | 0.08 | | | | |
| Sc | 44.72–45.48 | 45.10 | | | 41.54 | | 2.47 | | 12.63 | | | | | | |

[1] From Zhou *et al.* [6]; [2] from Taylor. [40]; AC, average concentration; EF, enrichment factor; ERL, effect range low value; ERM, effect range median value.

**Table 4.** Analytical concentrations of the selected trace elements in resulting leachates from 0–90 h in Experiment I (µg/L).

| Element | 0–9 h | 9–18 h | 18–27 h | 27–36 h | 36–45 h | 45–54 h | 54–63 h | 63–72 h | 72–81 h | 81–90 h | Trend Curve |
|---|---|---|---|---|---|---|---|---|---|---|---|
| As | 10.81 | 15.04 | 9.75 | 7.81 | 4.06 | 2.98 | 2.29 | 1.93 | 1.33 | 0.65 | |
| V | 33.18 | 35.79 | 24.74 | 19.16 | 14.44 | 12.28 | 9.20 | 8.24 | 5.97 | 3.97 | |
| Cr | 8.93 | 6.52 | 4.17 | 3.94 | 3.83 | 3.22 | 2.42 | 2.38 | 2.27 | 2.23 | |
| Co | 11.08 | 3.24 | 1.46 | 1.27 | 1.26 | 0.92 | 0.50 | 0.34 | 0.48 | 0.51 | |
| Ni | 21.64 | 8.40 | 4.09 | 2.48 | 2.16 | 1.36 | 1.34 | 0.58 | 0.63 | 0.58 | |
| Cu | 15.60 | 6.72 | 2.52 | 1.35 | 1.10 | 0.57 | 0.21 | 0.05 | 0.00 | 0.00 | |
| Zn | 55.93 | 44.25 | 40.47 | 27.60 | 25.87 | 18.37 | 19.73 | 18.09 | 19.19 | 21.83 | |
| Se | 28.85 | 0.67 | 0.44 | 0.48 | 0.24 | 0.30 | 0.18 | 0.30 | 0.02 | 0.28 | |
| Cd | 0.15 | 0.07 | 0.03 | 0.02 | 0.03 | 0.01 | 0.01 | 0.01 | 0.01 | 0.02 | |
| Sn | 0.37 | 0.22 | 0.11 | 0.09 | 0.09 | 0.07 | 0.02 | 0.01 | 0.01 | 0.01 | |
| Pb | 3.39 | 1.56 | 0.61 | 0.48 | 0.41 | 0.19 | 0.04 | 0.02 | 0.01 | 0.01 | |
| Hg | 0.19 | 0.16 | 0.11 | 0.09 | 0.07 | 0.23 | 0.12 | 0.08 | 0.09 | 0.09 | |

**Table 5.** Analytical concentrations of the selected trace elements in resulting leachates from 0–90 h in Experiment II (μg/L).

| Elements | 0–9 h | 9–18 h | 18–27 h | 27–36 h | 36–45 h | 45–54 h | 54–63 h | 63–72 h | 72–81 h | 81–90 h | Trend Curve |
|---|---|---|---|---|---|---|---|---|---|---|---|
| As | 10.72 | 10.45 | 7.78 | 6.88 | 4.43 | 3.49 | 2.35 | 2.54 | 2.09 | 1.62 | \ |
| V | 28.07 | 31.20 | 21.88 | 18.67 | 15.92 | 14.81 | 11.58 | 11.53 | 9.82 | 8.49 | ⌒\ |
| Cr | 6.87 | 7.44 | 4.46 | 4.07 | 4.30 | 3.79 | 2.71 | 2.58 | 2.48 | 2.42 | ⌒\ |
| Co | 8.09 | 3.45 | 1.52 | 1.26 | 1.37 | 0.91 | 0.55 | 0.36 | 0.32 | 0.30 | ∟ |
| Ni | 19.96 | 10.60 | 5.05 | 3.82 | 3.83 | 2.39 | 1.30 | 1.10 | 0.82 | 0.59 | ∟ |
| Cu | 13.45 | 8.71 | 3.79 | 2.14 | 1.81 | 1.26 | 0.44 | 0.44 | 0.13 | 0.01 | ∟ |
| Zn | 52.12 | 45.08 | 38.15 | 41.28 | 33.16 | 26.04 | 17.77 | 26.96 | 14.61 | 13.31 | ⋁\ |
| Se | 23.03 | 1.56 | 0.95 | 0.87 | 0.71 | 0.23 | 0.38 | 0.30 | 0.23 | 0.09 | ∟ |
| Cd | 0.12 | 0.05 | 0.02 | 0.06 | 0.06 | 0.02 | 0.01 | 0.01 | 0.01 | 0.01 | — |
| Sn | 0.30 | 0.29 | 0.13 | 0.11 | 0.12 | 0.10 | 0.01 | 0.01 | 0.01 | 0.0 | ⎯ |
| Pb | 2.10 | 2.37 | 0.96 | 0.71 | 0.54 | 0.29 | 0.09 | 0.06 | 0.03 | 0.01 | ⌒\ |
| Hg | 0.10 | 0.15 | 0.12 | 0.10 | 0.07 | 0.12 | 0.12 | 0.12 | 0.07 | 0.07 | ⌒⌒ |

**Table 6.** Analytical concentrations of the selected trace elements in resulting leachates from 0–90 h in Experiment III (μg/L).

| Elements | 0–9 h | 9–18 h | 18–27 h | 27–36 h | 36–45 h | 45–54 h | 54–63 h | 63–72 h | 72–81 h | 81–90 h | Trend Curve |
|---|---|---|---|---|---|---|---|---|---|---|---|
| As | 13.38 | 16.66 | 15.04 | 12.20 | 4.71 | 3.16 | 1.96 | 1.82 | 1.31 | 0.61 | ⌒\ |
| V | 38.07 | 43.29 | 31.48 | 26.78 | 13.76 | 12.73 | 7.80 | 6.50 | 5.26 | 4.05 | ⌒\ |
| Cr | 10.11 | 8.16 | 6.49 | 4.40 | 3.55 | 3.47 | 2.35 | 2.32 | 2.32 | 2.34 | \ |
| Co | 10.18 | 3.76 | 1.99 | 1.16 | 1.08 | 0.60 | 0.42 | 0.28 | 0.34 | 0.47 | ∟ |
| Ni | 22.59 | 11.02 | 8.14 | 3.08 | 2.34 | 1.39 | 0.82 | 0.75 | 0.56 | 0.67 | ∟ |
| Cu | 17.31 | 9.56 | 4.40 | 2.15 | 1.14 | 0.74 | 0.07 | 0.00 | 0.00 | 0.02 | \ |
| Zn | 68.14 | 49.93 | 59.40 | 43.74 | 38.47 | 18.92 | 21.12 | 25.11 | 14.86 | 19.76 | ⋁\ |
| Se | 25.08 | 1.71 | 0.57 | 0.55 | 0.54 | 0.50 | 0.19 | 0.43 | 0.17 | 0.21 | ∟ |
| Cd | 0.12 | 0.05 | 0.27 | 1.36 | 0.04 | 0.02 | 1.72 | 0.08 | 0.05 | 0.02 | ⌒⌒ |
| Sn | 0.39 | 0.28 | 0.16 | 0.10 | 0.07 | 0.06 | 0.00 | 0.00 | 0.00 | 0.00 | ⎯ |
| Pb | 4.14 | 2.82 | 1.67 | 0.84 | 0.55 | 0.38 | 0.05 | 0.02 | 0.03 | 0.03 | \ |
| Hg | 0.21 | 0.09 | 0.06 | 0.15 | 0.13 | 0.07 | 0.10 | 0.09 | 0.07 | 0.06 | ⌒⌒ |

**Table 7.** Analytical concentrations of the selected trace elements in resulting leachates from 0–90 h in Experiment IV (μg/L).

| Elements | 0–9 h | 9–18 h | 18–27 h | 27–36 h | 36–45 h | 45–54 h | 54–63 h | 63–72 h | 72–81 h | 81–90 h | Trend Curve |
|---|---|---|---|---|---|---|---|---|---|---|---|
| As | 12.79 | 11.13 | 11.36 | 11.49 | 8.35 | 7.51 | 6.79 | 6.55 | 6.00 | 4.01 | \ |
| V | 59.61 | 43.02 | 42.93 | 36.38 | 35.31 | 31.75 | 23.66 | 21.03 | 18.97 | 14.32 | \ |
| Cr | 31.75 | 17.67 | 18.27 | 12.18 | 14.46 | 11.33 | 4.95 | 4.27 | 4.49 | 3.54 | \ |
| Co | 34.68 | 14.79 | 8.69 | 5.66 | 5.15 | 3.63 | 1.68 | 1.20 | 1.22 | 0.89 | ∟ |
| Ni | 55.24 | 37.17 | 24.41 | 15.49 | 11.69 | 8.17 | 4.30 | 3.01 | 2.65 | 2.65 | \ |
| Cu | 36.32 | 27.21 | 20.27 | 12.42 | 8.49 | 5.80 | 2.55 | 1.79 | 1.58 | 1.40 | \ |
| Zn | 140.91 | 83.36 | 77.06 | 56.70 | 64.42 | 42.65 | 34.55 | 26.84 | 26.59 | 20.49 | \ |
| Se | 107.20 | 20.83 | 4.56 | 1.90 | 1.46 | 0.94 | 1.11 | 0.71 | 0.68 | 0.50 | ∟ |
| Cd | 0.51 | 0.15 | 0.12 | 0.06 | 0.07 | 0.04 | 0.03 | 0.02 | 0.02 | 0.02 | ∟ |
| Sn | 0.29 | 0.65 | 0.67 | 0.49 | 0.60 | 0.47 | 0.10 | 0.06 | 0.07 | 0.10 | ⌒\ |
| Pb | 17.87 | 8.16 | 7.73 | 4.45 | 4.83 | 3.02 | 0.94 | 0.68 | 0.79 | 0.43 | ∟ |
| Hg | 0.25 | 0.10 | 0.10 | 0.09 | 0.05 | 0.15 | 0.08 | 0.05 | 0.06 | 0.05 | ⎯ |

The concentrations of the 12 elements in the resulting leachates reached the maximum at different leaching time periods (Table 8). Approximately 70.83% of the leached concentrations reached the maximum at 0–3 h, especially for elements Co, Ni, Cu and Se. The maximum leach concentration occurred at 12–15 h, accounting for 10.42%. In the other periods of 6–9 h, 15–18 h, 21–24 h, 24–27 h, 27–30 h, 30–33 h, 51–54 h and 60–63 h, the resulting leachates reached the maximum concentration only once (2.08%). All 12 selected trace elements could be mostly leached out within no more than 30 h, except for elements Hg and Cd.

**Table 8.** The maximum leached concentrations ($C_{lm}$) of the 12 elements from the coal gangue of the Wulantuga open-cast coal mine area (μg/L).

| Elements | Experiment I | | Experiment II | | Experiment III | | Experiment IV | |
|---|---|---|---|---|---|---|---|---|
| | $C_{lm}$ | Tp (h) | $C_{lm}$ | Tp (h) | $C_{lm}$ | Tp (h) | $C_{lm}$ | Tp (h) |
| As | 18.19 | 15–18 | 11.63 | 3–6 | 17.64 | 27–30 | 13.99 | 0–3 |
| V | 40.07 | 6–9 | 35.13 | 12–15 | 45.69 | 12–15 | 81.15 | 0–3 |
| Cr | 11.23 | 0–3 | 10.21 | 12–15 | 13.47 | 0–3 | 49.10 | 0–3 |
| Co | 17.76 | 0–3 | 13.76 | 0–3 | 16.20 | 0–3 | 51.15 | 0–3 |
| Ni | 27.88 | 0–3 | 30.57 | 0–3 | 31.36 | 0–3 | 67.00 | 0–3 |
| Cu | 19.00 | 0–3 | 14.81 | 0–3 | 22.76 | 0–3 | 43.97 | 0–3 |
| Zn | 63.55 | 0–3 | 64.40 | 0–3 | 93.35 | 24–27 | 185.60 | 0–3 |
| Se | 53.50 | 0–3 | 64.70 | 0–3 | 49.73 | 0–3 | 196.90 | 0–3 |
| Cd | 0.26 | 0–3 | 0.24 | 0–3 | 5.87 | 30–33 | 0.79 | 0–3 |
| Sn | 0.48 | 0–3 | 0.44 | 12–15 | 0.56 | 0–3 | 1.07 | 21–24 |
| Pb | 4.83 | 0–3 | 3.72 | 12–15 | 5.85 | 0–3 | 29.50 | 0–3 |
| Hg | 0.41 | 51–54 | 0.22 | 60–63 | 0.27 | 0–3 | 0.33 | 0–3 |

Tp: time period of leaching.

### 3.2.2. Solution pH

Considering the natural rainfall conditions, three pH values, 6.0, 7.0 and 8.0, were examined in Experiments I, II and III in this study. In general, the trends and mean concentrations were similar in the three experiments for each individual element (Tables 4–7), indicating that there was no significant impact of pH on the leached concentrations and trends of the elements in the coal gangue of the Wulantuga open-cast coal mine area.

Figures 4 and 5 exhibit the changes of leachability of the 12 selected elements from the coal gangue over 90 h. The interrupted curve in the figures indicated that the element concentration in the resulting leachates was less than the black value. The leachability trends of most of the selected elements differed according to the acidity of the leaching solutions. The results of Experiment II, with a pH of approximately 8.0, differed from those of the other two experiments. For a pH of 6.0, the leachabilities of the elements from the coal gangue displayed a sharp drop in the initial leaching phase (0–18 h) and then a steady wave-like curve (in descending order of significance), whereas for a pH of 7.0, they showed a rise in the initial phase (0–18 h), followed by a sharp decline and, then, a steady decreasing order wave-like curve. Trace elements Co, Ni, Zn and Se showed almost the same trends of leachabilities in experiments with different pH values.

In terms of $L_{am}$ (μg/g) of the elements from the coal gangue in the study area, a slight reverse dependence with the pH of leaching solutions was observed, which suggested the leachability decreased with increasing pH (Table 6). Most of the selected elements displayed similar behavior for $L_{rm}$ and $L_{am}$ under the changing pH values. However, inconsistency was observed between the trends of $L_{rm}$ and $L_{am}$ for elements Co, Ni, Cu and Sn. The $L_{rms}$ values of these four elements remained constant at 0.001%, 0.003%, 0.002% and 0.001%, respectively, regardless of the pH of the leaching solutions (Table 9). As a whole, the pH values had little impact on the $L_{rm}$ of the 12 elements, but it could influence the $L_{am}$.

**Table 9.** The maximum leached amount ($L_{am}$) and leachability ($L_{rm}$) of metals from coal gangue in the Wulantuga open-cast coal mine area.

| Elements | Experiment I | | Experiment II | | Experiment III | | Experiment IV | |
|---|---|---|---|---|---|---|---|---|
| | $L_{am}$ (µg/g) | $L_{rm}$ (%) | $L_{am}$ (µg/g) | $L_{rm}$ (%) | $L_{am}$ (µg/g) | $L_{rm}$ (%) | $L_{am}$ (µg/g) | $L_{rm}$ (%) |
| As | 0.168 | 0.009 | 0.136 | 0.008 | 0.174 | 0.010 | 0.085 | 0.005 |
| V | 0.536 | 0.005 | 0.484 | 0.005 | 0.502 | 0.005 | 0.315 | 0.003 |
| Cr | 0.137 | 0.002 | 0.117 | 0.001 | 0.134 | 0.002 | 0.107 | 0.001 |
| Co | 0.060 | 0.001 | 0.043 | 0.001 | 0.046 | 0.001 | 0.059 | 0.001 |
| Ni | 0.120 | 0.003 | 0.116 | 0.003 | 0.116 | 0.003 | 0.127 | 0.003 |
| Cu | 0.072 | 0.002 | 0.071 | 0.002 | 0.069 | 0.002 | 0.089 | 0.002 |
| Zn | 1.015 | 0.007 | 0.860 | 0.006 | 1.096 | 0.008 | 0.519 | 0.004 |
| Se | 0.079 | 0.105 | 0.060 | 0.079 | 0.061 | 0.081 | 0.096 | 0.127 |
| Cd | 0.001 | 0.003 | 0.001 | 0.002 | 0.014 | 0.037 | 0.001 | 0.002 |
| Sn | 0.003 | 0.001 | 0.002 | 0.001 | 0.002 | 0.001 | 0.003 | 0.001 |
| Pb | 0.017 | 0.001 | 0.016 | 0.001 | 0.022 | 0.0014 | 0.038 | 0.001 |
| Hg | 0.005 | 0.037 | 0.003 | 0.025 | 0.004 | 0.027 | 0.001 | 0.007 |

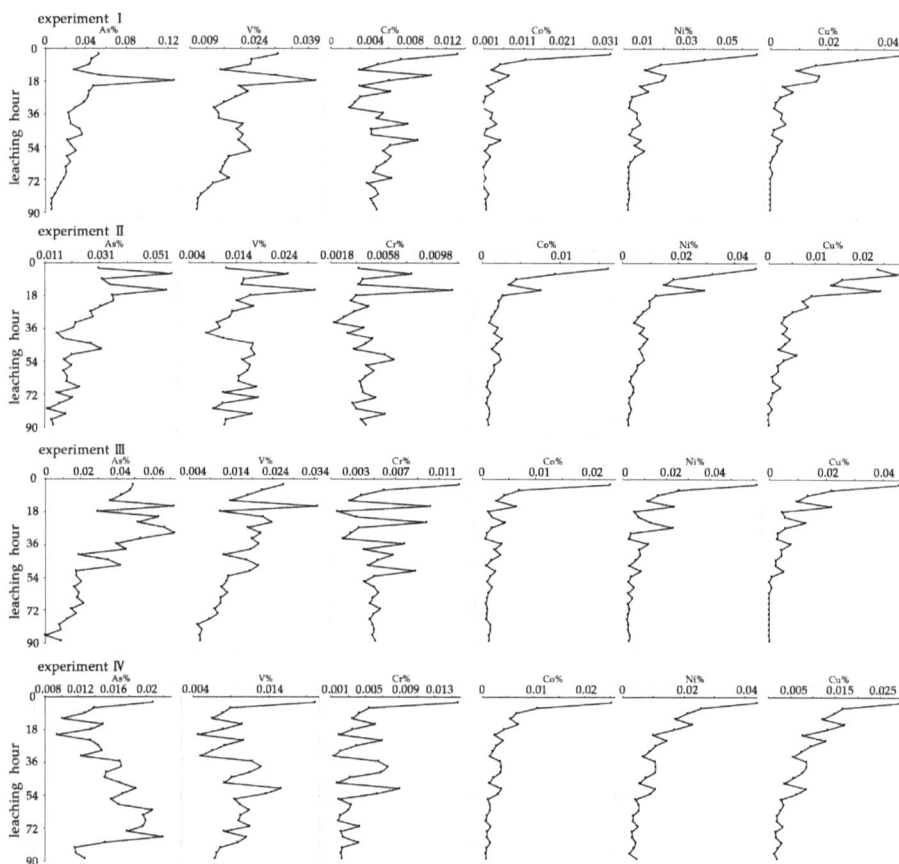

**Figure 4.** Leachabilities (%) of the elements As, V, Cr, Co, Ni and Cu from the coal gangue over 90 h in the four experiments.

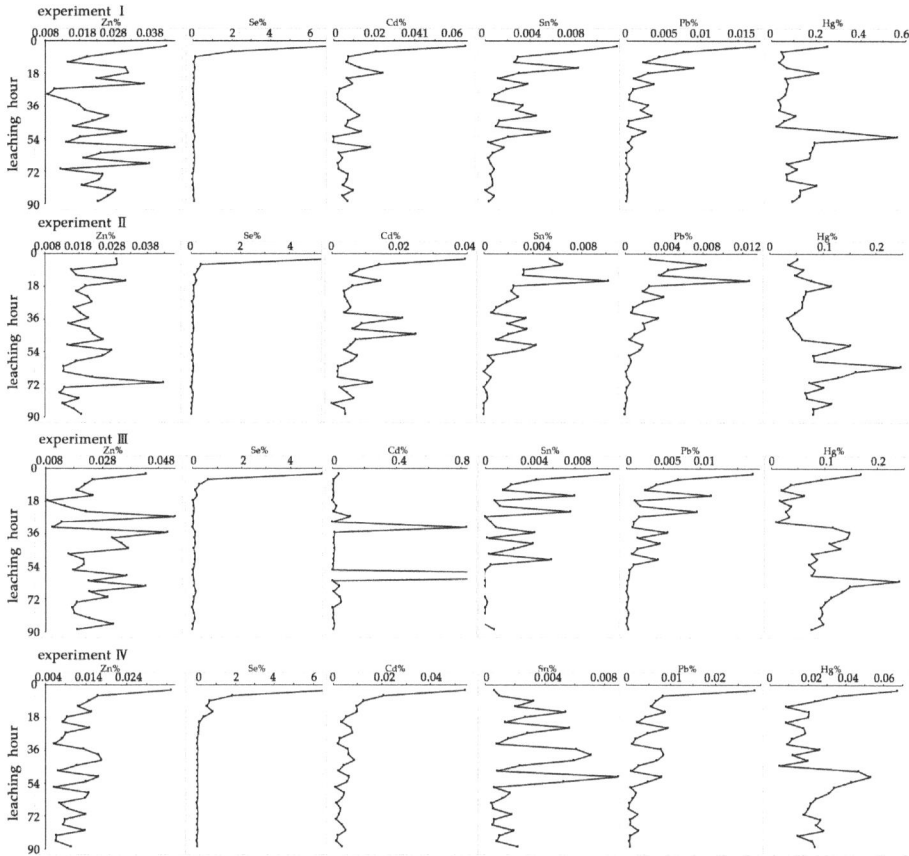

**Figure 5.** Leachabilities (%) of the elements Zn, Se, Cd, Sn, Pb and Hg from the coal gangue over 90 h in the four experiments.

### 3.2.3. Sample Amount

The effects of leaching time and pH values on the leaching behavior of the elements had previously been investigated [4,6,34]. However, the amount of samples was one of the impact factors for the leaching behavior of the elements, especially in column leaching tests. Therefore, the effect of sample amount was further discussed in this study. In the column leaching tests, sample amount should be considered and determined before leaching. Experiment III (30-g of sample) and Experiment IV (45-g of sample), under the same pH of solutions, were performed to evaluate the role of the mass of coal gangue samples in terms of leaching behavior of the elements, including leached concentrations, leachability, $L_{am}$ and $L_{rm}$.

All of the selected elements showed much higher leached concentrations with higher sample mass than those with less mass over 90 h, except for As, Cd, Sn and Hg (Tables 4–7). These four elements were found at almost the same leached concentrations in the two experiments. The leached concentrations reached maximum values in the 0–3-h phase in Experiment IV, with the exception of Sn. For Experiment III, there were four exceptions for the maximum leached concentration, namely As (27–30 h), V (12–25 h), Zn (24–27 h) and Cd (30–33 h).

According to the information displayed above, the sample amount showed little impact on the change trends of the element concentration over time, but a remarkable effect on the analytical concentrations of the elements in the resulting leachates. In terms of the changes in leachability of the elements from coal gangue over the leaching time, most of the elements showed little difference between Experiments III and IV (Figures 4 and 5). However, for Cd and Sn, the characteristics of the leachability trends differed significantly between the two experiments (Figure 5). Cd displayed a slow rise followed by a sharp rise and then a sudden wave curve under the condition of less sample mass. However, it showed a sharp initial drop followed by a steadily decreasing wave curve with higher sample mass. Sn exhibited adverse drastic fluctuations, beginning with a sharp decrease (30 g of sample) and increase (45 g of sample) in the two leaching tests (Figure 5). Furthermore, it also exhibited an obvious inverse leachability trend in the latter phase of leaching (45–90 h) for the element of As (Figure 5), which fluctuated with a general decrease in Experiment III and fluctuated with a general increase in Experiment IV. In this study, the $L_{am}$ and $L_{rm}$ of coal gangue showed no significant effect of sample mass in the leaching tests (Table 6). The $L_{rm}$ of most elements remained invariant, whereas the elements of V, Zn and Hg present a slight decrease with increasing sample mass (Table 9). This indicated that the sample amount perhaps affected the leaching behavior of some environment-sensitive elements.

### 3.3. Environmental and Ecological Risk Assessment

#### 3.3.1. Environmental Impact

In comparison with the Clarke values, the concentrations of the trace elements As, Co, Se and Pb were considerably higher in the coal gangue of the Wulantuga open-cast coal mine area (Table 3). The EF values for As and Se were 4.86 and 7.41, respectively, indicating a high pollution degree due to their high concentration levels (EF > 2).

The $L_{am}$ values of the selected trace elements provided information of the maximum potential environmental effect of the elements. The results indicated that all of the selected trace elements from coal gangue in the study area had no potential risk to soils and vegetation, based on comparing the $L_{am}$ values with the corresponding concentrations of the elements defined by the environmental quality standard for soils and limits of contaminants in foods (Tables 9 and 10). In contrast, the elements could perhaps have high potential impacts on the groundwater according to the comparison of $L_{am}$ with the corresponding concentration in the quality standard for groundwater (Tables 9 and 10). When compared to the quality standard for groundwater, the concentrations of the elements from the water samples were all at an acceptable level. All of the concentrations fell into Grade I (less than the natural background level) for the listing elements in the standard, except for element As, which fell into Grade II (natural background level) (Tables 9 and 10). According to the standard, water in Grade I and Grade II could be used for all kinds of purposes, including those of drinking water. For the concentrations of soil samples, elements Ni, Pb and Hg were found in Level 1 (natural background level), and As, Cr, Cu, Zn and Cd fell into Level 2 (no pollution to vegetation and the environment) (Tables 3 and 10). However, in terms of the vegetation samples, almost all of the element concentrations were higher than those defined in the limits of contaminants in foods (Tables 3 and 10).

#### 3.3.2. Ecological Risk

ERL and ERM values were employed to evaluate the biological toxicity of the selected trace elements. According to Table 3, it could be found that Ni was in the probable biological effects category (>ERMs value) for 50% of the coal gangue samples from the Wulantuga open-cast coal mine area. For As, Cr and Zn, 50% of samples were in the middle range, *i.e.*, occasional adverse biological effects (>ERL and <ERM) predicted. Therefore, the 12 sensitive environmental trace elements from coal gangue in the study produced moderate to high ecological hazard to the environment.

Table 10. Concentrations of trace elements defined in groundwater, soils and foods.

| Element | Quality Standard for Ground Water [1] (μg/L) | | | | | Environmental Quality Standard for Soils [2] (μg/L) | | | | | Limits of Contaminants in Foods [3] (μg/g) |
| | Grade I | Grade II | Grade III | Grade IV | Grade V | Level One | Level Two | | | Level Three | |
| | | | | | | Natural Background | pH < 6.5 | pH 6.5–7.5 | pH < 7.5 | pH > 6.5 | |
|---|---|---|---|---|---|---|---|---|---|---|---|
| As | ≤5 | ≤10 | ≤50 | ≤50 | >50 | ≤15 | ≤40 | ≤30 | ≤25 | ≤40 | ≤0.5 |
| V | - | - | - | - | - | - | - | - | - | - | - |
| Cr | ≤5 | ≤10 | ≤50 | ≤100 | >100 | ≤90 | ≤150 | ≤200 | ≤250 | ≤300 | ≤0.5 |
| Co | ≤5 | ≤50 | ≤50 | ≤1000 | >1000 | - | - | - | - | - | - |
| Ni | ≤5 | ≤50 | ≤50 | ≤100 | >100 | ≤40 | ≤40 | ≤50 | ≤60 | ≤200 | ≤1.0 |
| Cu | ≤10 | ≤50 | ≤1000 | ≤1500 | >1500 | ≤35 | ≤50 | ≤100 | ≤100 | ≤400 | - |
| Zn | ≤50 | ≤500 | ≤1000 | ≤5000 | >5000 | ≤100 | ≤200 | ≤250 | ≤300 | ≤500 | - |
| Se | ≤10 | ≤10 | ≤10 | ≤10 | >10 | - | - | - | - | - | - |
| Cd | ≤0.1 | ≤1 | ≤10 | ≤10 | >10 | ≤0.20 | ≤0.30 | ≤0.60 | ≤1.0 | - | ≤0.2 |
| Sn | - | - | - | - | - | - | - | - | - | - | ≤250 |
| Pb | ≤5 | ≤10 | ≤50 | ≤100 | >100 | ≤35 | ≤250 | ≤300 | ≤350 | ≤500 | ≤0.3 |
| Hg | ≤0.05 | ≤0.5 | ≤1 | ≤1 | >1 | ≤0.15 | ≤0.30 | ≤0.50 | ≤1.0 | ≤1.5 | ≤0.01 |

[1] GB/T14848-93 [41]; [2] GB15618-1995 [42]; [3] GB2762-2012 [43]. "-" means data are not available.

## 4. Discussion

### 4.1. Leaching Behavior and Experimental Procedures

The leaching behavior of the elements from the coal gangue is affected by various factors. Leaching time, pH values and sample amount were discussed to evaluate their roles in terms of the leaching behavior of the elements of the resulting leachates from the coal gangue in the Wulantuga open-cast coal mine area. The results of this study show that leaching time and sample mass have relatively obvious effects on the concentrations, leached amounts and leachability of elements. This result is consistent with the conclusions of other research [4,17,19,44,45].

Many previous studies suggested that the leaching behavior of the elements was closely associated with pH values in the leaching solutions [4,19,20,45]. However, it shows little role in the current study. Considering the natural conditions of rainfall, pH varies from 6.60–8.19 in the study area, and the pH values applied to evaluate its effect are 6.0, 7.0 and 8.0. These three values are consistent with the natural conditions, but could not demonstrate the effect of solution acidity or alkalinity. In this study, leaching behavior is considered to include the leached concentrations of the elements, leached amount, leachability, $L_{am}$ and $L_{rm}$. Fraction profiles of the elements could further assess the potential risks posed by the elements from coal gangue [6,46]. In future research, the fraction profiles, water-leachable, ion-changeable, carbonate-bound, organic-bound, silicate-bound and sulfide-bound particles of the selected 12 elements will be investigated.

### 4.2. Quantitative Analysis of Environmental Impacts

The assessment of the impact of trace elements from coal gangue on vegetation indicates that little potential risk exists to the plants around the open-cast coal mine area based on comparing $L_{am}$ values of the elements to corresponding concentrations in the limits of contaminants in foods, which could indirectly provide information of the impacts on human health. However, the trace element concentrations of the vegetation samples are higher than those of the limits for all of the selected elements (As, Cr, Ni, Cd, Pb and Hg) listed in the limits of contaminants in foods, except for Sn. This suggests that coal gangue contributed little to the concentrations of trace elements in vegetation in the study area, which are obviously not suitable for food consumption.

In addition to the method of contrasting standards, EF values, $L_{am}$, $L_{rm}$, ERL and ERM values were also applied to assess the trace element potential environmental and ecological hazards. These methods could be defined as semi-quantitative analyses for the trace element potential environmental and ecological impacts. Furthermore, the spatial extent and levels of pollution exposed by the trace element dispersal from coal gangue in coal mines are critical in determining the best measures and techniques for preventing environmental pollution and reclamation. The trace element transport pathways and the extent of potential environmental pollution should be identified by *in situ* sampling and analysis with the help of GIS in future research. Based on a large number of studies, a quantitative model of the trace element leaching behavior could be developed to evaluate environmental and ecological potential risks.

## 5. Conclusions

The changes of concentrations, leached amount, leachability, $L_{am}$ and $L_{rm}$ of the selected 12 trace elements from coal gangue in the Wulantuga open-cast coal mine area, Inner Mongolia, China, were investigated in this study. Based on the results of leaching behavior, the potential environmental and ecological hazards were also evaluated through different methods.

Leaching time and sample mass play important roles in determining the trace element concentrations, the leached amounts and leachabilities. pH values do not exhibit an obvious effect on the leaching behavior in this study. The coal gangue is enriched in As, Co, Se and Pb, and the EF values of As and Se indicate higher environmental pollution levels. All of the selected trace elements, namely, As, V, Cr, Co, Ni, Cu, Zn, Se, Cd, Sn, Pb and Hg, show no potential risk to soils and vegetation, but high potential risk to groundwater, based on the analysis of $L_{am}$ values. Simultaneously, according to the ERL and ERM values in the SQGs, Ni from the coal gangue is inclined to have high biological toxicity, and As, Cr and Zn show moderate ecological risk to the environment. Further research of the environment-sensitive trace element leaching behavior should be investigated through more quantitative methods with the aid of GIS to identify environmental pollution and effectively make decisions regarding prevention and reclamation.

**Acknowledgments:** This work is supported by the National Basic Research Program of China (973 Program, No. 2014CB238906). We thank Shifeng Dai at China University of Mining & Technology (Beijing) for his suggestions about the leaching tests and assisting in the ICP-MS analysis of the trace elements in coal gangue, the resulting leachates, soil, water and vegetation samples. We thank Xibo Wang at China University of Mining & Technology (Beijing) for his help with the XRF and XRD analysis of the chemical and mineral compositions of coal gangue. Two anonymous reviewers are especially thanked for their valuable comments, which greatly improved the paper quality.

**Author Contributions:** Liu Yang performed the analysis and designed the evaluation methods. Liu Yang, Jianfei Song and Bo Song helped to conceive of and design the experiments. Jianfei Song, Ruduo Wang and Bo Song performed the experiments. Liu Yang, Xue Bai and Jianfei Song analyzed the data. Xue Bai contributed to the standards and analysis. Jianfei Song and Tianhao Zhou developed the figures and tables. Jianli Jia and Haixia Pu provided the water and vegetation samples. Liu Yang and Xue Bai wrote the paper.

**Conflicts of Interest:** The authors declare no conflicts of interest.

## References

1. Dai, S.; Ren, D.; Chou, C.; Finkelman, R.; Vladimir, V.; Zhou, Y. Geochemistry of trace elements in Chinese coals: A review of abundances, genetic types, impacts on human health, and industrial utilization. *Int. J. Coal Geol.* **2012**, *94*, 3–21. [CrossRef]
2. Hower, J.C.; Robl, T.L.; Anderson, C.; Thomas, G.A.; Sakulpitakphon, T.; Mardon, S.M.; Clark, W.L. Characteristics of coal combustion products (CCP's) from Kentucky power plants, with emphasis on mercury content. *Fuel* **2005**, *84*, 1338–1350. [CrossRef]
3. Park, K.S.; Seo, Y.C.; Lee, S.J.; Lee, J.H. Emission and speciation of mercury from various combustion sources. *Powder Technol.* **2008**, *180*, 151–156. [CrossRef]
4. Wang, W.; Qin, Y.; Song, D.; Song, D.; Wang, K. Column leaching of coal and its combustion residues, Shizuishan, China. *Int. J. Coal Geol.* **2008**, *75*, 81–87. [CrossRef]
5. Finkelman, R.B.; Orem, W.; Castranova, V.; Tatu, C.A.; Belkin, H.E.; Zheng, B.; Lerch, H.E.; Maharaj, S.V.; Bates, A.L. Health impacts of coal and coal use: Possible solutions. *Int. J. Coal Geol.* **2002**, *50*, 425–443. [CrossRef]
6. Zhou, C.; Liu, G.; Wu, D.; Fang, T.; Wang, R.; Fan, X. Mobility behavior and environmental implications of trace elements associated with coal gangue: A case study at the Huainan Coalfield in China. *Chemosphere* **2014**, *95*, 193–199.
7. Bhattacharya, A.; Routh, J.; Jacks, G.; Bhattacharya, P.; Morth, M. Environmental assessment of abandoned mine tailings in Adak, Västerbotten district (northern Sweden). *Appl. Geochem.* **2006**, *21*, 1760–1780. [CrossRef]
8. Si, H.; Bi, H.; Li, X.; Yang, C. Environmental evaluation for sustainable development of coal mining in Qijiang, Western China. *Int. J. Coal Geol.* **2010**, *81*, 163–168. [CrossRef]
9. Querol, X.; Izquierdo, M.; Monfort, E.; Alvarez, E.; Font, O.; Moreno, T.; Alastuey, A.; Zhuang, X.; Lu, W.; Wang, Y. Environmental characterization of burnt coal gangue banks at Yangquan, Shanxi Province, China. *Int. J. Coal Geol.* **2008**, *75*, 93–104. [CrossRef]
10. Yue, M.; Zhao, F. Leaching experiments to study the release of trace elements from mineral separates from Chinese coals. *Int. J. Coal Geol.* **2008**, *73*, 43–51. [CrossRef]
11. Kang, Y.; Liu, G.; Chou, C.L.; Wong, M.; Zheng, L.; Ding, R. Arsenic in Chinese coals: Distribution, modes of occurrence, and environmental effects. *Sci. Total Environ.* **2011**, *412–413*, 1–13. [CrossRef] [PubMed]
12. Swaine, D.J. Why trace elements are important. *Fuel Process. Technol.* **2000**, *65–66*, 21–23. [CrossRef]
13. Dai, S.; Ren, D.; Tang, Y.; Yue, M.; Hao, L. Concentration and distribution of elements in Late Permian coals from western Guizhou Province, China. *Int. J. Coal Geol.* **2005**, *61*, 119–137. [CrossRef]

14. Chen, J.; Liu, G.; Kang, Y.; Wu, B.; Sun, R.; Zhou, C.; Wu, D. Atmospheric emissions of F, As, Se, Hg, and Sb from coal-fired power and heat generation in China. *Chemosphere* **2013**, *90*, 1925–1932. [CrossRef] [PubMed]

15. Praharaj, T.; Powell, M.A.; Hart, B.R.; Tripathy, S. Leachability of elements from sub-bituminous coal fly ash from India. *Environ. Int.* **2002**, *27*, 609–615. [CrossRef]

16. Hassett, D.; Pflughoeft-Hassett, D.; Heebink, L. Leaching of CCBs: Observations from over 25 years of research. *Fuel* **2005**, *84*, 1378–1383. [CrossRef]

17. Izquierdo, M.; Querol, X. Leaching behaviour of elements from coal combustion fly ash: An overview. *Int. J. Coal Geol.* **2012**, *94*, 54–66. [CrossRef]

18. Zhou, C.; Liu, G.; Yan, Z.; Fang, T.; Wang, R. Transformation behavior of mineral composition and trace elements during coal gangue combustion. *Fuel* **2012**, *97*, 644–650. [CrossRef]

19. Peng, B.; Wu, D. Leaching characteristics of bromine in coal. *J. Fuel Chem. Technol.* **2011**, *39*, 647–651. [CrossRef]

20. Huggins, F.E.; Seidu, L.B.A.; Shah, N.; Backus, J.; Huffman, G.P.; Honaker, R.Q. Mobility of elements in long-term leaching tests on Illinois #6 coal rejects. *Int. J. Coal Geol.* **2011**, *94*, 326–336.

21. Du, G.; Zhuang, X.; Querol, X.; Izquierdo, M.; Alastuey, A.; Moreno, T.; Font, O. Ge distribution in the Wulantuga high-germanium coal deposit in the Shengli coalfield, Inner Mongolia, northeastern China. *Int. J. Coal Geol.* **2009**, *78*, 16–26. [CrossRef]

22. Dai, S.; Liu, J.; Ward, C.R.; Hower, J.C.; Xie, P.; Jiang, Y.; Hood, M.M.; O'Keefe, J.M.K.; Song, H. Petrological, geochemical, and mineralogical compositions of the low-Ge coals from the Shengli Coalfield, China: A comparative study with Ge-rich coals and a formation model for coal-hosted Ge ore deposit. *Ore Geolo. Rev.* **2015**, *71*, 318–349. [CrossRef]

23. Jia, J.; Li, X.; Wu, P.; Liu, Y.; Han, C.; Zhou, L.; Yang, L. Human health risk assessment and safety threshold of harmful trace elements in the soil environment of the Wulantuga open-cast coal mine. *Minerals* **2015**, *5*, 837–848. [CrossRef]

24. Dai, S.; Wang, X.; Seredin, V.; Hower, J.C.; Ward, C.R.; O'Keefe, J.M.; Huang, W.; Li, T.; Li, X.; Liu, H.; *et al.* Petrology, mineralogy, and geochemistry of the Ge-rich coal from the Wulantuga Ge ore deposit, Inner Mongolia, China: New data and genetic implications. *Int. J. Coal Geol.* **2012**, *90–91*, 72–99. [CrossRef]

25. Zhuang, X.; Querol, X.; Alastuey, A.; Juan, R.; Plana, F.; Lopez-Soler, A.; Du, G.; Martynov, V.V. Geochemistry and mineralogy of the Cretaceous Wulantuga high germanium coal deposit in Shengli coal field, Inner Mongolia, Northeastern China. *Int. J. Coal Geol.* **2006**, *66*, 119–136. [CrossRef]

26. Dai, S.; Seredin, V.V.; Ward, C.R.; Jiang, J.; Hower, J.C.; Song, X.; Jiang, Y.; Wang, X.; Gornostaeva, T.; Li, X.; *et al.* Composition and modes of occurrence of minerals and elements in coal combustion products derived from high-Ge coals. *Int. J. Coal Geol.* **2014**, *121*, 79–97. [CrossRef]

27. Dai, S.; Hower, J.C.; Ward, C.R.; Guo, W.; Song, H.; O'Keefe, J.M.K.; Xie, P.; Hood, M.M.; Yan, X. Elements and phosphorus minerals in the middle Jurassic inertinite-rich coals of the Muli Coalfield on the Tibetan Plateau. *Int. J. Coal Geol.* **2015**, *144–145*, 23–47. [CrossRef]

28. Dai, S.; Wang, P.; Ward, C.R.; Tang, Y.; Song, X.; Jiang, J.; Hower, J.C.; Li, T.; Seredin, V.V.; Wagner, N.J.; *et al.* Elemental and mineralogical anomalies in the coal-hosted Ge ore deposit of Lincang, Yunnan, southwestern China: Key role of $N_2$–$CO_2$-mixed hydrothermal solutions. *Int. J. Coal Geol.* **2015**, *152*, 19–46. [CrossRef]

29. Li, X.; Dai, S.; Zhang, W.; Li, T.; Zheng, X.; Chen, W. Determination of As and Se in coal and coal combustion products using closed vessel microwave digestion and collision/reaction cell technology (CCT) of inductively coupled plasma mass spectrometry (ICP-MS). *Int. J. Coal Geol.* **2014**, *124*, 1–4. [CrossRef]

30. Dai, S.; Li, T.; Jiang, Y.; Ward, C.R.; Hower, J.C.; Sun, J.; Liu, J.; Song, H.; Wei, P.; Li, Q.; *et al.* Mineralogical and geochemical compositions of the Pennsylvanian coal in the Hailiushu Mine, Daqingshan Coalfield, Inner Mongolia, China: Implications of sediment-source region and acid hydrothermal solutions. *Int. J. Coal Geol.* **2015**, *137*, 92–110. [CrossRef]

31. Liu, K.; Wang, X.; Sun, H.; Chen, L. PH value and acid rain condition in Inner Mongolia area in recent 3 years. *Meteorol. J. Inn. Mongolia* **2010**, *3*, 32–33.

32. Duce, R.A.; Hoffmann, G.L.; Zoller, W.H. Atmospheric trace metals at remote northern and southern hemisphere sites: Pollution or natural? *Science* **1975**, *187*, 59–61. [CrossRef] [PubMed]

33. Bai, X.; Jia, H. The release of heavy metals in Gangue Leaching Process. *Environ. Sci. Tech.* **2009**, *22*, 5–8.

34. Wang, W.; Hao, W.; Bian, Z.; Lei, S.; Wang, X.; Sang, S. Effect of coal mining activities on the environment of Tetraenamongolica in Wuhai, Inner Mongolia, China—A geochemical perspective. *Int. J. Coal Geol.* **2014**, *132*, 94–102. [CrossRef]

35. Pekey, H.; Karakas, D.; Ayberk, S.; Tolun, L.; Bakoglu, M. Ecological risk assessment using trace elements from surface sediments of Izmit Bay (Northeastern Marmara Sea) Turkey. *Mar. Pollut. Bull.* **2004**, *48*, 946–953. [CrossRef] [PubMed]

36. Thomas, L. *Coal Geology*, 2nd ed.; John Wiley & Sons Inc.: Hoboken, NJ, USA, 2012; p. 454.

37. ASTM International. *Test Method for Moisture in the Analysis Sample of Coal and Coke*; ASTM Standard D3173–11; ASTM International: West Conshohocken, PA, USA, 2011.

38. ASTM International. Annual Book of ASTM Standards. In *Test Method for Ash in the Analysis Sample of Coal and Coke*; ASTM Standard D3174–11; ASTM International: West Conshohocken, PA, USA, 2011.

39. Dai, S.; Graham, I.T.; Ward, C.R. A review of anomalous rare earth elements and yttrium in coal. *Int. J. Coal Geol.* **2016**, *159*, 82–95. [CrossRef]

40. Tayor, S.R. Abundance of chemical elements in the continental crust: a new table. *Int. J. Geochim. Cosmochim. Acta.* **1964**, *28*, 1273–1285.

41. General Administration of Quality Supervision. *Quality Standard for Ground Water*; GB/T 14848-1993; General Administration of Quality Supervision: Beijing, China, 1993. (In Chinese)

42. General Administration of Quality Supervision. *Environmental Quality Standard for Soils*; GB 15618-1995; General Administration of Quality Supervision: Beijing, China, 1995. (In Chinese)

43. Ministry of Health P.R. China. *National Food Safety Standard of Maximum Levels of Contaminants in Foods*; GB 2762-2012. Ministry of Health P.R. China: Beijing, China, 2012. (In Chinese).

44. Guo, Y.; Huang, P.; Zhang, W. Leaching of heavy metals from Dexing copper mine tailings pond. *Trans. Nonferrous Met. Soc. China* **2013**, *23*, 3068–3075. [CrossRef]

45. Spears, D.A. The determination of trace element distributions in coals using sequential chemical leaching—A new approach to an old method. *Fuel* **2013**, *114*, 31–37. [CrossRef]

46. DeLemos, L.J.; Brugge, D.; Cajero, M.; Durant, J.L.; George, C.M.; Henio-Adeky, S.; Nez, T.; Manning, T.; Rock, T.; Seschillie, B.; *et al.* Development of risk maps to minimize uranium exposures in the Navajo Church rock mining district. *Environ. Health* **2009**, *8*, 29. [CrossRef] [PubMed]

*Article*

# Human Health Risk Assessment and Safety Threshold of Harmful Trace Elements in the Soil Environment of the Wulantuga Open-Cast Coal Mine

**Jianli Jia [1],\*, Xiaojun Li [1], Peijing Wu [1], Ying Liu [2], Chunyu Han [2], Lina Zhou [2] and Liu Yang [1]**

[1]   School of Chemical and Environmental Engineering, China University of Mining and Technology, Beijing 100083, China; 7lixiaojun9@gmail.com (X.L.); wpjhzh@gmail.com (P.W.); 108889@cumtb.edu.cn (L.Y.)

[2]   Yanqing Country Water Authority, Beijing 100083, China; 157liuying2015@gmail.com (Y.L.); hanchunyu2015@gmail.com (C.H.); zhoulinaqq2015@gmail.com (L.Z.)

\*   Correspondence: jjl@cumtb.edu.cn; Tel.: +86-10-6233-9289

Academic Editors: Shifeng Dai and David Cliff

Received: 6 September 2015; Accepted: 20 November 2015; Published: 30 November 2015

**Abstract:** In this study, soil samples were collected from a large-scale open-cast coal mine area in Inner Mongolia, China. Arsenic (As), cadmium (Cd), beryllium (Be) and nickel (Ni) in soil samples were detected using novel collision/reaction cell technology (CCT) with inductively-coupled plasma mass spectrometry (ICP-MS; collectively ICP-CCT-MS) after closed-vessel microwave digestion. Human health risk from As, Cd, Be and Ni was assessed via three exposure pathways—inhalation, skin contact and soil particle ingestion. The comprehensive carcinogenic risk from As in Wulantuga open-cast coal mine soil is 6.29–87.70-times the acceptable risk, and the highest total hazard quotient of As in soils in this area can reach 4.53-times acceptable risk levels. The carcinogenic risk and hazard quotient of Cd, Be and Ni are acceptable. The main exposure route of As from open-cast coal mine soils is soil particle ingestion, accounting for 76.64% of the total carcinogenic risk. Considering different control values for each exposure pathway, the minimum control value (1.59 mg/kg) could be selected as the strict reference safety threshold for As in the soil environment of coal-chemical industry areas. However, acceptable levels of carcinogenic risk are not unanimous; thus, the safety threshold identified here, calculated under a $1.00 \times 10^{-6}$ acceptable carcinogenic risk level, needs further consideration.

**Keywords:** carcinogenic risk; hazard quotient; open-cast coal mine; arsenic; soil; safety threshold; harmful trace elements

## 1. Introduction

Coal will continue to play an important role in the global energy supply, especially in China, for a long time to come [1], and will make significant contributions to the development of human society and the standards of living. However, some harmful trace elements, such as arsenic (As), cadmium (Cd), beryllium (Be) and nickel (Ni) are enriched in coal [2,3] with the accompanying minerals. Researchers observed that As and Hg (mercury) was hosted in pyrite, Be and U (uranium) adsorbed in clay minerals and, meanwhile, F (fluorine) enriched with kaolinite [4–6], through the effect of sedimentary diagenesis, microbial action, tectonism, magmatic hydrothermal activity or groundwater activity [7–9]. These trace harmful elements, in various forms may migrate into soil, groundwater, air and other environmental media [10] and negatively affect human health, through natural activities, such as hydrothermal activity, or human activities, like coal gasification or coal coking processes.

Chemicals, such as heavy metals, have been shown to cause human cancers [11]. As, Cd, Be, Ni and other harmful trace compounds found in coal, which conspicuously cause toxicity in humans, were documented and suggested by the U.S. Environmental Protection Agency (U.S. EPA) [12], as well as by the Ministry of Environmental Protection of the People's Republic of China [13]. Studies on the level of their risk to human health and corresponding risk control in the mining process are important for the safety and health of workers and residents in mining areas.

Health risk assessment [14] is a comprehensive evaluation method that links environmental pollution and human health [15]. Environmental risk assessment in China was started in the 1980s, and human health risk evaluation studies were developed in the 1990s. Based on the assessing processes and models used in different countries, software was developed for the assessment of health and the environmental risks of contaminated sites in China, named the Health and Environmental Risk Assessment (HERA) [16], and this software was applied to the assessment of contaminated sites, such as the areas surrounding oilfields or other chemical plants. In recent years, the human health risk caused by As, Cd, Be and other toxic trace elements in some sites was quantitatively evaluated using different methods of health risk assessment. Juhasz *et al.* [17] evaluated the human health risk of As in rice; the results indicated that different forms of As could cause different levels of risk to human health. Zhuang *et al.* [18] assessed the human health risk of Pb and Cd in the Huayuan mining area in China, and results indicated that Pb and Cd accumulated in vegetables had severe potential risks for human health. Ren *et al.* [19] evaluated the potential risk of Pb in the soil environment for children in Shenyang city, and Li *et al.* [20] calculated the health risk level caused by Cd, Cu and Se in rice grain in the Nanjing area.

Although there were several models and standards for human health risk assessment, both in China and globally, and several health risk assessments were carried out, research on health risk assessment of harmful trace elements in open-cast coal mines is still very limited. Considering the ecological system properties of the open-cast mining area in the northwest of China and the complex contamination characteristics of multiple trace elements, this study could be a useful complement in this field. Furthermore, this study aims to propose safety thresholds for harmful trace elements (As, Cd, Be and Ni) in the coal mine area, which has implications for the protection of workers and industry health. We comprehensively compared mainstream evaluation models and methods, such as CLEA (Contaminated Land Exposure Assessment [21,22]), RBCA (Risk-Based Corrective Action [23,24]) and HERA (Health and Environmental Risk Assessment [16]). This study used Chinese standard technical guidelines for risk assessment of contaminated sites (HJ25.3-2014) [25] to carry out human health risk assessment of harmful trace elements in the Wulantuga open-cast coal mine area.

## 2. Experimental Section

### 2.1. Sample Collection

Soil samples were collected from the Wulantuga coal mine area, which is located in Xilinhaote in Inner Mongolia (north latitude 43°56′57.86″ and east longitude 115°54′37.36″ in China) in July 2014. Soil samples were collected using a geotome for a 0–15-cm depth of each layer, and in each layer, three sampling points were set. The soil samples were stored in plastic sealing bags and stored in a portable freezer until they were returned to the laboratory. The Wulantuga open-cast coal mine is still in operation; the area where the coal mine is located has an annual average temperature of 0–3 °C. The average annual rainfall was less than 300 mm, with a perennial southwest wind. Proven coal reserves were 760 million tons; the annual output is 7.3 million tons, and 337 staff work here. Many scholars have studied the geochemistry and mineralogy of the coal deposit in this coal mine [26–29]. The open-cast coal mine and the sampling sites are illustrated in Figure 1, and the distribution of sampling points and soil profile information is shown in Figure 2. Background soil samples were taken from a grassland, which was about 15 km away from Xilinhaote city in the northeast direction.

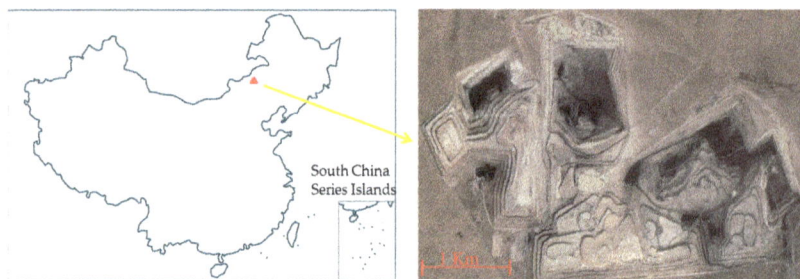

**Figure 1.** Location of the Wulantuga coal mine.

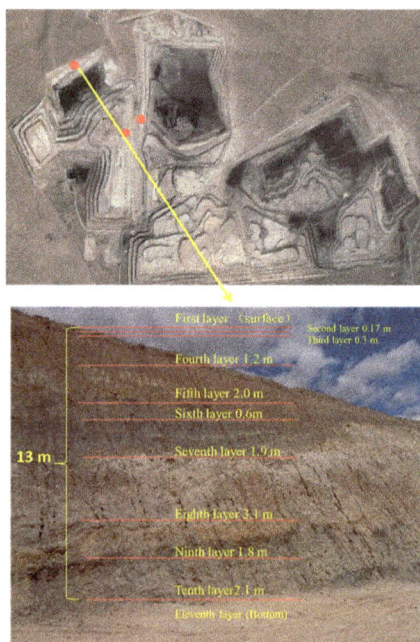

**Figure 2.** The distribution of sampling points and sections in the mining area.

*2.2. Sample Handling and Detection*

After drying the soil samples in an oven for 8 h at 105 °C [30], they were crushed to 200 mesh. The samples were digested in an UltraCLAVE microwave high-pressure reactor (Milestone, Milano, Italy) for 175 min [31]. Next, 50 mg of the soil sample were digested in 5 mL 40% HF, 2 mL 65% $HNO_3$ and 1 mL 30% $H_2O_2$. Initial nitrogen pressure was set at 50 bars. The heating process is: 12 min to 60 °C, 20 min to 125 °C, 8 min to 160 °C, 15 min to 240 °C, 60 min to 240 °C. [31]. Inductively-coupled plasma mass spectrometry (ICP-MS, ThermoScientific Xseries 2, Thermo Fisher Scientific, Waltham, MA, USA) was used to determine the amounts of the trace elements (plasma RF power set to 1400 W, sampling depth set to 130 steps, peristaltic pump speed set to 30 RPM, collision gas flow set to 4 mL/min, dwell time set to 10 ms, peak jumping acquisition mode, nebulizer gas flow set to 1.00 L/min, auxiliary gas flow set to 0.80 L/min, cool gas flow set to 13.00 L/min). The linearity of the calibration curves was considered acceptable in the range 0–100 μg/L with a determination coefficient $r^2 > 0.9999$. The method detection limit (MDL) of these elements was about 0.02 μg/L. As was determined using

ICP-MS with collision cell technology (CCT) due to its volatility. Polyfluoroalkoxy volumetric flasks were used without drying on an electric hot plate to avoid volatile loss. A laser particle size analyzer was used to determine the texture of the soil samples.

*2.3. Health Risk Assessment Methods*

2.3.1. Exposure Assessment

During the preliminary stage of this study, Co (cobalt), Hg, Cu (copper), Zn (zinc), Se (selenium) and U concentrations were found to be low and not considered to be potential human health risks, and there were no effective toxicity parameters of Cr (chromium) and Pb (plumbum). Therefore, we selected As, Cd, Be and Ni as the major elements to evaluate. Different land use patterns define the land type, for example residential, cultural and school land are defined as sensitive sites. Industrial lands are defined as non-sensitive sites. As the experimental site is a typical non-sensitive site, the ways in which human health could be influenced in this coal mining area were identified according to the recommended guidelines for human health risk assessment of contaminated sites [25]. Considering that there was no surface water in the area surrounding the mine, the groundwater was not used for drinking and based on published reports [32–35], three routes of exposure—inhalation of particles, skin contact and ingestion of soil particles—were selected to evaluate the human health risk of this mining area. The formulas by which corresponding soil exposure doses of the three exposure ways were calculated are listed in Table 1.

**Table 1.** Calculating models of soil exposure dose in three soil exposure pathways.

| Exposure Routes | Instruction | Formula for Calculation of Exposure Dose | Equation Number |
|---|---|---|---|
| Inhalation of particles | Carcinogenic risk | $OISER_{ca} = \dfrac{OSIR_a \times ED_a \times EF_a \times ABS_0}{BW_a \times AT_{ca}} \times 10^{-6}$ | (1) |
| | Non-carcinogenic risk | $OISER_{nc} = \dfrac{OSIR_a \times ED_a \times EF_a \times ABS_0}{BW_a \times AT_{nc}} \times 10^{-6}$ | (2) |
| Skin contact | Carcinogenic risk | $DCSER_{ca} = \dfrac{SAE_a \times SSAR_a \times EF_a \times ED_a \times E_V \times ABS_d}{BW_a \times AT_{ca}} \times 10^{-6}$ | (3) |
| | Non-carcinogenic risk | $DCSER_{nc} = \dfrac{SAE_a \times SSAR_a \times EF_a \times ED_a \times E_V \times ABS_d}{BW_a \times AT_{nc}} \times 10^{-6}$ | (4) |
| Ingestion of soil particles | Carcinogenic risk | $PISER_{ca} =$ $\dfrac{PM_{10} \times DAIR_a \times ED_a \times PIAF \times (fspo \times EFO_a + fspi \times EFI_a)}{BW_a \times AT_{ca}} \times 10^{-6}$ | (5) |
| | Non-carcinogenic risk | $PISER_{nc} =$ $\dfrac{PM_{10} \times DAIR_a \times ED_a \times PIAF \times (fspo \times EFO_a + fspi \times EFI_a)}{BW_a \times AT_{nc}} \times 10^{-6}$ | (6) |

The main parameters of the contaminated site risk-assessment model include concentration and toxicological parameters of the pollutants, site condition parameters and exposure parameters. The values of each concentration of the target pollutants and the site condition parameters were measured. The exposure factor parameters were applied without considering the exposure of children, based on the non-sensitive properties of the coal mining area in this paper (Table 2).

2.3.2. Toxicological Evaluation

Based on the parameter value selection and the calculation of the various exposure routes, the carcinogenic risk and hazard quotient were calculated using the formulas and parameters listed in Tables 2 and 3. Then, the comprehensive human health risk was summed up with the individual risk associated with each exposure route [25]. The specific level of human health risk for each sampling point thus obtained was compared to the acceptable level of human carcinogenic risk ($1.00 \times 10^{-6}$) and hazard quotient (with the standard value of 1.00) [25,35].

$CR_{ois}$ is the carcinogenic risk associated with the exposure route of the inhalation of particles (dimensionless); $CR_{dcs}$ is the carcinogenic risk associated with the exposure route of skin contact (dimensionless); $CR_{pis}$ is the carcinogenic risk associated with the exposure route of the ingestion of soil particles (dimensionless); $HQ_{ois}$ represents the hazard quotient associated with the exposure route of the ingestion of soil particles (dimensionless); $HQ_{dcs}$ is the hazard quotient associated with the exposure route of skin contact (dimensionless); $HQ_{pis}$ is the hazard quotient associated with the exposure route of the ingestion of soil particles (dimensionless). The remaining parameters are shown in Table 2.

**Table 2.** Major parameters in the exposure dose calculation models.

| Parameter | Implication | Value | Unit |
|---|---|---|---|
| $OSIR_a$ | Intake amount of soil per day | 100.00 | $mg \cdot day^{-1}$ |
| $ED_a$ | Exposure time | 25.00 | a |
| $EF_a$ | Exposure rate | 250.00 | $day \cdot a^{-1}$ |
| $BW_a$ | Weight of an adult | 56.80 | kg |
| $ABS_0$ | Absorption efficiency factor of inhaled particles | 26,280.00 | - |
| $AT_{ca}$ | Average carcinogenic effect time | 26,280.00 | day |
| $AT_{nc}$ | Average non-carcinogenic effect time | 91,280.00 | day |
| $SAE_a$ | Exposed skin area | 2854.62 | $cm^2$ |
| $SSAR_a$ | Soil adhesion coefficient of skin surface | 0.20 | $mg \cdot cm^{-2}$ |
| $ABS_d$ | Absorption efficiency factor of skin contact | 0.03 | - |
| $E_V$ | Frequency of skin contact per day | 1.00 | $time \cdot day^{-1}$ |
| $PM_{10}$ | Concentration of inhalable suspended particulate matter | 0.15 | $m^3 \cdot day^{-1}$ |
| $DAIR_a$ | Air intake per day | 14.50 | $m^3 \cdot day^{-1}$ |
| PIAF | Retention ratio of inhalable soil particles *in vivo* | 0.75 | - |
| fspi | Proportion of soil particles in indoor air | 0.80 | - |
| fspo | Proportion of soil particles in outdoor air | 0.50 | - |
| $EFI_a$ | Indoor exposure frequency | 187.50 | $day \cdot a^{-1}$ |
| $EFO_a$ | Outdoor exposure frequency | 62.50 | $day \cdot a^{-1}$ |
| $C_{sur}$ | Concentration of pollutants in the surface soil | Table 6 | $mg \cdot kg^{-1}$ |
| $SF_0$ | Oral intake slope factor of carcinogenic element | 1.50 | $(mg/kg \cdot day)^{-1}$ |
| $SF_d$ | Skin contact slope factor of carcinogenic element | 1.00 | $(mg/kg \cdot day)^{-1}$ |
| $SF_i$ | Breathing slope factor of carcinogenic element | 4.30 | $(mg/kg \cdot day)^{-1}$ |
| SAF | Reference dose distribution coefficient of soil exposure | 0.20 | - |
| $RfD_0$ | Reference dose for ingestion | $3.00 \times 10^{-4}$ | $mg \cdot kg^{-1} \cdot day^{-1}$ |
| $RfD_d$ | Reference dose for skin contact | $3.00 \times 10^{-4}$ | $mg \cdot kg^{-1} \cdot day^{-1}$ |
| $RfD_i$ | Reference dose for inhalation | $3.83 \times 10^{-6}$ | $mg \cdot kg^{-1} \cdot day^{-1}$ |

**Table 3.** Formulas for the calculation of carcinogenic risk and the hazard quotient.

| Exposure Routes | Instruction | Cancer Risk or Hazard Quotient Calculating Formulas | Equation Number |
|---|---|---|---|
| Inhalation of particles | Carcinogenic risk | $CR_{ois} = OISER_{ca} \times C_{sur} \times SF_o$ | (7) |
| | Hazard quotient | $HQ_{ois} = \dfrac{OISER_{nc} \times C_{sur}}{RfD_O \times SAF}$ | (8) |
| Skin contact | Carcinogenic risk | $CR_{dcs} = DCSER_{ca} \times C_{sur} \times SF_d$ | (9) |
| | Hazard quotient | $HQ_{dcs} = \dfrac{DCSER_{nc} \times C_{sur}}{RfD_d \times SAF}$ | (10) |
| Ingestion of soil particles | Carcinogenic risk | $CR_{pis} = PISER_{ca} \times C_{sur} \times SF_i$ | (11) |
| | Hazard quotient | $HQ_{pis} = \dfrac{PISER_{nc} \times C_{sur}}{RfD_i \times SAF}$ | (12) |

### 2.3.3. Calculation of Control Values

When carcinogenic risk exceeds the recommended safety value, the risk control value associated with the corresponding routes of exposure should be calculated (Table 4).

ACR refers to the acceptable level of human carcinogenic risk ($1 \times 10^{-6}$, dimensionless); AHQ is the acceptable level of the hazard quotient (1, dimensionless). The remaining parameters are listed in Table 2.

Table 4. Formulas for the calculation of the safety threshold.

| Exposure Routes | Instruction | Safety Threshold Formulas | Equation Number |
|---|---|---|---|
| Inhalation of particles | Carcinogenic risk | $RCVS_{ois} = \dfrac{ACR}{OISER_{ca} \times SF_0}$ | (13) |
| | Hazard quotient | $HCVS_{ois} = \dfrac{RfD_0 \times SAF \times AHQ}{OISER_{nc}}$ | (14) |
| Skin contact | Carcinogenic risk | $RCVS_{dcs} = \dfrac{ACR}{DCSER_{ca} \times SF_d}$ | (15) |
| | Hazard quotient | $HCVS_{dcs} = \dfrac{RfD_d \times SAF \times AHQ}{DCSER_{nc}}$ | (16) |
| Ingestion of soil particles | Carcinogenic risk | $RCVS_{pis} = \dfrac{ACR}{PISER_{ca} \times SF_i}$ | (17) |
| | Hazard quotient | $HCVS_{pis} = \dfrac{RfD_i \times SAF \times AHQ}{PISER_{nc}}$ | (18) |

## 3. Results and Discussion

### 3.1. Harmful Trace Elements' Concentrations and Exposure Levels

The concentrations of As, Cd, Be and Ni in each sample and carcinogenic and non-carcinogenic exposure, cancer risk and the hazard quotient under each exposure pathway are provided in Tables 6 and 7. The distribution of As was between 7.67 and 107.07 mg/kg, whereas that of Cd, Be and Ni was 0.27–0.70, 1.73–4.85 and 11.75–37.09 mg/kg, respectively. The concentrations of As, Cd, Be and Ni in raw coal were 14.08, 0.05, 0.01 and 75.50 mg/kg, respectively. Carcinogenic exposure level of As in this area under the exposure pathway of the inhalation of particles was $4.19 \times 10^{-7}$ m$^3$/day, whereas the non-carcinogenic exposure level of Cd, Be and Ni was $1.21 \times 10^{-6}$ m$^3$/day. Carcinogenic exposure levels of As and Cd under the exposure pathway of skin contact in this area were $7.17 \times 10^{-8}$ and $2.39 \times 10^{-9}$ m$^3$/day, respectively, whereas the non-carcinogenic exposure levels were $2.06 \times 10^{-7}$ and $6.88 \times 10^{-9}$ m$^3$/day, respectively. The carcinogenic exposure level of As under the exposure pathway of the ingestion of soil particles in this area was $4.95 \times 10^{-9}$ m$^3$/day, and the non-carcinogenic exposure level of Cd, Be and Ni was $1.43 \times 10^{-8}$ m$^3$/day. The particle size of the soil samples is shown in Table 5. The texture of the soil from "10 m to the edge of the mine" was silty loam and from "200 m to the edge of the mine" sandy clay loam, and the other twelve soil samples were all sandy loam soil.

Table 5. Particle size of each soil sample.

| Sampling Site | Percentage of Each Size (%) | | |
|---|---|---|---|
| | <0.002 mm | 0.02–0.002 mm | 2–0.02 mm |
| Grassland | 2.00 | 12.29 | 85.70 |
| 10 m to the edge of the mine | 5.64 | 45.76 | 48.58 |
| 200 m to the edge of the mine | 2.74 | 20.91 | 76.33 |
| First layer | 0.49 | 4.03 | 95.47 |
| Second layer | 1.15 | 10.15 | 88.68 |
| Third layer | 0.38 | 4.46 | 95.15 |
| Fourth layer | 1.32 | 8.56 | 90.11 |
| Fifth layer | 1.48 | 10.80 | 87.70 |
| Sixth layer | 0.97 | 6.94 | 92.07 |
| Seventh layer | 0.87 | 8.60 | 90.52 |
| Eighth layer | 1.09 | 8.11 | 90.79 |
| Ninth layer | 1.72 | 11.31 | 86.95 |
| Tenth layer | 0.52 | 7.04 | 92.43 |
| Eleventh layer | 0.92 | 7.38 | 91.69 |

**Table 6.** Concentrations and evaluation parameters of As under each exposure pathway.

| Description | Sampling Site | Concentration of As (mg/kg) | Inhalation of Particles | | Skin Contact | | Ingestion of Soil Particles | |
|---|---|---|---|---|---|---|---|---|
| | | | CR | HQ | CR | HQ | CR | HQ |
| Background soil | Grassland | 12.63 | $7.93 \times 10^{-6}$ | $2.60 \times 10^{-1}$ | $1.36 \times 10^{-6}$ | $4.00 \times 10^{-2}$ | $1.05 \times 10^{-6}$ | $2.40 \times 10^{-1}$ |
| Mine side soil | 10 m to the edge of the mine | 66.10 | $4.15 \times 10^{-5}$ | 1.34 | $7.14 \times 10^{-6}$ | $2.30 \times 10^{-1}$ | $5.51 \times 10^{-6}$ | 1.24 |
| | 200 m to the edge of the mine | 97.23 | $6.11 \times 10^{-5}$ | 1.96 | $1.05 \times 10^{-5}$ | $3.30 \times 10^{-1}$ | $8.11 \times 10^{-6}$ | 1.81 |
| Section soil | First layer | 13.67 | $8.59 \times 10^{-6}$ | $2.80 \times 10^{-1}$ | $1.48 \times 10^{-6}$ | $5.00 \times 10^{-2}$ | $1.14 \times 10^{-6}$ | $2.60 \times 10^{-1}$ |
| | Second layer | 10.56 | $6.63 \times 10^{-6}$ | $2.20 \times 10^{-1}$ | $1.14 \times 10^{-6}$ | $4.00 \times 10^{-2}$ | $8.81 \times 10^{-7}$ | $2.00 \times 10^{-1}$ |
| | Third layer | 7.67 | $4.82 \times 10^{-6}$ | $1.60 \times 10^{-1}$ | $8.29 \times 10^{-7}$ | $3.00 \times 10^{-2}$ | $6.40 \times 10^{-7}$ | $1.40 \times 10^{-1}$ |
| | Fourth layer | 47.97 | $3.01 \times 10^{-5}$ | $9.70 \times 10^{-1}$ | $5.18 \times 10^{-6}$ | $1.60 \times 10^{-1}$ | $4.00 \times 10^{-6}$ | $9.00 \times 10^{-1}$ |
| | Fifth layer | 107.07 | $6.72 \times 10^{-5}$ | 2.16 | $1.16 \times 10^{-5}$ | $3.70 \times 10^{-1}$ | $8.93 \times 10^{-6}$ | 2.00 |
| | Sixth layer | 47.45 | $2.98 \times 10^{-5}$ | $9.60 \times 10^{-1}$ | $5.12 \times 10^{-6}$ | $1.60 \times 10^{-1}$ | $3.96 \times 10^{-6}$ | $8.90 \times 10^{-1}$ |
| | Seventh layer | 11.39 | $7.15 \times 10^{-6}$ | $2.30 \times 10^{-1}$ | $1.23 \times 10^{-6}$ | $4.00 \times 10^{-2}$ | $9.50 \times 10^{-7}$ | $2.10 \times 10^{-1}$ |
| | Eighth layer | 32.94 | $2.07 \times 10^{-5}$ | $6.70 \times 10^{-1}$ | $3.56 \times 10^{-6}$ | $1.10 \times 10^{-1}$ | $2.75 \times 10^{-6}$ | $6.20 \times 10^{-1}$ |
| | Ninth layer | 20.71 | $1.30 \times 10^{-5}$ | $4.20 \times 10^{-1}$ | $2.24 \times 10^{-6}$ | $7.00 \times 10^{-2}$ | $1.73 \times 10^{-6}$ | $3.90 \times 10^{-1}$ |
| | Tenth layer | 14.21 | $8.92 \times 10^{-6}$ | $2.90 \times 10^{-1}$ | $1.54 \times 10^{-6}$ | $5.00 \times 10^{-2}$ | $1.19 \times 10^{-6}$ | $2.70 \times 10^{-1}$ |
| | Eleventh layer | 25.92 | $1.63 \times 10^{-5}$ | $5.20 \times 10^{-1}$ | $2.80 \times 10^{-6}$ | $9.00 \times 10^{-2}$ | $2.16 \times 10^{-6}$ | $4.80 \times 10^{-1}$ |

Annotation: CR represents carcinogenic risk; HQ represents hazard quotient.

**Table 7.** Concentrations and evaluation parameters of each exposure pathway for different elements.

| Sample Description | Sampling Site | Concentration (mg/kg) | | | HQ of Inhalation of Particles | | | HQ of Skin Contact | HQ of Ingestion of Soil Particles | | |
|---|---|---|---|---|---|---|---|---|---|---|---|
| | | Cd | Be | Ni | Cd | Be | Ni | Cd | Cd | Be | Ni |
| Background soil | Grassland | 0.73 | 2.02 | 25.92 | $4.44 \times 10^{-3}$ | $6.11 \times 10^{-3}$ | $7.85 \times 10^{-3}$ | $7.34 \times 10^{-1}$ | $2.06 \times 10^{-2}$ | $2.83 \times 10^{-2}$ | $8.08 \times 10^{-2}$ |
| Mine side soil | 10 m to the edge of the mine | 0.40 | 3.71 | 24.13 | $2.42 \times 10^{-3}$ | $1.12 \times 10^{-2}$ | $7.31 \times 10^{-3}$ | $4.00 \times 10^{-1}$ | $1.12 \times 10^{-2}$ | $5.20 \times 10^{-2}$ | $7.52 \times 10^{-2}$ |
| | 200 m to the edge of the mine | 0.38 | 3.08 | 22.71 | $2.32 \times 10^{-3}$ | $9.32 \times 10^{-3}$ | $6.88 \times 10^{-3}$ | $3.83 \times 10^{-1}$ | $1.07 \times 10^{-2}$ | $4.32 \times 10^{-2}$ | $7.08 \times 10^{-2}$ |
| Section soil | First layer | 0.43 | 1.73 | 22.92 | $2.61 \times 10^{-3}$ | $5.25 \times 10^{-3}$ | $6.94 \times 10^{-3}$ | $4.31 \times 10^{-1}$ | $1.21 \times 10^{-2}$ | $2.43 \times 10^{-2}$ | $7.14 \times 10^{-2}$ |
| | Second layer | 0.51 | 2.28 | 20.11 | $3.08 \times 10^{-3}$ | $6.92 \times 10^{-3}$ | $6.09 \times 10^{-3}$ | $5.09 \times 10^{-1}$ | $1.42 \times 10^{-2}$ | $3.20 \times 10^{-2}$ | $6.27 \times 10^{-2}$ |
| | Third layer | 0.33 | 3.85 | 20.61 | $1.98 \times 10^{-3}$ | $1.17 \times 10^{-2}$ | $6.25 \times 10^{-3}$ | $3.28 \times 10^{-1}$ | $9.18 \times 10^{-3}$ | $5.40 \times 10^{-2}$ | $6.42 \times 10^{-2}$ |
| | Fourth layer | 0.70 | 2.10 | 11.75 | $4.21 \times 10^{-3}$ | $6.38 \times 10^{-3}$ | $3.56 \times 10^{-3}$ | $6.96 \times 10^{-1}$ | $1.95 \times 10^{-2}$ | $2.95 \times 10^{-2}$ | $3.66 \times 10^{-2}$ |
| | Fifth layer | 0.54 | 3.18 | 37.09 | $3.30 \times 10^{-3}$ | $9.63 \times 10^{-3}$ | $1.12 \times 10^{-2}$ | $5.45 \times 10^{-1}$ | $1.53 \times 10^{-2}$ | $4.46 \times 10^{-2}$ | $1.16 \times 10^{-1}$ |
| | Sixth layer | 0.32 | 4.44 | 23.65 | $1.92 \times 10^{-3}$ | $1.35 \times 10^{-2}$ | $7.17 \times 10^{-3}$ | $3.17 \times 10^{-1}$ | $8.87 \times 10^{-3}$ | $6.23 \times 10^{-2}$ | $7.37 \times 10^{-2}$ |
| | Seventh layer | 0.27 | 2.70 | 18.52 | $1.61 \times 10^{-3}$ | $8.17 \times 10^{-3}$ | $5.61 \times 10^{-3}$ | $2.66 \times 10^{-1}$ | $7.46 \times 10^{-3}$ | $3.78 \times 10^{-2}$ | $5.77 \times 10^{-2}$ |
| | Eighth layer | 0.41 | 4.85 | 31.26 | $2.48 \times 10^{-3}$ | $1.47 \times 10^{-2}$ | $9.47 \times 10^{-3}$ | $4.11 \times 10^{-1}$ | $1.15 \times 10^{-2}$ | $6.80 \times 10^{-2}$ | $9.74 \times 10^{-2}$ |
| | Ninth layer | 0.45 | 2.50 | 18.55 | $2.75 \times 10^{-3}$ | $7.58 \times 10^{-3}$ | $5.62 \times 10^{-3}$ | $4.54 \times 10^{-1}$ | $1.27 \times 10^{-2}$ | $3.51 \times 10^{-2}$ | $5.78 \times 10^{-2}$ |
| | Tenth layer | 0.49 | 2.04 | 26.27 | $2.94 \times 10^{-3}$ | $6.19 \times 10^{-3}$ | $7.96 \times 10^{-3}$ | $4.86 \times 10^{-1}$ | $1.36 \times 10^{-2}$ | $2.87 \times 10^{-2}$ | $8.19 \times 10^{-2}$ |
| | Eleventh layer | 0.35 | 3.19 | 28.45 | $2.15 \times 10^{-3}$ | $9.65 \times 10^{-3}$ | $8.62 \times 10^{-3}$ | $3.55 \times 10^{-1}$ | $9.93 \times 10^{-3}$ | $4.47 \times 10^{-2}$ | $8.87 \times 10^{-2}$ |

### 3.2. Health Risk Assessment

#### 3.2.1. Carcinogenic Risk

Regarding the harmful trace elements, carcinogenic risk of As was the most significant, whereas no obvious carcinogenic effect was observed for other elements. The variation of carcinogenic risk of As in each soil profile layer is illustrated in Figure 3. In the first few soil section layers, the carcinogenic risk level of As was lower, but still exceeded the recommended safety value ($1 \times 10^{-6}$). Overall, it did not show an obvious change with increasing depth. A high carcinogenic risk value was observed at a depth of 1–7 m. The highest carcinogenic risk value observed was $8.77 \times 10^{-5}$, which is 87.70-times the recommended safety value. Therefore, it could be concluded that the carcinogenic risk level of As is high, which suggests that it is not safe for workers or other people to stay in this area for a long period. Therefore, it is necessary to adopt effective safety measures for the staff working in this open-cast coal mining area.

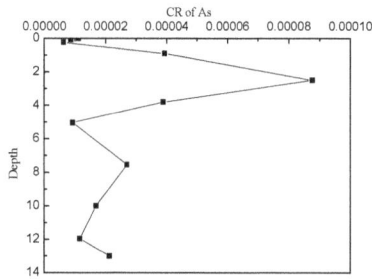

**Figure 3.** The carcinogenic risk level of As in each section layer.

#### 3.2.2. Hazard Quotient

The variation in the hazard quotient value of As, Cd, Be and Ni in each soil profile layer is illustrated in Figure 4. Among these, the hazard quotient of As was most prominent. Samples from three sampling points exceeded the recommended safety value under the exposure pathways of the inhalation of particles and the ingestion of soil particles. The highest value was 2.16-times the acceptable risk level, and the total hazard quotient of each exposure pathways was up to 4.70-times the acceptable risk level. Generally, the hazard quotients of other elements in each soil section layer were much lower; even the maximum value did not exceed the recommended safety value. Therefore, this study did not investigate the hazard quotients levels of those elements that were acceptable. However, because exploration of coal has been carried out for a long time in this area, the possibility of an increase in the hazard quotient with coal mine excavation should be studied.

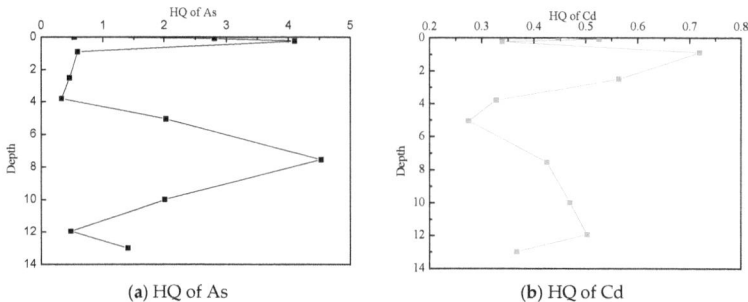

(a) HQ of As    (b) HQ of Cd

**Figure 4.** *Cont.*

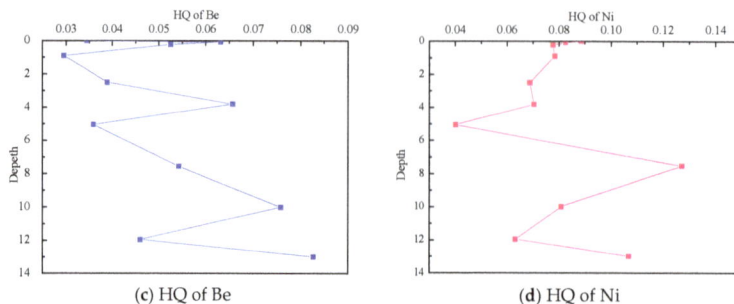

(c) HQ of Be            (d) HQ of Ni

**Figure 4.** The hazard quotient (HQ) value of As, Cd, Be and Ni in each section layer and the changing trend.

### 3.2.3. Contribution of Different Exposure Pathways

In order to devise strategies for the mitigation and prevention of human health risk in coal mines, the contribution of different exposure pathways to human risk was calculated in this paper (Figure 5).

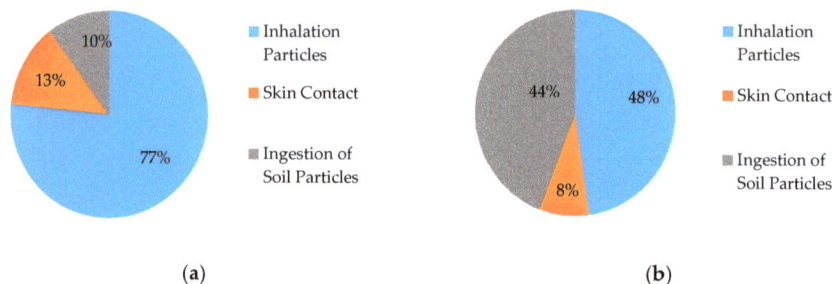

(a)                   (b)

**Figure 5.** The contribution of different exposure pathways to human health risk. (**a**) Carcinogenic risk; (**b**) Non-carcinogenic risk.

The carcinogenic risk of As by the inhalation of particles exposure pathway could reach 76.64% in this open-cast coal mining area (Figure 5). Inhalation of particles was also the most important exposure pathway for non-carcinogenic risk; the contribution of the ingestion of soil particles increased to 44.22%, and it can be concluded that different exposure pathways can have different contribution ratios when the damage type (carcinogenic risk or hazard quotient) is different. Additionally, in order to control and decrease the human health risk in the open-cast coal mining area, risk control should be aimed at blocking the main exposure pathway, specifically to prevent the inhalation of particles by workers, by advising them to wear safety masks.

### 3.3. Safety Threshold Identification

According to the human health risk assessment of the open-cast coal mine area, only the carcinogenic risk of As in each sampling point exceeded the acceptable standard level, so in this research, the risk control values of As under the corresponding routes of exposure were calculated, according to the method provided in Table 4; the calculation results are shown in Table 8. There are still three sampling points that exceed the recommended safety value under the exposure pathways of the inhalation of particles and the ingestion of soil particles, respectively. The risk control values of the two exposure pathways were also calculated (Table 2).

The risk control values of As in these open-cast coal mine soils varied among different exposure pathways. The lowest risk control value of arsenic is 1.59 mg/kg.

**Table 8.** Risk control value of arsenic in open-cast coal mine area soil.

| Exposure Route | Type of Risk | Control Value (mg/kg) |
|---|---|---|
| Inhalation of particles | carcinogenic | 1.59 |
| Skin contact | carcinogenic | 9.26 |
| Ingestion of soil particles | carcinogenic | 11.99 |
| Inhalation of particles | non-carcinogenic | 49.50 |
| Ingestion of soil particles | non-carcinogenic | 55.55 |

However, it should be noted that this open-cast coal mine is located in the northwest of China, which is windy and dry in most seasons; this leads to an abundance of dust and light soil particles. As a result, frequent inhalation of soil particles is unavoidable. Therefore, considering the principle of strict management for risk control and taking into consideration the natural weather conditions, the concentration value of 1.59 mg/kg As could be selected as the reference safety threshold for As in this area, in order to protect the health of personnel working in this coal mine and to ensure sustainable development of this regional environment. However, the acceptable levels of carcinogenic risk vary (the United States usually uses $10^{-6}$, whereas $10^{-5}$ is usually used in the UK, and The Netherlands recommends a more relaxed $10^{-4}$ [36]), suggesting that the value of 1.59 mg/kg, calculated under a $10^{-6}$ acceptable carcinogenic risk level, as the safety threshold for As in the soil environment in a coal chemical industry area needs further discussion. Then, the final feasible threshold of As in the soil environment should be determined holistically by considering the background value, geological conditions, biological parameters, regional climatic characteristics and regional development planning.

## 4. Conclusions

Among the harmful trace elements in the Wulantuga open-cast coal mine area, the carcinogenic risk of As is most significant. High carcinogenic risk was found at a depth of 1–7 m. The highest carcinogenic risk value achieves $8.77 \times 10^{-5}$, which is 87.70-times the recommended safety value. It is necessary to adopt effective safety protection measures for personnel working in this coal mine area.

In the soil environment of the Wulantuga open-cast coal mining area, the main route of exposure of As is the inhalation of particles, which contributes to 68.64% of the carcinogenic risk. Therefore, in order to mitigate and prevent human health risk from the coal mine, blocking the inhalation particle exposure route appears to be the best method.

Considering the different control values in each exposure pathway, the minimum control value (1.59 mg/kg) in the pathway of the ingestion of soil particles can be selected as the strict reference safety threshold for As in the soil environment in the coal chemical industry area, which would provide a basis for the protection of the operators working in the area. However, the acceptable levels of carcinogenic risk vary, suggesting that the value of 1.59 mg/kg, calculated under a $10^{-6}$ acceptable carcinogenic risk level, as the safety threshold for As in soil environment in the coal chemical industry area needs further discussion.

**Acknowledgments: Acknowledgments:** The authors would like to thank the National Basic Research Program of China (973 Program, No. 2014CB238906) for financial support. We thank Shifeng Dai of China University of Mining and Technology (Beijing) for assisting in the ICP-MS analysis of heavy metals in soil samples.

**Author Contributions: Author Contributions:** Jianli Jia determined the evaluation method; Ying Liu and Xiaojun Li performed the experiments; Chunyu Han and Lina Zhou analyzed the data; Peijing Wu contributed materials/analysis tools; Liu Yang provided the soil samples; Jianli Jia and Xiaojun Li wrote the paper.

**Conflicts of Interest:** The authors declare no conflict of interest.

## References

1. Longwell, J.P.; Rubin, E.S.; Wilson, J. Coal: Energy for the future. *Prog. Energy Combust. Sci.* **1995**, *21*, 269–360. [CrossRef]
2. Qi, H.W.; Hu, R.Z.; Zhang, Q. Concentration and distribution of trace elements in lignite from the Shengli Coalfield, Inner Mongolia, China: Implications on origin of the associated Wulantuga Germanium Deposit. *Int. J. Coal Geol.* **2007**, *71*, 129–152. [CrossRef]
3. Dai, S.F.; Seredin, V.V.; Ward, C.R.; Jiang, J.H.; Hower, J.C.; Song, X.L.; Jiang, Y.F.; Wang, X.B.; Gornostaeva, T.; Li, X.; *et al.* Composition and modes of occurrence of minerals and elements in coal combustion products derived from high-Ge coals. *Int. J. Coal Geol.* **2013**, *121*, 79–97. [CrossRef]
4. Chen, J.; Chen, P.; Liu, W.Z. The occurrences and environmental effects of 12 kinds of trace elements in Huainan coal-mining area. *Coal Geol. Explor.* **2009**, *37*, 47–52. (In Chinese)
5. Huggins, F.E.; Huffman, G.P. Modes of occurrence of trace elements in coal from XAFS spectroscopy. *Int. J. Coal Geol.* **1996**, *32*, 31–53. [CrossRef]
6. Finkelman, R.B. Modes of occurrence of environmentally-sensitive trace elements in coal. *Environ. Asp. Trace Elem. Coal Springer Neth.* **1995**, *2*, 24–50.
7. Ren, D.Y.; Zhao, F.H.; Wang, Y.Q.; Yang, S.J. Distributions of minor and trace elements in Chinese coals. *Int. J. Coal Geol.* **1999**, *40*, 109–118. [CrossRef]
8. Ren, D.Y.; Xu, D.W.; Zhao, F.H. A preliminary study on the enrichment mechanism and occurrence of hazardous trace elements in the Tertiary lignite from the Shenbei coalfield, China. *Int. J. Coal Geol.* **2004**, *57*, 187–196. [CrossRef]
9. Dai, S.F.; Li, T.; Seredin, V.V.; Ward, C.R.; Hower, J.C.; Zhou, Y.P.; Zhang, M.Q.; Song, X.L.; Song, W.J.; Zhao, C.L. Origin of minerals and elements in the Late Permian coals, tonsteins, and host rocks of the Xinde mine, Xuanwei, eastern Yunnan, China. *Int. J. Coal Geol.* **2014**, *121*, 53–78. [CrossRef]
10. Wang, J.N.; Liu, Z.X.; Rong, L.I.; Qiang, L.I.; Deng, Y.F. Analysis of element content and discussion on migration of coal and flyash in coal-fired power plant. *J. Henan Univ. Urban Constr.* **2014**, *23*, 27–31.
11. Matés, J.M.; Segura, J.A.; Alonso, F.J.; Márquez, J. Roles of dioxins and heavy metals in cancer and neurological diseases using ROS-mediated mechanisms. *Free Radic. Biol. Med.* **2010**, *49*, 1328–1341. [CrossRef] [PubMed]
12. Soil Screening Guidance: Technical Background Document. Available online: http://www2.epa.gov/ superfund/soil-screening-guidance-technical-background-document (accessed on 12 November 2015).
13. Environmental Quality Standard for Soils (Amendment). Available online: http://kjs.mep.gov.cn/hjbhbz/ bzwb/trhj/trhjzlbz/199603/W020070313485587994018.pdf (accessed on 12 November 2015).
14. Johnson, B.B.; Slovic, P. Presenting uncertainty in health risk assessment: Initial studies of its effects on risk perception and trust. *Risk Anal. Off. Publ. Soc. Risk Anal.* **1995**, *15*, 485–494. [CrossRef]
15. Fei, C. *Health Risk assessment of Heavy Metals in Multimedia Environment in Shen-Fu Irrigation Area in Liaoning Province*; Chinese Research Academy of Environmental Sciences: Beijing, China, 2009.
16. Chen, M.F.; Luo, Y.M.; Song, J.; Li, C.P.; Wu, C.F.; Luo, F.; Wei, J. Theory and commonly used models for the derivation of soil generic assessment criteria for contaminated sites. *Adm. Techn. Environ. Monit.* **2011**, *23*, 19–25.
17. Juhasz, A.L.; Smith, E.; Weber, J.; Rees, M.; Rofe, A.; Kuchel, T.; Sansom, L.; Naidu, R. *In vivo* assessment of arsenic bioavailability in rice and its significance for human health risk assessment. *Environ. Health Perspect.* **2007**, *114*, 1826–1831. [CrossRef]
18. Zhuang, P.; McBride, M.B.; Xia, H.P.; Li, N.Y.; Li, Z.A. Health risk from heavy metals via consumption of food crops in the Vicinity of Dabaoshan mine, South China. *J. Sci. Total Environ.* **2009**, *407*, 1551–1561. [CrossRef] [PubMed]
19. Ren, M.F.; Wang, J.D.; Wang, G.P.; Zhang, X.L.; Wang, C.M. Influence of soil lead upon children blood lead in Shenyang city. *J. Environ. Sci.* **2005**, *26*, 153–158.
20. Li, Z.W.; Zhang, Y.L.; Pan, G.X.; Li, J.H.; Huang, X.M.; Wang, J.F. Grain Contents of Cd, Cu and Se by 57 rice cultivars and the risk significance for human dietary uptake. *J. Chin. J. Environ. Sci.* **2003**, *21*, 112–115.
21. Swartjes, F.A.; Cornelis, C. Human health risk assessment. *Deal. Contam. Sites* **2011**, *283*, 107–172.
22. Walden, T. Risk assessment in soil pollution: Comparison study. *Rev. Environ. Sci. Biotechnol.* **2005**, *4*, 87–113. [CrossRef]

23. McNeel, P.J., IV; Atwood, C.J.; Dibley, V. *Case Study Comparisons of Vapor Inhalation Risk Estimates: ASTM RBCA Model Predictions vs Site Specific Soil Vapor Data*; Ground Water Publishing Co.: Westerville, OH, USA, 1997.

24. Yang, M.M.; Sun, L.N.; Luo, Q.; Bing, L.F. Health risk assessment of polycyclic aromatic hydrocarbons in soils of the tiexi relocated old industrial area, shenyang, china. *Chin. J. Ecol.* **2013**, *32*, 675–681.

25. Ministry of Environmental Protection of the People's Republic of China. *Technical Guidelines for Risk Assessment of Contaminated Sites (HJ25.3-2014)*; National Environmental Protection Standards of the People's Republic of China: Beijing, China, 2014.

26. Zhuang, X.; Querol, X.; Alastuey, A.; Juan, R.; Plana, F.; Lopez-Soler, A. Geochemistry and mineralogy of the Cretaceous Wulantuga high-germanium coal deposit in Shengli coal field, Inner Mongolia, Northeastern China. *Int. J. Coal Geol.* **2006**, *66*, 119–136. [CrossRef]

27. Dai, S.F.; Wang, X.B.; Seredin, V.V.; Hower, J.C.; Ward, C.R.; O'Keefe, J.M.K.; Huang, W.H.; Li, T.; Li, X.; Liu, H.D.; *et al.* Petrology, mineralogy, and geochemistry of the Ge-rich coal from the Wulantuga Ge ore deposit, Inner Mongolia, China: New data and genetic implications. *Int. J. Coal Geol.* **2011**, *90*, 72–99. [CrossRef]

28. Dai, S.F.; Liu, J.J.; Colin, R.W.; Hower, J.C.; Xie, P.P.; Jiang, Y.F.; Hood, M.M.; O'Keefe, J.M.K.; Song, H.J. Petrological, geochemical, and mineralogical compositions of the low-Ge coals from the Shengli coalfield, China: A comparative study with Ge-rich coals and a formation model for coal-hosted Ge ore deposit. *Ore Geol. Rev.* **2015**, *71*, 318–349. [CrossRef]

29. Qi, H.W.; Hu, R.Z.; Zhang, Q. REE Geochemistry of the Cretaceous lignite from Wulantuga Germanium Deposit, Inner Mongolia, Northeastern China. *Int. J. Coal Geol.* **2007**, *71*, 329–344. [CrossRef]

30. Jia, J.L.; Li, G.H.; Zhong, Y. The Relationship between abiotic factors and microbial activities of microbial eco-system in contaminated soil with petroleum hydrocarbons. *Environ. Sci.* **2004**, *25*, 110–114.

31. Li, X.; Dai, S.F.; Zhang, W.G.; Li, T.; Zheng, X.; Chen, W.M. Determination of As and Se in coal and coal combustion products using closed vessel microwave digestion and collision/reaction cell technology (CCT) of inductively coupled plasma mass spectrometry (ICP-MS). *Int. J. Coal Geol.* **2014**, *124*, 1–4. [CrossRef]

32. Khan, N.I.; Owens, G.; Bruce, D.; Naidu, R. Human arsenic exposure and risk assessment at the landscape level: A review. *Environ. Geochem. Health* **2009**, *31*, 143–166. [CrossRef] [PubMed]

33. Dong, J.Y.; Wang, J.Y.; Zhang, G.X.; Wang, S.G.; Shang, K.Z. Population exposure to PAHs and the health risk assessment in Lanzhou city. *Ecol. Environ. Sci.* **2012**, *21*, 327–332.

34. Yang, J.K.; Wang, Q.S.; Lu, J.F.; Wang, L.K.; Bei, X.Y.; Zhu, W.S. Human health risk assessment on trihalomethanes with multiple exposure pathways in drinking water. *China Resour. Compr. Util.* **2009**, *27*, 27–30.

35. Yost, L.J.; Shock, S.S.; Holm, S.E.; Lowney, Y.W.; Noggle, J.J. Lack of complete exposure pathways for metals in natural and fgd gypsum. *Hum. Ecol. Risk Assess.* **2010**, *16*, 317–339. [CrossRef]

36. Hua, Y.P.; Luo, Z.J.; Cheng, S.G.; Xiang, R. Analysis of Soil Remediation Limits in Site Based on Health Risk. *Ind. Saf. Environ. Prot.* **2012**, *38*, 68–71.

*Article*

# The Transformation of Coal-Mining Waste Minerals in the Pozzolanic Reactions of Cements

Rosario García-Giménez [1,*], Raquel Vigil de la Villa Mencía [1], Virginia Rubio [2] and Moisés Frías [3]

[1]   Dpto de Geología y Geoquímica, Unidad Asociada CSIC-UAM, Universidad Autónoma de Madrid, 28049 Madrid, Spain; raquel.vigil@uam.es

[2]   Dpto de Geografía, Unidad Asociada CSIC-UAM, Universidad Autónoma de Madrid, 28049 Madrid, Spain; virginia.rubio@uam.es

[3]   Eduardo Torroja Institute (CSIC), Spanish National Research Council, 28033 Madrid, Spain; mfrias@ietcc.csic.es

*   Correspondence: rosario.garcia@uam.es; Tel.: +34-914-974-819

Academic Editors: Shifeng Dai, Xibo Wang and Lei Zhao
Received: 22 January 2016; Accepted: 22 June 2016; Published: 30 June 2016

**Abstract:** The cement industry has the potential to become a major consumer of recycled waste materials that are transformed and recycled in various forms as aggregates and pozzolanic materials. These recycled waste materials would otherwise have been dumped in landfill sites, leaving hazardous elements to break down and contaminate the environment. There are several approaches for the reuse of these waste products, especially in relation to clay minerals that can induce pozzolanic reactions of special interest in the cement industry. In the present paper, scientific aspects are discussed in relation to several inert coal-mining wastes and their recycling as alternative sources of future eco-efficient pozzolans, based on activated phyllosilicates. The presence of kaolinite in this waste indicates that thermal treatment at 600 °C for 2 h transformed these minerals into a highly reactive metakaolinite over the first seven days of the pozzolanic reaction. Moreover, high contents of metakaolinite, together with silica and alumina sheet structures, assisted the appearance of layered double hydroxides through metastable phases, forming stratlingite throughout the main phase of the pozzolanic reaction after 28 days (as recommended by the European Standard) as the reaction proceeded.

**Keywords:** metakaolinite; micas; coal-mining waste; LDH (layered double hydroxides); pozzolanic reaction

## 1. Introduction

A key vector of opportunity in the construction sector, in general, and for manufacturers of cement-based materials, in particular, is the efficient use of materials and energy resources, which moderates the carbon footprint of the final products. Our production of cements, mortars, and concretes, the most widely manufactured materials on the planet, involves energy-intensive exploitation of raw materials that remains a source of extremely high $CO_2$ emissions.

The research, validation, and enhancement of new mineral additives for cement should ensure their availability in sufficient quantities for profitable investments, thereby reducing the environmental impacts of Portland-cement-clinker production, the overall volume of waste products, and the energy consumption of the final product. These additives also assist reactivity that densifies the hydration products.

The coal industry, more than any other sector, has one of the most negative effects on the environment. Coal waste (in abundant amounts from various extraction processes) is disposed of in landfills or is incinerated [1]. The use of coal waste to prevent environmental impacts has been explored in China [2] and elsewhere [3], especially in relation to building [4].

Coal-mining waste from the extraction and washing of debris from mines contains kaolinite, illite, and quartz of varying composition according to geological conditions and methods of extraction and purification. Above all, the composition of claystones generally consists of illite and kaolinite. Ferrous minerals, quartz, and carbonic matter may also be found in small amounts. All carboniferous rocks may also contain dolomitic veins, pyritic encrustations, and extensive quantities of plant detritus [4].

Pozzolans found in industrial waste have been linked to environmental and technological advantages that have driven research into their use. The incorporation of these industrial by-products and wastes together with natural materials in various production stages of blended cements was first prioritized in the cement industry. Experimentation with rice husk, fly ash, and palm oil has all been reported in the literature [5–8]. Materials with "pozzolanic" properties are linked to siliceous/aluminous materials. When added to water as fines, they can form cementitious properties in reaction with $Ca(OH)_2$ [9]. Another class of pozzolans is from natural raw sources (volcanic material, limestone) or calcined materials (burnt shale, calcined kaolinite) with pozzolanic properties [10,11]. Blended with lime, their use in construction projects has been documented throughout history; natural pozzolans from magma deposits following volcanic activity have been added to mortars since classical antiquity. Natural pozzolans continue to be applied in many cement manufacturing processes to this day.

Calcined clays with pozzolanic activity draw structural water from the layers of crystalline clay, leaving amorphous or semi-amorphous materials with high surface areas and chemical reactivity.

Thermal activation of kaolinite at controlled temperatures produces metakaolinite, a product with highly pozzolanic properties. The investigation of thermal activation, within the ranges of between 650 and 750 °C with kaolinite-containing waste, resulted in products with a high latent pozzolanicity [12–18].

The works of Li et al. [19] and Beltramini et al. [20] pioneered the study of cement matrices containing activated coal-mining waste, providing useful results for future research. However, performance criteria at the percentages of coal-mining waste that they added and its activation conditions still require further research. Currently, studies in this investigative field are multiplying [21–23].

In this paper, the mineralogical transformations of coal waste are studied across a range of temperatures (500–900 °C) for the establishment of optimum calcination conditions that yield products with sufficient pozzolanic properties to be used as additives in the manufacture of cements and related materials.

## 2. Materials and Methods

### 2.1. Materials

Mining generates high volumes of waste that are currently dumped in slag heaps of no apparent utility. Our study concerns coal-mining waste supplied by a coal-mining company (Sociedad Anónima Hullera Vasco-Leonesa) in Santa Lucía (Province of León, Spain). The mine supplies coal to the power plant in the region of La Robla, and the mine waste employed in this study is used in a cement factory in the same region.

The geological materials from La Robla vary greatly. They are separated by easily identifiable irregularities and contain a variety of rocks: metamorphic shales and sedimentary limestone and dolomite with *Facies Utrillas* (detrital deposit formed in a sedimentary environment of river systems). Igneous activity is very low and only found in small porphyry dikes [24].

The waste products from that area are reasonably uniform and contain white quartzite with highly recrystallized micro-conglomerates, feldspar mixed with sandstone, foliated slate (containing phyllite with mica and quartz, sericite, chlorite, zircon, pyrite, monazite, apatite, and tourmaline). The structure of dolomite changes in contact with St Lucia limestones and develops reddish hues.

There are grey limestones, found in marine environments, interbedded in massive reefs, with fauna in the form of bryozoans, crinoids, and brachiopods.

The mining waste material under study (coal, ore, and gangue with charcoal remains) was heat treated at different temperatures for activation, converting the kaolinite into metakaoline, among others chemical reactions.

Following its thermal activation at temperatures of 500, 600, 700, 800 and 900 °C for 2 h in an electric laboratory furnace, the best activation conditions were selected for use as a pozzolan in the manufacture of cement. The activated samples were placed in an agate mortar and pestle and crushed to particle sizes of less than 63 μm.

### 2.2. Methods

Chemical characterization was performed with a Philips PW 1404 X-ray Fluorescence (XRF) Spectrometer (Philips, Eindhoven, The Netherlands), Loss on ignition (LOI) was calculated in accordance with the method specified in the European standard (at 950 °C/1 h). Bulk sample mineralogical compositions were analyzed by random powder X-ray diffraction (XRD), using the oriented film method for the <2 μm fraction, employing a Cu anode and a Siemens D-5000, X-ray diffractometer (Siemens, Madrid, Spain) in both instances. The operating conditions were, respectively, 30 mA and 40 kV, with a divergence slit of 2 mm and a receiving slit of 0.6 mm. Each sample was scanned (2θ) at steps of 0.041 and a count time of 3-s. Bulk sample characterization and semi-quantification used the random-powder method, between 3° to 65° 2θ at a rate of 2°/min. Determination of phyllosilicates in the <2 μm fraction employed the oriented-slides method, operating from 2° to 40° at a scan rate of 1°/min. The Rietveld method was used for quantitative determination of the mineralogical composition.

A Thermo Scientific NICOLET 6700 spectrometer fitted with a DGTS CsI detector (Thermo Fisher, Waltham, MA, USA) performed the FTIR (Fourier Transform Infrared Spectroscopy) analyses, recording 64 scans on the samples. Specimen preparation was done by mixing 1 mg of the sample in 300 mg of KBr. Spectral analyses within the range of 4000–400 cm$^{-1}$ was performed at a spectral resolution of 4 cm$^{-1}$.

A Renishaw Raman RM2000 Microscope System (Renishaw, Wotton-under-Edge, UK) fitted with a Leica microscope (Leica, Wetzlar, Germany), an electrically refrigerated charge-coupled device camera (CCD), a 785 nm diode laser, and a 633 nm He–Ne Renishaw RL633 laser (Renishaw) were used to record dispersive Raman spectra at 633 nm. Frequency calibration used a 520 cm$^{-1}$ silicon line with a spectral resolution at 4 cm$^{-1}$. A 50× lens was used for triplicate spectra recordings over wave ranges of 4000–100 cm$^{-1}$ at 10 s exposure times with 10 accumulations for each spectrum.

An Inspect FEI Company Electron Microscopy (Hillsboro, OR, USA) fitted with an energy dispersive X-ray analyzer (W source, DX4i analyzer and Si/Li detector) performed SEM/EDX (Scanning Electron Microscopy/Energy Dispersive X-Ray Spectroscopy) a morphological observation and microanalysis for each sample. The average chemical composition of each sample was based on ten analyses.

An accelerated chemical method was applied to solid waste in order to study the pozzolanic behavior of a pozzolan/calcium hydroxide (lime) system. The test involves leaving the material (1 g) in a lime-saturated solution (75 mL) and analyzing the CaO concentrations at 1, 7, and 28 days into the reaction time. The difference between its concentration in the lime-saturated control solution (17.69 mmol/L) and the Cao content of the solution in contact with the sample gave the combined CaO (mmol/L) content. The hydrated solid sample was then filtered, washed in ethanol, and heated at 60 °C for 24 h until the hydration reaction ended [25].

## 3. Discussion and Results

### 3.1. Raw-Carbon Wastes

The XRF results in relation to the carbon waste (initial sample without treatment) showed that the main oxides were $SiO_2$ (57%), $Al_2O_3$ (25%), $Fe_2O_3$ (5%), CaO (4%), and $K_2O$ (3%), with some $SO_3$ (0.29%) as well. Loss on ignition (LOI) was 15% for initial waste and 3% for material that had been calcined at 500 °C; this change is related to the dehydroxylation processes of kaolinite and organic matter (carbon). Minor elements or traces were also present, such as chromium (120 ppm), vanadium (139 ppm), nickel (53 ppm), and cobalt (21 ppm).

XRD analysis showed the mineralogical composition of the carbon waste (room temperature), revealing the existence of mica (25%) and kaolinite (14%), quartz (37%), calcite (17%), dolomite (5%), and feldspars (2%) (Figures 1 and 2). Both mica and kaolinite contribute to pozzolanic activity, as it is well established that atmospheric thermical activation of a variety of clay minerals at 600 °C/900 °C following dehydroxylation causes the (partial) destruction of the crystal lattice structure, leading to a transitional and highly reactive phase. Mica requires temperatures over 930 °C for its activation and, as a result, it usually results in a weak pozzolan; tending to dissipate at temperatures over 900 °C.

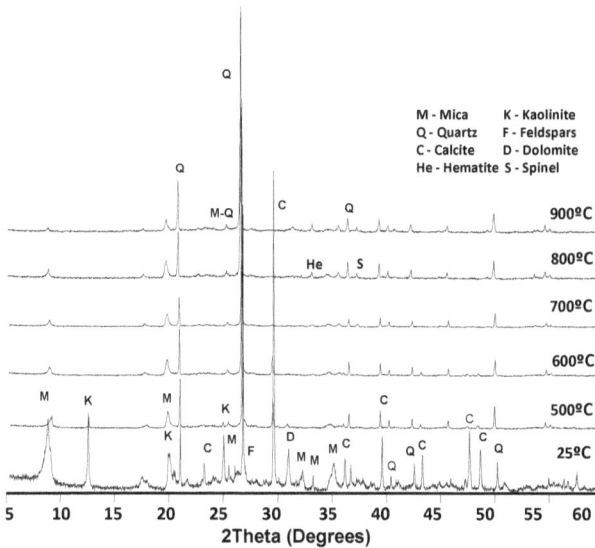

**Figure 1.** Mineralogical composition by X-ray diffraction (XRD) for the coal mining waste and the coal mining activated waste.

The XRD analysis also showed the formation of anorthite, the presence of which might be due to metakaolinite (from clay dehydration) reacting with calcium carbonate. The formation of mullite that the clay would otherwise generate during firing can be reduced by metakaolinite in reaction with calcium carbonate [26].

In comparison to the samples of coal-mining waste used in this study, in general terms, coal contains aluminum-bearing minerals that include kaolinite, illite, montmorillonite, chlorite, sanidine, albite, plagioclase, biotite, hornblende, and muscovite [27–30]. South African coal samples, for instance, contain quartz, kaolinite, illite I/S, calcite, dolomite, siderite, pyrite, analcite, basanite, bohemite, anatase, diaspore, and jarosite [31].

Even though it was not identified in the unheated sample in this study, the prevalent clay mineral in most South African coals is kaolinite [32–34] and would therefore be found in the raw coal. However, anhydrite and hematite, identified in the ash produced at 350 °C, have not been observed in South African coals. Their presence in the heated samples used in the present study might suggest an alternative origin.

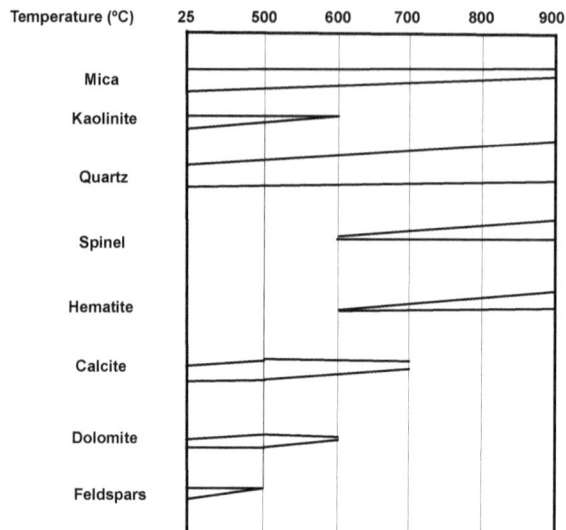

**Figure 2.** Mineralogical evolution by XRD for the coal mining waste and the coal mining activated waste.

In contrast, the main compounds reported in Australian coals are quartz, kaolinite, illite/smectite, calcite, dolomite, albite, siderite, pyrite, albite, apatite, gypsum, hematite, and anatase [35].

Calcite is the main carbonate mineral in Taoshuping coals with an average content of 9%, and lower amounts of siderite, anchorite, and dolomite [36].

*3.2. Activated Coal-Mining Waste*

After analyzing the samples of coal waste, they were then thermically activated by calcination at between 500 and 900°C. In Figures 1 and 2, the XRD patterns of the activated products at different temperatures are shown. The spectra reflects low levels of quartz and kaolinite dissipated at 600 °C/2 h of thermical activation, due to their transformation into metakaolinite. Detrital quartz can also be cell-, cleat-, and fracture-fillings that are epigenetic in nature or syngenetic, such as aluminosilicates in coal [33,34,37].

The temperature at which kaolinite is calcinated to produce metakaolinite in the active state lies within the range of 600–800 °C. The activation of mica requires temperatures over 930 °C, although it usually produces poor pozzolans. Temperature tests (at 500, 600, 700, 800 and 900 °C) determined both phyllosilicate dehydroxylation and new phase-formation temperatures. Further studies were conducted after this test at 600 and at 900 °C, showing that 600 °C was the total dehydroxylation temperature of kaolinite and its neoformation into spinel-like phases took place at 900 °C.

Quasi-stable dehydroxylated mica was formed following dehydroxylation after thermical treatment between 700 and 900 °C. The quasi-stable dehydroxylated phases of these dioctahedral micas were unlike the trioctahedral mica phases, which tended to dehydroxylate and recrystallize more or less simultaneously. Dehydroxylation of the 2:1 mica layers needed higher temperatures. At 600 °C/2 h, a new characteristic peak appeared, explained by the appearance of hematite. As the heating increased from 700 to 900 °C, both hydrous oxides and ferrous hydroxides in the coal-mining waste crystallized in the form of hematite as they lost water. The presence of hematite in all likelihood reflects pyritic oxidation of the coal. Calcite, habitually found in the four coal cleaning residues, is easily differentiated from other calcium carbonates, due to its Raman spectrum, which showed bands of 1085, 711 and 280 cm$^{-1}$ [30] in similar studies of a range of American coal samples. As observed in Brazilian materials, the Fe-oxide mineral hematite is commonly found in most coals and coal-cleaning residues.

XRD analysis between 600 and 900 °C showed reflections at 2.43 Å (36.98° 2θ) and 2.85 Å (31.38° 2θ), revealing spinel-like phases, resulting from the heating of aluminous clay minerals. Hematite was present as a pyritic product in minor concentrations in the original coal waste and its presence increased at higher temperatures. At temperatures of 700, 800 and 900 °C, well crystallized hematite appeared following loss of water in the hydrous oxides and the ferrous hydroxides of the original coal waste. Finally, dolomite dissipated at 600 °C and subsequently, calcite at 700 °C.

As oxidation continued, ferric sulfate was formed, eventually producing sodium, calcium, magnesium, and potassium sulfate.

The principal mineralogical compounds identified from the XRD data were subjected to FTIR analysis. Kaolinite was identified in the (O–H) bands at 3696, 3656 and 3620 cm$^{-1}$, mica at 3628 and 3545 cm$^{-1}$ (both overlapping with kaolinite and water), and at 3423 cm$^{-1}$. Standard carbonate group bands were detected at 1426 and 874 cm$^{-1}$. With regard to silicates, Si–O vibrations revealed the presence of kaolinite at 1032 and 1007 cm$^{-1}$ and quartz at 1090 cm$^{-1}$ at tetrahedral sites. At lower frequencies (<1000 cm$^{-1}$), the Al–O–H vibration, mainly associated with kaolinite (also at 751 cm$^{-1}$) and mica, was detected at 912 cm$^{-1}$. Vibrations of the Si–O quartz bond were identified at bands of 798, 778 (doublet), 694, and 472 cm$^{-1}$. OH absorption of mica is suggested by the presence of absorption bands in the OH-stretching region at 3432 cm$^{-1}$. The original coal residues revealed a further absorption band at 3656 cm$^{-1}$, also explained by OH group stretching frequencies, which disappeared in the thermally activated waste. This observation points to the appearance of impure, low crystalline mica, and further isomorphous substitution in the crystalline structure.

Characterization of all tectosilicates, feldspars, and quartz was within the absorption band ranges of 950 to 1200 cm$^{-1}$, reflecting stretching vibrations of Si–O–Si, and between 400 and 550 cm$^{-1}$, reflecting O–Si–O bending vibrations. Two hematite bands were observed at 535 and 469 cm$^{-1}$, both overlapping the main band of mica. IR spectra (465 and 614 cm$^{-1}$) of the thermally activated waste revealed $AB_2O_4$ spinel.

A Raman spectrum in the thermally activated coal waste revealed two bands at 1597 and 1346 cm$^{-1}$, corresponding to graphite and disordered peak bands, respectively, from carbon. The Raman spectrum of calcite, which is commonly found in all four coal-waste composites, may easily be differentiated from other calcium carbonates by its bands at 1085, 711 and 280 cm$^{-1}$.

### 3.3. Pozzolanic Reactivity of Activated Coal-Mining Wastes

The removal of structural water from the crystalline clay layers of calcined clay waste is a consequence of its pozzolanic reactivity, leaving semi-amorphous products with high surface areas and chemical reactivity. The calcining temperature that is required is dependent upon the nature of the clay mineral and the thermal energy required for dehydroxylation of the clays. The calcining temperature required to produce this active state is usually in the range 600–1000 °C [9,12,38]. Crystallization occurs above this temperature and activity declines. Even higher firing temperatures lead to the formation of a liquid phase that cools into a solid amorphous glass phase, also showing pozzolanic activity. Reactive metakaolinite is formed at 600 °C and hematite and spinel is formed between 600 and 900 °C that contains mica with a low crystallinity, at 900 °C; all of which contribute to the pozzolanic activity of the coal-mining wastes calcined to 600 and 900 °C (Figure 3). Clay minerals were shown to have pozzolanic activity that influenced the reaction kinetics.

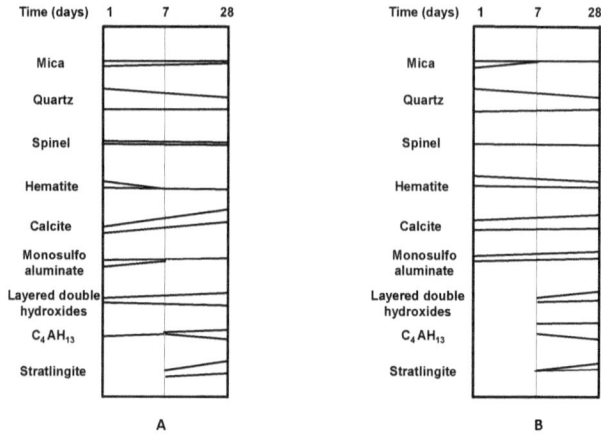

**Figure 3.** Mineralogical evolution of coal mining waste by XRD; (**A**) activated at 600 °C/2 h; (**B**) activated at 900 °C/2 h.

The pozzolanic activity of the coal-mining waste that was activated at 600 °C and 900 °C/2 h had values, at one day, of 65.2%, at 600 °C, and 3.0%, at 900 °C; at seven days, these values were 67.2%, at 600 °C, and 33.1%, at 900 °C, and at 28 days into the reaction reached 72.4%, at 600 °C, and 65.8%, at 900 °C. These percentile results reflect a higher reaction rate at 600 °C than at 900 °C, over short reaction times (one to seven days); both activated products showed a high pozzolanic activity in terms of the fixed lime results over longer reaction times (28 days).

Recommended activation conditions from both an energetic and an economic point of view were therefore set at 600 °C/2 h retention time for the research.

### 3.4. Evolution of Hydrated Phases in Activated Coal Waste/Lime Systems

Characterization of the hydrated phases by XRD and SEM-EDX in both cases confirmed evidence of a pozzolanic reaction between the metakaolinate (from the coal waste) and the lime.

XRD analysis of the activated coal waste at 600 °C and 900 °C-2 h/Ca(OH)$_2$ at one, seven, and 28 days into the reaction revealed the following crystalline hydrated phases: stratlingite (C$_2$ASH$_8$), tetracalcium aluminate hydrate (C$_4$AH$_{13}$), monosulfoaluminate hydrate (C$_3$A·SO$_4$Ca·12H$_2$O), and layered double hydroxides. All the crystalline phases in the XRD analysis with their evolution and reaction times are shown in Figure 2. The formation of monosulfoaluminate hydrate (C$_3$A·SO$_4$Ca·12H$_2$O) in the thermally activated coal was evident at day one, as a consequence of the reaction between the reactive alumina of pozzolan, sulfate ions, and portlandite. Traces of tetracalcium aluminate hydrate (C$_4$AH$_{13}$) were identified at short reaction times; slight increases in its content were observed in the coal waste activated at 600 °C/2 h at day 28 of the reaction. Activation at 900 °C was joined by the formation of layer double hydroxides; stratlingite C$_2$ASH$_8$ was formed after seven days of reaction at 600 °C/2 h. Calcination at 900 °C prevented any further formation of stratlingite.

Alkaline solutions of these samples had pH values of ≈12. The most prevalent were Al$^{3+}$, Al(OH)$_2$$^+$, Al(OH)$^{2+}$, Al(OH)$_3$, and Al(OH)$_4$$^-$, without other ligands. The most prevalent species in the solution at high pH (pH > 10) was Al(OH)$_4$$^-$. These ions combined with readily available Ca$^{2+}$ ions through metastable phases yielding C$_3$AH$_6$. At high pH values (pH > 10), total dissolved silica concentrations in equilibrium with quartz and amorphous silica increased. At pH values of ≈12, the sum of the ionized and un-ionized species (H$_4$SiO$_4$, H$_3$SiO$_4$$^-$, and H$_2$SiO$_4$$^{2-}$) was equal to the total concentration of dissolved silica. If supersaturation of the total silica concentration in the solution with respect to amorphous silica occurs, polymers form that combine with the Ca$^{2+}$ ions to form CSH

(Calcium silicate hydrate) gels and with $(CO_3)^{2-}$ ions via metastable phases to form LDH (Layered double hydroxides) [39].

The appearance of $C_4AH_{13}$ is explained by supersaturation of calcium hydroxide and low metakaolinite contents in the aqueous phase. Increased temperatures of coal waste at 900 °C/2 h reduced the presence of alumina and metakaolinite and silica sheet structures, assisting the formation of the main hydration product, $C_4AH_{13}$.

Traces of portlandite and tetracalcium aluminate hydrate were identified at short reaction times [40,41]. However, tetracalcium aluminate hydrate was the prevalent phase in the pozzolanic reaction of coal waste at 900 °C at seven and 28 days; layered double hydroxides formed at day one of the reaction in coal waste at 600 °C to become the prevalent phase at day seven of the pozzolanic reaction. Stratlingite was evident at day seven of the reaction in the activated wastes and was the prevalent phase in the pozzolanic reaction after 28 days.

Following SEM analysis (Figure 4), an enriched composition of laminar microaggregates of silica were observed, with very porous surfaces covered with CSH gels, more abundantly at 600 °C/2 h, having in all cases a spongy appearance. The Ca/Si ($CaO/SiO_2$) ratio varied between 1.26 at 600 °C over 2 h (I-type gels) [10] and at all times was 1.73 in 900 °C/2 h (II-type gels) [10]. A laminar phase, interlaced with the gels, consisted of layered double hydroxide, stratlingite $C_2ASH_8$, monosulfoaluminate hydrate ($C_3A \cdot SO_4Ca \cdot 12H_2O$), and tetracalcium aluminate hydrate ($C_4AH_{13}$) of widely varying sizes. SEM imagery showed a gradual layered double hydroxide with stratlingite crystallization throughout the hydrothermal treatment and an increase in crystal size that depended on the duration of the temperature. Both the SEM and the XRD analyses yielded similar results.

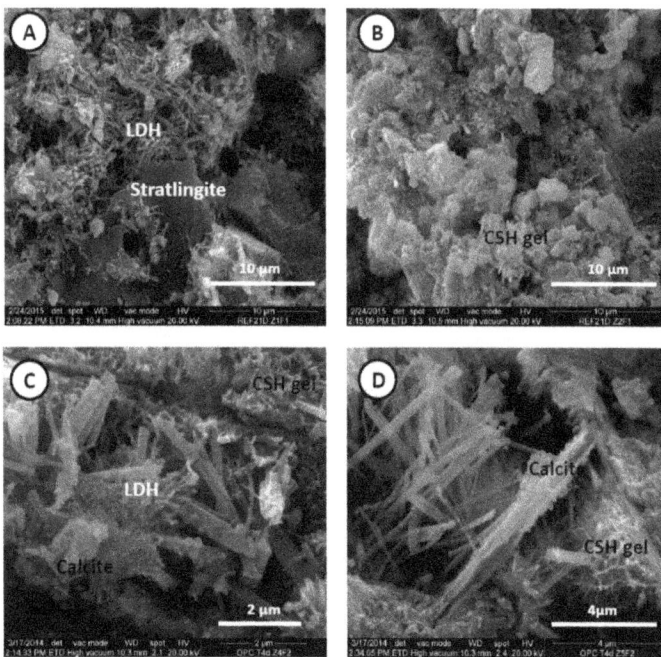

**Figure 4.** Morphological aspect of the CSH (Calcium silicate hydrate) gel, C4ASH12, and C2ASH8 phases at 600 °C/2 h (**A** and **B**). Morphological aspect of the CSH gel, C4AH13 and LDH (Layered double hydroxide) structures at 900 °C/2 h (**C** and **D**). (**A**) Layered double hydroxides with stratlingite; (**B**) Calcium silicate hydrate with spongy appearance; (**C**) Calcite crystal with typical morphology; (**D**) Calcite fibbers and layered double hydroxides.

## 4. Conclusions

The coal waste that has been studied had a mineralogical composition of kaolinite, micas, quartz, calcite, dolomite, and feldspars.

The presence of kaolinite indicated that thermal treatment at 600 °C/2 h transformed kaolinite in metakaolinite, a highly reactive component of the pozzolanic reaction.

Unlike hydrated phases obtained from a natural kaolinite (CSH gels, $C_4AH_{13}$, and $C_2ASH_8$), metakaolinite from thermal activation of the coal-mining waste in the stable phase of the pozzolanic reaction assisted the appearance of metastable layered double hydroxide compounds and stratlingite. All these compounds indicate that the products resulting from thermal activation at 600 °C/2 h from coal-mining waste contain highly pozzolanic properties over the first seven days into the reaction.

Following the activation of the coal-mining waste at 900 °C/2 h, $C_4AH_{13}$ was the stable phase in the pozzolanic reaction, following supersaturation of the aqueous phase in the presence of calcium hydroxide and low metakaolinite and tetrahedral and octahedral layers resulting from the dehydroxylation of the mica content. SEM/EDX analysis pointed to CSH gels among the main hydrated phases of the pozzolanic reaction in activated/lime systems during.

The use of this waste generates benefits and limits environmental damage such as (1) land occupation, and both soil and groundwater contamination, are reduced; (2) less exploitation of natural resources (kaolin deposits) and reduced emissions of greenhouse gases per unit of cement produced ($CO_2$ in the clinker production process); (3) coal waste containing residues of fossil carbon give the starting material a heating value which can be used in the alignment process (heating to the required temperature).

**Acknowledgments:** This research has been supported by the Spanish Ministry of Economy and Competitiveness (Project Ref. MAT2012-37005-CO3-01/02/03 and BIA2015-65558-C3-1-2-3R (MINECO/FEDER)). The authors are also grateful to the Sociedad Anónima Hullera Vasco-Leonesa and to the Spanish Cement Institute (IECA) for their assistance with this research.

**Author Contributions:** Moisés Frías and Raquel Vigil de la Villa Mencía conceived and designed the experiments; Virginia Rubio performed the experiments; Moisés Frías, Raquel Vigil de la Villa Mencía and Rosario García-Giménez analyzed the data; Virginia Rubio contributed materials and analysis tools; Rosario García-Giménez wrote the paper.

**Conflicts of Interest:** The authors declare no conflict of interest.

## References

1. Zhao, Y.; Zhang, J.; Zheng, C.G. Transformation of aluminum-rich minerals during combustion of a bauxite-bearing Chinese coal. *Int. J. Coal Geol.* **2012**, *94*, 182–190. [CrossRef]
2. Liu, H.; Liu, Z. Recycling utilization patterns of coal mining waste in China. *Res. Cons. Recycl.* **2010**, *54*, 1331–1340.
3. Bian, Z.; Dong, J. The impact of disposal and treatment of coal mining wastes on environment and farmland. *Environ. Geol.* **2009**, *58*, 625–634. [CrossRef]
4. Skarzynska, K. Reuse of coal mining wastes in civil engineering—Part 2: Utilization of minestone. *Waste Manag.* **1995**, *15*, 83–126. [CrossRef]
5. Nehdi, M.; Duquette, J.; El Damatty, A. Performance of rice husk ash produced using a new technology as a mineral admixture in concrete. *Cem. Concr. Res.* **2003**, *33*, 1203–1210. [CrossRef]
6. Chindaprasirt, P.; Homwuttiwong, S.; Jaturapitakkul, C. Strength and water permeability of concrete containing palm oil fuel ash and rice husk–bark ash. *Constr. Build. Mater.* **2007**, *21*, 1492–1499. [CrossRef]
7. García Giménez, R.; Vigil de la Villa, R.; Goñi, S.; Frías, M. Fly ash/paper sludge as constituents of cements: Hydration phases. *J. Environ. Eng. Sci.* **2015**, *10*, 46–52. [CrossRef]
8. Frías, M.; Villar-Cociña, E.; Savastano, H. Brazilian sugar bagasse ashes from the cogeneration industry as active pozzolans for cement manufacture. *Cem. Concr. Comp.* **2011**, *33*, 490–496. [CrossRef]
9. Sabir, B.B.; Wild, S.; Bai, J. Metakaolin and calcined clays as pozzolans for concrete: A review. *Cem. Concr. Comp.* **2001**, *23*, 441–454. [CrossRef]
10. Taylor, H.F.W. *Cement Chemistry, 1997*; Thomas Telford Services Ltd.: London, UK, 1997.

11. Siddique, R.; Klaus, J. Influence of metakaolin on the properties of mortar and Concrete: A review. *Appl. Clay Sci.* **2009**, *43*, 392–400. [CrossRef]

12. Ambroise, J.; Murat, M.; Pera, J. Hydration reaction and hardening of calcined clays and related minerals: V. Extension of the research and general conclusions. *Cem. Concr. Res.* **1985**, *15*, 261–268. [CrossRef]

13. De la Vigil Villa, R.; Frías, M.; Sánchez de Rojas, M.I.; Vegas, I.; García, R. Mineralogical and morphological changes of calcined paper sludge at different temperatures and retention in furnace. *Appl. Clay Sci.* **2007**, *36*, 279–286. [CrossRef]

14. Frias, M.; García, R.; Vigil, R.; Ferreiro, S. Calcination of art paper sludge waste for the use as a supplementary cementing material. *Appl. Clay Sci.* **2008**, *42*, 189–193. [CrossRef]

15. Banfill, P.F.G.; Rodríguez, O.; Sánchez de Rojas, M.I.; Frías, M. Effect of activation conditions of a kaolinite based waste on rheology of blended cement pastes. *Cem. Concr. Res.* **2009**, *39*, 843–848. [CrossRef]

16. Rodríguez Largo, O.; de la Vigil Villa, R.; de Sánchez Rojas, M.I.; Frías, M. Novel use of kaolin wastes in blended cements. *J. Am. Ceram. Soc.* **2009**, *92*, 2443–2446. [CrossRef]

17. Vegas, I.; Urreta, J.; Frías, M.; García, R. Freeze-thaw resistance of blended cements containing calcined paper sludge. *Constr. Build. Mater.* **2009**, *23*, 2862–2868. [CrossRef]

18. Frías, M.; Rodríguez, O.; Nebreda, B.; García, R.; Villar-Cocina, E. Influence of activation temperature of kaolinite based clay wastes on pozzolanic activity and kinetic parameters. *Adv. Cem. Res.* **2010**, *22*, 135–142. [CrossRef]

19. Li, D.; Song, X.; Gong, C.; Pan, Z. Research on cementitious behaviour and mechanism of pozzolanic cement with coal gangue. *Cem. Concr. Res.* **2006**, *36*, 1752–1759. [CrossRef]

20. Beltramini, L.B.; Suárez, M.L.; Guilarducci, A.; Carrasco, M.F.; Grether, R.O. Aprovechamiento de residuos de la depuración del carbón mineral: Obtención de adiciones puzolánicas para el cemento Portland. *Revista Técnica de Ciencias. Universidad Tecnológica Nacional de Argentina* **2010**, *3*, 7–18. (In Spanish)

21. Frías, M.; de Sánchez Rojas, M.I.; García, R.; Juan, A.; Medina, C. Effect of activated coal mining wastes on the properties of blended cement. *Cem. Concr. Comp.* **2012**, *34*, 678–683. [CrossRef]

22. Vigil de la Villa, R.; Frías, M.; García-Giménez, R.; Martínez-Ramírez, S.; Fernández-Carrasco, L. Chemical and mineral transformations that occur in mine waste and washery rejects during pre-utilization calcination. *Int. J. Coal Geol.* **2014**, *132*, 123–130. [CrossRef]

23. García, R.; Vigil de la Villa, R.; Frías, M.; Rodríguez, O.; Martínez-Ramírez, S.; Fernández-Carrasco, L.; de Soto, I.S.; Villar-Cociña, E. Mineralogical study of calcined coal waste in a pozzolan/Ca(OH)$_2$ system. *Appl. Clay Sci.* **2015**, *108*, 45–54. [CrossRef]

24. Leyva, F.; Matas, J.; Rodríguez Fernández, L.R. *Memoria y Hoja del Mapa Geológico de España, Escala 1:50.000, No. 129 La Robla, 2ª Serie Magna*; IGME: Madrid, Spain, 1984. (In Spanish)

25. Frías, M.; Vigil de la Villa, R.; García, R.; Sánchez de Rojas, M.I.; Baloa, T.A. Mineralogical evolution of kaolin-based drinking water treatment waste for use as pozzolanic material. The effect of activation temperature. *J. Am. Ceram. Soc.* **2012**, *96*, 3188–3195. [CrossRef]

26. Zimmer, A.; Bergmann, C.P. Fly ash of mineral coal as ceramic tiles raw material. *Waste Manag.* **2007**, *27*, 59–68. [CrossRef] [PubMed]

27. Dai, S.; Ren, D.; Chou, C.L.; Li, S.; Jiang, Y. Mineralogy and geochemistry of the No. 6 Coal (Pennsylvanian) in the Junger Coalfield, Ordos Basin, China. *Int. J. Coal Geol.* **2006**, *66*, 253–270. [CrossRef]

28. Vassilev, S.V.; Vassileva, C.G. Mineralogy of combustion wastes from coal-fired power stations. *Fuel Process. Technol.* **1996**, *47*, 261–280. [CrossRef]

29. Vassilev, S.V.; Vassileva, C.G. A new approach for the classification of coal fly ashes based on their origin, composition, properties, and behaviour. *Fuel* **2007**, *86*, 1490–1512. [CrossRef]

30. Ward, C.R.; Bocking, M.A.; Ruan, C. Mineralogical analysis of coals as an aid to seam correlation in the Gloucester Basin, New South Wales, Australia. *Int. J. Coal Geol.* **2001**, *47*, 31–49. [CrossRef]

31. Pinetown, K.L.; Ward, C.R.; van der Westhuizen, W.A. Quantitative evaluation of minerals in coal deposits in the Witbank and Highveld Coalfields, and the potential impact on acid mine drainage. *Int. J. Coal Geol.* **2007**, *70*, 166–183. [CrossRef]

32. Gaigher, J.L. The Mineral Matter in Some South African coals. Master Thesis, University of Pretoria, Pretoria, South Africa, 1980.

33. Ward, C.R. Analysis and significance of mineral matter in coal seams. *Int. J. Coal Geol.* **2002**, *50*, 135–168. [CrossRef]

34. Dai, S.; Tian, L.; Chou, C.L.; Zhou, Y.; Zhang, M.; Zhao, L.; Wang, J.; Yang, Z.; Cao, H.; Ren, D. Mineralogical and compositional characteristics of Late Permian coals from an area of high lung cancer rate in Xuan Wei, Yunnan, China: Occurrence and origin of quartz and chamosite. *Int. J. Coal Geol.* **2008**, *76*, 318–327. [CrossRef]

35. Ruan, C.-D.; Ward, C.R. Quantitative X-ray powder diffraction analysis of clay minerals in Australian coals using Rietveld methods. *Appl. Clay Sci.* **2002**, *21*, 227–240. [CrossRef]

36. Wang, X.; Dai, X.; Chou, C.; Zhang, M.; Wang, J.; Song, X.; Wang, W.; Jiang, Y.; Zhou, Y.; Ren, D. Mineralogy and geochemistry of Late Permian coals from the Taoshuping Mine, Yunnan Province, China: Evidences for the sources of minerals. *Int. J. Coal Geol.* **2012**, *96*, 49–59. [CrossRef]

37. Tian, L.; Dai, S.; Wang, J.; Huang, Y.; Ho, S.C.; Zhou, Y.; Lucas, D.; Koshland, C.P. Nanoquartz in Late Permian C1 coal and the high incidence of female lung cancer in the Pearl River Origin area: A retrospective cohort study. *BMC Public Health* **2008**, *8*, 398. [CrossRef] [PubMed]

38. Ambroise, J.; Martín Calle, S.; Pera, J. Pozzolanic behavior of thermally activated kaolin. In Proceedings of the Fourth CANMET/ACI International Conference on Fly Ash, SF, Slag and Natural Pozzolans in Concrete, Istambul, Turkey, 3–8 May 1992; Malhotra, V.M., Ed.; Volume 1, pp. 731–741.

39. De Wintd, L.; Deneele, D.; Maubec, N. Kinetic of lime/bentonite pozzolanic reaction at 20 and 50 °C: Batch tests and modeling. *Cem. Concr. Res.* **2014**, *59*, 34–42. [CrossRef]

40. Kaminskas, R.; Cesnauskas, V.; Kulibiute, R. Influence of different artificial additives on Portland cement hydration and hardening. *Constr. Build. Mater.* **2015**, *95*, 537–544. [CrossRef]

41. De Azeredo, A.F.N.; Azeredo, G.; Carneiro, A.M.P. Performance of lime-metakaolin pastes and mortars in two curing conditions containing kaolin wastes. *Key Eng. Mater.* **2016**, *668*, 419–432. [CrossRef]

*minerals*

MDPI

*Article*

# Brown Coal Dewatering Using Poly (Acrylamide-Co-Potassium Acrylic) Based Super Absorbent Polymers

**Sheila Devasahayam [1],\*, M. Anas Ameen [1], T. Vincent Verheyen [1] and Sri Bandyopadhyay [2]**

[1] School of Applied and Biomedical Sciences, Federation University, Gippsland Campus, Northways Rd, Churchill Vic. 3842, Australia; anasameen@students.federation.edu.au (M.A.A.); vince.verheyen@federation.edu.au (T.V.V.)

[2] School of Materials Science and Engineering, University of New South Wales, Sydney 2052, Australia; s.bandyopadhyay@unsw.edu.au

\* Correspondence: s.devasahayam@federation.edu.au; Tel.: +61-3-5327-6466

Academic Editor: Kota Hanumantha Rao

Received: 27 July 2015 ; Accepted: 17 September 2015 ; Published: 30 September 2015

**Abstract:** With the rising cost of energy and fuel oils, clean coal technologies will continue to play an important role during the transition to a clean energy future. Victorian brown coals have high oxygen and moisture contents and hence low calorific value. This paper presents an alternative non evaporative drying technology for high moisture brown coals based on osmotic dewatering. This involves contacting and mixing brown coal with anionic super absorbent polymers (SAP) which are highly crossed linked synthetic co-polymers based on a cross-linked copolymer of acryl amide and potassium acrylate. The paper focuses on evaluating the water absorption potential of SAP in contact with 61% moisture Loy Yang brown coal, under varying SAP dosages for different contact times and conditions. The amount of water present in Loy Yang coal was reduced by approximately 57% during four hours of SAP contact. The extent of SAP brown coal drying is directly proportional to the SAP/coal weight ratio. It is observed that moisture content of fine brown coal can readily be reduced from about 59% to 38% in four hours at a 20% SAP/coal ratio.

**Keywords:** brown coal; osmotic dewatering; super absorbent polymers; FTIR; dewatering mechanism

## 1. Introduction

Victorian brown coal is a cost effective fuel for power generation. It is very cheap to mine and low in sulphur and ash yield but the high moisture content and hence low calorific value result in high $CO_2$ emission intensity relative to bituminous coal. This high moisture has ensured that power stations are located adjacent to their mines to minimise handling, transportation and environmental problems. Run-of-mine brown coal will require upgrading before its use in new generation thermal power plants.

Practical and economic advantages in reducing the moisture content in brown coals include enhanced handling characteristics and reduced transportation costs. Additional benefits include reduced boiler capital costs, higher combustion efficiency, lower water consumption and waste disposal costs.

### 1.1. Victorian Brown Coals

Brown coal is at an intermediate stage in the geochemical conversion of accumulated vegetable debris from peat into hard or bituminous coal [1]. Brown coals typically have high moisture contents, in the 30%–70% range, with Victorian brown coal at the extreme end of this range. This high moisture content has a negative impact on every thermal application for brown coal. There is an uninterrupted converse relationship between the moisture content of Victorian brown coal and useful heat accessible from combustion of the coal (the net wet specific energy).

According to Mackay [2], Victorian brown coals occur in seams of a few centimetres to over 100 m thick, extending laterally from a few metres to over 50 km. Within the seams there are bands of coal which vary in appearances and properties. These bands are formed of coals belonging to different lithotypes (rock types) which reflect different depositional environments. The samples for current study are supplied from Loy Yang brown coal mines, Gippsland, Victoria (Figure 1).

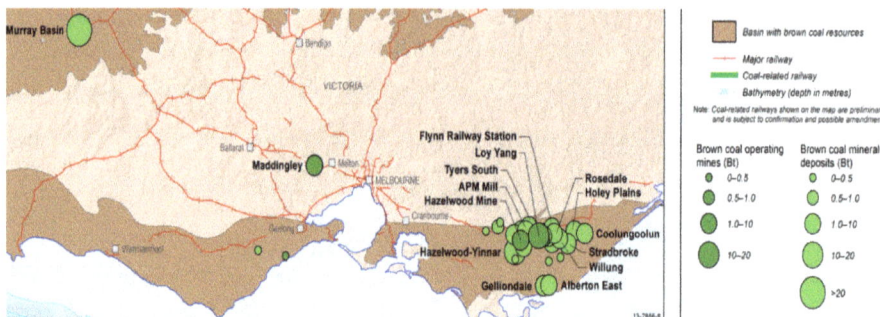

**Figure 1.** Current operating brown coal mines located in Victoria website [3].

The freshly exposed surface of as mined brown coal changes in colour from red brown to dark brown, as drying or oxidation occurs. When brown coal is air dried (to moisture 10%–15%) the colour varies from yellow to black.

Verheyen and Perry [4] reported Victorian brown coals to be physically complex and heterogeneous due to their detrital origin. Their relatively low carbon aromaticity, presence of residual carbohydrates (in woody macerals), along with methoxy phenols, unsaturated diterpenoids and fatty acids all confirm that Victorian brown coals have only been exposed to very mild (diagenetic) conditions. Structural heterogeneity tends to decrease as coalification continues due to condensation and cleavage reactions.

Iyengar, Sibal and Lahiri [5] reported that the water sorbed as a monolayer on coal is attached to hydrophilic sites on the coal surface. These sites were identified as oxygen-containing functional groups. They have been confirmed for a variety of coal ranks including Victorian brown coals [6,7].

Molecular simulation techniques were applied by Kumagai, Chiba and Nakamura [8] to define a model for coal structure. They modelled the structure of Yallourn brown coal as two oligomers, namely a tetramer (molecular weight (MW) 1540) and a pentamer (MW 1924) based on a monomer of composition $C_{21}H_{20}O_7$ as represented in Figure 2. The unit structure was modelled on the basis of combined data from elemental analysis (C: 65.6, H: 5.2, O: 29.2 wt %) by Schafer [7] and $^{13}$C-NMR spectroscopy. First two molecules were joined with 360 water molecules matching to 65.3% moisture content (wet basis).

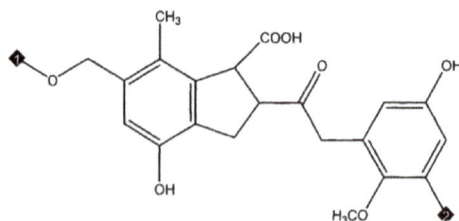

**Figure 2.** Monomer structure of brown coal used as basis of Yallourn brown coal molecular model (Kumagai, Chiba and Nakamura 1999) [8] corresponds to C: 65.6, H: 5.2, O: 29.2 wt % (Molecular Weight (MW) = 384.4).

## 1.2. Brown Coal Drying

Processes for drying or dewatering brown coals are grouped into three wide categories: (a) evaporative (thermal); (b) non-evaporative (thermal) and (c) other non-evaporative drying processes.

The common commercial procedures include evaporative drying where heat is applied to evaporate the water from coal at atmospheric pressure. Non evaporative drying processes are attractive due to their enhanced energy effectiveness, since water is separated in liquid form and the latent heat of vaporisation is not expended.

Studies have been conducted to investigate feasibility of non-evaporative drying of brown coal such as the Fleissner process [9] which incorporates pressurised steam treatment at temperatures above 200 °C. The Evans-Siemon process is the updated Fleisssner process where pressurised hot water is used to improve thermal efficiency by avoiding a pressure cycle [10]. The Koppelman (K-fuel) process [11] involves high pressure with low operating temperature; however, this process was not demonstrated commercially. The mechanical thermal expression dewatering process [12] involves use of elevated temperatures, up to 250 °C and pressure <12.7 MPa, resulting in significant reductions in residual moisture content which may be largely attributed to the destruction of brown coal porosity. Attempts at electro dewatering brown coals [13] did not demonstrate encouraging results. Our current research aims at developing alternative technologies to dewater brown coals.

Dzinomwa, Wood and Hill [14] employed super absorbents to de-water fine black coal particles, revealed some advantages over the alternative non evaporative drying technologies, and offered an alternative lower energy osmotic water removal approach to the thermal technologies outlined above.

## 1.3. Super Absorbent Polymers (SAP)

SAPs comprise high molecular weight crossed linked hydrophilic polymers which absorb several tens to hundred times their individual mass of water as they expand in size, but still preserve distinct particle identity [15–17]. SAP can be cationic and anionic. The amount of water that a specific SAP can absorb depends on its chemical composition and morphology as well as the quality of absorbed water [18], particularly with respect to occurrence of ionic salts. The recycled water used in mining plant operations generally contains a significant concentration of salts [19] and these need to be managed to maximize the polymeric absorption potential.

Factors affecting the capacity of a SAP to absorb water are as follows:

1.  Swelling properties (attributed to presence of hydrophilic groups in the network).
2.  Cross linking density (generally higher molecular weight with lower cross linking densities exhibits higher absorption capacities).
3.  Structural integrity (high cross linking density is crucial to retain the structural integrity of the polymer loaded with moisture as the high cross-linking density offers high mechanical strength).
4.  The "availability" of target water *i.e.*, how tightly it is bound to the drying substrate.

Figure 3 shows the mechanism of water absorption in SAP. The initial diffusion of water inside hydrophilic SAP causes ionization of neutralized acrylate groups into negative carboxylate ions and positive sodium ions. Negative electrical charges along the SAP backbone cause mutual repulsion of carboxylate ions, increasing the osmotic pressure inside the gel, thereby resulting in expansion and swelling of the SAP chains due to absorbed water. Finally, cross links between chains inhibit solubilisation of SAP (in water) thus governing the extent of swelling or absorption by restricting infinite swelling.

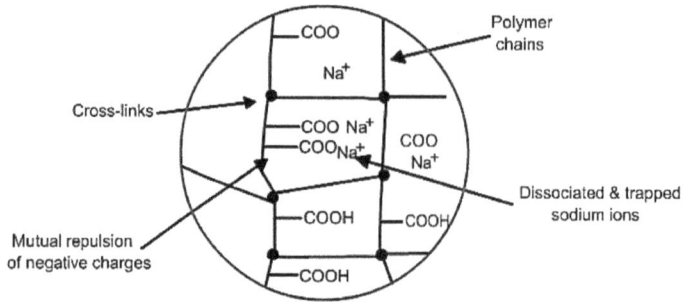

**Figure 3.** Mechanism of water absorption in super absorbent polymers (SAP).

*1.4. Standard Method of Moisture Determination*

According to Allardice [1], determination of moisture content of brown coals is complicated by the lack of an inadequate definition for what constitutes moisture in coal. The most widely accepted definition is that the moisture content is the water present in the coal as water molecules ($H_2O$), which can be released at 105–110 °C. This is not intended to include water from the decomposition of functional groups or chemically adsorbed water.

Allardice [1] reported on two basic types of standard moisture determination methods *i.e.*, (a) azeotropic distillation, in an immiscible liquid such as toluene and (b) oven drying at 105 °C. In azeotropic distillation moisture content is determined directly from the volume of water collected in the condenser, in contrast to most oven drying methods where water is estimated indirectly by net weight loss. The standard method for determining moisture content of brown coals does not discriminate between water of decomposition at a temperature up to 105 °C and molecular water present in it.

## 2. Methodology of the SAP Dewatering Process

Figure 4 depicts an outline of the osmotic dewatering process, using SAP to decrease moisture content of brown coal. This is achieved by direct contact between SAP and sized brown coal using an end-over-end tumbler, tumbling through 360° from top to bottom providing complete mixing and a sieve shaker to affect final separation. High moisture brown coal is intimately mixed with dry SAP by shaking in air tight bottles to prevent evaporation. This batch storage approach also enabled the SAP to passively draw and absorb water from the surface of brown coal particles. Mechanical end-over-end tumbling (vertical bottle rotation at 60 rpm) for a set time at constant speed was used to maximize surface contact between SAP and brown coal particles thereby reducing the equilibration time. The apparatus provided reproducible and gentle particle contact conditions. Water absorption causes SAP to swell thus permitting the separation of swollen SAP from shrunken brown coal particles by sieving.

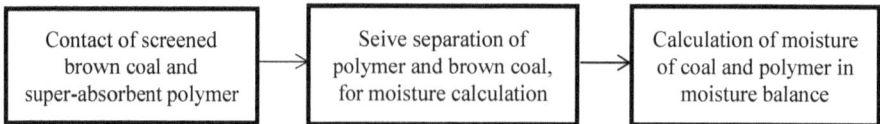

**Figure 4.** Proposed processes for reducing water content of brown coal by treatment with SAP.

## 3. Experimental

### 3.1. Materials

#### 3.1.1. Brown Coal

The brown coal (<80 mm) was provided by Omnia Specialities from a bulk sample collected from the AGL Loy Yang mine in 2014. It was stored in sealed plastic pails prior to analysis. Its proximate and ultimate analysis is presented in Table 1. According to [20] brown coals of Victoria, particularly from Latrobe have low ash yields, but as would be predicted from such huge volumes of coal, even separate seams have slight to substantial differences in physical properties and chemical composition.

**Table 1.** Loy Yang Brown coal analysis.

| Coal Properties | Percentage (% Dry Basis) |
| --- | --- |
| Moisture | 59.3 * |
| Ash | 2.2 |
| Volatile matter | 50.5 |
| Carbon | 68.4 |
| Hydrogen | 5.1 |
| Sulphur | 0.3 |

\* % as received.

#### 3.1.2. Super Absorbent Polymers (SAP) Used in this Work

Aquasorb 3005 (<1000 μm) supplied by SNF (Australia) Pty Ltd. (Lara, Victoria, Australia), is a highly crossed linked, synthetic co-polymer of acryl amide and potassium acrylate. It is water insoluble with maximum water absorption of 150% w/w in 1000 ppm NaCl solution.

The polymer consists of a set of polymeric chains that are parallel to each other and regularly linked to each other by cross-linking agents, forming a network. When water comes in contact with one of these chains, it is drawn into the molecule by osmosis. Water rapidly migrates into the interior of the SAP network where it is stored. The pH of SAP is alkaline at 8.1 with a density of $1.10 \text{ g/cm}^3$. The maximum water absorption of 400 wt % occurs for deionized water. SAP particles were sieved and those >600 μm and <850 μm were selected for use. Figure 5 presents a molecular structural representation of Aquasorb.

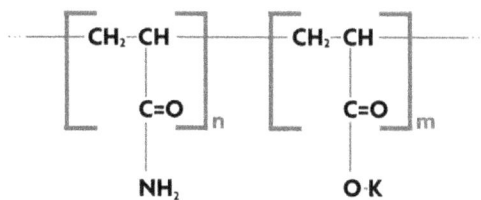

**Figure 5.** Structure of Aquasorb.

### 3.2. Experimental Methods

#### 3.2.1. Fourier Transform Infrared (FTIR)

Fourier transform infrared (FTIR) spectroscopy is widely used to determine/evaluate functional groups in coal structure. The reference [21] attempted structural assignments via deconvolution of particular absorbance bands associated functional groups. FTIR spectral comparison of dewatered and run of mine brown coal samples has the potential to elucidate changes in hydrogen bonding and oxygen functional groups occurring during drying.

In this project, the FTIR spectra were recorded with a FTS1000 FTIR (Varian, MA, USA) fitted with a ATR (attenuated total reflectance) accessory (Specac, Golden Gate Mark II, Orpington, UK); 128 scans were accumulated with spectral resolution of 4 cm$^{-1}$, over the range of 4000–650 cm$^{-1}$. The spectrometer was equipped with ATR diamond anvil cell with single reflection. After measurement of spectra, the remaining coal sample on the crystal was removed with soft paper soaked in methanol and allowed to dry. Regular background spectra (air) were collected to ensure no cross contamination. The raw reflectance spectra were corrected for frequency distortion and converted to absorbance by standard software.

Figure 6 presents the ATR-FTIR spectrum of run-of-mine moisture Loy Yang brown coal. The ATR technique probes only the surface layers of the brown coal structure. Peaks identified at 3365, 2900, 1700, 1621 and 1440 cm$^{-1}$ correspond to OH groups [22], aliphatic groups, carboxylic and ketone groups, carbon oxygen double bonded aromatic rings and bending mode of H-bonded O–H groups [23] respectively (Table 2). Water in brown coal constitutes a progressive series of forms—each is more difficult to remove from apparent bulk water to that resulting from thermal decomposition of hydroxyl groups in brown coal and water of hydration of impure minerals.

The pH of the Loy Yang coal sample was found to be pH 4.05 suggesting that carboxylic groups will mainly be involved in dewatering.

**Table 2.** Structural assignments for absorption bands observed in infrared spectra of brown coal.

| Frequency (cm$^{-1}$) | Functional Groups | Reference |
|---|---|---|
| 3350–3600 | Stretching vibration of OH groups (water, alcohol, phenol, carbohydrates, peroxides) as well as amides (3650 cm$^{-1}$) | (Liu *et al.*, 2006) [22] Vijayalakshmi and Ravindhran, 2012 [24] |
| 2750–3000 | CH$_3$ and CH$_2$ (aliphatic) | (Chandarlal *et al.*, 2014) [23] |
| 1700 | Carboxylic acid and ketone groups | (Chandarlal *et al.*, 2014) [23] |
| 1649 | Adsorbed water | (Li *et al.*, 2001) [24] |
| 1618–1622 | C=O, aromatic rings | (Chandarlal *et al.*, 2014) [23] |
| 1440 | CH$_2$, C=O bending mode of H-bonded O–H groups | (Chandarlal *et al.*, 2014) [23] |
| 1000–1300 | Phenoxy structure, aliphatic ethers, alcohols | (Chandarlal *et al.*, 2014) [23] |
| 1355 | Benzene or condensed benzene rings | (Chandarlal *et al.*, 2014) [23] |
| 1180 | Sp-3 rich structure COOH and OH | (Chandarlal *et al.*, 2014) [23] |
| 997–1130 | Stretching vibration of C–O of mono, oligo and carbo-hydrates | (Vijayalakshmi and Ravindhran, 2012) [25] |
| <1000 | C–H bending vibration from isoprenoids | (Vijayalakshmi and Ravindhran, 2012) [25] |
| 870–750 | Weak absorption due to C–H bending vibrations of olefinic and aromatic structures | (Verheyen and Perry 1991) [4] |

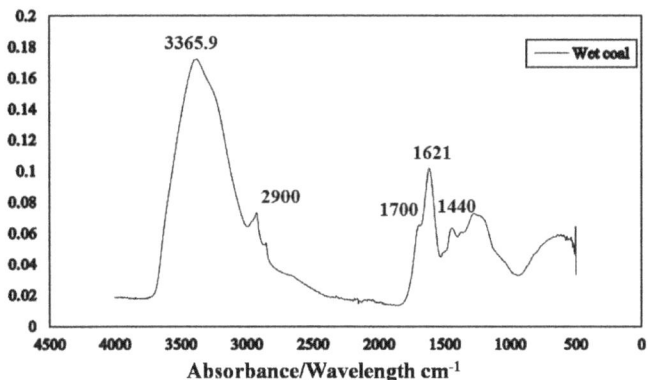

**Figure 6.** Attenuated total reflectance Fourier transform infrared (ATR FTIR) spectrum of as received Loy Yang coal.

### 3.2.2. Dewatering Tests

A series of tests were conducted to optimise the SAP dosage and brown coal/SAP contact time, to identify the effective operating conditions to maximise the water removal.

Loy Yang coal (as received) was crushed and screened <600 μm (to ensure a smaller size than SAP) and stored in sealed containers prior to use. 20 g coal sub samples were mixed with predetermined amounts of SAP to achieve target loadings *i.e.*, 0%, 5%, 10%, 15%, 20%, 25% and 30% SAP/coal. The SAP/coal mixtures were sealed in air tight glass 600 mL lab bottles fitted with plastic screw caps. Up to twelve bottles (replicates) were placed inside the sample box of a vertical tumbler for gentle "top over bottom" bottle agitation at 60 rpm. The sample was weighed before and after the predetermined tumbling time period to ensure no evaporative losses occurred. Contact time was monitored by determining the weight loss of brown coal (after SAP separation) until there was no further change at which point the moisture was considered to be at equilibrium. The mixture was readily separated into dewatered fine brown coal and swollen polymer via a sieve due to their size difference—SAP swells and increases in size, whilst the coal shrinks as it loses moisture. Changes in moisture were measured by a moisture balance programmed for constant weight at a temperature of 105 °C. Figure 7 shows loaded SAP after dewatering of brown coal. It was found the coal lost to the SAP was <0.2%.

**Figure 7.** SAP after dewatering brown coal. Note increase in size and the dark colouration afforded by traces of humic material adsorbed onto the alkaline SAP.

## 4. Results and Discussion

According to Li *et al.* [26] there is presence of significant amount of alkali and alkaline earth metals associated with the carboxylic and phenolic functionalities in the structure of low rank coal. At low pH, ion exchange mainly takes place with carboxylic groups to form carboxylates. According to Schafer [27], acidic groups and phenolic, are primarily responsible for ion exchange properties of brown coals. At the *in-situ* pH prevalent in wet brown coals, carboxyl groups may interchange cations with pore water to form carboxylates. Cations normally related with these groups are calcium, magnesium, sodium and iron. Meanwhile, phenolic groups do not interchange cations to any degree until system pH is greater than pH 8. The mean acid group content of the brown coal from Loy Yang field is shown in Table 3.

**Table 3.** Mean acidic group content of Loy Yang brown coal (Schafer, 1991) [27].

| Phenolic OH (Dry Basis) | COOH (Dry Basis) | COO (Dry Basis) | Phenolic Oxygen (Dry Basis) | Carboxylic Oxygen (Dry Basis) | Acidic Oxygen | Total Oxygen (Dmif) | Acidic Oxygen |
|---|---|---|---|---|---|---|---|
| meq/g | | | Percentage (%) | | | | |
| 3.04 | 2.39 | 0.10 | 4.86 | 7.97 | 13.0 | 24.7 | 53 |

Dmif—dry mineral and inorganic free basis.

### 4.1. Moisture Results

SAP dosage rates ranging from 5 to 30 wt % SAP/brown coal, and contact times between 1 to 6 h were evaluated; to determine the optimum drying conditions for our experimental setup.

Test samples of fine brown coal prior to drying had a water content of 59.3%. Figure 8 and Table 4 show the final coal moisture for different SAP doses and contact times. It was considered that dosage of 5% of SAP by weight of brown coal and a contact time of six hours would give an adequate moisture reduction while avoiding excessive cost of SAP and holding capacity. The brown coal was observed to lose maximum of 53.25% of its original moisture after six hours contact with 20% w/w SAP.

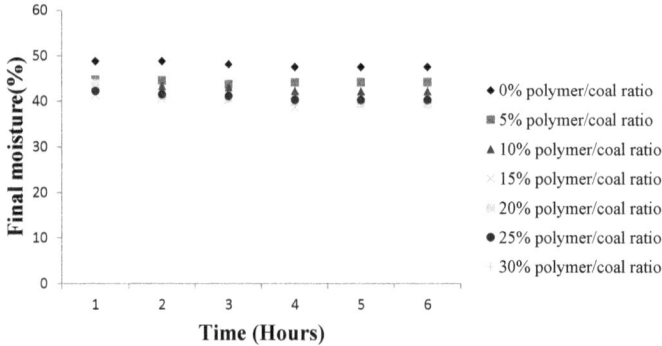

**Figure 8.** Variation in Loy Yang moisture content with contact time and Super absorbent Polymer, SAP/coal.

A SAP/coal of 25 wt % caused a further decrease in moisture level. However at 30 wt % SAP dosage, there was no further reduction in coal moisture instead a slight increase was observed. This implies that higher weight percentage SAP loadings may release water from SAP which is reabsorbed on to the brown coal.

Figure 8 reveals the final Loy Yang moisture values with 20% SAP/coal achieving the optimal drying for moisture reduction. Static osmotic dewatering of brown coal *i.e.*, without the tumbling, produced a final coal moisture of 47.3%, compared to the 38.6% with tumbling for the same contact time. This result illustrates the importance of continuously exposing fresh surfaces for maximum contact osmotic drying rates. The higher convective mass transfer and increased surface contact interactions afforded by tumbling make it possible to reduce the brown coal drying time. It was ascertained that a contact time of four hours was the optimal treatment time. Further drying via SAP contact could be achieved by increasing the pH of coal as the SAP has higher moisture affinity under basic conditions and the Loy Yang coal is acidic in nature.

**Table 4.** Effect of anionic super absorbent polymers (SAP) dosage and contact time on reduction of water content of fine brown coal (initial water content of brown coal = 59.3%).

| | Polymer/Coal Ratio (%) | | | | | | |
|---|---|---|---|---|---|---|---|
| - | 0 | 5 | 10 | 15 | 20 | 25 | 30 |
| Hours | Final Moisture Content (%) | | | | | | |
| 0 | 59.3 | 59.3 | 59.3 | 59.3 | 59.3 | 59.3 | 59.3 |
| 1 | 58.7 | 48.8 | 44.8 | 43.6 | 41.3 | 44 | 42.2 |
| 2 | 58.7 | 48.8 | 44.6 | 43.4 | 40.4 | 40.8 | 41.5 |
| 3 | 58.4 | 48.2 | 43.7 | 43.1 | 40.2 | 40.4 | 41.7 |
| 4 | 58 | 48.5 | 44.2 | 42.3 | 38.6 | 39.7 | 40.3 |
| 5 | 58.2 | 48.6 | 44.2 | 42.2 | 39.7 | 39.3 | 40.3 |
| 6 | 58.2 | 47.6 | 44.3 | 42.2 | 39.2 | 39.3 | 40.2 |

*4.2. Dewatering Kinetics*

According to Szekely *et al.* [28] analysis of a heterogeneous reaction system must start from the recognition that reaction takes place at the interface and hence mass must be transported to and from this interface. It follows that structure, pertinent to mass transfer during and after contact plays a significant role on the overall rate of water transfer. The fraction of moisture removed from brown coal was calculated from the following equation:

$$\alpha = \frac{\text{weight loss of brown coal due to moisture removal at any time (t)}}{\text{Total moisture content}}$$

where, $\alpha$ = fraction of moisture removed.

Figure 9 shows the fraction of moisture removed with respect to the time. It can be seen that the bulk of the drying occurs within the first two hours (Regions 1 and 2) and then attains a steady state/equilibrium (Region 3). Figure 9 indicates the maximum removal of moisture by SAP within the first two hours.

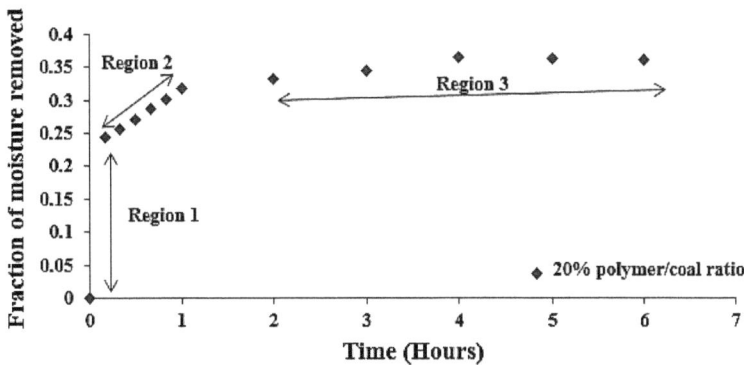

**Figure 9.** Fraction of moisture removed ($\alpha$) *vs.* time at 20% SAP to coal ratio.

The three different regions identified in Figure 9 are as follows:

1. Region 1: Dissociation and breaking of chemically adsorbed and surface moisture (chemical process).
2. Region 2: Desorption of water (bound and free water)/and breaking of liquid bridges in coal (chemical and physical process).
3. Region 3: Diffusion/mass transfer process (physical process).

*4.3. Surface Chemical and Diffusion Control Model*

Accordingly the dewatering kinetics were tested for surface chemical model ($R_3$ ($\alpha$)) and diffusion control model ($D_3$ ($\alpha$)). Where,

$$D_3(\alpha) = 1 - \frac{2}{3}\alpha - (1 - \alpha)^{2/3}$$

$$R_3(\alpha) = 1 - (1 - \alpha)^{1/3}$$

It follows from Figure 9 that Region 1 and parts of Region 2 should follow the surface chemical model and parts of Region 2 and Region 3 should follow the diffusion model. Figure 10 shows the plots for surface chemical shrinking core model, $R_3$ ($\alpha$) *vs.* time and diffusion control model, $D_3$ ($\alpha$) *vs.* time, for 20% SAP to coal ratio. Plots for the pore diffusion model (Figure 10) indicate a better linear trend until the end of the reaction, indicating that the mass transfer-diffusion process controls the reactions. A linear trend in the Region 2 for $R_3$ ($\alpha$) plots indicates a competing chemical, and mass transfer-diffusion process controlling the dewatering process.

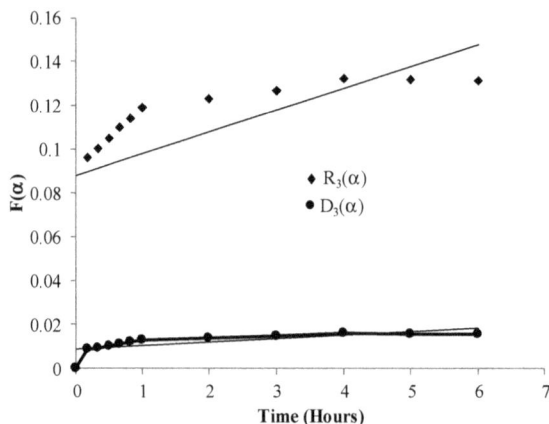

**Figure 10.** Plot of diffusion control model [D3 ($\alpha$)] *vs.* time and surface control model [R3 ($\alpha$)] *vs.* time for 20% SAP to coal ratio.

*4.4. Fourier Transform Infrared (FTIR ATR)*

The Fourier transform infrared spectroscopy of the brown coal sample and SAP sample before and after dewatering over the wave number ranges 4000–600 $cm^{-1}$ are presented in Figures 11 and 12 respectively. The amplified difference spectrum in Figure 12 reveals that the as received and SAP treated brown coal samples have only minimal difference in absorptions due to inherent moisture *i.e.,* positive broad H-bonding band centred at approx. 3365 $cm^{-1}$ and the minor contribution to the band near 1621 $cm^{-1}$. Removal of all the non-bound water at 105 °C resulted in the typical IR absorbance spectrum (Figure 11) for dried brown coal with the O–H and C–H of aromatic and aliphatic stretching,

(3700–2400 cm$^{-1}$) [4,29] and more pronounced absorptions due to C–H stretch near 2900 cm$^{-1}$ [4,23] and bend near 1450 cm$^{-1}$; carbonyl 1700 cm$^{-1}$ [23] and aromatic ring (breathing mode enhanced by ring substitution) at 1621 cm$^{-1}$. The fingerprint region 1440–949.2 cm$^{-1}$ which incorporates a multitude of functional group contributions [23,25], including etheric oxygen and mineral salts which is also enhanced in the oven dried sample. The sensitivity (IR extinction coefficients) changes for the organic bands in the coal on removal of moisture with slightly higher band intensities observed for most organic functional groups. For the SAP treated sample as seen in Figures 11 and 12 the doublet band centred near 2380 cm$^{-1}$ is assigned to adsorbed $CO_2$ on the surface of the brown coal. The variation in this band in SAP treated coal is thought to relate to both atmospheric $CO_2$ adsorption by the sample and $CO_2$ transfer from the SAP itself. Figures 13 and 14 provide an IR spectral comparison of SAP before and after contact with Loy Yang coal. Differences in absorption in the 1440–1600 cm$^{-1}$ region (Figure 14) corresponds to fine structural bands that are lost from the polymer on wetting and the spectrum is dominated by positive water bands. There is an expected increase in peaks at approximately 3335 and 1650 cm$^{-1}$ for SAP after coal contact indicative of adsorbed water.

The SAP would need to be recycled in a commercial coal drying application by a non-evaporative dehydration process. This can be achieved by pH shock wherein the SAP shrinks on contact with acids such as HCl [14]. The low pH water self-drains from the SAP thereby reducing the need for thermal evaporation. Further research into SAP recycling is warranted given the encouraging results presented here for removing bulk moisture osmotically from Loy Yang coal.

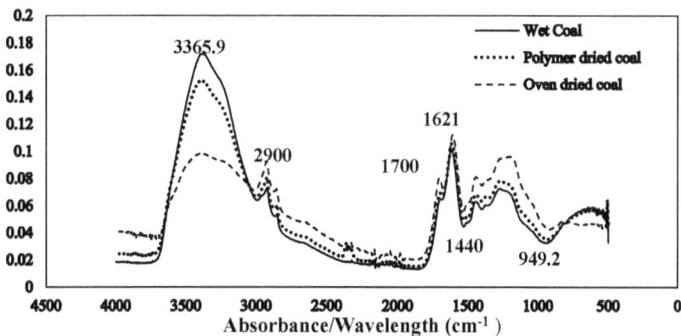

**Figure 11.** Stacked ATR–IR spectra of as received (59.3% moisture) SAP treated (38.6% moisture) and oven dried (105 °C in air) Loy Yang coal. Spectra presented at the same scale.

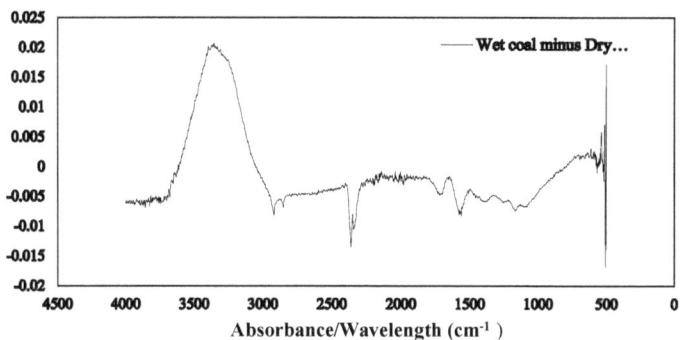

**Figure 12.** ATR–IR difference spectrum for as received moisture Loy Yang coal minus its SAP treated product.

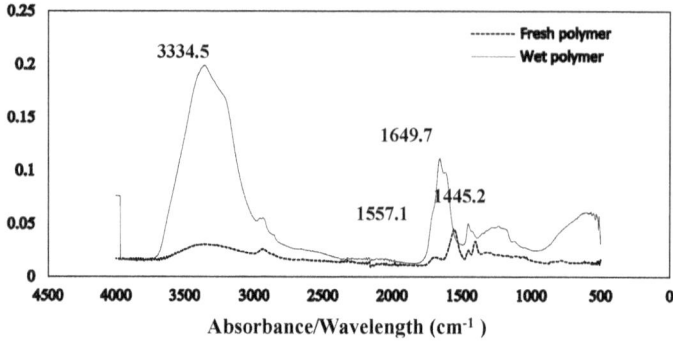

**Figure 13.** Stacked ATR–IR spectra of fresh SAP and its water loaded equilibrium (with coal) moisture product. Spectra presented at the same scale.

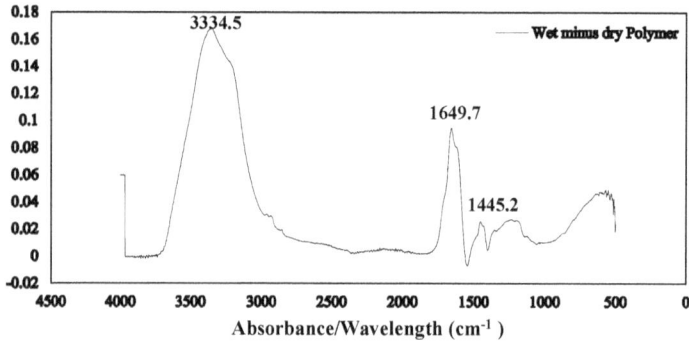

**Figure 14.** ATR–IR difference spectrum for wet minus fresh SAP Note the lack of any coal derived absorbance bands despite the brown colouration of the loaded SAP.

## 5. Conclusion

Osmotic dewatering of brown coal using SAP indicates that it is possible to achieve reduction in Loy Yang moisture from 59.3% to 38.6% equivalent to removing 57% (on dry basis) of the initial water present. Under the tumbling contact conditions employed mixing for four hours is necessary to attain equilibrium moisture loadings between polymer and brown coal. This process is safer than direct thermal drying and will produce fine brown coal with a consistent moisture content. Laboratory tests revealed that the rate of moisture absorption remains constant between 4–6 h.

**Acknowledgments: Acknowledgments:** We acknowledge the CRN seed funding awarded to Sheila Devasahayam by the federation University Australia and David Hill's (University of Queensland, Australia) technical input and discussions during the course of the project. Andrew Stranieri's, (FoST, Federation University Australia), interest, encouragement, facilitating of the project are very much appreciated.

**Author Contributions: Author Contributions:** Sheila Devasahayam is the supervisor of this student's project. Anas Ameen, the master's student in Mining Engineering at the Federation University, Australia carried out this project in partial fulfilment of his master's degree, under the supervision of Sheila Devasahayam, co-supervised by Vince Verheyen (Fed Uni) and Sri Bandyopadhyay, UNSW.

**Conflicts of Interest:** The authors declare no conflict of interest.

# References

1. Allardice, D.J.; Newell, B.S. Industrial imlication of the properties of brown coal. In *The Science of Victorian Brown Coal: Structure Properties and Consequences for Utilization*; Durie, R.A., Ed.; Butterworh-Heinmann Ltd.: Oxford, UK, 1991.

2. Mackay, A.M.; George, G.H. Petrology. In *The Science of Victorian Brown Coal: Structure, Properties and Consequences for Utilization*; Durie, R.A., Ed.; Buterworth-Heinemann Ltd.: Oxford, UK, 1991.

3. Australia's Coal Industry. Available online: http://www.minerals.org.au/resources/coal/coal_mines_by_state (accessed on 24 September 2015).

4. Verheyen, T.V.; Perry, G.J. Chemical structure of Victorian brown coal. In *The Science of Victorian Brown Coal: Structure Properties and Consequences for Utilization*; Durie, R.A., Ed.; Butterworth-Heinmann Ltd.: Oxford, UK, 1991.

5. Iyengar, M.S.; Sibal, D.H.; Lahiri, A. Role of hydrogen bonds in the briquetting of lignite. *Fuel* **1957**, *36*, 76–84.

6. Allardice, D.J.; Evans, D.G. The brown coal/water system: Part 2. Water sorption isotherms on bed moisture Yallourn brown coal. *Fuel* **1971**, *50*, 236–253. [CrossRef]

7. Schafer, H.N.S. Factors effecting the equilibrum moisture content of low rank coals. *Fuel* **1972**, *51*, 4–9. [CrossRef]

8. Kumagai, H.; Chiba, T.; Nakamura, K. Change in physical and chemical characteristics of brown coal along with a progress of moisture release. In *Abstracts of Papers of the American Chemical Society*; American Chemical Society: Washington, DC, USA, 1999; Volume 218.

9. Fleissner, H. The Drying of Fuels And The Australian Coal Industry. Sonderdruck Spartwirtschaft, 1927; Nos.10 and 11.

10. Higgins, R.S.; Allardice, D.J. *Development of Brown Coal Dewatering*; Reserach and Deveopment Report; State Electricity Commission of Victoria (SECV): Victoria, Australia, 1973.

11. Murray, R.G. Stable high enery solid fuel from lignite. *Coal Process Tech.* **1979**, *5*, 211–214.

12. Hulston, J.; Favas, G.; Chaffee, A.L. Physico-chemical properties of Loy Yang lignite dewatered by mechanical thermal expression. *Fuel* **2005**, *84*, 1940–1948. [CrossRef]

13. Reuter, F.; Molek, H.; Kockert, W.; Lange, W. Laboratory desalination of brown coal by elecrto-osmosis. *Neue Bergbautechnik* **1981**, *11*, 186–187.

14. Dzinomwa, G.P.T.; Wood, C.J.; Hill, D.J.T. Fine coal dewatering using pH-sensitive and temeprature sensitive super absorbent polymers. *Polym. Adv. Technol.* **1997**, *8*, 767–772. [CrossRef]

15. Masuda, K.; Iwata, H.; Masuda, K.; Iwata, H. Dewatering of particulate materials utilizing highly water-absorptive polymer. *Powder Technol.* **1990**, *63*, 113–119. [CrossRef]

16. Moody, G.M. Role of polyacrylamides and related products in treatment of mineral processing effluent. *Trans. Inst. Min. Metall.* **1990**, *99*, C136.

17. SadeghiI, M.; Hosseinzadeh, H. Synthesis and properties of collagen-g-poly(sodium acrylate-co-2-hydroxyethylacrylate) superabsorbent hydrogels. *Braz. J. Chem. Eng.* **2013**, *30*, 379–389. [CrossRef]

18. Jiang, S.Q.; Sun, X.W.; Xie, Z.X.; Qin, L. Study on synthesis and property of anti-salt super absorbent resin. *Adv. Mater. Res.* **2013**, *873*, 683–688. [CrossRef]

19. Laftha, W.A.; Hashim, S.; Ibrahim, A.N. Polymers hydrogel: A review. *Polym. Plast. Technol. Eng.* **2011**, *50*, 1475–1486. [CrossRef]

20. Gloe, C.S.; Holdgate, G.R. Geology and Resources. In *The Science of Victorian Brown Coal: Structure, Properties and Consequences for Utilization*; Durie, R.A., Ed.; Butterworh-Heinmann Ltd.: Oxford, UK, 1991.

21. Painter, P.C.; Starsinic, M.; Coleman, M.M. *Fourier Transform Infrared Spectroscopy Applications to Chemical Systems*; Ferraro, J.R., Basile, J.L., Eds.; Academic Press: Waltham, MA, USA, 1985.

22. Liu, H.; Sun, S.Q.; Lv, G.H.; Chan, K.K. Study on Angelica and its different extracts by Fourier transform infrared spectroscopy and two-dimensional correlation IR spectroscopy. *Spectrochim. Acta Part A* **2006**, *64*, 321–326. [CrossRef] [PubMed]

23. Chandarlal, N.; Mahapatra, D.; Shome, D.; Dasgupta, P. Behaviour of low rank high moisture coal in small stockpile under controlled ambient condition—A approach. *Am. Int. J. Res. Form. Appl. Nat. Sci.* **2014**, *6*, 98–108.

24. Li, H.; Khor, K.A.; Cheang, P. HVOF Sprayed Hydroxyapatite Coatings: Powders' Smelting State and Mechanical Properties. In *Thermal Spray 2001: New Surface for a New Millenium*; Berndt, C.C., Khor, K.A., Lugscheider, E.F., Eds.; ASM International: Geouga County, OH, USA, 2001; pp. 99–104.

25. Vijayalakshmi, R.; Ravindhran, R. Comparitive fingureprint and extraction yield of diospyrus ferrea(wild.) Bakh. root with phenol compounds (gallic acid), as determined by UV-Vis and FT-IR spectroscopy. *Asian Pac. J. Trop. Biomed.* **2012**, *2*, 1367–1371. [CrossRef]

26. Hayashi, J.; Li, C.-Z. Structure and properties. In *Advances in the Science of Victorian Brown Coal*, 1st ed.; Li, C.Z., Ed.; Elsevier Ltd.: Amsterdam, The Netherlands, 2004; pp. 11–78.

27. Schafer, H.N.S. Functional groups and ion exchange prooperties. In *The Science of Victorian Brown Coal*; Durie, R.A., Ed.; Butterworh-Heinmann Ltd.: Oxford, UK, 1991.

28. Szekely, J.; Evans, J.W.; Sohn, H.Y. *Gas-Solid Reaction*; Academic Press: New York, NY, USA, 1976.

29. Meekum, U. Study of the molecular srain of polymerizable cyclic oligocarbonates using the spectroscopic techique. *Suranaree J. Sci. Technol.* **2005**, *12*, 107–113.

*minerals*

MDPI

Article

# The Fate of Trace Elements in Yanshan Coal during Fast Pyrolysis

Jiatao Dang, Qiang Xie *, Dingcheng Liang, Xin Wang, He Dong and Junya Cao

Department of Chemical Engineering, School of Chemical and Environmental Engineering, China University of Mining and Technology (Beijing), D-11 Xueyuan Rd., Haidian District, Beijing 100083, China; dangjt1988@163.com (J.D.); 13810562597@163.com (D.L.);18010125892@163.com (X.W.); donghe_cumtb@126.com (H.D.); caojy@cumtb.edu.cn (J.C.)
* Correspondence: dr-xieq@cumtb.edu.cn; Tel.: +86-10-6233-1014

Academic Editor: Alireza Somarin
Received: 8 January 2016; Accepted: 31 March 2016; Published: 6 April 2016

**Abstract:** In this study, a high-sulfur and high-ash yield coal sample obtained from the Yanshan coalfield in Yunnan, China was analyzed. A series of char samples was obtained by pyrolysis at various temperatures (300, 400, 500, 600, 700, 800, and 900 °C) and at a fast heating rate (1000 °C/min). A comprehensive investigation using inductively coupled plasma mass spectrometry (ICP-MS), a mercury analyzer, ion-selective electrode (ISE) measurements, X-ray diffraction (XRD) analysis, and Fourier transform infrared (FTIR) spectroscopy was performed to reveal the effects of the pyrolysis temperature on the transformation behavior of trace elements (TEs) and the change in the mineralogical characteristics and functional groups in the samples. The results show that the TE concentrations in the raw coal are higher than the average contents of Chinese coal. The concentrations of Be, Li, and U in the char samples are higher than those in raw coal, while the opposite was observed for As, Ga, Hg, and Rb. The F and Se concentrations are initially higher but decrease with pyrolysis temperature, which is likely caused by associated fracturing with fluoride and selenide minerals. Uranium shows the highest enrichment degree, and Hg shows the highest volatilization degree compared to the other studied TEs. As the temperature increases, the number of OH groups decreases, and the mineral composition changes; for example, pyrite decomposes, while oldhamite and hematite occur in the chars. It is suggested that the behavior and fate of TEs in coal during fast pyrolysis are synergistically influenced by self-characteristic modes of occurrence and mineralogical characteristics.

**Keywords:** trace elements; minerals in coal; transformation behavior; coal fast pyrolysis

---

## 1. Introduction

Approximately two-thirds of the basic chemical materials in China have been derived from coal or down-stream products. China has been the world's largest coal producer and consumer for several years. The abundance of coal makes it a stable, reliable, and basic fossil energy source for the sustained and rapid development of the Chinese economy and society; for example, approximately 74% of the total primary energy consumption is met by coal [1]. It is anticipated that the utilization of coal will remain stable for a few years in China while a new generation of clean coal technology, especially entrained-flow coal gasification, is being developed to meet global energy demands and to address environmental issues [2].

Coal is a complex flammable rock composed of inorganic and organic material and incorporates almost all of the elements present in the earth's crust. Some potentially toxic and hazardous trace elements (TEs) in coal, such as arsenic, mercury, fluorine, beryllium, and uranium, are thought to pose a potential risk for public health and the natural environment, even if present only at the parts-per-million level. Although the TE concentrations in coal are low, significant emissions of these pollutants into the environment due to large quantities of anthropogenic coal consumption causes several problems and attracts attention from regulatory authorities and scientists. Coal consumption and related environmental regulations are increasing. Therefore, a detailed investigation of the concentration, distribution, occurrence, and transformation behavior of TEs during coal production, preparation, utilization, and waste disposal is necessary to obtain comprehensive information [3].

There are many studies on the concentration, geochemical, and mineralogical characteristics of TEs in coal [4–10]. For example, Dai *et al.* [11] investigated mineral composition and the TE concentrations in coal from the Huayinshan coalfield. The authors concluded that rare-earth elements mainly occur in rhabdophane and silicorhabdophane, while mercury and selenium are incorporated in pyrite and marcasite. A study by Liu *et al.* [12] showed that U, Mo, Re, Se, Cr, and V are mainly associated with organic components and are less associated with illite or mixed-layer illite/smectite. Extensive work on the partitioning behavior of TEs during coal combustion has been carried out [13,14], such as on the concentration and distribution of TEs in combustion products [15–17], the volatilization or enrichment behavior and chemical composition of TEs in ash [18,19], and the morphology and control technologies of TEs [20,21]. However, information about the behavior and the physical and chemical forms of TEs emitted during coal pyrolysis and gasification is scarce [22]. Several studies address the behavior of TEs during fixed-bed gasification [23,24], fluidized-bed gasification [25], and using other gasification technologies [26–28] employing thermodynamic simulations [29–31]. However, few studies focus on the TE concentration during fast coal pyrolysis [32–34] and entrained-flow coal gasification because simulation experiments are difficult and expensive. Moreover, data are extremely limited with respect to TE behavior during different reaction stages due to the difficulty of analyzing the fate of TEs at various temperatures.

Pyrolysis is an important thermochemical process that can be regarded as both the initial step and/or the accompanying process in most coal conversion processes, such as combustion, gasification, and liquefaction of coal. Furthermore, pyrolysis is a coal-cleaning technology producing fuel or basic chemical materials. Consequently, it is of great importance to analyze the transformation and behavior of TEs during coal pyrolysis. Many of the previous studies on pyrolysis processes were based on low heating rates, which are very different from those in modern coal combustion or gasification processes that are characterized by fast heating.

Hence, the goal of this work is to study coal pyrolysis under fast heating conditions, up to 1000 °C/min, similar to the heating process of modern coal pyrolysis and entrained-flow gasification. We collected a series of Yanshan coal samples from the province of Yunnan in Southwest China, which is enriched with several TEs, such as arsenic, fluorine, gallium, mercury, lithium, rubidium, selenium, and uranium, compared with the average contents of Chinese coal [35]. Experiments were conducted at various temperatures with a heating rate of 1000 °C/min, and TE concentrations were analyzed in detail. This research aims to determine the concentration and partitioning of TEs, and to characterize the mineralogical and chemical compositions during the fast pyrolysis of coal at different temperatures, which can provide insights into possible transformation mechanisms and predict the emission characteristics of TEs during gasification. The experimental operating conditions of the present work may not be fully representative of modern coal pyrolysis and entrained-flow gasification. However, the results provide basic theories or suggestions for TE emission and control.

## 2. Experiments

### 2.1. Samples

Ganhe meager coal from the Yanshan coalfield, located in Yunnan, was collected and used in the present study. The high TE, sulfur, and ash content in this coal make it atypical. The sampling procedure followed the Chinese Standard Method GB 474-2008 [36]. After the coal was air-dried, the samples were crushed and pulverized to obtain particle sizes < 0.074 mm. Fine-grained coal samples were then stored hermetically for experimental use. The characteristics of the Yanshan coal are shown in tab:minerals-06-00035-t001. The coal ash was sampled at 815 °C, in accordance with the procedure described in the Chinese Standard Method GB/T 212-2008 [37]. X-ray fluorescence spectrometry (PANalytical Axios-Max, Almelo, The Netherlands) was employed to measure the concentrations of major element oxides in the coal ash sample [38]. The results are shown in tab:minerals-06-00035-t002.

**Table 1.** Proximate and ultimate analysis of the coal sample.

| Sample | Proximate Analysis % | | | | Ultimate Analysis % | | | | |
|---|---|---|---|---|---|---|---|---|---|
| | $M_{ad}$ | $A_d$ | $V_{daf}$ | $FC_{daf}$ | $C_{daf}$ | $H_{daf}$ | $O_{daf}$ | $N_{daf}$ | $S_{t,daf}$ |
| Yanshan Coal | 3.32 | 40.98 | 17.32 | 82.68 | 77.97 | 3.25 | 1.72 | 1.08 | 15.98 |

M, moisture; A, ash yield; V, volatile matter; FC, fixed carbon; ad, air-dry basis; $S_t$, total sulphur; d, dry basis; daf, dry and ash-free basis.

**Table 2.** Concentrations of major element oxides in the coal ash sample.

| Sample | $Na_2O$ | $MgO$ | $Al_2O_3$ | $SiO_2$ | $SO_3$ | $K_2O$ | $CaO$ | $TiO_2$ | $V_2O_5$ | $Fe_2O_3$ |
|---|---|---|---|---|---|---|---|---|---|---|
| Mass Percentage (%) | 0.36 | 1.60 | 26.98 | 54.43 | 4.15 | 3.41 | 3.34 | 0.78 | 0.15 | 4.60 |

## 2.2. Pyrolysis Procedure

Approximately 1 g of fine coal samples was weighed and placed into a quartz crucible. Highly pure nitrogen (99.999%) was used to purge the crucible, and the reaction zone was closed. Subsequently, char samples were obtained from pyrolysis at a fast heating rate (1000 °C/min) and at the different pyrolysis temperatures of 300, 400, 500, 600, 700, 800, and 900 °C, in accordance with the procedure by Xie *et al.* [39]. The crucible and reaction zone were heated by a microwave transduction cavity, which can transform the microwave from the workstation into general thermal radiation and can avoid the potential structural influence from the microwave. The final temperature was held for approximately 5 min to complete the pyrolysis. Char samples were weighed again and stored hermetically after the quartz crucible cooled to ambient temperature. The average cooling rate was approximately 45 °C/min before the temperature of the reaction zone was below 200 °C. Char samples were labelled as YCGR-300–900 according to the pyrolysis temperature, while the raw coal sample was labelled as YCGR-000.The char yields of the coal sample at different pyrolysis temperatures are shown in tab:minerals-06-00035-t003.

**Table 3.** Char yields of the coal sample at different pyrolysis temperatures.

| Sample | YCGR-300 | YCGR-400 | YCGR-500 | YCGR-600 | YCGR-700 | YCGR-800 | YCGR-900 |
|---|---|---|---|---|---|---|---|
| Char Yield (%) | 98.41 | 97.62 | 93.25 | 91.95 | 89.40 | 87.31 | 86.58 |

## 2.3. X-Ray Diffraction (XRD) Analysis

Mineralogical characteristics of all of the samples were determined using XRD analysis. The raw coal and char samples were analyzed using a Japanese Rigaku D/max-2500PC X-ray diffractometer (Rigaku Corporation, Tokyo, Japan) with a Cu tube. The XRD pattern was recorded over a 2θ range of 2.5°–70° with a step size of 0.02°. The scanning speed was 4°/min. The accelerating voltage and the tube current of the X-ray diffractometer were 40 kV and 100 mA, respectively.

## 2.4. Fourier Transform Infrared (FTIR) Spectroscopy

Functional groups and transition characteristics of the raw coal sample as well as the char samples were analyzed using a NICOLET iS-10 FTIR spectrometer (ThermoFisher, Waltham, MA,

USA). The KBr disc method was used, with a 1:160 mixture of sample and KBr. Spectra were recorded in the range of 400–4500 cm$^{-1}$ with an accuracy of 1.929 cm$^{-1}$. Thirty-two scans were taken with a 4 cm$^{-1}$ spectral resolution.

*2.5. TE Analysis*

Concentrations of Hg in coal and char samples were determined using a mercury analyzer (DMA-80, Milestone, Sorisole, Italy). Concentrations of F in all of the samples were measured by pyrohydrolysis with an ion-selective electrode (ISE) following the processes stipulated in the ASTM standard. Other TEs, such as arsenic, beryllium, gallium, lithium, rubidium, selenium, and uranium, were determined using inductively coupled plasma mass spectrometry (ICP-MS) (ThermoFisher, X-II) in a three points per peak pulse counting mode. For ICP-MS analysis, a Milestone UltraClave microwave high-pressure reactor was used to digest the samples. The digestion reagents consisted of 5 mL of 65% HNO$_3$ and 2 mL of 40% HF for each 50 mg of coal, while the solutions consisted of 2 mL of 65% HNO$_3$ and 5 mL of 40% HF for each 50 mg of non-coal samples. Sub-boiling distillation was used to further purify HF (the Guaranteed Reagent) and HNO$_3$ (the Guaranteed Reagent) to reduce interferences. Selenium and arsenic contents were measured by ICP-MS using collision-cell technology to eliminate interference due to polyatomic ions. Multi-element standards, such as U in CCS-1, and Be, Ga, and Rb in CCS-4, were used to calibrate the TE concentrations [11,12,40].

## 3. Results and Discussion

*3.1. Effect of Temperature on TE Concentration During Pyrolysis*

The TE concentrations of the raw coal sample and the char samples derived from pyrolysis at different temperatures at a fast heating rate are shown in tab:minerals-06-00035-t004. The concentrations of Be, Li, and U in all of the chars obtained over the studied temperature range are higher than those of raw coal, while the concentrations of As, Ga, Hg, and Rb of all of the chars obtained are lower than those of raw coal. In contrast, the concentrations of F and Se of some of the char samples are higher than those of raw coal, and others are lower than those of raw coal. The data indicate that Be, Li, and U tend to be enriched in the solid products, whereas As, Ga, Hg, and Rb tend to volatilize during pyrolysis. F and Se are initially enriched in the solid products but then volatilize with increasing temperature. At temperatures lower than 600 °C, F does not show volatilization during fast experimental pyrolysis. The volatility of Hg increases rapidly with increasing temperature in the temperature range below 400 °C and increases slowly above 400 °C because of a surplus shortage. As exhibits a lower volatility than Hg, which may be due to the association with sulfide and silicate minerals. Minimal amounts of volatile TEs, such as As, Hg, and Se, were found in chars compared to raw coal, which indicates that these TEs escape into the gas phase during pyrolysis.

**Table 4.** Trace element (TE) concentrations in raw coal and chars obtained from pyrolysis at different temperatures.

| Sample | As (µg/g) | Be (µg/g) | F (µg/g) | Ga (µg/g) | Hg (µg/kg) | Li (µg/g) | Rb (µg/g) | Se (µg/g) | U (µg/g) |
|---|---|---|---|---|---|---|---|---|---|
| YCGR-000 | 13.3 | 1.73 | 1153 | 16.1 | 165 | 55.9 | 34.8 | 25.7 | 67.6 |
| YCGR-300 | 11.7 | 1.95 | 1194 | 13.9 | 98.0 | 63.6 | 28.4 | 26.4 | 76.2 |
| YCGR-400 | 11.5 | 1.95 | 1252 | 14.0 | 19.7 | 63.7 | 27.9 | 24.2 | 86.6 |
| YCGR-500 | 12.3 | 2.04 | 1270 | 14.4 | 4.60 | 65.8 | 29.0 | 26.0 | 105 |
| YCGR-600 | 11.9 | 2.01 | 1254 | 14.4 | 2.87 | 64.0 | 29.2 | 24.5 | 93.3 |
| YCGR-700 | 12.5 | 2.09 | 1195 | 15.0 | 6.84 | 67.8 | 30.1 | 22.9 | 101 |
| YCGR-800 | 12.1 | 2.19 | 1183 | 15.0 | 3.31 | 69.1 | 26.1 | 20.3 | 124 |
| YCGR-900 | 11.7 | 2.32 | 1141 | 15.2 | 4.09 | 69.8 | 29.1 | 19.6 | 119 |

A relative parameter ($\Delta C$) is used to clearly and quantitatively describe the changing behavior of TE concentrations in the samples. The value of $\Delta C$ was calculated based on the data from tab:minerals-06-00035-t003 and is summarized in Figure 1.

$$\Delta C = \frac{c_{chars} - c_{coal}}{c_{coal}} \tag{1}$$

Here, $\Delta C$ is the changing rate of concentrations; $C_{chars}$ is the TE concentration in the char samples; and $C_{coal}$ is the TE concentration in the raw coal sample.

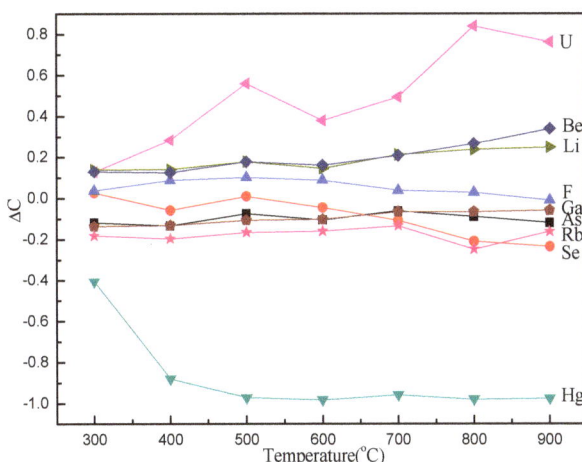

**Figure 1.** $\Delta C$ of trace elements (TEs) in raw coal and char samples obtained from pyrolysis at different temperatures at a fast heating rate.

The enrichment and volatilization of TEs were investigated by introducing a relative parameter ($\Delta C$), which compares the TE content in chars with that in raw coal. Figure 1 presents the $\Delta C$ of raw coal and char samples obtained from pyrolysis at different temperatures at a fast heating rate, showing that the $\Delta C$ of U, Be, Li, and F is positive, except for F at 900 °C, while the $\Delta C$ of Ga, As, Rb, Se, and Hg is negative, except for Se at 300 °C. Uranium shows the highest increase of $\Delta C$, and Hg reveals the largest decrease. It is notable that the enrichment order from high to low is U, Be, Li, and F, and the enrichment of U, Be, and Li increases with increasing pyrolysis temperature, except for F. Mercury almost completely volatizes at 500 °C and has a much higher volatility than Se, Rb, As, and Ga, even at low temperatures. The volatility of Se is unnoticeable at low temperatures and then increases gradually. The enrichment of Li is very similar to that of Be, and the partitioning of As is similar to that of Se in the samples (Figure 1). The volatility of As, Ga, Hg, Rb, and Se is much higher than that of F, suggesting that a large concentration of these elements is vaporized to the gas phase during pyrolysis.

Studies suggest that the elements associated with the organic matter are more easily vaporized than those associated with minerals, which can hinder TE partitioning into the gas phase. However, there has been no evidence to support that Hg is associated with organic matter in Yanshan coal. The phenomenon of Hg being one of the most volatile TEs in Yanshan coal is likely caused by the Hg characteristics. The conclusion can be drawn that Se, Rb, As, and Ga are volatile TEs during fast pyrolysis, while F has the potential for volatilization at high temperatures, and U, Be, and Li are non-volatile TEs in Yanshan coal. In addition, the behavior of TEs is also influenced by technological parameters, such as reaction atmosphere, mixing status, and product properties. Because of increased

TE emission and the potentially serious risk for the ecosystem environment, more effective measures and countermeasures should be investigated.

### 3.2. Effect of Temperature on Functional Groups During Pyrolysis

The FTIR spectra of raw coal and char samples obtained from pyrolysis at various temperatures at a fast heating rate are presented in Figure 2. The spectra show well-resolved peaks as IR bands.

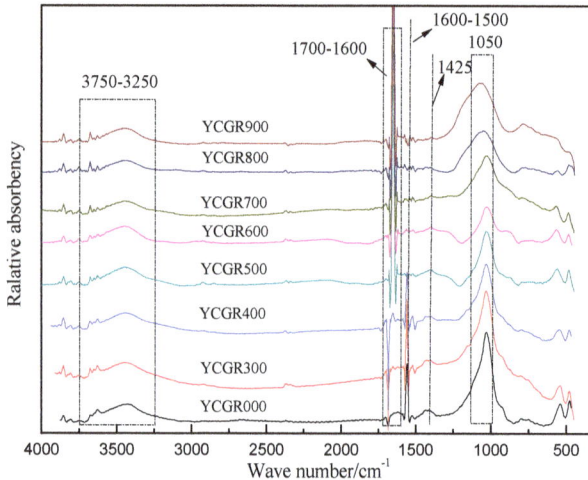

**Figure 2.** Fourier transform infrared (FTIR) spectra of raw coal and char samples obtained from pyrolysis at different temperatures at a fast heating rate.

It has been reported that some TEs are associated with organic matter or are influenced by organic matter in coal [8,12,41]. Therefore, helpful information may be obtained through an investigation of the changing behavior of the functional groups or organic structure during fast pyrolysis. A band in the region between 3750–3250 cm$^{-1}$ assigned to OH stretching vibrations was observed in all of the samples. However, the band intensity in this region tends to decrease with increasing pyrolysis temperature (Figure 2). The absorbance in this region is attributed to free OH groups, alcohol OH groups, carboxyl OH groups, and the associated OH of aromatic clusters of samples. It is suggested that the OH concentration decreases during pyrolysis. Moreover, a stretching vibration-induced band appears in the region of 1700–1600 cm$^{-1}$, which is assigned to the aliphatic C=O and –COOH. A small band was observed in the region of 1600–1500 cm$^{-1}$. This feature, which is assigned to C=C aromatic stretching vibrations of aromatic rings or an aromatic nucleus [42], gradually disappears with the increasing pyrolysis temperature. A possible explanation for these changes is that some aromatic rings in the macromolecular structure of coal or char samples decompose, and a few aliphatic C=O groups are generated from pyrolysis reactions during fast pyrolysis. A notable peak is found in all of the samples at 1050 cm$^{-1}$ due to kaolinite [43]. The peak is sharp in YCGR-000 and then weakens gradually, but it becomes stronger in YCGR-900 compared to the samples obtained at lower temperatures. Calcite stretching vibrations cause a peak at 1425 cm$^{-1}$[42]. A weak peak at this wave number is found in YCGR-000–600 and disappears in YCGR-700. The possible reason is that calcite decomposes because of the fast heating rate. Note that this also demonstrates the variation of kaolinite and calcite compared to XRD analyses and possibly is a useful source of information about TE migration mechanisms, especially some TEs associated with the organic matter. In general, the peak positions in the FTIR spectra of all of the samples were similar. The difference in absorbance indicates that the quantity and structures of functional groups or minerals changed.

### 3.3. Effect of Temperature on Mineralogical Characteristics During Pyrolysis

To elucidate the mineralogical characteristics of the samples, XRD patterns of the raw coal and char samples obtained from pyrolysis at various pyrolysis temperatures at a fast heating rate are presented in Figure 3. In raw coal, quartz, mica, gypsum, and pyrite have been identified. The mineralogical characteristics of the char samples from pyrolysis show a minor but interesting variation. The common mineral is quartz, exclusively composed of $SiO_2$ ( tab:minerals-06-00035-t002).    With increasing temperature, illite, oldhamite, and hematite appear in the chars, which may be due to the transformation of mica, gypsum, and pyrite [44].

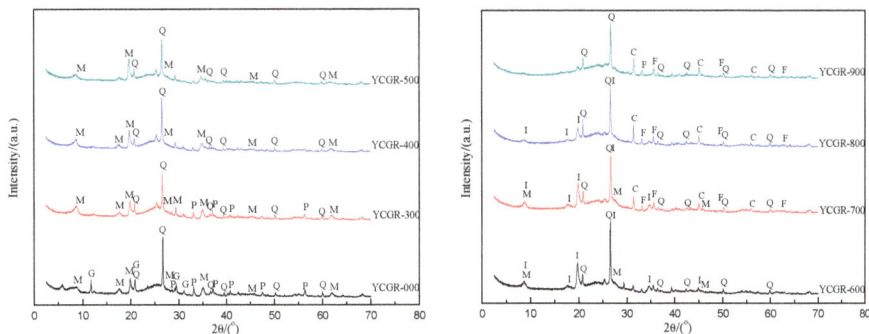

**Figure 3.** X-ray diffraction (XRD) patterns of raw coal and char samples obtained from pyrolysis at different temperatures at a fast heating rate. Q: quartz; P: pyrite; M: mica (muscovite, polylithionite, fluorphlogopite); G: gypsum; I: illite; C: oldhamite; F: hematite.

The TE contents are below the detection limit of XRD but are included in many minerals. The results obtained by XRD analysis are in agreement with the behavior of some TEs in the pyrolysis temperatures range. The concentration of F is related to silicate associations [41] and is low in YCGR-900, coincidentally, at the location where fluorphlogopite disappeared in the XRD pattern. It has been reported that most As is associated with pyrite [45,46], and pyrite is found in coal and not in chars, as Figure 3 shows, which provides a possible explanation for the volatilization of As. Minerals identified in chars are different from those in coal, which provides a possible interpretation according to the phenomena of TE enrichment or volatilization. The behavior of TEs during pyrolysis is also related to their characteristics—e.g., Hg expresses higher volatilization than As, although both TEs show similar primary modes of occurrence in coal. Some fluorides and selenides are found in raw coal and char samples, which indicates that there is a competing mechanism of partitioning behavior of F and Se, such as the volatility of F and Se and the hindering influence of minerals. In addition, inorganic cationic matrices, such as iron and calcium, also influence TE atomization [47], which suggests that the partitioning behavior of TEs, such as that of As in Figure 1, is different from the anticipated volatilization degree. Hence, a calcium-based absorber may be useful for the TE solidification in solid products. Research on this phenomenon may be useful to develop environmental pollution control methods and technologies for volatile TEs.

### 4. Summary and Conclusions

This paper presents a comprehensive study of the influence of different pyrolysis temperatures on TE behavior during coal pyrolysis at a heating rate of 1000 °C/min by analyzing the concentrations of As, Be, F, Ga, Hg, Li, Rb, Se, and U in raw coal and char samples. In addition, the effects of various pyrolysis temperatures on functional groups are investigated to interpret the behavioral characteristics of TEs during fast pyrolysis. The main conclusions drawn are as follows:

(1) The concentrations of As, F, Ga, Hg, Rb, Se, and U are much higher, and the concentration of Li in Yanshan coal is only slightly higher than the average TE values of Chinese coal. However, the raw coal is also very rich in sulfur and ash.

(2) The concentrations of Be, Li, and U in the char samples obtained from pyrolysis are higher than those in raw coal samples. Contrary results are obtained for As, Ga, Hg, and Rb, except for F and Se. The concentrations of F and Se in chars are initially higher but are lower than those in raw coal. The concentration of F in the char sample at 900 °C is lower than that in coal, while a similar phenomenon can be observed when the temperature is above 500 °C for Se, which is likely caused by the association fracturing with fluoride and selenide minerals at this specific temperature.

(3) The enrichment degree of U is much higher than that of the other selected non-volatile TEs, and the volatilization degree of Hg is much higher than that of the other studied volatile TEs. The enrichment of U and the volatilization of Hg increase with increasing pyrolysis temperature. The Hg content changes only slightly above 500 °C, whereas chars show the highest U concentration at 800 °C.

(4) The number of OH groups decreases during fast coal pyrolysis. Interesting changes were also observed regarding the strength of C=O and aromatic rings or aromatic nuclei above 500 °C.

(5) The minerals identified in chars are different from those in coal, which is caused by the pyrolysis reaction and may explain some of the TE behaviors. Pyrite has not been identified in chars, which indicates that As and Hg, associated with or related to pyrite, are being released during pyrolysis. In contrast, oldhamite and hematite are found with increasing pyrolysis temperature. TE partitioning is affected by not only the mode of occurrence in minerals but also the temperature and the element itself. To effectively control TE emission during fast pyrolysis or entrained-flow gasification, further research is necessary.

**Acknowledgments:** The authors gratefully acknowledge the financial and other support from the National Key Basic Research Program of China (NO. 2014CB238905). The authors also express special and sincere thanks to Shifeng Dai for the kind assistance in the analysis of the concentrations of TEs and to Yuegang Tang for his assistance in the sampling.

**Author Contributions:** Qiang Xie contributed to the conception of the work. Jiatao Dang, Dingcheng Liang, and Xin Wang performed the pyrolysis experiments. Jiatao Dang, He Dong, and Xin Wang performed the FTIR and XRD experiments. Qiang Xie, Junya Cao, Jiatao Dang, Dingcheng Liang, Xin Wang, and He Dong contributed to the manuscript preparation. Jiatao Dang wrote the manuscript with constructive help from Qiang Xie. All of the authors read and approved the manuscript.

**Conflicts of Interest:** The authors declare no conflict of interest.

## References

1. Dai, S.F.; Ren, D.Y.; Chou, C.L.; Finkelman, R.B.; Seredin, V.V.; Zhou, Y.P. Geochemistry of trace elements in Chinese coals: A review of abundances, genetic types, impacts on human health, and industrial utilization. *Int. J. Coal Geol.* **2012**, *94*, 3–21. [CrossRef]
2. Li, C.Z. Some recent advances in the understanding of the pyrolysis and gasification behavior of Victorian brown coal. *Fuel* **2007**, *86*, 1664–1683. [CrossRef]
3. Swaine, D.J. Why trace elements are important. *Fuel Process. Technol.* **2000**, *65*, 21–33. [CrossRef]
4. Riley, K.W.; French, D.H.; Farrel, O.P.; Wood, R.A.; Huggins, F.E. Modes of occurrence of trace and minor elements in some Australian coals. *Int. J. Coal Geol.* **2012**, *94*, 214–224. [CrossRef]
5. Dai, S.F.; Ren, D.Y.; Ma, S.M. The cause of endemic fluorosis in western Guizhou Province, Southwest China. *Fuel* **2004**, *83*, 2095–2098. [CrossRef]
6. Diehl, S.F.; Goldhaber, M.B.; Koening, A.E.; Lowers, H.A.; Ruppert, L.F. Distribution of arsenic, selenium, and other trace elements in high pyrite Appalachian coals: Evidence for multiple episodes of pyrite formation. *Int. J. Coal Geol.* **2012**, *94*, 238–249. [CrossRef]
7. Zhao, L.; Ward, C.R.; French, D.; Graham, I.T. Major and trace element geochemistry of coals and intra-seam claystones from the Songzao Coalfield, SW China. *Minerals* **2015**, *5*, 870–893. [CrossRef]

8. Finkelman, R.B. Modes of occurrence of potentially hazardous elements in coal: Levels of confidence. *Fuel Process. Technol.* **1994**, *39*, 21–34. [CrossRef]
9. Wang, X.B.; Wang, R.X.; Wei, Q.; Wang, P.P.; Wei, J.P. Mineralogical and geochemical characteristics of Late Permian coals from the Mahe Mine, Zhaotong Coalfield, Northeastern Yunnan, China. *Minerals* **2015**, *5*, 380–396. [CrossRef]
10. Vejahati, F.; Xu, Z.H.; Gupta, R. Trace elements in coal: Associations with coal and minerals and their behavior during coal utilization—A review. *Fuel* **2010**, *89*, 904–911. [CrossRef]
11. Dai, S.F.; Luo, Y.B.; Seredin, V.V.; Ward, C.R.; Hower, J.C.; Zhao, L.; Liu, S.D.; Zhao, C.L.; Tian, H.M.; Zou, J.H. Revisiting the Late Permian coal from the Huayingshan, Sichuan, Southestern China: Enrichment and occurrence modes of minerals and trace elements. *Int. J. Coal Geol.* **2014**, *122*, 110–128. [CrossRef]
12. Liu, J.J.; Yang, Z.; Yan, X.Y.; Ji, D.P.; Yang, Y.C.; Hu, L.C. Modes of occurrence of highly-elevated trace elements in superhigh-organic-sulfur coals. *Fuel* **2015**, *156*, 190–197. [CrossRef]
13. Xu, M.H.; Yan, R.; Zheng, C.G.; Qiao, Y.; Han, J.; Sheng, C.D. Status of trace element emission in a coal combustion process: A review. *Fuel Process. Technol.* **2003**, *85*, 215–237. [CrossRef]
14. Wang, G.M.; Luo, Z.X.; Zhang, J.Y.; Zhao, Y.C. Modes of occurrence of fluorine by extraction and SEM method in a coal-fired power plant from Inner Mongolia, China. *Minerals* **2015**, *5*, 863–869. [CrossRef]
15. Swanson, S.M.; Engle, M.A.; Ruppert, L.F.; Affolter, R.H.; Jones, K.B. Partitioning of selected trace elements in coal combustion products from two coal-burning power plants in the United States. *Int. J. Coal Geol.* **2013**, *113*, 116–126. [CrossRef]
16. Tang, Q.; Liu, G.J.; Zhou, C.C.; Sun, R.Y. Distribution of trace elements in feed coal and combustion residues from two coal-fired power plants at Huainan, Anhui, China. *Fuel* **2013**, *107*, 315–322. [CrossRef]
17. Roy, B.; Choo, W.L.; Bhattacharya, S. Prediction of distribution of trace elements under oxy-fuel combustion condition using Victorian brown coals. *Fuel* **2013**, *114*, 135–142. [CrossRef]
18. Oboirien, B.O.; Thulari, V.; North, B.C. Major and trace elements in coal bottom ash at different oxy coal combustion conditions. *Appl. Energy* **2014**, *129*, 207–216. [CrossRef]
19. James, D.W.; Krishnamoorthy, G.; Benson, S.A.; Seames, W.S. Modeling trace elements partitioning during coal combustion. *Fuel Process. Technol.* **2014**, *126*, 284–297. [CrossRef]
20. Quick, W.J.; Irons, R.M.A. Trace element partitioning during the firing of washed and untreated power station coals. *Fuel* **2002**, *81*, 665–672. [CrossRef]
21. Yi, H.H.; Hao, J.M.; Duan, L.; Tang, X.L.; Ning, P.; Li, X.H. Fine particle and trace elements emissions from an anthracite coal-fired power plant equipped with a bag-house in China. *Fuel* **2008**, *87*, 2050–2057. [CrossRef]
22. Clarke, L.B. The fate of trace elements during coal combustion and gasification: An overview. *Fuel* **1993**, *72*, 731–736. [CrossRef]
23. Bunt, J.R.; Waanders, F.B. Trace element behavior in the Sasol-Lurgi MK IV FBDB gasifier. Part 1—The volatile elements: Hg, As, Se, Cd and Pb. *Fuel* **2008**, *87*, 2374–2387. [CrossRef]
24. Bunt, J.R.; Waanders, F.B. Trace element behavior in the Sasol-Lurgi MK IV FBDB gasifier. Part 2—The semi-volatile elements: Cu, Mo Ni and Zn. *Fuel* **2009**, *88*, 961–969. [CrossRef]
25. Huang, Y.J.; Jin, B.S.; Zhong, Z.P.; Rui, X.; Zhou, H.C. The relationship between occurrence of trace elements and gasification temperature. *Proc. CSEE* **2006**, *26*, 10–15. (In Chinese).
26. Liu, S.Q.; Wang, Y.T.; Yu, L.; Oakey, J. Volatilization of mercury, arsenic and selenium during underground coal gasification. *Fuel* **2006**, *85*, 1550–1558. [CrossRef]
27. Yoshiie, R.; Taya, Y.; Ichiyanagi, T.; Ueki, Y.; Naruse, I. Emissions of particles and trace elements from coal gasification. *Fuel* **2013**, *108*, 67–72. [CrossRef]
28. Liu, S.Q.; Wang, Y.T.; Yu, L.; Oakey, J. Thermodynamic equilibrium study of trace element transformation during underground coal gasification. *Fuel Process. Technol.* **2006**, *87*, 209–215. [CrossRef]
29. Duchesne, M.A.; Hall, A.D.; Hughes, R.W.; Mccalden, D.J.; Anthony, E.J.; Macchi, A. Fate of inorganic matter in entrained-flow slagging gasifiers: Fuel characterization. *Fuel Process. Technol.* **2014**, *118*, 208–217. [CrossRef]
30. Gibbs, B.M.; Thompson, D.; Argent, B.B. A thermodynamic equilibrium comparison of the mobilities of trace elements when washed and unwashed coals are burnt under Pffiring conditions. *Fuel* **2004**, *83*, 2271–2284. [CrossRef]
31. Diaz-Somoano, M.; Martinez-Tarazona, M.R. Trace element evaporation during coal gasification based on a thermodynamic equilibrium calculation approach. *Fuel* **2003**, *82*, 137–145. [CrossRef]

32. Guo, R.X.; Yang, J.L.; Liu, D.Y.; Liu, Z.Y. Transformation behavior of trace elements during coal pyrolysis. *Fuel Process. Technol.* **2002**, *77*, 137–143. [CrossRef]
33. Guo, R.X.; Yang, J.L.; Liu, D.Y.; Liu, Z.Y. The fate of As, Pb, Cd, Cr and Mn in a coal during pyrolysis. *J. Anal. Appl. Pyrolysis* **2003**, *70*, 555–562. [CrossRef]
34. Guo, R.X.; Yang, J.L.; Liu, Z.Y. Behavior of trace elements during pyrolysis of coal in a simulated drop-tube reactor. *Fuel* **2004**, *83*, 639–643. [CrossRef]
35. Dai, S.F.; Wang, X.B.; Chen, W.M.; Li, D.H.; Chou, C.L.; Zhou, Y.P.; Zhu, C.S.; Li, H.; Zhu, X.W.; Xing, Y.W.; et al. A high-pyrite semianthracite of Late Permian age in the Songzao Coalfield, Southwestern China: Mineralogical and geochemical relations with underlying mafic tuffs. *Int. J. Coal Geol.* **2010**, *83*, 430–445. [CrossRef]
36. *Standardization Administration of the People's Republic of China. Method for Preparation of Coal Sample*; Chinese Stand GB/T 474-2008; Standardization Administration of the People's Republic of China: Beijing, China, 2008. (In Chinese)
37. *Standardization Administration of the People's Republic of China. Proximate Analysis of Coal*; Chinese Stand GB/T 212-2008; Standardization Administration of the People's Republic of China: Beijing, China, 2008. (In Chinese)
38. Hower, J.C.; Eble, C.F.; O'Keefe, J.M.; Dai, S.F.; Wang, P.P.; Xie, P.P.; Liu, J.J.; Ward, C.R.; French, D. Petrology, palynology, and geochemistry of Gray Hawk Coal (Early Pennsylvanian, Langsettian) in Eastern Kentucky, USA. *Minerals* **2015**, *5*, 592–622. [CrossRef]
39. Xie, Q.; Liang, D.C.; Tian, M.; Dang, J.T.; Liu, J.C.; Yang, M.S. Influence of heating rate on structure of chars derived from pyrolysis of Shenmu coal. *J. Fuel Chem. Technol.* **2015**, *43*, 798–908. (In Chinese)
40. Dai, S.F.; Wang, X.B.; Zhou, Y.P.; Hower, J.C.; Li, D.H.; Chen, W.M.; Zhu, X.W.; Zou, J.H. Chemical and mineralogical compositions of silicic, mafic, and alkali tonsteins in the Late Permian coals from the Songzao Coalfield, Chongqing, Southwest China. *Chem. Geol.* **2011**, *282*, 29–44. [CrossRef]
41. Wang, X.B.; Dai, S.F.; Sun, Y.Y.; Li, D.; Zhang, W.G.; Zhang, Y.; Luo, Y.B. Modes of occurrence of fluorine in the Late Paleozoic No.6 coal from the Haerwsu Surface Mine, Inner Mongolia, China. *Fuel* **2011**, *90*, 248–254. [CrossRef]
42. Okolo, G.N.; Neomagus, H.W.; Everson, R.C.; Roberts, M.J.; Bunt, J.R.; Sakurovs, R.; Mathews, J.P. Chemical-structural properties of South African bituminous coals: Insights from wide angle XRD-carbon fraction analysis, ATR-FTIR, solid state $^{13}$C NMR, and HRTEM techniques. *Fuel* **2015**, *158*, 779–792. [CrossRef]
43. Qin, Z.H.; Chen, H.; Yan, Y.J.; Li, C.S.; Rong, L.M.; Yang, X.Q. FTIR quantitative analysis upon solubility of carbon disulfide/N-methyl-2-pyrrolidinone mixed solvent to coal petrographic constituents. *Fuel Process. Technol.* **2015**, *133*, 14–19. [CrossRef]
44. Tian, C.; Zhang, J.Y.; Zhao, Y.C.; Gupta, R. Understanding of mineralogy and residence of trace elements in coals via a novel method combining low temperature ashing and float-sink technique. *Int. J. Coal Geol.* **2014**, *13*, 162–171. [CrossRef]
45. Huggins, F.E.; Huffman, G.P. Modes of occurrence of trace elements in coal from XAFS spectroscopy. *Int. J. Coal Geol.* **1996**, *32*, 31–53. [CrossRef]
46. Dai, S.F.; Zeng, R.S.; Sun, Y.Z. Enrichment of arsenic, antimony, mercury, and thallium in a Late Permian anthracite from Xingren, Guizhou, Southwest China. *Int. J. Coal Geol.* **2006**, *66*, 217–226. [CrossRef]
47. Raeva, A.A.; Dongari, N.; Artemyeva, A.A.; Kozliak, E.I.; Pierce, D.T.; Seames, W.S. Experimental simulation of trace element evolution from the excluded mineral fraction during coal combustion using GFAAS and TGA-DSC. *Fuel* **2014**, *124*, 28–40. [CrossRef]

MDPI AG

St. Alban-Anlage 66

4052 Basel, Switzerland

Tel. +41 61 683 77 34

Fax +41 61 302 89 18

http://www.mdpi.com

*Minerals* Editorial Office

E-mail: minerals@mdpi.com

http://www.mdpi.com/journal/minerals